Foundations of Physical Chemistry

This book tries to lighten the difficulties of learners by focalizing in few areas of Physical Chemistry and approaching these areas with extensive and in-depth explanations combined with an easy-to-read style of presentation. This book deals only with quantum mechanics, atomic structure, thermodynamics, and symmetry because these are both at the basis of the whole Chemistry and are the topics requiring most attention.

- Formulas are extensively accompanied by clarifications of their roles and consequences followed by discussions and comments, prioritizing the explanation of the line of thinking over the technicalities, but also preferring to address directly the most challenging points, rather than conceal them.
- The selection of very few topics allows following a *colloquial style with a simple language*, while using an in-depth treatment of the subjects.
- Some topics pertinent to Mathematics (linear spaces and matrices, two necessary tools for quantum mechanics and symmetry) are treated separately with a more extensive presentation in appendixes, avoiding formalization, but exploiting the analogy with the vectors of ordinary 3D space.
- Strong emphasis is given to the connections between different areas of Physical Chemistry, such as the laws of motion (mechanics), symmetry, and macroscopic behavior (thermodynamics).

Prospected readers are not only students of Physical Chemistry courses in the first years for a degree in Chemistry, but also more advanced students willing to strengthen their previous knowledge, *i.e.,* graduate or Ph.D. students in various areas of Chemistry or related fields such as Biotechnology, Materials Chemistry or Energy and Environmental Technologies.

This is a reference book for a deeper understanding of the most fundamental parts of Physical Chemistry. It has been written both for making the study of the beginner easier and helping the advanced learner to clarify the most critical points.

To Milena, Laura and Anna (GS),
To Luisella and Federico (PG),
To our students

Foundations of Physical Chemistry

Giorgio Spinolo and Paolo Ghigna

CRC Press
Taylor & Francis Group
Boca Raton London New York

CRC Press is an imprint of the
Taylor & Francis Group, an **informa** business

Designed cover image: Designed by the Authors.

First edition published 2027
by CRC Press
2385 NW Executive Center Drive, Suite 320, Boca Raton FL 33431

and by CRC Press
4 Park Square, Milton Park, Abingdon, Oxon, OX14 4RN

CRC Press is an imprint of Taylor & Francis Group, LLC

ISBN: 9781041082330 (hbk)
ISBN: 9781041082354 (pbk)
ISBN: 9781003644422 (ebk)

DOI: 10.1201/9781003644422

Typeset in Times
by codeMantra

Contents

Preface ix
About the Authors xi

Chapter 1 Quantum Mechanics 1

1 The Starting Point: Classical Mechanics 1
1.1 Degrees of Freedom and Configuration Space 2
1.2 Newton Law 2
1.3 Generalized Coordinates 5
1.4 Hamilton Function 6
2 The First Postulates of QM 7
2.1 Quantum Mechanics *vs* Classical Mechanics 7
2.2 Postulate I: The Wavefunction 8
2.3 Postulate II: Expectation Values 11
2.4 Postulate III: QM Operators 15
2.5 Postulate IV: Schrödinger Equation 18
2.6 Postulate V: Eigenvalues and Eigenvectors 18
3 Some Results of the First Five Postulates 19
3.1 Stationary States 19
3.2 Degeneracy 21
3.3 Commuting Operators 21
3.4 Eigenvalues of Angular Moments 23
3.5 Separable Hamiltonian Operators 24
4 Spin 26
4.1 Postulate VI. Spin: Meaning and Properties 27
4.2 Pauli Principle 28
4.3 Quantum Statistics 28
5 Linear Spaces and Matrices in QM 28
5.1 Linearity and QM 29
5.2 Matrices in QM 32
6 Perturbation Methods 35
6.1 Time-Independent Perturbations without Degeneracy 36
6.2 Time-Independent Perturbations with Degeneracy 36
6.3 Interaction with the Electromagnetic Radiation 38
6.4 Spectroscopy 39
6.5 Polarizability 42
7 Remarks 44
7.1 The Turnover Rule 44
7.2 Commutation of Angular Moments 44
7.3 Algebraic Solution of the Angular Moment Problem 45
7.4 The Dipole Transition Operator 47
8 Exercises 47
Notes 48

Chapter 2 The Electronic Structure of Atoms 49

1 Introduction and Antisymmetry 49
1.1 Interactions in Atoms with Many Electrons 49
1.2 The Simplest Model: Independent Electrons 50
1.3 Antisymmetry 53

2 Charge Interactions among Electrons 59
2.1 Outline 59
2.2 Degeneracy Removal in Open Shell Configurations 62
2.3 Exchange Interaction and Magnetism 66
2.4 The Vector Model 69
2.5 Summary 73
3 Spin-Orbit Interaction and LS (Russel Saunders) Coupling 73
3.1 Spin-Orbit Interaction 73
3.2 ***LS*** (Russell-Saunders) Coupling 75
3.3 Selection Rules in Atomic Spectroscopy 76
4 Remarks 77
4.1 Details for the $1s^12s^1$ Example 77
4.2 Counting the States 78
5 Exercises 79
Notes 79

Chapter 3 Thermodynamics 80

1 Classical Thermodynamics and Statistical Thermodynamics 80
1.1 Introduction 80
1.2 Basic Definitions 81
1.3 The Laws of Classical Thermodynamics 85
1.4 The Postulates of Statistical Thermodynamics 87
2 Classical Thermodynamics 89
2.1 Energy, Heat, and Work 89
2.2 General Equilibrium Conditions 90
2.3 Thermodynamic Potentials 92
3 Equilibria in Chemical Systems 95
3.1 A Pure Substance in a Single Phase 95
3.2 Mixtures 97
3.3 Phase Equilibrium 102
3.4 Chemical Reactions 109
3.5 Heterogeneous Equilibrium 113
3.6 Reactions of Charged Species 113
3.7 Electrochemical Equilibrium 115
3.8 Quasi-Chemical Equilibria 117
3.9 Thermodynamic Degrees of Freedom 119
4 Statistical Thermodynamics 119
4.1 Partition Functions 119
4.2 Analytic Approach and Computational Approach 121
5 The Analytic Approach to ST 122
5.1 Independent Sub-Systems 122
5.2 Distinguishable and Indistinguishable Sub-Systems 123
5.3 Occupancy of One-Electron States 127
5.4 High-Temperature Limit and ***Classical*** Results 127
6 Some Analytic Results 128
6.1 Energy Equipartition 128
6.2 Independent and Equal Sub-Systems 129
6.3 An Experimental Macroscopic Validation 134
6.4 Non-Ideal Systems: A Simple Model 134
7 Short Notes on the Computational Approach 136
7.1 Molecular Dynamics (MD) 136
7.2 The Monte Carlo (MC) Approach 137
8 Remarks 139
8.1 Some Properties of Thermodynamic Functions 139
8.2 The Partition Function of the Canonical Ensemble 140

8.3 Quantum Statistics and the Grand-Canonical Partition Function 144
8.4 The Ideal Mixture in ST 144
8.5 Maxwell-Boltzmann Statistics 145
8.6 Temperature Dependence of the Chemical Potential of the Electrons 147
9 Exercises 147
Notes 148

Chapter 4 Symmetry in Chemistry 149

1 Introduction and Examples 149
1.1 Symmetry Groups 149
1.2 Symmetry and Wave Functions 159
1.3 Symmetry and Degeneracy 162
1.4 Symmetry Ascent and Symmetry Descent 164
1.5 Spectroscopic Selection Rules 164
1.6 Summary 165
2 The Simple Results 166
2.1 Finding the Symmetry Species of Commutative Groups 166
2.2 Symmetry Adapted Functions (SAFs) 169
2.3 Spectroscopic Transitions 175
3 Translational Symmetry 178
3.1 The Crystal Structure 179
3.2 Bloch Theorem 181
3.3 Crystal Orbitals as LCAO 185
3.4 "Lattice" Vibrations 187
3.5 Conclusions 190
4 Substructures and Superstructures of Groups 190
4.1 Inner Structures 190
4.2 Group Products and Product Group 192
5 Representations and Characters 192
5.1 Matrix Representations 192
5.2 Reducible and Irreducible Representations 200
5.3 Characters 201
5.4 Orthogonality Theorems 203
5.5 Character Tables 206
5.6 Operations on Representations 206
5.7 Summary 207
6 Applications 207
6.1 Symmetry and QM 207
6.2 Building Symmetry Adapted Functions 211
6.3 Relations between Groups 222
6.4 Continuous Groups 224
6.5 Spectroscopic Transitions in Molecules 225
6.6 The Solid State beyond the Translational Symmetry 227
7 Remarks 232
7.1 The Basic Property of the Cayley Table 232
7.2 The Regular Representation 232
7.3 Another Orthogonality Theorem for Characters 234
7.4 Orthogonality of SAF of Different IRs and Different Partners 234
7.5 Filled Shells 234
8 Exercises 234
Notes 236

Chapter 5 Appendices 237

1 The Hamiltonian Version of Classical Mechanics 237
1.1 The Method 237
1.2 Central Field Motion 239
1.3 Molecular Degrees of Freedom and Equipartition of Energy 245
2 Linear Spaces (Vector Spaces) 250
2.1 Basic Operations 250
2.2 Components, Projections and Dot Product 252
3 Matrix Calculus 254
3.1 Definitions 254
3.2 Matrices and Linear Spaces 257
3.3 Other Matrix Operations 263
4 Biographical Notes 264
Notes 268

Index 269

Preface

This book deals with the most fundamental parts of Physical Chemistry using, as far as possible, an easy to follow and tutorial style.

There are several reasons for the difficulties encountered in studying Physical Chemistry. Chemistry has developed as an experimental science, and various areas of Physical Chemistry provide its basic theory using the methods and language of Physics, with a significant role for Mathematics. It is a fact that the interest in Physical Chemistry takes on strongly vocational connotations and must deal with a negative attitude toward Mathematics that is so often reported by the teachers, and sometimes openly displayed by the students. This attitude can lead to a complete dismissal of Physical Chemistry and a lack of interest in the foundations of Chemistry. Another reason is that education must pay attention to both the topics that are parts of Physical Chemistry in its strict sense and those that play an important role in other areas of Chemistry. Finally, Physical Chemistry is deeply divided into distinct areas, and each requires adequate coverage by textbooks: from classical thermodynamics to kinetics and transport phenomena, from theoretical chemistry (including quantum chemistry and the foundation of computational chemistry) to many spectroscopies, from statistical thermodynamics to solid-state science and physical chemistry of materials, to name just a few broad areas that can be further subdivided into autonomous sectors.

In addition to the specific difficulties of Physical Chemistry, there are common challenges shared with many scientific disciplines due to the explosive increase in teaching content required by the advancement of knowledge, often combined with a reduction in the time available for teaching and individual study. Seemingly, to keep within reasonable limits the workload requested to the students, the dilemma is whether to totally delete some topics or to compact as far as possible every single topic deleting the parts that are considered less important. However, compacting too much a topic reduces the possibility of a wider and deeper discussion and severely affects the effectiveness of a presentation. On the other hand, the topics selected for removal, thinking that they have only marginal or historical interest, frequently are at the borderline of different sections of a discipline or different disciplines. Then, the removal enlarges the separation, contrary to the current demand for a more interdisciplinary approach and better connections between different areas of knowledge.

What are really needed are teaching solutions that go out of the horns of this dilemma. We think that an appropriate approach is to focus on the most problematic and fundamental parts, using more extensive and in-depth explanations and an easy-to-read presentation for these parts only.

This approach makes this book completely different from typical textbooks in terms of purpose, content, and presentation style.

The *purpose* is not to replace but instead to complement a textbook. Essentially, we are willing to offer in a single book all the learning material required for achieving the sound understanding that learners of previous generations were able to grasp with a long and hard work by examining in parallel specialized books characterized by different contents, styles, and presentation levels.

Concerning the *content*, this book *does not cover* the whole area of physical chemistry as most textbooks do but deals with only a few parts. Essentially, these are Quantum Mechanics, Thermodynamics, and Symmetry. Not only these are at the basis of whole Chemistry, but also are the parts requiring most attention, according to our personal experience as teachers. In addition, within these parts, the book does not treat all the topics that a textbook typically treats, but focuses on the most difficult and problematic topics, and those that are most effective for clarifying other parts of Physical Chemistry. While most emphasis is given to the fundamental aspects, more advanced topics are also discussed, when their explanation deserves particular attention, and the underlying basic arguments are already available. On the other hand, as the book is planned to complement teacher's lectures and textbooks, many topics are assumed to be already known and are freely recalled for illustrating the consequences of an argument or for establishing connections between various areas of Physical Chemistry.

The chapter on *Quantum Mechanics* essentially gives an ample and easy to read presentation of the fundamental postulates. It starts with a short summary of classical mechanics that introduces the Hamiltonian function but also outlines the separation of degrees of freedom as a general tool for facing many complex problems in both classical and quantum mechanics. The chapter does not discuss the exactly solvable problems (the particle in a box, the harmonic oscillator, the hydrogen atom) or the LCAO approach to the molecular orbitals: these topics are treated in a satisfactory manner in most textbooks and do not require deeper insight. Instead, it discusses the perturbation method for both time-independent and time-dependent fields: this topic generally receives lesser attention but is enormously important for better understanding of many chemical applications. Finally, the chapter is supplemented (i) by an Appendix that gives a more complete account of the Hamiltonian formulation of classical mechanics and discusses the problem of two particles bonded by a central force and (ii) by a whole chapter on the electronic structure of atoms, that essentially aims to foster familiarity with the technicalities of quantum mechanics and to clarify features of more complex systems with many electrons, such as molecules and solids.

The chapter on *Thermodynamics*, follows an integrated approach, treating in a parallel way Classical and Statistical Thermodynamics. In comparison to the traditional approach, where the two topics are discussed in different chapters of a textbook, or taught in different classes, maybe in different semesters, this hopefully maximizes synergies and makes possible to use procedures, concepts, and examples of one theory for giving insights into the other. A more advanced argument is the extension of the formal treatment of chemical reactions and the introduction of quasi-chemical reactions, which is so useful to present in a unifying way many Materials Science topics.

Finally, the chapter on *Symmetry* first develops a treatment of symmetry that does not require matrix representations, orthogonality theorems and character tables. This simplified treatment holds only for commutative groups but is indeed effective in showing the power of the methods based on group theory. This might be sufficient when a short presentation of the role of symmetry in Chemistry is preferred. The more advanced part discusses, clarifies and explains results that are typically shown in textbooks as black box procedures.

Concerning the *style of presentation*, the book always follows a colloquial style with a simple language. Formulas are extensively accompanied by clarifications of their roles and consequences, followed by discussions on their validity and limitations, along with comments. When possible and appropriate, derivations are done informally, prioritizing the explanation of the line of thinking over the technicalities. When encountering particularly challenging points, the approach has consistently been to address them directly rather than conceal them. Also, we strive to explicitly emphasise the critical aspects, even in cases where, for various reasons, complete solutions are not given. Repetitions have not been avoided, preferring the risk of repeating a well-assimilated point rather than missing an opportunity for further clarifying possible sources of misunderstanding. To keep the presentation easily readable, sometimes the details of an argument are not fully developed and left to exercises, while some more difficult derivations are reported in Remarks at the end of each chapter. Finally, some parts pertinent to Mathematics (linear spaces and matrices, two necessary tools for Quantum Mechanics and Symmetry) are treated separately with a more extensive presentation in the Appendices, avoiding strong formalization, but exploiting the analogy with the vectors of ordinary 3D space. Hopefully, this approach can help to reduce the above said negative attitude toward Mathematics.

The potential users of this book are all students with some curiosity. We have planned the content, level, and style of presentation to meet the needs of undergraduate Chemistry students wishing to achieve a deeper understanding of the fundamentals of Physical Chemistry than usually required in preliminary courses. The book could be used also by more advanced students wishing to strengthen and deepen their previous knowledge. Graduate students or doctoral students in various areas of Chemistry or in related areas such as Biotechnology, Materials Chemistry or Energy and Environmental Technologies may then benefit also from some more in-depth sections not usually covered in introductory textbooks.

This book has been reviewed by the IUPAC project "Introducing the IUPAC Seal of Approval for a wider adoption of IUPAC-recommended symbols, terminology, and nomenclature: Stage 1 – Symbols" (https://iupac.org/project/2022-008-4-500/), and it was found that the symbols used in this book are consistent with the IUPAC Green Book (https://iupac.org/what-we-do/books/greenbook/).

This book has been written while receiving inspiration from a number of books that we found fundamental as learners and teachers:

H.B. Callen. *Thermodynamics and an Introduction to Thermostatics*. John Wiley & Sons, New York, 1985. ISBN 0-471-86256-8.

M.S. Dresselhaus, G. Dresselhaus, and A. Jorio. *Group Theory - Application to the Physics of Condensed Matter*. Springer-Verlag, Berlin, Heidelberg, 2008. ISBN 978-3-540-32897-1.

H. Goldstein. *Classical Mechanics*. Addison-Wesley Publishing Company, Reading, 1950. ISBN 978-0-201-65702-9.

T.L. Hill. *An Introduction to Statistical Thermodynamics*. Dover Publications, Inc., New York, 1986, first published by the Addison-Wesley Publishing Company, Inc., Reading, 1960. ISBN-13: 978-0-486-65242-9, ISBN-10 0-486-65242-4.

F.L. Pilar. *Elementary Quantum Chemistry*. Dover Publications, Inc., New York, 2001, first published by McGraw-Hill Publishing Company, New York, 1990. ISBN 978-0-486-41464-5.

J. von Neumann. *Mathematical Foundations of Quantum Mechanics*. Translated from the German edition by Bobert T. Beyer. Princeton University Press, Princeton, 1971. ISB 0-691-08003-8.

J.C. Slater. *Quantum Theory of Matter*. McGraw-Hill Book Company, New York, 1986. ISBN 0070580405, 9780070580404.

About the Authors

Giorgio Spinolo was a full Professor of Physical Chemistry at the Department of Chemistry of the University of Pavia. He made several research contributions and two books in various fields of Physical Chemistry of Materials. He has over thirty years of teaching experience on basic Physical Chemistry courses.

Paolo Ghigna is full Professor of Physical Chemistry in the Department of Chemistry at the University of Pavia. He is author of more than 205 research paper in various fields of Physical Chemistry of Materials, and several book chapters in the field.

1 Quantum Mechanics

When compared to classical mechanics (CM), quantum mechanics (QM) clearly appears as a *difficult* topic to the beginner. However, some knowledge of the basics of QM is a real need also for learners with predominant interest in experimental chemistry. Indeed, QM is the physical theory explaining most fundamental aspects of Chemistry, such as the periodic table of the elements and the chemical bond now called covalent, found by 19th century chemists and for a long time recognized as incompatible with the physical theories of the time (CM and classical electromagnetism). Going beyond the historical point of view, many peculiar aspects of QM and its very language permeate all modern chemistry so that a trustworthy understanding is required to deal with the matter in a sensible way.

Basic ideas, procedures, and results of QM are surely *less familiar* than those of CM: when added to the inevitable conciseness of a textbook, this explains the feeling of a larger *difficulty*. So, this chapter aims at reducing this feeling by following a more colloquial style and a wider presentation for explaining, clarifying, and discussing topics deserving deeper attention.

The presentation of QM must obviously follow a compromise between conflicting goals: achieving as soon as possible the most important results, but also establishing a solid foundation, also capable of rewarding the learners that are more interested in basic aspects, that are planning to deepen their knowledge later, or that are simply looking for sound explanations.

QM is introduced here as a theory derived from postulates. Some textbooks rather do *not* state the postulates explicitly: this approach is possibly seen as too abstract, so that passing over these aspects is a choice both of simplicity and didactic efficacy. On the contrary, we here follow the line of thinking that an explicit statement improves the understanding, when adequately explained.

As occurs with any theory, the technical language must be clarified before discussing principles, derivations, and results. Indeed, this seems to be another source of difficulty when approaching QM the first time, so that the purpose of this chapter is also to make the student more familiar with the language of QM.

This chapter does not deal with some topics that may already be known in some way from previous courses or do not require in our opinion deeper insights, being regularly treated in standard textbooks. When useful, however, these topics are freely referenced to clarify some aspects of the basic theory or to show examples of applications. This regards for instance the historical introduction to the crisis of CM and the first formulations of QM, the so-called exactly solvable problems (the particle in a box and in a potential well, the harmonic oscillator, the free rotor and the hydrogen-like atom) and the variational method with the LCAO approach to molecular orbitals, particularly in the simplest version due to Hückel.

The line of presentation that is followed here requires some previous knowledge of CM. QM can be considered an extension of CM and its most fundamental tool, the Hamiltonian operator, derives from the Hamiltonian function of CM. Moreover, the reader should be acquainted with the idea that in most cases a mechanical problem is better treated using a set of coordinates different from the cartesian coordinates of each particle. For these reasons, the first part of this chapter summarizes these prerequisites with a very simple language. We understand that this approach is unusual for Chemistry students, who are customarily taught to use the Hamiltonian operator as a postulate. We preferred to give a detailed explanation, as this book is focused on the foundations of Physical Chemistry: of course, the non-interested reader can skip this part without compromising the understanding of the following. Moreover, an Appendix on CM gives a deeper account on the Hamiltonian formulation of CM and other useful results that are easily transferred to the QM treatment.

Concerning the mathematical prerequisites, these essentially boil down to complex numbers and linear spaces, including matrices. The first topic is very shortly *summarized* here in connection with the first postulate, assuming that it has been already taught the student previously but rarely dealt with later. Concerning linear spaces, this chapter includes at point 5 a short outline of what is most important for QM. Readers that are less familiar with the topic, or prefer a deeper insight, can find a more detailed presentation in the Appendix, which anyway follows a less formal approach than most specialized textbooks on QM, and essentially exploits the analogy with the vectors of ordinary three-dimensional space.

1 THE STARTING POINT: CLASSICAL MECHANICS

Starting a presentation of the principles of physical chemistry with classical mechanics (CM) may seem surprising, as it is well known that the mechanical theory for studying atoms and molecules is quantum mechanics (QM).

It is important to show how a problem is faced in CM, because the treatment in QM - in a sense that will be clarified later - is the *translation* into the language of QM of the treatment that is followed in CM for the same problem. For this purpose, however, the approach in CM must be developed differently from the most elementary approach based on Newton laws as typically presented in preparatory courses. A different formulation of CM is needed. This

DOI: 10.1201/9781003644422-1

is the Hamiltonian formulation and is here outlined in a strongly simplified way.

The focal point is the analysis of the degrees of freedom of a system of many particles and the exact or approximate separation of the pertinent motions.

1.1 Degrees of Freedom and Configuration Space

To specify a point in space we must specify three numbers. We say that we need three *coordinates* to describe the position of a particle in space. These can be the Cartesian coordinates x, y, z - this is the simplest case - but various other possibilities are frequently used, for example the spherical polar coordinates or the cylindrical coordinates. Sometimes, systems with reduced dimensionality are also discussed (one-dimensional - 1D - or two-dimensional - 2D motions): these cases generally work as an introduction to three-dimensional (3D) systems.

The spherical polar coordinates are usually denoted by r, ϑ, ϕ and are the distance from the origin of the axes and two angles (see also Section 2.2.3). The cylindrical coordinates are denoted by r, φ, z, where r is again the distance from the z axis, z is the same as the cartesian coordinate, and φ is a longitude. For 2D systems, plane polar coordinates are frequently used: these are the same as the r and φ cylindrical coordinates. Also, ϑ is sometimes used instead of ϕ to denote cylindrical or plane polar coordinates.

There are two aspects to keep in mind in this regard: *number* and *nature* of the coordinates. As for the number, it is obvious that we need three numbers for each particle. In fact, with a smaller number of coordinates, we would not be able to completely specify the position of the particles. For example, if the value of a Cartesian coordinate for a particle were not specified, it would indicate not a point in 3D space but an entire line parallel to the axis of the missing coordinate. On the other hand, with a higher number of coordinates some coordinates can be written as a function of the others and therefore can be dropped out.

As for the nature of the coordinates, the essential point is to be sure that they are *independent* of each other. The meaning is that one coordinate does not have to be a function of the others: if it were, it would be redundant, it would have to be eliminated and it would be necessary to find another coordinate to specify the position in space of the various particles. Cartesian coordinates are an independent set, just as planar (2D) or spherical (3D) polar coordinates, or cylindrical coordinates are. The final part of 1.3 reports further explanations.

We here consider only systems consisting of **point particles**, also called *material points* or *dimensionless masses*. Considering a point particle means ignoring the extension of the particle and its consequences: we essentially ignore the fact that the particle can rotate or deform and therefore has rotational or deformation energy. In many cases, however, these energies must be considered for a correct balance, for example when discussing the collision between polyatomic molecules. Then, if necessary, we will consider systems made up of many point masses and therefore characterized by rotation and deformation energies in addition to translation energy. Incidentally, we note that, unlike applications of CM to various areas of engineering, in chemical problems it is rarely necessary to resort to the treatment of *extended* bodies, *i.e.* to a continuous mass distribution.

We say that a system of n point particles has $3n$ **degrees of freedom** to highlight that $3n$ quantities are needed, regardless of what these quantities are in detail, if they are independent of each other.

We call **configuration** the arrangement in space (at a given time) of the particles of a system. Changing even just one coordinate of a single particle changes the entire configuration. The configuration of a system can be visualized not only as a set of n points in the usual 3D space but also using a $3n$-dimensional *abstract* space. This is called the **configuration space** and a point in this space represents the whole system. In the case of one point particle, the configuration space clearly is the same as the ordinary 3D space. The abstraction involved in considering the configuration space instead of the ordinary 3D space is justified both by the simplicity in extending the algebraic treatment and by the greater freedom that the configuration space offers in choosing the variables used to describe a system of n particles. There are, in fact, cases in which it may be useful to use coordinates that do not refer to one or another particle but describe a quantity related to two or more particles such as the distance between two particles or the centre of mass of two or more particles (see, for example, Section 1.3). The set of points that in the configuration space represent the system - time by time - is called the **system trajectory**. The time function that describes the distance travelled along the trajectory is called the **time law of motion**.

1.2 Newton Law

CM and QM strongly differ in the way they describe the motion of a system and obtain the desired information. In CM, describing the motion of a system essentially means describing how the coordinates change over time. Ultimately, it means obtaining the **functions** that describe the **time dependence** of each coordinate. To derive these functions, we write and solve the **equations of motion**. As well known, in the elementary version of CM the equations of motion are given by **Newton law**. For a point particle of mass m moving along a single direction in space (a 1D motion) and a force F, the law is:

$$F = ma \tag{1.1}$$

In Eq. (1.1), the acceleration (a) is the second total derivative with respect to time of the Cartesian coordinate x:

$$a = \mathrm{d}^2x/\mathrm{d}t^2 \tag{1.2}$$

If the point particle we are considering can move along the three directions of ordinary three-dimensional space (3D), we have:

$$\vec{F} = m\vec{a} \tag{1.3}$$

Force ($\vec{F}$) and acceleration ($\vec{a}$) are now vectors, *i.e.*, quantities made of three components, and Eq. (1.3) is the short notation for three scalar equations:

$$F_x = ma_x \quad F_y = ma_y \quad F_z = ma_z \tag{1.3a}$$

These equations express the simplest relation (a proportionality) between two vector quantities. A proportionality relation (also called a linear relation) between vector quantities assumes *in the general case* a slightly more complicated form, where each component of the first quantity is proportional to all components of the other quantity. If we denote the two vector quantities by $\boldsymbol{c}$ and $\boldsymbol{d}$, we have:

$$c_x = m_{xx}d_x + m_{xy}d_y + m_{xz}d_z$$

$$c_y = m_{yx}d_x + m_{yy}d_y + m_{yz}d_z$$

$$c_z = m_{zx}d_x + m_{zy}d_y + m_{zz}d_z$$

Thus, nine coefficients are needed to express linearity between two 3D vectors in the general case. The quantity with two indices that groups the nine coefficients is called a tensor. Examples of such a relationship between vector quantities are given by Eq. (6.27) of this chapter and by Eq. (1.30) of the appendix on the Hamiltonian formulation of CM. There is a particular significance of the fact that Newton law requires only one coefficient. It means, among other things, that $\vec{F}$ and $\vec{a}$ are parallel to each other and that the mass of a particle is a scalar.

The Eq. (1.3a) are second order differential equations, where the unknowns are the three functions: $x(t)$, $y(t)$, $z(t)$, while the forces are given functions, and the masses are constant. They are called differential equations because the unknowns to be found are not numbers, but functions and are derived from relations that involve their derivatives. They are called second order because in each equation the unknown functions appear along with their derivatives up to the second derivatives, that is the accelerations.

Finally, the extension to a system of n point particles is easily written:

$$\vec{F_1} = m_1\vec{a_1} \tag{1.4a}$$

$$\vec{F_2} = m_2\vec{a_2} \tag{1.4b}$$

$$\vec{F_n} = m_n\vec{a_n} \tag{1.4c}$$

1.2.1 Some Examples

Let us outline some extremely simple cases.

1.2.1.1 Inertial 1D Motion

For a 1D motion with no force, the equation of motion is:

$$m\mathrm{d}^2x/\mathrm{d}t^2 = 0 \tag{1.5}$$

Remembering that zero derivative means constant function, we immediately obtain that the first derivative of $x(t)$, that is the velocity (along the given x direction) is constant and therefore equal to the velocity at initial time that we denote with v_0. With a second integration we obtain that the position x changes in time according to the function:

$$x = x_0 + v_0 t \tag{1.6}$$

where x_0 is another constant, the initial value of x.

1.2.1.2 Uniformly Accelerated 1D Motion

Still considering a 1D case, if a *constant* force (F) were to act, we would have:

$$\mathrm{d}^2x / \mathrm{d}t^2 = F / m \tag{1.7}$$

and, writing $g = F/m$, we would finally obtain the formula for uniformly accelerated motion:

$$x = x_0 + v_0 t + gt^2 / 2 \tag{1.8}$$

This function describes the motion in a field that is uniform (F does not depend on the position in space, *i.e.* it does not depend on x) and constant (F does not depend on time). This is the case of the falling motion of a heavy object, if the fall occurs in a space short enough to neglect the change of F or g with height.

1.2.1.3 Harmonic 1D Motion

A final example of 1D motion concerns a particle that is tied to a point by a spring that exerts a force proportional to the deformation of the spring and with opposite sign

$$F = -kx \tag{1.9}$$

where $k>0$ is a constant and the spring is fixed is at the origin.

Then: $\mathrm{d}^2x / \mathrm{d}t^2 = -kx / m$.

Knowing that the sine or cosine functions are equal to their second derivative multiplied by a negative constant, we directly obtain the solution of the differential equation as

$$x = A\sin(\omega x) + B\cos(\omega x) \tag{1.10}$$

where $\omega = (k/m)^{1/2}$, A and B are constants specified by the initial conditions. An alternative writing, that can be obtained with the usual trigonometric formulas is

$$x = C\sin(\omega x + \varphi) \quad (1.11)$$

where C (the amplitude of the oscillation) and φ (the initial phase) are two new constants. We note that F can be derived from a potential energy of the form $V=kx^2/2$ by $F=-(\partial V/\partial x)$.

1.2.1.4 A 2D Motion

Consider now the case of a point particle on which only the gravitational force acts along the direction that we can call y. Starting from the position (x_0, y_0), the particle is set in motion with a velocity v_0 along the horizontal direction (which we call x) and zero velocity along y. Then, the motion is described by the two equations:

$$\mathrm{d}^2x/\mathrm{d}t^2 = 0 \text{ with initial conditions}$$
$$x(t=0) = x_0; v_x(t=0) = v_0 \quad (1.12a)$$

$$\mathrm{d}^2y/\mathrm{d}t^2 = g \text{ with initial conditions}$$
$$y(t=0) = y_0; v_y(t=0) = 0 \quad (1.12b)$$

The gravitational force $F=mg$ and the corresponding gravitational acceleration g are assumed to be constant and uniform, *i.e.* independent of time and position. The solution is simple, thanks to the fact that it is a system of two *uncoupled* equations, each of which is solved independently of the other. Looking at the solutions of the previous examples Sections 1.2.1.1 and 1.2.1.2, we immediately have:

$$x = x_0 + v_0 t \quad (1.13a)$$

$$y = y_0 + gt^2/2 \quad (1.13b)$$

1.2.2 Derived Quantities

From the solution of Newton equations (Eq. 1.4), that is from the knowledge of the functions that describe the time dependence of the coordinates, CM derives all the other mechanical quantities as functions of time.

For example, for each particle i the modulus of the velocity vector is given by:

$$v_i = \sqrt{v_{x,i}^2 + v_{y,i}^2 + v_{z,i}^2}; \quad v_{x,i} \stackrel{\text{def}}{=} \frac{\mathrm{d}x_i}{\mathrm{d}t}$$

From the velocities we can obtain the **kinetic energy** as sum of the products $m_i v_i^2$ for all the n particles:

$$T = \frac{1}{2}\sum_{i=1}^{n} m_i v_i^2$$

The **potential energy** (V) has the important property of describing synthetically the forces. As said, the partial derivative of potential energy with respect to the Cartesian coordinate of a particle, with reversed sign, gives the Cartesian component (in the direction of the same coordinate) of the force acting on the particle:

$$F_{x,i} = -\frac{\partial V}{\partial x_i} \quad F_{y,i} = -\frac{\partial V}{\partial y_i} \quad F_{z,i} = -\frac{\partial V}{\partial z_i}$$

In most cases, the potential energy (V) is a function only of the coordinates: then, it can be shown that the sum of potential energy and kinetic energy does not depend on time: this is the **principle of conservation of energy**. The systems with this property are called **conservative.** Unless explicitly stated otherwise, only conservative systems will be discussed hereafter.

There are cases, however, in which the potential energy of a particle and therefore the force to which it is subjected depend on its velocity. These are called *dissipative* systems. Simple examples of velocity-dependent forces are the Lorentz force acting on a charged particle moving in a magnetic field or the mechanical systems with friction or viscosity.

Other important mechanical quantities are the **moments.** *Linear* moments are the product of the mass of a particle and a Cartesian component of the velocity vector of the particle:

$$p_{x,i} = m_i v_{x,i} \quad (1.14)$$

and play a central role in discussing linear motions and collisions.

Angular moments are important for discussing rotational motions. These are defined as:

$$l_{x,i} = m_i d_i v_i \quad (1.15)$$

where $d_i=|\vec{d}_i|$, $\vec{d}_i$ is the vector orthogonal to the axis of rotation from particle i to the x axis, and v_i is the velocity along the direction orthogonal to $\vec{d}_i$ and to the axis.

All these auxiliary quantities are important because they better explain some aspects of a particular motion, follow conservation laws and sometimes directly provide the solution of a mechanical problem. Let us consider, for example, the motion of a mass falling in a uniform and constant gravitational field and suppose we want to know the velocity of the body after a vertical fall h. Instead of solving the equations of motion (see Section 1.2.1.2) we can calculate the potential energy corresponding to h, which is equal to the final value of the kinetic energy, in the case of zero velocity at the initial time. From the kinetic energy we derive directly the velocity that the body must have at the end of the downward jump. Denoting the initial value of y by $h>0$, the initial value of potential energy is mgh and the final value is 0; for kinetic energy the initial value is 0 and the final value is $\frac{1}{2}mv^2$. Then we directly have: $v = \sqrt{2gh}$.

1.3 Generalized Coordinates

As already said, to describe a mechanical system it is possible to use different coordinates (Cartesian, polar, cylindrical or others). However, *Newton equations explicitly refer to a particular choice*, that is, to the *three Cartesian coordinates of each point particle* of the system. To emphasize this situation, it is expedient to call these coordinates **Newtonian coordinates**.

The use of coordinates other than Newtonian coordinates has essential advantages, that are better introduced with an example. Let us consider the **1D** motion of **two** particles with the same mass m and linked by a force proportional to their distance (the so-called harmonic force). There are two degrees of freedom. Using Newtonian coordinates the equations of motion are:

$$k(x_2 - x_1) = ma_1 = m\mathrm{d}^2x_1 / \mathrm{d}t^2 \quad (1.16a)$$

$$k(x_1 - x_2) = ma_2 = m\mathrm{d}^2x_2 / \mathrm{d}t^2 \quad (1.16b)$$

These differential equations are coupled to each other. Indeed, according to Eq. (1.16a), we can obtain x_1 by integrating its second derivative a_1 only if we preliminarily know x_2. Similarly, Eq. (1.16b) indicates that we can obtain x_2 by integrating a_2 if we preliminarily know x_1.

Now, let us use as coordinates the quantities defined by:

$$X = (x_1 + x_2)/2 \quad \text{and} \quad x = (x_2 - x_1) \quad (1.17)$$

apply the inverse transformation:

$$x_1 = X - x/2 \quad \text{and} \quad x_2 = X + x/2 \quad (1.18)$$

take the second derivatives:

$$\mathrm{d}^2x_1 / \mathrm{d}t^2 = \mathrm{d}^2X / \mathrm{d}t^2 - \left(\mathrm{d}^2x / \mathrm{d}t^2\right)/2 \quad \text{and}$$
$$\mathrm{d}^2x_2 / \mathrm{d}t^2 = \mathrm{d}^2X / \mathrm{d}t^2 + \left(\mathrm{d}^2x / \mathrm{d}t^2\right)/2$$

substitute in Eq. (1.16) and finally take the sum and difference of the results. The equations of motion then become:

$$m\mathrm{d}^2X / \mathrm{d}t^2 = 0 \quad \text{and} \quad (1.19)$$

$$m\mathrm{d}^2x / \mathrm{d}t^2 + 2kx = 0 \quad (1.20)$$

We now discuss these equations from two different points of view.

From a technical point of view, that is, considering how the differential equations can be solved, the result is that with the new quantities X and x **the differential equations are decoupled**. They can therefore be discussed separately from each other and are both easily solved. By comparison with the first example (Section 1.2.1), we see that the first equation is in fact an inertial motion characterized by a constant velocity:

$$\mathrm{d}^2X / \mathrm{d}t^2 = 0 \quad \rightarrow X = X_0 + V_0t \quad (1.21)$$

Here X_0 and V_0 are two constants depending on the initial conditions and are precisely the initial values of X and its total derivative with respect to time (V_0=dX/dt, that is the speed) that remains equal to its initial value. Comparing the second equation with the example of the harmonic motion (Section 1.2.1.3), we obtain the solution:

$$\mathrm{d}^2x / \mathrm{d}t^2 = -2kx / m \rightarrow x = R\cos(\omega t + \varphi) \quad \text{or}$$
$$x = A\sin(\omega t) + B\cos(\omega t) \quad (1.22)$$

Here $\omega = (2k/m)^{1/2}$ includes the specific mechanic parameters, while R and φ (or A and B) depend on the initial conditions. We understand that the Newtonian equations are coupled because of the expression of the forces, which mix x_1 with x_2. If the forces depended on only one coordinate (or none), the coupling would not exist. Since the forces are usually derived from a potential energy, the same consideration applies to the potential energy: the potential energy mixes the Newtonian coordinates and therefore gives coupled equations of motion (Eq. 1.16). The $(x_1, x_2) \rightarrow (X, x)$ transformation (Eq. 1.17) decouples the equations of motion because the forces and the potential energy depend only on the difference x=(x_2-x_1). Using x as a coordinate, we have that the forces and therefore also the potential energy depend on only one independent variable (x).

Going beyond the example, we can understand that a system of differential equations is certainly **easier** to solve **when the equations are decoupled**, because each equation constitutes a separate problem and does not influence the solution of the other equations. Another simple example of decoupled equations is given in Section 1.2.1.4. As a matter of fact, however, one rarely deals with uncoupled equations of motion. Usually, the equations are coupled because some particles of the system interact in some way with other particles. We say that the system has internal forces: the force acting on a given particle is the vector sum of the forces due to each of the other particles with which it interacts, possibly added to the forces due to external fields (gravitational, electric, magnetic, …). In these cases, it is necessary to modify the equations of motion in some way to solve them exactly (*if and when this can be done*).

From **the point of view of physical meaning**, the two results (Eqs. 1.21 and 1.22) also provide the best **explanation** of the solution. In plain words, Eqs. (1.21) and (1.22) tell us that the original problem of the two equal masses connected by a harmonic force (Eq. 1.16, a problem with two degrees of freedom), is **equivalent** to two independent motions: an inertial motion and a simple (1 degree of freedom) harmonic motion (see Figure 1.1).

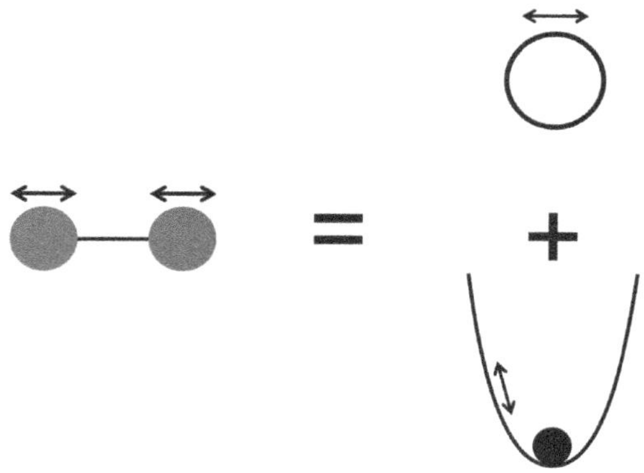

FIGURE 1.1 The problem of two masses (grey) linked by a harmonic force is equivalent to two independent motions: an inertial motion of the whole system (white circle) and a harmonic oscillation of the bond.

In conclusion, the above example shows that

> to discuss and understand the motion of a mechanical system, the Newton coordinates are not necessarily the best choice.

On the contrary,

> the flexibility offered by generalized coordinates is necessary for finding (if and when possible) a coordinate system that allows to separate a complex motion into independent motions.

As said, when looking for a suitable set of coordinates, one must always pay attention to their independence, that is, to the fact that a coordinate must *not* be a function of the others. It is therefore necessary to distinguish between the *independence* between the coordinates and the *separation* between the motions:

a. both the Newtonian coordinates $\{x_1, x_2\}$ and the generalized coordinates $\{X, x\}$ are independent of each other: x_1 is not a function of x_2 and x_2 is not a function of x_1 as well as X is not a function of x and x is not a function of X, but
b. due to Hooke law, both cartesian coordinates $\{x_1, x_2\}$ appear in each of the original equations of motion (Eq. 1.16), while only one coordinate appears with its derivatives in each of the equations for X and x (Eqs. 1.19 and 1.20) and in their solutions (Eqs. 1.21 and 1.22) and therefore the motions are separated.

1.4 Hamilton Function

To profit from the above considerations, we must know how to write the equations of motion for generalized coordinates. Differently from what the example of the previous point might suggest, *the changes of coordinates are not simple tricks to be sought from time to time but involve an analysis of the conceptual bases of CM* and a true *reformulation*. The branch of mathematics that deals with this problem is Rational Mechanics, also known as Analytical Mechanics. Rational Mechanics offers more than one solution to the problem of writing the equations of motion for generalized coordinates, but only some very general considerations about the approach named after **Hamilton** are requested to introduce QM.

The Hamiltonian formulation of MC describes a mechanical system with n degrees of freedom using $2n$ *independent* variables: n generalized coordinates q_j and n corresponding moments p_j, that are linked to each other only by the equations of motion. A moment is defined as the partial derivative of the kinetic energy T with respect to a generalized velocity ($\dot{q}_j$):

$$p_j \overset{\text{def}}{=} \frac{\partial T}{\partial \dot{q}_j} \qquad (1.23)^1$$

where a generalized velocity is the total derivative with respect to time of a generalized coordinate q_j: $\dot{q}_j \overset{\text{def}}{=} \frac{dq_j}{dt}$. From Eq. (1.23) we see that each q_j – however chosen, even a Newtonian coordinate – always corresponds to a (unique) moment p_j. We say that it is the **conjugate moment** of the q_j coordinate.

Finally, the **Hamiltonian function** is the sum of potential energy V and kinetic energy T both written as functions of coordinates and moments:

$$H(q_j, p_j) = V(q_j) + T(q_j, p_j) \qquad (1.24)$$

Now, we need

a. to find the generalized moments for a given mechanical problem and a given choice of generalized coordinates (see Eq. 1.23),
b. to write all physical quantities as functions of coordinates and moments and, in particular,
c. to write the Hamiltonian function (see Eq. 1.24).

This is what is strictly required about Hamiltonian Mechanics to introduce the basic ideas of QM. For the sake of completeness, we may also quote the equations of motion of the Hamiltonian formulation of CM:

$$\dot{q}_j = \frac{\partial H}{\partial p_j} \quad \text{and} \quad \dot{p}_j = \frac{-\partial H}{\partial q_j} \quad \text{(Hamilton equations)} \qquad (1.25)$$

With the introduction of moments, which are independent variables at the same level as coordinates, we also introduce the $2n$-dimensional **phase space** in place of the n-dimensional *configuration* space. Equation (1.25) shows that a point in phase space fully specifies the motion, while a point in configuration space does not provide enough information because also the velocities must be known to specify the motion.

We do not derive Hamilton equations from Newton law: for the interested reader, however, the appendix on Hamiltonian Mechanics shows that Hamilton equations (Eq. 1.25) do indeed provide the same results provided by Newton equation. To become acquainted with all the machinery of Hamiltonian Mechanics, the same appendix also reports, as an example, the treatment according to the Hamiltonian formulation of CM of the motion in ordinary space of two particles linked by a central force, that is a force directed along their joining line. This model is the extension to the 3D ordinary space and to other forces of the simple harmonic model described by Eq. (1.16) and is important in itself because it is directly related to two fundamental problems for Chemistry, the hydrogen atom and the diatomic molecule.

As a final comment, it can be noted that Newtonian mechanics in some cases "discovers" - let us say - that the energy $(V+T)$ is conserved by solving the equations of motion. Instead, the Hamiltonian approach uses since the beginning the fundamental role of energy in determining the evolution in time of a mechanical system

> the Hamiltonian function
> is the characteristic function of the system
> and implicitly contains its time evolution.

The equations of motion are obtained by "operating" on the Hamiltonian function.

This point of view is shared also by QM.

2 THE FIRST POSTULATES OF QM

2.1 Quantum Mechanics *vs* Classical Mechanics

Quantum mechanics (QM) is a theory of wider validity than classical mechanics (CM): we say that **QM extends CM**. Broadly speaking, the extension concerns the atomic-molecular world, *i.e.*, particles having size and mass that are orders of magnitude lower than those of the macroscopic world.[2] In other words, the relation between CM and QM is that QM applies to all problems for which CM holds, but also to problems for which CM does not hold.

What concerns atoms and molecules is sometimes referred to as the *microscopic world*, an expression that is generally used also here despite some ambiguity. CM *can* be applied to some problems of the microscopic world, but important approximations must then be made, and due care must be taken to verify the correctness of the approach. CM generally provides a kind of asymptotic limit of the results of QM: it can be shown, for example, that the results of CM for a given mechanical problem approach the QM results as the quantum numbers approach infinity.

Being an extension of CM, QM cannot be derived from CM and requires a foundation of its own. For example, one cannot derive the Schrödinger equation from Newton law or Hamilton equations. We can then ask how QM is justified. The answer is the same that must be given for any fundamental theory: QM is justified by experiments. QM is developed as a deductive theory from some unproven statements, called axioms or postulates. Then, the results derived from the postulates are either validated by experiments (and therefore further assess the theory without fully proving it) or else are shown false (and therefore the whole theory, starting from the postulates, is rejected, or significantly modified).

To some extent, the statements used as postulates of QM can be selected in an arbitrary way. The scheme here used is based on six postulates. Initially, however, only five postulates are presented, because several results developed from them are a prerequisite for the sixth postulate.

A significant difference between QM and CM is the **intrinsically probabilistic character** of QM: QM does not say, for example, *where* a particle is at a given time, but says *what is the probability of finding* the particle in a given region of space at a given time. On the contrary, the solution of a mechanical problem in CM is a set of functions describing how each coordinate depends on time: we know exactly where the system is at the given time because there is a single value for each coordinate. CM is then said to be a *deterministic* theory. For the sake of completeness, note that a probabilistic treatment is possible also within CM. Some applications of CM deal with problems having starting conditions that are not exactly known, and a probability distribution of starting conditions is introduced. So, a statistical treatment is superimposed over the intrinsically deterministic treatment of CM, and the solutions of the mechanical problem assign probabilities to different configurations of the system at a given time. Such a probabilistic superstructure is an essential part, for example, of the kinetic theory of gases. The intrinsically probabilistic character of QM is a different matter because it is an inevitable feature. This difference between the two mechanical theories has strong philosophical impact and has also shown wide appeal with the public.

A second difference is that - *in some cases* - the product of two different physical quantities is **non-commutative**. In CM, for instance, $xp_x=p_xx$ but the equality does not hold in QM (see Section 2.4.2). This is a clear sign of the higher complexity of the mathematical foundation of QM. On the other side, the commutativity of two specific physical quantities, when it occurs, is an important property that is frequently used to discuss a mechanical problem (see Section 3.3).

Another fundamental difference between CM and QM is that in QM **a measurement changes the state of the system**: this will be explained when dealing with the fifth postulate (2.6).

Despite these large differences, QM shares with CM most of the very basic ideas. A general way of treating a mechanical problem in QM is indeed to start with a preliminary approach in CM and then, let us say, *translate* the approach into the language and the procedures of QM, as explained in Section 2.4. To this purpose, however, the simplest version of CM, *i.e.,* the version based on Newton law, is not appropriate and another version (the Hamiltonian version) must be used, as shortly discussed previously (see Section 1.4). A significant feature of the Hamiltonian version of CM that is shared by QM, is the possibility of using coordinate systems different from the cartesian coordinates of each particle and writing the fundamental equations of mechanics using these *generalized* coordinates.

Finally, it is worth noting that QM can be formulated in various ways. This chapter presents a simple and limited version which should be adequate to illustrate the basic concepts and most applications to Chemistry.

2.2 Postulate I: The Wavefunction

Postulate I
The state of an insulated physical system is represented, for any time, by a *wavefunction*.

The first postulate essentially assesses the wavefunction as the fundamental tool for describing the state of motion of a system (for short: the **state**), discusses its meaning and summarizes its properties. As for CM, different states are possible for a problem. Obviously, different states are described by different wavefunctions. However, the *same state* is possibly described by *different wavefunctions*, if these are in some way *equivalent* to each other. This point will be clarified later.

2.2.1 Meaning and Properties of the Wavefunction

The wavefunction contains all the physical information that it is possible to obtain for the state of a system. The second and third postulates jointly specify **how** to obtain the values of the physical quantities we are interested in. The fifth postulate clarifies a more specific point in that regard, while the fourth postulate states the fundamental equation that a wavefunction must satisfy and that must be solved to obtain the wavefunctions of a system.

The wavefunction is a **function of coordinates and time.** For example, the wavefunction of a particle moving along a line is $\varphi(x, t)$ if x denotes the space coordinate and t denotes time. For a particle in ordinary 3D space, the wavefunction is $\varphi(x, y, z, t)$ if cartesian coordinates are selected. For a system of two particles in 3D space, the wavefunction is $\varphi(x_1, y_1, z_1, x_2, y_2, z_2, t)$, ... In the following, it will be convenient to write a wavefunction as $\varphi(q, t)$ where q indicates either a *single* coordinate or *the set of all* coordinates. As said, QM keeps the possibility of using different coordinate systems, as in CM, and one coordinate system may prove more suitable than another. Generally, the same coordinate system is used in CM and QM, because the simplification realized by a given coordinate system in CM is also valid in QM.

Concerning time, the reader is surely familiar with wavefunctions independent of time. In principle, this is not correct: the wavefunction *does* depend on time. Sometimes, however: (i) it *can* be written as a product of a factor depending only on coordinates and a factor depending only on time, *and* (ii) many important results are obtained by considering only the time-independent factor. Also, to solve a QM problem, one typically goes through the determination of the space-dependent part of the wavefunction (see Section 3.1).

In CM, the solutions of the equations of motion have a *direct physical meaning*; in QM, on the contrary, the wavefunction does not (need to) have a *physical meaning* by itself. What is meaningful is *what is obtained from the wavefunction.*

A usual line of presentation (**Born interpretation**) assigns the wavefunction $\varphi(q, t)$ the meaning of *probability amplitude.* So, the squared modulus of the wavefunction [*i.e.* $|\varphi(q, t)|^2 = \varphi(q, t) \cdot \varphi^*(q, t)$, which is a positive or zero real quantity] has the meaning of a *probability density* of finding the system at point q, and its integral over a region of configuration space is the *probability* of the system being somewhere in this region. Finally, the integral over the whole configuration space is the probability of the system being somewhere in the whole configuration space and therefore *must be equal to one.* Born interpretation is correct and useful, but it is not strictly necessary as part of the postulate: what the theory requires is only that the integral of the squared modulus of the wavefunction over the whole configuration space exists and is $=1$ (normalization condition, see Section 2.2.3).

The wavefunction is *(generally)* **a complex function.** This means that the wavefunction has both a real and an imaginary part (see Section 2.2.2).

Some introductions to QM emphasize that the wavefunction must also be **single valued**. This is a possible source of confusion. Given the n-tuple of coordinate values specifying a point in configuration space, the wavefunction surely takes *only one (generally complex) value.* This is the very meaning of *function*, and it is hardly necessary to specify. Confusion can arise with angular coordinates. Let us consider a longitude (say ϑ), which needs to be specified in the $(0-2\pi)$ interval and $\varphi(\theta) = \varphi(\theta + 2\pi) = \varphi(\theta + 4\pi) = \cdots = (\theta + 2n\pi)$ for any positive, negative or zero integer n. This must occur because these different values of ϑ refer the same point in physical space but is *not* a particular or strange feature of QM.

The wavefunction must be **continuous**. Generally, also the derivatives of the wavefunction are continuous. In some QM models, however, there are singular discontinuities in the first derivative(s) of the wavefunction (points, lines, or surfaces where a derivative shows a jump). An example is the *particle in a box* model, where the box-shaped potential (somehow artificial indeed) features discontinuous first derivative at the boundaries and causes a similar behaviour of the derivative of the wavefunction.

2.2.2 Complex Numbers

Having to deal with **complex functions** is not a trouble: the mathematical difficulties would have been much greater in dealing with functions of complex *variables*. As far as we deal with *real independent variables*, we don't need to tackle more difficult mathematics, because the required tools are already acquired. To be acquainted with complex functions of real variables one must just remember a few points that are briefly outlined below.

A. A complex number z is *uniquely* written as the sum of a real part, and the product of a real coefficient by the imaginary unit (i): $z=a+ib$ (a and b are real numbers)

B. Then, a complex function of real variables [$f(q, t)$] can be uniquely written as the sum *of two real functions of real variables* (the second one multiplied by the imaginary unit):

$$f(q,t) = g(q,t) + ih(q,t)$$

(g and h are *real* functions, q and t are real variables)

C. The derivatives can be directly written recalling the property that the derivative of a sum is the sum of the derivatives (and treating the imaginary unit i as a usual constant):

$$df / dq = (dg / dq) + i \cdot (dh / dq)$$

D. Analogously, the integral of a complex function is the sum of the integrals of the real part and of the coefficient of the imaginary (the second integral multiplied by the constant i):

$$\int_{\Omega} f(q,t)\, dq = \int_{\Omega} g(q,t)\, dq + i \int_{\Omega} h(q,t)\, dq$$

E. The square of i is $i^2=-1$.

F. Some important transformations of the complex number $z=a+ib$ are:
   a. The **complex conjugate**: $z^* = a - ib$
   b. The **modulus**: $|z| \stackrel{\text{def}}{=} \sqrt{z \cdot z^*}$
   Note: $|z|$ is a positive real number, but $|z| = 0$ if and only if z is the unique complex number with zero real and imaginary parts. If $|z| = 0$, z is identified with the real number 0.
   c. The **inverse**
   $$(z)^{-1} = z^* / (zz^*) = (a - ib) / (a^2 + b^2) = \frac{a}{(a^2 + b^2)} - i \frac{b}{(a^2 + b^2)}$$

G. Some algebraic manipulations are greatly simplified switching from the additive notation, $z = a + ib$:
   a. to the **exponential notation**: $z = \rho \cdot e^{i\theta}$, $\rho = |z| = \sqrt{a^2 + b^2}$, $\tan\theta = \frac{b}{a}$
   b. or to the **trigonometric notation**: $z = \rho \cdot (\cos\theta + i \cdot \sin\theta)$.
   As already said, the real variable ϑ (the *argument* of z, or arg(z)) is not uniquely specified: using a different argument (ϑ') we indicate the same complex number z if ϑ and ϑ' differ by an integer multiple of 2π.

H. It is worth recalling that complex numbers have a very useful *graphic representation* as points in an abstract plane (known as **Gauss** or **Argand plane**) where the abscissa represents the real part (a) and the ordinate represents the coefficient (b) of the imaginary part of $z = a + ib$. A complex number can then be represented as a vector in the Argand 2D space. Correspondingly, also algebraic operations on complex numbers have a useful geometric representation
   a. The *sum* (z_1+z_2) *of two complex numbers* has the meaning of vector addition, in the sense that the vector (z_1+z_2) has real (a_1+a_2) and imaginary ($b_1 + b_2$) components given by the sums of the respective components of vectors $z_1=a_1+ib_1$ and $z_2=a_2+ib_2$: $(z_1+z_2)=(a_1+a_2)+i(b_1+b_2)$.
   b. The *product* (z_1z_2) of two complex numbers is a vector with modulus $\rho=\rho_1 \cdot \rho_2$, with $\rho_1 = |z_1|$ and $\rho_2 = |z_2|$ and argument $\vartheta=\vartheta_1+\vartheta_2$, with ϑ_1 =arg(z_1) and ϑ_2 =arg(z_2): when multiplied by z_2, the vector z_1 is then *expanded* (or *contracted*) by ρ_2 and *rotated* by ϑ_2.
   c. In the Argand plane, the *complex conjugate* ($z^*=a-ib$) of $z=a+ib$ has the same abscissa, but opposite ordinate with respect to z, *i.e.*, it is *rotated* around the real axis.

We often deal with complex numbers z with $\rho=1$. While there are only two *real* numbers with $\rho=1$ ($z=+1$ and $z=-1$), an *infinity* of *complex numbers* with $|z| = \rho = 1$ exists. These are also known as *phase factors*, are analytically represented by $e^{i\vartheta}$ (ϑ is any real number) and correspond to the points on the circumference of radius = 1 and centre on the origin of the Argand plane.

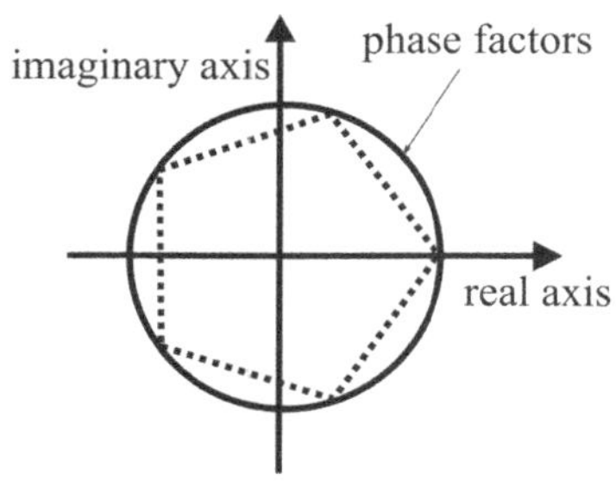

FIGURE 1.2 The Argand plane and the phase factors; the vertices of the dotted pentagon are the fifth roots of unity.

Among the phase factors, there are all the nth roots of 1. These are geometrically identified by the vertices of the regular polygon with n sides and radius = 1 if the polygon is rotated so that one vertex is on the real axis at $z = 1$. For instance, the vertices of the dotted pentagon of Figure 1.2 mark the five fifth roots of unity. Analogously, the four fourth roots of unity are 1, i, −1 and $-i$: $1^4 = i^4 = (-1)^4 = (-i)^4 = 1$.

2.2.3 Normalizing the Wavefunction

A wavefunction of QM must comply with the **normalization condition**:

$$\int_{\Omega} \varphi(q,t) \cdot \varphi^*(q,t) \cdot \mathrm{d}q = \int_{\Omega} |\varphi(q,t)|^2 \cdot \mathrm{d}q = 1 \qquad (2.1)$$

With one, two, three, … n variables, we obviously have simple, double, triple, … integrals: in Eq. (2.1), q is the set of all coordinates, while dq is the corresponding *volume element*. For example, $\mathrm{d}q = \mathrm{d}x\mathrm{d}y\mathrm{d}z$ if we are dealing with ordinary 3D space, or $\mathrm{d}q = \mathrm{d}x_1\mathrm{d}y_1\mathrm{d}z_1\mathrm{d}x_2\mathrm{d}y_2\mathrm{d}z_2$ for a system of two particles and Cartesian coordinates. Ω is the whole configuration space. **In cases other than Cartesian coordinates**, the volume element is *not* necessarily a simple product as dxdydz. For example, if we use spherical polar coordinates instead of Cartesian coordinates for a 3D problem:

> from a function $\varphi(x, y, z, t)$ we must switch to a *different function* $\chi(\rho, \vartheta, \varphi, t)$ where ρ is the radial coordinate (running from 0 to +∞), ϑ is the angle (0 to $+\pi$) between the radial line through the point and the z axis (0 to $+\pi$), and φ is the longitude (0 to $+2\pi$);
>
> (a) the integral: $\int_{-\infty}^{+\infty}\int_{-\infty}^{+\infty}\int_{-\infty}^{+\infty} |\varphi(x,y,z,t)|^2 \cdot \mathrm{d}x\mathrm{d}y\mathrm{d}z$ changes
>
> to: $\int_0^{+\infty}\left(\int_0^{\pi}\left(\int_0^{2\pi}|(\rho,\vartheta,\varphi)|^2 \cdot \rho^2 \cdot \sin\vartheta \cdot \mathrm{d}\varphi\right)\mathrm{d}\vartheta\right)\mathrm{d}\rho$,
>
> where $\rho^2 \cdot \sin\vartheta$ is due to the transformation from Cartesian coordinates to polar coordinates $(x, y, z) \to (\rho, \vartheta, \varphi)$. Technically, it is the *Jacobian* of the coordinate transformation.

Two aspects of the condition are better considered separately:

a. The **integral** in Eq. (2.1) **exists and is finite**: this means that the squared modulus of the wavefunction *can be integrated* over the whole configuration space, or – in other words – it is meaningful and possible to calculate the integral.
b. The value of the **integral is 1**.

The **important aspect is the first one** and it must be clear that this is not a **trivial property**.

The integration domain is *typically* infinite. In the case of only one variable (x), examples directly show that there are functions $\varphi(x)$ that cannot be integrated over the interval from −∞ to +∞. Obviously, the function must approach zero as x tends to infinity ($x \to -\infty$ and $x \to +\infty$), but this is not enough, if the trend is not *fast enough. The request for integrability strongly constrains the analytic forms of the wavefunction of a QM problem*. For example, this is why the radial part of the hydrogenic atomic orbitals (1s, 2s, 2p, …, 3d, …, 4f, …, 5g, …), possibly after some oscillations, approaches zero with an exponential trend (e^{-r}) as r tends to infinity.

If the integral of the square modulus exists, we say that the function is *square-summable* or that it can be *normalized*. We also see that the integral *must be a positive real number*: this comes from the fact that the integrand is a non-negative real function. While not strictly required, a function that is not square-summable generally is not considered as a possible wavefunction. However, it is not necessary to discuss this topic because it has scarce interest in this presentation of QM.

The second point (integral = 1) is less important as it can always be solved. In fact, after having obtained a solution $\varphi(q,t)$ that is square-summable, its normalization integral can be calculated, and a positive real value is obtained:

$$\int_{\Omega} \varphi(q,t) \cdot \varphi^*(q,t) \cdot \mathrm{d}q = a$$

Then, if $a \neq 1$, we define a new function $\psi(q,t) \overset{\text{def}}{=} \frac{\varphi(q,t)}{\sqrt{a}}$, thus obtaining:

$$\int_{\Omega} \psi(q,t) \cdot \psi^*(q,t) \cdot \mathrm{d}q$$

$$= \int_{\Omega} \frac{1}{\sqrt{a}}\varphi(q,t)\frac{1}{\sqrt{a}}\varphi^*(q,t) \cdot \mathrm{d}q = a\frac{1}{\sqrt{a}}\frac{1}{\sqrt{a}} = 1$$

This is known as the **normalization of the wavefunction**.

Some results from the following postulates must be here disclosed to clarify *why* QM is required to normalize a wavefunction. Briefly, when considering the different procedures that can be done on a wavefunction to obtain physically significant quantities, we will see that:

a. in some cases (discussed when dealing with the 5th postulate) these procedures provide numerical values that do **not** depend on the scale factor of the wavefunction,
b. in other cases (discussed when dealing with 2nd and 3rd postulates) these procedures provide numerical values that **are proportional** to the square of the scale factor of the wavefunction, so that the scale factor must then be specified to get the result.

Essentially, normalized wavefunctions are required for completeness of the theory. Obviously, in the Born interpretation, the normalization procedure appears in a more direct way as the tool for obtaining a value = 1 for the integral of the probability density over the whole configuration space. From a practical point of view, we must deal with the normalization procedure because the methods of solving the Schrödinger equation usually provide non-normalized functions. When a wavefunction is obtained in a not (yet) normalized form, sometimes it can be used as it is, sometimes it must be used only after *normalization*: these are precisely the cases (a) and (b) discussed above.

There is another ambiguity concerning the analytic form of a wavefunction. Let us replace in the previous example the normalizing factor $1/\sqrt{a}$ with its opposite, $-1/\sqrt{a}$: then the values of the physical quantities we would obtain from the wavefunction wave according to the second and fifth postulates would not change. *The sign of the wavefunction does not matter.* Even more, given that in general the wavefunction is a complex quantity, *a wavefunction (even after normalization) is specified except for a multiplication factor of modulus* = 1, *i.e.*, in the form $\alpha \equiv e^{i\theta}$ where ϑ is any real number (see Section 2.3.2 for a deeper discussion). As said, these are the phase factors graphically placed in the Argand plane on the circumference centred at the origin and with radius = 1 (see Figure 1.2).

Coming back to the discussion on the real or complex form of a wavefunction, we can now understand that if a state can be described by a wavefunction with a real space-dependent part, it can also be described in an equivalent manner by a wavefunction with a complex space-dependent part: to make clear the point, just multiply the original function by a suitable phase factor, for example *i*. The argument also explains the fact - already said - that the same state can be represented by different wavefunctions.

Finally, among the solutions of the equation of motion of QM (Schrödinger equation) there is always **the function that is everywhere = 0**. This function *cannot be normalized* because the integrand is zero and it is obviously impossible to find a scale factor capable of producing a result other than zero. For this reason, the function that is zero everywhere is **explicitly excluded** from the set of wavefunctions that have *physical meaning*. Again, Born interpretation is particularly effective: the everywhere null function is not a possible wavefunction for a QM problem, because it would give a probability density everywhere = 0 and a null probability of finding somewhere the system.

2.3 Postulate II: Expectation Values

Postulate II
The expectation value A_ψ of a measurable quantity A in the state $\psi(q,t)$
is given by $\langle A \rangle_\psi = \int\limits_\Omega \psi^*(q,t) \cdot \left[\hat{A}\psi(q,t)\right] \cdot dq$
where $\hat{A}$ is a **linear Hermitian** operator.

The second postulate regards **the mean value of a physical quantity**, *and to understand the meaning of this postulate we need to specify what is meant by* **expectation value**, **measurable quantity** *and* **linear Hermitian** *operator.*

The point must be discussed step by step. Let us start by considering a mechanical problem. For instance, we can consider the "particle in a box" problem, or the hydrogen atom, or the motion of the electrons in the CH_4 molecule with the nuclei in a fixed position, or the motion of electrons and nuclei in the benzene molecule, etc....

For that problem, let us consider a particular state, described by wavefunction $\psi(q,t)$. For instance, we can consider the 1s or the $2p_x$ state of the hydrogen atom, ...

Let us finally consider a particular physical quantity (*A*) that can be measured: in QM it is said to be an observable. For instance, we can consider the total energy, the angular moment, ...

Then, the second postulate allows us to calculate the value of the quantity *A* in the state $\psi(q,t)$ of the system. More specifically, remembering the intrinsically probabilistic character of QM, the value given by the second postulate is the mean value of the quantity *A* in the state $\psi(q,t)$. It is known as the expectation value of the *A* observable in the $\psi(q,t)$ state and is written $\langle A_\psi \rangle$ where the angular parenthesis and the subscript respectively denote the mean value and the state.

This value is obtained as follows.

a. In QM, each observable (*A*) is assigned a **Hermitian operator** ($\hat{A}$). Meaning and properties of *operators* are dealt with later (Section 2.3.1): meanwhile it is enough to say that an operator is a *rule* that can be applied to a QM wavefunction $\psi(q,t)$ and produces a function $\chi(q,t)$. We write $\hat{A}\psi(q,t) = \chi(q,t)$. Generally, but not necessarily, $\psi(q,t) \neq \chi(q,t)$.
b. Then, $\hat{A}\psi(q,t)$ is multiplied by the complex conjugate $\psi^*(q,t)$ of the original function.
c. The final step is to integrate the product $\psi^*(q,t) \cdot \chi(q,t) = \psi^*(q,t) \cdot \hat{A}\psi(q,t)$ over the whole configuration space (Ω).

The second postulate *identifies the value of the integral with the expectation value*:

$$\langle A\rangle_\psi = \int_\Omega \psi^*(q,t)\cdot\left[\hat{A}\psi(q,t)\right]\cdot \mathrm{d}q \tag{2.2}$$

Several logical steps are needed also to clarify what the experimental counterpart of the result of Eq. (2.2) is.

1. We start by experimentally preparing the system in the state corresponding to $\psi(q,t)$.
2. Then, we measure the physical quantity A. The result can be equal to the expectation value but can also be different: the second postulate does not provide this information.
3. We repeat the measurement several times *always starting with the system in the same state* $\psi(q,t)$. So, we obtain several results $a_1, a_2, a_3, a_4, a_5, \ldots a_n$, that can be all different from each other, or only partially different, or else all equal.
4. Ideally making an infinite series of measurements, the mean value of these measurements equals the expectation value

$$\langle A\rangle_\psi = \lim_{n\to\infty}\left(\sum_{i=1}^{n} a_n\right)/n \tag{2.3}$$

The expectation value of a measurable quantity corresponds to the limit (for an infinite number of experiments) of the average of the measurements.

An essential point is that, after each measurement, the system must be put again in state $\psi(q,t)$ because, in QM, **a measurement changes the state of the system**. As said, this is another fundamental difference between CM and QM and will be discussed in more detail when dealing with the fifth postulate (see Section 2.6). Note that the second postulate does *not* say

- how to write the operator of the physical quantity A (this is the content of third postulate);
- what are the values obtained in the measurements (this is the content of fifth postulate).

2.3.1 Operators

Let us now clarify the technical terms *operator*, *linear* operator and *Hermitian* operator.

An operator is a particular case of the set-theoretic concept of **map**.

In set theory, a **map** R is a rule that assigns to every element (x, y, z, …) of a set X an element (a, b, c, …) of a set A: $R(x)=a$. In other words, (i) every element of X corresponds to an element of A (no element of X is left without its corresponding element in A); (ii) the corresponding element is unique. The set X is called the *domain* of the map, the set A is called the *codomain* of the map, the set of all elements of A that are related to an element of X is called the *image* of the map.

The reader is surely familiar with some typical cases that are known with specific names when domain and codomain are sets with particular properties as shown in the table below.

Map	Domain	Codomain
Sequence	N (integers)	R (real numbers)
Function	C (real or complex numbers)	C (real or complex numbers)
Operator	Function	Function
Functional	Function	C (real or complex numbers)

We speak of an **operator** when the **domain X is a set of functions** (f) of one or more variables (q), and **the codomain A is also a set of functions** (g): then we write $g(q)=\hat{R}\left[f(q)\right]$. In other words, an operator transforms a function (taken from a prefixed set of functions, for example functions of one, or two, … variables; continuous, differentiable, integrable functions; functions which are solutions of a particular differential equation, …) into another function, generally different and belonging to an equal or different set of functions.

The language may seem cumbersome, but there are clearly no difficulties in the idea of operator. For example, "to take the derivative with respect to q" of a function of variable q, means "to apply the differentiation operator d/dq" (or possibly the partial differentiation operator) to $f(q)$:

$$g(q)=\hat{O}\left[f(q)\right]=\frac{\mathrm{d}f(q)}{\mathrm{d}q};\quad \hat{O}\overset{def}{=}\frac{\mathrm{d}}{\mathrm{d}q}$$

Surely, many operators are already well known. Examples of operators are

- multiplying the function by a constant, by a variable or by a given function, …
- taking the square of the function, or the squared modulus, …
- taking the square root of the function, …
- … as well as all the various combinations of these operations and many others that need not be written here.

QM uses operators with two properties, namely being **linear** (always) and being **Hermitian** (in most cases). The development of the theoretical apparatus of QM essentially concerns an in-depth discussion of these properties. This is what many textbooks expressly dedicated to Quantum Chemistry typically report. On the other hand, most Physical Chemistry textbooks introduce the basic aspects of QM without explicit reference to these properties.

We here prefer to follow an intermediate line of presentation by clarifying the meaning of linear operators and Hermitian operators using a style as discursive as possible. The aim is, first, to make familiar a language that is consolidated in the area and therefore it is useful to know even without necessarily deriving in detail all its implications. Another aim is to serve as an introduction to following studies. Finally, some knowledge of the subject is also useful to obtain important results (see, for example, Section 2.3.5).

2.3.2 Linear Operators

Going back to the example of the differentiation operation, we note that

- the derivative of a sum of functions ($f+g$) is the sum of the derivatives and that
- the derivative of the product of a function (f) by a constant (k) is the product of the constant by the derivative of the function.

So, for any pair of functions (f and g) and for any number k

$$\hat{O}[f+g]=\hat{O}[f]+\hat{O}[g] \text{ and } \hat{O}[k\cdot f]=k\cdot\hat{O}[f]\left(\hat{O}\overset{def}{=}\frac{\mathrm{d}}{\mathrm{d}q}\right)$$

These are *not trivial* properties and therefore deserve a specific name: the operator is a *linear operator.* It is easy to see that the property of linearity is frequently encountered: many of the operators that we know have this property, but there are certainly many usual operators that are *not* linear. For example, the square root and square operators are not linear: the square root of a sum is not the sum of the square roots, and the square of a sum is not the sum of the squares.

The linearity property is often written by saying that an operator $\hat{R}$ is linear if (and only if) for any pair of functions (f and g) in the domain of the operator and for any pair of numbers (a and b)

$$\hat{R}[a\cdot f+b\cdot g]=a\cdot\hat{R}[f]+b\cdot\hat{R}[f] \tag{2.4}$$

Note that the numbers (a, b) we are talking about here are complex numbers (and that the functions are complex).

It is an essential part of the theory of QM (of the second postulate) that

QM uses only linear operators.

In this regard it may be simply said that **QM is a linear theory**. We will come back to this aspect later (Section 5.1).

N. B: By looking at the expression of the expectation values (Eq. 2.2) and using the linearity property of the operators representing observables, one can now understand why it was said (with regard to the first postulate, see Section 2.2.3) that *a (normalized) wave function is specified up to a numerical multiplicative factor of modulus = 1.*

Indeed, the value of the integral of the expectation value does not change by multiplying a wave function (ψ) by any complex number with modulus = 1. Although cumbersome, it is useful to follow all the steps in detail to make this property evident. If: $\langle A\rangle_\psi=\int_\Omega \psi^*(q,t)\cdot\left[\hat{A}\psi(q,t)\right]\cdot\mathrm{d}q$ and if we consider a second wave function, equal to the first one multiplied by a complex number (a), the integral becomes $\int_\Omega a^*\cdot\psi^*(q,t)\cdot\left\{\hat{A}\left[a\cdot\psi(q,t)\right]\right\}\cdot\mathrm{d}q$ (the conjugate of a product is the product of the conjugates). The fact that *the operator is linear* allows us to write $\hat{A}\left[a\cdot\psi(q,t)\right]=a\cdot\hat{A}\left[\psi(q,t)\right]$ and we obtain: $\int_\Omega a^*\cdot\psi^*(q,t)\cdot\left\{a\cdot\hat{A}\left[\psi(q,t)\right]\right\}\cdot\mathrm{d}q$. The constants can be taken outside the integral so that we have: $(a^*\cdot a)\cdot\int_\Omega \psi^*(q,t)\cdot\left\{\hat{A}\left[\psi(q,t)\right]\right\}\cdot\mathrm{d}q$. Now, if the complex number a has modulus = 1, the product outside the integral is $(a^* a)=1$ and therefore the expectation value calculated with the new wave function is equal to that calculated with the old wave function.

2.3.3 Hermitian Operators

The second property that QM operators can have (but some important operators do not) is Hermiticity. This is not a simple and familiar property like linearity and requires an introduction. The second postulate tells us how to obtain the mean (statistical) value of an **observable** (*i.e.* of a measurable quantity). Then, the operation must always yield a real number. More precisely, the theory must be formulated in such a way that

- whatever the QM problem we are discussing,
- whatever the wavefunction (if it is an acceptable wavefunction, *i.e.* if it has all the properties required by the first and following fourth postulates),
- whatever the observable A we are interested in,

the integral

$$\langle A\rangle_\psi=\int_\Omega \psi^*(q,t)\cdot\left[\hat{A}\psi(q,t)\right]\cdot\mathrm{d}q \quad \text{must yield a } \textbf{real \textit{value}} \tag{2.5}$$

It is obvious that this request, because of its generality, must translate into a sort of constraint on the analytical form of an operator, but it is not easy to understand what this constraint is in detail. Let us get to the point step by step by first considering a linear operator ($\hat{A}$), not necessarily corresponding to an observable. Let us then consider another operator (also linear), abstractly defined as the operator that provides as expectation value a number that is the complex conjugate of the number that is provided by $\hat{A}$ for the same QM problem and for the same state. To make clear how the operators are related, let us indicate the second operator with the same letter and adding a dagger: $\hat{A}^\dagger$ and call it the **adjoint** or *Hermitian conjugate* of $\hat{A}$.

Thus, the definition of $\hat{A}^\dagger$ is:

$$\left[\int_\Omega \psi^*(q,t)\cdot\left[\hat{A}^\dagger\psi(q,t)\right]\cdot\mathrm{d}q\right]=\left[\int_\Omega \psi^*(q,t)\cdot\left[\hat{A}\psi(q,t)\right]\cdot\mathrm{d}q\right]^* \tag{2.6}$$

whatever the QM problem we are discussing, whatever the wavefunction, and whatever the *linear* operator $\hat{A}$.

Finally, the original request, that the mean value of an observable is always a real number, corresponds to the request that *an operator corresponding to an observable equals its adjoint*:

$$\hat{A}^{\dagger} = \hat{A}. \tag{2.5}$$

This surely gives a real value of the integral in Eq. (2.7)

An operator that is equal to its adjoint is simply called **Hermitian**. In conclusion, we say that

an **observable** corresponds to a **Hermitian** operator

As already said, the QM operators are *not* necessarily Hermitian. There are QM operators for which we do not *require* the property of being Hermitian because they do *not* correspond to observables. Important examples are the symmetry operators. Other non-Hermitian operators are used as tools to derive some important results: for examples on this regard, see Section 7.3.

Being Hermitian is a property with very rich consequences on the algebraic level and therefore it is an important tool for the axiomatic development of QM theory (which is *not* presented here, as already said). A particular difficulty in understanding these aspects is surely the abstraction. The definition does not say how to construct the Hermitian conjugate of an operator: we are indeed talking about operators without having yet specified what analytical form they have, because this is specified by the third postulate. Instead, if the analytical form is known, the definition can be applied and the property of an operator to be Hermitian can be verified directly in a surprisingly simple way.

Consider, for example, a mechanical problem described by a single coordinate (q) and consider $\hat{q}$ as the operator consisting of "multiplying what comes after" by the independent variable q. As we will see when discussing the third postulate, this is, in fact, an operator that corresponds to an observable. Let us therefore evaluate

$$\int_{\Omega} \varphi^{*}(q) \cdot \hat{q} \cdot \varphi(q) \cdot \mathrm{d}q$$

It is immediately clear that the integrand is real (since the product of $\varphi^*\varphi$ is real for any point q in the configuration space and the same is true for the product $\varphi^* q \varphi$, since q is real and the integral must also be real. It then follows that the *operator* $\hat{q}$ is equal to its adjoint, because it provides a real result. Note that we have not made any assumptions about the function φ: the proof is valid whatever it is and therefore it is *a property of the operator.*

It is not even very difficult to verify that an operator is *not* Hermitian. Let us now consider the derivative operator and then the integral:

$$\int_{\Omega} \varphi^{*}(q) \cdot \frac{\partial \varphi(q)}{\partial q} \cdot \mathrm{d}q$$

Assuming that all the necessary conditions are satisfied, we integrate by parts and indicate with a, b the extremes of the interval Ω, obtaining:

$$\int_{\Omega} \varphi^{*}(q) \cdot \frac{\partial \varphi(q)}{\partial q} \cdot \mathrm{d}q = \left[\varphi^{*} \cdot \varphi\right]_{b}^{a} - \int_{\Omega} \varphi(q) \cdot \frac{\partial \varphi^{*}(q)}{\partial q} \cdot \mathrm{d}q$$

Let us now assume that, as is frequently found, the wavefunctions cancel at the extremes of the intervals of definition. Since we are only giving an example, we can safely accept this limitation, which is frequently true and has the advantage of making things simpler. In this hypothesis, the first term to the right cancels and therefore:

$$\int_{\Omega} \varphi^{*}(q) \cdot \frac{\partial \varphi(q)}{\partial q} \cdot \mathrm{d}q = -\int_{\Omega} \varphi(q) \cdot \frac{\partial \varphi^{*}(q)}{\partial q} \cdot \mathrm{d}q$$
$$= \left[-\int_{\Omega} \varphi^{*}(q) \cdot \frac{\partial \varphi(q)}{\partial q} \cdot \mathrm{d}q\right]^{*}$$

It follows that (whatever the function φ) the operator $-\mathrm{d}/\mathrm{d}q$ gives as expectation value the complex conjugate of the value given by the operator $\mathrm{d}/\mathrm{d}q$. We then say that $(\mathrm{d}/\mathrm{d}q)^{\dagger} = -\mathrm{d}/\mathrm{d}q$. In fact, we can anticipate that, according to the third postulate, this operator *does not* correspond to an observable.

Continuing with the same example, we see how it is possible to obtain a Hermitian operator. Let us consider not the simple derivative operator, but the "derivative multiplied by $-i$" operator. Then, from the previous relation.

$$\int_{\Omega} \varphi^{*}(q) \cdot \frac{\partial \varphi(q)}{\partial q} \cdot \mathrm{d}q = -\int_{\Omega} \varphi(q) \cdot \frac{\partial \varphi^{*}(q)}{\partial q} \cdot \mathrm{d}q$$

we obtain:

$$\int_{\Omega} \varphi^{*}(q) \cdot (-i) \cdot \frac{\partial \varphi(q)}{\partial q} \cdot \mathrm{d}q = i \cdot \left[\int_{\Omega} \varphi(q) \cdot \frac{\partial \varphi^{*}(q)}{\partial q} \cdot \mathrm{d}q\right].$$

Developing further we get:

$$i \cdot \int_{\Omega} \varphi(q) \cdot \frac{\partial \varphi^{*}(q)}{\partial q} \cdot \mathrm{d}q = (-i)^{*} \cdot \left[\int_{\Omega} \varphi^{*}(q) \cdot \frac{\partial \varphi(q)}{\partial q} \cdot \mathrm{d}q\right]^{*}$$

and finally:

$$\int_{\Omega} \varphi^{*}(q) \cdot (-i) \cdot \frac{\partial \varphi(q)}{\partial q} \cdot \mathrm{d}q =$$
$$(-i)^{*} \cdot \left[\int_{\Omega} \varphi^{*}(q) \cdot \frac{\partial \varphi(q)}{\partial q} \cdot \mathrm{d}q\right]^{*} = \tag{2.8}$$
$$= \left[\int_{\Omega} \varphi^{*}(q) \cdot (-i) \cdot \frac{\partial \varphi(q)}{\partial q} \cdot \mathrm{d}q\right]^{*} \tag{2.9}$$

The comparison between Eqs. (2.8) and (2.9) shows that the operator "derivative multiplied by $-i$", gives a result equal to its conjugate and therefore a real result: the operator is Hermitian, and by discussing the third postulate we will see that it actually corresponds to an observable.

2.3.4 Bra-ket Notation

A notation introduced by P. A. M. Dirac is frequently used instead of Eq. (2.2):

$$\langle A\rangle_{\psi}=\left[\int_{\Omega}\psi^{*}(q,t)\cdot\left[\hat{A}\psi(q,t)\right]\cdot\mathrm{d}q\right]\overset{def}{=}\left\langle\psi\middle|\hat{A}\middle|\psi\right\rangle \quad (2.10)$$

and is known as *bra-ket notation*. The conciseness of the notation is well evident, as it does not require writing everything that must be present and is implicit (the integral, the integration variables, the configuration space, the complex conjugate sign of the antecedent function).

The notation also applies when the first position (the *bra*) contains a function other than the third position (the *ket*):

$$\left\langle\varphi\middle|\hat{A}\middle|\psi\right\rangle\overset{def}{=}\left[\int_{\Omega}\varphi^{*}(q,t)\cdot\left[\hat{A}\psi(q,t)\right]\cdot\mathrm{d}q\right] \quad (2.11)$$

Finally, when no operator appears, we have:

$$\langle\varphi|\psi\rangle\overset{def}{=}\left[\int_{\Omega}\varphi^{*}(q,t)\cdot\left[\psi(q,t)\right]\cdot\mathrm{d}q\right] \quad (2.12)$$

When dealing with functions taken from a list, instead of $\left\langle\varphi_i\middle|\hat{A}\middle|\varphi_j\right\rangle$ we write: $\left\langle i\middle|\hat{A}\middle|j\right\rangle$.

The notation is also used to denote a single wavefunction: for example, $|\varphi\rangle$, or $|i\rangle$ to denote φ_i. Moreover, a ket can denote a function using a complete set (see Section 3.3.1) of eigenvalues or quantum numbers. For example, the well-known wavefunctions of the hydrogen atom that are usually written $\varphi_{n,l,m_l}(\rho,\vartheta,\varphi)$, in Dirac notation are concisely denoted as $|n,l,m_l\rangle$.

2.3.5 The Turnover Rule

A relation holds that plays a fundamental role in the development of the theory:

$$\left\langle\varphi\middle|\hat{A}\middle|\psi\right\rangle=\left[\left\langle\psi\middle|\hat{A}^{\dagger}\middle|\varphi\right\rangle\right]^{*} \quad (2.13)$$

The rule essentially says that the following three operations

- exchange of the bra with the ket, *i.e.* φ with ψ,
- exchange of the operator with its Hermitian conjugate, *i.e.* $\hat{A}$ with $\hat{A}^{\dagger}$,
- exchange of the complex number [...] on the left with its conjugate [...]*,

leave the result unchanged provided that *all three* are performed. The proof is reported at the end of the chapter (see Section 7.1) mainly to show that many derivations of the basic theory of QM are actually much easier than one might imagine, even if they are long and boring.

When the operator is Hermitian, it is evidently not necessary to perform the second exchange, and the rule becomes more simply:

$$\left\langle\varphi\middle|\hat{A}\middle|\psi\right\rangle=\left[\left\langle\psi\middle|\hat{A}\middle|\varphi\right\rangle\right]^{*},\ \text{if}\ \hat{A}=\hat{A}^{\dagger} \quad (2.14)$$

2.4 Postulate III: QM Operators

Postulate III
The observables of QM are obtained
from the quantities of CM in its Hamiltonian
formulation using **correspondence rules**

2.4.1 The Correspondence Rules

Explicit reference is made to the **Hamiltonian formulation** of **CM,** which uses two sets of independent variables conjugated to each other: coordinates and moments.

The QM operator corresponding to a coordinate is the coordinate itself, meaning that the coordinate multiplies the wavefunction: $\hat{q}f(q,t)\overset{\text{def}}{=}q\cdot f(q,t)$. For problems with two or more coordinates: $(q_1,q_2,\ldots,q_n)$: $\widehat{q_r}f(q_1,q_2,\ldots,q_n,t)\overset{\text{def}}{=}q_r\cdot f(q_1,q_2,\ldots,q_n,t)$ where r indicates one of the n coordinates.

The QM operator corresponding to a moment is the derivative with respect to the conjugate coordinate multiplied by $-i\hbar$ (as usual, i is the imaginary unit and $\hbar$ is the reduced Planck constant, *i.e.* h divided by 2π): $\widehat{p_s}f(q_s,t)\overset{\text{def}}{=}-i\cdot\hbar\cdot\dfrac{\partial f(q_s,t)}{\partial q_s}$.

For problems with two or more coordinates $(q_1,q_2,\ldots,q_n)$: $\widehat{p_s}f(q_1,q_2,\ldots,q_n,t)\overset{\text{def}}{=}-i\cdot\hbar\cdot\left(\dfrac{\partial f(q_1,q_2,\ldots,q_n,t)}{\partial q_s}\right)_{q_{j\neq s}}$ where the index s specifies one of the n moments and its conjugate coordinate.

If the observable is not a moment or a coordinate, but a composite quantity, it must be first expressed as a function of coordinates and moments, according to Hamilton formulation of the problem in CM, and then the corresponding operator is obtained as follows.

- The *multiplicative constants* remain as they are.
- The *sum of two quantities* becomes the *sum operator*, that is, the operator that (when applied to a function) produces the result that would be obtained by applying the two operators and then adding the results. If $\hat{C}=\hat{A}+\hat{B}$, then $\hat{C}f(x,t)\overset{\text{def}}{=}\hat{A}f(x,t)+\hat{B}f(x,t)$
- The *product of two quantities* corresponds to the sequential application of the two operators. In

other words, the *product operator* of $\hat{A}$ and $\hat{B}$ is defined by: $\hat{C}f(x,t) \stackrel{\text{def}}{=} \hat{B}\left[\hat{A}f(x,t)\right]$. Note that we apply $\hat{A}$ first and then $\hat{B}$ and that the order in which we apply the operators is important (see note 4 below and more details in Section 3.3).

When one wants to use **coordinates other than the Cartesian coordinates** of the individual particles (*i.e.* the coordinates of Newton law), it is also possible (and sometimes more convenient) to write the operators in Newton coordinates and derive the expression in the generalized coordinates using the usual rules of mathematical analysis (see note 2 below).

2.4.2 Examples and Comments

N. B. 1: The kinetic energy operator $\hat{T}$ for a 3D motion of a particle is an example that shows various rules. In the Hamiltonian formulation of the CM, the kinetic energy must be written: $T = \frac{1}{2m}\left(p_x^2 + p_y^2 + p_z^2\right)$, where p_x, p_y and p_z are the conjugate moments of the Cartesian coordinates. In QM we have:

$$\hat{T} = \frac{1}{2m}\left\{-i\hbar\frac{\partial}{\partial x}\left(-i\hbar\frac{\partial}{\partial x}\right) - i\hbar\frac{\partial}{\partial y}\left(-i\hbar\frac{\partial}{\partial y}\right) - i\hbar\frac{\partial}{\partial z}\left(-i\hbar\frac{\partial}{\partial z}\right)\right\}$$

Recall that $i^2 = -1$ and therefore $(-i\hbar)^2 = -\hbar^2$. Products such as p^2 correspond to second derivatives, since the "product" of two derivatives means performing one derivative after the other.

$$\text{Then}: \hat{T} = \frac{1}{2m}\left\{-\hbar^2\frac{\partial^2}{\partial x^2} - \hbar^2\frac{\partial^2}{\partial y^2} - \hbar^2\frac{\partial^2}{\partial z^2}\right\}$$
$$= \frac{-\hbar^2}{2m}\left\{\frac{\partial^2}{\partial x^2} + \frac{\partial^2}{\partial y^2} + \frac{\partial^2}{\partial z^2}\right\} = \frac{-\hbar^2}{2m}\nabla^2$$

where ∇^2 is the Laplacian operator. The meaning becomes perhaps clearer by writing the function (χ) to which the operator is applied:

$$\hat{T}\chi(x,y,z,t) = \frac{-\hbar^2}{2m}\left\{\frac{\partial^2\chi}{\partial x^2} + \frac{\partial^2\chi}{\partial y^2} + \frac{\partial^2\chi}{\partial z^2}\right\}$$
$$= \frac{-\hbar^2}{2m}\nabla^2\chi(x,y,z,t)$$

N. B. 2: Let us see how to go from Newton coordinates to generalized coordinates (see Section 1.3). For example, if we want to use spherical coordinates, we first write the new coordinates as functions of the old ones:

$$r = \sqrt{x^2 + y^2 + z^2}; \quad \theta = \arccos\left(\frac{z}{\sqrt{x^2 + y^2 + z^2}}\right); \quad \varphi = \arccos\left(\frac{x}{\sqrt{x^2 + y^2}}\right) \tag{2.15}$$

Then, the expressions of the operators in spherical coordinates can be obtained using the usual rules of transformation from one system of independent variables to another.

For example, $\frac{\partial\chi}{\partial x} = \frac{\partial\chi}{\partial r}\frac{\partial r}{\partial x} + \frac{\partial\chi}{\partial\theta}\frac{\partial\theta}{\partial x} + \frac{\partial\chi}{\partial\varphi}\frac{\partial\varphi}{\partial x}$, so that $\frac{\partial\chi}{\partial x}$ can be replaced by $\frac{\partial\chi}{\partial r}, \frac{\partial\chi}{\partial\theta}, \frac{\partial\chi}{\partial\varphi}$ multiplied by the derivatives $\frac{\partial r}{\partial x}, \frac{\partial\theta}{\partial x}, \frac{\partial\varphi}{\partial x}$ obtained from Eq. (2.15).

With long but trivial calculations (see Exercise 1.10) we obtain the Laplacian operator ∇^2:

$$\nabla^2 = \frac{\partial^2}{\partial x^2} + \frac{\partial^2}{\partial y^2} + \frac{\partial^2}{\partial z^2} = \frac{1}{r^2}\frac{\partial}{\partial r}\left(r^2\frac{\partial}{\partial r}\right) + \frac{\hat{\Lambda}}{r^2}, \tag{2.16}$$

$$\hat{\Lambda} = \frac{1}{\sin\vartheta}\left[\frac{\partial}{\partial\vartheta}\left(\sin\vartheta\frac{\partial}{\partial\vartheta}\right) + \frac{1}{\sin\vartheta}\frac{\partial^2}{\partial\varphi^2}\right] \tag{2.17}$$

From Eqs. (2.16) and (2.17) we obtain the expressions in spherical coordinates:

$$\hat{T} = -\frac{\hbar^2}{2m}\nabla^2 = -\frac{\hbar^2}{2m}\left\{\frac{1}{r^2}\frac{\partial}{\partial r}\left(r^2\frac{\partial}{\partial r}\right) + \frac{1}{r^2}\frac{1}{\sin\vartheta}\left[\frac{\partial}{\partial\vartheta}\left(\sin\vartheta\frac{\partial}{\partial\vartheta}\right) + \frac{1}{\sin\vartheta}\frac{\partial^2}{\partial\varphi^2}\right]\right\} \tag{2.18}$$

and the square modulus of the angular moment (see Eq. 2.24):

$$\left|\widehat{L^2}\right| \stackrel{\text{def}}{=} \hbar^2\hat{\Lambda} = \frac{\hbar^2}{\sin\vartheta}\left[\frac{\partial}{\partial\vartheta}\left(\sin\vartheta\frac{\partial}{\partial\vartheta}\right) + \frac{1}{\sin\vartheta}\frac{\partial^2}{\partial\varphi^2}\right] \tag{2.19}$$

Warning: in the evaluation of the integrals (see Section 2.2.3) it is necessary to include the *Jacobian* of the transformation. See also Exercise 1.10.

N. B. 3: It is useful to also report the analytical expressions of the **angular moment** operators that play an important role in the next part. In CM and Cartesian coordinates, the three components of the angular moment L related to the motion of a particle are:

$$L_x \stackrel{\text{def}}{=} y \cdot p_z - z \cdot p_y; \quad L_y \stackrel{\text{def}}{=} z \cdot p_x - x \cdot p_z; \quad L_z \stackrel{\text{def}}{=} x \cdot p_y - y \cdot p_x \tag{2.20}$$

Here x, y, z, p_x, p_y and p_z are the Cartesian components of position and linear momentum. It is convenient to write the three Eq. (2.20) with a concise notation that uses the rule for calculating the determinant of a 3×3 matrix:

$$\vec{L} \stackrel{\text{def}}{=} \det\begin{vmatrix} \vec{i} & \vec{j} & \vec{k} \\ x & y & z \\ p_x & p_y & p_z \end{vmatrix}$$

$$= \left[(y\cdot p_z)-(z\cdot p_y)\right]\cdot\vec{i} + \left[(z\cdot p_x)-(x\cdot p_z)\right]\cdot\vec{j} + \left[(x\cdot p_y)-(y\cdot p_x)\right]\cdot\vec{k} \quad (2.21)$$

The vectors in the first row are the unit vectors along x, y, z and the determinant has been developed according to the usual Sarrus rule:

$$\begin{vmatrix} a & b & c \\ d & e & f \\ g & h & i \end{vmatrix} = aei + bfg + cdh - ceg - bdi - afh$$

For short:

$$L_i \stackrel{\text{def}}{=} j\cdot p_k - k\cdot p_j \quad (2.22)$$

Here, i, j, k denote the triplet of Cartesian components in one of the **cyclic** permutations (x, y, z or y, z, x, or z, x, y, but **not** x, z, y nor z, y, x nor y, x, z).

The correspondence rules give the Cartesian expression of the QM operators:

$$\widehat{L_x} \stackrel{\text{def}}{=} -i\cdot\hbar\cdot\hat{y}\cdot\frac{\partial}{\partial z} + i\cdot\hbar\cdot\hat{z}\cdot\frac{\partial}{\partial y};$$

$$\widehat{L_y} \stackrel{\text{def}}{=} -i\cdot\hbar\cdot\hat{z}\cdot\frac{\partial}{\partial x} + i\cdot\hbar\cdot\hat{x}\cdot\frac{\partial}{\partial z};$$

$$\widehat{L_z} \stackrel{\text{def}}{=} -i\cdot\hbar\cdot\hat{x}\cdot\frac{\partial}{\partial y} + i\cdot\hbar\cdot\hat{y}\cdot\frac{\partial}{\partial x} \quad (2.23)$$

Here too, there is a compact version:

$$\widehat{L_i} \stackrel{\text{def}}{=} -i\cdot\hbar\cdot\hat{j}\cdot\frac{\partial}{\partial k} + i\cdot\hbar\cdot\hat{k}\cdot\frac{\partial}{\partial j}$$

Finally, the operator of the square modulus of the angular momentum is:

$$\left|\widehat{L^2}\right| \stackrel{\text{def}}{=} \left|\widehat{L_x^2}\right| + \left|\widehat{L_y^2}\right| + \left|\widehat{L_z^2}\right| \quad (2.24)$$

In spherical coordinates, we can use the expressions already obtained for the Laplace and the kinetic energy operators (Eqs. 2.16 and 2.18):

$$\left|\widehat{L^2}\right| = \hbar^2\hat{\Lambda};\quad \hat{\Lambda} = \frac{1}{\sin\vartheta}\left[\frac{\partial}{\partial\vartheta}\left(\sin\vartheta\frac{\partial}{\partial\vartheta}\right) + \frac{1}{\sin\vartheta}\frac{\partial^2}{\partial\varphi^2}\right] \quad (2.19)$$

Also, if φ is the spherical coordinate corresponding to the angle ($0 \le \varphi < 2\pi$) of the rotation around the z-axis, (see Eq. 2.15):

$$\widehat{L_z} = -i\cdot\hbar\cdot\frac{\partial}{\partial\varphi} \quad (2.25)$$

The result is consistent with the fact that in CM the conjugate moment of a rotation coordinate (φ) is the component of the angular moment (L) along the rotation axis (z).

The properties of angular momentum operators are discussed more extensively in Section 3.4.

N. B. 4: We need to be careful with the combination of operators when a CM quantity corresponds to the **product of different quantities**. Consider the product of a coordinate and its conjugate moment, for example the product (xp_x). In QM, this becomes the operator $(-i\hbar)x\frac{\partial}{\partial x}$:

$$(-i\hbar)\left(x\frac{\partial}{\partial x}\right)f(x,\ldots,t) = -i\hbar\cdot x\cdot\frac{\partial f(x,\ldots,t)}{\partial x}$$

Let us now consider when the two operations are performed in the **reverse order**:

$$(-i\hbar)\left(\frac{\partial}{\partial x}x\right)f(x,\ldots,t) = -i\hbar\cdot\frac{\partial\left[x\cdot f(x,\ldots,t)\right]}{\partial x}$$

$$= -i\hbar\cdot\left\{f(x,\ldots,t) + x\cdot\frac{\partial f(x,\ldots,t)}{\partial x}\right\}$$

The result is different! (In CM, obviously, there is no difference).

The product of two QM operators (in general) is NOT commutative.

When two operators do *not* commute, it is necessary to distinguish multiplication "on the right" from multiplication "on the left" and to remember that it is not possible to simplify some expressions. For example, in the algebra of operators, if $a=bc$ and $b=ef$, then $a=efc$, but you cannot write $a=ecf$ or $a=cef$ (and neither one of the other three forms in which e, f are in the order fe). Furthermore, if $ab=bc$, then $a=c$ is NOT necessarily true, but you must multiply both expressions by b^{-1} (the inverse of b) from the right or from the left, obtaining $c=b^{-1}ab$ or $a=bcb^{-1}$. The fact that the algebra of operators is non-commutative has been a significant difficulty in the initial development of QM theory. Another category of mathematical objects whose multiplication is non-commutative are matrices: it is well known that one of the two historical formulations of QM (the one due to Heisenberg) is based directly on the matrix expression of QM operators.

Important properties related to the commutation of QM operators are discussed in Section 3.3. When two operators ($\hat{A}$ and $\hat{B}$) do not commute, it makes sense to define their **commutator** $\left[\hat{A},\hat{B}\right]$:

$$\left[\hat{A},\hat{B}\right] \stackrel{\text{def}}{=} \hat{A}\hat{B} - \hat{B}\hat{A}$$

For the relation between commutators and uncertainty relations, see Exercises 1.1–1.4.

2.5 Postulate IV: Schrödinger Equation

Postulate IV
A QM wavefunction must satisfy the

$$\text{Schrödinger equation}: \hat{H}\psi(q,t) = i\cdot\hbar\cdot\frac{\partial\psi(q,t)}{\partial t} \tag{2.26}$$

It is also known as the **time-dependent Schrödinger equation**. $\hat{H}$ denotes the Hamiltonian operator, which is the operator corresponding to the Hamiltonian function of CM, *i.e.* the total energy of the system. As already said, not only in CM, but also in QM (although in different ways) energy is the quantity that drives the temporal evolution of the system.

There is another equation which also goes by the name of Schrödinger:

$$\hat{H}\chi(q,t) = E\cdot\chi(q,t) \tag{2.27}$$

This is the **time-independent Schrödinger equation** and is preferably written as:

$$\hat{H}\varphi(q) = E\cdot\varphi(q) \tag{2.28}$$

to make it clear that it does not depend on time.

The connection between these equations and their solutions (ψ, χ and φ) will be discussed later (see Section 3.1). For now, it is only necessary to remember that between the two Schrödinger equations (Eq. 2.26 and Eq. 2.28), the one that fully plays the role of the *equation of motion of QM* is the first one. The second one can, in certain cases, be used in place of the first one. Also, deriving the solutions of the second one is a step towards deriving the solutions of the first one.

Warning: a wave function of a given QM problem not only is a solution of the pertinent Schrödinger equation, but must also satisfy the other conditions (continuity, differentiability, boundary conditions) and be different from the function that is zero everywhere (see Section 2.2.1).

2.6 Postulate V: Eigenvalues and Eigenvectors

Postulate V
The possible results of a measurement
are obtained by solving an **eigenvalue equation**.

The fifth postulate addresses a problem similar to that addressed by the second postulate. The second postulate gives the expectation value of an observable when the system is in a specified state. The fifth postulate instead tells us that the results (A_i) of a measurement of the same observable and for the same system are given by:

$$\hat{A}\varphi_i(q,t) = A_i\cdot\varphi_i(q,t) \tag{2.29}$$

In Eq. (2.29), φ_i is called the **eigenfunction** and A_i is called the **eigenvalue** of the operator $\hat{A}$. A_i is the eigenvalue corresponding to the eigenfunction φ_i.

N. B. 1: The meaning of Eq. (2.29) is that, given a certain operator, there are functions, *i.e.* elements of the operator domain, on which the effect of the operator is simply to *multiply* them by a *constant.*

The simplest example is given by the derivative operator and exponential functions. The exponential function e^x, as we know, has the characteristic of being equal to its derivative. If we then consider a slightly more complicated exponential as: e^{cx}, (c being a constant) then $\frac{d(e^{cx})}{dx} = c\cdot e^{cx}$ and this can be written as an eigenvalue equation $\hat{A}\varphi = A\cdot\varphi$, with $\hat{A} \overset{\text{def}}{=} \frac{d}{dx}$. The eigenfunction of the operator is $\varphi = e^{cx}$ and the eigenvalue is $A = c$. A function that is NOT an exponential is NOT an eigenfunction of the derivative operator. The eigenvalue equation for the derivative operator therefore identifies an exponential function of the type e^{cx} (including the case $c = 1$) as the solution. Warning: the example is not relevant to QM problems, and the derivative operator is *not* an observable as already said in Section 2.3.3.

N. B. 2: An eigenvalue equation has two unknowns: the *eigenfunction* (*i.e.*, the function that has the characteristic that its transformation with the operator is a simple multiplication by a number) and the *eigenvalue* (*i.e.*, the number that multiplies it). By solving the eigenvalue equation, you get both together.

N. B. 3: A QM eigenvalue equation is typically a partial differential equation, because many QM operators contain partial derivatives (see, *e.g.*, energy). Later, we will see that the differential equation is associated with a linear algebraic equation.

N. B. 4: It makes sense to talk about eigenvalues even for *linear non-Hermitian* operators. Obviously, these are not observables, there is no reference to a measure, and the eigenvalue can be complex. Instead, if the operator corresponds to an observable and therefore is a *Hermitian operator*, then the *eigenvalues* are *real*. This is easily shown by the consideration that an eigenvalue is only a particular case of an expectation value, which must be real by the second postulate.

N. B. 5: The **number** of eigenfunctions of an observable depends on the problem: of course, when we count eigenfunctions, we must consider as a single eigenfunction all the functions that differ from each other by a simple multiplicative constant. More generally, we must count *how many linearly independent solutions there are.* Usually, we are dealing with an infinite set of eigenfunctions. As an example, consider the eigenfunctions and eigenvalues of the Hamiltonian for the H atom. The principal quantum number takes the possible values $n=1, 2, \ldots$ (any integer) and for each n there are n^2 states (or twice, if one considers the spin of the electron).

However, there are cases in which the number is finite and small (for example, the eigenfunctions of the spin operators of an electron are only two).

The problems concerning the correspondence between eigenfunctions and eigenvalues and the **degeneracy** are discussed in Section 3.2.

N. B. 6: As already said, the expectation value of the second postulate refers to a particular state, that is the state in which the system has been prepared. This state, however, does *not* appear in the eigenvalue equation (Eq. 2.29). Both the eigenfunctions and the eigenvalues are specified once and for all and are independent of the previous history through which the system was brought to a particular state. The previous history instead determines the expectation value of the second postulate. There is no contradiction, and the explanation is very important:

a. Already said: in QM the measurement modifies the state.
b. To be precise: the effect of the measurement is to *bring the system from the initial state to one of the states described by the eigenfunctions of the operator.* The result of the measurement is then the eigenvalue corresponding to this eigenfunction.
c. The eigenstate to which the system is brought by the measurement is selected at random, according to a probability **that depends on the initial state**.
d. It is certainly possible that the state of the system after the measurement is the same as before. This happens if and only if the state before the measurement is described by an eigenfunction of the operator of the quantity being measured. Then, the result of the measurement is the eigenvalue, the expectation value is the same as the eigenvalue, and the dispersion of the results around the mean value is zero.
e. Anticipating later considerations (see Section 5.1), the fact is that one can always write the initial state as a finite or infinite linear combination (*i.e.* as a *series*) of the eigenfunctions (φ_i) of the operator: for any state (ψ) one can always write $\psi = \sum_i a_i \cdot \varphi_i$. Then, the measurement randomly selects one of the components (φ_i) with probability given by $|a_i|^2$ (if the starting state is normalized).
f. *Different* expectation values therefore arise as the result of averages made with different *statistical weights*. The *possible values* used to calculate the averages *do not depend on the starting state*, the weights instead *depend on the starting state*.

N. B. 7: The time-independent Schrödinger equation (Eq. 2.27) is the eigenvalue equation for the Hamiltonian operator.

N. B. 8: By looking at the expression of an eigenvalue equation, one can now understand why it has been said (for the first postulate) that the operations that can be performed on the wave function to obtain physically significant quantities provide in some cases values that are independent of the scaling factors of the wave function. The cases are those considered in the fifth postulate: it is immediately clear, in fact, that if $\hat{A}\varphi_i(q,t) = A_i \cdot \varphi_i(q,t)$ holds for a normalized wave function, then it also holds for a non-normalized wave function, since both the right-hand and left-hand terms are multiplied by the same complex factor of modulus $\neq 1$.

N. B. 9: The eigenvalues of an operator can be a **discrete**, **continuous**, or **mixed** set.

Example 1: The eigenvalues of energy for the harmonic oscillator model are a discrete set: there is one energy level $E_n = \left(n + \frac{1}{2}\right)\hbar\omega$, and one state for every integer $n \geq 0$ (see Section 3.2).

Example 2: The energy eigenvalues for the bound states of the hydrogen atom are a discrete set, because: $E_n = \frac{-m \cdot e^4}{\hbar^2}\left(\frac{1}{2 \cdot n^2}\right)$ and for every integer $n>0$ there is an energy level with negative energy. The zero of the energy scale corresponds to the situation in which the nucleus and electron are at infinite distance. For any positive energy, the spectrum of eigenvalues is continuous. In fact, the Schrödinger equation of the hydrogen atom also describes non-bound states, usually ignored in elementary treatments. These states concern the ionized atom. The problem of the hydrogen atom is therefore characterized by a spectrum of energy eigenvalues of a mixed type: discrete for $E<0$, continuous for $E \geq 0$.

Note 10: The eigenvalues of an operator can be bounded (upper, lower, or both) or unbounded.

Examples. The energy eigenvalues for the harmonic oscillator model are a lower bounded, but upper unbounded set. The energy eigenvalues for the bound states of the hydrogen atom are a lower bounded ($E_n \geq -\frac{m \cdot e^4}{2 \cdot \hbar^2}$) and upper bounded ($E_n < 0$) set. Adding the states of the ionised atom, the spectrum is instead only lower bounded.

3 SOME RESULTS OF THE FIRST FIVE POSTULATES

3.1 STATIONARY STATES

We can now delve deeper into the relationship between the **time-dependent** Schrödinger equation:

$$\hat{H}\psi(q,t) = i \cdot \hbar \cdot \frac{\partial \psi(q,t)}{\partial t} \tag{2.26}$$

and the **time-independent** Schrödinger equation:

$$\hat{H}\chi(q,t) = E \cdot \chi(q,t) \quad (2.27)$$

Equaton (2.26) is the equation of motion of QM, that is, the **fundamental equation** that every wavefunction must satisfy (4th postulate, 2.5). Equation (2.27) is the **eigenvalue equation** of the Hamiltonian operator of the system. Equations (2.26) and (2.27) use different symbols for the wavefunctions because we must clarify whether they are the same functions or not.

The most frequent and most important situation for chemistry is when the Hamiltonian operator **does not explicitly contain time**: the exceptions essentially concern the interaction with electromagnetic radiation, which is treated separately (see Section 6.3). If the Hamiltonian does not explicitly contain time, Eq. (2.24) obviously does not allow us to evaluate the time dependence of the eigenfunction $\chi(q,t)$ but can only determine its dependence $[\varphi(q)]$ on the spatial coordinates. Therefore, Eq. (2.27) must be written:

$$\hat{H}\varphi(q) = E \cdot \varphi(q) \quad (2.28)$$

The time dependence of the eigenfunction must appear in the form of a multiplicative coefficient $f(t)$ that is independent of the coordinates:

$$\chi(q,t) = f(t)\cdot\varphi(q) \quad \text{so that}$$
$$\hat{H}\left[f(t)\cdot\varphi(q)\right] = f(t)\cdot\hat{H}\left[\varphi(q)\right] = f(t)\cdot E\cdot\varphi(q) \quad (3.1)$$

The time dependence $f(t)$ can be determined by considering that $X(q,t)$ must also satisfy the time-dependent Schrödinger equation:

$$f(t)\cdot E\cdot\varphi(q) = \hat{H}\left[f(t)\cdot\varphi(q)\right] = i\cdot\hbar\cdot\frac{\partial\psi(t)}{\partial t}$$
$$= i\cdot\hbar\cdot\varphi(q)\cdot\frac{\partial f(t)}{\partial t}$$

Then: $\frac{\partial f(t)}{\partial t} = \frac{E}{i\cdot\hbar}\cdot f(t) = -i\cdot\frac{E}{\hbar}\cdot f(t)$ and therefore: $f(t) = e^{-iEt/\hbar}$

This finally gives:

$$\chi(q,t) = e^{-iEt/\hbar}\cdot\varphi(q) \quad (3.2)$$

The Xs of Eq. (3.2) are the eigenfunction of the Hamiltonian with their time dependence. They are also solutions of the time-dependent Schrödinger equation but do not represent *all* solutions of (Eq. 2.26). To understand this, it is enough to look at the linear combination of two (or more) functions of this kind (Eq. 3.2) and with different eigenvalues E:

$$\psi(q,t) = a_1\cdot e^{-iE_1t/\hbar}\cdot\varphi_1(q) +$$
$$a_2\cdot e^{-iE_2t/\hbar}\cdot\varphi_2(q) \quad \text{with} \quad E_1 \neq E_2$$

This is a solution of Eq. (2.26), but is not of the kind described by Eq. (3.2), as can be seen by substitution:

$$\hat{H}\psi(q,t) = \hat{H}\left[a_1\cdot e^{-iE_1t/\hbar}\cdot\varphi_1(q) + a_2\cdot e^{-iE_2t/\hbar}\cdot\varphi_2(q)\right]$$
$$= a_1\cdot e^{-iE_1t/\hbar}\cdot\hat{H}\left[\varphi_1(q)\right] + a_2\cdot e^{-iE_2t/\hbar}\cdot\hat{H}\left[\varphi_2(q)\right]$$
$$= a_1\cdot e^{-iE_1t/\hbar}\cdot E_1\cdot\varphi_1(q) + a_2\cdot e^{-iE_2t/\hbar}\cdot E_2\cdot\varphi_2(q)$$

to be compared with the right side of Eq. (2.26):

$$i\cdot\hbar\cdot\frac{\partial\left[a_1\cdot e^{\frac{-iE_1t}{\hbar}}\cdot\varphi_1(q) + a_2\cdot e^{\frac{-iE_2t}{\hbar}}\cdot\varphi_2(q)\right]}{\partial t}$$
$$= i\cdot\hbar\cdot\left[a_1\cdot\varphi_1(q)\cdot\frac{\partial e^{\frac{-iE_1t}{\hbar}}}{\partial t} + a_2\cdot\varphi_2(q)\cdot\frac{\partial e^{\frac{-iE_2t}{\hbar}}}{\partial t}\right]$$
$$= i\cdot\hbar\cdot\left[a_1\cdot\varphi_1(q)\cdot\left(\frac{-iE_1}{\hbar}\right)\cdot e^{\frac{-iE_1t}{\hbar}} + a_2\cdot\varphi_2(q)\cdot\left(\frac{-iE_2}{\hbar}\right)\cdot e^{\frac{-iE_2t}{\hbar}}\right]$$
$$= a_1\cdot\varphi_1(q)\cdot E_1\cdot e^{\frac{-iE_1t}{\hbar}} + a_2\cdot\varphi_2(q)\cdot E_2\cdot e^{\frac{-iE_2t}{\hbar}}$$

The two expressions are, in fact, equal.

Wavefunctions of the type

$$\chi(q,t) = e^{-iEt/\hbar}\cdot\varphi(q) \quad (3.2)$$

where $\varphi(q)$ is an eigenfunction of the Hamiltonian, play a fundamental role in QM and deserve a specific name: they are called **stationary states**. The name comes directly from the fact that for any stationary state the expectation values of any observable that does not explicitly contain time are constant (independent of time).

The proof is straightforward. Consider the expectation value of the observable A in state ψ:

$$A_\chi = \int_\Omega \psi^*(q,t)\cdot\left[\hat{A}\psi(q,t)\right]\cdot dq \quad (\textit{see}\ \text{Eq. 2.2})$$

If ψ is a stationary state, that is $\psi(q,t) = \chi(q,t) = e^{-iEt/\hbar}\cdot\varphi(q)$, then:

$$A_\chi = \int_\Omega e^{+iEt/\hbar}\cdot\varphi^*(q)\cdot\hat{A}\left[e^{-iEt/\hbar}\cdot\varphi(q)\right]\cdot dq$$

If the operator does not contain time:

$$\hat{A}\left[e^{\frac{-iEt}{\hbar}}\cdot\varphi(q)\right] = e^{-iEt/\hbar}\cdot\hat{A}\left[\varphi(q)\right] \quad \text{and finally:}$$

$$A_\chi = \int_\Omega e^{+iEt/\hbar}\cdot e^{-iEt/\hbar}\cdot\varphi^*(q)\cdot\hat{A}\left[\varphi(q)\right]\cdot dq$$
$$= \int_\Omega \varphi^*(q)\cdot\hat{A}\left[\varphi(q)\right]\cdot dq$$

which is independent of time.

An expectation value is a linear combination of eigenvalues, as already said on the comments to the 5th postulate (see Section 2.6). Now, we can understand why the expectation value of a time-independent observable in a steady state is independent of time: this is *not* due to a compensation between time-varying weights of different eigenvalues, but to the fact that both the eigenvalues and their weights are independent of time.

The stationary states are important also because **the time-dependent Schrödinger equation is typically solved first by solving the time-independent Schrödinger equation**.

More in detail and still anticipating a result that will be better explained when discussing the structure of linear spaces (Section 5.1), *a linear combination of two or more solutions of a time-dependent Schrödinger equation has all the properties required by the first postulate to be a valid wavefunction* (it is continuous, it can be normalized, …), if the functions that combine have these properties in turn and *the set of all linear combinations of stationary states is the set of all possible wavefunctions, i.e.* the set of *all solutions* of the time-dependent Schrödinger equation.

So, if the Hamiltonian operator is time-independent, what needs to be done is to solve the time-independent Schrödinger equation and obtain the spatial parts $\varphi_i(q)$ with the pertinent E_i. Then,

- either the summation: $\sum_{i=1}^{n} a_i \cdot e^{-iE_i t/\hbar} \cdot \varphi_i(q)$ or
- the series expansion $$\sum_{i=1}^{\infty} a_i \cdot e^{-iE_i t/\hbar} \cdot \varphi_i(q) = \sum_{i=1}^{\infty} a_i \cdot \chi_i(q,t)$$

are the most general wavefunctions for the given problem (described by the given Hamiltonian).

3.2 Degeneracy

There is **not one-to-one correspondence** between eigenfunctions and eigenvalues

- for each eigenfunction there is - quite obviously - only one eigenvalue, while
- an eigenvalue is possibly shared by several linearly independent eigenfunctions.

In the latter case the eigenvalue is said to be *degenerate*, and another index (or another quantum number) is needed to specify a particular eigenfunction within the set of degenerate eigenfunctions. If an eigenvalue is n times degenerate, this does not mean that there are n eigenfunctions. There are infinitely many of them, because any linear combination of eigenfunctions with that eigenvalue is still an eigenfunction with the same eigenvalue. Therefore, **the set of eigenfunctions** chosen for that level is not fixed but **is the result of a choice**. For example:

if: $\hat{A}\varphi_{i,1}(q,t) = A_i \cdot \varphi_{i,1}(q,t)$ and $\hat{A}\varphi_{i,2}(q,t) = A_i \cdot \varphi_{i,2}(q,t)$ where A_i is the same while $\varphi_{i,1}(q,t)$ and $\varphi_{i,2}(q,t)$ are two linearly *independent* wavefunctions,
then: any linear combination of $\varphi_{i,1}(q,t)$ and $\varphi_{i,2}(q,t)$ is also an eigenfunction *with the same eigenvalue*.

In fact, using the linearity property:

$$\hat{A}\{c \cdot \varphi_{i,1}(q,t) + d \cdot \varphi_{i,2}(q,t)\}$$
$$= c \cdot \hat{A}\varphi_{i,1}(q,t) + d \cdot \hat{A}\varphi_{i,2}(q,t)$$
$$= A_i \cdot \{c \cdot \varphi_{i,1}(q,t) + d \cdot \varphi_{i,2}(q,t)\}$$

What matters is to specify *a set of n linearly independent eigenfunctions* of the level.

The degeneracy of an eigenvalue is the number of *linearly independent* eigenfunctions that can be found for that eigenvalue.

When the degeneracy is $=1$ it is usually said that the level is **non-degenerate**. The definition of degeneracy correctly applies also to the case of degeneracy $=1$, that is absence of degeneracy. The set of all functions that differ from a given function by an arbitrary (complex) multiplicative constant ($\neq 0$) is the set of all linear combinations of the given function.

An important property (see Section 5.1.3) is that

Eigenfunctions with different eigenvalues are always linearly independent of each other.

3.3 Commuting Operators

A remarkable difference between QM and CM is that the product of two different physical quantities of CM is always commutative, while the corresponding operators of QM possibly do not commute (for examples of operators that commute or do not commute, see last note in Section 2.4.2). We here discuss an important property of the eigenfunctions of two commuting operators of QM. *Generally*, there are no relations between the eigenfunctions of two different operators of QM. *When two operators do commute*, however, they *can* share the eigenfunctions.

If two operators commute, it is always possible to find a set of functions such that each function is eigenfunction of **both** operators. In other words it is always possible to find a set of **common eigenfunctions** of two commuting operators

It is not difficult to prove a *particular case* of this property. When two operators (A and B) related by

$$\hat{A} \cdot \hat{B} = \hat{B} \cdot \hat{A} \tag{3.3}$$

an eigenfunction of the latter operator (B) that belongs to a *non-degenerate* level is also eigenfunction of the former operator (A). Indeed, let us consider an eigenfunction of $\hat{B}$: $\hat{B}\varphi_i(q,t) = B_i \cdot \varphi_i(q,t)$. Then:

$$\hat{A}\left(\hat{B}\varphi_i(q,t)\right) = \hat{A}B_i\varphi_i(q,t) = B_i\hat{A}\varphi_i(q,t) \tag{3.4}$$

where the second equality holds because a number always commutes with an operator. Moreover, by Eq. (3.3):

$$\hat{A}\left(\hat{B}\varphi_i(q,t)\right) = \hat{B}\left(\hat{A}\varphi_i(q,t)\right) \text{ so that}$$

$$\hat{B}\left(\hat{A}\varphi_i(q,t)\right) = B_i\hat{A}\varphi_i(q,t)$$

which means that $\hat{A}\varphi_i(q,t)$ is also an eigenfunction of $\hat{B}$ with the same eigenvalue B_i as the original function φ_i. Now, if B_i is a non-degenerate level, $\hat{A}\varphi_i(q,t)$ can differ from the original eigenfunction φ_i only by a multiplicative factor a: $\hat{A}\varphi_i = a \cdot \varphi_i$. This shows that φ_i is an eigenfunction also of operator $\hat{A}$, and the constant a is its eigenvalue.

It is important to note that *it is possible to find* a set of common eigenfunctions: the eigenfunctions of a degenerate level are not uniquely specified, different sets of independent linear combinations can be chosen, and a set possibly contains degenerate eigenfunctions of one operator that are *not* eigenfunctions of the other operator. Then, a specific procedure is required to obtain the set of common eigenfunctions.

When the degeneracies do not overlap, *the common set represents a physically sensible way of selecting the eigenfunctions of a degenerate level*, because they can be *classified* using a *precise value* of *both quantities*. For a reader already familiar with atomic (hydrogenic) orbitals, we can show the meaning of this property referring to the operators corresponding to: (A) the squared modulus of angular moment (quantum number l) and (B) its component along z (quantum number m). The eigenvalues are $l\,(l+1)$ and m in dimensionless units. Let us consider for instance the three degenerate 2p eigenfunctions of the squared modulus ($l = 1$, eigenvalue $= 2$). The eigenfunctions of the component operator are non-degenerate (the eigenvalues are 1, 0, −1) and directly give the common set, because each of them is also an eigenfunction of the squared modulus. For the squared modulus other eigenfunctions are possibly selected: the set $\{2p_x, 2p_y, 2p_z\}$ is an example, but then two eigenfunctions of the squared modulus ($2p_x$ and $2p_y$) are *not* characterized by a precise value of the component along z.

3.3.1 Complete Systems of Commuting Operators

Let us now consider the case of *more than two* commuting operators. If *all* these operators commute with each other, it can be shown that it is possible to find a set of functions that are eigenfunctions of *all* these operators. So, we can use the eigenvalues of all these operators to classify the common eigenfunctions. If there are enough commuting operators the classification is *complete*, meaning that each different set of eigenvalues uniquely identifies a function of the common set. Then, the set of operators is called a *complete set of commuting observables* (CSCO).

A simple example is again given by the atomic orbitals. Without considering the spin, the eigenvalues of energy (H), squared modulus of the angular moment (L^2) and z component (L_z) uniquely identify a state. These three operators indeed commute with each other: H commutes with L^2 and with L_z, L^2 commutes with L_z. When the spin is considered, these three operators are *not* a CSCO, and the spin operator (quantum number s_z) must be added to the set.

3.3.2 Commutation and Symmetry

Symmetry is a wide topic with many different consequences and is more deeply discussed in the specific chapter of this book. We can here anticipate a point that is in itself very important and frequently used.

Let us start by clarifying that with symmetry we mean a set of linear transformations of coordinates that leave the equations of motion unchanged. A simple example is 1D harmonic motion, where $\hat{H} = \frac{1}{2m}\frac{\partial^2}{\partial x^2} + \frac{1}{2}k(x)^2$. Let's consider the symmetry operation represented by the transformation $x \to -x$. Then, $\frac{\partial}{\partial(-x)} = -\frac{\partial}{\partial x}$, the Hamiltonian operator becomes $\frac{1}{2m}\frac{\partial^2}{\partial(-x)^2} + \frac{1}{2}k(-x)^2 = \frac{1}{2m}\frac{\partial^2}{\partial x^2} + \frac{1}{2}k(x)^2$ and equals the operator before transformation. It is not difficult to prove that the result holds also for the Hamiltonian function of CM.

The important point is that

in a symmetric system, every symmetry operator commutes with the Hamiltonian

This is not difficult to prove. Let us start by applying a coordinate transformation described by the *linear operator* $\hat{R}$ to a Schrödinger equation $\hat{H}\psi(q,t) = E \cdot \psi(q,t)$. Then: $\widehat{H'}\left[\hat{R}\psi(q,t)\right] = E \cdot \hat{R}\psi(q,t)$, where $\widehat{H'}$ is the Hamiltonian operator as modified by $\hat{R}$. $\widehat{H'}$ an be specified by considering a new operator (let us write it $\hat{R}^{-1}$) that cancels the effect of $\hat{R}$ and brings everything back to the beginning. By applying $\hat{R}^{-1}$ to the previous relation we get $\hat{R}^{-1}\widehat{H'}\left[\hat{R}\psi(q,t)\right] = \hat{R}^{-1}E \cdot \hat{R}\psi(q,t)$ and $\hat{R}^{-1}E \cdot \hat{R}\psi(q,t) = E \cdot \hat{R}^{-1}\hat{R}\psi(q,t) = E \cdot \psi(q,t)$ because a linear operator commutes with a number (E). Finally, $\hat{R}^{-1}\widehat{H'}\left[\hat{R}\psi(q,t)\right] = \hat{H}\left[\psi(q,t)\right]$ that is $\hat{R}^{-1}\widehat{H'}\hat{R} = \hat{H}$ or $\widehat{H'} = \hat{R}\hat{H}\hat{R}^{-1}$.

Now, let us assume that $\hat{R}$ (and therefore also $\hat{R}^{-1}$) are *symmetry operations* of the system. Then we must require that $\hat{H}$ *remains unchanged, i.e.* $\widehat{H'} = \hat{H}$: therefore $\hat{R}^{-1}\widehat{H'}\hat{R} = \hat{R}^{-1}\hat{H}\hat{R} = \hat{H}$ and $\hat{R}\hat{R}^{-1}\hat{H}\hat{R} = \hat{R}\hat{H}$ but $\hat{R}\hat{R}^{-1}\hat{H}\hat{R} = \hat{H}\hat{R}$, so that $\hat{H}\hat{R} = \hat{R}\hat{H}$ as required.

Care must be taken to avoid improper implications of this result. Indeed, it is *not true in general* that all symmetry operators commute with each other in pairs, even if each one commutes with the Hamiltonian.[3] *This fact makes far from trivial the applications of symmetry to the solution of QM and CM problems.*

It is worth discussing here a simple application of the commutation between a symmetry operator and the Hamiltonian. It concerns the symmetry operators that are equal to their inverse ($\hat{R}^{-1} = \hat{R}$): let us call them *self-inverse*. By repeating their application, we go back to the top. It is not difficult to find examples: inversion, reflection, rotation of half a turn, exchange of coordinates between two equal particles. Considering an eigenfunction of a self-inverse operator $\hat{R}$ and applying two times the transformation, we obtain again the original function, that is we multiply it by 1. Therefore, the eigenvalue (r) is one of the two square roots of unity ($r=1$ or $r=-1$).

$$\text{If } \hat{R}\psi(q,t) = r\cdot\psi(q,t) \text{ and } \hat{R} = \hat{R}^{-1}(\text{self-inverse}), \text{ then } r = \begin{cases} 1 \\ -1 \end{cases} \tag{3.5}$$

Indeed: $\hat{R}\left[\hat{R}\psi(q,t)\right] = \hat{R}\left[r\cdot\psi(q,t)\right] = r\cdot\hat{R}\left[\psi(q,t)\right] = r^2\cdot\left[\psi(q,t)\right] = \left[\psi(q,t)\right]$(the operator is self-inverse). Then $r^2 = 1$.

Warning: the eigenvalues of symmetry operators *are not necessarily real numbers*, since a symmetry operator is not necessarily Hermitian (it does not correspond to an observable). However, the square roots of unity (+1 and −1) are both real numbers.

Using the commutativity of $\hat{R}$ with $\hat{H}$, the eigenfunctions of the Hamiltonian can then be classified into two groups: those for which $r=1$ and those for which $r=-1$.

Depending on the nature of the operation $\hat{R}$, the two cases take various names, for example

- even/odd (when speaking of inversion or reflection) or
- symmetric/antisymmetric (when speaking of exchange symmetry).

3.4 Eigenvalues of Angular Moments

We complete the presentation of the angular moment operators (see Section 2.4.2) here. Eigenvalues and eigenfunctions of these operators can obviously be obtained by solving the eigenvalue equations, after having translated the CM expressions of the corresponding quantities into operators according to the correspondence rules of the third postulate (see Section 2.4.1).

Before discussing the results, however, it is worth noting some properties of angular moments, and their consequences. The operators concerning the angular moments obey **commutation rules**: the components do not commute among themselves and of each of them commute with the modulus or the squared modulus. The first set of commutation rules is that:

$$\begin{aligned} \left[\hat{L}_x,\hat{L}_y\right] &\overset{\text{def}}{=} \hat{L}_x\cdot\hat{L}_y - \hat{L}_y\cdot\hat{L}_x = i\hbar\hat{L}_z \\ \left[\hat{L}_y,\hat{L}_z\right] &\overset{\text{def}}{=} \hat{L}_y\cdot\hat{L}_z - \hat{L}_z\cdot\hat{L}_y = i\hbar\hat{L}_x \\ \left[\hat{L}_z,\hat{L}_x\right] &\overset{\text{def}}{=} \hat{L}_z\cdot\hat{L}_x - \hat{L}_x\cdot\hat{L}_z = i\hbar\hat{L}_y \end{aligned} \tag{3.6}$$

The derivation is long and boring and can be found by the interested reader at the end of this chapter (see Section 7.2). These three relations can be summarized in one: $\left[\hat{L}_i,\hat{L}_j\right] \overset{\text{def}}{=} \hat{L}_i\cdot\hat{L}_j - \hat{L}_j\cdot\hat{L}_i = i\hbar\hat{L}_k$, where the triplet ($i$, j, k) indicates the variables x, y, z in one of their three *cyclic* permutations x, y, z or y, z, x or z, x, y).

The second set of commutation rules is that:

$$\left[\widehat{L^2},\hat{L}_x\right] = \left[\widehat{L^2},\hat{L}_y\right] = \left[\widehat{L^2},\hat{L}_z\right] = 0 \tag{3.7}$$

as can be obtained similarly by the rules of Eq. 3.6 and the definition:

$$\left|\widehat{L^2}\right| \overset{\text{def}}{=} \left|\widehat{L_x^2}\right| + \left|\widehat{L_y^2}\right| + \left|\widehat{L_z^2}\right| \quad (\text{see Eq. 2.23}) \tag{2.24}$$

According to what already said about commutating operators (3.3) this means that

- (*in general*) it is not possible to know exactly two or three of the quantities $\hat{L}_x$, $\hat{L}_y$, $\hat{L}_z$, but
- it is possible to know exactly and simultaneously both the value of the square modulus and that of a single component $\hat{L}_x$ or $\hat{L}_y$ or $\hat{L}_z$.

Note, however, that when $L=0$ we know exactly the value $=0$ of all three quantities.

So, we can choose eigenfunctions such that they are (at the same time) eigenfunctions of

$\widehat{L^2}$ and $\hat{L}_x$, or $\widehat{L^2}$ and $\hat{L}_y$ or $\widehat{L^2}$ and $\hat{L}_z$

Now, writing the operators according to the correspondence rule of the third postulate (Section 2.4.1) we can solve the eigenvalue equations (for example: for $\hat{L}^2$ and $\hat{L}_z$) thus obtaining that the eigenvalues depend on two quantum numbers (L and M) and are given by:

$$\widehat{L^2} \Rightarrow L\cdot(L+1)\cdot\hbar^2; \quad \widehat{L_z} \Rightarrow M\cdot\hbar \tag{3.8}$$

Here: $L=0$, or 1, or 2, …, (any integer positive number). For a given L, M can take all $(2L+1)$ integer values ranging from $-L$ to L (including 0).

3.4.1 Generalized Angular Moments

In applications of QM many different angular moments are used. All these are vector quantities that group various operators (the three components, their squares and their sum, *i.e.* the square of the modulus) and are always characterized by the particular behaviour specified by the commutation rules: Eqs. 3.6 and 3.7. The last postulate of QM (see Section 4.1) will introduce analogous quantities (the spins, and particularly the electron spin) that do not have a classical analogue. The corresponding operators, therefore, cannot be derived from CM according to the rules of the third postulate, as can be done for the operators seen so far. Instead, they are introduced by a definition that refers to the commutation rules of Eqs. 3.6 and 3.7.

As shown in detail in Section 7.3, *it is possible to use directly the commutation rules to find the eigenvalues*: this realises a generalisation of the angular moment quantities and leads to results that include as a special case those obtained by directly solving the eigenvalue equation of the angular moments that have a classical analogue (Eq. 2.23). The generalisation, however, also gives other results. It is therefore convenient to look directly at the general results.

Let us consider a generic angular moment, which we can indicate with $\vec{j}$.

For the various operators (components, modulus, etc.) related to $\vec{j}$ we consider only the commutation rules already seen (Eqs. 3.6 and 3.7) and we deliberately ignore any reference to an analytical form that can be derived from CM. Obviously, the properties that can be derived from the commutation rules still hold. We still have wavefunctions that are simultaneously eigenfunctions of the operators related to:

- the square modulus of the vector: $\left|\vec{j}\right|^2$ and
- its component in a predetermined direction.

Then, **on the sole basis of the commutation rules**, the following results are obtained (see Section 7.3 for the derivation).

- The eigenvalue of the **square modulus** is: $\left|\vec{j}\right|^2 = j\cdot(j+1)\cdot\hbar^2$, where j is the quantum number relative to the modulus and can only be 0, ½, 1, 3/2, 2, ….
- The eigenvalue of the component is: $\hat{j}_z = m_j\cdot\hbar$, where the quantum number m_j takes all the values between j and $-j$ in steps of 1 (and therefore $2j+1$ different values).

For the angular moments that have a classical analogue, only the integer values of j are possible. Semi-odd values can be found only for angular moments that do *not* have a classical analogue, but these moments can also have integer values for j. The fact that a quantum number can only be 0, ½, 1, 3/2, 2, … is stated in various ways, *e.g.* saying that it must be a *half-integer*, or that it must be *either an integer or a half-integer*. To show these features of the components and the modulus of an angular moment of QM, a convenient graphical representation is frequently used: an angular moment vector is represented as a cone of which we know (in units of $\hbar$)

- the side $\sqrt{j\cdot(j+1)}$, that is, the modulus of the vector, and
- the height (m_j), that is, the component in the "vertical" direction that we identify with z,

while the indeterminacy of the x and y components translates into the freedom to choose any point of the base circle of the cone. Figure 1.3 reports the three simplest cases, after the trivial case $j=0$.

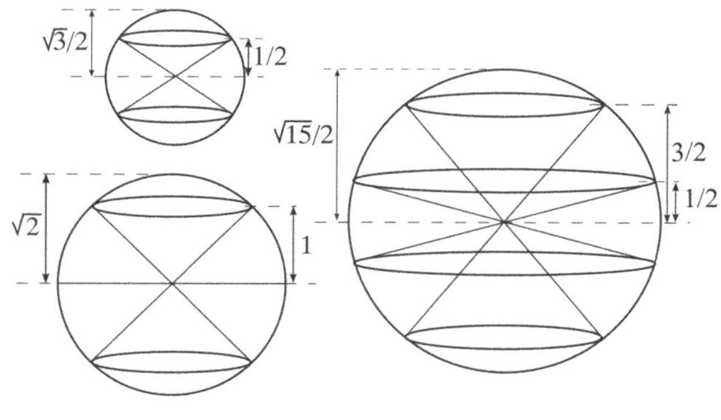

FIGURE 1.3 Graphical representation of the angular moments in QM: $j=1/2$, $j=1$, and $j=3/2$.

3.5 Separable Hamiltonian Operators

In Section 1 we discussed an important simplification of the equations of motion in CM. This occurs when it is possible to choose coordinates and moments that allow to write the Hamiltonian function as a sum of **disjoint parts**. We recall that a function f of several variables (x_1, …, x_n) is the sum of (for example) three *disjoint* parts f_1, f_2 and f_3 if $f=f_1+f_2+f_3$, if each of these functions depend on a particular set of variables and if the different sets do not overlap (a variable belongs to only one set). In QM the simplification can be stated as follows.

if: it is possible to find a set of coordinates such that the Hamilton operator is separated into two disjoint parts

$$\hat{H}(q_1,\ldots,q_n) = \widehat{H_1}(q_1,\ldots,q_{r-1}) + \widehat{H_2}(q_r,\ldots,q_n) \quad (3.9)$$

then: the solutions of the eigenvalue equation can be obtained with the following steps.

A. We solve separately the two eigenvalue equations:

$$\widehat{H_1}\chi_a(q_1,\ldots,q_{r-1}) = u_a\cdot\chi_a(q_1,\ldots,q_{r-1})$$

$$\widehat{H_2}\varphi_b(q_r,\ldots,q_n) = v_b \cdot \varphi_b(q_r,\ldots,q_n)$$

where a and b indicate quantum numbers (or sets of quantum numbers), while (u, χ) and (v, φ) are eigenvalues and eigenfunctions corresponding to the two disjoint parts.

B. We write eigenfunctions and eigenvalues of the sum operator as products of eigenfunctions and sums of eigenvalues of the disjoint parts, respectively:

$$\psi_{a,b}(q_1,\ldots,q_n) = \chi_a(q_1,\ldots,q_{r-1}) \cdot \varphi_b(q_r,\ldots,q_n) \quad \text{and}$$

$$\hat{H}\psi_{a,b}(q_1,\ldots,q_n) = (u_a + v_b) \cdot \psi_{a,b}(q_1,\ldots,q_n)$$

It is an easy task to verify the result by substitution, and it is equally easy to verify by iteration the extension of the procedure to Hamiltonian operators made of **more than two disjoint parts**.

Possibly, the simplest **example** of this procedure is the 3D "particle in a box" model where the Hamiltonian operator contains only the kinetic energy operator:

$$\hat{H} = \hat{T} = -\frac{\hbar^2}{2m}\nabla^2 = -\frac{\hbar^2}{2m}\left(\frac{\partial^2}{\partial x^2} + \frac{\partial^2}{\partial y^2} + \frac{\partial^2}{\partial z^2}\right) \tag{3.10}$$

and the solutions are products of the solutions for the three 1D cases, *e.g.*: $\widehat{H_x} = -\frac{\hbar^2}{2m}\left(\frac{\partial^2}{\partial x^2}\right)$

Separability holds also in a broader form: a part of the Hamiltonian can contain also the other coordinates, but only as a simple multiplication factor:

$$\hat{H}(q_1,\ldots,q_n) = \widehat{H_1}(q_1,\ldots,q_{r-1}) + f(q_1,\ldots,q_{r-1}) \cdot \widehat{H_2}(q_r,\ldots,q_n) \tag{3.11}$$

Starting once again from the separate equations

$$\widehat{H_1}\chi_a(q_1,\ldots,q_{r-1}) = h_{1,a} \cdot \chi_a(q_1,\ldots,q_{r-1}) \quad \text{and}$$

$$\widehat{H_2}\varphi_b(q_r,\ldots,q_n) = h_{2,b} \cdot \varphi_b(q_r,\ldots,q_n)$$

we find that the products:

$$\psi_{a,b}(q_1,\ldots,q_n) = \chi_a(q_1,\ldots,q_{r-1}) \cdot \varphi_b(q_r,\ldots,q_n)$$

are the desired solutions because:

$$\hat{H}\psi_{a,b}(q_1,\ldots,q_n) = \left[\widehat{H_1} + f \cdot \widehat{H_2}\right]\left[\chi_a(q_1,\ldots,q_{r-1}) \cdot \varphi_b(q_r,\ldots,q_n)\right] = (h_{1,a} + f \cdot h_{2,b}) \cdot \left[\chi_a(q_1,\ldots,q_{r-1}) \cdot \varphi_b(q_r,\ldots,q_n)\right] \tag{3.12}$$

It is not difficult to prove Eq. (3.12):

$$\widehat{H_1}\left[\chi_a(q_1,\ldots,q_{r-1}) \cdot \varphi_b(q_r,\ldots,q_n)\right] + \left[f \cdot \widehat{H_2}\right]\left[\chi_a(q_1,\ldots,q_{r-1}) \cdot \varphi_b(q_r,\ldots,q_n)\right]$$

$$= \varphi_B(q_r,\ldots,q_n) \cdot \widehat{H_1}\chi_a(q_1,\ldots,q_{r-1}) + \left[f \cdot \chi_a(q_1,\ldots,q_{r-1})\right] \cdot \widehat{H_2}\varphi_b(q_r,\ldots,q_n)$$

$$= \varphi_B(q_r,\ldots,q_n) \cdot h_{1,a} \cdot \chi_a(q_1,\ldots,q_{r-1}) + f \cdot \chi_a(q_1,\ldots,q_{r-1}) \cdot h_{2,b} \cdot \varphi_b(q_r,\ldots,q_n)$$

$$= (h_{1,a} + f \cdot h_{2,b}) \cdot \left[\chi_a(q_1,\ldots,q_{r-1}) \cdot \varphi_b(q_r,\ldots,q_n)\right]$$

3.5.1 An example: The Hydrogen Atom

There is a very important application of these procedures. It corresponds to a system made of two point-particles linked by a force directed along their joining line and introduces the whole chapter on atomic structure.

This is both the problem of the (spinless) hydrogen atom (nucleus+one electron), but also the diatomic molecule, when considered as made of two dimensionless masses.

As described in part 1.2 of the Appendices, the Hamilton function of CM can be split into several disjoint parts by transforming the coordinates with a two steps procedure. The first step replaces the Newton coordinates (x_1, y_1, z_1, x_2, y_2, z_2) with the cartesian coordinates (X, Y, Z) of the centre of mass of the system and the cartesian components (x, y, z) of the vector that connects a particle to the other. With this transformation, the original H depending on six degrees of freedom splits into three 1D parts H(X), H(Y), H(Z) plus one 3D part $H(x, y, z)$:

$$H(x_1, y_1, z_1, x_2, y_2, z_2) \Rightarrow H(X) + H(Y) + H(Z) + H(x, y, z)$$

$H(X)$, $H(Y)$, and $H(Z)$ describe the **free motion of the centre of mass** of a system with a mass $M = m_1 + m_2$ equal to the sum of the masses of the original particles. $H(x, y, z)$ describes a particle with mass $\mu = m_1 m_2/(m_1 + m_2)$ moving under the effect of a force directed towards the origin of the (x, y, z) coordinates.

The second step replaces the (x, y, z) coordinates in $H(x, y, z)$ with their spherical counterpart (r, ϑ, φ). Given that the potential energy depends only on the radial coordinate (r), this transformation eventually gives an (approximate) separation between radial and angular coordinates, which, for a molecule, corresponds to an approximate separation into vibrational and rotational motions.

Both coordinate transformations can be applied also in QM. The $\{x_1, y_1, z_1, x_2, y_2, z_2\} \Rightarrow \{X\}, \{Y\}, \{Z\}, \{x, y, z\}$ transformation is easily done by directly applying the correspondence rules (Section 2.4.1) on the Hamiltonian functions $H(X)$, $H(Y)$, $H(Z)$, $H(x, y, z)$ of the CM treatment. The first three parts correspond to the particle in a box model (Eq. 3.10), so that, for example:

$$\hat{H}(X) = -\frac{\hbar^2}{2M}\left(\frac{\partial^2}{\partial X^2}\right)$$

The second transformation concerns $H(x, y, z)$ and is easily achieved by writing the Laplacian operator in spherical coordinates (see Eq. 2.18):

$$\nabla^2 = \frac{1}{r^2}\frac{\partial}{\partial r}\left(r^2\frac{\partial}{\partial r}\right) + \frac{1}{r^2}\frac{1}{\sin\vartheta}\left[\frac{\partial}{\partial\vartheta}\left(\sin\vartheta\frac{\partial}{\partial\vartheta}\right) + \frac{1}{\sin\vartheta}\frac{\partial^2}{\partial\varphi^2}\right]$$

$$\hat{T} = -\frac{\hbar^2}{2m}\nabla^2, \quad \hat{\Lambda} = \frac{1}{\sin\vartheta}\left[\frac{\partial}{\partial\vartheta}\left(\sin\vartheta\frac{\partial}{\partial\vartheta}\right) + \frac{1}{\sin\vartheta}\frac{\partial^2}{\partial\varphi^2}\right], \quad \text{and}$$

$$\hat{H} = \hat{T} + \hat{V} = -\frac{\hbar^2}{2\mu}\left\{\frac{1}{r^2}\frac{\partial}{\partial r}\left(r^2\frac{\partial}{\partial r}\right) + \frac{\hat{\Lambda}}{r^2}\right\} + V(r)$$

The Hamiltonian operator is the type of Eq. (3.11)

$$\widehat{H_1}(r) \Leftrightarrow -\frac{\hbar^2}{2\mu}\left\{\frac{1}{r^2}\frac{\partial}{\partial r}\left(r^2\frac{\partial}{\partial r}\right)\right\} + V(r), \quad \text{and}$$

$$\widehat{H_2}(\vartheta,\varphi) \Leftrightarrow -\frac{\hbar^2}{2\mu}\left\{\frac{1}{r^2}\frac{1}{\sin\vartheta}\left[\frac{\partial}{\partial\vartheta}\left(\sin\vartheta\frac{\partial}{\partial\vartheta}\right) + \frac{1}{\sin\vartheta}\frac{\partial^2}{\partial\varphi^2}\right]\right\}$$

$$= -\frac{\hbar^2}{2\mu}\left\{\frac{\hat{\Lambda}}{r^2}\right\}$$

because the first (radial) part contains only the variable r and its derivatives, while the second part (the angular part) contains the variables ϑ, φ multiplied by $1/r^2$ (there are no operators such as $\partial/\partial r$). Moreover, the angular part of the Laplace operator gives the angular moment operator already seen:

$$\hat{\Lambda} = \hbar^2\widehat{L^2} \tag{2.19}$$

Now, it is possible to follow the procedure of Eq. (3.12) starting with the eigenvalue equation for the angular moment (the result is derived in Section 7.3):

$$\widehat{L^2}Y_{lm}(\vartheta,\varphi) = \hbar^2 \cdot l \cdot (l+1)Y_{lm}(\vartheta,\varphi) \quad \text{or} \quad \hat{\Lambda}Y_{lm}(\vartheta,\varphi) = l \cdot (l+1)Y_{lm}(\vartheta,\varphi)$$

where $\hbar^2 \cdot l \cdot (l+1)$ is the eigenvalue, l, m are the well-known quantum numbers (3.4), and the eigenfunctions $Y_{lm}(\vartheta,\varphi)$ are known as spherical harmonics.

Then, we write $\psi(r,\vartheta,\varphi) = R(r) \cdot Y(\vartheta,\varphi)$ and we replace the left-hand side of the above equation with its right-hand side in

$$\hat{H} = \hat{T} + \hat{V} = \left\{-\frac{\hbar^2}{2m}\left[\frac{1}{r^2}\frac{\partial}{\partial r}\left(r^2\frac{\partial}{\partial r}\right) + \frac{\hat{\Lambda}}{r^2}\right] + V(r)\right\}\psi(r,\vartheta,\varphi),$$

$$= \left\{-\frac{\hbar^2}{2\mu}\left[\frac{1}{r^2}\frac{\partial}{\partial r}\left(r^2\frac{\partial}{\partial r}\right) + \frac{l(l+1)}{r^2}\right] + V(r)\right\} R(r) \cdot Y(\vartheta,\varphi)$$

$$= E \bullet R(r) \cdot Y(\vartheta,\varphi)$$

Finally, we are left with the radial equation:

$$\hat{H}(r;l) = \left\{-\frac{\hbar^2}{2\mu}\left[\frac{1}{r^2}\frac{\partial}{\partial r}\left(r^2\frac{\partial}{\partial r}\right) + \frac{l(l+1)}{r^2}\right] + V(r)\right\}$$

$$R(r) = E \cdot R(r)$$

where the operator also depends on the quantum number l of the angular moment. This is a differential equation for only one degree of freedom. Multiplied by the spherical harmonics, the solutions give the well-known AOs $R(r;l) \cdot Y(\vartheta,\varphi)$ for the H atom.

4 SPIN

Spin is a separate issue in the elementary presentation of QM followed here. Using the axiomatic apparatus presented so far, that is the correspondence rules of the third postulate, it is indeed impossible to introduce the electron spin as the intrinsic angular moment of a charged particle rotating on itself as usually done in the naive approaches.

In fact, if one were to follow this procedure one would obtain results that are quantitatively at odds with the experiments, and one would not explain a whole series of experimental phenomena that have been recognized as spin dependent (see also Section 4.2). In particular, from a classical angular moment operator one cannot obtain the value ½ (in units of $\hbar = h/2\pi$) which characterizes the modulus of the spin angular moment of an electron, because if an angular moment has a classical analogue its possible values are only integers (see Section 3.4.1). It is not correct, also, to say that spin describes the rotation of an electron on itself.

On the other hand, the intrinsic angular moment of the electron has a solid experimental basis, which has also played a fundamental role in the historical development of QM. We certainly cannot consider a QM approach in which the electron spin is not included. Therefore, the spin requires an *ad hoc* postulate, which is added to the others.

The situation certainly is far from optimal from a theoretical point of view and is fundamentally a choice of simplicity. In fact, there is a formulation of QM in which it is not necessary to specifically introduce spin, because spin emerges automatically, and with the desired properties, from a different theoretical framework. This different formulation of QM is based on the Dirac equation that is the relativistic version of

Schrödinger equation. As said, there are cases in which relativistic effects are important also in Chemistry. However, for most applications, the Dirac equation, with the complications that its use entails, can be avoided precisely thanks to the ad hoc introduction of spin with the VI postulate.

4.1 Postulate VI. Spin: Meaning and Properties

> Postulate VI
> Spin ($\hat{s}$) is an angular moment.

Spin is both a *property* of a particle and a *coordinate*. As a property, it characterizes the particle in a fundamental way, like mass and charge. As a coordinate, spin is added to the other independent variables of the wave function (the spatial coordinates and time). Contrarily to spatial coordinates and time, the spin coordinate has only discrete values, and these are specific to the particle.

4.1.1 Spin as an Intrinsic Property of a Particle

We have the operators in the usual form: the three components x, y, z, *i.e.* $\widehat{s_x}, \widehat{s_y}, \widehat{s_z}$, the squares of the components $(\widehat{s_x^2}, \widehat{s_y^2}, \widehat{s_z^2})$, the square of the modulus $(\widehat{s^2} = \widehat{s_x^2} + \widehat{s_y^2} + \widehat{s_z^2})$ and the modulus (its square root). The usual commutation rules for angular moments also hold:

$$\left[\widehat{s_x}, \widehat{s_y}\right] = i\hbar\widehat{s_z};\ \left[\widehat{s_y}, \widehat{s_z}\right] = i\hbar\widehat{s_x};\ \left[\widehat{s_z}, \widehat{s_x}\right] = i\hbar\widehat{s_y};\ \left[\widehat{s^2}, \widehat{s_z}\right] = 0$$

As with all angular moments, the spin is specified by two quantum numbers. One is related to the modulus, is specific to the particle, and is denoted by s. The other one is related to the z component, is denoted by m_s, and can take all values from $-s$ to $+s$ in steps of 1.

The spin operators commute with the operators corresponding to spatial coordinates.

The quantum number (s) specific to the particle is:

- for the electron, proton and neutron: $s = ½$,
- for the photon: $s = 1$,
- for the κ meson and the π meson: $s = 0$, etc.

4.1.2 The Spin Coordinate

The spin coordinate takes the discrete values given by the quantum number m_s, that is, all the values from $-s$ to s in steps of 1.

From now on (unless otherwise said), it is convenient to limit ourselves to considering only electrons.

For an **electron**, the **spin coordinate**, σ, can assume only two values: $\sigma = +½$ and $\sigma = -½$. A function dependent on the spin variable, $f(\sigma)$, is specified once the two values $f(+½)$ and $f(-½)$ have been specified.

When there are no spin-dependent interactions, the Schrödinger equation cannot determine the part of the wave function that depends on spin. This part then appears in the solution of the equation as a multiplicative constant to be specified. In conclusion, in the absence of spin-dependent interactions, a mono-electronic wave function for an electron in an atom can be written as the product

- of an **orbital** (a function of the **spatial coordinates** only)
- and a **spin function** (a function of the **spin variable** only).

The **spin function** is written $\omega(\sigma)$. The two values [ω (+½) and ω (−½)] of the spin function are constrained by the requirement that the sum of their squared moduli = 1. This is the normalization of the spin functions. Usually the two simplest (normalized) spin functions are used:

$$\alpha(\sigma): \left\{ \begin{matrix} \alpha = 1,\ \sigma = +½ \\ \alpha = 0,\ \sigma = -½ \end{matrix} \right\} \quad \text{and} \quad \beta(\sigma): \left\{ \begin{matrix} \beta = 0,\ \sigma = +½ \\ \beta = 1,\ \sigma = -½ \end{matrix} \right\} \tag{4.1}$$

In common language, these are the two situations in which the electron has spin +½ or −½.

This pair of functions represents a complete set, because a linear combination of the two allows us to express any spin function, that is, any pair of values ω (+½) and ω (−½). This is easy to verify: if we need a generic spin function, for example $\omega(\sigma): \left\{ \begin{matrix} \sigma = +½ \rightarrow \omega = a \\ \sigma = -½ \rightarrow \omega = b \end{matrix} \right\}$ with $|a|^2 + |b|^2 = 1$ because of the normalization, we just consider the linear combination $a\alpha + b\beta$.

In fact, we have:

$$\omega(\sigma): \left\{ \begin{matrix} \omega = a\alpha(+½) + b\beta(+½) = a & \sigma = +½ \\ \omega = a\alpha(-½) + b\beta(-½) = b & \sigma = -½ \end{matrix} \right\}$$

Finally, it is convenient to define the **spin-orbitals (SO)**, that is, functions that are precisely the product of an orbital by one spin function. In formulas

$$\varphi(\tau) \overset{\text{def}}{=} \chi(\vec{r}) \cdot \omega(\sigma) \tag{4.2}$$

where χ is an orbital, ω is α or β and, for simplicity, the spatial coordinates ($\vec{r}$) and the spin coordinate (σ), have been grouped in a single symbol, that is, $\tau = \{\vec{r}, \sigma\}$.

4.1.3 The Magnetic Moment Related to Spin

Consistent with the fact that the electron is a charged particle (with charge $= -e$, $e > 0$), the angular moment of spin ($\vec{s}$) corresponds to a magnetic moment:

$$\vec{\mu} = -g\frac{e}{2mc^2}\vec{s} \tag{4.3}$$

The g coefficient ($g = 2.0023$) is of relativistic origin. This is another aspect of the lack of a classical analogue of spin: Eq. (4.3) does not agree with the value $\vec{\mu} = \frac{-e}{2mc^2}\vec{s}$ that would be classically obtained for the rotation of the electron on itself.

4.2 Pauli Principle

The introduction of spin has two important consequences for all cases in which two or more equal particles are treated. One concerns the mechanical treatment, the other the statistical treatment.

From the point of view of the mechanical treatment, it must be noted first that every transformation of coordinates consisting in **the exchange of the coordinates of two equal particles** is a **symmetry operation** of the system.

To put it very simply: electrons are indistinguishable. If we assume that we have a system consisting of two electrons, the case where electron 1 is *here* and electron 2 is *there* is identical to the case where electron 1 is there and electron 2 is here. The mechanical treatment *must* therefore *produce formulas that do not distinguish between the two cases.*

Considering the case of N electrons, the coordinate transformation we are talking about **must** produce a new function that has the same analytical expression as the function to be transformed, with the only difference that now the coordinates of electron k appear where the coordinates of electron l previously were and vice versa:

$$\hat{P}_{kl}\psi(\tau_1, \tau_2, \ldots, \tau_k, \ldots, \tau_l, \ldots, \tau_N) = \psi(\tau_1, \tau_2, \ldots, \tau_l, \ldots, \tau_k, \ldots, \tau_N) \tag{4.4}$$

The fact that this transformation of coordinates is a symmetry operation of the system implies its commutativity with the Hamiltonian. Finally, we use the fact that repeating the exchange brings us back to the beginning, as seen for the other self-inverse operators, Eq. (3.5). Therefore, we obtain that the eigenfunctions of the Hamiltonian of a system with two or more equal particles are automatically to be classified as symmetric or antisymmetric.

- For **symmetric functions** the eigenvalue of each exchange operator is = 1 and therefore no exchange operator changes the wavefunctions in any way.
- For **antisymmetric functions** the eigenvalue of each exchange operator is = −1 and therefore the exchanges only change the sign of the wavefunction.

Generally, wavefunctions do not follow the rule and therefore it is necessary to **symmetrize** or **anti-symmetrize** a wavefunction.

There is another aspect, which is completely general and distinguishes permutation symmetry from geometric symmetries. **Symmetry and antisymmetry** are *not* simultaneously possible.

> (Pauli Principle) Systems of indistinguishable particles are described only by symmetric or anti-symmetric states under any exchange between two particles.
>
> It is not possible to have **both** cases.
> **Symmetry** is valid for particles with **integer spin**, **antisymmetry** for particles with **semi-odd spin**.

In the case of electrons, we then have the

> (Principle of antisymmetry) For systems with more than one electron, only antisymmetric states are possible.

In the QM approach that is followed here, the Pauli principle is a principle, that is, an unproven statement, which finds its justification on an empirical basis. In essence, it is experimentally verified that in nature there are no symmetric states with respect to the exchange of two electrons. In other words, QM treatments that do not include the effect of the Pauli principle produce results that disagree with experience. Instead, the disagreement disappears after having introduced the effect of the Pauli principle into the same treatments.

The comparison with experiment is particularly direct here since the Pauli principle simply means cancelling out the states that are not antisymmetric (in the case of electrons, for example) or the states that are not symmetric (if we are dealing with particles with integer spin).

The consequences of the antisymmetry principle regarding the states of electrons in an atomic, molecular, etc. system are - as is well known - very important (an example is what is known as the **exclusion principle**) and are discussed in the chapter dealing with atomic structure.

4.3 Quantum Statistics

For the consequence of statistical nature that the introduction of spin entails, we refer to the part concerning statistical thermodynamics. Here it is sufficient to mention that the two cases (symmetric/integer spin or antisymmetric/semi-odd spin) correspond to two different statistical treatments known as

Bose-Einstein statistics (for particles with integer spin)
Fermi-Dirac statistics (for particles with semi-odd spin).

5 LINEAR SPACES AND MATRICES IN QM

For the elementary problems of QM, the usual approach looks for the *analytical* solution of the time-independent Schrödinger equation. Eigenfunctions and eigenvalues are obtained with an *exact* procedure in the form of explicit *formulas*. These are known as the *exactly solvable problems* and include, *e.g.*, the models of the particle in a box, the harmonic oscillator, the free rotor, the hydrogen atom.

The analytic solutions clearly show several basic features of QM. They directly explain how and when a quantization arises, clarify the relation between eigenvalues and expectation values, provide examples of spectra of eigenvalues that are discrete or continuous, bounded or unbounded, and allow to relate the QM results with the CM results of the same problem.

For example, the eigenvalue equation for the Hamiltonian operator of the 1D harmonic oscillator is:

$$\hat{H}\psi_n = E_n \cdot \psi_n, \quad \text{with: } \hat{H} = \frac{-\hbar^2}{2m}\frac{\partial^2}{\partial x^2} + \frac{kx^2}{2} \quad (5.1)$$

This is a second order differential equation, and the solutions are:

$$\psi_n = N_n \cdot H_n(\alpha \cdot x) \cdot e^{\frac{-\alpha^2 \cdot q^2}{2}}; \quad E_n = \left(\frac{1}{2} + n\right) \cdot \hbar \cdot \omega \quad (5.2)$$

In Eqs. (5.1) and (5.2), x is the space coordinate, $n=0$, 1, 2, … is a quantum number, ψ_n is a stationary state, E_n is its energy (the eigenvalue), and the constants ω and α are given by:

$$\omega \overset{\text{def}}{=} \sqrt{\frac{k}{m}}; \quad \alpha^2 \overset{\text{def}}{=} \frac{m \cdot \omega}{\hbar} = \frac{\sqrt{m \cdot k}}{\hbar}$$

where m and k are the two characteristic mechanical parameters of the harmonic oscillator, the mass of the point particles and the "spring constant", *i.e.*, the constant value of the second derivative of the potential energy with respect to the coordinate ($\frac{\partial^2 V}{\partial x^2} = k$). N_n is a normalization factor, while $H_n(\alpha x)$ is one of the so-called Hermite polynomials.

There are expressions as function of n for the eigenvalues (Eq. 5.2), but also for the Hermite polynomials and normalization factors. For example, the first cases are: $H_0(q)=1$, $H_1(q)=q$, $H_2(q)=4 \cdot q^2 - 2$, $H_3(q)=8 \cdot q^3 - 12 \cdot q$ using $q=\alpha x$. It can be verified by substitution that the ψ_n of Eq. (5.2) are indeed solutions of the Schrödinger equation (5.1). The ψ_n are more exactly the spatial parts of the stationary solutions of the QM equation of motion. The time dependent factors, as usual, are the imaginary exponentials of the energies:

$$\varphi_n(x,t) = \psi_n(x) \cdot e^{-iE_n t/h} = N_n \cdot H_n(\alpha \cdot x) \cdot e^{\frac{-\alpha^2 \cdot q^2}{2}} \cdot e^{-iE_n t/h}$$

However, once we have examined the few exactly solvable problems things become much more intricate: there are no exact solutions, and it is inevitable to discuss approximate solution methods. Compared to the solvable problems, the treatment of these problems and the related solution methods are different also from the mathematical point of view. The key point is that we typically deal with wave functions written as:

$$\psi_a(x) = \sum_i a_i \cdot \varphi_i \quad (5.3)$$

where the φ_i are pre-assigned functions, ψ_a is the function we are looking for and the coefficients a_i are - in general - complex numbers.

Surely, the reader is already familiar with expressions of this type for the MO (molecular orbitals) of a molecular system: the φ_i are suitable AO (atomic orbitals) and the method is known as the LCAO method (linear combination of atomic orbitals).

Note that Eq. (5.3) can imply either

- a finite number n of indices ($i=1, \ldots, n$), *i.e.* it can be a *summation*: $\psi_a = \sum_{i=1}^{n} (a_i \cdot \varphi_i)$, or
- an infinite number of indices ($i=1, \ldots, n, \ldots$), *i.e.* a *series*: $\psi_a = \sum_{i=1}^{\infty} (a_i \cdot \varphi_i)$.

Since the basis functions are pre-assigned, by specifying the coefficients a_i, the ψ_a is also specified and we can say that we *know* the wavefunction. The a_i are obtained by solving algebraic equations, *not* differential equations, and the mathematics underlying this way of approaching QM is the mathematics of **linear spaces**. Finally, all the procedures concerning linear spaces have their counterparts in **matrix calculations**, which provide the tools for solving the algebraic equations with automatic numerical calculations, that is with a computer. For these reasons, linear spaces and matrices are part of the basic language of QM and need to be known.

What follows is a simple *summary* of the essential features of these topics, with focus on the applications to QM. A reader lacking the basic knowledge of linear spaces and matrices can find in the Appendices a wider presentation that does not follow the usual mathematical style but uses an easier language rich in explanations and insights.

5.1 Linearity and QM

5.1.1 Wavefunctions as Linear Spaces

Let us come back to the basic properties of the QM and some **operations** that can be done on the wavefunctions

- **adding** two wavefunctions and
- **multiplying** a wavefunction by a number,

that is making **a linear combination**. The fundamental point is that **a linear combination of solutions is also a solution** of the same equation.

In detail

- if $\hat{H}\psi = i\hbar\partial\psi/\partial t$ and $\hat{H}\varphi = i\hbar\partial\varphi/\partial t$ (that is: if ψ and φ are solutions of a Schrödinger equation)
- then any linear combination of them (for example: $\chi=a\psi+b\varphi$, where a and b are arbitrary complex numbers) is a solution of the same Schrödinger equation.

So, we must demonstrate that $\hat{H}\chi = i\hbar\partial\chi/\partial t$ also holds. It is not difficult. On the left we have

$\hat{H}\chi = \hat{H}(a\psi + b\varphi) = a\hat{H}\psi + b\hat{H}\varphi$ since the Hamiltonian operator is linear. On the right we have: $i\hbar\,\partial\chi/\partial t = i\hbar\,\partial(a\psi + b\varphi)/\partial t = i\hbar a\,\partial\psi/\partial t + i\hbar b\,\partial\varphi/\partial t$ since also $\partial/\partial t$ is linear. Now, substituting $\hat{H}\psi = i\hbar\,\partial\psi/\partial t$ and $\hat{H}\varphi = i\hbar\,\partial\varphi/\partial t$, we see that the expressions on the right and left are the same.

In fact, much more can be proved. The linear combinations also have the other features (see Section 2.2.1) that a QM wave function must have. They are continuous on the appropriate regions of configuration space, they are differentiable, … and in particular they **can be normalized, if** the given functions (ψ and φ) **also have these features**. Some proofs are trivial (the sum of two continuous functions is continuous, etc.), except for normalization (which is not presented here). To summarise,

- we have a set of functions that are wavefunctions of a QM problem and
- by making all possible linear combinations we **remain within the set**.

We say that the set has an **internal structure**, *i.e.* the algebraic structure corresponding to the operation of making linear combinations and to the properties of the operation. The structure is called **linear space** (or **vector space**).

> The wavefunctions of a QM problem are vectors of a linear space.

The linear space structure is a **fundamental part of QM**. As already said (Section 2.3.2), QM is a linear theory, and the operators of QM are linear operators. According to the different presentations of QM, the operators are explicitly declared linear in a postulate, or their linearity is, let us say, discovered by developing the consequences of postulates formulated in another way.

The linear spaces of QM wavefunctions share many of their properties with the ordinary 3D space or its 1D or 2D subspaces, line and plane. With respect to the spaces of elementary geometry, the linear spaces of QM wavefunctions are characterized by

1. the use of **complex** numbers in the linear combinations that define the elements of the linear space,
2. the **dimension** of the linear space and the problem (clarified later) of the **completeness** of a linear space,
3. the **metric properties**, *i.e.* the meaning of quantities such as the length of a vector and the angle between two vectors (from which the property of orthogonality between vectors follows).

Concerning the first point, the QM wavefunctions are a **linear space over the field of complex numbers**, while the ordinary 3D space is over the field of *real* numbers.

Concerning the second point, and referring again to the Appendices for deeper explanations, we say that:

> the dimension of a QM wavefunction space is *usually* infinite

The meaning is that (generally) Eq. (5.3) must be written as a series: $\psi_a = \sum_{i=1}^{\infty}(a_i \cdot \varphi_i)$.

An example and a warning.

Let us recall the example of the harmonic oscillator (see Eqs. 5.1 and 5.2) and assume that the eigenfunctions of the Hamiltonian are linearly independent (this can be proved). These functions are infinite in number, because there is an expression that produces an independent function for every natural number. So, there is not an upper limit to the i index in Eq. (5.3). We say then that the space of wavefunctions of the harmonic oscillator has infinite dimension.

However, there are QM problems for which a finite-dimensional space can be used. For example, in discussing the state of a particle with spin ½ (and neglecting its position in 3D space) a two-dimensional space of wave functions is used, with a basis constituted by the two states normally denoted by α and β or "spin up" and "spin down". A system with two spin ½ particles is similarly described with a four-dimensional space and so on. A quantum mechanical treatment of this type is for example sufficient to describe the most elementary features of NMR and EPR spectroscopies.

As the Appendices shows, the infinite dimensional nature of most linear spaces of QM show additional difficulties for what concerns the mathematical foundation. On the other side, the translation - let us say - into matrix calculus and numerical algorithms obviously refers to a finite number of basis functions. For simplicity, in the following, we omit the upper limit of the indices when a single formula holds for both cases.

Concerning the third point, the linear spaces of QM require the introduction of an inner product.

5.1.2 Inner Product and Orthonormality of Wavefunctions

In the ordinary 3D space, one can define the *scalar product* or *dot product* of two vectors. This is *different* from the operations that define a linear space (addition and multiplication by a scalar) but is *compatible* with them. In general, the scalar product makes possible to introduce **metric properties**, that is, makes possible to define the *projection* of a vector on another vector, the *length* of a vector, the *distance* and *angle* between two vectors, and also the fact that a vector may have length = 1 or that two vectors are *orthogonal*, in the sense that they are in directions that have nothing to do with each other. The scalar product $(\vec{a},\vec{b})$ of two vectors of the ordinary 3D space is defined as:

$$(\vec{a},\vec{b})\cdot|\vec{a}|\cdot|\vec{b}|\cdot\cos\left(\widehat{\vec{a},\vec{b}}\right)$$

The essential properties are that:

1. it is defined for every pair of vectors (as well as for a vector with itself),
2. it is *symmetric* with respect to the exchange of the two vectors $(\vec{a},\vec{b})=(\vec{b},\vec{a})$,
3. it is *linear* with respect to the two vectors: both the relation $(\vec{a},(\beta\cdot\vec{b}+\gamma\cdot\vec{c}))=\beta\cdot(\vec{a},\vec{b})+\gamma\cdot(\vec{a},\vec{c})$ and the relation $((\beta\cdot\vec{b}+\gamma\cdot\vec{c}),\vec{a})=\beta\cdot(\vec{b},\vec{a})+\gamma\cdot(\vec{c},a)$ hold,
4. it is *positive definite*, that it is always positive, except for the scalar product by the null vector.

A closely related operation can be introduced also on the spaces of QM wavefunctions. For the purpose, the operation

- must be meaningful for the wavefunctions of QM,
- must yield a number for each ordered pair of functions and
- must be characterized by the formal properties that the abstract algebraic theory requires for a scalar product (the properties (1) … (4) reported above).

On the other side, on the wavefunctions of QM we introduced integrals of the type:

$$\left[\int_\Omega \psi_a^*(q,t)\cdot\psi_a(q,t)\cdot \mathrm{d}q\right] \quad \text{and} \quad \left[\int_\Omega \psi_a^*(q,t)\cdot\hat{G}\psi_a(q,t)\cdot \mathrm{d}q\right]$$

and we now want to find the meaning of these integrals in the context of linear spaces.

We must also consider the cases where these integrals contain two different functions:

$$\left[\int_\Omega \psi_a^*(q,t)\cdot\psi_b(q,t)\cdot \mathrm{d}q\right] \quad \text{and} \quad \left[\int_\Omega \psi_a^*(q,t)\cdot\hat{G}\psi_b(q,t)\cdot \mathrm{d}q\right]$$

For all these integrals we had introduced, for convenience, the Dirac (bra-ket) notation:

$$\left[\int_\Omega \psi_a^*(q,t)\cdot\psi_b(q,t)\cdot \mathrm{d}q\right] \Leftrightarrow \langle a|b\rangle \quad \text{and}$$

$$\left[\int_\Omega \psi_a^*(q,t)\cdot\hat{G}\psi_b(q,t)\cdot \mathrm{d}q\right] \Leftrightarrow \langle a|\hat{G}|b\rangle$$

The two integrals $\langle a|b\rangle$ and $\langle a|\hat{G}|b\rangle$ can be treated in a unified way, if we consider that an operator produces a new function ($|c\rangle \overset{\text{def}}{=} \hat{G}|b\rangle$) from the given function and therefore $\langle a|\hat{G}|b\rangle \overset{\text{def}}{=} \langle a|c\rangle$.

Directly jumping to conclusions,[4] the scalar product, and more generally the metric properties of a linear space, are topics strictly connected with the integrals of functions of QM. In summary

- there is an operation defined for two QM wavefunctions that
  - gives as a result a number and
  - can be considered as the analogue of the scalar product of the ordinary vectors of 3D space, meaning that it is characterized by the same formal properties, and
- this is the operation between wavefunctions for which we were looking for a meaning in the context of the theory of linear spaces, namely it is the integral:

$$\left[\int_\Omega \psi_a^*(q,t)\cdot\psi_b(q,t)\cdot \mathrm{d}q\right] \Leftrightarrow (\vec{a},\vec{b}) \tag{5.4}$$

- which is generally known as the **inner product** of two wavefunctions.

It is easily verified that this relation satisfies the requirements of the theory for a scalar product given above, with **two extensions** due to the use of complex numbers instead of real numbers.

Unlike the scalar product between vectors in ordinary 3D space, this product (Eq. 5.4) is *not* necessarily a real number but can be a complex number. It is also **conjugate-symmetric**:

$$(\vec{a},\vec{b}) = \int_\Omega \psi_a^*(q,t)\cdot\psi_b(q,t)\cdot \mathrm{d}q = \left[\int_\Omega \psi_b^*(q,t)\cdot\psi_a(q,t)\cdot \mathrm{d}q\right]^* = (\vec{b},\vec{a})^*$$

However, the inner product of a state with itself is always real. Indeed, it is positive or exceptionally zero: the integrand is everywhere real and is$=0$ if and only if $\psi_a(q,t)$ is the function that is everywhere$=0$.

The other extension concerns the **linearity** (property (3) above).

As can be easily verified (see Section 2.3.5), the inner product is *linear* in the second factor. If $\vec{b}=(\gamma\cdot\vec{c}+\delta\cdot\vec{d})$, then the inner product is $(\vec{a},\vec{b})=(\vec{a},(\gamma\cdot\vec{c}+\delta\cdot\vec{d}))=\gamma\cdot(\vec{a},\vec{c})+\delta\cdot(\vec{a},\vec{d})$.

However, for the first factor a different relation holds. If $\vec{a}=(\gamma\cdot\vec{c}+\delta\cdot\vec{d})$, the inner product is

$$(\vec{a},\vec{b})=((\beta\cdot\vec{b}+\gamma\cdot\vec{c}),\vec{a})=\beta^*\cdot(\vec{b},\vec{a})+\gamma^*\cdot(\vec{c},\vec{a})$$

and it is said that the scalar product of Eq. (5.4) is ***antilinear* in the first factor**.

These two differences with respect to the scalar product of ordinary 3D space are obviously due to the nature of QM wavefunctions as a linear space over the field of *complex* numbers.

The integral $\int_{\Omega} \psi_a^*(q,t) \cdot \psi_b(q,t) \cdot dq$ has the meaning of **inner product** between the states ψ_a and ψ_b.

The notation $\langle a|b \rangle$ visually highlights that it is the inner product (a, b) between vectors a, b: this further shows how useful is Dirac notation. Moreover,

if $\int_{\Omega} \psi_a^*(q,t) \cdot \psi_a(q,t) \cdot dq = 1$ then the state is said to be **normalized**.

If each vector has zero projection along the other, the two vectors are said to be orthogonal:

In QM, if the inner product of two states (a and b) is 0, then the states are said to be **orthogonal**.

Considering the inner product between different functions, we recall that, in elementary geometry, the scalar product between two vectors serves to quantify the **projection** of one vector along the other, that is, **how much of one vector is contained in the other**.

A linear space over the complex field, with an inner product, and the property of being complete[5] is called a **Hilbert** space:

the states of a QM problem are a Hilbert space.

5.1.3 Eigenfunctions and Subspaces

Subspaces are subsets of a linear space that are themselves vector spaces. They can be related to the eigenvalue equation (postulate no. 5). Let us consider for example the operator corresponding to a rotation of 120° around a given axis of the vectors of the ordinary 3D space. A generic vector is modified in a way that depends on its orientation with respect to the axis. A vector *collinear to the axis* does not change direction nor magnitude: it is an eigenvector with eigenvalue = 1. There are infinitely many vectors with this characteristic. Once a unit vector is chosen, all the other vectors can be obtained as multiples of this vector, *i.e.* they are its linear combinations. The line containing the axis of rotation is a (1D) **subspace** of the global space.

Instead, a vector *perpendicular to the axis* of rotation does not change its magnitude but changes direction while remaining in the plane perpendicular to the axis of rotation. Each vector of this plane can be obtained as a linear combination of two (linearly independent) vectors of the plane itself. Therefore, all the vectors of this plane are a (2D) subspace of the global space.

The rotation operator then identifies two subspaces, and we see that

- an eigenfunction that belongs to one subspace is transformed by the operator into an eigenfunction of the same *subspace* and with the same *eigenvalue*,
- eigenfunctions that belong to different subspaces are orthogonal to each other.

Coming back to the linear spaces of QM, we see that (a) different eigenvalues identify different subspaces and (b) **eigenfunction with different eigenvalues are orthogonal to each other**.

To derive the property b), let us consider $\langle a|\hat{G}|b \rangle = \int \varphi_a^* \cdot \hat{G}\varphi_b d\tau$ and the assumption that both functions are eigenfunctions of the Hermitian operator: $\hat{G}\varphi_a = G_a \cdot \varphi_a$, $\hat{G}\varphi_b = G_b$ with $G_a \neq G_b$. Then: $\int \varphi_a^* \cdot \hat{G}\varphi_b d\tau = G_b \int \varphi_a^* \cdot \varphi_b d\tau$ and, analogously, $\int \varphi_b^* \cdot \hat{G}\varphi_a d\tau = G_a \int \varphi_b^* \cdot \varphi_a d\tau$. According to the turnover rule (2.3.5): $\int \varphi_a^* \cdot \hat{G}\varphi_b d\tau = \left[\int \varphi_b^* \cdot \hat{G}\varphi_a d\tau \right]^*$ because $\hat{G}$ is Hermitian. So: $\int \varphi_a^* \cdot \hat{G}\varphi_b d\tau - \left[\int \varphi_b^* \cdot \hat{G}\varphi_a d\tau \right]^* = 0$ and therefore: $G_b \int \varphi_a^* \cdot \varphi_b d\tau - G_a \left[\int \varphi_b^* \cdot \varphi_a d\tau \right]^* = 0$.

From: $\int \varphi_a^* \cdot \varphi_b d\tau = \left[\int \varphi_b^* \cdot \varphi_a d\tau \right]^*$, we finally obtain $(G_b - G_a) \cdot \int \varphi_a^* \cdot \varphi_b d\tau = 0$. By hypothesis, the two eigenvalues are different, and this proves $\int \varphi_a^* \cdot \varphi_b d\tau = 0$.

Note however that **different subspaces *can* have the same eigenvalue**.

5.2 Matrices in QM

The correspondence between vectors and wavefunctions and the possibility of writing a wavefunction as a linear combination of a given set of functions

$$\psi_a = \sum_i a_i \varphi_i \qquad (5.3)$$

also allows us to replace the operations on the functions with corresponding operations on the coefficients.

Given a complete set of linearly independent basis functions, each function of a linear space is uniquely associated with a list of coefficients that are the components of the function with respect to the given basis.

For example, let us use φ_1 and φ_2 as basis functions and assume that they are linearly independent, *i.e.* that one cannot be written as a multiple of the other. For any two functions (χ and ψ) let us consider the linear combinations

$$\chi = a_1 \cdot \varphi_1 + a_2 \cdot \varphi_2 \quad \text{and} \quad \psi = b_1 \cdot \varphi_1 + b_2 \cdot \varphi_2$$

The numbers (a_1 and a_2) or (b_1 and b_2) are the *components* of the functions χ or ψ in the given basis φ_1 and φ_2). Technically, the components are numbers belonging to the scalar field on which the vector space is defined: for the wavefunctions of QM they are (in general) complex numbers. The one-to-one correspondence between vectors and components with respect to a given basis is valid for every vector space and, in particular, it is valid for both the ordinary 3D space and for the spaces of QM wave functions. This is technically called an isomorphism and every operation on the vectors of a space has its corresponding operation on the components.

The components and operations on the components then become the elements of an alternative formulation of QM. This alternative "language" is *matrix calculus*. It is no coincidence, therefore, that two mathematical formulations are at the origin of QM. These are due to Heisenberg, Jordan and Born, and to Schrödinger and are known as *matrix mechanics* and *wave mechanics*. They were later recognized as *equivalent* formulations of the same theory.

The appendix on matrix calculus provides a brief introduction to the matrices and the operations that can be performed on them concerning the applications to QM. Here, it is again assumed that the most important points are already known, and we simply recall them quickly.

5.2.1 Basic Matrices

A matrix is a rectangular pattern of real or complex numbers on n rows and m columns ($n \times m$ matrix):

$$\begin{pmatrix} A_{11} & A_{12} & \cdots & A_{1m} \\ A_{21} & A_{22} & \cdots & A_{2m} \\ \vdots & \vdots & \ddots & \vdots \\ A_{n1} & A_{n2} & \cdots & A_{nm} \end{pmatrix}$$

Particularly important are **square** matrices ($n \times n$), **column** matrices ($n \times 1$) and **row** matrices ($1 \times n$), **diagonal** matrices (square matrices that have non-zero elements only along the main diagonal, that is elements A_{ii}) and, among the diagonal matrices, **scalar** matrices, in which all the elements along the main diagonal are equal.

Among the ($n \times m$) matrices there is one (unique) matrix that plays the role of a zero element with respect to addition and has all elements$=0$. This is called the **null matrix** and is written $\mathbf{0}_{nm}$.

The transpose of an $n \times m$ matrix ($\boldsymbol{A}$) is the $m \times n$ matrix (written $\boldsymbol{A}^{\mathrm{T}}$) that contains the same elements as the original matrix but rearranged by exchanging rows and columns: $(\boldsymbol{A}^{\mathrm{T}})_{ij} = A_{ji}$.

With complex matrices (*i.e.* matrices whose elements are complex numbers) instead of the transpose we are typically interested in the **adjoint** (or **Hermitian conjugate**) written with the superscript dagger †. Here, in addition to the row/column exchange we also switch to the complex conjugate: $(\boldsymbol{A}^{\dagger})_{ij} = A^*_{ji}$.

5.2.2 Sum and Product of Matrices

Matrix **sum** and **scalar multiplication** are simple and intuitive operations because they are performed independently on each element of the matrices. These operations correspond exactly to the fundamental operations (vector addition and multiplication by a scalar) that are the basis of the notion of linear space.

For instance, if the basis functions are $\{\varphi_i\}$ and if $\psi_a = \sum_{i=1}^{n} (a_i \cdot \varphi_i)$ and $\psi_b = \sum_{i=1}^{n} (b_i \cdot \varphi_i)$, $\psi_c = \psi_a + \psi_b$ is given by $\psi_c = \sum_{i=1}^{n} (c_i \cdot \varphi_i)$. The coefficients of the sum are: $c_1 = a_1 + b_1$, $c_2 = a_2 + b_2$, ...; $c_n = a_n + b_n$, ... and analogously holds for the product with a number: $\psi_d = k \sum_{i=1}^{n} (a_i \cdot \varphi_i)$. In matrix notation we have:

$$\text{if } \psi_a \Leftrightarrow \begin{pmatrix} a_1 \\ a_2 \\ \vdots \end{pmatrix} = \boldsymbol{a};\ \psi_b \Leftrightarrow \begin{pmatrix} b_1 \\ b_2 \\ \vdots \end{pmatrix} = \boldsymbol{b};\ \psi_c \Leftrightarrow \begin{pmatrix} c_1 \\ c_2 \\ \vdots \end{pmatrix}$$

$$= \boldsymbol{c};\quad \psi_d \Leftrightarrow \begin{pmatrix} d_1 \\ d_2 \\ \vdots \end{pmatrix} = \boldsymbol{d};$$

then, $\psi_c = \psi_a + \psi_b$ translates to $\boldsymbol{c} = \boldsymbol{a} + \boldsymbol{b}$

$$= \begin{pmatrix} a_1 \\ a_2 \\ \vdots \end{pmatrix} + \begin{pmatrix} b_1 \\ b_2 \\ \vdots \end{pmatrix} = \begin{pmatrix} a_1 + b_1 \\ a_2 + b_2 \\ \vdots \end{pmatrix} = \begin{pmatrix} c_1 \\ c_2 \\ \vdots \end{pmatrix}$$

and $\psi_d = k\psi_a$ translates to $\boldsymbol{d} = k \cdot \boldsymbol{a}$

$$= k \cdot \begin{pmatrix} a_1 \\ a_2 \\ \vdots \end{pmatrix} = \begin{pmatrix} ka_1 \\ ka_2 \\ \vdots \end{pmatrix} = \begin{pmatrix} d_1 \\ d_2 \\ \vdots \end{pmatrix}$$

Therefore, in the application of matrix calculus to QM, the **components** of a function (with respect to a given basis) are written in a **column matrix**.

NB. The choice to write the components of a vector with respect to a pre-assigned basis in a column matrix is conventional. It would be equally correct to choose to write them in a row matrix. Clearly, the following formulas should therefore be modified accordingly.

There is another important basic matrix operation. This is the **matrix-by-matrix multiplication** and is used for:

a. **linear operators**,
b. **changes of basis**, *i.e.* transformations for going from one set of basis vectors to another set of basis vectors, where the "new" basis vectors are linear combinations of the "old" ones,
c. and **scalar product** of vectors.

The matrix-by-matrix multiplication operation **is not always possible**. For matrix-by-matrix multiplication to make sense, the second dimension of the first factor must be equal to the first dimension of the second factor. A matrix $\boldsymbol{A}$ $(n \times m)$ and a matrix $\boldsymbol{B}$ $(r \times s)$ can be multiplied together (in this order) only if $m=r$ and the result $\mathbf{C}=\boldsymbol{A} \cdot \boldsymbol{B}$ is a matrix $(n \times s)$. $\boldsymbol{A} \cdot \boldsymbol{B}$ is also written as $\boldsymbol{AB}$.

Matrix-by-matrix multiplication is also a more complicated operation than matrix addition and scalar by matrix multiplication. Each of the elements of $\boldsymbol{C}=\boldsymbol{A} \cdot \boldsymbol{B}$ is obtained by a row by column product according to the recipe given in Appendix 3.1.3:

$$\text{Matrix-by-matrix product: } \boldsymbol{C} = \boldsymbol{A} \cdot \boldsymbol{B} \Leftrightarrow C_{ij} = \sum_{k=1}^{m} A_{ik} \cdot B_{kj} \tag{5.5}$$

Recall that matrix multiplication is **not** (necessarily) **commutative**. Diagonal matrices are a special case, and this is obviously valid also for scalar matrices *i.e.* diagonal matrices with all the elements of the main diagonal equal (see again Appendix 3.1.3).

The scalar matrix of order n with all non-zero elements$=1$ is the **unit matrix** of order n, and it is written as $\mathbf{1}_n$. For any $(n \times n)$ matrix $\boldsymbol{A}$: $\boldsymbol{A} \cdot \mathbf{1}_n = \mathbf{1}_n \cdot \boldsymbol{A}$. Given the existence of a unit matrix, it makes sense to define the **inverse** ($\boldsymbol{B}$) of an $(n \times n)$ matrix $\boldsymbol{A}$: we write $\boldsymbol{B}=\boldsymbol{A}^{-1}$ and $\boldsymbol{A}=\boldsymbol{B}^{-1}$ if $\boldsymbol{AB}=\boldsymbol{BA}=\mathbf{1}_n$. **Not all** square $(n \times n)$ matrices are **invertible**: the inverse does not always exist but, **if it exists, it is unique.**

A square matrix is said to be **symmetric** if it is equal to its transpose; it is said to be **Hermitian** if it is equal to its adjoint.

An **orthogonal** matrix is a real invertible matrix whose transpose is equal to the inverse. A **unitary** matrix is a complex invertible matrix whose adjoint (the transpose of the conjugate) is equal to the inverse. Orthogonal matrices and unitary matrices are important for **similarity transformations**, that is, for the transformations that linear operators undergo when the basis functions are changed.

5.2.3 Matrix Representation of QM States

Let us directly discuss some results useful for QM, referring once again to the Appendix on matrix calculus for more details.

Given a set of basis functions $\{\varphi_i\} \stackrel{\text{def}}{=} \{\varphi_1, \varphi_2, \ldots, \varphi_i, \ldots\}$: a QM state is represented by a column-matrix

$$\psi = \sum_{i=1}^{n} (a_i \cdot \varphi_i) \Leftrightarrow \boldsymbol{a} = \begin{pmatrix} a_1 \\ a_2 \\ \vdots \end{pmatrix}$$

A linear operator is represented by a square matrix.

$$\hat{G} \Leftrightarrow \boldsymbol{G} = \begin{pmatrix} G_{11} & G_{12} & \cdots \\ G_{21} & G_{22} & \cdots \\ \vdots & \vdots & \ddots \end{pmatrix}$$

The elements of $\boldsymbol{G}$ are given by: $\boldsymbol{G}_{rs} = \langle r|\hat{G}|s \rangle = \int \varphi_r^* \cdot \hat{G} \varphi_s \mathrm{d}\tau$ (φ_r and φ_s are basis functions).

A Hermitian operator is represented by a Hermitian matrix.

This is easily derived from the turnover rule (Section 2.3.5): $\langle i|\hat{A}|j \rangle = \left[\langle j|\hat{A}|i \rangle\right]^*$ for a Hermitian operator ($\hat{A} = \hat{A}^\dagger$). This gives $A_{ij} = (A_{ji})^*$ which is the definition of a Hermitian matrix.

If $\psi_a = \sum_i (a_i \cdot \varphi_i)$ (that is $\psi_a \Leftrightarrow \boldsymbol{a}$) and $\psi_a \Leftrightarrow \sum_i (b_i \cdot \varphi_i)$ (that is $\psi_b \Leftrightarrow \boldsymbol{b}$) then: the integrals of an operator $\hat{\mathbf{G}}$ between two states are given by:

$$\langle a|\hat{G}|b \rangle = \sum_{j=1} \left(\sum_{i=1} (a_j^* \cdot G_{ji} \cdot b_i) \right) = \boldsymbol{a}^\dagger \cdot \boldsymbol{G} \cdot \boldsymbol{b}$$

The expectation value for the observable $\boldsymbol{G}$ in state $\boldsymbol{a}$ is

$$\langle a|\hat{G}|a \rangle = \sum_{j=1} \left(\sum_{i=1} (a_j^* \cdot a_i \cdot G_{ji}) \right) = \boldsymbol{a}^\dagger \cdot \boldsymbol{G} \cdot \boldsymbol{a}$$

The integrals between φ_i and φ_j are given by:

$$(\boldsymbol{S})_{ij} = \langle \varphi_i^* | \varphi_j \rangle = \int_{\Omega} \varphi_i^* \cdot \varphi_j \cdot \mathrm{d}\tau$$

When going from one basis set to another, the coordinates of a vector are transformed ($\boldsymbol{a} \Rightarrow \boldsymbol{b}$) according to $\boldsymbol{b} = \boldsymbol{T}^{-1} \cdot \boldsymbol{a}$ and $\boldsymbol{a} = \boldsymbol{T} \cdot \boldsymbol{b}$.

The columns of the transformation matrix $\boldsymbol{T}$ are the "new" basis vectors ($\boldsymbol{t}_i$) as a function of the "old" ones: $\boldsymbol{T} = \begin{pmatrix} t_1 & t_2 & \cdots \end{pmatrix}$.

A change of basis transforms the operator matrices according to a similarity transformation:

$$\boldsymbol{G} \Rightarrow \boldsymbol{F} \quad \boldsymbol{F} = \boldsymbol{T}^{-1} \cdot \boldsymbol{G} \cdot \boldsymbol{T} \quad \boldsymbol{G} = \boldsymbol{T} \cdot \boldsymbol{F} \cdot \boldsymbol{T}^{-1}$$

Also:

Solving an eigenvalue equation $\hat{G}\psi_i = E_i \cdot \psi_i$ corresponds to diagonalizing the operator matrix

Let us now see the matrix form of the **eigenvalue equation with an orthonormal basis**. Given a basis of functions, we write the matrix $\boldsymbol{G}$ of an operator. Then, we diagonalize $\boldsymbol{G}$

$$\boldsymbol{A}^{-1} \cdot \boldsymbol{G} \cdot \boldsymbol{A} = \boldsymbol{\lambda}$$

$\boldsymbol{A}$ is the matrix that transforms by similarity $\boldsymbol{G}$ into a diagonal matrix $\boldsymbol{\lambda}$. $\boldsymbol{\lambda}$ contains the eigenvalues of the operator, $\boldsymbol{A}$ contains the eigenvectors. In each column of $\boldsymbol{A}$, we find the coefficients that give an eigenfunction as a linear combination of the basis vectors.

Concerning the **eigenvalue equation with a non-orthonormal basis**:

$$\boldsymbol{G} \cdot \boldsymbol{A} = \boldsymbol{\lambda} \cdot \boldsymbol{S} \cdot \boldsymbol{A} \;\; (\textit{NON}\text{-orthonormal basis})$$

$\boldsymbol{S}$ is the matrix of integrals between basis functions as defined above; $\boldsymbol{\lambda}$ contains the eigenvalues; $\boldsymbol{A}$ contains the eigenvectors. Operationally, one diagonalizes ($\boldsymbol{S}^{-1} \bullet \boldsymbol{G}$) instead of $\boldsymbol{G}$.

Other properties:

The eigenvalues of an observable are always real.
The expectation values of an observable are always real (whatever the state).
The eigenfunctions of a Hermitian matrix are a **complete** and **orthogonal set**.

The *completeness* of a basis set essentially means that **any state** can be represented as a linear combination of the functions of that basis set. This property must be specified for infinite-dimensional linear spaces. Then, it means that every function of linear space can be written as a series expansion on the basis set, *i.e.* that the series converges.

As for **orthogonality, eigenvectors related to different eigenvalues are automatically orthogonal**, as shown in Section 5.1.3. The choice of an orthogonal set for a degenerate eigenvalue is discussed at point 3.2.5.3 of the Appendices.

Finally, we remind that an eigenvalue equation is satisfied both by normalized eigenvectors and by non-normalized eigenvectors (*i.e.* multiplied by an arbitrary complex factor).

6 PERTURBATION METHODS

In basic QM textbooks, three approximate methods are traditionally presented: the variational method, the perturbative method and the WKB method (Wentzel-Kramers-Brillouin). Concerning the WKB method, it is enough to mention that it essentially develops the asymptotic value of the CM, using its results to approximate those of QM. It is not even necessary to discuss the variational method, since in textbooks for Chemistry courses it is usually reported in detail, in the version that leads to the calculation of MOs as LCAO. It is perhaps worth emphasizing, however, that the MO-LCAO method is only one of the many forms of the variational method, both from the methodological point of view (a variational method does not necessarily use linear combinations) and from the point of view of applications.

However, it is useful to dedicate a few pages to the perturbative method because it is reported less frequently in Chemistry textbooks and because it is very widely used in different situations and presents features that are in some respects complementary to the variational method. To compare the two methods and to summarize as much as possible, the variational method is based on a hypothesis - about the analytical form of the solution - suggested by considerations external to the method itself. The method leaves the widest freedom in the choice of this hypothesis, indicates how to use it best and provides - we can say - a grade for it, but it is not useful for formulating the hypothesis. That a mono-electronic wave function for a molecule can be written as a linear combination of orbitals of the constituent atoms is an idea that does not arise from the variational method. However, once this assumption is accepted, the variational method provides (under certain conditions) the best coefficients, in the specific sense that they are those that provide the best estimate of the energy. It should also be emphasized that these coefficients may provide less than optimal information for other observables.

The perturbation methods, on the contrary, *operationally* indicate the form of the solution of a quantum mechanical problem because, essentially, they encode a method of successive approximations. To be precise, the perturbation methods are applied to the problem of a Hamiltonian operator $\hat{H} = \widehat{H^0} + \hat{V}$ that differs by a 'small' part ($\hat{V}$, called perturbation) from the Hamiltonian ($\widehat{H^0}$) of a problem whose solutions $\{\varphi_n, E_n^0\}$ are already known.

The energies and wave functions are obtained as "corrections" to the unperturbed problem and there are subsequent corrections (to the first order $\delta^{(1)}$, to the second order $\delta^{(2)}$, …) as needed:

$$E_n = E_n^0 + \delta E_n = E_n^0 + \delta^{(1)} E_n + \delta^{(2)} E_n + \delta^{(3)} E_n + \cdots \quad \text{and} \tag{6.1}$$

$$\psi_n = \varphi_n + \delta\varphi_n = \varphi_n + \delta^{(1)}\varphi_n + \delta^{(2)}\varphi_n + \delta^{(3)}\varphi_n + \cdots \tag{6.2}$$

A **general condition** for the applicability of the perturbation methods is that the **perturbation** is **small** (in the sense that the theory itself specifies) and that the subsequent corrective terms ($\delta^{(1)}$, $\delta^{(2)}$, $\delta^{(3)}$, …) on the functions and on the energy are *progressively smaller*, in a sort of convergence.

There are no special requirements as to which interaction can be treated as a perturbation: it can be either an external field acting on the particles of a material system, or an additional interaction between the particles of the system. In the opposite case in which - for example - the interaction of an external field with the particles that constitute the material system of interest is sufficiently "strong", it is possible to apply the perturbative method by reversing the point of view, *i.e.* by treating the interaction between the particles of the system as a "small" perturbation of their interactions with the external field.

The key point is that the corrections are written as *series expansions over the unperturbed functions*, thanks to the fact that the unperturbed functions are a complete and orthonormalized basis set ($\langle n|m\rangle = 0$ if $n \neq m$ or $=1$ if $n = m$)

$$\delta^{(1)}\varphi_n = \sum_{j\neq n} a_j^{(1)} \cdot \varphi_j \tag{6.3}$$

and similarly for the subsequent corrections $\delta^{(2)}, \delta^{(3)}, \ldots$

6.1 Time-Independent Perturbations without Degeneracy

To begin with, it is convenient to consider the case when both the unperturbed Hamiltonian and the perturbation **do not contain time**, and the **unperturbed system** is **non-degenerate**.

The detailed derivation of the resolution formulas is not particularly significant and can safely be left to technical books: basically, Eq. (6.3) is replaced in Eqs. (6.1) and (6.2), orthonormality between the unperturbed basis functions is considered, the different orders of corrections are grouped together and are equalized one by one. In the end, we arrive at:

First order correction for energy:

$$\delta^{(1)}E_n = \langle n|\hat{V}|n\rangle \quad . \tag{6.4}$$

First order correction for functions:

$$\delta^{(1)}\varphi_n = \sum_{j\neq n} \frac{\langle j|\hat{V}|n\rangle}{E_j^0 - E_n^0} \cdot \varphi_j \tag{6.5}$$

Secondorder correction for energy:

$$\delta^{(2)}E_n = \sum_{j\neq n} \frac{|\langle j|\hat{V}|n\rangle|^2}{E_n^0 - E_j^0} \tag{6.6}$$

Expressions can be developed to any order. Those in Eqs. (6.4)–(6.6) are just the first and most frequently used ones. Often, one simply wants to understand what the effect of the perturbation is. Then, it is sufficient to stop at Eq. (6.4) alone (first-order correction for E, no correction for the functions). If Eq. (6.4) provides no correction, one moves on to consider Eq. (6.5) together with Eq. (6.6) (second-order correction for E with first-order correction for the functions) and so on as necessary.

Finally, it is possible to apply more perturbations (let us say $\widehat{V_1}$ and $\widehat{V_2}$) to an unperturbed problem ($\widehat{H^0}$), either in the most trivial way (*i.e.* working with a global perturbation $\hat{V} = \widehat{V_1} + \widehat{V_2}$), or by applying the perturbations one after the other: in this last case it is necessary to start from the most important one (the one that produces the greatest variation in energy).

6.2 Time-Independent Perturbations with Degeneracy

Let us now move on to the case in which the unperturbed Hamiltonian and the perturbation do not yet contain time, but the unperturbed system has some degenerate levels. Then, the same formulas as above apply for the non-degenerate levels, but the treatment must be revised for the degenerate levels. It is obvious, in fact, that Eqs. (6.5) and (6.6) cannot be applied if $E_n^0 = E_j^0$, since the denominator vanishes, and the normal perturbation series would diverge.

The essential point is related to the choice of the eigenfunctions of a degenerate level. As Figure 1.4 shows by resorting once again to the analogy between vectors of the 3D ordinary space and the function spaces of QM, it is necessary to specify which are the functions - within a set of degenerate eigenfunctions - to which the correction algorithm has to be applied. In other words, it is necessary to find which are the independent linear combinations (of the initial functions) that are to be corrected. We say that these are the functions *adapted to the perturbation.*

Let us consider a level (say the nth level) and assume that it is m-times degenerate with eigenfunctions $(\varphi_{n1},\ldots,\varphi_{nm})$. As said many times before, any set of independent linear combinations of these eigenfunctions $(\chi_{n1},\ldots,\chi_{nm})$ would be equally acceptable as a basis set. The second set is obtained by applying a unitary linear operator on the first set.

The problem has a simple solution, because the perturbation itself provides us with the appropriate unperturbed functions. Given an arbitrary set $\varphi_{n1},\ldots,\varphi_{nm}$ of eigenfunctions of the unperturbed Hamiltonian relative to the nth m times degenerate level, we first build the matrix representation of the perturbation operator in this basis:

$$V \Leftrightarrow \hat{V}_{ij} = \langle i|\hat{V}|j\rangle \tag{6.7}$$

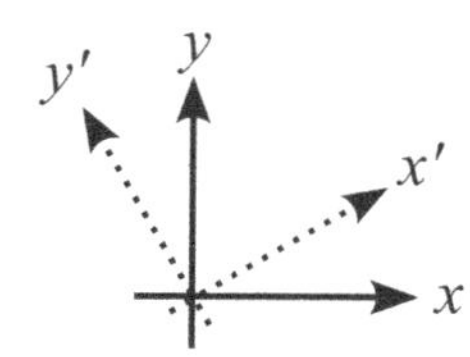

FIGURE 1.4 Scheme of the functions adapted to the perturbation, in case of twofold degeneracy. Full Lines: unperturbed functions of the degenerate energy level, dotted lines: perturbation-adapted functions for the same level.

where i and j denote one of the $\varphi_{n1},\ldots,\varphi_{nm}$. Then we diagonalize $\boldsymbol{V}$ to obtain (a) the eigenfunctions $(\chi_{n1},\ldots,\chi_{nm})$ adapted to the perturbation and (b) the eigenvalues of the perturbation, which are also the so-called zero-order corrections to the energies. **If** the eigenvalues of $\boldsymbol{V}$ originating from the diagonalization are **all different from each other** and therefore the modified energies are also all different from each other, we say that **the perturbation has completely removed the degeneracy**. Then, if we want to proceed further, we regularly apply the perturbation procedure seen before (Eqs. 6.5 and 6.6 and those that follow) adding the eigenfunctions of $\boldsymbol{V}$ thus found to the unperturbed eigenfunctions of the non-degenerate levels. Instead, if the eigenvalues originating from the diagonalization are **partially equal**, it is said that the perturbation has **partially removed the degeneracy**. The eigenfunctions of $\boldsymbol{V}$ so found are added to the eigenfunctions of the non-degenerate levels and we continue by applying the perturbative procedure for non-degenerate levels, ignoring the corrective terms between the functions that remained degenerate. The eigenvalues of the perturbation **can** also **all be equal** (but not necessarily zero): in this case we say that **the perturbation has not removed the degeneracy**.

The degeneracies of $\boldsymbol{H^0}$ and $\boldsymbol{V}$ (and hence of $\boldsymbol{H}$) are essentially due to symmetry, as discussed much more extensively in the chapter dedicated to symmetry. The (total or partial) removal of degeneracy is then predictable knowing the symmetry of $\boldsymbol{H^0}$ and $\boldsymbol{V}$ and is due to the lower symmetry of $\boldsymbol{V}$ compared to $\boldsymbol{H^0}$.

6.2.1 The Stark Effect in the Hydrogen Atom

The example of the perturbation of the electronic levels of the *hydrogen atom* due to a static electric field (Stark effect) is particularly interesting because it allows us to illustrate various aspects of the perturbation methods. Let us consider, therefore, the interaction of an electric field $\vec{E}$ with the electric dipole moment $\vec{\mu}$:

$$\hat{W} = -\vec{\mu}\cdot\vec{E} = -\mu_z\cdot E = -z\cdot e\cdot E \quad E \overset{\text{def}}{=} \left|\vec{E}\right| \tag{6.8}$$

assuming that the z direction is that of the external field.

The **first degenerate level** is the $n=2$ level with the four states $|nlm\rangle = |200\rangle$[6], *i.e.* $|2s\rangle$, $|21-1\rangle$, $|210\rangle$, *i.e.* $|2p_z\rangle$, and $|211\rangle$, keeping in mind that the two states $|21-1\rangle$ and $|211\rangle$ are linear combinations with equal weights of the states $|2p_x\rangle$ and $|2p_y\rangle$, and vice versa. To study the effect of the perturbation, it is therefore necessary to write the 4×4 matrix of the operator W between these four states, that is, to calculate the integrals:

$$\int \varphi_a^*\cdot z\cdot\varphi_b \mathrm{d}V \tag{6.9}$$

It is not difficult to see by simple symmetry considerations that **all diagonal elements of the perturbation operator are zero**. The result is due, in fact, to the symmetry by inversion with respect to the centre of the atom. All the hydrogen functions are a) *even* (the functions s, d, g, …) that is, they do not change when going from x, y, z to $-x$, $-y$, $-z$ or b) they are *odd* (the functions p, f, …) that is, they change sign when going from x, y, z to $-x$, $-y$, $-z$. We must calculate the integrals on a product of three functions that - by inversion with respect to the centre of the atom - are characterized by a specific symmetry. As z is odd, in the *diagonal* elements of the perturbation operator, *i.e.*when $a=b$ in Eq. (6.9), the two wave functions either remain unchanged or both change sign, the integrand is odd, and the integral is zero.

Going on to the **non-diagonal elements**, we see that the only non-zero matrix elements are those between $|200\rangle = |2s\rangle$ and $|210\rangle = |2p_z\rangle$, where the integrand is: $|2s\rangle\, z$. In fact, one function changes sign on inversion but the other does not, the product changes sign, the integrand of Eq. (6.9) does not change sign and finally the integral *can* be different from zero. The only function that does not change sign *on inversion* is $|200\rangle = |2s\rangle$ and so we should only consider the integrals between this and each of the others, but not the integrals of the 2p functions between themselves. Now, it is easy to see that the products $|z\cdot 2p_x\rangle$ and $|z\cdot 2p_y\rangle$ give a null integral while the product $|z\cdot 2p_z\rangle$ is everywhere positive and therefore gives an integral $\neq 0$

Arranging the functions in the order $|200\rangle = |2s\rangle$, $|210\rangle = |2p_z\rangle$, $|2p_x\rangle$, $|2p_y\rangle$ for convenience and denoting with k the value of the integral that symmetry does not cancel out, the matrix of the interaction operator becomes:

$$\begin{pmatrix} 0 & k & 0 & 0 \\ k & 0 & 0 & 0 \\ 0 & 0 & 0 & 0 \\ 0 & 0 & 0 & 0 \end{pmatrix}$$

The 2×2 top right partition is directly diagonalized with two eigenvalues k and $-k$ and a matrix of eigenvectors:

$$\frac{1}{\sqrt{2}}\begin{pmatrix} 1 & 1 \\ 1 & -1 \end{pmatrix}$$

that is, the sum and difference normalized with $1/\sqrt{2}$. In conclusion, the level $n=2$ which is 4 times degenerate undergoes partial removal of degeneracies: the states $|200\rangle = |2s\rangle$ and $|210\rangle = |2p_z\rangle$ combine (sum and difference with equal weights) to give two *sp* hybrids which have energies $+keE$ and $-keE$ with respect to the unperturbed level, while the energies of the states $|2p_x\rangle$ and $|2p_y\rangle$ (or $|21-1\rangle$ and $|211\rangle$) are not modified by the perturbation.

Generalizing the considerations made for the two combinations $|2s\rangle \pm |2p_z\rangle$ and given that the diagonal elements are in any case zero, we see that the presence of a degenerate level with different l ($n=3, 4, \ldots$) is an essential condition for having degeneracy removal and therefore for a Stark effect at the first order of the perturbation.

Extending the discussion to the AOs $|nlm\rangle$ of poly-electronic atoms, we see that the Stark effect seen so far (and called **linear**) does not appear because now the degeneracy between AOs with the same n but different l does not occur. Instead, there are **second-order** corrections (Eq. 6.6) that are proportional to the square of the field E: this is the **quadratic Stark effect**.

6.3 Interaction with the Electromagnetic Radiation

Among the applications of the perturbation method to time-dependent external fields, by far the most important case is the interaction of a material system with electromagnetic radiation.

This is a very broad problem with different applications and is approached with different methods and different levels of complexity depending on the properties that one wants to derive and the properties of the field, *i.e.* the frequency range and the intensity. The problem is also intricate and difficult to explain here, because some prerequisites may not be known from previous Physics courses. These include some topics of classical electromagnetism (Maxwell equations, the vector potential), the quantization of the electromagnetic field, Planck distribution law, and finally some hints on relativity. For these reasons, instead of an exhaustive treatment, here we essentially give an illustrative presentation of the most important results.

The V field of electromagnetic (hereafter: e.m.) radiation is described by a distribution of sinusoidal oscillations (monochromatic waves) that depend on space and time through two characteristic parameters:

- the **frequency** ν, often replaced in the formulas by the *angular frequency* $\omega=2\pi\nu$ and
- the **wave vector**, usually indicated by $\vec{k}$, which embodies both the direction and the sense of propagation of the monochromatic wave, and its wavelength: $|\vec{k}| = 2\pi/\lambda$.

In vacuum: $\lambda\nu=c$. In matter, λ and the speed of light decrease and $|\vec{k}|$ increases. Furthermore, $\vec{k}$ can acquire a complex component that describes the attenuation of the intensity of radiation as it passes through matter.

The analytical expression of a monochromatic wave is:

$$V = V_0 \cdot \cos\left(\vec{k}\cdot\vec{r} - \omega t\right) \tag{6.10}$$

V_0 contains (i) the amplitude of the wave, whose square modulus is the intensity of the radiation or, more precisely, its spectral density, and (ii) some parameters which describe the detail of the interaction of the material system with the electric or magnetic field of the radiation, as discussed later. For simplicity, it is convenient for now to leave V_0 unspecified and to write the time- and space-dependent part of V in complex exponential form:

$$\cos\left(\vec{k}\cdot\vec{r} - \omega t\right) = \left[e^{i\vec{k}\cdot\vec{r}}e^{-i\omega t} + e^{-i\vec{k}\cdot\vec{r}}e^{i\omega t}\right]/2 \tag{6.11}$$

Many techniques, of course, involve polychromatic e.m. radiation. However, thanks to the linearity of the interactions, the results may be derived for a single frequency. At the end, if necessary, a result is integrated over all component frequencies using the appropriate spectral distribution of the radiation.

The wave function of the system that interacts with the e.m. radiation is written in the most general form:

$$\psi = \sum_j a_j(t)\cdot\varphi_j(r,t) \tag{6.12}$$

as a series expansion over the stationary states - $\varphi_j(r,t)$ - of the unperturbed system, which are assumed to be completely known. Since **the perturbation depends on time**, unlike the previous cases, we have that:

a. the coefficients a_j are functions of time,
b. the time-dependent part of the stationary states must be inserted

$$\varphi_j(r,t) = |j\rangle e^{-i\cdot E_j^0\cdot t/\hbar} \text{ using } |j\rangle$$

to indicate the space dependent part (6.13)

c. and one must resort to the time-dependent Schrödinger equation:

$$\hat{H}\psi(q,t) = i\cdot\hbar\cdot\frac{\partial\psi(q,t)}{\partial t} \tag{6.14}$$

Substituting Eqs. (6.11) and (6.12) into (6.14), we have:

$$\left(\widehat{H^0} + \hat{V} - i\cdot\hbar\cdot\frac{\partial}{\partial t}\right)\left(\sum_j a_j(t)\cdot\varphi_j(r,t)\right) = 0$$

Let us consider a particular index (k), multiply by $\left[\varphi_k(r,t)\right]^*$ and integrate. Due to the orthogonality and normalization of the eigenfunctions of the unperturbed H, the unperturbed part becomes:

$$\int\left[\varphi_k(r,t)\right]^*\cdot\widehat{H^0}\left(\sum_j a_j(t)\cdot\varphi_j(r,t)\right)d\tau = a_k(t)\cdot\left(E_k^0\right)$$

The part containing the perturbation operator becomes:

$$\sum_j a_j(t)\cdot\left\langle k\left|\hat{V}\right|j\right\rangle\cdot e^{-i\left(E_j^0-E_k^0\right)\cdot t/\hbar}$$

while the part containing the derivative with respect to time becomes:

$$\int [\varphi_k(r,t)]^* \cdot \left(-i\cdot\hbar\cdot\frac{\partial}{\partial t}\right)\left(\sum_j a_j(t)\cdot\varphi_j(r,t)\right)d\tau$$

$$= -i\cdot\hbar\cdot\left[\frac{\partial a_k(t)}{\partial t} + a_k(t)\cdot\left(-i\frac{E_k^0}{\hbar}\right)\right]$$

Finally, summing the three parts, $a_k(t)\cdot(E_k^0)$ in the first and third parts cancel out and we get:

$$i\cdot\hbar\cdot\frac{\partial a_k(t)}{\partial t} = \sum_j a_j(t)\cdot\langle k|\hat{V}|j\rangle\cdot e^{-i(E_j^0-E_k^0)\cdot t/\hbar} \quad (6.15)$$

This equation is exact, but difficult to solve. In fact, by integration:

$$\text{For each } f\neq n,\ a_f^{(1)}(t) = \frac{-i}{\hbar}\int_0^t \langle f|V|n\rangle\cdot e^{-i\cdot(E_f^0-E_k^0)\cdot t'/\hbar}\,dt'. \quad (6.16)$$

The difficulty of solving Eq. (6.15) in the general case should now become apparent from the consideration that to obtain an $a_k(t)$ it is necessary to know all the other coefficients and their time dependence. To obtain useful results, it is necessary to consider some additional conditions consistent with the specific applications of interest.

6.4 Spectroscopy

A first result of interest concerns spectroscopic transitions. This is the probability that a system makes a transition **from one stationary state to another** due to the interaction with radiation.

For this purpose, let us consider a system that is initially in the stationary state $\varphi_n(r,t)=|n\rangle\cdot e^{-i\cdot E_n^0\cdot t/\hbar}$. At time $t=0$ we therefore have $a_n=1$ and $a_{j\neq n}=0$. With time and due to the interaction with the external field, a_n changes and becomes $|a_n|<1$, the various $a_{j\neq n}$ become (in general) ¹ 0, vary in time as described by Eq. (6.15) and at each instant express **how much of each stationary state *j*** there is in the wave function of the system which is interacting with the e.m. radiation and was initially φ_n. Thus, the coefficient $a_f\,(f\neq n)$ is the probability amplitude of transition between the **initial state** *n* and a specific **final state** *f*. The square modulus of this amplitude integrated from the initial time:

$$P_{nf}(t) = \int_0^t |a_f(t')|^2\,dt' \quad (6.17)$$

is the probability that the system undergoes a transition due to radiation between the initial time and time *t*. Dividing by *t*, we obtain the transition probability per unit time.

According to the perturbative method, for each $a_k(t)$ we write a series of corrections: $a_k^{(0)}+a_k^{(1)}(t)+a_k^{(2)}(t)+a_k^{(3)}(t)+\cdots$ where the unperturbed and time-independent term is $a_k(0)=1$ (if $k=n$) or $=0$ (if $k\neq n$). We derive the time functions $a_k^{(r)}(t)$ that express the corrections at the various orders by following an iterative scheme that uses the corrections of order $r-1$ at time *t* in Eq. (6.16) to derive the corrections of order *r* ($r=1, 2, \ldots$) at time *t*:

$$a_k(t) = a_k(0) - \frac{i}{\hbar}\Sigma_j \int_0^t a_j(t')\cdot\langle k|\hat{V}|j\rangle\cdot e^{-i(E_j^0-E_k^0)\cdot t'/\hbar}dt'.$$

For each $f\neq n$, $a_k^{(r)}(t) = a_k^{(r-1)}(t=0) - \frac{i}{h}\sum_j \cdot\int_0^t a_j^{(r-1)}(t')\cdot \langle k|V|j\rangle\cdot e^{-i\cdot(E_j^0-E_k^0)\cdot t'/\hbar}\,dt'$, because there is only one coefficient of order zero at $t=0$ and is $a_n^{(0)}=a_n(0)=1$. Stopping at first order, at the end $a_f(t)=a_f^{(1)}(t)$.

6.4.1 Frequency Selection

To begin with, it is convenient to focus on the time dependence only. From Eq. (6.18) below we see that the quantity to be integrated to obtain the transition probability amplitude between the initial state *n* and the state *f* contains three time-dependent factors:

$$\int_0^t e^{+i\cdot E_f^0\cdot t'/\hbar}\cdot e^{i\omega t'}\cdot e^{-i\cdot E_n^0\cdot t'/\hbar}\cdot dt' = \int_0^t e^{i(\Delta E/\hbar+\omega)\cdot t'}\cdot dt'$$
$$\text{with: } \Delta E \overset{\text{def}}{=} E_f^0 - E_n^0 \quad (6.18)$$

The term containing $e^{i\omega t}$ will be accompanied by a similar term containing the exponentials with exponents changed in sign. Then, we immediately derive that $P_{nf}(t)$ is different from zero if and only if $\Delta E/h+\omega=0$ (or $\Delta E/h-\omega=0$ for the term with the other sign). Let us write $\alpha=(\Delta E/\hbar+\omega)$. Then, $\int_0^t e^{i\alpha t'}\cdot dt' = \left|\frac{e^{i\alpha\cdot t'}}{i\alpha}\right|_{t'=0}^{t'=t} = \frac{e^{i\alpha\cdot t}-1}{i\alpha}$.

Writing the modulus $\left|\frac{e^{i\alpha\cdot t}-1}{i\alpha}\right|^2 = \frac{e^{i\alpha\cdot t}-1}{i\alpha}\frac{e^{-i\alpha\cdot t}-1}{-i\alpha}$, we get: $\frac{1-e^{i\alpha\cdot t}-e^{-i\alpha\cdot t}+1}{\alpha^2} = \frac{4\cdot\sin^2(\alpha\cdot t/2)}{\alpha^2}$. This is an interference

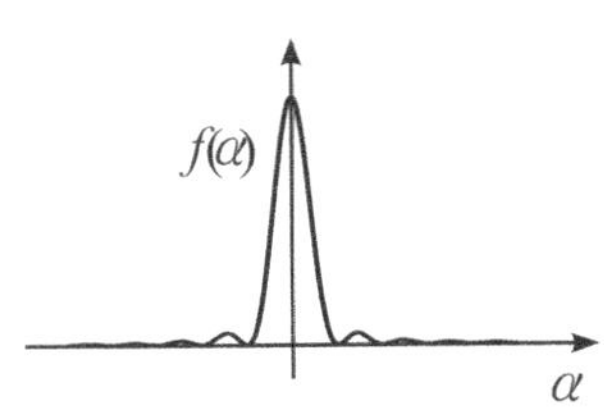

FIGURE 1.5 Interference function: $f(\alpha)$ at a given *t* (see text). The principal maximum is proportional to t^2, width proportional to $1/t$ and area proportional to *t*, so that the transition probability per unit of time $P_{nf}(t)/t$ is constant.

function: as shown in Figure 1.5 the principal maximum is at $\alpha=0$, that is $\omega = -\Delta E / \hbar$.

This is the well known

frequency selection rule:

there is a **transition induced by e.m. radiation** when $|\Delta E| = \hbar\omega = h\upsilon$

that is the difference in energy between the initial state (n) and another state (f) is:

$|\Delta E| = |E_f^0 - E_n^0|$

In detail: $\Delta E / \hbar + \omega = 0 \Rightarrow \Delta E < 0 \Rightarrow E_f^0 < E_n^0$: (stimulated) emission of radiation

$\Delta E / \hbar - \omega = 0 \Rightarrow \Delta E > 0 \Rightarrow E_f^0 > E_n^0$: absorption of radiation

6.4.2 Spontaneous Emission and Einstein Coefficients

It is evident that there is an important gap in the discussion carried so far. It is a conceptual gap and is not due to the approximations used in the derivations: the previous approach does not describe **spontaneous emission**, *i.e.* the decay to a lower energy state occurring in the *absence* of an e.m. field.

Historically, the need to include spontaneous emission was derived by Einstein from balance considerations between the kinetics of absorption and emission transitions considering the simple case of material systems in two non-degenerate levels (1 and 2). For N_1+N_2 systems:

$$\left[\frac{-\mathrm{d}N_1(t)}{\mathrm{d}t}\right]_{\mathrm{ass}} = \left[\frac{\mathrm{d}N_2(t)}{\mathrm{d}t}\right]_{\mathrm{ass}} = B_{12}(t)\cdot\rho(\nu_{12})\cdot N_1(t) \quad (\text{absorption}) \tag{6.19}$$

$$\left[\frac{-\mathrm{d}N_2(t)}{\mathrm{d}t}\right]_{\mathrm{stim}} = B_{21}(t)\cdot\rho(\nu_{12})\cdot N_2(t) \quad (\text{stimulated decay}) \tag{6.20}$$

$$\left[\frac{-\mathrm{d}N_2(t)}{\mathrm{d}t}\right]_{\mathrm{spont}} = A_{21}(t)\cdot N_2(t) \quad (\text{spontaneous decay}) \tag{6.21}$$

where $h\nu_{12} = E_2 - E_1 > 0$, $\rho(\nu)$ is the spectral density of the e.m. energy[7] and the three coefficients A_{21}, B_{12}, B_{21} are known as **Einstein coefficients**. Combining the three kinetic laws we have

$$\frac{-\mathrm{d}N_1(t)}{\mathrm{d}t} = \frac{\mathrm{d}N_2(t)}{\mathrm{d}t} = \left[\frac{\mathrm{d}N_2(t)}{\mathrm{d}t}\right]_{\mathrm{ass}} + \left[\frac{\mathrm{d}N_2(t)}{\mathrm{d}t}\right]_{\mathrm{spont}} + \left[\frac{\mathrm{d}N_2(t)}{\mathrm{d}t}\right]_{\mathrm{stim}} \tag{6.22}$$

so that $$\frac{\mathrm{d}N_2(t)}{\mathrm{d}t} = B_{12}(t)\cdot\rho(\nu_{12})\cdot N_1(t) - A_{21}(t)\cdot N_2(t) - B_{21}(t)\cdot\rho(\nu_{12})\cdot N_2(t)$$

At equilibrium, the time-dependent quantities become constant, and the following relation holds:

$$B_{12}\cdot\rho(\nu_{12})\cdot N_1 - A_{21}\cdot N_2 - B_{21}\cdot\rho(\nu_{12})\cdot N_2 = 0$$

while the thermal equilibrium of material systems requires

$$\frac{N_1}{N_2} = \mathrm{e}^{-(E_2-E_1)/kT} \overset{\mathrm{def}}{=} \mathrm{e}^{-h\nu_{12}/kT}. \quad \text{Then:}$$

$$\rho(\nu_{12}) = \frac{A_{21}\cdot N_2}{B_{12}\cdot N_1 - B_{21}\cdot N_2} = \frac{A_{21}}{\left(\frac{N_1}{N_2}\right)B_{12} - B_{21}} = \frac{A_{21}}{B_{12}}\frac{1}{\mathrm{e}^{h\nu_{12}/kT} - \left(\frac{B_{21}}{B_{12}}\right)} \tag{6.23}$$

Finally, from the comparison with Planck energy distribution law of e.m. radiation

$$\rho(\nu_{12}) = \frac{8\pi h}{c^3}\nu_{12}^3 f_{BE}(T,\nu_{12})$$

with the Bose-Einstein distribution of photons $f_{BE}(T,\nu) = \left(\mathrm{e}^{h\nu/kT} - 1\right)^{-1}$ we get:

$$\rho(\nu_{12}) = \frac{A_{21}}{B_{12}}\frac{1}{\mathrm{e}^{h\nu_{12}/kT} - \left(\frac{B_{21}}{B_{12}}\right)} = \frac{8\pi h}{c^3}\nu_{12}^3\frac{1}{\left(\mathrm{e}^{h\nu_{12}/kT} - 1\right)}$$

Then, to balance the dependence on T of the expressions on the right and left:

$$B_{12} = B_{21} \text{ and} \tag{6.24}$$

$$A_{21} = \frac{8\pi h}{c^3}\cdot\nu_{12}^3\cdot B_{12} \text{ to match the scale factors} \tag{6.25}$$

Note from Eq. (6.25) that spontaneous emission rapidly becomes favoured over stimulated emission as the frequency increases.

It may be surprising to know that the direct derivation of spontaneous emission - a seemingly simpler phenomenon than absorption and stimulated emission - requires a theoretically much more difficult treatment: it requires relativistic considerations with the quantization of the e.m. field and is based on the interaction of the material system with the zero-point vibrations of the quantized field.

6.4.3 Electric Dipole and Multipole Transitions

Coming back to the e.m.-stimulated transitions,

the probability of transition per unit time from state n to state f is proportional to the square of the matrix element of electric dipole:

$$\frac{P_{nf}(t)}{t} \propto \left|\langle n|\hat{\mu}|f\rangle\right|^2 \qquad (6.26)$$

This is called the **Fermi golden rule**. In Eq. (6.26) $\hat{\mu}$ is the **electric dipole moment** operator of the system, *i.e.* of its charged particles (*i*: electrons and nuclei):

$$\hat{\mu} = \sum_i q_i \vec{r}_i \quad q_i \text{ is the charge of the particle}$$

Equation (6.26) is derived from Eq. (6.10) with several considerations and some further approximations.

A. A first point is that dealing with the interaction with the e.m. field requires replacing $\vec{p}^2$ with $\left[\vec{p}-(q/2m)\vec{A}\right]^2$ for each particle of the system in the unperturbed Hamiltonian. Here q and m are the charge and mass of each interacting particle, while $\vec{A}$ is the **vector potential** of the radiation.

   In the theory of electromagnetism, the vector potential $\vec{A}$ of an e.m. field is specified up to an additional condition called gauge condition. If we choose the gauge condition that cancels the divergence of the vector potential:

$$\text{div}\,\vec{A} = \frac{\partial A_x}{\partial x} + \frac{\partial A_y}{\partial y} + \frac{\partial A_z}{\partial z} = 0$$

   then we get that (i) the linear moment operator $\hat{p}$ commutes with $\hat{A}$ (and therefore it does not matter whether the operator to be inserted is $\hat{A}\cdot\hat{p}$ or $\hat{p}\cdot\hat{A}$) and (ii) the product is Hermitian.

B. Another point is that we usually neglect the quadratic term in $\vec{A}$ leaving only the linear term:

$$\left[\hat{p} - \frac{q}{2m}\hat{A}\right]^2 \Rightarrow \widehat{p^2} - \frac{q}{m}\hat{A}\cdot\hat{p}$$

   This approximation restricts the validity of the result to the case of **low fields**, i.e. low radiation intensities. In discursive terms, by neglecting the quadratic term we ignore the fact that while absorbing or emitting a photon, the system may interact with another photon, *i.e.* we only deal with *one-photon* processes. The quadratic term is also important in deriving diamagnetism.

C. We also make the approximation of considering $\vec{A}$ uniform in space; that is, we ignore its variation within the material system that is being considered. This approximation goes in the same direction as the *long wavelength approximation* (see following point *E*).

D. The matrix elements of $\hat{p}$ between unperturbed stationary states can finally be replaced by the matrix elements of $\vec{r}$ multiplied by the difference in energy (the derivation is given in Section 7.4)

$$\langle m|\hat{p}|n\rangle = \left\langle m\left|\hat{r}\widehat{H^0} - \widehat{H^0}\hat{r}\right|n\right\rangle\left(E_n^0 - E_m^0\right)\langle m|\hat{r}|n\rangle$$

   or:

$$e\left(E_n^0 - E_m^0\right)\langle m|\hat{r}|n\rangle = \left(E_n^0 - E_m^0\right)\langle m|\hat{\mu}|n\rangle$$

   Where e is the charge of the electron.

E. Another point is that the two terms $\exp(i\vec{k}\cdot\vec{r})$ and $\exp(-i\vec{k}\cdot\vec{r})$ of Eq. (6.11) can be drastically simplified. In fact, if we consider (i) systems of atomic dimensions and (ii) radiation in the near UV, visible, IR or even at longer wavelengths, the exponent $\vec{k}\cdot\vec{r} \ll 1$ and therefore the two exponentials can be assumed = 1 by truncating their series expansions at the first term. This is the well-known *approximation of long wavelengths*. Indeed, for an atomic system, r is of the order of 0.1 nm. If we consider radiation in the near UV (e.g. 100 nm), $k = 2\pi/\lambda < 10/(100\,\text{nm}) = 0.1\ (\text{nm}^{-1})$ and $\vec{k}\cdot\vec{r}$ is of the order of 10^{-2} and becomes much lower for longer wavelength radiations (visible, IR, …).

F. Finally, one must consider the mutual orientation between $\vec{\mu}$ and the unit vector in the direction of the oscillation of the electric field component of e.m. radiation.

As regards the approximation of long wavelengths, it should be noted that the subsequent term of the series expansions: $e^{i\vec{k}\cdot\vec{r}} = 1 + \vec{k}\cdot\vec{r} + \ldots$ or $e^{-i\vec{k}\cdot\vec{r}} = 1 - \vec{k}\cdot\vec{r} + \ldots$, *i.e.* $(\vec{k}\cdot\vec{r})$, correspond to the **electric quadrupole moment** and **magnetic dipole moment** operators. Going to the successive powers of the series expansions, one would obtain the successive terms (electric octupole, magnetic quadrupole, …) of the so-called multipole series, *i.e.* the terms that according to electromagnetism express in the most general case the field generated by a distribution of charges and currents. The corresponding operators contain the quadratic functions x^2, y^2, z^2, xy, yz, zx (for the electric quadrupole and magnetic dipole moments), the cubic functions (for the electric octupole and magnetic quadrupole) and so on.

The general consideration is that

- going up in the series expansion, the terms become progressively less important, but
- the trend is less and less marked as the radiation frequency increases.

For near UV or softer radiations, … (visible, IR, …) the ratio between the probability of an electric quadrupole or magnetic dipole transition and that of an electric dipole transition is typically 10^{-5} or less and then it is acceptable to neglect the magnetic dipole and electric quadrupole transitions.

Of course, it becomes important to consider these transitions when the electric dipole transitions have zero probability (a case that typically occurs for symmetry reasons).

6.5 Polarizability

In setting up a solution of Eq. (6.15), the goal in (6.4) was to find the probability of transition between a given initial state (n) and one (k) among the possible final states. For this purpose, the wave function of the evolving system was written as linear combination of stationary states, the condition was imposed that the system initially was in the given stationary state and the time behaviour of the coefficients $a_{k\neq n}(t)$ relative to the other stationary states was obtained. In other words, the effect of the interaction between the material system and radiation was sought only as the transition between stationary states, *assuming* that these states *remained unchanged.*

In addition to this result and always under the hypothesis of 'small perturbations', Eq. (6.15) can be used also to understand how the interaction with e.m. radiation can modify the properties of the stationary states of a material system. In particular, the **generation of induced dipoles** is of interest here.

The assumptions to be used to find the solutions of Eq. (6.15) suitable for this purpose are:

- (similarly to the transition problem) that the system is initially in a stationary state (n) and
- (unlike the transition problem) that it remains, at least approximately, in that same stationary state.

As for the time-dependent coefficients it is then assumed that $a_n=1$ at time zero and always remains "close to 1", while the other stationary states contribute to the wave function of the evolving state with coefficients ($a_{k\neq n}$) that always remain sufficiently "small", and the charge distribution produced by these contributions generates an (induced) electric dipole moment.

We then introduce a quantity called polarizability ($\vec{\alpha}$) which expresses a proportionality between the *induced* electric dipole moment ($\vec{\mu}$) and the electric field ($\vec{E}$) of the electromagnetic radiation:

$$\vec{\mu} = \vec{\alpha} \cdot \vec{E} \tag{6.27}$$

Polarizability is a *two-index tensor, i.e.* a set of coefficients. Each coefficient specifies the proportionality between a component (x, y, z) of the induced dipole and an equal or different component of the electric field of the radiation, *e.g.* a relation such as: $\mu_y=\alpha_{yz}\cdot E_z$. The polarizability tensor obviously has properties consistent with the symmetry of the system (it behaves like second-order functions in the coordinates: x^2, y^2, z^2, xy, yz, zx). Moreover, it is symmetric, meaning that $\alpha_{ij}=\alpha_{ji}$.

Proportionality is evidently the first and most important term in the relationship between the cause (the field) and the effect (the induced dipole), often the only one that is of interest. When higher terms are significant (quadratic, ...) we speak of **non-linear** effects.

The final result, for example for the xx component is:

$$\vec{\alpha} = \frac{2e^2}{\hbar}\sum_k \frac{E_k - E_n}{\omega_{kn}^2 - \omega^2}\left|\langle n|\hat{x}|k\rangle\right|^2 \text{ using } \omega^2 = \frac{E_k - E_n}{\hbar} \tag{6.28}$$

Prior to the introduction of QM, a result close to that of Eq. (6.28) had been obtained by combining electromagnetism with CM in a model known as the **Lorentz model**. The model essentially considers an electron bound by a centrosymmetric potential that has a minimum at a distance r_0 from the nucleus and adds the interaction with an electric field oscillating at angular frequency ω. Using x to indicate the deviation from the equilibrium position in the direction of the electric field vector and $\omega_0 = \sqrt{k/m}$ to indicate the frequency of the harmonic oscillation, the equation to be solved in CM is:

$$m\frac{\partial^2 \vec{x}}{\partial t^2} + m\omega_0^2\vec{x} = -e\vec{E}_0 \cdot e^{-i\omega t}$$

and is known as the forced harmonic oscillator model. This is a standard problem of CM and the solution, here shown without proof, gives:

$$\frac{e^2}{m}\frac{1}{\omega_0^2 - \omega^2} \tag{6.29}$$

Then, the trend as a function of ω of the quantum polarizability (Eq. 6.28) agrees with the semiclassical model (Eq. 6.29) with the remarkable difference that various ω_{kn} appear instead of the single ω_0.

In resonance conditions, *i.e.* when the external force has the same frequency as the free oscillation (ω_0), the solution (Eq. 6.29) diverges, and the Lorenz model is modified by introducing a viscous force proportional to the velocity through a constant γ. In complex form, the solution then becomes:

$$x = -\frac{e}{m}\cdot\frac{1}{\omega_0 - \omega - i\gamma\omega}\cdot E_0 \cdot \cos(\omega t) \text{ with}$$

$$\frac{1}{\omega_0 - \omega - i\gamma\omega} = \frac{\omega_0 - \omega}{(\omega_0^2 - \omega^2) + (\gamma\omega)^2} + i\frac{\gamma\omega}{(\omega_0^2 - \omega^2) + (\gamma\omega)^2} \tag{6.30}$$

where the real part and the imaginary part indicate the responses respectively in phase and out of phase with respect to the external force.

In the quantum treatment of Eq. (6.28), the divergence is removed at each of the $n \to k$ transitions by explicitly considering the lifetime of the excited states (Γ):

$$\frac{e^2}{\hbar}\left\{\sum_k (E_k - E_n)\left|\langle n|\hat{x}|k\rangle\right|^2\left[R_k(\omega) + i\cdot I_k(\omega)\right]\right\}, \text{where} \tag{6.31}$$

$$R_k(\omega) = \frac{[\omega_{kn} - \omega]}{(\omega_{kn} - \omega)^2 + \Gamma^2/4} \quad \text{and}$$

$$I_k(\omega) = \frac{\Gamma/2}{(\omega_{kn} - \omega)^2 + \Gamma^2/4}$$

Figure 1.6 shows the frequency dependence of the real and imaginary parts $R_k(\omega)$ and $I_k(\omega)$.

6.5.1 Frequency Dispersion

As a first application, Eqs. (6.28) and (6.31) allow us to derive the polarization vector $\vec{P}$ that appears in Maxwell equations, and which is simply the total *induced* dipole moment:

$$\vec{P} = N\vec{\mu}; \quad (\text{Number of induced dipoles})$$

$$\vec{D} \overset{\text{def}}{=} \vec{E} + \vec{P}; \quad \vec{D} = \epsilon\vec{E} \quad \text{and then} \quad \vec{P} = (\epsilon - 1)\vec{E};$$

$$\text{and finally:} \quad (\epsilon - 1) = N\alpha$$

With the quantum expression - or with the Lorenz model Eq. (6.30) - we then obtain the dependence of the refractive index on the frequency of the radiation. This is characterized by a drastic change near the resonances (the ω_0 of the Lorenz model, or the ω_{nf} transitions of the quantum treatment). The dielectric constant contains the refractive index (n) in the forms:

$$(\epsilon - 1) = n^2 (\text{in the gas phase}) \text{ or}$$

$$(\epsilon - 1) = 3\frac{n^2 - 1}{n^2 + 2}(\text{in condensed matter})$$

Compared to the semiclassical model, the quantum mechanical treatment also explains the so-called negative dispersion (refractive index decreasing before resonance and increasing afterwards) as caused by transitions from excited states ($E_n > E_f$).

6.5.2 Elastic and Inelastic Scattering

Let us now consider as a second application of the results for polarizability, namely the emission of e.m. radiation from an oscillating dipole.[8] In terms of balance of photons, scattering is an interaction with *two* photons, one provided by the exciting radiation and absorbed by the material system and the other emitted by the material system. We speak of *elastic* scattering when the incident radiation, the induced dipole and the diffused radiation have the same frequency because there is no gain or loss of energy from the material system.

Of great historical interest is the formula, obtained with the semiclassical model, for the ratio between the elastically diffused intensity (I_D) at the angle ϑ and the intensity of the incident radiation (I_0):

$$\frac{I_D}{I_0} \propto \frac{1 + \cos^2\vartheta}{r^2}\left[\frac{\omega^2}{\omega_0^2 - \omega^2}\right]^2$$

where r is the distance between the scattering centre and detector. This is usually known in two limiting forms as Rayleigh scattering (for "large" wavelengths, $\omega \ll \omega_0$):

$$\frac{I_D}{I_0} \propto \frac{1 + \cos^2\vartheta}{r^2}\frac{\omega^4}{\omega_0^4}$$

and Thomson scattering (for hard radiation, especially X-rays, $\omega \gg \omega_0$):

$$\frac{I_D}{I_0} \propto \frac{1 + \cos^2\vartheta}{r^2}$$

Going into more detail, if we write the electric field vector of a monochromatic incident radiation as $\vec{E}_i \cdot \cos\omega t$ and if the frequency of a proper vibration of the system is ω_v, we obtain that the induced dipole is given by:

$$\begin{aligned}\vec{\mu} &= \vec{\vec{\alpha}} \cdot \vec{E} = \left(\vec{\vec{\alpha}}_0 + \Delta\vec{\vec{\alpha}} \cdot \cos\omega_v t\right) \cdot \vec{E}_i \cdot \cos\omega t \\ &= \vec{\vec{\alpha}}_0 \cdot \vec{E}_i \cdot \cos\omega t + \Delta\vec{\vec{\alpha}} \cdot \cos\omega_v t \cdot \vec{E}_i \cdot \cos\omega t\end{aligned} \tag{6.32}$$

where the parenthesis contains the static part of the polarizability ($\vec{\vec{\alpha}}_0$) and its component at the frequency $\omega_v (\Delta\vec{\vec{\alpha}})$. We then obtain:

$$\begin{aligned}\vec{\mu} &= \vec{\vec{\alpha}} \cdot \vec{E} \\ &= \vec{\vec{\alpha}}_0 \cdot \vec{E}_i \cdot \cos\omega t + \frac{\Delta\vec{\vec{\alpha}}}{2} \cdot \vec{E}_i \cdot \{\cos(\omega - \omega_v) \cdot t + \cos(\omega + \omega_v) \cdot t\}\end{aligned} \tag{6.33}$$

Equation (6.33) is derived from Eq. (6.32) with the trigonometric formulas for $\cos(\omega + \omega_v)$ and $\cos(\omega - \omega_v)$ and shows, therefore, that in addition to the interaction at the frequency (ω) of the exciting radiation, two other induced dipoles oscillating at $\omega - \omega_v$ and $\omega + \omega_v$ must be considered. There are three diffuse radiations at three different frequencies (ω_s):

- *Rayleigh scattering* ($\omega_s = \omega$, elastic scattering) which consists of radiation with the same frequency as the incident radiation ω,

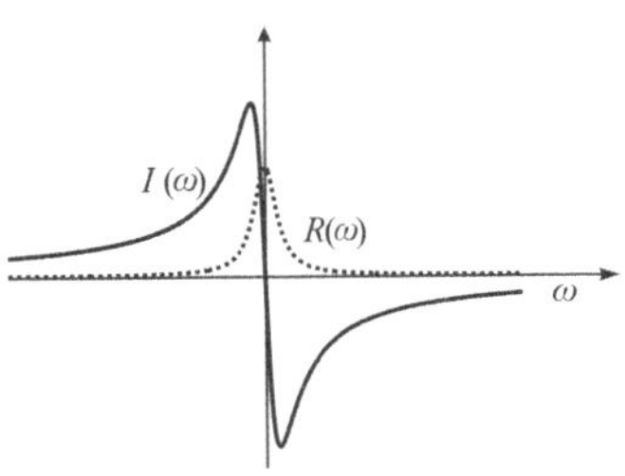

FIGURE 1.6 Real part (dotted) and imaginary part (solid line) of the frequency dependence of a contribution (k) to polarizability. The peak of the imaginary part and the zero of the real part are at $\omega = \omega_{kn}$.

- *Stokes scattering* ($\omega_s=\omega+\omega_v$) at higher frequency $\omega_s>\omega$ and
- *anti-Stokes scattering* ($\omega_s=\omega-\omega_v$) at lower frequency $\omega_s<\omega$.

As far as the material system is concerned, a Stokes line is globally an **uphill** transition from the ground state (typically) to an excited state, while an anti-Stokes line is the **downhill** transition from a higher energy state to a lower energy state, typically to the ground state.

The Stokes ($\omega_s=\omega+\omega_v$) and anti-Stokes ($\omega_s=\omega-\omega_v$) effects are called *inelastic* scattering but more frequently *Raman scattering* after one of its first discoverers.

7 REMARKS

7.1 The Turnover Rule

The basic relation: $\langle\varphi|\hat{A}|\psi\rangle=\left[\langle\psi|\hat{A}^{\dagger}|\varphi\rangle\right]^{*}$ (see Section 2.3.5) can be proved in four steps.

7.1.1 Step 1

Let us consider two wavefunctions, φ and χ, and a linear operator $\hat{A}$. Then:

1a) $\langle\varphi|\hat{A}|\varphi\rangle=\left[\langle\varphi|\hat{A}^{\dagger}|\varphi\rangle\right]^{*}$ and:

1b) $\langle\chi|\hat{A}|\chi\rangle=\left[\langle\chi|\hat{A}^{\dagger}|\chi\rangle\right]^{*}$

These come from the definitions of $\hat{A}^{\dagger}$.

7.1.2 Step 2

Now, let us consider the wavefunction $|a\rangle=\varphi+\chi$. Then:

$$\langle a|\hat{A}^{\dagger}|a\rangle=\langle\varphi+\chi|\hat{A}^{\dagger}|\varphi+\chi\rangle=\langle\varphi+\chi|\hat{A}^{\dagger}|\varphi\rangle+\langle\varphi+\chi|\hat{A}^{\dagger}|\chi\rangle$$
$$=\langle\varphi|\hat{A}^{\dagger}|\varphi\rangle+\langle\varphi|\hat{A}^{\dagger}|\chi\rangle+\langle\chi|\hat{A}^{\dagger}|\varphi\rangle+\langle\chi|\hat{A}^{\dagger}|\chi\rangle$$

because of the linearity of the operator. Analogously:

$$\langle a|\hat{A}|a\rangle=\langle\varphi+\chi|\hat{A}|\varphi+\chi\rangle=\langle\varphi+\chi|\hat{A}|\varphi\rangle+\langle\varphi+\chi|\hat{A}|\chi\rangle$$
$$=\langle\varphi|\hat{A}|\varphi\rangle+\langle\varphi|\hat{A}|\chi\rangle+\langle\chi|\hat{A}|\varphi\rangle+\langle\chi|\hat{A}|\chi\rangle$$

The results are related by: $\langle a|\hat{A}^{\dagger}|a\rangle=\left[\langle a|\hat{A}|a\rangle\right]^{*}$, so that:

$$\langle\varphi|\hat{A}^{\dagger}|\varphi\rangle+\langle\varphi|\hat{A}^{\dagger}|\chi\rangle+\langle\chi|\hat{A}^{\dagger}|\varphi\rangle+\langle\chi|\hat{A}^{\dagger}|\chi\rangle$$

2a) $$=\langle\varphi|\hat{A}|\varphi\rangle^{*}+\langle\varphi|\hat{A}|\chi\rangle^{*}+\langle\chi|\hat{A}|\varphi\rangle^{*}+\langle\chi|\hat{A}|\chi\rangle^{*}$$

7.1.3 Step 3

Let us consider a different wavefunction: $|b\rangle=\varphi+i\chi$. Then:

$$\langle b|\hat{A}^{\dagger}|b\rangle=\langle\varphi+i\chi|\hat{A}^{\dagger}|\varphi+i\chi\rangle=\langle\varphi+i\chi|\hat{A}^{\dagger}|\varphi\rangle$$
$$+(i)^{*}\langle\varphi+\chi|\hat{A}^{\dagger}|\chi\rangle$$
$$=\langle\varphi|\hat{A}^{\dagger}|\varphi\rangle+i\langle\varphi|\hat{A}^{\dagger}|\chi\rangle-i\langle\chi|\hat{A}^{\dagger}|\varphi\rangle+\langle\chi|\hat{A}^{\dagger}|\chi\rangle$$
$$(\text{using } i^{*}=-i \text{ and } -i\cdot i=1)$$

Analogously:

$$\langle b|\hat{A}|b\rangle=\langle\varphi|\hat{A}|\varphi\rangle+i\langle\varphi|\hat{A}|\chi\rangle-i\langle\chi|\hat{A}|\varphi\rangle+\langle\chi|\hat{A}|\chi\rangle$$

The results are related by: $\langle b|\hat{A}^{\dagger}|b\rangle=\left[\langle b|\hat{A}|b\rangle\right]^{*}$, so that:

2b) $$\langle\varphi|\hat{A}^{\dagger}|\varphi\rangle+i\langle\varphi|\hat{A}^{\dagger}|\chi\rangle-i\langle\chi|\hat{A}^{\dagger}|\varphi\rangle+\langle\chi|\hat{A}^{\dagger}|\chi\rangle$$
$$=\langle\varphi|\hat{A}|\varphi\rangle^{*}-i\langle\varphi|\hat{A}|\chi\rangle^{*}+i\langle\chi|\hat{A}|\varphi\rangle^{*}+\langle\chi|\hat{A}|\chi\rangle^{*}$$

7.1.4 Step 4

Using 1a) and 1b), 2a) becomes:

3a) $$\langle\varphi|\hat{A}^{\dagger}|\chi\rangle+\langle\chi|\hat{A}^{\dagger}|\varphi\rangle=\langle\varphi|\hat{A}|\chi\rangle^{*}+\langle\chi|\hat{A}|\varphi\rangle^{*}$$

and 2b) becomes:

3b) $$i\langle\varphi|\hat{A}^{\dagger}|\chi\rangle-i\langle\chi|\hat{A}^{\dagger}|\varphi\rangle=-i\langle\varphi|\hat{A}|\chi\rangle^{*}+i\langle\chi|\hat{A}|\varphi\rangle^{*}$$

then:

4b) $$-\langle\varphi|\hat{A}^{\dagger}|\chi\rangle+\langle\chi|\hat{A}^{\dagger}|\varphi\rangle=\langle\varphi|\hat{A}|\chi\rangle^{*}-\langle\chi|\hat{A}|\varphi\rangle^{*}$$

Now, adding 3a) and 4b):

5a) $\langle\chi|\hat{A}^{\dagger}|\varphi\rangle=\langle\varphi|\hat{A}|\chi\rangle^{*}$ This is the **turnover rule**.

7.2 Commutation of Angular Moments

The Cartesian expressions of the various operators are obtained from the determinant:

$$\hat{L}\overset{\text{def}}{=}-i\hbar\begin{vmatrix}\vec{i} & \vec{j} & \vec{k}\\ x & y & z\\ \dfrac{\partial}{\partial x} & \dfrac{\partial}{\partial y} & \dfrac{\partial}{\partial z}\end{vmatrix}$$

$\left[\vec{i},\vec{j},\vec{k}\text{ are the unit vectors along } x, y, x\right]$

For example, $\widehat{L_x}=-i\hbar\left(y\dfrac{\partial}{\partial z}-z\dfrac{\partial}{\partial y}\right)$.

Let us consider the commutator:

$$\left[\widehat{L_x},\widehat{L_y}\right]_- \overset{\text{def}}{=} \widehat{L_x}\cdot\widehat{L_y}-\widehat{L_y}\cdot\widehat{L_x}$$

The first product is:

$$\widehat{L_x}\cdot\widehat{L_y}=-\hbar^2\left(y\frac{\partial}{\partial z}-z\frac{\partial}{\partial y}\right)\cdot\left(z\frac{\partial}{\partial x}-x\frac{\partial}{\partial z}\right)$$

$$=-\hbar^2\left(y\frac{\partial}{\partial z}z\frac{\partial}{\partial x}-y\frac{\partial}{\partial z}x\frac{\partial}{\partial z}-z\frac{\partial}{\partial y}z\frac{\partial}{\partial x}+z\frac{\partial}{\partial y}x\frac{\partial}{\partial z}\right)$$

and the second one:

$$\widehat{L_y}\cdot\widehat{L_x}=-\hbar^2\left(z\frac{\partial}{\partial x}-x\frac{\partial}{\partial z}\right)\cdot\left(y\frac{\partial}{\partial z}-z\frac{\partial}{\partial y}\right)$$

$$=-\hbar^2\left(z\frac{\partial}{\partial x}y\frac{\partial}{\partial z}-x\frac{\partial}{\partial z}y\frac{\partial}{\partial z}-z\frac{\partial}{\partial x}z\frac{\partial}{\partial y}+x\frac{\partial}{\partial z}z\frac{\partial}{\partial y}\right)$$

Some terms may be deleted. In fact:

$$y\frac{\partial}{\partial z}x\frac{\partial}{\partial z}=yx\frac{\partial}{\partial z}\frac{\partial}{\partial z}=x\frac{\partial}{\partial z}y\frac{\partial}{\partial z} \quad \text{and}$$

$$z\frac{\partial}{\partial y}z\frac{\partial}{\partial x}=z^2\frac{\partial}{\partial y}\frac{\partial}{\partial x}=z\frac{\partial}{\partial y}z\frac{\partial}{\partial x}$$

The difference $y\frac{\partial}{\partial z}z\frac{\partial}{\partial x}-z\frac{\partial}{\partial x}y\frac{\partial}{\partial z}$ is $y\frac{\partial}{\partial x}$. In fact, writing a function f for clarity, we have:

$$y\frac{\partial}{\partial z}z\frac{\partial f}{\partial x}=yz\frac{\partial}{\partial z}\frac{\partial f}{\partial x}+y\frac{\partial f}{\partial x} \quad \text{and}$$

$$z\frac{\partial}{\partial x}y\frac{\partial f}{\partial z}=zy\frac{\partial}{\partial x}\frac{\partial f}{\partial z} \quad \text{then}$$

$$y\frac{\partial}{\partial z}z\frac{\partial f}{\partial x}-z\frac{\partial}{\partial x}y\frac{\partial f}{\partial z}=y\frac{\partial f}{\partial x}$$

Similarly,

$$z\frac{\partial}{\partial y}x\frac{\partial f}{\partial z}-x\frac{\partial}{\partial z}z\frac{\partial f}{\partial y}=zx\frac{\partial}{\partial y}\frac{\partial f}{\partial z}-x\frac{\partial f}{\partial y}-zx\frac{\partial}{\partial y}\frac{\partial f}{\partial z}=-x\frac{\partial f}{\partial y}$$

Finally: $\widehat{L_x}\cdot\widehat{L_y}-\widehat{L_y}\cdot\widehat{L_x}=-\hbar^2\left(y\frac{\partial}{\partial x}-x\frac{\partial}{\partial y}\right)$.

This is $-i\cdot\hbar\widehat{L_z}$.

In fact: $\widehat{L_z}=i\cdot\hbar\left(y\frac{\partial}{\partial x}-x\frac{\partial}{\partial y}\right)$.

7.3 Algebraic Solution of the Angular Moment Problem

Solvable QM problems are typically treated by finding solutions to the corresponding differential equations and are greatly simplified by using the separability of the Hamiltonian as shown for the example in Section 3.5.

It may be of interest to know that - in addition to the analytical techniques for solving differential equations - there is also a solution method of an essentially algebraic type, which is available for the cases of the harmonic oscillator, of the angular momentum, of the radial part of the hydrogen atom problem and for a generalization of these problems.

It may be useful to see the algebraic solution of the eigenvalue equations of the angular momentum as an example of this approach and, above all, because of its importance in the generalization to angular momenta that have no classical analogue, With reference to what was anticipated in Section 3.4.1, we want to solve the problem of the eigenvalues of the angular momentum operators *using only the commutation rules* already seen:

$$\left[\widehat{L_x},\widehat{L_y}\right]\overset{\text{def}}{=}\widehat{L_x}\cdot\widehat{L_y}-\widehat{L_y}\cdot\widehat{L_x}=i\hbar\widehat{L_z}$$

$$\left[\widehat{L_y},\widehat{L_z}\right]\overset{\text{def}}{=}\widehat{L_y}\cdot\widehat{L_z}-\widehat{L_z}\cdot\widehat{L_y}=i\hbar\widehat{L_x} \tag{3.6}$$

$$\left[\widehat{L_z},\widehat{L_x}\right]\overset{\text{def}}{=}\widehat{L_z}\cdot\widehat{L_x}-\widehat{L_x}\cdot\widehat{L_z}=i\hbar\widehat{L_y}$$

as proved in (7.2) and the rule

$$\left[\widehat{L^2},\widehat{L_x}\right]=\left[\widehat{L^2},\widehat{L_y}\right]=\left[\widehat{L^2},\widehat{L_z}\right]=0 \tag{3.7}$$

based on the definition:

$$\widehat{L^2}\overset{\text{def}}{=}\widehat{L_x^2}+\widehat{L_y^2}+\widehat{L_z^2} \tag{2.24}$$

In this way, we obtain results that are valid even for angular moment operators that have no classical analogue and therefore cannot be derived from the rules of the third postulate (see Section 3.4). While not particularly difficult, the derivation is long and intricate and is better to proceed step by step. From the commutation rules we know that we can have functions that are common eigenfunctions of $\widehat{L^2}$ and $\widehat{L_z}$: let us indicate them with the respective dimensionless eigenvalues α and β:

$$\widehat{L^2}|\alpha\beta\rangle=\left(\alpha\hbar^2\right)|\alpha\beta\rangle \quad \widehat{L_z}|\alpha\beta\rangle=\beta\hbar|\alpha\beta\rangle$$

First, let us introduce two new operators

$$\widehat{L_+}\overset{\text{def}}{=}\widehat{L_x}+i\widehat{L_y} \quad \text{and} \quad \widehat{L_-}\overset{\text{def}}{=}\widehat{L_x}-i\widehat{L_y}$$

We can verify that these are not Hermitian operators but are one the Hermitian conjugate of the other:

$$\left(\widehat{L_+}\right)^\dagger=\widehat{L_-} \quad \text{and} \quad \left(\widehat{L_-}\right)^\dagger=\widehat{L_+}$$

Let us consider the products $\widehat{L_z}\widehat{L_+}$ and $\widehat{L_z}\widehat{L_-}$ and derive the relations:

$$\hat{L}_z\hat{L}_+ = \hat{L}_+\left(\hat{L}_z + \hbar\right) \qquad (7.1)$$

and $$\hat{L}_z\hat{L}_- = \hat{L}_-\left(\hat{L}_z - \hbar\right) \qquad (7.2)$$

In fact: $\hat{L}_z\hat{L}_+ = \hat{L}_z\hat{L}_x + i\hat{L}_z\hat{L}_y$. Replacing both products with the commutator: $\hat{L}_z\hat{L}_x + i\hat{L}_z\hat{L}_y = \left[\hat{L}_z,\hat{L}_x\right] + \hat{L}_x\cdot\hat{L}_z\cdot$ $+\hat{L}_x\cdot\hat{L}_z + i\left\{-\left[\hat{L}_y,\hat{L}_z\right] + \hat{L}_y\cdot\hat{L}_z\right\}$. Making the commutators explicit we have: $i\hbar\hat{L}_y + \hat{L}_x\cdot\hat{L}_z + i\hat{L}_y\cdot\hat{L}_z - (i)^2\hbar\hat{L}_x$. Rearranging: $\hat{L}_x\cdot\left(\hat{L}_z+\hbar\right) + i\hat{L}_y\cdot\left(\hat{L}_z+\hbar\right)$, from which the result. We proceed analogously for the other product.

Let us also consider the products

$$\hat{L}_+\cdot\hat{L}_- = \hat{L}_x^2 + \hat{L}_y^2 + \hbar\hat{L}_z \quad \text{and} \quad \hat{L}_-\cdot\hat{L}_+ = \hat{L}_x^2 + \hat{L}_y^2 - \hbar\hat{L}_z \qquad (7.3)$$

(derived using the commutator $\left[\hat{L}_x,\hat{L}_y\right] = i\hbar\hat{L}_z$) and the relations:

$$\hat{L}^2 - \hat{L}_z^2 = \hat{L}_x^2 + \hat{L}_y^2 = \hat{L}_-\cdot\hat{L}_+ + \hbar\hat{L}_z = \hat{L}_+\cdot\hat{L}_- - \hbar\hat{L}_z \qquad (7.4)$$

Now, let us apply these products to the common eigenfunctions of $\hat{L}^2$ and $\hat{L}_z$:

$$\langle\alpha\beta|\hat{L}_-\cdot\hat{L}_+|\alpha\beta\rangle = \hat{L}^2 - \hat{L}_z^2\big|\langle\alpha\beta - \hbar\langle\alpha\beta\rangle\hat{L}_z|\alpha\beta\rangle$$
$$= \hbar^2\left[\alpha - \beta^2 - \beta\right] \qquad (7.5)$$

$$\langle\alpha\beta|\hat{L}_+\cdot\hat{L}_-\big|\langle\alpha\beta = \langle\alpha\beta\big|\hat{L}^2 - \hat{L}_z^2\big|\langle\alpha\beta + \hbar\langle\alpha\beta\,\hat{L}_z|\alpha\beta\rangle$$
$$= \hbar^2\left[\alpha - \beta^2 + \beta\right] \qquad (7.6)$$

We also consider the two products $\hat{L}^2\hat{L}_+$ and $\hat{L}^2\hat{L}_-$ and prove that they are both commutative:

$$\hat{L}^2\hat{L}_+ = \hat{L}^2\hat{L}_+ \quad \text{and} \quad \hat{L}^2\hat{L}_- = \hat{L}_-\hat{L}^2 \qquad (7.7)$$

In fact: let us start with $\hat{L}^2\hat{L}_+ = \left(\hat{L}^2 - \hat{L}_z^2\right)\hat{L}_+ - \hat{L}_z^2\hat{L}_+$ and develop separately the two parts of the right-hand side.

A. $\left(\hat{L}^2 - \hat{L}_z^2\right)\hat{L}_+$ becomes $\left(\hat{L}_+\hat{L}_- - \hbar\hat{L}_z\right)\hat{L}_+$ by Eq. (7.4) and then $\hat{L}_+\hat{L}_-\hat{L}_+ - \hbar\hat{L}_z\hat{L}_+ = \hat{L}_+\left(\hat{L}_+\hat{L}_- - 2\hbar\hat{L}_z\right)$ $-\hbar\hat{L}_+\left(\hat{L}_z + \hbar\right)$ using the second equation of Eq. (7.4) for the product $\hat{L}_-\hat{L}_+$ and Eq. (7.1) for $\hat{L}_z\hat{L}_+$.

Eventually, we come to $\hat{L}_+\left(\hat{L}^2 - \hat{L}_z^2 - \hbar\hat{L}_z\right) - \hbar\hat{L}_+\hat{L}_z$ $-\hbar^2\hat{L}_+ = \hat{L}_+\hat{L}^2 - \hat{L}_+\hat{L}_z^2 - \hbar\hat{L}_+\hat{L}_z - \hbar\hat{L}_+\hat{L}_z - \hbar^2\hat{L}_+$.

B. $\hat{L}_z^2\hat{L}_+$ becomes $\hat{L}_z\left(\hat{L}_+\hat{L}_z + \hbar\hat{L}_+\right) = \hat{L}_z\hat{L}_+\hat{L}_z + \hbar\hat{L}_z\hat{L}_+$ using the commutator and then $\left(\hat{L}_+\hat{L}_z + \hbar\hat{L}_+\right)\hat{L}_z$ $+\hbar\left(\hat{L}_+\hat{L}_z + \hbar\hat{L}_+\right) = \hat{L}_+\hat{L}_z^2 + 2\hbar\hat{L}_+\hat{L}_z + \hbar^2\hat{L}_+$.

Adding the two parts: $\hat{L}_+\hat{L}^2 - \hat{L}_+\hat{L}_z^2 - \hbar\hat{L}_+\hat{L}_z$ $-\hbar\hat{L}_+\hat{L}_z - \hbar^2\hat{L}_+ + \hat{L}_+\hat{L}_z^2 + 2\hbar\hat{L}_+\hat{L}_z + \hbar^2\hat{L}_+ = \hat{L}_+\hat{L}^2$ so that $\hat{L}^2\hat{L}_+ = \hat{L}_+\hat{L}^2$ and the commutativity is proved. For the other product we proceed analogously.

Now let us apply the products of Eqs. (7.7) and (7.1) to the common eigenfunctions of $\hat{L}^2$ and $\hat{L}_z$:

$$\hat{L}^2\hat{L}_+|\alpha\beta\rangle = \hat{L}_+\hat{L}^2|\alpha\beta\rangle = \hat{L}_+\alpha\hbar^2|\alpha\beta\rangle = \left(\alpha\hbar^2\right)\hat{L}_+|\alpha\beta\rangle; \text{ and:}$$

$$\hat{L}_z\hat{L}_+|\alpha\beta\rangle = \hat{L}_+\left(\hat{L}_z + \hbar\right)|\alpha\beta\rangle = \hat{L}_+(\beta+\hbar)\hbar|\alpha\beta\rangle = (\beta\hbar)\hbar\hat{L}_+|\alpha\beta\rangle$$

We see that $\hat{L}_+$ produces a function that is still an eigenfunction of both $\hat{L}^2$ and $\hat{L}_z$ with eigenvalues that are respectively unchanged (α) and increased by 1: $\hbar(\beta+1)$:

$$\hat{L}_+|\alpha\beta\rangle = |\alpha(\beta+1)\rangle \qquad (7.8)$$

With obvious modifications, we obtain similarly from the other Eq. (7.7) that:

$$\hat{L}_-|\alpha\beta\rangle = |\alpha(\beta-1)\rangle \qquad (7.9)$$

We highlight these properties by saying that the two operators $\hat{L}_+$ and $\hat{L}_-$ are **step-up** and **step-down operators**, respectively (they are also known as **ladder operators**).

Obviously, one should not draw the conclusion that the sequence $|\alpha\beta\rangle \rightarrow |\alpha(\beta+1)\rangle \rightarrow |\alpha(\beta+2)\rangle, \ldots$ with the step-up operator must continue upwards to infinity, nor that the sequence with the step-down operator must continue downwards to infinity.

The solution for the angular momentum operators that have a classical analogue (which is anyway included in the general case that we want to find) imposes the existence of boundary values for the step-up and step-down sequences.

The meaning instead is (for example, for the step-up case) that, once the maximum eigenvalue ($\beta_{\max}$) compatible with the given value α is reached, the subsequent application of the step-up operator produces the null function:

$$\hat{L}_+|\alpha\beta_{\max}\rangle = 0 \qquad (7.10)$$

If one prefers a more 'physical' argument, it is possible to use the consideration that the square of a component cannot exceed the square of the modulus, β^2 cannot exceed α and therefore (for a given α) there *must* be a β_{max}.

Similarly, for the step-down operator the minimum eigenvalue corresponds to:

$$\hat{L}_-|\alpha\beta_{\min}\rangle = 0 \qquad (7.11)$$

Once the limit value is reached (β_{max} for step-up and β_{min} for step-down), if we continued to apply the operator, we would have that by operating on the null function we still obtain the null function. When applied to $|\alpha\beta_{\max}\rangle$ and $|\alpha\beta_{\min}\rangle$ Eqs. (7.5) and (7.6) become:

$$\langle\alpha\beta|\widehat{L_-}\cdot\widehat{L_+}|\alpha\beta\rangle=\hbar^2\left[\alpha-\beta_{\max}^2-\beta_{\max}\right]=0 \qquad (7.12)$$

$$\langle\alpha\beta|\widehat{L_+}\cdot\widehat{L_-}|\alpha\beta\rangle=\hbar^2\left[\alpha-\beta_{\min}^2+\beta_{\min}\right]=0 \qquad (7.13)$$

since (in Eq. 7.12) Eq. (7.10) now holds and (in Eq. 7.13) Eq. (7.11) holds. Let us rewrite:

$$\alpha-\beta_{\max}(\beta_{\max}+1)=0 \quad \text{and} \quad \alpha-\beta_{\min}(\beta_{\min}-1$$

Then: $\beta_{\max}(\beta_{\max}+1)=\beta_{\min}(\beta_{\min}-1)$, using $\beta_{\max}^2-\beta_{\max}^2+\beta_{\max}-\beta_{\min}=0$. Finally, $(\beta_{\max}+\beta_{\min})(\beta_{\max}-\beta_{\min}+1)=0$. There are two solutions: $\beta_{max}=-\beta_{min}$ or $\beta_{max}=\beta_{min}-1$. Obviously, the second one must be rejected. We obtain:

$$\beta_{\max}=-\beta_{\min} \quad \text{and} \quad \alpha=\beta_{\max}(\beta_{\max}+1)$$

The usual notation uses the letter l (or L) instead of β_{max} and m (or M) instead of β. Finally, recalling Eqs. (7.10) and (7.11), we understand that m ranges from $-l$ to $+l$ (extremes included) in steps of 1, l can only take integer or semi-odd values, $\alpha=l(l+1)$ and in conclusion, all the results anticipated in Section 3.4.1 are proved.

7.4 The Dipole Transition Operator

For a particle of mass m, the unperturbed Hamiltonian $(\widehat{H^0})$ has the form:

$$\widehat{H^0}=\frac{1}{2m}\widehat{p^2}+U(r).$$

Let us consider the commutator between $\widehat{H^0}$ and r: $\hat{r}\widehat{H^0}-\widehat{H^0}\hat{r}$.

The operator corresponding to the coordinate r commutes with the function $U(r)$, so we only need the commutator with the kinetic energy operator and, preliminarily, the commutator between $\hat{r}$ and $\widehat{p_r}=-i\hbar\cdot\partial/\partial r$. For a generic function χ we have:

$$\hat{r}\widehat{p_r}\chi-\widehat{p_r}\hat{r}\chi=(-i\hbar)r\cdot\frac{\partial\chi}{\partial r}-(-i\hbar)\frac{\partial}{\partial r}r\chi$$

$$=(-i\hbar)r\cdot\frac{\partial\chi}{\partial r}-(-i\hbar)\left[r\cdot\frac{\partial\chi}{\partial r}+\chi\right]$$

$$=(i\hbar)\chi \quad \text{and:} \quad \hat{r}\widehat{p_r}-\widehat{p_r}\hat{r}=i\hbar$$

Also: $\hat{r}\frac{1}{2m}\widehat{p_r^2}-\frac{1}{2m}\widehat{p_r^2}\hat{r}=\left(\frac{1}{2m}\right)\left[\hat{r}\widehat{p_r}\widehat{p_r}-\widehat{p_r}\widehat{p_r}\hat{r}\right].$

The triple products can be simplified using the commutator $\hat{r}\widehat{p_r}-\widehat{p_r}\hat{r}=i\hbar$

$$\left(\hat{r}\widehat{p_r}\right)\widehat{p_r}=\left[i\hbar+\widehat{p_r}\hat{r}\right]\widehat{p_r} \quad \text{and} \quad \widehat{p_r}\left(\widehat{p_r}\hat{r}\right)$$

$$=\widehat{p_r}\left[\hat{r}\widehat{p_r}-i\hbar\right]$$

$$\Rightarrow\left[\hat{r}\widehat{p_r}\widehat{p_r}-\widehat{p_r}\widehat{p_r}\hat{r}\right]=i\hbar\widehat{p_r}+\widehat{p_r}\hat{r}\widehat{p_r}-\widehat{p_r}\hat{r}\widehat{p_r}$$

$$+i\hbar\widehat{p_r}=2i\hbar\widehat{p_r}$$

and $$\left(\frac{1}{2m}\right)\left[\hat{r}\widehat{p_r}\widehat{p_r}-\widehat{p_r}\widehat{p_r}\hat{r}\right]=\left(\frac{i\hbar}{m}\right)\widehat{p_r}$$

So, we are left with: $\hat{r}\widehat{H^0}-\widehat{H^0}\hat{r}=\left(\frac{i\hbar}{m}\right)\widehat{p_r}$

Then the matrix element of the moment operator between two eigenfunctions (s, t) of the unperturbed Hamiltonian is:

$$\left(\frac{i\hbar}{m}\right)\langle s|\widehat{p_r}|t\rangle=\langle s|\hat{r}\widehat{H^0}-\widehat{H^0}\hat{r}|t\rangle=\langle s|\hat{r}\widehat{H^0}|t\rangle-\langle s|\widehat{H^0}\hat{r}|t\rangle$$

The first term is $\langle s|\hat{r}\widehat{H^0}|t\rangle=\langle E_n^0 s|\hat{r}|t\rangle$. The second term must be processed with the turnover rule (7.1), considering that the adjoint of a product is the product of the adjoints multiplied in reverse order and that the two operators - $\widehat{H^0}$ and $\hat{r}$ - are Hermitian:

$$\langle s|\widehat{H^0}\hat{r}|t\rangle=\left\langle t\left|\left(\widehat{H^0}\hat{r}\right)^{\dagger}\right|s\right\rangle^{*}=\left\langle t|\hat{r}\widehat{H^0}|s\right\rangle^{*}=E_s^0\left[\langle t|\hat{r}|s\rangle\right]^{*}=E_s^0\langle s|\hat{r}|t\rangle$$

Finally: $\left(\frac{i\hbar}{m}\right)\langle s|p|t\rangle=\langle s|\hat{r}\widehat{H^0}-\widehat{H^0}\hat{r}|t\rangle=\left(E_t^0-E_s^0\right)\langle s|\hat{r}|t\rangle.$

The result can be extended to 3D and to a set of equal particles: in the Hamiltonian instead of one moment there is a sum of moments and instead of $U(r)$ there is a function U that depends on several coordinates.

8 EXERCISES

Exercise 1.1

Prove that the commutator between position and moment operators is $\left[\hat{x},\widehat{p_x}\right]=-\frac{\hbar}{i}$.

Exercise 1.2

The deviations $\widehat{\Delta x}$ and $\widehat{\Delta p_x}$ are $\widehat{\Delta x}=\sqrt{\widehat{x^2}-\hat{x}^2}$ and $\widehat{\Delta p_x}=\sqrt{\widehat{p_x}^2-\widehat{p_x}^2}$. Introduce the operators $\hat{\alpha}=\hat{x}-x$ and $\hat{\beta}=\widehat{p_x}-p_x$, and show that their commutator is $\left[\hat{\alpha},\beta\right]=i\hbar$. In addition, show that $\widehat{\alpha x}=\sqrt{\widehat{\alpha^2}}$ and $\widehat{\alpha p_x}=\sqrt{\widehat{\beta^2}}$.

Exercise 1.3

Take a normalized wavefunction φ, and write the expectation values for $\widehat{\Delta x}$ and $\widehat{\Delta p_x}$ in the Dirac bra-ket notation. Do the same for $\widehat{\Delta x^2}$ and $\widehat{\Delta p_x^2}$ in terms of the operators $\hat{\alpha} = \hat{x} - x$ and $\hat{\beta} = \widehat{p_x} - p_x$. Use the Cauchy-Schwarz inequality $|x||y| \geq |\langle x|y\rangle|$ to prove that $\widehat{\Delta x^2}\widehat{\Delta p_x^2} \geq \left|\langle \varphi | \widehat{\Delta\beta} | \varphi \rangle\right|^2$.

Exercise 1.4

Write the product $\widehat{\alpha\beta}$ in terms of $\frac{\widehat{\alpha\beta} + \widehat{\beta\alpha}}{2}$ and $\frac{\widehat{\alpha\beta} - \widehat{\beta\alpha}}{2}$. Use the results of Exercises 1.1 – 1.3 to prove that $\widehat{\Delta x^2} \cdot \widehat{\Delta p_x^2} \geq \left|\left\langle \varphi | \frac{\widehat{\alpha\beta} + \widehat{\beta\alpha}}{2} | \varphi \right\rangle\right|^2 + \frac{\hbar^2}{4}$ and then, even more so, that $\widehat{\Delta x} \cdot \widehat{\Delta p_x} \geq \frac{\hbar}{2}$, which is the Heisenberg uncertainty principle for position and moment.

Exercise 1.5

Take an eigenfunction φ of operator $\hat{x}$, with eigenvalue x_0. Use the commutator $[\hat{x}, \hat{p}_x] = -\frac{\hbar}{i}$ to prove that φ cannot be an eigenfunction of p_x. int: apply the commutator to φ, and then use the reduction ad impossible argument supposing that φ eigenfunction of $\hat{p}_x$ with eigenvalue $p_{x,0}$.

Exercise 1.6

Given the matrices:

$$\sigma_x = \begin{pmatrix} 0 & 1 \\ 1 & 0 \end{pmatrix} \; \sigma_y = \begin{pmatrix} 0 & -i \\ i & 0 \end{pmatrix} \; \sigma_z = \begin{pmatrix} 1 & 0 \\ 0 & -1 \end{pmatrix},$$

show that the commutators are $[\sigma_x, \sigma_y] = 2i\sigma_z$, $\sigma_y, \sigma_z \; = 2i\sigma_x$ and $[\sigma_z, \sigma_x] = 2i\sigma_y$.

Exercise 1.7

Show that the matrices of Exercise 1.5 are Hermitian, and that their determinant is –1.

Exercise 1.8

Refer to Appendix 3.2.5, where the procedure for finding the eigenvalues of a matrix is explained. Analytically, for finding the eigenvalues of the matrix $\boldsymbol{G}$, one should solve the equation $det(\boldsymbol{G} - \lambda\mathbf{1}) = 0$, where $\mathbf{1}$ is the unit matrix, and λ the eigenvalue. Show that the eigenvalues of the matrices of Exercise 1.6 are +1 and –1.

Exercise 1.9

The matrices of Exercise 1.5 are known as Pauli matrices. Use the results of Exercises 1.5–1.7 to prove that, neglecting a factor $\frac{\hbar}{2}$, the matrices behave as angular moment operators. Hint: construct the σ^2 matrix and prove that σ^2 commutes with each of the σ_i. Indeed, Pauli matrices can be used for describing the spin observable for a fermionic particle.

Exercise 1.10

Use the relations:

$$\begin{cases} x = r\sin\theta\cos\varphi \\ y = r\sin\theta\sin\varphi \\ z = r\cos\theta \end{cases} \quad \text{and}$$

$$\begin{cases} \dfrac{\partial}{\partial x} = \sin\theta\cos\varphi\dfrac{\partial}{\partial r} + \dfrac{1}{r}\cos\theta\cos\varphi\dfrac{\partial}{\partial\theta} + \dfrac{1}{r}\dfrac{\sin\varphi}{\sin\theta}\dfrac{\partial}{\partial\varphi} \\ \dfrac{\partial}{\partial y} = \sin\theta\sin\varphi\dfrac{\partial}{\partial r} + \dfrac{1}{r}\cos\theta\sin\varphi\dfrac{\partial}{\partial\theta} + \dfrac{1}{r}\dfrac{\cos\varphi}{\sin\theta}\dfrac{\partial}{\partial\varphi} \\ \dfrac{\partial}{\partial z} = \cos\theta\dfrac{\partial}{\partial r} - \dfrac{1}{r}\sin\theta\dfrac{\partial}{\partial\theta} \end{cases}$$

to express the angular moment operators and the ∇^2as a function of the spherical polar coordinates (see Section 2.4.2).

NOTES

1 For non-conservative systems, the potential energy V depends also on velocities. Then, Eq. (1.23) becomes $p_j \overset{\text{def}}{=} \frac{\partial(T-V)}{\partial \dot{q}_j}$ and Eq. (1.24) becomes $H(q_j, p_j) = V(q_j, p_j) + T(q_j, p_j)$.

2 It is well known that QM was developed around the same years relativistic mechanics was being developed. Relativistic mechanics extends CM towards problems characterized by high speeds with respect to the speed of light. Here, the two extensions of the CM are taken as fully independent of each other, and only QM is dealt with. Note however that there are chemical problems where relativistic aspects play an important role.

3 In other words, if the symmetry operators $\hat{R}$ and $\hat{S}$ commute with $\hat{H}$: $\hat{H}\hat{R} = \hat{R}\hat{H}$ and $\hat{H}\hat{S} = \hat{S}\hat{H}$, it is not necessarily true that $\hat{S}\hat{R} = \hat{R}\hat{S}$.

4 We make again reference to the part of the Appendices concerning linear spaces (point 2.2) for deeper explanations.

5 What is a *complete* linear space is explained in the Appendix.

6 $|nlm\rangle$ is an eigenfunction of the hydrogenic atom with quantum numbers n, l, and m.

7 The integral of $\rho(\nu)$ between two frequencies is the total e.m. energy in that frequency range.

8 This is sometimes called light *diffusion*, but *scattering* is more frequently used.

2 The Electronic Structure of Atoms

The approach to the electronic structure of atoms and its main results are a well-established topic: nature, shape, and energy order of the atomic orbitals (AOs), meaning of electronic configuration, and the exclusion and *aufbau* principles are well assessed. Seemingly, there is no need to enter the details of the theoretical foundation, such as explaining when and why the wavefunction of a many-electron system can be written as a product of single-electron wavefunctions, why these are different from the hydrogenic orbitals, and where the exclusion principle comes from. It seems that, once "the electrons are put in the orbitals" there is not much more to say.

There are good reasons, however, for a chapter on the electronic structure of atoms. One point concerns the open shells where the configuration corresponds to more than one quantum state. According to the simplest account, these states belong to a single energy level. On the contrary, experiments show - in some cases - a partial removal of degeneracy: the energy splitting can be measured and the states belonging to different sub-levels are possibly characterized by different spectroscopic and chemical properties. To deal with these aspects, one must go beyond the simple approach of "putting electrons into orbitals" and understand why, when, and how it can be modified. The treatment of open shell systems also clarifies the basics of magnetism: it is then worth mentioning at least the most basic aspects of this wide field of science.

Another point is that a deeper understanding allows us to better distinguish what strictly concerns atoms from what holds also for other systems with many electrons. For instance, when the energy difference between the highest occupied orbital and the lowest unoccupied orbital of a fundamental configuration is low (in the order of kT or below) the exclusion principle must be revised as it is not anymore true that one is fully occupied, and the other one is empty. This case occurs frequently in condensed matter, where it gives rise to a whole class of materials (metals) with unique properties.

More generally, the electronic structure of atoms shows an ample set of problems, and its treatment is important also from a didactic point of view for gaining more familiarity with the methods and language of quantum mechanics (QM). Approaching systems with many electrons is not easy because of the effects of various interactions combined with antisymmetry. Not only each interaction is approached with its own methods, but the methods are combined differently when dealing with different systems, for instance for atoms with lower or higher number of electrons.

Nowadays, the outstanding advancements of computational quantum chemistry make possible to investigate systems with a huge number of electrons with a high degree of accuracy using methods capable of addressing in a unifying approach the various interactions and their complex relations. From a didactic point of view, however, all these aspects are better introduced step by step and discussed separately. The aim of this chapter is then to introduce and clarify what the computational approaches quantify accurately. To this aim, it is reasonable to discuss through examples topics that can be addressed theoretically only with a complex and difficult derivation, and it is useful to exploit an approximate and empirically justified approach (the vectorial model).

1 INTRODUCTION AND ANTISYMMETRY

1.1 Interactions in Atoms with Many Electrons

The most important contributions to the energy of an atomic system of n electrons are:

1. **The Kinetic Energy of Each Electron**: $\hat{T}_i = \frac{-\hbar^2}{2m_e}\nabla_i^2$, where ∇_i^2 is the Laplacian operator, which in Cartesian coordinates has the form: $\nabla_i^2 \overset{\text{def}}{=} \frac{\partial^2}{\partial x_i^2} + \frac{\partial^2}{\partial y_i^2} + \frac{\partial^2}{\partial z_i^2}$, (x_i, y_i, z_i) being the coordinates of electron i, and m_e the electron mass.
2. **The Potential Energy of Coulomb Attraction between Each Electron and the Nucleus**: $V_i = \frac{Z \cdot e^2}{r_i}$, where r_i is the distance of the i-th electron from the nucleus ($r_i = \sqrt{x_i^2 + y_i^2 + z_i^2}$ and Z is the charge of the nucleus which is placed at the origin of the coordinate system.

   The precise form of this relation (*e.g.*, with or without a factor $4\pi\varepsilon_0$ in the denominator) depends on the system of units. In poly-nuclear cases, this term becomes the *sum* of the attractive interactions of each electron (i) with each nucleus. It is convenient to include the two operators relating to a single electron in a single Hamiltonian operator:

$$\hat{H}_i^0 \overset{\text{def}}{=} \hat{T}_i + \hat{V}_i \tag{1.1}$$

and define the sum of the H_i^0 (one for each electron)

$$\hat{H}^0 \overset{\text{def}}{=} \sum_{i=1}^{n} \hat{H}_i^0 \tag{1.2}$$

We call it the **independent particle Hamiltonian** because, if we consider only these energy terms,

DOI: 10.1201/9781003644422-2

the position of an electron does not modify the interactions that act on another electron.

3. **The Potential Energy of Coulomb Repulsion between Electrons:**

$$V^{ee} \overset{\text{def}}{=} \sum \frac{e^2}{r_{ij}} \tag{1.3}$$

where r_{ij} is the distance between electrons i and j, and the summation extends to all electron *pairs* ij. It is an (electrostatic or Coulomb) interaction between charges and is repulsive: it increases the energy with respect to the independent particle Hamiltonian. The sum of the independent particle Hamiltonian and the electron-electron interaction

$$\hat{H}_{NR} \overset{\text{def}}{=} \hat{H}_i^0 + \hat{V}^{ee} \tag{1.4}$$

is called (somewhat improperly) **the non-relativistic** (NR) **Hamiltonian.**

4. **The Spin-Orbit Interaction**: that is, the interaction between the magnetic moments related to the orbital motions and the spins of the electrons.

As a matter of fact, these four interactions are sufficient to account for the most important effects, and it is therefore convenient to stop here. Other interactions are at work, but their effect is quantitatively much smaller than that of interactions 1–4, while the qualitative consequences, for our purposes, can be neglected. The analysis of the first interactions (Eqs. 1.2 and 1.3) immediately clarifies the difficulties in solving the Schrödinger equation in a poly-electronic system. In fact, the *e-e* charge interaction term (Eq. 1.3) "mixes" the spatial coordinates of the electrons and therefore prevents the resolution through the separation of variables. The fourth energy term mixes the spatial coordinates with the spin coordinates and adds further difficulties.

In principle, the inclusion of electrostatic repulsion between electrons would force us to write "intrinsically" poly-electronic wave functions, that is, functions of variables relating to the coordinates of the several electrons and not to each electron separately. A useful term of comparison to explain what is meant could be what has already been done in the treatment in classical mechanics (CM) of the degrees of freedom of two particles linked by a harmonic potential. In that case, we no longer speak of the coordinates of one or the other particle, but of two "global" coordinates, *i.e.* an average value (position of the centre of mass) and a distance between the two particles. The choice of these new coordinates produces an exact separation of those two degrees of freedom. A solution of this kind for the electrons in an atom was, in fact, tried in the 1950s, for the He atom, by writing wave functions of the type:

$$\psi = \left(1 + au + bt^2 + cs + ds^2 + fu^2\right) \cdot e^{(-\eta s)}$$

where a, b, c, d, f, η are parameters to be determined and, instead of the positions ($\vec{r_1}$ and $\vec{r_2}$) of the two electrons, the global coordinates are used:

$$s \overset{\text{def}}{=} \left|\vec{r_1}\right| + \left|\vec{r_2}\right|, \quad t \overset{\text{def}}{=} \left|\vec{r_1}\right| - \left|\vec{r_2}\right|, \quad u \overset{\text{def}}{=} \left|\vec{r_2}\right| - \left|\vec{r_1}\right|$$

Despite the excellent results obtained (in relation to the time), this method (due to Hylleraas) was abandoned because its generalization to cases with more than two electrons is not known.

1.2 The Simplest Model: Independent Electrons

To better understand the various aspects of the problem and their relationships, it is useful to analyse simplified systems in which only part of the various interactions is considered. In particular, it is useful to start with a system in which both the charge interaction between electrons and the spin-orbit interaction are completely ignored. This simplified system is particularly informative because it can be solved exactly. In fact, if we limit ourselves to the contributions of the kinetic energy and the electron-nucleus interactions, we have the *independent particle model* (Eq. 1.2). The solution of the poly-electronic problem can be obtained by preliminarily solving a mono-electronic problem: $\hat{H}_i^0 \varphi_i^0 = E_i^0 \cdot \varphi_i^0$. Then, the general lines of the solution of a *separate* Hamiltonian, as described at Section 3.5.1 of Chapter dealing with QM, allow us to write the wave functions of a system with 2, 3, … electrons as *products* of 2, 3, … one-electron functions (φ_i^0) and to express the energies of the poly-electronic system as the *sum* of the corresponding eigenvalues (E_i^0). The treatment would then be conceptually analogous to the treatment of the independent translational, rotational and vibrational degrees of freedom of a molecule, which is based on the (exact) separation between external and internal degrees of freedom and (approximate) separation of the internal degrees into rotational and vibrational degrees.

The independent particle model is an approximation, because the charge interaction (V^{ee}) couples the motions of the electrons. Discussing this model is however very important because, first, it allows us to analyse in the simplest and most direct way some general characteristics that any poly-electronic wave function must have due to antisymmetry. Secondly, the model is in any case a first step towards the solution of the problem of the poly-electronic atom because - as we will see later - it is the basis of an important approximate solution, according to which one still resorts to *monoelectronic functions* and takes into account in an indirect and approximate way the interactions by *modifying* the monoelectronic functions and their energies with respect to the functions and energies of the mono-electronic system. Following this line of presentation, the problems to be discussed are:

1. the anti-symmetrization of poly-electronic wavefunctions, both in general and in the particular case in which they are written as products of mono-electronic functions,
2. the treatment of interactions added to the interactions of each electron with the nucleus (or with the nuclei, in the case of poly-nuclear systems), in particular the charge repulsion between electrons and the spin-orbit interaction.

The **combined effect** of these interactions not only produces important consequences but also affects the very way in which the study of a poly-electronic system is approached and its results are presented.

1.2.1 Orbital Functions

To begin with, it is useful to analyse in some detail the characteristics of monoelectronic functions. The *spatial part* of a monoelectronic wavefunction, that is, a function (generally complex) of the three spatial coordinates of a single electron, is called *orbital function* or simply **orbital**. In fact, various types of orbitals are of interest and are used.

1.2.1.1 Hydrogen-Like Orbitals

The quantum mechanical model known as the *hydrogen atom* describes the motion of a single electron in the Coulomb field of the nucleus, neglecting the spin-orbit interaction (discussed in Section 3.1). The model applies to hydrogen and to ions such as He^+, Li^{2+}, …. The solutions are called **hydrogenic** or **hydrogen-like** orbitals. Using spherical polar coordinates (ρ, ϑ, φ) the Hamiltonian is separable (as shown in Section 3.5 of Chapter 1) into a part containing only the radial coordinate ρ and a part in which the angular coordinates ϑ, φ appear in operators while the radial coordinate appears only in a multiplicative function. Then, the solutions φ come up as products of a radial part (R) which is a function of only ρ, multiplied by an angular part (Y, called *spherical harmonic*) which is a function only of the coordinates ϑ and φ and are identified by three quantum numbers (n, l, m):

$$\Phi_{nlm}(\rho,\vartheta,\varphi) = R_{nl}(\rho)\cdot Y_{lm}(\vartheta,\varphi) \quad (1.5)$$

As it is well known, the solutions are denoted in Dirac notation as $|nlm\rangle$.

- The quantum number n (called principal) takes the values: $n = 1, 2, \ldots$
- The other two quantum numbers take the values: $l = 0, \ldots, n-1$; $m = -l, \ldots, +l$ in steps of 1: in fact, they are the quantum numbers of an angular moment, which is called *orbital* to distinguish it from the moment relative to spin.
- The energy depends on the nuclear charge (Z) and on the quantum number n, but does not depend on the quantum numbers l and m

$$E_n = \frac{-m\cdot e^4}{\hbar^2}\left(\frac{Z^2}{2\cdot n^2}\right),$$ in the so-called atomic units, or,

In SI units: $$E_n = \frac{-m\cdot e^4}{(4\pi\epsilon_0)^2\hbar^2}\left(\frac{Z^2}{2\cdot n^2}\right) = \frac{-m\cdot e^4}{8\epsilon_0{}^2h^2}\left(\frac{Z^2}{n^2}\right) \quad (1.6)$$

The functions of Eq. (1.5) are eigenfunctions of three operators: the energy, the magnitude of the angular moment $\left(\left|\vec{l}\right|\right)$ and the z-component of the same moment (l_z). The energy is specified by Eq. (1.6), the magnitude of the angular moment by $\left|\vec{l}\right| = \sqrt{l\cdot(l+1)}\cdot\hbar$ and the z-component of the angular moment by $\vec{l_z} = m\cdot\hbar$. Neglecting the electron spin, these three are a complete system of commuting operators: eigenfunctions and eigenvalues are completely specified by the three quantum numbers nlm. The eigenfunctions are conveniently denoted as $|nlm\rangle$ or, if one prefers, as 1s, 2s, $2p_1$, $2p_0$, $2p_{-1}$, 3s, $3p_1$, $3p_0$, $3p_{-1}$, $3d_2$, $3d_1$, $3d_0$, $3d_{-1}$, $3d_{-2}$, … where 1, 2, 3, … denote the quantum number n; the letters s, p, d, … refer to the quantum number l (s ⇔ 0; p ⇔ 1, d ⇔ 2, f ⇔ 3, g ⇔ 4 and so on along the alphabetic order), while the subscript is the quantum number m.

There is ambiguity in the choice of hydrogen-like AOs. The well-known functions 1s, 2s, $2p_x$, $2p_y$, $2p_z$, 3s, $3p_x$, $3p_y$, $3p_z$, $3d_{xy}$, $3d_{yz}$, $3d_{zx}$, … are linear combinations of functions $|nlm\rangle$ that have *the same n and the same l, but different m* and are degenerate. The three functions $2p_x$, $2p_y$, $2p_z$ are *independent* linear combinations of the three functions $2p_1$, $2p_0$, $2p_{-1}$, that is $|211\rangle$, $|210\rangle$ and $|21-1\rangle$. Similarly for the three functions 3p, the three functions 4p, the three functions 5p, …. The five functions $3d_{z^2}$, $3d_{xy}$, $3d_{yz}$, $3d_{zx}$, $3d_{x^2+y^2}$ are *independent* linear combinations of $3d_2$, $3d_1$, $3d_0$, $3d_{-1}$, $3d_{-2}$. Similarly for the five functions 4d, the five functions 5d, …. For the seven functions 4f, 5f, 6f, …, the nine functions 5g, … analogous considerations apply. The two sets of hydrogen-like functions are alternative and equivalent. For each pair n, l, one goes from one set to the other through *independent linear combinations* within a set of degenerate eigenfunctions of E and l. The analytic expressions of the AOs and the precise relations between the functions of the two sets are reported in all the specialized texts. Here it is rather interesting to explain the reasons why two sets are used.

- The $|nlm\rangle$ functions are used when one needs functions characterized by a precise value of the z component of the angular moment and of the related magnetic moment, in particular in problems of magnetism. They are in general complex functions, those with $m = 0$ are real.
- The functions with indices x, y, z are not (always) characterized by a precise value of the z component of the angular moment, have a directional shape and therefore are particularly suitable for describing the chemical bonds. They are real functions and therefore are also simpler to represent graphically.

1.2.1.2 Central Field Atomic Orbitals, CF-AOS

The considerations on the different ways of expressing the hydrogen-like AOs (*i.e.* $2p_x$, $2p_y$, $2p_z$, or $2p_1$, $2p_0$, $2p_{-1}$, ...) hold for *all centrosymmetric atomic orbitals*, since the radial part $R_{nl}(\rho)$ has no influence. There are other atomic orbitals (different from the previous ones, but also indicated with the notations 1s, 2s, $2p_x$, $2p_y$, $2p_z$, ...) characterized by central field (CF) symmetry. An example are the so-called **Slater orbitals**. The difference from the Coulomb field of the hydrogen-like system is that the dependence on *r* is *not* of the type $-Z/r$. Central field symmetry (CF) implies that there are three 2p functions (or three 3p functions, or five 3d functions, ...) *at* the same energy. These degeneracies are due to symmetry. The degeneracy according to which (for hydrogen-like systems) the 2s and 2p functions or 3s, 3p, 3d, or 4s, 4p, 4d, 4f, ... have the same energy is instead specific to the Coulomb field, does not hold for a generic CF potential and in particular does not hold for a poly-electronic atom.

1.2.1.3 Other Atomic Orbitals

Other well-known AOs are the **hybrids** sp, sp^2, d^2sp^3, ... These AOs *cannot* be written as products of a radial part and a spherical harmonic. A monoelectronic function is called *orbital* even in the case of *polynuclear* systems (molecules, crystals, ...).

Warning: the graphical representations of orbitals sometimes are incorrectly commented on. For example, it is frequently written that an orbital is "the region of space in which there is an electron" or "electrons in fact inhabit regions of space known as orbitals". Sentences like these contain various inaccuracies that can lead to misunderstanding: an orbital is **not** a **region** of space but is a function of spatial coordinates, and an orbital usually extends over the whole of ordinary 3D space and *is not limited* to a particular region.

Graphical representations of hydrogen-like AOs are readily available on many websites.[1]

1.2.1.4 Orbital Shells

The word **shell** specifies a set of *degenerate* orbitals: in the case of CF-AOs we are therefore talking about shells 1s, 2s, 2p, 3s, 3p, 3d, ... Just from the definition, we distinguish between 2s and 2p shells, between 3*s*, 3*p* and 3d, etc.... because we want to consider the general case where the degeneracy between AOs with the same *n* but different *l* is lacking. We speak of shells also in polynuclear systems (molecules, solids).

1.2.2 Spin Functions, Spin Variables, Spin-Orbital Functions

Now, let us remind that among the *coordinates* that describe the motion of an electron there is also the spin (see Section 1.4 of Chapter 1). The *spin coordinate of an electron*, denoted by σ, can assume the two values $\sigma = +½$ and $\sigma = -½$. The non-relativistic Hamiltonian (NR, Eq. 1.4) does not contain energy terms dependent on the spin variables. That means that the Schrödinger equation does not determine (we better say: it *cannot* determine) the part of the wave function that depends on the spin. This part then appears as a multiplicative constant to be specified. In conclusion, a monoelectronic wave function, complete with spin dependence, can be written as the **product of**:

- a function of the spatial coordinates only (the orbital) multiplied by
- a function of the spin variable only, which we denote by $\omega(\sigma)$.

Usually the two simplest (normalized) spin functions are used:

$$\alpha(\sigma):\left\{\begin{array}{l}\alpha=1;\ \sigma=+½\\ \alpha=0;\ \sigma=-½\end{array}\right\}\quad \beta(\sigma):\left\{\begin{array}{l}\beta=0;\ \sigma=+½\\ \beta=1;\ \sigma=-½\end{array}\right\} \tag{1.7}$$

These are the two functions corresponding, in common language, to the two situations in which the electron has spin +½ or spin −½. This pair of functions is a complete set. Also, it is convenient to group the spatial coordinates ($\vec{r}$) and the spin coordinate (σ) under a single symbol (τ) and introduce, alongside the orbitals, the spin-orbitals (SOs), that is, functions which are the product of an orbital (χ) by one of the two usual spin functions (ω) of Eq. (1.7):

$$\varphi(\tau) \stackrel{\text{def}}{=} \chi(\vec{r})\cdot\omega(\sigma) \tag{1.8}$$

All subsequent discussion is greatly simplified by the use of SOs.

1.2.3 Hartree Product

In the independent electron model, as already mentioned, the wavefunctions of a poly-electronic system are written - at least in a first step - as products of mono-electronic functions, *i.e.* SOs:

$$\psi(\tau_1,\tau_2,\ldots,\tau_n) \stackrel{\text{def}}{=} \varphi_a(\tau_1)\cdot\varphi_b(\tau_2)\cdot\ldots\cdot\varphi_z(\tau_n) \tag{1.9}$$

The product of SOs plays a central role in the treatment of both atoms and poly-electronic systems in general, and it is called *Hartree product*. A Hartree product is *a product of spinorbitals*, as many SOs as the electrons of the system.

N. B. 1: The name Hartree product does not imply that the mono-electronic functions (the SOs) have a particular shape or are obtained with a particular computational method. They can be hydrogen-like functions, if we are talking about a model with truly independent electrons, but – anticipating the discussion that will be carried out later – they can also be functions that account in a more or less accurate way for the charge repulsion between electrons.

N. B. 2: We speak of Hartree products also in the molecular case: obviously, an SO will then be the product of an MO (molecular orbital) by one of the two spin functions. For crystalline systems we will similarly speak of Hartree products of crystalline SOs.

1.3 Antisymmetry

1.3.1 Permuting the Coordinates

Poly-electronic functions – by the very fact that they describe the motion of equal particles – have particular properties with important consequences. For clarity, these issues are addressed by discussing poly-electronic functions written as Hartree products, keeping in mind that most considerations we now discuss also hold when the functions are *not* Hartree products.

Using the same SOs $\varphi_a,\varphi_b,\ldots,\varphi_z$ (cfr. Eq. 1.9), we *can write more than one Hartree product*, because we can assign electrons to the spin orbitals in different ways. For example, if we consider a system of three electrons and three distinct spin orbitals S_a, S_b, S_c, we have the Hartree product $S_a(1) \times S_b(2) \times S_c(3)$ [$S_a(1)$ is a concise writing for $S_a(\tau_1)$]. Other Hartree products with the same spin orbitals are, for example: $S_a(1) \times S_b(2) \times S_b(3)$, or $S_a(1) \times S_b(2) \times S_a(3)$, $S_b(1) \times S_b(2) \times S_a(3)$, etc. Note: even using only two spin orbitals (S_a, S_b) we could still have more Hartree products, for example: $S_a(1) \times S_b(2) \times S_b(3)$, or $S_a(1) \times S_b(2) \times S_a(3)$, or $S_b(1) \times S_b(2) \times S_a(3)$, etc. .... We will see later that these products are forbidden by antisymmetry. It is possible to write several functions starting from the same analytical expressions, even when these are **not** Hartree products. For example, one can consider the function $\psi_a = \psi(\tau_1,\tau_2,\tau_3)$ and $\psi_b = \psi(\tau_1,\tau_3,\tau_2)$ which depends on the second coordinate in the same way as ψ_a depends on the third coordinate. The "exchange" operation (P_{kl}) between the k and l electrons is described by the exchange operator:

$$\hat{P}_{kl}\psi(\tau_1,\tau_2,\ldots,\tau_k,\ldots,\tau_l,\ldots,\tau_n) \overset{\text{def}}{=} \psi(\tau_1,\tau_2,\ldots,\tau_l,\ldots,\tau_k,\ldots,\tau_n) \tag{1.10}$$

In the particular case of a **Hartree product**:

$$\hat{P}_{kl}\left[\varphi_a(\tau_1)\cdot\varphi_b(\tau_2)\cdot\ldots\varphi_m(\tau_k)\cdot\ldots\cdot\varphi_n(\tau_l)\cdot\ldots\cdot\varphi_z(\tau_n)\right]$$
$$\overset{\text{def}}{=} \varphi_a(\tau_1)\cdot\varphi_b(\tau_2)\cdot\ldots\varphi_m(\tau_l)\cdot\ldots\cdot\varphi_n(\tau_k)\cdot\ldots\cdot\varphi_z(\tau_n)$$

When there are more than two electrons, it is necessary to consider the exchanges between the coordinates of three, four, ... electrons in addition to the *simple* exchanges between the coordinates of only two electrons. These more complex transformations correspond to the successive applications of two or more simple exchanges. All together, they are called **permutations**, and the set of all permutations is the **permutation group**. For algebraic consistency, the identity must also be counted among these transformations: this is the initial sequence or else the permutation that "leaves everything as it is". A simple method of listing the permutations of the group is to list the *sequences of electron indices* as they are modified by the permutations. For example, with *two* electrons we have only one exchange and therefore the set of permutations is made up of only two elements, because there are only two possible sequences for the electron indices: 12 and 21 and the permutation group is made up of only two operations: the identity [12 → 12] and the simple exchange [12 → 21]. With *three* electrons, we have six sequences: 123, 132, 213, 231, 312, 321. The sequence 132 is obtained simply by exchanging 2 and 3, which we can indicate with (2,3), while the sequences 213 and 321 are obtained similarly from the basic sequence 123 by performing the simple exchanges (1,2) and (1,3) respectively. For the others, at least two successive simple exchanges are needed, for example the sequence 312 is obtained by first performing the exchange (1,3): 123 → 321 and then the exchange (2,1): 321 → 312. The *number* of permutations of n equal objects is $n!$ This is easily proved. It is basically a matter of writing down all the possible sequences of the numbers 1, 2, ... n using them all only once. In first place we can put any of the n numbers, in second place any of the remaining $(n - 1)$ numbers, ... So, there are $n \times (n - 1) \times (n - 2) \times \ldots \times 2$ possibilities.

For example, with $n = 3$, there are $3! = 2 \times 3 = 6$ permutations (see above), with $n = 4$, there are $4! = 2 \times 3 \times 4 = 4 \times 3! = 24$ permutations, with $n = 5$, there are $5! = 2 \times 3 \times 4 \times 5 = 5 \times 4! = 120$ permutations, with $n = 6$, there are $2 \times 3 \times 4 \times 5 \times 6 = 6 \times 5! = 720$ permutations, ... A given permutation can be realized with different sequences of simple exchanges, and different *numbers* of exchanges. However, in any case, the *number of simple exchanges* to be performed is *always even* or *always odd*.[2] For example, the sequence 132, corresponding to the simple exchange of 2 with 3 that we write (2,3), can be obtained also by performing three simple exchanges: (1,3) → 321; (1,2) → 312; (1,3) → 132 or (1,2) → 213; (1,3) → 231; (1,2) → 132. If a permutation requires an even number of simple exchanges, then it is said to be **even**; if the number of simple exchanges required is odd, the permutation is said to be **odd**.

1.3.2 Eigenvalues and Symmetry Operators

In QM, as is well known, the symmetry that characterizes a system translates into a symmetry of the eigenfunctions of the Hamiltonian operator. Let us consider an example that does not concern electrons in atoms. In the harmonic oscillator model, the potential energy is proportional to the square of the coordinate and therefore does not change if "one goes from x to $-x$". The kinetic energy does not change either: differentiating twice with respect to x or $-x$ makes no difference. This symmetry of the system is called **parity**. Due to this feature, each stationary wavefunction of the harmonic oscillator must have a defined behaviour if "one goes from x to $-x$". The wavefunctions must have "even" parity (the function does not change if "one goes from x to $-x$") or "odd" parity (the function changes sign if "one goes from x to $-x$"). Only these behaviours are compatible with parity. In particular, it is impossible for a stationary wave function to have no symmetric behaviour (neither even nor odd). It should be noted that the energy operator and the energy function have even parity, but some stationary solutions are "even" and some are "odd". An "odd" function is still a function with parity symmetry, because it has a defined behaviour in "going from x to $-x$" and is therefore fundamentally different from a function that does not have a defined behaviour.

All the previous discussion about permutations has a simple reason. No matter how we write a poly-electronic function, not necessarily as a Hartree product, every transformation that we apply by permuting the coordinates of the various electrons is *a symmetry operation of the system.* "Putting one electron in place of another" cannot, quite obviously, change the energy of the system. So, we have eigenvalues and eigenfunctions of the operators of permutation symmetries, and we can derive the eigenvalues starting from the simple consideration that performing a simple exchange (p_{kl}) twice means bringing the system back to the top and therefore necessarily corresponds to the eigenvalue $(p_{kl})^2 = 1$. Then, the eigenvalue of any simple exchange (p_{kl}) can only be one of the two square roots of unity: +1 or −1.

- Functions for which every simple exchange has eigenvalue (p_{kl}) = 1 are called **symmetric**.
- Functions for which every simple exchange has eigenvalue (p_{kl}) = −1 are called **antisymmetric**.

1.3.3 Pauli Principle

This permutation symmetry is the basis of a very general property of QM wavefunctions known as the **antisymmetry principle** (or **Pauli principle**). In its best known and electron-specific form, the Pauli principle states that

> In a system of many electrons, only wave functions that change sign for *every* exchange of the coordinates of *two* electrons are possible:

$$\hat{P}_{kl}\psi(\tau_1,\tau_2,\ldots,\tau_k,\ldots,\tau_l,\ldots,\tau_n)$$
$$= -\psi(\tau_1,\tau_2,\ldots,\tau_k,\ldots,\tau_l,\ldots,\tau_n) \quad (1.11)$$

> Wavefunctions that change sign for every exchange of the coordinates of two electrons are called **antisymmetric.**

In other words, with poly-electronic systems (atoms, molecules, crystals, …) we can *only use antisymmetric wavefunctions* and we must neglect functions that are not, even if in other respects they seem possible. Moreover, an even number of simple exchanges (exchange of the coordinates of *two* electron) corresponds to an even number of changes of sign, In the end:

- an *antisymmetric* function is an eigenfunction of every *permutation* operator:
- its eigenvalue (p_{kl}) is +1 (if the permutation is even) or −1 (if the permutation is odd).

Although much less important in chemistry, it can be useful for comparison to consider the case of two or more independent and equal particles whose wavefunctions must be *symmetric*, rather than antisymmetric. Obviously, these are not electrons. If a function is symmetric, a permutation operator corresponding to a simple exchange multiplies it by +1. Applying a permutation operator that does not correspond to a simple exchange means applying a sequence of simple exchanges and then multiplying by +1 several times. In the end, we get that

- a symmetric function is an eigenfunction of every permutation operator:
- its eigenvalue (p_{kl}) is always +1.

1.3.4 Slater Determinants

Now, we see that a *Hartree product* has no special behaviour with respect to permutations (it is neither antisymmetric nor symmetric). Let's consider for example the case of two electrons and two SOs (1sα and 1sβ), that is the Hartree product

$$\varphi_{1s\alpha}(\tau_1)\times\varphi_{1s\beta}(\tau_2)$$

By exchanging electrons, we obtain: $\varphi_{1s\alpha}(\tau_2)\times\varphi_{1s\beta}(\tau_1)$. Given two sets of values for the two variables τ, the two expressions assume different and unpredictable values if the two functions are not specified previously. To clarify what instead **symmetric function** means, let's consider a slightly more complicated function, for example, the sum of the two Hartree products built with the same spin orbitals:

$$\left[\varphi_{1s\alpha}(\tau_1)\times\varphi_{1s\beta}(\tau_2)+\varphi_{1s\beta}(\tau_1)\times\varphi_{1s\alpha}(\tau_2)\right]$$

Exchanging 1 with 2 we obtain

$$\left[\varphi_{1s\alpha}(\tau_2)\times\varphi_{1s\beta}(\tau_1)+\varphi_{1s\beta}(\tau_2)\times\varphi_{1s\alpha}(\tau_1)\right]$$
$$=\left[\varphi_{1s\beta}(\tau_2)\times\varphi_{1s\alpha}(\tau_1)+\varphi_{1s\alpha}(\tau_2)\times\varphi_{1s\beta}(\tau_1)\right]$$
$$=\left[\varphi_{1s\alpha}(\tau_1)\times\varphi_{1s\beta}(\tau_2)+\varphi_{1s\beta}(\tau_1)\times\varphi_{1s\alpha}(\tau_2)\right]$$

thus, it is a **symmetric wavefunction** (with respect to the permutation between 1 and 2). Now consider the difference instead of the sum:

$$\left[\varphi_{1s\alpha}(\tau_1)\times\varphi_{1s\beta}(\tau_2)-\varphi_{1s\beta}(\tau_1)\times\varphi_{1s\alpha}(\tau_2)\right]$$

Applying the permutation of 1 with 2 we get:

$$\left[\varphi_{1s\alpha}(\tau_2)\times\varphi_{1s\beta}(\tau_1)-\varphi_{1s\beta}(\tau_2)\times\varphi_{1s\alpha}(\tau_1)\right]$$
$$=\left[-\varphi_{1s\beta}(\tau_2)\times\varphi_{1s\alpha}(\tau_1)+\varphi_{1s\alpha}(\tau_2)\times\varphi_{1s\beta}(\tau_1)\right]$$
$$=\left[-\varphi_{1s\alpha}(\tau_1)\times\varphi_{1s\beta}(\tau_2)-\varphi_{1s\beta}(\tau_1)\times\varphi_{1s\alpha}(\tau_2)\right]$$

This linear combination is an **antisymmetric function**. The result does not depend on any particular property of the functions φ but is *the direct consequence* of using *a particular linear combination* in place of the two original functions. The example also shows that, in order to have functions in agreement with the Pauli principle, it is necessary an operation that makes the Hartree products *anti-symmetric*. Essentially, we have transformed the product $\varphi_{1s\alpha}(\tau_1)\times\varphi_{1s\beta}(\tau_2)$ into an *antisymmetric function* by making its linear combination (with coefficients +1 and −1) *with its transform* $\varphi_{1s\alpha}(\tau_2)\times\varphi_{1s\beta}(\tau_1)$. Note that the transform is also one of the Hartree products that can be made using the same SO. To generalize the example to the case of any number of electrons, the **anti-symmetrization** of a poly-electronic function requires in detail:

- to perform all possible permutations obtaining all the transformed functions and
- to linearly combine the original function with the transformed functions using the coefficients +1 (for even permutations, including the case of the original untransformed function) or −1 (for odd permutations).

The fact that this procedure is correct can be rigorously proved by considerations based on symmetry (group theory). *The fundamental difference between parity and other kinds of symmetry* is the *principle of antisymmetry* that allows *only one particular behaviour instead of both.*

In the special case of poly-electronic functions written as Hartree products, the mathematical tool for anti-symmetrization is the **Slater determinant**. So, the Hartree product:

$$\psi(\tau_1,\tau_2,\ldots,\tau_n)=\varphi_a(\tau_1)\cdot\varphi_b(\tau_2)\cdot\ldots\cdot\varphi_z(\tau_n) \quad (1.12)$$

is made anti-symmetric with the:

$$\frac{1}{\sqrt{n!}}\begin{vmatrix}\varphi_a(\tau_1) & \cdots & \varphi_z(\tau_1)\\ \vdots & \ddots & \vdots\\ \varphi_a(\tau_n) & \cdots & \varphi_z(\tau_n)\end{vmatrix} \quad \text{(Slater determinant)} \quad (1.13)$$

For example, for a three-electronic system with SOs S_a, S_b and S_c:

$$\Psi=\frac{1}{\sqrt{6}}\begin{vmatrix}S_a(\tau_1) & S_b(\tau_1) & S_c(\tau_1)\\ S_a(\tau_2) & S_b(\tau_2) & S_c(\tau_2)\\ S_a(\tau_3) & S_b(\tau_3) & S_c(\tau_3)\end{vmatrix}$$

$$=\frac{1}{\sqrt{6}}\left[S_a(\tau_1)\cdot S_b(\tau_2)\cdot S_c(\tau_3)+S_b(\tau_1)\cdot S_c(\tau_2)\cdot S_a(\tau_3)+S_c(\tau_1)\cdot S_a(\tau_2)\cdot S_b(\tau_3)\right]$$

$$-\frac{1}{\sqrt{6}}\left[S_c(\tau_1)\cdot S_b(\tau_2)\cdot S_a(\tau_3)+S_b(\tau_1)\cdot S_a(\tau_2)\cdot S_c(\tau_3)+S_a(\tau_1)\cdot S_c(\tau_2)\cdot S_b(\tau_3)\right]$$

Without *deriving* the result, let us see the main features and properties of the Slater determinants and verify that they are what is required to agree with the Pauli principle.

A. The determinant has n rows and n columns (n is the *number of electrons* to be treated). On a column there is always the same mono-electronic function, *i.e.* the same SO. On a row there is always the same variable, that is the set of spatial and spin coordinates of the same electron.
B. Each matrix element of the Slater determinant is a particular SO written as a function of the coordinates of a particular electron and thus corresponds to "a particular electron in a particular SO".
C. The determinant is basically the *linear combination of products* of different matrix elements with the following features. Each product is obtained by multiplying n elements of the matrix. These elements are always taken from different rows and columns. Each product contains all the SOs and is different from another product because some particular SO depends on the coordinates of a different electron. Reciprocally, each product contains the coordinates of all electrons and is different from another product because the coordinates of one or more electrons appear as the argument of a different SO. Summarizing, each Hartree product within the determinant has a different permutation of the electrons on the various SOs and *every product linearly combined in the determinant is one of the various Hartree products that can be made with the same SO.*
D. In the determinant, the electrons are permuted on the various SOs in all possible ways: we leave out no Hartree product and no Hartree product appears more than once. In fact, in a determinant there are $n!$ Hartree products to be combined linearly, just as there are $n!$ permutations between n electrons. The main diagonal corresponds to the Hartree product that we can call "original" or prototype (Eq. 1.12), that is, the Hartree product that we want to anti-symmetrize using the Slater determinant.
E. In the linear combinations of the Hartree products, a product appears with coefficient +1 (or −1) if, with respect to the order specified by the main diagonal, the electrons are exchanged an even (or

odd) number of times. Therefore, the exchange of two rows (or two columns) of the entire square matrix only changes the sign of the determinant (= simply multiplies it by −1). This is the property that "realizes" the anti-symmetrization. The $1/\sqrt{n!}$ factor in Eq. (1.13) is due to the normalization of the $n!$ linear combinations with coefficients = +1 or −1, i.e. with modulus = 1. Thanks to this factor the Slater determinant is a normalized function, if the individual SOs already are.

We note that the calculation of the determinant is an operation of complexity rapidly increasing with n, since $n!$ Hartree products must be combined, each consisting of n factors. For example, with six electrons we must consider 720 Hartree products, each of which is the product of six SOs. Simple formulas are available for the cases $n = 2$ and $n = 3$ (Sarrus' rule), in the other cases the calculation of a determinant of order n is generally solved by linearly combining n determinants of order $(n - 1)$, and so on recursively.

1.3.5 Exclusion Principle

The effects of anti-symmetrization can now be derived from the properties of determinants. Among these, the most interesting one is that by exchanging two rows (or two columns) the determinant changes sign. It follows, then, that a determinant with two equal rows or two equal columns is zero, since it must simultaneously be equal to itself and to its opposite.

$$\Psi = \frac{1}{\sqrt{N!}}\begin{vmatrix} S_a(\tau_1) & S_b(\tau_1) & \cdots & S_b(\tau_1) \\ S_a(\tau_2) & S_b(\tau_2) & \cdots & S_b(\tau_2) \\ \vdots & \vdots & \ddots & \vdots \\ S_a(\tau_n) & S_b(\tau_n) & \cdots & S_b(\tau_n) \end{vmatrix} = 0$$

Hence the fundamental and well-known **exclusion principle**, which is usually stated by saying that it is not possible to place more than one electron in an SO or it is possible to place a maximum of two electrons in an orbital but, if there are two, they must have different spins. More correctly, this means that it is not possible to use the same SO more than once or else that it is not possible to use an orbital more than twice and if you use it twice you must use two different spin functions.

1.3.6 The Electrons Are Indistinguishable

Slater determinant clearly highlights an important and "new" aspect, that it is not found in CM. It is expressed by saying that electrons are *indistinguishable. It is not possible to say in which SO a given electron is and, vice versa, it is not possible to say which electron "occupies" a given SO.* Let us consider, for example, a generic state for a three-electron system:

$$\frac{1}{\sqrt{6}}\begin{vmatrix} \psi_a(\tau_1) & \psi_b(\tau_1) & \psi_c(\tau_1) \\ \psi_a(\tau_2) & \psi_b(\tau_2) & \psi_c(\tau_2) \\ \psi_a(\tau_3) & \psi_b(\tau_3) & \psi_c(\tau_3) \end{vmatrix}$$

It is not possible to say which is the SO where electron 1 is: its coordinates appear as arguments of all three SOs and appear with the same statistical "weight", which is one-sixth, as given by the square modulus of the coefficient. It is not even possible to say which electron is in a given SO: in fact, among the arguments of an SO the coordinates of all three electrons appear with the same weight. It is worth pointing out that some frequently used expressions may be misleading. It is not correct to say, for example, "I place the *first* electron in SO *a*, the *second* in SO *b*, the *third* in SO *c*". Instead, one should say: "I place *one* electron in SO a, *another* in SO *b*, and yet another in SO *c*" or, better, "I use three SO, *a*, *b*, and *c*" and there is no need to specify their order. The meaning, anyway, is that a Slater determinant is made using SOs *a*, *b*, and *c*.

1.3.7 Electronic Configuration and *aufbau*

Since the energy of a system of *independent electrons* is the sum of the energies of the various SOs, what matters in determining the energy of the poly-electronic system is the occupancy of the electronic shells (see Section 1.2.1.4), that is, the occupancy of the *energy levels* of the orbitals. With the **electronic configuration,** we indicate how many SOs of each shell we use. The principle of antisymmetry forbids the use of the same SO more than once. Then, to find the state or states with minimum energy of systems with two, three, four, …, *n*, electrons, we progressively use the SOs of the various shells, starting from the shell at lowest energy, and going up to another shell when all the SOs of a lower energy shell have been completely used. This is the **aufbau principle**.

The configuration that agrees with the aufbau principle is the **fundamental** electronic configuration or the **ground state**. Configurations at higher energies are **excited** configurations.

In some configurations all the shells are either **completely occupied** or **completely empty**. These are said to be **closed shells**. An electronic configuration that is *not a closed shell* is an **open shell**. Examples are: (i) $1s^2$ is the fundamental electronic configuration of He and is a closed shell configuration; (ii) $1s^1 2s^1$ is an excited configuration of He and is an open shell configuration; (iii) $2s^2$ is an excited configuration of He and is a closed shell configuration; (iv) $1s^2 2s^1$ is the fundamental configuration of Li and is an open shell configuration; etc. A trivial observation: due to the spin, each shell is made up of an even number of SOs. Then, a system with an odd number of electrons has only open shell configurations.

1.3.8 Counting the States

We now want to count *how many distinct* non-zero *Slater determinants* there are for an electronic configuration. This is not a trivial task, because

Antisymmetry places very strong constraints on the number of states of an electronic configuration.

This is easily seen for *closed-shell configurations*. If the number of spin orbitals to be used (*i.e.* the number of electrons, let us say *m*) is equal to the number of available SOs, we must use them all, since none of them can be used more than once. Then, we have *m*! Hartree products corresponding to *m*! different permutations of the electrons assigned to each SO. These are a single Slater determinant and there is **only this possibility**. See Exercise 2.1

A closed-shell electronic configuration always has only one state.

Open-shell electronic configurations always have *more than one state*. For example, the $1s^2 2s^1$ configuration of a three-electronic system (*e.g.* the fundamental electronic configuration of Li) has two states. In fact, in the first and second columns of the determinant we place the SOs $1s\alpha$ and $1s\beta$: for the third column we can instead choose between two possibilities: $2s\alpha$ or $2s\beta$:

$$\frac{1}{\sqrt{6}}\begin{vmatrix} \psi_{1s\alpha}(\tau_1) & \psi_{1s\beta}(\tau_1) & \psi_{2s\alpha}(\tau_1) \\ \psi_{1s\alpha}(\tau_2) & \psi_{1s\beta}(\tau_2) & \psi_{2s\alpha}(\tau_2) \\ \psi_{1s\alpha}(\tau_3) & \psi_{1s\beta}(\tau_3) & \psi_{2s\alpha}(\tau_3) \end{vmatrix}$$

$$\text{and} \quad \frac{1}{\sqrt{6}}\begin{vmatrix} \psi_{1s\alpha}(\tau_1) & \psi_{1s\beta}(\tau_1) & \psi_{2s\beta}(\tau_1) \\ \psi_{1s\alpha}(\tau_2) & \psi_{1s\beta}(\tau_2) & \psi_{2s\beta}(\tau_2) \\ \psi_{1s\alpha}(\tau_3) & \psi_{1s\beta}(\tau_3) & \psi_{2s\beta}(\tau_3) \end{vmatrix}$$

Of course, any permutation of columns would not change the result. To approach the most general case, let us first consider a system with *n* electrons and a larger number (*m*: $m > n$) of distinct SOs, *all degenerate*. Then there are m^n different Hartree products, but the states are only:

$$\binom{m}{n} = \frac{m!}{n!\cdot(m-n)!} \quad \text{(also known as binomial coefficient)} \tag{1.14}$$

This is the number of *simple combinations* of *m* objects taken *n at a time* (*simple* means *without repetition*). As a special case, if $m = n$, that is, if the shell is closed, we see that $\binom{m}{n} = 1$ as it must be. This number (Eq. 1.14) answers the question: "in how many ways can I arrange *n* equal objects in *m* distinct boxes with at most one object per box?". The objects are equal because we do not want or are not able to distinguish, as is the case of the electrons. The boxes are the SOs.

N. B. 1: Electrons are equal and indistinguishable, while the SOs are all different and therefore distinguishable. The number of electrons enters the calculation only because it indicates how many SOs to use: if we have *n* electrons, we must use *n* SOs, we can choose them in all possible ways from a set of *m* SOs ($m \geq n$) and there are $\binom{m}{n}$ different ways.

N. B. 2: The calculation to be done and the formula are a direct consequence of the Slater determinant and Pauli principle. If we have *n* electrons to deal with, the Slater determinants have dimension $n \times n$. In each determinant there are *n* columns that contain a different SO each, otherwise the determinant would cancel. If we have *m* degenerate SOs, then the number of distinct determinants is $\binom{m}{n}$ and the order does not matter. Changing the order would change at most the sign which has no relevance in the counting.

N. B. 3: The coefficient $\binom{m}{n}$ is called binomial because it represents the *n*th coefficient of the expansion of the binomial $(x + y)^m$, that is, the coefficient that multiplies the monomial $x^n y^{m-n}$.

It is clear that, in *the fundamental* electronic configuration of a poly-electronic system, *only the last shell can be open*. Then, the number of states is still given by the formula (Eq. 1.14), where however the numbers *m* and *n refer only to the uppermost shell*.

It may be useful to consider the example of the $2p^2$ electronic configuration. Without anti-symmetrization and without the exclusion principle, we should say that having six SOs and having to use two of them, the possible states are $6^2 = 36$. This is not what we need, because (i) it contains repetitions (which are forbidden by the Pauli principle) and (ii) it distinguishes the order. From the 36 products, we must remove six products that correspond to using the same SO twice. The remaining 30 must be combined two by two (by exchanging two SOs we do not have a new Slater determinant) and therefore $(36 - 6)/2 = 15$. This is exactly what the binomial formula counts.

Let us finally consider **excited configurations with more than one open shell**. To count the states of these configurations, we look at *each open shell and multiply the results*. An example: an excited electronic configuration of C is $1s^2 2s^1 2p^3$. For the 1s shell we have two SOs and we must use two of them so there is only one possibility. For the 2s shell we have two SOs and we must use one: two possibilities. For the 2p shell we have six SOs and we must use three: 20 possibilities. Each possibility for one shell must be associated with each possibility for any other shell. So:

$$\binom{2}{1} = 2; \quad \binom{6}{3} = 20.$$

$$\text{Overall, } 1 \times 2 \times 20 = 40 \text{ determinants}$$

It is important to specify here that the calculations must be done *separately for each shell*. If, instead, the calculations are carried out "globally", the result is wrong because it counts the states of more than one electronic configuration. As an example, let us again take the fundamental electronic configuration of C: $1s^2 2s^2 2p^2$. It would seem obvious to say that, having a total of ten SOs (two SOs 1s, two SOs 2s and six SOs 2p) and wanting to "arrange" six electrons (better: wanting to use six different SOs), one should count:

$$\binom{10}{6} = \frac{10!}{6! \cdot 4!} = \frac{10 \cdot 9 \cdot 8 \cdot 7 \cdot 6!}{6! \cdot 4 \cdot 3 \cdot 2} = \frac{10 \cdot 9 \cdot 8 \cdot 7}{4 \cdot 3 \cdot 2}$$
$$= 10 \cdot 3 \cdot 7 = 210$$

That's wrong! The reason is that in this way we count something else, namely the number of states of all electronic configurations in which six SOs are used by taking them in all possible ways from the 1s, 2s and 2p shells. These configurations are:

$$(1s)^2 (2s)^1 (2p)^3 : 2 \times \binom{6}{3} : \quad 40 \text{ states}$$

$$(1s)^1 (2s)^2 (2p)^3 : 2 \times 20 : \quad 40 \text{ states}$$

$$(1s)^1 (2s)^0 (2p)^5 : 2 \times 6 : \quad 12 \text{ states}$$

$$(1s)^0 (2s)^1 (2p)^5 : 2 \times 6 : \quad 12 \text{ states}$$

$$(1s)^2 (2s)^2 (2p)^2 : \quad \binom{6}{2} = \frac{6!}{2! \cdot 4!} : \quad 15 \text{ states,}$$

$$(1s)^2 (2s)^0 (2p)^4 : \quad \binom{6}{4} = \frac{6!}{2! \cdot 4!} : \quad 15 \text{ states,}$$

$$(1s)^1 (2s)^1 (2p)^4 : \quad 2 \times 2 \times 15 : \quad 60 \text{ states,}$$

$$(1s)^0 (2s)^2 (2p)^4 : \quad 15 \text{ states,}$$

$$(1s)^0 (2s)^0 (2p)^6 : \quad 1 \text{ state}$$

The total is exactly 210 states, but it is clear that most of them do not pertain to the $1s^2 2s^2 2p^2$ configuration.

Finally, it is important to note that the usual energy schemes for AO occupancy provide the correct solution to counting the states. This graphical representation automatically shows the different Slater determinants since the horizontal line indicates the AO, the arrow indicates the spin function and both globally specifies an occupied SO. The diagrams count the determinants and not the Hartree products because (i) the arrows are anonymous, do not say whether it is one electron or another, and (ii) drawing two arrows with the same direction on the same AO is explicitly forbidden.

If we label the 2p AO as p_x, p_y, p_z, the diagram of Figure 2.1 faithfully represents the Slater determinant:

$$\Psi = \frac{1}{\sqrt{2}} \begin{vmatrix} \varphi_{1s\beta}(\tau_1) & \varphi_{2p_x\alpha}(\tau_1) \\ \varphi_{1s\beta}(\tau_2) & \varphi_{2p_x\alpha}(\tau_2) \end{vmatrix}$$

1.3.9 Beyond the Atomic Systems

In elementary treatments of the electrons in systems with many nuclei, the degrees of freedom of the nuclei are typically ignored, because they are already considered "separated". One usually resorts to the *Born-Oppenheimer separation*. Basically, this approximation first treats the motion of electrons by considering the positions of all nuclei to be fixed (either in the equilibrium configuration or in deformed configurations), then solves the vibrational problem by inserting the dependence of the energy of the electronic system on the nuclear positions as an additional contribution to the potential energy of interaction between the nuclei.

That said, most of the considerations made so far and their consequences apply to any poly-electronic system, even poly-nuclear (for example, to molecules). Here too we speak of mono-electronic functions (MO, molecular orbitals, and CO, crystalline orbitals), orbital shells, electronic configuration and counting of the states of a configuration.

A **first example** concerns the π-type MOs of *benzene* built on the six $2p_z$ AOs of the C atoms. There are six MOs and four energy levels with degeneracies 1, 2, 2, 1 in order of increasing energy, let us call them *g*, *h*, *i*, *l*. The other MOs can be ignored because they are outside the energy interval of the four MO-π: those at lower energies are completely occupied, while those at higher energies are completely empty. In the neutral molecule, the four levels are occupied by six electrons. Then, the fundamental electronic configuration is: g^2, h^4, i^0, l^0, it is a closed-shell

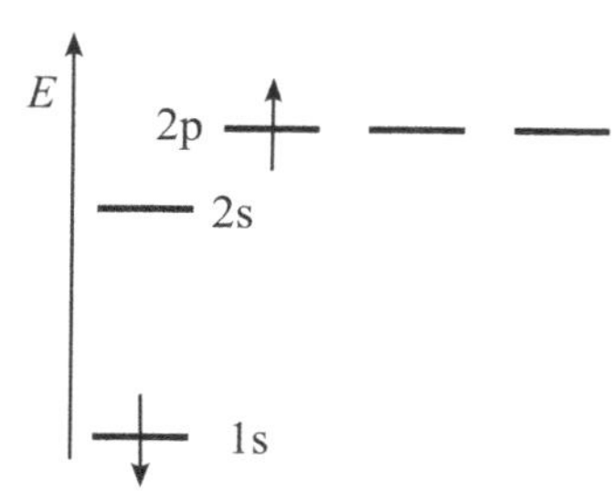

FIGURE 2.1 A determinant of the $1s^1 2p^1$ configuration.

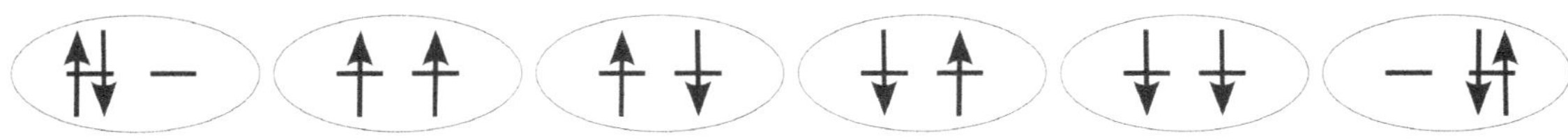

FIGURE 2.2 The states of the electronic configuration $(\pi^*)^2$ of O_2.

configuration and there is only one state. The lowest excited configurations involve only MO-π shells: the other MO shells remain either completely full or completely empty. Thus, the first excited configuration is: g^2, h^3, i^1, l^0 with: $1 \times \binom{4}{3} \times \binom{4}{1} = 1 \times 4 \times 4 = 16$ states.

Another simple **example** concerns the O_2 molecule. In the neutral molecule, after having filled the shells at lower energies, two degenerate (antibonding) π* MOs remain, *i.e.* four SOs, and two electrons. Figure 2.2 shows the six determinants of the electronic configuration $(\pi^*)^2$.

The combinatorial formula indeed gives: $\binom{4}{2} = \frac{4!}{2!\cdot 2!} = \frac{4\cdot 3\cdot 2!}{2!\cdot 2} = \frac{4\cdot 3}{2} = 6$. For further considerations on the oxygen molecule and the degeneracy removal in this electronic configuration, see Section 2.3.1.

2 CHARGE INTERACTIONS AMONG ELECTRONS

2.1 Outline

Now, we must understand how to insert, into the treatment followed so far, the additional energy term related to the *charge interactions between electrons*:

$$V^{ee} \overset{\text{def}}{=} \sum \frac{e^2}{r_{ij}} \tag{2.1}$$

where r_{ij} is the distance between electron i and electron j and the summation is over *all electron pairs*. In detail, it is written $V^{ee} \overset{\text{def}}{=} \sum_{i=1}^{n-1} \sum_{j=i+1}^{n} \frac{e^2}{r_{ij}}$ or else $\frac{1}{2} \sum_{i=1}^{n} \sum_{j=1,\, j\neq i}^{n} \frac{e^2}{r_{ij}}$ where the factor 1/2 is used to avoid counting the same pair of different indices twice. The interaction in Eq. (2.1) is the Coulomb law of electrostatics and is repulsive because the interacting charges have the same sign, so that it increases the energy with respect to the independent-particle Hamiltonian. As already mentioned, the *e-e* electrostatic interaction prevents the exact solution through separation of variables because it "mixes up" the spatial coordinates of the various electrons.

Let us try to get a very rough estimate of the effect of the electrostatic repulsion between electrons. We can get an idea, for example, by considering the $1s^2$ configuration of He and calculating the expectation values of the *e-e* repulsion for hydrogen-like functions, which are certainly very rough approximations to the correct functions. If for simplicity we use the a.u. (atomic unit) of energy, which is precisely the energy of a 1s orbital of the H atom, for the two-electronic system He we obtain: $\hat{H}^0 = -4$ u.a. because we use two 1s AOs, instead of just one, and each has double energy, since the atomic number Z is 2, and not 1, and thus the nuclear charge is doubled. The expectation value of the repulsion operator between two electrons, each in a 1s AO, is

$$\hat{V}^{ee} \approx +1.25 \text{ u.a.} \Rightarrow \hat{H} = \hat{H}^0 + \hat{V}^{ee} \approx -2.75 \text{ u.a.}$$

This is about 30%. Even if the calculation is rough, there is no doubt that it is an important contribution: the fact can also be verified experimentally, for example by spectroscopy. This means that it cannot be evaluated through successive corrections *i.e.* with perturbative methods but requires specialized methods. Generally speaking, the method that has prevailed is based on the division of the V^{ee} into two parts. A **first part**, the most quantitatively relevant, is treated in the same way as the Coulomb potential due to the nucleus. In essence, the potential acting on each single electron now includes an *additional* static potential due to the distributed electric charge of all the other electrons. Consequently,

- the main part of the total Coulomb energy, that is the independent particle Hamiltonian + this first part of the V^{ee} interaction, still follows a separable particle model,
- the motion of each electron can be solved independently of the others,
- the poly-electronic wavefunction can still be written as a product of mono-electronic functions which, obviously, are *not the hydrogen-like functions*, because the potential has a different dependence on the coordinates,
- the total energy is given by the sum of the energies obtainable by solving the Schrödinger equation for each separate part of the new Hamiltonian.

The **second part** of V^{ee} essentially quantifies the error made by considering only the first part: it is less relevant and is addressed in a following step.

2.1.1 The Screen Constant Approach

Historically, the first way used to address the most important part of the V^{ee} is based on the consideration that, for each electron, the electrostatic attraction due to the nucleus is decreased by the repulsive effect of the other electrons. This is modelled by assuming that each electron interacts with a nuclear charge reduced by a quantity called the **screen constant**. We then consider orbitals that are solutions of the problem of motion of a single electron in a modified hydrogen-like potential (that is, which includes only kinetic energy + Coulomb interaction with a nucleus of charge Ze), in the sense that for most orbitals the charge is $(Z-\sigma)\cdot e$ instead of Ze:

$$V_i = \frac{(Z-\sigma)\cdot e^2}{r_i}$$

The screen constant σ is different for different orbital shells because it describes the effect of the **inner occupied** orbitals.

This first naive treatment later developed into the Hartree method (better known as the Hartree–Fock method for its version that takes antisymmetry into account).

2.1.2 Hartree Approach

A poly-electronic problem is treated as a sum of mono-electronic problems. Only one electron is considered at a time. The complex of two-body Coulomb interactions (left diagram of Figure 2.3) is dealt with by considering, one by one, the various interactions of a given electron (star in the middle diagram) with the nucleus (large white circle) and with all the other electrons (small black circles). For each electron, a *one-electron Schrödinger problem* is then solved: the potential $f(\mathbf{r}_i)$ acting on the electron (i) is determined by the distance from the nucleus, by the nuclear charge Z and by the *charge densities of the other* $(N-1)$ *electrons* (right part of Figure 2.3). The charge densities are in turn obtained from the *squared moduli of the corresponding wavefunctions.*

2.1.2.1 The SCF Method

To evaluate the charge density due to a given electron (i), it is necessary to know its wave function, which is obtained by solving a problem requiring a knowledge of the charge densities of the other electrons. However, to evaluate the charge densities of each of these other electrons, it is necessary to know the wave function of the i-th electron as well. So, the Hartree approach seemingly has *no operational validity.* We then resort to an *iterative procedure* that step by step improves the mono-electronic wave functions and the related eigenvalues. *It is a fact* that the procedure *actually converges,* in the sense that each iteration modifies the result of the previous iteration by increasingly smaller quantities, getting closer and closer to an asymptotic value and the comparison with experiments shows that this asymptotic value is reliable. The method is called **self-consistent field** (SCF) and is outlined in the scheme of Figure 2.4.

A wise choice of the starting AOs obviously influences both the speed (*i.e.* the number of iterations) of the "convergence" and quality of the result obtained with a given number of iterations. As regards the operations to be performed at each cycle, Hartree's original scheme has been later modified to account for anti-symmetrization. Finally, the calculation cycle is stopped after as many iterations as necessary, checking the results at each iteration and comparing them with those of the previous iteration. For this purpose, appropriate numerical indicators are used, such as the values of the energies and the coefficients expressing the improved AOs as linear combinations of the basis functions.

2.1.2.2 Correlation Energy

It is important to point out a critical aspect of the method: the *e-e* interaction is treated only approximately by the SCF approach, which cannot evaluate part of the interaction energy. In fact, the scheme evaluates the interaction energy of an electron (star) with the others considered *fixed*, that is, it treats the *e-e* charge interaction in a "static" way (Figure 2.5, scheme on the left), while it is to be expected that the motion of the electron under consideration causes a shift in the distribution of the other electrons (Figure 2.5, scheme on the right).

The part of the interaction energy that the SCF treatment does not evaluate is called *correlation energy*. An important part of the research for the development of accurate quantum chemical methods has historically been devoted to the evaluation of correlation energy.

2.1.3 The HF-SCF-CF Approach

As with any poly-electronic function, a product of SCF AOs must also be made compatible with the antisymmetry principle. The original Hartree's approach has been therefore modified to take this aspect into account and the method that incorporates Hartree's SCF approach and anti-symmetrization is known as **Hartree-Fock** (HF) method. Briefly, the iterative SCF calculations are done on Slater determinants

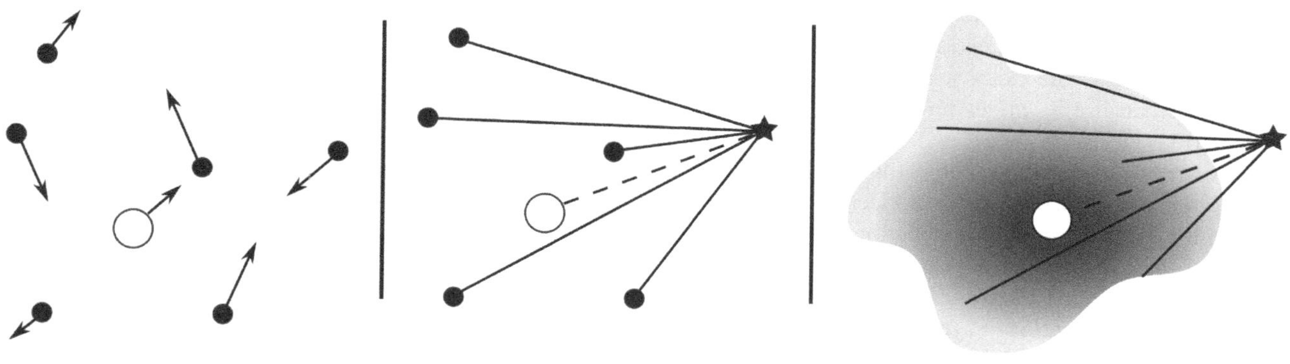

FIGURE 2.3 In a system (left part) with nucleus (white) and electrons (black) the charge interactions (middle part) are evaluated considering an electron at a time (star) and the charge density of the other electrons (right part).

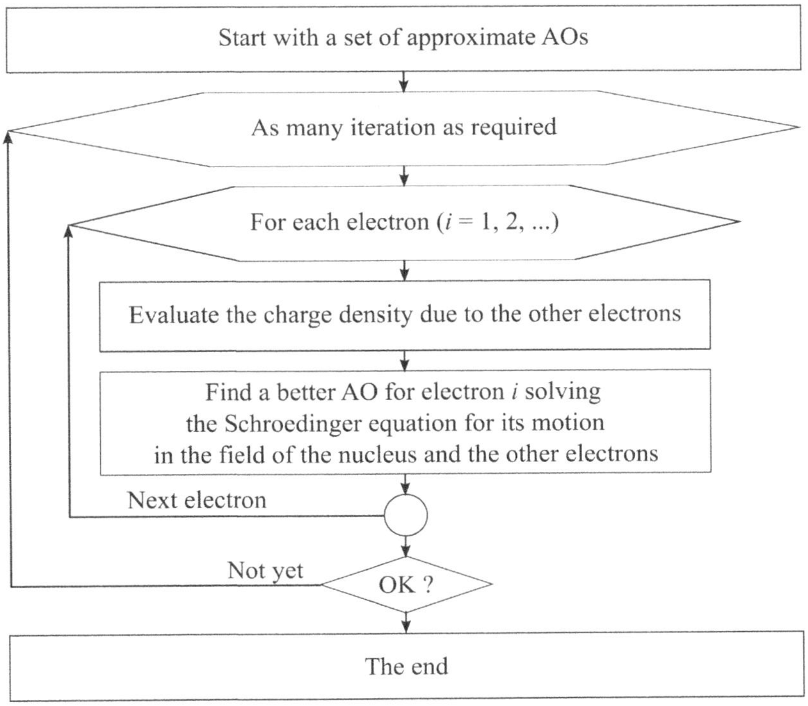

FIGURE 2.4 Flow chart of the self-consistent field method.

FIGURE 2.5 The Hartree method is a *static* approach to the *e-e* interaction.

instead of simple Hartree products. Furthermore, various versions of the HF-SCF method have been developed and widely used. The most accurate versions are not the most suitable to show in a simple way the most interesting results and the most important consequences for chemistry. One aspect to consider is how the SCF SOs are evaluated from the charge densities of the other electrons. When the cumulative charge density of the other electrons is centrosymmetric, also the potential felt by the i-th electron is centrosymmetric and the numerical integrations involve only the radial coordinate. For interpretative purposes, it is then convenient to use *a less accurate version of the SCF method*, called SCF-CF (self-consistent field – central field), in which the potential felt by the i-th electron is made centrosymmetric, even when it is not, *by averaging over the various directions* the cumulative charge density of the other electrons. With this assumption it is possible to write each AO as the product of a radial part (R) to be determined with 1D numerical calculations, multiplied by an angular part that is again a spherical harmonic (Y).

$$\Phi_{nlm}(\rho,\vartheta,\varphi) = R_{nl}(\rho)\cdot Y_{lm}(\vartheta,\varphi)$$

Figure 2.6 shows the trend of the energies.

Thanks to the CF approximation, many aspects of the electronic structure of atoms can also be discussed using hydrogen-like AOs as a comparison. The first important result is that

> The e-e charge interaction removes the degeneracy between AOs having the same n but different l.

The qualitative result is obvious: we must consider only the degeneracy that can be derived from the CF symmetry (3 for p-type AOs, 5 for d AOs, 7 for f AOs, …). The additional degeneracy of the hydrogen-like case is the effect of

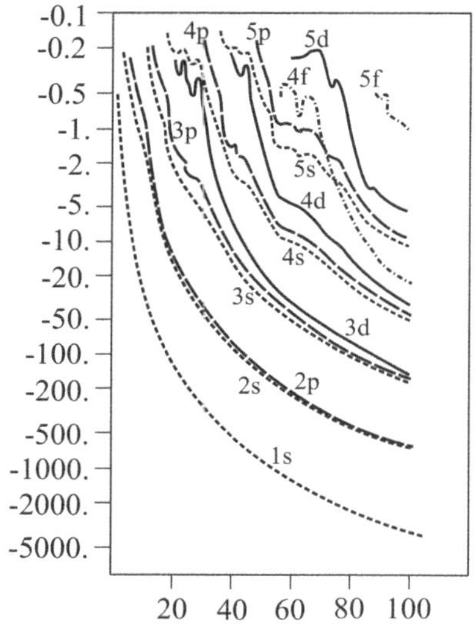

FIGURE 2.6 HF-SCF-CF energies for the various CF shells of the elements of the periodic table. The energy is in atomic units (a.u.), also called Hartree units, and in logarithmic scale. The abscissae are the atomic numbers.

an additional symmetry that depends on the fact that the potential acting on an electron depends on r as $-1/r$. When the potential, while remaining centrosymmetric, does not depend on r as $-1/r$, the additional degeneracy does not occur.

Compared to hydrogen-like AOs, in HF-SCF-CF SOs:

- the angular part is the same and there are still AOs s, p, d, f, …
- the *radial* part is different and is the one that must be evaluated numerically and
- the quantum number n *does not uniquely determine the energy* which now depends also on l. In particular, it is not true that the dependence of the energy on n is $\propto -\frac{1}{2 \cdot n^2}$.

The quantum number n loses the meaning it has in the hydrogen-like case, but we can still speak of AOs 1s, 2s, 2p, 3*s*, … in the more limited sense whereby 1s is the AO with the lowest energy among the AOs of s-type symmetry, the 2p AO is the AO with the lowest energy among the p-type AOs, etc.…

2.1.4 Summary of the Preliminary Results

In the independent particles model, the basic assumption is that the electrons do not interact with each other. This makes it possible to write the wave function of a system with m electrons as a Hartree product of m wave functions called spin-orbitals (SO), each dependent on the coordinates of a single electron.

Anti-symmetry essentially requires that any exchange between the coordinates of two different electrons produces the same wave function with changed sign. A simple Hartree product cannot show this behaviour but can be anti-symmetrized using its linear combination with *all* the other Hartree products containing the same set of SOs but with all permutations of the electronic coordinates. The linear combination is the *Slater determinant*. A Slater determinant one-to-one correspond to the well-known energy schemes where the horizontal lines specify the AOs and up or down arrows specify the spin state.

A fundamental consequence of anti-symmetry is that the electrons are *indistinguishable*. A simple Hartree product is not anti-symmetric because each electron (i) is placed in a particular SO (φ_a): $\varphi_a(i)$, while a Slater determinant is anti-symmetric because each electron - let us say - *inhabits* all SOs with equal weight. Anti-symmetry is also the basis of the *exclusion principle*, which does not allow to use more than once a given SO because the corresponding Slater determinant is zero.

In the independent particles model, the energy of an electronic state is the sum of the energies of the constituent AOs and depends only on the *electronic configuration*. The *aufbau principle* allows to distinguish the *fundamental* electronic *configuration* (at lowest energy) from the *excited* configurations (at higher energies); another important difference is between *closed shell* and *open shell* configurations.

The charge interactions between electrons can be approximately treated with the HF (Hartree-Fock) Self-Consistent Field (SCF) approach by including in the electric potential sensed by an electron the static effect of the charge distribution of the other electrons and following an iterative procedure until the desired accuracy is achieved. Within this approximation, *it is still possible to use SOs and Hartree products, and other results obtained with the independent particles model.*

This applies to all systems with more than one electron: atoms, molecules, and condensed matter. For *atomic* systems a further approximation is to resort to central-field (CF) orbitals, where spherical symmetry is assumed for the electric potential sensed by each electron. These HF-SCF-CF AOs can still be grouped into the well-known (1s, 2s, 2p, …) shells, because they differ from the hydrogenic AOs only in the dependence on the radial coordinate, while the dependence on the angular coordinates is the same as the hydrogenic AOs. Another difference is that the energy of these AOs also depends on the l quantum number, thus removing the degeneracy of the hydrogenic AOs with the same n but different l.

2.2 Degeneracy Removal in Open Shell Configurations

Going on in our approach, we find *another important effect* that is not obvious nor trivial but comes out – quite unexpectedly – from the detailed calculations, and only for *open shells*. The effect is a **removal** of **degeneracy** of an **electronic configuration**: it does *not* occur for *all* open shell configurations, but just in some cases and *it is never complete*. For example, in the fundamental configuration of the Li atom ($1s^2 2s^1$) there are two states because we can use the 1sα, 1sβ and 2sα SOs, or the 1sα, 1sβ and 2sβ SOs, and we can build two Slater determinants. These two states are expected to be degenerate, as the 2sα and 2sβ SOs belong to the same degenerate shell. In this case, the calculations we are going to describe confirm that these two states have the same energy. On the contrary, in the fundamental configuration of the C atom ($1s^2 2s^2 2p^2$: 15 states), the detailed calculations show that there are three energy levels, and the degeneracies are 9, 1, 5 in order of increasing energy. This is unexpected, as the difference between the 15 Slater determinants is only in the different occupations of the same six *degenerate* SOs. In addition, the 15 states *are not* the "original" 15 Slater determinants, but 15 linear combinations that are obtained from the solution of the matrix form of the Schrödinger equation in the basis of the 15 Slater determinants. For the sake of completeness, we anticipate here a point that will be discussed in the following: including another contribution to the total energy, *i.e.* the spin-orbit coupling, this ninefold degenerate level further splits into three levels with 1, 3, 5 degeneracies, giving a total of five energy levels with 1, 3, 5, 1, 5 degeneracies. There are particularly important effects of this effect in chemistry and spectroscopy. The proof, however, is not easy and it is convenient to start with an example.

2.2.1 The $1s^1 2s^1$ Example

Let us consider the $1s^1 2s^1$ atomic configuration of an atomic system with two electrons (*e.g.*: the He atom). This is the simplest case: we have the minimum number of electrons and an open shell, and we will se that the effect we are going to study does occur. This is an excited configuration, but it does not matter; also, this case is easier to discuss than an open shell *fundamental* configuration: the *fundamental* configuration with the smallest number of electrons showing the effect we are going to discuss is $1s^2 2s^2 2p^2$, a problem with six electrons.

There are four states for the $1s^1 2s^1$ configuration, as we have four Slater determinants (D_A, D_B, D_C and D_D) that are faithfully described by the schemes *A–D* of Figure 2.7.

The Slater determinants are:

$$D_A = \frac{1}{\sqrt{2}} \begin{vmatrix} \chi_{1s} \cdot \alpha(\tau_1) & \chi_{1s} \cdot \alpha(\tau_2) \\ \chi_{2s} \cdot \alpha(\tau_1) & \chi_{2s} \cdot \alpha(\tau_2) \end{vmatrix} = \frac{1}{\sqrt{2}} \left[|ac\rangle - |ca\rangle \right] \tag{2.2}$$

$$D_B = \frac{1}{\sqrt{2}} \begin{vmatrix} \chi_{1s} \cdot \alpha(\tau_1) & \chi_{1s} \cdot \alpha(\tau_2) \\ \chi_{2s} \cdot \beta(\tau_1) & \chi_{2s} \cdot \beta(\tau_2) \end{vmatrix} = \frac{1}{\sqrt{2}} \left[|ad\rangle - |da\rangle \right] \tag{2.3}$$

$$D_C = \frac{1}{\sqrt{2}} \begin{vmatrix} \chi_{1s} \cdot \beta(\tau_1) & \chi_{1s} \cdot \beta(\tau_2) \\ \chi_{2s} \cdot \alpha(\tau_1) & \chi_{2s} \cdot \alpha(\tau_2) \end{vmatrix} = \frac{1}{\sqrt{2}} \left[|bc\rangle - |cb\rangle \right] \tag{2.4}$$

$$D_D = \frac{1}{\sqrt{2}} \begin{vmatrix} \chi_{1s} \cdot \beta(\tau_1) & \chi_{1s} \cdot \beta(\tau_2) \\ \chi_{2s} \cdot \beta(\tau_1) & \chi_{2s} \cdot \beta(\tau_2) \end{vmatrix} = \frac{1}{\sqrt{2}} \left[|bd\rangle - |db\rangle \right] \tag{2.5}$$

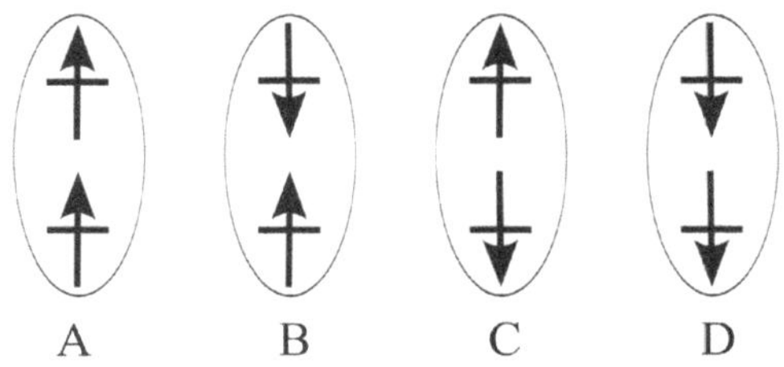

FIGURE 2.7 The four Slater determinants of the $1s^1 2s^1$ configuration.

In Eqs. (2.2)–(2.5) *a*, *b*, *c*, *d* respectively indicate the $\chi_{1s}\alpha, \chi_{1s}\beta, \chi_{2s}\alpha, \chi_{2s}\beta$ SOs. On the right, we use Dirac notation and agree that the order specifies the arguments: $\cdots |bc\rangle \stackrel{\text{def}}{=} \chi_{1s}\beta(\tau_1) \cdot \chi_{2s}\alpha(\tau_2)$, while $|cb\rangle \stackrel{\text{def}}{=} \chi_{1s}\beta(\tau_2) \cdot \chi_{2s}\alpha(\tau_1) = \chi_{2s}\alpha(\tau_1) \cdot \chi_{1s}\beta(\tau_2)$.

For the sake of simplicity, let us consider AOs deriving from a SCF-CF procedure, which approximately incorporates in the CF the charge interaction between electrons. We also assume that $\{a, b, c, d\}$ is an orthonormal set. Let us now verify if the "true" states of this configuration are represented by the four Slater determinants of Eqs. (2.2)–(2.5) and are degenerate. To this aim, we set up the matrix of the Hamiltonian in the basis of the four determinants, specifically considering the so-called non-relativistic Hamiltonian (Eq. 1.4), which includes: (i) the kinetic energy of each electron; (ii) the potential energy of the interaction between each electron with the nucleus; (iii) the charge interaction in each electron pair. We can disclose that:

- **If** we find that, in the basis of the four determinants, the Hamiltonian matrix is **diagonal**, then the four determinants D_A, …, D_D are the true states. If the four diagonal elements are also equal, these states are degenerate.
- **If**, on the contrary, we find that the matrix is **non-diagonal**, then:
  - the D_A, …, D_D determinants are not the true states of the system, but we need to find four linear combinations, corresponding to the eigenvectors of the matrix, and
  - the energies of these linear combinations are given by the eigenvalues and are not necessarily all equal.

It is worth noting that, also in this simple example, the calculations are quite intricate, as we need: ten matrix elements for the whole Hamiltonian (*i.e.*, the *independent* elements of a *Hermitian* 4 × 4 matrix), for each of these ten elements, three operators, two (one for each electron) for the kinetic energy + the energy of interaction with the nucleus: $\hat{H}_i^0 \stackrel{\text{def}}{=} \hat{K}_i + \hat{V}_i \left[i = 1,2 \right]$ and one for the electrostatic interaction between the two electrons: $V^{ee} \stackrel{\text{def}}{=} \sum e^2 / r_{12}$ and, at the end, for each of the 10 × 3 = 30 operators, four integrals, as both bra and ket are differences of **two** Hartree products.

Then, we need to calculate 120 integrals, combine them into the matrix elements, and finally diagonalize the resulting Hamiltonian matrix.

Actually, the calculations are simplified by the orthogonality of AOs or the spin functions or both. The detailed calculations are lengthy and tedious, but not particularly difficult: Section 4.1 reports some parts of the derivation, the rest is left as an exercise to the reader. Finally, we have the following Hamiltonian

$$\boldsymbol{H} = \begin{pmatrix} \alpha + J - K & 0 & 0 & 0 \\ 0 & \alpha + J & -K & 0 \\ 0 & -K & \alpha + J & 0 \\ 0 & 0 & 0 & \alpha + J - K \end{pmatrix} \tag{2.6}$$

where: $\alpha = \alpha_{1s} + \alpha_{2s}$, $\alpha_{1s} = 1s\left|H_\mu^0\right|1s = \int \chi_{1s}^* H_\mu^0 \chi_{1s} \mathrm{d}r_\mu$ ($\boldsymbol{\mu}$ = 1, 2 specifies the electron) and analogously for α_{2s}, while J and K are respectively given by:

$$J = \left\langle 1s2s \left| \frac{1}{r_{12}} \right| 1s2s \right\rangle = \int \chi_{1s}^*(\vec{r}_1)\chi_{2s}^*(\vec{r}_2)\frac{1}{r_{12}}\chi_{1s}(r_1)\chi_{2s}(\vec{r}_2)\mathrm{d}\vec{r}_1\mathrm{d}\vec{r}_2$$

$$K = \left\langle 1s2s \left| \frac{1}{r_{12}} \right| 2s1s \right\rangle = \int \chi_{1s}^*(\vec{r}_1)\chi_{2s}^*(\vec{r}_2)\frac{1}{r_{12}}\chi_{2s}(\vec{r}_1)\chi_{1s}(\vec{r}_2)\mathrm{d}\vec{r}_1\mathrm{d}\vec{r}_2$$

These integrals are *only over space coordinates* ($\vec{r}_1$ and $\vec{r}_2$); α_{1s} and α_{2s} account for the kinetic energy of one electron in addition to the *attractive* interaction between electron and nucleus. α_{1s} and α_{2s} are both < 0, which is obvious, as we are dealing with bound electrons. On the other hand, J and K are > 0, as they account for the *repulsive* interaction between two electrons.

Integral J describes the electrostatic interaction between the charge densities due to electron 1 (in the 1s orbital, regardless of spin up or down) and electron 2 (in the 2s orbital, also in this case regardless of spin). This is called **Coulomb integral**. Considering electron 1 in the 2s orbital and electron 2 in the 1s orbital would not make any difference (the functions and the operator $1/r$ are multiplied to each other and therefore the order is irrelevant, and the electrons are identical and indistinguishable). Therefore, we use just one J symbol for

$$\left\langle 1s2s \left| \frac{1}{r_{12}} \right| 1s2s \right\rangle = \left\langle \chi_{1s}(\vec{r}_1)\cdot\chi_{2s}(\vec{r}_2) \left| \frac{1}{r_{12}} \right| \chi_{1s}(\vec{r}_1)\cdot\chi_{2s}(\vec{r}_2) \right\rangle \quad \text{and}$$

$$\left\langle 2s1s \left| \frac{1}{r_{12}} \right| 2s1s \right\rangle = \left\langle \chi_{2s}(\vec{r}_1)\cdot\chi_{1s}(\vec{r}_2) \left| \frac{1}{r_{12}} \right| \chi_{2s}(\vec{r}_1)\cdot\chi_{1s}(\vec{r}_2) \right\rangle$$

This integral is basically what we would have written for evaluating the energy of repulsion between the two electrons, using the given orbitals. We say that ***J* has a classical analogue**.

The K integral looks somewhat surprising: seemingly, it is like J, but a closer inspection shows that it is different (its numerical value is different as well) as electrons are *exchanged* between the two orbitals in the bra and the ket. For this reason, K is called the **exchange integral**. Moreover, the turnover rule shows that there is only one exchange integral:

$$\left\langle \chi_{1s}(\vec{r}_1)\cdot\chi_{2s}(\vec{r}_2) \left| \frac{1}{r_{12}} \right| \chi_{1s}(\vec{r}_2)\cdot\chi_{2s}(\vec{r}_1) \right\rangle$$

$$= J = \left\langle \chi_{1s}(\vec{r}_2)\cdot\chi_{2s}(\vec{r}_1) \left| \frac{1}{r_{12}} \right| \chi_{1s}(\vec{r}_1)\cdot\chi_{2s}(\vec{r}_2) \right\rangle$$

The reasons that give rise to the exchange integral are not obvious. The real explanation relies on the detailed calculations (see Section 4.1) of the expressions with the Slater determinants and with the charge interaction between electrons. We say that there is **not** a **classical analogue** of the exchange integral. The introduction of the exchange integral is essentially due to anti-symmetry: this is made obvious by the fact that, with respect to the Coulomb integral with the same electrons and SO pairs, electrons are exchanged between bra and ket.

The diagonalization of $\boldsymbol{H}$ (see Eq. 2.6) requires only the diagonalization of the 2 × 2 block in the middle of the matrix and gives:

State 1: $E = \alpha + J - K$ eigenfunction D_A (2.7)

State 2: $E = \alpha + J - K$ eigenfunction $(D_B + D_C)/\sqrt{2}$ (2.8)

State 3: $E = \alpha + J + K$ eigenfunction $(D_B - D_C)/\sqrt{2}$ (2.9)

State 4: $E = \alpha + J - K$ eigenfunction D_D (2.10)

Then, we find that the "true" states only in part correspond to the Slater determinants: two Slater determinants (D_A and D_D) are still eigenstates, the other two (D_B and D_C) must be linearly combined as *sum and difference with equal weights*. In addition, we have a partial removal of degeneracy: instead of one level with four states, we have two levels ($E = \alpha + J + K$ and $E = \alpha + J - K$). There is one state (the third one, given by $[D_B - D_C]/\sqrt{2}$) at energy $E = \alpha + J + K$; it is called a **singlet**. At energy $E = \alpha + J - K$ we have three degenerate states, forming the so-called **triplet**. As K is positive, the triplet level is at lower energy.

This example is noteworthy also for what concerns the **properties** of the new eigenfunctions with respect to the original Slater determinants. Indeed, it can be shown that

- Concerning *spectroscopy*, the triplet states [A, (*B* + *C*) and *D*] have zero probability of transition to the fundamental state ($1s^2$ configuration): this is a “spin-forbidden transition”. On the contrary, the transition from the singlet state (*B*–*C*) to the fundamental state is “allowed” and its probability is “large”.
- Concerning *chemical properties*, the triplet states describe what is essentially a *diradical* (as it is well apparent by looking at D_A and D_D) and show a reactivity which is typical of these chemical species. Instead, the singlet has chemical properties like those of a *carbene*.

It is worth noting that the result of Figure 2.8 *can be obtained by considering the commutation* between the Hamilton operator and the spin operators without writing and diagonalizing the ***H*** matrix (Eqs. 2.6–2.10). The Hamiltonian operator here considered does not contain spin-dependent interactions and therefore commutes with spin-only operators (see point 2.3 of the Chapter 1). This allows us to classify its eigenfunctions using eigenfunctions and eigenvalues of the spin operators. As the electrons are indistinguishable, these spin operators cannot be single-electron operators ($\vec{s_1}$ and $\vec{s_2}$). They must concern the global spin operator $\vec{S} = \vec{s_1} + \vec{s_2}$, that is: (i) $\vec{S}^2$ with quantum number *S* and eigenvalue $S(S + 1)$; together with (ii) its component along a particular direction (let us say *z*), that is S_z, with quantum number M_s. Therefore, the eigenstates are classified as $|S, M_S\rangle$ states and are $|1,1\rangle$, $|1,0\rangle$, $|1,-1\rangle$ and $|0,0\rangle$. This approach can be extended to more than two electrons and essentially is how the vector model works (see Section 2.4).

Moreover, in the case of two electrons, the wave functions can be written as a product of a space-dependent part and a spin-dependent part. The spin dependent parts are

$$\alpha(1)\cdot\alpha(2),\quad \beta(1)\cdot\beta(2),\quad \text{and}\quad \alpha(1)\cdot\alpha(2)+\alpha(1)\cdot\alpha(2)$$

respectively for the $|1,1\rangle$, $|1,-1\rangle$, and $|1,0\rangle$, states and

$$\alpha(1)\cdot\alpha(2)-\alpha(1)\cdot\alpha(2)\quad \text{for the } |0,0\rangle \text{ state)}$$

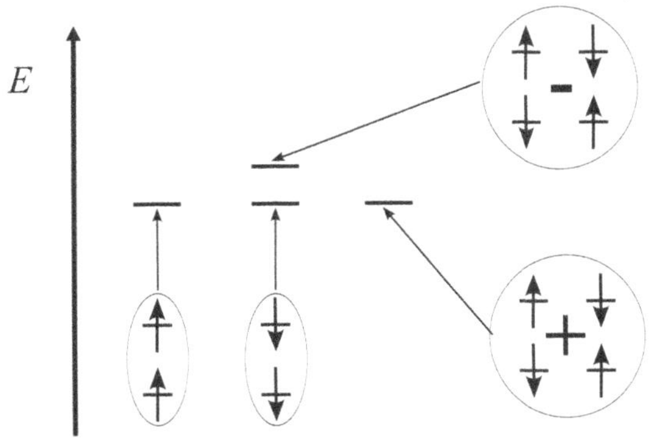

FIGURE 2.8 Energies of the states obtained by diagonalizing the Hamiltonian matrix in the basis of the determinants D_A, D_B, D_C, D_D (Eqs. 2.16–2.19).

As the spin functions $\alpha(1)\cdot\alpha(2), \beta(1)\cdot\beta(2)$, and $\alpha(1)\cdot\alpha(2)+\alpha(1)\cdot\alpha(2)$ are symmetric, the space dependent part of $|1,1\rangle$, $|1,-1\rangle$, and $|1,0\rangle$ must be anti-symmetric and therefore is $1s(t_1)2s(t_2) - 2s(t_1)1s(t_2)$. The spin dependent part of $|0,0\rangle$ is antisymmetric so that its space dependent part must be the symmetric combination $1s(t_1)2s(t_2) + 2s(t_1)1s(t_2)$. This derivation, however, cannot be generalized because the wave functions of a system with more than two electrons cannot be written as simple products of a space dependent part and a spin dependent part.

2.2.1.1 The Meaning

It is not easy to give an explanation in plain words of the result obtained for the $1s^1 2s^1$ example and it is even more difficult for the general case of an open shell configuration.

A possible *illustration* can be obtained by calculating numerically the *joint probability* density of finding an electron at a distance r_1 from the nucleus, and the other electron at a distance r_2. *For the triplet states* the *probability* of finding both electrons at the same distance from the nucleus, independently of the *angular* coordinates of the electrons, *is always zero*: this situation is strictly forbidden for the triplet states but is not forbidden for the singlet.

A similar result can be found analytically. Considering the Slater determinant D_A (see Eq. (2.21)) and carrying on the calculations, we can factor out the spin functions:

$$|ac\rangle - |ca\rangle = \left[\chi_{1s}(\vec{r_1})\cdot\chi_{2s}(\vec{r_2}) - \chi_{2s}(\vec{r_1})\cdot\chi_{1s}(\vec{r_2})\right] \cdot\left[\alpha(\sigma_1)\cdot\alpha(\sigma_2)\right] \tag{2.11}$$

It can be immediately seen that the first expression in square brackets vanishes when both electrons are in the same point ($\vec{r_1} = \vec{r_2}$). Then, the function and the probability density vanish as well. The essential point is that **in the triplet states**, contrary to the singlet state, *the probability density is zero* where $\vec{r_1} = \vec{r_2}$, that is *where the electron-electron repulsion diverges to infinity*.

Also, it is worth noting that eigenvalues and eigenvectors given by Eqs. (2.21)–(2.24) are obtained without changing the AO basis: a calculation done without this constraint would show also differences in the inter-electron repulsion and the attractive electron - nucleus interactions.

Finally, the degeneracy removal is sometimes reported by saying that anti-symmetrization causes an additional repulsion between electrons with the same spin, and that this repulsion is different from and stronger than the electrostatic repulsion. Instead, it must be stressed that *the lower energy of triplet states, compared to singlet states, is electrostatic in nature*. This is a result which is made evident by Eqs. (2.21)–(2.24) and should be explicitly pointed out, as it is possibly mistaken as a **magnetic** interaction (which occurs, but is not what is under discussion here). For a comparison between magnetic and exchange interactions, see Sections 2.3.1 and 2.3.2.

2.2.2 Generalizing the Example

Let us now analyse what we have obtained from the example. By using the Slater determinants as states of an electronic configuration, one might assume that all the states of an electronic configuration are degenerate, as they correspond to using the *same number* of SOs for each *shell,* and that all the SOs in a shell are degenerate. What we have seen above is a *counterexample* to this conclusion and – as such – is *valid in general.* This means that in open shell systems *degeneracy may be (at least partially) removed.* The actual outcome must be analysed case-by case (see Section 2.4). In detail

a. for some electronic configurations degeneracy removal does not occur (an example is the $1s^2 2s^1$ configuration) and
b. the *possible* degeneracy removal is due to the combined effect of electrostatic repulsion and anti-symmetry *is never complete.*

We can disclose here that full removal of degeneracy requires an external magnetic field. The following conclusions can be generalized.

1. In a system with more than one electron, the states of an open shell configuration do not necessarily correspond to Slater determinants. At least in some cases, we need to consider linear combinations of Slater determinants.
2. Analogously, an open shell electron configuration generally (but not necessarily) corresponds to more than one energy level, that is it shows degeneracy removal. In our example, states A, $(B + C)$ and D are degenerate, with energy $\alpha + J - K$. The $(B–C)$ state lies at higher energy $(\alpha + J + K)$.
3. Both results (1) and (2) are due to the presence of off-diagonal elements in the $\boldsymbol{H}$ matrix in the basis of the Slater determinants.
4. In turn, the $\boldsymbol{H}$ matrix is non-diagonal due to the exchange integral K (without the K integral, the matrix would be diagonal with all eigenvalues alike).
5. The exchange integral is a combined effect of charge repulsion between electrons and anti-symmetry with respect to exchange of electronic coordinates. Without charge repulsion, we would have neither the J nor the K integrals: the four Slater determinants of our example would have been the "true" states, with energies all equal to α, and, in general, each electronic configuration would have been made by all-degenerate states. Without anti-symmetry, the integral J would be present, but not K: the four Slater determinants of our example would have been the "true" states, all with energy equal to $\alpha + j$.
6. A classical analogue of the exchange integral does not exist. It is a direct consequence of anti-symmetrization of wave-functions written as Hartree products (the Coulomb J integral *does* have a classical analogue).

We finally point out that *the exchange integral does not* represent a *new interaction* (*i.e.,* an interaction *not* already included in the non-relativistic Hamiltonian, as is the case - for example – of the spin-orbit interaction that will be discussed in the following).

The $1s^1 2s^1$ example, on the contrary, cannot be directly generalized to predict which electronic configurations show degeneracy removal and in how many levels the energy is split. These points will be discussed in Section 2.4.

What the example of the $1s^1 2s^1$ configuration shows in a most general way is that:

Putting electrons into orbitals does not solve the problem of energy levels in many-electron systems.

The degeneracy removal between states of the same configuration cannot be neglected, nor can be neglected the requirement of considering states that are not simple Slater determinants, but are linear combinations and, therefore, have different properties. Indeed, degeneracy removal:

- can be experimentally detected and quantitatively measured by spectroscopy,
- identifies states that are different both for what concerns **spectroscopic** and **chemical properties**,
- gives a direct explanation of some **magnetic properties** of matter.

Considering the reasons behind the phenomenon, we can understand that, with obvious changes, the result is not limited to atoms and is valid for any poly-electronic system.

2.3 Exchange Interaction and Magnetism

Before going back to the electronic structure of atoms, it is interesting to digress on another important consequence of the same quantum phenomenon: magnetism.

2.3.1 Molecular Paramagnetism

An interesting case is that of the oxygen molecule, in which there is a shell of two MOs occupied by two electrons, a situation that shows *some* similarities with the $1s^1 2s^1$ atomic configuration (see Section 1.3.9). There are two π^* (antibonding) MOs as LCAO of the $2p_x$ and $2p_y$ of the two atoms, A and B (x and y are orthogonal to the molecular axis). From the two $2p_x$, we obtain two MOs that we can denote as π_x and $(\pi_x)^*$ but are preferably denoted with the symmetry symbols such as π_u^x (the bonding MO) and π_g^x (the antibonding MO):

$$\pi_x = \pi_u^x = \frac{1}{\sqrt{2}}(2p_{x,A} + 2p_{x,B}) \quad (\pi_x)^*$$

$$= \pi_g^x = \frac{1}{\sqrt{2}}(2p_{x,A} - 2p_{x,B})$$

The symbols u - ungerade, *i.e.* odd - and g - gerade, *i.e.* even - refer to the behaviour with respect to the inversion centre. Two MOs π_y similarly built on the $2p_y$ are denoted as π_u^y (the bonding MO) and π_g^y (the antibonding MO):

$$\pi_y = \pi_u^y = \frac{1}{\sqrt{2}}(2p_{y,A} + 2p_{y,B}) \quad (\pi_y)^*$$

$$= \pi_g^y = \frac{1}{\sqrt{2}}(2p_{y,A} - 2p_{y,B})$$

It is convenient to set up linear combinations that are eigenfunctions of the z-component of the orbital angular moments. The *real* AOs $2p_x$ and $2p_y$ can be combined into two *complex* AOs:

$$2p_{+1,A} = \frac{1}{\sqrt{2}}(2p_{x,A} + i \cdot 2p_{y,A}) \quad \text{and} \quad 2p_{-1,A}$$

$$= \frac{1}{\sqrt{2}}(2p_{x,A} - i \cdot 2p_{y,A})$$

$$\text{and} \quad 2p_{+1,B} = \frac{1}{\sqrt{2}}(2p_{x,B} + i \cdot 2p_{y,B}) \quad \text{and} \quad 2p_{-1,A}$$

$$= \frac{1}{\sqrt{2}}(2p_{x,B} - i \cdot 2p_{y,B})$$

that are eigenfunctions of $\widehat{l_{z,A}}$ (or $\widehat{l_{z,B}}$) with eigenvalues $m_l =$ +1 and $m_l = -1$:

$$\widehat{l_{z,A}} 2p_{+1,A} = \hbar \cdot 2p_{+1,A} \quad \text{and} \quad \widehat{l_{z,A}} 2p_{-1,A} = -\hbar \cdot 2p_{-1,A}$$

$$\text{and} \quad \widehat{l_{z,B}} 2p_{+1,B} = \hbar \cdot 2p_{+1,B} \quad \text{and} \quad \widehat{l_{z,B}} 2p_{-1,B} = -\hbar \cdot 2p_{-1,B}$$

Similarly, the two antibonding MO are combined into two *complex* MOs:
$\pi_g^+ = \frac{1}{\sqrt{2}}(\pi_g^x + i\pi_g^y)$ and $\pi_g^- = \frac{1}{\sqrt{2}}(\pi_g^x - i\pi_g^y)$ with analogous properties. Finally introducing the global angular momentum L as the sum of the two atomic l, we find that π_g^+ and π_g^- are eigenfunctions of its z-component $\widehat{L_z} = \widehat{l_{z,A}} + \widehat{l_{z,B}}$ with eigenvalues $m_l = 1$ and $m_l = -1$. Indeed:

$$\widehat{L_z}\pi_g^+ = \widehat{L_z}\left[\frac{1}{\sqrt{2}}(\pi_g^x + i\pi_g^y)\right]$$

$$= \frac{1}{2}\widehat{L_z}\left\{\left[(2p_{x,A} - 2p_{x,B}) + i(2p_{y,A} - 2p_{y,B})\right]\right\}$$

$$= \frac{1}{2}\widehat{L_z}\left\{(2p_{x,A} + i \cdot 2p_{y,A}) - (2p_{x,B} + i \cdot 2p_{y,B})\right\}$$

$$= \frac{1}{2}\left(\widehat{l_{z,A}} + \widehat{l_{z,B}}\right)\left\{2p_{+1,A} - 2p_{+1,B}\right\} = \frac{\hbar}{2}\left\{2p_{+1,A} - 2p_{+1,B}\right\}$$

$$= \frac{\hbar}{\sqrt{2}}\left\{\frac{1}{\sqrt{2}}(2p_{x,A} + i \cdot 2p_{y,A}) - \frac{1}{\sqrt{2}}(2p_{x,B} + i \cdot 2p_{y,B})\right\}$$

$$= \frac{\hbar}{\sqrt{2}}\left\{(2p_{x,A} - 2p_{x,B}) + i \cdot (2p_{y,A} - 2p_{y,B})\right\}$$

$$= \frac{\hbar}{\sqrt{2}}\left\{\pi_g^x + i \cdot \pi_g^y\right\} = \pi_g^+$$

This proves that π_g^+ is eigenfunction of $\widehat{L_z}$ with eigenvalue = +1. Analogously, π_g^- is an eigenvalue of the same operator, with eigenvalue = −1.

The two bonding MOs are fully occupied. The two antibonding MOs are occupied by only two electrons. Compared to the $1s^1 2s^1$ case, the main difference is that the two partially occupied antibonding orbitals (π_g^+ and π_g^-) now are at the same energy. Therefore, the degeneracy of the configuration is 6 instead of 4:

$$\binom{4}{2} = \frac{4!}{2! \cdot 4!} = \frac{4 \cdot 3 \cdot 2!}{2! \cdot 2!} = 6$$

There are *three triplet states* analogous to those in the $1s^1 2s^1$ case, namely:

$$\left|\pi_g^+\alpha(1)\pi_g^-\alpha(2)\right|, \quad \left|\pi_g^+\beta(1)\pi_g^-\beta(2)\right|$$

and the additive combination

$$\frac{1}{\sqrt{2}}\left[\left|\pi_g^+\alpha(1)\pi_g^-\beta(2)\right| + \left|\pi_g^-\beta(1)\pi_g^+\alpha(2)\right|\right]$$

where 1 and 2 indicate the electrons and the vertical bars indicate the Slater determinants. These states have $M_L = 0$, $S = 1$ and, respectively, $M_S = 1, -1, 0$.

The *singlet states are three* because the two cases of double occupancy of one MO

$$\left|\pi_g^+\alpha(1)\pi_g^+\beta(2)\right| \quad \text{and} \quad \left|\pi_g^-\alpha(1)\pi_g^-\beta(2)\right|$$

are now added to the subtractive combination:

$$\frac{1}{\sqrt{2}}\left[\pi_g^+\alpha(1)\pi_g^-\beta(2) - \pi_g^-\beta(1)\pi_g^+\alpha(2)\right]$$

as already seen for $1s^1 2s^1$

The first two states are degenerate and have different energy than the third one. The six states are shown in Figure 2.9.

The angular moment of a charged particle corresponds to a magnetic moment. Considering them as completely independent, that is ignoring their interactions, the magnetic moments of molecules in the gas phase can be aligned along a particular direction by an external magnetic field

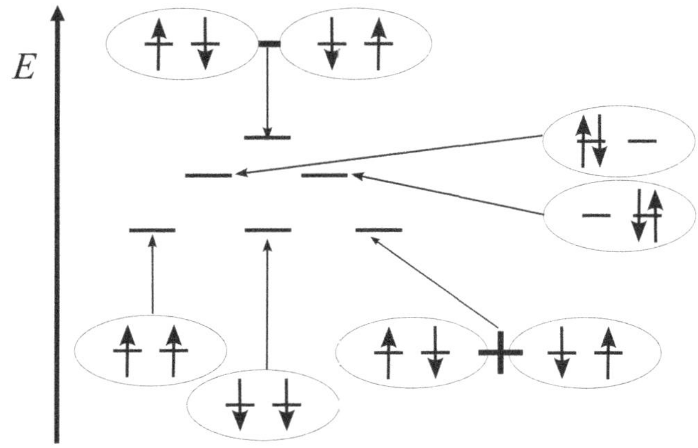

FIGURE 2.9 The states of the $(\pi^*)^2$ configuration of molecular oxygen.

and therefore show a global magnetic moment ≠ 0 and dependent on the intensity of the external magnetic field. This is the essence of the simplest magnetic phenomenon, **paramagnetism.** Triplet oxygen is paramagnetic because it has an angular moment that can assume the values −1, 0, +1. Singlet states are not paramagnetic.

Systems with an odd number of electrons obviously have non-zero angular and magnetic moment and are therefore paramagnetic. The theoretical interest in the paramagnetism of molecular oxygen is also in its *even* number of electrons.

The thermodynamic treatment of paramagnetism, reported at point 6.2.6 of Chapter 3, describes the dependence of the total magnetic moment on the applied field and *temperature*. Here it is sufficient to recall, for small fields, the linear relation between field and total moment at a given temperature.

2.3.2 Magnetostatic Interaction and Exchange Interaction

The example of molecular oxygen is also interesting because it gives experimental evidence that *the energy split corresponding to different spin alignments is **not** due to a magnetic interaction.*

For this purpose, let us consider two magnetic dipoles μ_1 and μ_2 placed at a distance r from each other. From magnetostatics we know that their interaction energy is (see Figure 2.10):

$$E = \frac{|\vec{\mu}_1| \cdot |\vec{\mu}_2|}{r^3}\left(\cos\alpha - 3\cdot\cos\beta\cdot\cos\gamma\right)$$
$$= A\left(\cos\alpha - 3\cdot\cos\beta\cdot\cos\gamma\right)$$

For the simplest case (two dipoles perpendicular to the connecting line: $\cos\beta = \cos\gamma = 0$) this relation predicts a *lower energy* for the *antiparallel* alignment ($\cos\alpha = -1$) with respect to the parallel alignment ($\cos\alpha = 1$), contrary to what shown in Figures 2.8 and 2.9.

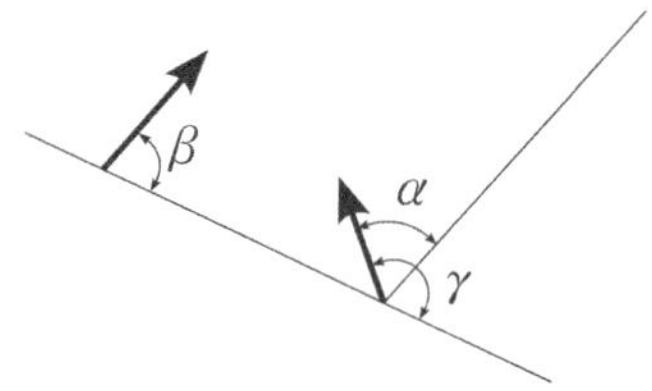

FIGURE 2.10 Magnetostatic interaction between two magnetic moments $\vec{\mu}_1$ (left) and $\vec{\mu}_2$ (right). α is the angle between the two moments, β is the angle between $\vec{\mu}_1$ and the vector joining the dipoles, and γ is the corresponding angle for $\vec{\mu}_2$.

Even more important is the comparison of the *values* of energy splits. Indeed, by inserting a reasonable value for a magnetic moment on the atomic scale, such as the quantity known as Bohr magneton, and a distance of the order of 0.1 nm, we obtain that the magnetic interaction energy is of the order of **a few J/mol**. On the contrary, the spectroscopic determination of the energy difference between the triplet and the three singlet states of molecular oxygen gives values of the order of **several tens of kJ/mol**.

This is decisive experimental evidence *against* the hypothesis that this removal of degeneracy is due to magnetic interactions. Despite a possible misunderstanding, the effect of spin alignments on the charge interaction between electrons is typically approached by introducing an **exchange energy**, a name that well reminds both the role of the antisymmetry and the *exchange* integral.

2.3.3 Ferromagnetism

In condensed matter, the exchange energy "acts" between AOs of atoms located at distances of the order of a fraction of a nanometre. This *generally* produces a *parallel* alignment of the magnetic moments: this is **ferromagnetism**. However, when we look more closely at the various contributions to be considered for the case of **two adjacent** identical **atoms**, the energetic positions of the triplet and singlet can change significantly compared to the case of molecular oxygen and the change is reflected in a change in the value of the energy separation $E(S = 1) - E(S = 0)$, which ends up depending on various factors and *can change sign* thus favouring *antiparallel* alignment.

For explaining, let us consider the overlap between adjacent AOs (let's call them p and q), then the singlet and triplet energies become respectively:

$$E(S=0) = \frac{\alpha + J + K + S_0 V}{1 + S_0} \quad \text{and}$$

$$E(S=1) = \frac{\alpha + J - K - S_0 V}{1 - S_0}, \quad \text{where:}$$

$$S_0 = \langle p|q\rangle \quad \text{and} \quad V = \langle p|\widehat{H_\mu}|q\rangle$$

In these expressions, $\widehat{H_\mu}$ is the one-electron Hamiltonian and V is the resonance integral between the two AOs: it is an attractive term, and its value can reverse the relative position of the two levels. Other considerations work in the same direction, in particular the inclusion of the two configurations now called ionic and corresponding to two electrons in one AO, *p* or *q*, and none in the other. Their combination with the singlet modifies the energy.

Last but not least, the exchange interaction is strongly dependent on the relative orientation of the *p* and *q* AOs and on their nature, in particular when the atoms are different.

The exchange interaction, although strongly prevalent at short distance over the dipolar interaction, as just seen, decays rapidly (exponentially) with distance. Then, the magnetic interaction, which depends on r^{-3} (see Section 2.3.2), ends up prevailing at large distances over the parallel exchange interaction. The conflict between the two trends explains the formation of **magnetic domains**. Within a domain the dipoles are oriented according to the exchange interaction (except for thin boundary regions), but different domains are oriented anti-parallel to each other to minimize the overall magnetostatic energy.

2.4 The Vector Model

The general problem of degeneracy of electronic configurations in atomic systems (see Section 2.2.2) is much more intricate and difficult than the $1s^12s^1$ example may suggest. To give an example, the fundamental configuration of C ($1s^22s^22p^2$) requires the calculation of a 15 × 15 Hermitian matrix, and therefore 15 × 16/2 = 120 distinct matrix elements, each element of $\boldsymbol{H}$ in turn involves calculations concerning six mono-electronic operators and 15 operators for the charge interactions of 15 electron pairs. Each integral between Slater determinants corresponds to 6! × 6! integrals between the Hartree products of a single determinant, …. And this is, as already said, the *simplest* case of a *fundamental* electronic configuration that exhibits the effect we are discussing. Of course, there are numerical or symbolic procedures to deal with specific cases in detail. However, if one wants to discuss all electronic configurations in a general way, it becomes necessary to look for a more direct and simpler procedure that allows at least to grasp the most significant results. This procedure exists and essentially involves the use of **symmetry**. In the case of atomic systems and assuming the use of AOs with *central field symmetry* (CF), the procedure for classifying the sublevels of an electronic configuration is particularly simple and represents what is known as the **vector model of the atom**.

The approach is based on the very general considerations that, if the Hamiltonian operator commutes with another operator, then we can use common eigenfunctions of *both operators*. If the other operator is a symmetry operator, then what we do is to classify the eigenfunctions of the Hamiltonian on the basis of their symmetry or, more precisely, find the *linear combinations adapted to symmetry*. This is similar to what has already been done to make Hartree products compatible with antisymmetry. Linear combinations that are differently adapted to symmetry (or belong to different symmetry species, as it is said technically) do *not necessarily have the same energy*: there is *no* reason why they should have the same energy. Also, in the general case we are not dealing with a single symmetry operation, but with an entire symmetry group and this makes the approach more complex.

However, the vector model of the atom is based *only on the quantum numbers of the angular moments*. As a matter of fact, these quantum numbers *do express* symmetry properties, but the approach of the vector model *does not use* these properties *directly*. We could say that the symmetry is present in the vector model, but it does not *appear explicitly*. We refer to the chapter dealing with symmetry for a more in-depth discussion of *central field symmetry*.

Operationally, *the vector model says*

- *how many different energy levels* there are in an electronic configuration,
- what is the *degeneracy* of each of these levels,
- what properties distinguish all the states of a level from those of another level.

On the contrary, it *does not specify the energies* of these levels. It does not specify their order nor the extent of their separation in energy. In this regard, there are empirical rules (known as **Hund's rules**) that essentially predict the order in energy but still do not predict the extent of the energy differences. Quantitative information can be obtained with perturbative calculations.

The vector model of the atom is *very simple* because it is based only on the *processing of quantum numbers* relating to angular moments and on the addition of angular moments. Due to the necessary assumption of central field symmetry, it must essentially be considered as *an empirical model*: it originates from long experimental research in atomic spectroscopy and essentially is a powerful *interpretative scheme*, proposed well before the development of QM. While various aspects of the model can be derived using appropriate approximations, here we essentially show how the model works. Without searching for completeness, we justify some aspects of the model, when it is useful and possible to do so in a simple way.

Of course, modern computational chemistry does not require the vector model and in particular the central field hypothesis. As already said - the numerical procedures address the complex interactions of a poly-electronic atom providing results that, depending on the case, are more or less in agreement with what the vector model predicts with much lower accuracy.

2.4.1 Using the Angular Moments

As said, the vector model of the atom elaborates CF-type AOs and its essential ingredient is the calculation of *total angular moments* relative to both all the electrons and the sum of orbital and spin angular moments.

- A CF SO is characterized (in addition to n) by a total of *three* quantum numbers: two for orbital (l, m_l) and one for spin (m_s) angular moment, while the quantum number s is always = 1/2.
- In a Hartree product, a specific electron is assigned to a specific SO and therefore each electron has specific values of (n, l, m_l and m_s).
- In a Slater determinant different quantum numbers (n, l, m_l and m_s) are assigned to the same electron but the *sum* over all the electrons *does not change* because they have always the same permuted values.

In other words, an anti-symmetrized poly-electronic state **cannot** be characterized by precise quantum numbers (n, l, m_l and m_s) for each electron because then it would not agree with the Pauli principle. However, it is characterized by global quantum numbers related to the angular moments of all the electrons and, more precisely, the quantum numbers related to the **sum** of the corresponding **angular moments**.

The possibility of classifying the states of poly-electronic atoms by the sum of angular moments is based on the commutation rules between the Hamiltonian and these moments. In this regard, any *spin moment* (whether of a single electron or global) commutes with the non-relativistic Hamiltonian, because the latter does not contain interactions that depend on the spin. Concerning the *angular moments related to the orbital motions*, the important point is that the operators related to the *total* angular moment commute with the electron-electron electrostatic interaction operator. The proof of this important result is left to more specialized texts.

As already seen in QM, an angular moment is a vector quantity. It is not possible to know all the three components of such vectors, but only (i) the square modulus and (ii) its component in a predetermined direction, usually z. An angular moment (let us write $\vec{j}$) is characterized by two quantum numbers (j and m_j) and by two eigenvalues of these observables: $\left|\vec{j}\right|^2 \Rightarrow j\cdot(j+1)\cdot\hbar^2$ and $\hat{j}_z \Rightarrow m_j\cdot\hbar$. The quantum number relating to a modulus can only be j = 0, 1/2, 1, 3/2, 2, … that is, it can only be integer or semi-odd and in any case non-negative For angular moments that have a classical analogue, only the integer values of j are valid, the semi-odd values are found only for angular moments that do *not* have a classical analogue. Angular moments that do *not* have a classical analogue, however, can have integer values. In both cases, the quantum number relating to the z component takes all the values between j and $-j$ in steps of 1 (and therefore $2j + 1$ possible values). When dealing with a poly-electronic system, it is also necessary to distinguish between the angular moments (orbital and spin) of *a single electron*, conventionally indicated with *lowercase letters*, and those related to all the electrons in the system, *i.e.* the *sum of the moments of all the electrons*. These moments are conventionally indicated – for atomic systems – with the corresponding *capital letters*.

In summary, for the *square modulus* of all these angular moments the following applies, remembering once again that it is always s = ½ because this is a fundamental property of the electron:

$$\left|\vec{l}\right|^2 = l\cdot(l+1)\cdot\hbar^2; \qquad \left|\vec{s}\right|^2 = s\cdot(s+1)\cdot\hbar^2 = ¾\cdot\hbar^2;$$

$$\left|\vec{L}\right|^2 = L\cdot(L+1)\cdot\hbar^2; \qquad \left|\vec{S}\right|^2 = S\cdot(S+1)\cdot\hbar^2$$

For the z *components* we have instead:

$$\hat{l}_z = m_l\cdot\hbar; \qquad s_z = m_s\cdot\hbar = \pm ¾\cdot\hbar;$$

$$\hat{L}_z = M_L\cdot\hbar;\ \ S_z = M_S\cdot\hbar.\ (m \text{ and } M \text{ are also used instead of } m_l \text{ and } M_L)$$

Note: in the vector model scheme, atomic systems are treated as centrosymmetric, an assumption that holds only approximately. Another important point is that a closed shell has a centrosymmetric charge distribution and therefore it is only the external shells that cause a deviation from the central field approximation.

The *use of spin* angular moments *holds even in molecular systems*.

2.4.2 Sum of Angular Moments

The *addition* rules for the angular moments of QM are not the same as those used in CM to add two or more vector quantities. Obviously, even for the sum of angular moments it is possible to know only one component and the modulus, so the problem is to determine what the possible values for the quantum numbers for the **sum** of two angular moments are, knowing the quantum numbers relating to the moduli of the angular moments that are added.

For example, let us consider the addition of an angular moment with l_1 = 1 and another with l_2 = 2. We then have for the first moment three (l_1, m_1) pairs: [(1, 1), (1, 0), (1, –1)] and for the second one 5 (l_2, m_2) pairs: (2, 2), (2, 1), (2, 0), (2, –1), (2, –2) and therefore in total 3 × 5 = 15 (l_1, m_1, l_2, m_2) states. Let us consider the possible values for M, which are easily obtained by adding the components (respectively 1, 0, –1 and 2, 1, 0, –1, –2) of the addend moments:

$$3 = 1+2 \qquad -3 = -1-2$$

$$2 = \begin{cases} 2+0 \\ 1+1 \end{cases} \qquad -2 = \begin{cases} -2+0 \\ -1-1 \end{cases}$$

$$1 = \begin{cases} 1+0 \\ 2-1 \\ 0+1 \end{cases} \qquad -1 = \begin{cases} -1+0 \\ -2+1 \\ -1+0 \end{cases}$$

$$0 = \begin{cases} 1-1 \\ -1+1 \\ 0+0 \end{cases} \quad \text{For a total of 15 possibilities, as it should be}$$

Since the sum is also an angular moment, the fact that the extreme values are +3 and –3 immediately tells us that L *can* be 3 and that this is its maximum value. Then, by removing the components (3, 2, 1, 0, −1, −2, −3) of the case $L = 3$, we are left with the values 2 and −2 (once), 1, −1 and 0 (all twice). Reasoning similarly, we see that $L = 2$ is also possible (with components 2, 1, 0, −1, −2) and that the three remaining M (1, 0, −1) are the components of an angular moment with $L = 1$.

The procedure of the example can be converted into a constructive proof. It is not necessary to present it here and it can be found in specialized textbooks, but we easily understand that L is at most $(l_1 + l_2)$ and at least $|l_1 - l_2|$. The rule of addition of angular moments is therefore as follows. Let $\vec{l_1}$ and $\vec{l_2}$ be the two angular moments to be added and $\vec{L}$ be the *sum vector* ($\vec{L} = \vec{l_1} + \vec{l_2}$). The result of interest is:

Rule of addition of angular moments L (quantum number of the square module of the sum) takes all values between $l_1 + l_2$ and $|l_1 - l_2|$ at intervals of 1. The same is true for spin moments.

(**Note**: the quantity $|l_1 - l_2|$ is sometimes called *positive difference*).

Of course, for each L, M then takes all values (spaced by 1) between L and $-L$ as it must be for an angular moment. Applying the rule, for the case $l_1 = 1$ and $l_2 = 2$, we have: $L = 3$, or 2, or 1. For $L = 3$, we have seven M values: $M = 3$, 2, 1, 0, −1, −2, −3. For $L = 2$, five M values: $M = 2$, 1, 0, −1, −2. For $L = 1$, three M values: $M = 1$, 0, −1. There are 15 pairs of L, M values.

The total does not change, as it should be because we are considering linear combinations of the 15 original states and the independent linear combinations are still 15. This is a general result that is not difficult to prove (4.2). Finally:

When more than two angular momenta must be added, we proceed in sequence: we add the first two, then we add the third to each result of the sum, etc….

Example: Let us consider $l_1 = 1$, $l_2 = 2$ and $l_3 = 2$. The sum of the first two gives as a result $L_{12} = 3, 2, 1$. For the total quantum number L we obtain: $3 + l_3$, giving 5, 4, 3, 2, 1; $2 + l_3$, giving 4, 3, 2, 1, 0; and finally $1 + l_3$, giving 3, 2, 1.

2.4.3 Spectroscopic Terms

An electronic configuration can correspond to several different values of the quantum numbers L, M_L, S and M_S. It is essential to group the states according to the possible pairs of values of L and S (each of which corresponds to one or more values of M_L and M_S) because **the energy depends on L and S but does not depend on M_L and M_S**:

- different values of S correspond to states with different energy values, independently of the values of M_S,
- and, similarly, different values of L correspond to states with different energy values, independently of the values of M_L.

The fact that the energy does not depend on M_L and M_S is an obvious consequence of the fact that different orientations cannot be associated with different energies, in the absence of external fields favouring a specific direction. The fact that the energy depends on both L and S is the subject of Hund's first two (empirical) rules.

Each pair (L, S) of a configuration is called a **spectroscopic term**.

The states of a term have the same energy, which is not necessarily the same for the states of another term. It is then convenient to find a symbol that univocally denotes a term, the pair (L, S) and the corresponding energy level. The symbology used is a bit intricate, because it is based on symbols introduced before the "discovery" of QM, that is, before the tools required to understand the phenomenology. In a term symbol:

- the quantum number L is specified by an upright **capital letter** following the same notation used for hydrogen-like AOs: $L = 0 \Rightarrow$ S, $L = 1 \Rightarrow$ P, $L = 2 \Rightarrow$ D, $L = 3 \Rightarrow$ F, $L = 4 \Rightarrow$ G and then in alphabetical order (H, I, …).
- The quantum number S is specified by writing the **spin multiplicity**. This is **$2S + 1$** and is written to the top left of the letter specifying L.

For example, the pair $L = 0$, $S = 0$ is written ^{1}S and is read S singlet. The pair $L = 1$, $S = ½$ is written ^{2}P and is read P doublet. The pair $L = 4$, $S = 3/2$ is written ^{4}G and is read G quadruplet. Note: be careful not to confuse **spin multiplicity** with **M_S**. Spin multiplicity is the *number of M_S values* corresponding to the given value of S. For example, a term with $S = 3/2$ has spin multiplicity $= 2S + 1 = 4$, meaning that there are four states, with $M_S = 3/2, 1/2, -1/2, -3/2$.

2.4.4 Counting the States

The number of states of a configuration is given by *the number of Slater determinants* compatible with the configuration. Even if the determinants must be combined linearly, the count does not change, because the number of independent linear combinations is equal to the number of determinants to be combined. We find that:

The number of states of a configuration is NOT necessarily equal to the number of different results of angular moment addition.

It is sufficient to give a counterexample of the opposite statement. Let us examine the $2p^2$ electronic configuration of C which, as we have already seen, has 15 states. L is 2, 1, 0, S is 1, 0: this corresponds to the terms ^{3}D, ^{1}D, ^{3}P, ^{1}P, ^{3}S, ^{1}S. Using the data of the previous example, the number of states is $15 + 5 + 9 + 3 + 3 + 1 = 36$. *Why this difference?*

When we list the terms and then count the states of each term, we are simply counting in a different order the sums of the angular quantum numbers of the spin orbitals of the shell. *This procedure can lead to a result incompatible with antisymmetry, i.e.* it possibly counts combinations of quantum numbers that are forbidden because they correspond to *null Slater determinants.* Considering again the $2p^2$ configuration, we see that the value 36 obtained before is 6^2. Six is the number of SOs and 36 is exactly the number of states that we should assign to the configuration if we did not consider the anti-symmetrization. In fact, it means six possible SOs for "the first" electron, *i.e.* the second to last column of the Slater determinant and another six SOs for the "other electron", *i.e.* the last column of the Slater determinant. As already seen, the result is wrong because: (i) it also counts the six Slater determinants in which the same SO is used twice and (ii) the remaining 36 – 6 = 30 cases must be grouped in pairs: in each pair the same SOs are used and therefore each pair counts as a single poly-electronic state. As a check: (36 – 6)/2 = 15.

It is easily seen that some states must necessarily be excluded. For example, the ^3D term of the $2p^2$ configuration would correspond to $L = 2$ (so $M_L = 2, 1, 0, -1, -2$) and $S = 1$ (so $M_S = 1, 0, -1$). Of these, $M_L = 2$ can only come from $m_{l1} + m_{l2} = 1 + 1$, *i.e.* using the same orbital twice; $M_S = 1$ can only come from $m_{s1} + m_{s2} = ½ + ½$, *i.e.* "two spins up": so, the hypothetical state with $M_L = 2$ and $M_S = 1$ is certainly not allowed because it would require using twice the same AO with the same spin.

A particularly clear example is that of **closed shells**, where there is only one state, but vector additions give numbers > 1. For example, the vector sum over an ns^2 configuration would give $L = 0$ and $S = 1, 0$ (*i.e.* ^3S and ^1S), an np^6 would give $L = 6, 5, 4, 3, 2, 1, 0$ and $S = 3, 2, 1, 0$. For a closed shell we must necessarily use both the SO with an m_l and the one with $-m_l$, so the sum over the m_l is necessarily zero. Similarly for spin. Therefore, the only term compatible with the Pauli principle has $L = 0$, $S = 0$: ^1S.

The last consideration is also important in general for all configurations with many occupied shells (even if only partially). *Closed shells* (as well as the completely empty shells) *must not be considered.* The fact that a closed-shell electronic configuration corresponds to only one state holds well outside the vector model, for example for molecular systems.

2.4.4.1 The Hole Formalism

When a p, d, f, … shell is occupied by many electrons, the rule for the addition of the angular moments says that the first two are added, then the third is added to the various results of the first sum, and so on. This is a long and intricate procedure because it requires a large number of vector additions. It is easy to see that, in reality, these calculations are largely useless because we can already exclude many terms from the start and therefore simplify the calculation. Let us consider the example of the $2p^5$ configuration. We would add the first two giving $L = 2, 1, 0$, then the third, giving 3, 2, 1, 0, then the fourth, giving 4, 3, 2, 1, 0, and finally the fifth, giving 5, 4, 3, 2, 1, 0. Similarly, for S we would obtain the values 5/2, 3/2, and 1/2. It is immediately clear that many values of L and S can easily be excluded. Two orbitals must be doubly occupied and only one can have occupancy = 1. Therefore, S cannot be anything other than 1/2, while L can only be 1: ^2P.

For a configuration with a **shell more than half filled**, the hole formalism reduces the calculation of the terms to the calculation for the corresponding *less than half filled configuration.* In detail, if the shell can contain p electrons (p even), the configurations with m electrons ($m < p$) and with $p - m$ electrons, *i.e.* with m **holes**, have *the same terms.* The result derives directly from the formula of the binomial coefficients: $\binom{p}{m} = \frac{p!}{m!(p-m)!} = \binom{p}{p-m}$.

For example, the configurations $2p^1$ and $2p^5$ have the same terms, as do $2p^2$ and $2p^4$, or $3d^2$ and $3d^8$. A further consequence of the hole formalism is Hund's third rule (see Section 3.2.2).

Therefore, the difficulty of counting the terms and states of a configuration *does not arise*:

- for *closed shells* since these have only one state with term symbol ^1S,
- for *open shells filled with only one electron*, since there is no possibility of using the same SO more than once,
- for "almost filled" *shells, i.e. with only one empty SO* since these correspond to the previous case using the hole formalism.

In conclusion: this difficulty never occurs for s shells, occurs for 2, 3, 4 electrons in a p shell, for 2, 3, 4, 5, 6, 7, 8 electrons in a d-shell, for 2, …, 12 electrons for an f-shell, etc.…

2.4.4.2 What Are the Terms to Be Excluded

There is an analytical procedure to find, configuration by configuration, which states to exclude, but it is particularly intricate and of a purely technical nature. We can safely ignore it because knowing it does not improve our understanding of the general aspects of the problem. There is, however, a result of the analytic procedure that is particularly useful: the procedure *excludes* all the states of a term.

For open shells with more than one electron (or shells missing more than one electron) the Pauli principle requires eliminating some **terms**.

It is therefore convenient to refer directly to tables already available. Here is a short list of the allowed terms for some electronic configurations, referring to specialized texts for other cases.

np^2, np^4:	^1S, ^1D, ^3P	⇒ 15 states
np^3:	^2P, ^2D, ^4S	⇒ 20 states
nd^2, nd^8:	^1S, ^1D, ^1G, ^3P, ^3F	⇒ 45 states
nd^3, nd^7:	^2D(2), ^2P, ^2F, ^2G, ^2H, ^4P, ^4F	⇒ 120 states
nd^4, nd^6:	^1S(twice), ^1D(twice), ^1F, ^1G(2), ^1I, ^3P(twice), ^3D, ^3F(twice), ^3G, ^3H, ^5D	⇒ 210 states
nd^5:	^2S, ^2P, ^2D(three times), ^2F(twice), ^2G(twice), ^2H, ^2I, ^4P, ^4D, ^4F, ^4G, ^6S	⇒252 states

2.4.5 The Procedure

In conclusion, to find the terms and classify the states of a configuration, we proceed as follows.

A. We find all the possible values for S.
B. We find all the possible values for L.
C. We consider all their combinations. *i.e.* the possible terms.
D. (If necessary) we remove the terms forbidden by the Pauli principle, using the tables.
E. We check that the total number of states is equal to what is obtained from the combinatorial calculation for the number of Slater determinants of the configuration.

Example 1. We already know that the $2p^2$ configuration has $\binom{6}{2} = 15$ states. The list above reports three terms: ^1S, ^1D, ^3P. Now: ^1S ($S = 0$, $L = 0$) has only one state, ^1D ($S = 0$, $L = 2$) 5 states, ^3P ($S = 1$, $L = 1$) 9 states. So, $1 + 5 + 9 = 15$.

Example 2. The $3d^2$ configuration has $\binom{10}{2} = 45$ states. Then, ^1S has only one state, ^1D 5 states, ^1G 9 states, ^3P 9 states, ^3F 21 states. So, $1 + 5 + 9 + 9 + 21 = 45$.

2.5 Summary

The whole discussion of the charge interaction between electrons can be summarized by highlighting three important consequences.

- *There is a significant additional positive contribution to the energy* of each electronic configuration, and the energy *increases* with respect to the independent particle Hamiltonian. This is obvious since the interaction between charges of the same sign is positive.
- The *degeneracy between AOs with the same n but different l is removed.*
- The *degeneracy between states of the same electronic configuration can be partially removed.*

Concerning the last point, the electron-electron interaction mixes the various Slater determinants of an electronic configuration into linear combinations that differ by a greater or lesser effect of charge repulsion. The energy levels into which an electronic configuration is split are called *terms* and can be labelled with the quantum numbers L and S of the *vector model.*

The procedure for identifying the terms of a configuration is made complex by *the need to exclude some terms to comply with the Pauli principle.* The problem arises only for configurations that are open shells and have two or more electrons in an open shell, except the shells in which only one electron is missing. The allowed terms are reported in appropriate tables.

The vector model does not specify the energy order of the terms, nor does it quantify energy differences. This is done either by direct calculations of the perturbation on each term or by Hund empirical rules that will be presented later.

However, before going into these details it is worth considering another interaction. In fact, *the difference in energy between terms of the same electronic configuration is smaller:*

- both of the full contribution of the *e-e* interaction,
- and (generally) of the difference in energy between electronic configurations.

and can become comparable with other interactions that we have ignored so far, having considered only the so called"non-relativistic" Hamiltonian. As a matter of fact, the additional interaction that needs to be considered is the **spin-orbit interaction**.

3 SPIN-ORBIT INTERACTION AND LS (RUSSEL SAUNDERS) COUPLING

3.1 Spin-Orbit Interaction

The spin-orbit interaction is *a magnetic interaction* and describes the interaction between the spin and orbital magnetic moments. Orbital magnetic moments arise because of angular orbital moments that have a classical analogue, and the rotation of a charge classically corresponds to a current flowing in a wire arranged in a loop and is therefore analogous to a magnetic dipole. As for spin, which has no classical analogue, the postulate establishes the existence of a magnetic moment and its relation to the spin angular moment which is near twice the classical value of a rotating charged particle. In the general case of a poly-electronic system, the spin-orbit interaction fits into a framework that is already complex due to the charge interactions and antisymmetry.

For better clarity, it is therefore convenient to begin with the simpler problem of an atom with a single electron, *i.e.* a hydrogen-like system. Note that the treatment of the hydrogen atom usually given in elementary textbooks does *not* include the spin-orbit interaction.

3.1.1 The One Electron Case

In a mono-electronic system, the spin-orbit interaction is particularly simple because there is only one orbital magnetic moment and one spin magnetic moment, and both are proportional to their respective angular moments. Then, the magnetic interaction is proportional to the scalar product of the magnetic moments

$$E^{sl} \propto \vec{l} \cdot \vec{s} \tag{3.1}$$

An alternative way to get the same result is to consider the interaction between

- the spin magnetic moment of the electron $\vec{\mu}$ and
- the magnetic field ($\vec{H}$) generated by the orbital motion of the electron, in formulas:

$$E^{sl} = -\vec{H} \cdot \vec{\mu} \quad (3.2)$$

As said, $\vec{\mu} \propto \vec{s}$. The magnetic field $\vec{H}$ produced by the orbital motion turns out to be proportional to the orbital angular momentum through a constant that does not depend on the angular coordinates and is a function of the radial coordinate only. So, we have:

$$E^{sl} = f(r) \cdot \vec{l} \cdot \vec{s} \quad (3.3)$$

The derivation is reported in most textbooks, which also specify the exact form of the function *f(r)*. However, we can ignore it and pay attention only to the angular moments.

3.1.1.1 The Angular Moment $\vec{j} = \vec{l} + \vec{s}$

To deal with an interaction proportional to the scalar product of the orbital and spin angular moments requires introducing a new angular moment so defined: $\vec{j} = \vec{l} + \vec{s}$.

In fact, if $\vec{j} = \vec{l} + \vec{s}$, the squared modulus of this new angular moment is:

$$\left|\vec{j}\right|^2 = \vec{j} \cdot \vec{j} = \left(\vec{l} + \vec{s}\right) \cdot \left(\vec{l} + \vec{s}\right)$$

$$= \vec{l} \cdot \vec{l} + \vec{s} \cdot \vec{s} + 2 \cdot \vec{l} \cdot \vec{s} = \widehat{l^2} + \widehat{s^2} + 2 \cdot \vec{l} \cdot \vec{s}$$

$$\text{and} \quad \vec{l} \cdot \vec{s} = \frac{1}{2}\left(\left|\vec{j}\right|^2 - \widehat{l^2} - \widehat{s^2}\right) \quad (3.4)$$

From the correspondence rule (third postulate of Quantum Mechanics, see Section 2.4.1 of Chapter 1) this becomes:

$$\widehat{\vec{l} \cdot \vec{s}} = \frac{1}{2}\left(\widehat{j^2} - \widehat{l^2} - \widehat{s^2}\right) \quad (3.5)$$

As for any angular moment, we can write:

$$\widehat{j^2}\left|j, m_j, \ldots\right\rangle = j \cdot (j+1) \cdot \hbar^2 \cdot \left|j, m_j, \ldots\right\rangle \quad (3.6)$$

$$\widehat{j_z}\left|j, m_j, \ldots\right\rangle = m_j \cdot \hbar \cdot \left|j, m_j, \ldots\right\rangle \quad (3.7)$$

The dots indicate that the eigenfunctions require additional quantum numbers to be specified.

The reason why we must introduce the sum of the two angular moments is directly linked to the properties of this new quantity. Equation (3.4) can be immediately translated into the corresponding equations for the operators. Now, let us consider functions (φ) that are simultaneously eigenfunctions of $\widehat{j^2}$, $\widehat{l^2}$ and $\widehat{s^2}$. For these functions, the following holds:

$$\left(\widehat{\vec{l} \cdot \vec{s}}\right)\varphi = \frac{1}{2}\left(\widehat{j^2} - \widehat{l^2} - \widehat{s^2}\right)\varphi$$

$$= \frac{1}{2}\left[(j+1) - l(l+1) - s(s+1)\right]\hbar^2\varphi \quad (3.8)$$

Equation (3.8) is the fundamental result: the eigenfunctions of $\widehat{j^2}$, $\widehat{l^2}$ and $\widehat{s^2}$ are also eigenfunctions of the spin-orbit interaction operator.

3.1.1.2 The Energy Split Due to the Spin-Orbit Interaction

The problem, essentially, is to add the spin-orbit interaction to an unperturbed Hamiltonian. In the poly-electronic case, this is the so-called non-relativistic Hamiltonian $\hat{H}_{NR}$ but, in the case of a hydrogen-like problem, it is only the Coulomb interaction between nucleus and electron plus the kinetic energy of the electron itself. The unperturbed Hamiltonian is much simpler as it does not include charge interactions between electrons and does not present anti-symmetrization problems. Its energy levels are the usual shells 1s, 2s, 2p, 3s, 3p, 3d, 4s, … and a shell is made up of various states which are all eigenfunctions of the $\widehat{l^2}$ operator with the same eigenvalue. An energy level or shell is always degenerate (at least twice). For the problem we are dealing with, the shells 1s, 2s, 2p, 3s, 3p, 3d, … of the hydrogen-like problem correspond to the *terms* of the poly-electronic atom. In the poly-electronic atom, where the non-relativistic Hamiltonian includes charge interactions between electrons and antisymmetry must be considered, an energy level (*i.e.* a term) is indeed characterized by a specific eigenvalue of $\widehat{L^2}$ and a specific eigenvalue of $\widehat{S^2}$. For uniformity and to facilitate the subsequent generalization, it is therefore useful to say that, even in the particular case of any hydrogen-like problem, the levels of the unperturbed Hamiltonian without spin-orbit interaction are characterized by precise values of both $\widehat{l^2}$ and $\widehat{s^2}$.

The degenerate states for a given level (*nl*) of the hydrogen-like Hamiltonian can be chosen in various ways. The most obvious way is to use the different eigenvalues of the *z* components of the orbital angular moment and the spin momentum *i.e.* m_l and m_s. Alternatively, from the value of l one or two possible values of j are obtained according to the rules of vector addition.

If $l = 0$, $j = s = ½$. If $l > 0$, $j = l + ½$ or $l - ½$.

We can then classify the degenerate states of the shell using j and m_j instead of m_l and m_s. With this choice, Eq. (3.8) allows us to know *how many energy sublevels the shell is split into* by the spin-orbit interaction and *what are the degeneracies of these sublevels*.

For example, in the $3d^1$ configuration, the quantum number j can be 5/2 (from 2 + 1/2) or 3/2 (from 2 − 1/2). We have six states for the first case (m_j = 5/2, 3/2, 1/2, −1/2, −3/2, −5/2) and four states for the second case (m_j = 3/2, 1/2, −1/2, −3/2). The total number of states is still ten, but they are new linear combinations of the ten states $\left|n, l, m_l, m_s\right\rangle$ or of the ten states $3d_{xy}\alpha$, $3d_{xy}\beta$, $3d_{yz}\alpha$, $3d_{yz}\beta$, ... According to Eq. (3.8) the two j correspond to different energies for the spin-orbit interaction. So, we have a split into two sublevels, and their degeneracies are 6 and 4.

Another example: the $2p^1$ configuration. It is no longer six times degenerate but is "split" into two sublevels ($j = 1 + 1/2 = 3/2$ and $j = 1 - 1/2 = 1/2$) with degeneracies 4 and 2.

The simple calculation of the quantum numbers j and m_j does not specify the eigenfunctions, that is, the coefficients of the linear combinations of the functions $|n,l,m_l,m_s\rangle$ in a $|n,l,s,j,m_j\rangle$ state and the energy of the various sublevels. For the latter point, the third Hund's empirical rule is used.

3.1.2 Many-Electron Systems

In the general case of the poly-electronic atom, the spin-orbit interaction is made up of various components and is particularly intricate because there are many orbital moments (angular and magnetic) and just as many spin moments. Before tackling the problem, however, it is necessary to consider that, generally, we have three types of energy contributions:

a. the independent-particle Hamiltonian,
b. the charge interactions between electrons and
c. the spin-orbit interactions.

In the hydrogen-like case, where (b) is missing, the spin-orbit interaction is addressed with an approximate perturbative treatment on the exact results for the hydrogenic Hamiltonian, whose energy levels are corrected and possibly split as we have just seen.

For the poly-electronic case, essentially, two corrections must be applied, and the quantitatively more relevant correction must be applied first. Then, the quantitatively less important correction is applied to the results of the first calculation. However, the relative importance of the various energy terms varies with the number of electrons or the nuclear charge Z: (a) "goes as" Z^2, (b) as Z, and (c) as Z^4. For heavy atoms the spin-orbit interaction (c) becomes more important than the charge (Coulombic) interaction between electrons (b). For light atoms, however, the spin-orbit interaction is less important, with the exception of the H atom where the (b) contribution is zero.

As a matter of fact, the two cases are treated in the following ways:

- At low values of Z we add the spin-orbit interaction to the treatment seen so far for the Coulomb charge interactions between electrons, that is to the treatment in which the removal of degeneracy of an electronic configuration is due to the different results of the coupling of the orbital moments within a total moment $\vec{L}$ and of the spin moments within a total moment $\vec{S}$. The spin-orbit interaction is evaluated with the coupling between $\vec{L}$ and $\vec{S}$ in a total moment $\vec{J}$. We therefore speak of ***LS* coupling**.
- At high values of Z (essentially: after the lanthanides) we **first** treat the spin-orbit interactions by evaluating the coupling of the orbital angular momentum $\vec{l}$ and the spin angular momentum $\vec{s}$ of each electron into a total single-electron total angular momentum $\vec{j}$ and then we combine various single-electron moments into the total angular momentum $\vec{J}$. We therefore speak of ***jj* coupling**.

For the intermediate cases, treatments have been developed that somehow interpolate between the two extreme cases. Here, only the first case, which is really the most important, is considered.

3.2 *LS* (Russell-Saunders) Coupling

3.2.1 Outline

The *LS* coupling scheme for a poly-electronic atom essentially combines in sequence what has been seen so far for the charge interaction between electrons with a treatment of the spin-orbit interaction analogous to that of the mono-electronic case.

Let us find the linear combinations of the states $|L, S, M_L, M_S\rangle$ of a term (*i.e.* keeping both L and S fixed) which are eigenfunctions of the two operators ($\widehat{J^2}$ and $\widehat{J_z}$) relative to $\vec{J} = \vec{L} + \vec{S}$. These linear combinations are characterized by the quantum numbers J and M_J as well as by the L and S of the term. The spin-orbit interaction is now approached as for the one-electron case:

$$\widehat{J^2} = \vec{J}\cdot\vec{J} = \left(\vec{L}+\vec{S}\right)\cdot\left(\vec{L}+\vec{S}\right) = \vec{L}\cdot\vec{L}+\vec{S}\cdot\vec{S}+2\cdot\vec{L}\cdot\vec{S}$$

$$= \widehat{L^2}+\widehat{S^2}+2\cdot\vec{L}\cdot\vec{S} \quad \text{From which we get:} \tag{3.9}$$

$$\widehat{\vec{L}\cdot\vec{S}} = \frac{1}{2}\left(\widehat{J^2}-\widehat{L^2}-\widehat{S^2}\right) \tag{3.10}$$

Example: ^{3}P, that is $L = 1$, $S = 1$ and $J = 2, 1, 0$. Then $\vec{L}\cdot\vec{S} = (4-1-1)/2 = 1$ (for $^3P_2$), or $\vec{L}\cdot\vec{S} = (1-1-1)/2 = -1$ (for $^3P_1$), or else $\vec{L}\cdot\vec{S} = (0-1-1)/2 = -1(^3P_0)$.

For each term (L, S) the spin-orbit contribution to the energy depends on the scalar product $\vec{L}\cdot\vec{S}$ and therefore on J (see Eq. (3.10)). The energy level of the term is therefore "split" into as many sublevels as there are different values for J according to the vector addition rules. These sublevels are called **multiplets**.

The *complete removal of degeneracy* of an electronic configuration, *i.e.* the removal of degeneracy of a multiplet, is ultimately achieved by applying a magnetic field which introduces an additional energy contribution proportional to M_J: it is the energy of a magnetic dipole of component M_J in the direction of the external field.

3.2.2 Hund Rules

The discussion is finally complemented with empirical rules derived from spectroscopic data about the size of the various energy splits of an electronic configuration.

(First rule)
The quantum number that plays the most important role is S and **the energy increases as S decreases**.
(Second rule)
The second most important quantum number is L and **the energy increases as L decreases.**

The split produced by different L is less than the separation produced by different S

(Third rule)
The least important quantum number is J.

We must then distinguish between configurations with a shell that is more than half filled or less than half filled.

If the shell is less than half filled, the energy increases as J increases.
If the shell is more than half filled, the energy increases as J decreases.

The third rule does not apply for configurations with a single hole: the highest energy is for the highest J.

3.2.3 The 2p² Example (Carbon Atom)

The general energy scheme is shown in Figure 2.11 and the procedure is as follows.

1. Add the spin moments of single electrons. There are two unpaired electrons, so $S = 1$ (½ + ½), *i.e.* spin multiplicity = 3, or $S = 0$ (½ − ½) *i.e.* spin multiplicity = 1.
2. Add the orbital moments of single electrons. Both l's are 1, so L ranges from 2 (1 + 1) to 0 (1 − 1): L = 2, 1, 0. The possible terms are then ^{3}D, ^{3}P, ^{3}S, ^{1}D, ^{1}P, ^{1}S.
3. The configuration requires finding terms to be removed. In fact, there is more than one electron in the 2p shell (and the case of only one electron missing to complete the shell is excluded). Using the tables we see that the terms that survive are: ^{3}P, ^{1}D, ^{1}S.
4. For each term, combine the quantum numbers L and S to obtain J.
   a. For ^{3}P ($L = 1, S = 1$), $J = 2$ (=1 + 1) ($^3P_2$), $J = 1$ ($^3P_1$), $J = 0$ (=1 − 1) ($^3P_0$);
   b. For ^{1}D ($L = 2, S = 0$) there is only $J = 2$ (2 + 0) *i.e.* $^1D_2$.
   c. For ^{1}S ($L = 0, S = 0$) there is only $J = 0$ (0 + 0) *i.e.* $^1S_0$.
5. By Hund's first rule, $E(^3P) < E(^1D), E(^1S)$.
6. By Hund's second rule, $E(^1D) < E(^1S)$ and hence the energy order of the terms is: $E(^3P) < E(^1D) < E(^1S)$ and the second separation is smaller than the first.
7. By Hund's third rule, $E(^3P_0) < E(^3P_1) < E(^3P_2)$.
8. The degeneracies are $^3P_0$: 1; $^3P_1$: 3; $^3P_2$: 5; $^1D_2$: 5; $^1S_0$: 1. In total there are 15 states, as confirmed by the counting of Slater's determinants: $\binom{6}{2} = \frac{6!}{2!4!}$.
9. For any multiplet, the degeneracy can be completely removed by an external magnetic field $\vec{B}$ (Zeeman effect) which produces a shift in energy proportional to $M_J|\vec{B}|$ and hence does not change the average energy of the multiplet.

3.3 Selection Rules in Atomic Spectroscopy

Spectroscopic transitions occur between the states resulting from all the previous discussion. Two clarifications are necessary. First of all, as is known, the operators that control the intensity of the transitions are different. In order of importance, they are:

a. electric dipole transitions,
b. magnetic dipole transitions,
c. electric quadrupole transitions,
d. magnetic quadrupole transitions,

For radiation in the visible range, the intensities of the first three types (a, b, c) are typically in the ratio 1: 10^{-5}: 10^{-7}, while the other types they are even less intense. For this reason, we can limit ourselves to considering only the **electric dipole transitions**, whose probabilities are controlled by the square modulus of the matrix element between the initial state and the final state of the electric dipole operator:

$$p_{f\leftarrow i} \propto |\mu_{if}|^2; \quad \mu_{if} \stackrel{\text{def}}{=} \langle i|\vec{\mu}|f\rangle \stackrel{\text{def}}{=} e\cdot\langle i|\vec{r}|f\rangle \tag{3.11}$$

Second, what we want to obtain are **selection rules**, that is, results that establish which transitions are allowed (regardless of their probability) and which are forbidden, that is, have **identically zero** probability. To explain where

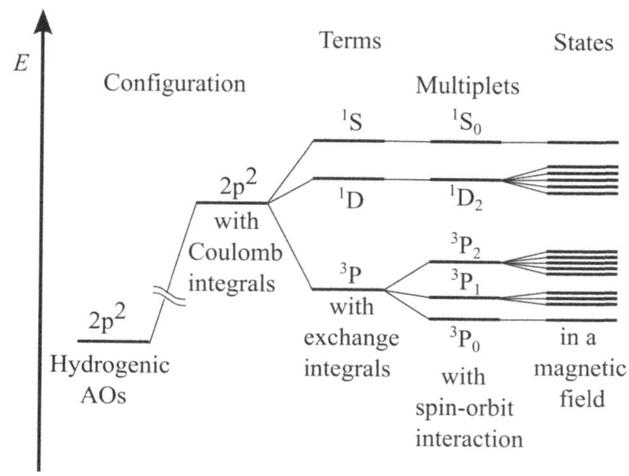

FIGURE 2.11 Scheme of the effect of various interactions for the fundamental configuration of the C atom.

a selection rule comes from and the difference between a value very close to zero and one that is identically zero, one can examine the calculation of the integral $\langle i|\hat{x}|f\rangle$. If the states $|i\rangle$ and $|f\rangle$ both have odd parity, *i.e.* $\varphi_i(x) = -\varphi_i(-x)$ and $\varphi_f(x) = -\varphi_f(-x)$ then for the integrand of $\langle i|\hat{x}|f\rangle$ we have

$$\varphi_i s(-x)\cdot(-x)\cdot\varphi_f(-x) = -[\varphi_i(x)\cdot x\cdot\varphi_f(x)]$$

This means that in the integration from $-\infty$ to 0 we obtain the same result (with a change of sign) as in the integration from 0 to $+\infty$. Therefore, the integral from $-\infty$ to $+\infty$ is zero *regardless of the specific form of the functions* φ_i and φ_f. To express this result, we say that the integral is identically zero. Incidentally, the same result is obtained even if both functions have even symmetry, that is, $\varphi_i(x) = \varphi_i(-x)$ and $\varphi_f(x) = \varphi_f(-x)$. So, the vanishing rule of the integral would be that *the initial and final states must have the same parity*: the transition between these two states has identically zero probability, and it is said to be forbidden by a selection rule.

The elementary example makes it clear that a selection rule is essentially due to symmetry.

The electric dipole operator has three Cartesian components (μ_x, μ_y, μ_z). If, for a pair of states (i, f), all three components have a null matrix element, the transition is forbidden. If, for at least one component, the matrix element is not identically zero, the transition is allowed.

To be precise, if the matrix element is zero for only one component, the absorbed or emitted radiation is polarized in the plane of the other two components (it is said to be σ-polarized) and appears circularly polarized if observed along an axis perpendicular to the polarization plane.

If the matrix element is zero for two components, the radiation is polarized along the corresponding axis (π-polarized).

The results *(exact, approximate or empirical)* in the ***LS* coupling scheme** are as follows.

1. **Only one electron jumps at a time.** This is not a true selection rule in the sense specified above, but rather allows us to interpret many experimental results well.
2. For the jumping electron, **the quantum number l must change by 1 ($\Delta l = \pm 1$).**
3. For the atom as a whole **$\Delta S = 0$ (conservation of spin multiplicity)**. This is a selection rule valid as long as the spin-orbit interaction "counts for little".
4. For the atom as a whole **$\Delta L = 0$ or ± 1.** This rule is also valid as long as the spin-orbit interaction "counts for little".
5. For the atom as a whole **$\Delta J = 0$ or ± 1**, but **the transition "0 ↔ 0" is forbidden**.
6. For the atom as a whole **$\Delta M_J = 0$ or ± 1**, but **the transition "0 ↔ 0" is forbidden if $\Delta J = 0$.**

4 REMARKS

4.1 Details for the 1s¹2s¹ Example

To understand the essential aspect that is hidden by the details of the calculation, let us look at the expression of two particular elements ($\boldsymbol{H}_{33}$ and $\boldsymbol{H}_{23}$) of the Hamiltonian matrix for the $1s^1 2s^1$ configuration (2.2.1). To simplify the notation a bit, it is convenient to regularly follow the convention of using the order to specify the two electrons: for example,

$$\left\langle 1s2s\left|\frac{1}{r_{12}}\right|1s2s\right\rangle \text{ means: } \left\langle \chi_{1s}(\vec{r}_1)\cdot\chi_{2s}(\vec{r}_2)\left|\frac{1}{r_{12}}\right|\chi_{1s}(\vec{r}_1)\cdot\chi_{2s}(\vec{r}_2)\right\rangle,$$

while

$$\left\langle 2s1s\left|\frac{1}{r_{12}}\right|1s2s\right\rangle \text{ means: } \left\langle \chi_{1s}(\vec{r}_2)\cdot\chi_{2s}(\vec{r}_1)\left|\frac{1}{r_{12}}\right|\chi_{1s}(\vec{r}_1)\cdot\chi_{2s}(\vec{r}_2)\right\rangle$$

Also: $a \Leftrightarrow \chi_{1s}\alpha$; $b \Leftrightarrow \chi_{1s}\beta$; $c \Leftrightarrow \chi_{2s}\alpha$; $d \Leftrightarrow \chi_{2s}\beta$.

4.1.1 The H_{33} Element

$$\boldsymbol{H}_{33} = \frac{1}{2}bc - \left\langle cb\left|\hat{H}\right|bc\right\rangle - cb$$

The constant ½ comes from the product of the two factors $1/\sqrt{2}$ (see Eqs. (2.16)–(2.19)). For the linearity of the operator:

$$\boldsymbol{H}_{33} = \frac{1}{2}\left[\left\langle bc\left|\hat{H}\right|bc\right\rangle - \left\langle cb\left|\hat{H}\right|bc\right\rangle - \left\langle bc\left|\hat{H}\right|cb\right\rangle + \left\langle cb\left|\hat{H}\right|cb\right\rangle\right]$$

For the $\left\langle bc\left|\hat{H}\right|bc\right\rangle$ **and** $\left\langle cb\left|\hat{H}\right|cb\right\rangle$ **integrals**, the operator is the sum of three operators:

$$\left\langle bc\left|\hat{H}\right|bc\right\rangle = \left\langle bc\left|\widehat{H_1^0}\right|bc\right\rangle + \left\langle bc\left|\widehat{H_2^0}\right|bc\right\rangle + \left\langle bc\left|\frac{1}{r_{12}}\right|bc\right\rangle$$

In all three cases, the operators do not act on the spin functions which therefore need to be treated separately, *e.g.*:

$$\left\langle bc\left|\widehat{H_1^0}\right|bc\right\rangle = \left\langle 1s2s\left|\widehat{H_1^0}\right|1s2s\right\rangle\cdot\langle\beta\alpha|1|\beta\alpha\rangle$$

$$= \left\langle 1s2s\left|\widehat{H_1^0}\right|1s2s\right\rangle\cdot\langle\alpha\,|\,\alpha\rangle\cdot\langle\beta\,|\,\beta\rangle = \left\langle 1s2s\left|\widehat{H_1^0}\right|1s2s\right\rangle$$

Then, observing that the operator $\widehat{H_1^0}$ acts only on the coordinates of the first electron, the second orbitals behave as multiplicative constants and can be factored separately: $\left\langle 1s2s\left|\widehat{H_1^0}\right|1s2s\right\rangle = \left\langle 1s\left|\widehat{H_1^0}\right|1s\right\rangle\cdot\langle 2s|1|2s\rangle = \left\langle 1s\left|\widehat{H_1^0}\right|1s\right\rangle \stackrel{\text{def}}{=} \alpha_{1s}$.

Similarly, the operator H_2^0 acts only on the coordinates of the second electron. Therefore:

$$\left\langle 1s2s\left|\widehat{H_2^0}\right|1s2s\right\rangle = \langle 1s|1|1s\rangle\cdot\left\langle 2s\left|\widehat{H_2^0}\right|2s\right\rangle = 1\cdot\left\langle 2s\left|\widehat{H_2^0}\right|2s\right\rangle \stackrel{\text{def}}{=} \alpha_{2s}$$

The operator $\frac{1}{r_{12}}$ acts on the coordinates of both electrons and therefore it is not possible to factor it into an integral that contains the coordinates of only one electron: the **integral is bi-electronic**:

$$\left\langle 1s2s \left| \frac{1}{r_{12}} \right| 1s2s \right\rangle \stackrel{\text{def}}{=} J_{1s2s}$$

Then: $\left\langle bc \left| \hat{H} \right| bc \right\rangle = \langle \alpha_{1s} + \alpha_{2s} + J_{1s2s} \rangle$. As expected, the result is the sum of:

a. the kinetic energy of an electron plus the interaction energy of the electron with the nucleus (α_{1s}) plus
b. the corresponding term for another electron in the 2s orbital (α_{2s}) plus
c. the interaction energy between an electron in the 1s orbital and another in the 2s orbital.

To simplify, we write: $\left\langle bc \left| \hat{H} \right| bc \right\rangle = \alpha + J$, incorporating the two integrals ($\alpha \stackrel{\text{def}}{=} \alpha_{1s} + \alpha_{2s}$) into a single factor and ignoring the indices of C_{1s2s} because it is the only integral of this type.

This result also holds for $\left\langle cb \left| \hat{H} \right| cb \right\rangle$.

For **integrals** $\left\langle cb \left| \hat{H} \right| bc \right\rangle$ and $\left\langle bc \left| \hat{H} \right| cb \right\rangle$, the first operator gives: $\left\langle cb \left| \widehat{H_1^0} \right| bc \right\rangle = \left\langle 2s1s \left| \widehat{H_1^0} \right| 1s2s \right\rangle \cdot \langle \alpha\beta |1| \beta\alpha \rangle = \left\langle 2s1s \left| \widehat{H_1^0} \right| 1s2s \right\rangle \cdot \langle \alpha \mid \beta \rangle \cdot \langle \beta \mid \alpha \rangle = 0$ because the spin functions are orthogonal. So, the exchange of the electrons between the two SOs cancels the integral and we see that this is also true for the other two operators ($\widehat{H_2^0}$ and $\frac{1}{r_{12}}$). Then the whole integral $\left\langle cb \left| \hat{H} \right| bc \right\rangle$ is zero. For $\left\langle bc \left| \hat{H} \right| cb \right\rangle$, the same result holds.

In conclusion, $\boldsymbol{H}_{33} = \alpha + J$.

4.1.2 The H_{23} VWW Element

$$\boldsymbol{H}_{23} = \left\langle B \left| \hat{H} \right| C \right\rangle = \frac{1}{2} \left\langle ad - da \left| \hat{H} \right| bc - cb \right\rangle$$

$$= \frac{1}{2} \left[\left\langle ad \left| \hat{H} \right| bc \right\rangle - \left\langle da \left| \hat{H} \right| bc \right\rangle - \left\langle ad \left| \hat{H} \right| cb \right\rangle + \left\langle da \left| \hat{H} \right| cb \right\rangle \right]$$

For the $\left\langle ad \left| \hat{H} \right| bc \right\rangle$ **and** $\left\langle da \left| \hat{H} \right| cb \right\rangle$ **integrals.**

$$\left\langle ad \left| \hat{H} \right| bc \right\rangle = \left\langle ad \left| \widehat{H_1^0} \right| bc \right\rangle + \left\langle ad \left| \widehat{H_2^0} \right| bc \right\rangle + \left\langle ad \left| \frac{1}{r_{12}} \right| bc \right\rangle$$

As before, all three operators do NOT act on spin functions, which must therefore be treated separately. Now, however, they do NOT produce a simplification but cancel the entire integral. Consider, for example $\left\langle ad \left| \widehat{H_1^0} \right| bc \right\rangle = \left\langle 1s2s \left| \widehat{H_1^0} \right| 1s2s \right\rangle \cdot \langle \alpha\beta |1| \beta\alpha \rangle = \left\langle 1s2s \left| \widehat{H_1^0} \right| 1s2s \right\rangle \cdot \langle \alpha \mid \beta \cdot \beta \mid \alpha \rangle = \left\langle 1s2s \left| \widehat{H_1^0} \right| 1s2s \right\rangle \cdot 0 \cdot 0 = 0$. At the end, one can see that both $\left\langle ad \left| \hat{H} \right| bc \right\rangle$ and $\left\langle da \left| \hat{H} \right| cb \right\rangle$ are identically zero.

For the $\left\langle da \left| \hat{H} \right| bc \right\rangle$ and $\left\langle ad \left| \hat{H} \right| cb \right\rangle$ integrals, the exchange of electrons between antecedent and consequent leads to a different result, *e.g.*:

$$\left\langle da \left| \widehat{H_1^0} \right| bc \right\rangle = \left\langle 2s1s \left| \widehat{H_1^0} \right| 1s2s \right\rangle \cdot \langle \alpha\beta \mid \alpha\beta \rangle$$

$$= \left\langle 2s1s \left| \widehat{H_1^0} \right| 1s2s \right\rangle \cdot \langle \alpha \mid \alpha \rangle \cdot \langle \beta \mid \beta \rangle = \left\langle 2s1s \left| \widehat{H_1^0} \right| 1s2s \right\rangle$$

and similarly for the other two operators. For the two one-electron operators this means that the integrals cancel each other out because they act only on the coordinates of the first electron, the second orbitals must be factored separately and their integral is zero since they are different and orthogonal orbitals, *e.g.*: $\left\langle 2s1s \left| \widehat{H_1^0} \right| 1s2s \right\rangle = \left\langle 2s \left| \widehat{H_1^0} \right| 1s \right\rangle \cdot \langle 1s \mid 2s \rangle = 0$.

The same conclusion does *not* hold for the operator $\frac{1}{r_{12}}$:

$$\left\langle 2s1s \left| \frac{1}{r_{12}} \right| 1s2s \right\rangle = \left\langle 2s1s \left| \frac{1}{r_{12}} \right| 2s1s \right\rangle$$

$$= \left\langle \chi_{2s}(r_1) \cdot \chi_{1s}(r_2) \left| \frac{1}{r_{12}} \right| \chi_{2s}(r_1) \cdot \chi_{1s}(r_2) \right\rangle$$

In fact, the two-electronic operator acts on the coordinates of both electrons and therefore there is no possibility of factorizing apart the 1s orbitals or the 2s orbitals. The integral that remains is precisely the exchange integral:

$$\left\langle 2s1s \left| \frac{1}{r_{12}} \right| 1s2s \right\rangle \stackrel{\text{def}}{=} J_{1s2s}$$

In conclusion, we see that in some matrix elements it is necessary to insert the exchange integral.

Without needing to develop the calculations, it is immediately seen that the exchange integral must also appear in $\mathbf{H}_{32}$, since the matrix is Hermitian. The same sequence of calculations would also show that the exchange integral does **not** appear in the other matrix elements.

4.2 Counting the States

We show that the total number of states classified by two distinct angular moments is equal to the number of states classified by all possible values of the quantum numbers of the sum of the angular moments.

We have $(2l_1 + 1)$ pairs of (l_1, m_1) quantum numbers and $(2l_2 + 1)$ pairs of (l_2, m_2) numbers: the states are in total $(2l_1 + 1)(2l_2 + 1) = 4l_1l_2 + 2(l_1 + l_2) + 1$. Since the result does not change by exchanging l_1 and l_2, we can safely assume $l_2 \leq l_1$ and, applying the addition rule, we have L ranging from $l_1 + l_2$ to $l_1 - l_2$, and for each L we have $2L + 1$ different values of M_L. The number of states is then:

$$\sum_{L=l_1-l_2}^{L=l_1+l_2}(2L+1)=2\sum_{L=l_1-l_2}^{L=l_1+l_2}(L)+(2l_2+1)$$

$$=2\left(\frac{l_1+l_2+l_1-l_2}{2}\right)(2l_2+1)+2l_2+1$$

$$=(2l_1)(2l_2+1)+2l_2+1=4l_1l_2+2(l_1+l_2)+1$$

Where we used the facts that the terms in the summation are $2l_2+1$, and that the sum of n terms of an arithmetic sum is the number of terms multiplied by the mean between the first and last term.

5 EXERCISES

Exercise 2.1

Consider four electrons and four different spin orbitals that we call S_a, S_b, S_c, S_d and assume that they are a closed shell (although the example is not realistic for atomic systems). How many Hartree products are there? Write down all of them. Write down the Slater determinant made starting from the four S_a, S_b, S_c, S_d SOs. *What is the relationship between the Hartree products and the Slater determinant? How many states are in this electronic configuration?*

Exercise 2.2

For the fundamental Li configuration ($1s^22s^1$) write down the two Slater determinants. Check with the binomial formula that the number of states is 2. How many states are in the fundamental electronic configuration of B: $1s^22s^22p^1$? How many states are in the fundamental electronic configuration of C: $1s^22s^22p^2$?

Exercise 2.3

Considering a nitrogen atom: (i) What is the electronic configuration of the fundamental state, and how many states are in there? (ii) Write down a Slater determinant for this configuration. How many Hartree products are in there? (iii) What are the L values for this electronic configuration? What are the S and J values?

Exercise 2.4

Take a system in the $^3P_0$ fundamental state and suppose that the first excited state is ^3D. Draw an energy diagram with all the levels and find the allowed transitions.

Exercise 2.5

Show that in an S term there is just one state, that in a P term there are at most three states, that for a D term there are at most five states, etc.

Exercise 2.6

The magnetic dipole operator is basically the angular moment operator multiplied by a constant: $\hat{m}=\gamma_e\hat{L}$. Show that $\Delta l=0$ magnetic dipole transitions are allowed, that is magnetic dipole transitions couple states with the same parity.

Exercise 2.7

In core level spectroscopies, such as x-ray absorption spectroscopy (XAS), an electron from a core level is ejected to unbound continuum states giving rise to absorption edges that are named K, L, M, etc. if the principal quantum number n of the core level is 1, 2, 3 etc. Show that there is just one K edge, but three L edges (L_I, L_{II}, L_{III}, in order of decreasing energy), five M edges (M_I, M_{II}, M_{III}, M_{IV}, M_V, in order of decreasing energy), etc. Hint: consider the final state electronic configuration, i.e. $1s^1$, $2s^1$, $2p^5$ etc. Explain why all the other electrons can be ignored.

Exercise 2.8

Referring to Exercise 2.7, in the XAS spectrum of Br, nine absorption edges are found. Explain why, then draw an energy diagram of the levels involved. What should be the intensity ratio between the L_{II} and L_{III} edges? Hint: consider the multiplicity of J. Explain the intensity ratio between the M_{IV} and M_V edges.

NOTES

1 Here is a very short list: http://en.wikibooks.org/wiki/High_School_Chemistry/Shapes_of_Atomic_Orbitals; http://en.wikipedia.org/wiki/Atomic_orbital, https://commons.wikimedia.org/wiki/Hydrogen_orbitals_3D; http://chemwiki.ucdavis.edu/Physical_Chemistry/Atomic_Theory

2 This can be rigorously proven.

3 Thermodynamics

A well-known comment by Arnold Sommerfeld well summarizes the difficulties typically faced in learning thermodynamics: "Thermodynamics is a funny subject. The first time you go through it, you don't understand it at all. The second time you go through it, you think you understand it, except for one or two small points. The third time you go through it, you know you don't understand it, but by that time you are so used to it, it doesn't bother you anymore."[1]

Both Classical Thermodynamics (CT) and Statistical Thermodynamics (ST) *can be* discussed as a series of statements axiomatically deduced from the *laws* of CT or the *postulates* of ST. However, it is not easy to specify clearly at the beginning the nature and properties of the objects treated by these theories, in particular for what concerns the macroscopic approach of CT. So, instead of a purely deductive line of presentation, such as that followed for QM, we here prefer to gradually introduce the different aspects and reach a sensible treatment by frequently restating and better specifying points previously outlined in a simpler way.

In the traditional approach, CT and ST are treated as different subjects and taught in different courses. We here adopt an integrated approach, to maximize synergies and allow us to use procedures, ideas, and examples of one theory for giving insights into the other.

Concerning the content, this book does not deal with topics that are surely well known to the reader from previous studies or that are widely spread across the general public, such as the meaning and implications of the second law. The book also skips some more technical sections such as the colligative properties and the thermodynamics of diluted solutions, surfaces, colloids, and nano-systems that are regularly treated in general textbooks. Analogously, electrochemistry is treated only in its most general thermodynamic aspects. Finally, only a simplified derivation of the partition function of the canonical ensemble[2] is reported in a Note because of its technical difficulties.

On the other hand, some space is spent to clarify various basic topics that are skipped in most textbooks. One of these concerns the comparison of metastability and instability and the meaning and onset of critical points. Another is the connection between the ideal solution model in CT and ST and the indistinguishability of microscopic particles. We also think it important to show how the idea of a chemical reaction is possibly generalized and how the so-called quasi-chemical approach treats in an effective manner various topics of Solid-State and Materials Sciences.

1 CLASSICAL THERMODYNAMICS AND STATISTICAL THERMODYNAMICS

1.1 Introduction

Classical thermodynamics (CT) and statistical thermodynamics (ST) are two theories that study the equilibrium properties of macroscopic systems and energy transformations. They differ in their theoretical framework and partly also in their objectives. In some respects, they are alternative theories, but in many other respects they are complementary to each other.

CT and ST both refer to other physical theories, essentially of a mechanical type: continuum mechanics, in particular fluid mechanics, and classical or quantum mechanics of atomic and molecular systems. When necessary, other physical theories are also involved, such as electrostatics or magnetism. Thermodynamics (CT and ST) is a sort of superstructure of these theories and deals with many quantities that these theories deal with. Among these we find, for example, volume (V), mass (m), amount of substance (n), pressure (p) and various forms of energy and work (kinetic energy, potential energy, expansion work - here often referred to briefly as pV work). When relevant, electrical and magnetic quantities are added.

In addition, there are **quantities** that are **specific to thermodynamics** and are better called **thermal** quantities. Among these, there are first of all *temperature* and *entropy*, then other quantities that contain temperature or entropy or both: for example, Gibbs and Helmholtz energies and chemical potentials.

CT was formulated first: historically, it was introduced to deal with the transformations between work and heat and heat engines and was then systematically applied, among other things, to the study of chemical processes, such as reactions and phase transformations. CT is based on some "laws" from which the results are deduced in an axiomatic way. The laws are assumed to be true without proof and are *justified* as a form of generalization of experimental facts.

A fundamental aspect of CT that is useful to anticipate here and that is shared by ST, concerns the introduction of appropriate **thermodynamic potentials**, for example: entropy, Gibbs energy, Helmholtz energy. These are characterized by the property that their minimum (or the maximum in the case of entropy) **identifies the equilibrium** of a system under certain conditions imposed from the outside.

Another important aspect of CT is best introduced by an example. In CT it is shown (see Section 3.1.1.1) that the

DOI: 10.1201/9781003644422-3

heat capacity at constant volume (C_V) and the heat capacity at constant pressure (C_p) of a pure substance are related by the relation:

$$C_p - C_V = \alpha^2 VT\beta^{-1}$$

where α is the isobaric thermal expansion coefficient and β is the isothermal compressibility. The relation, which is a typical example of the numerous exact relations derived from CT, has great direct experimental utility because it allows us to determine C_V, that is usually difficult to measure directly for condensed phases, by measuring C_p, α, β and V at temperature T. The point that is important to underline is that the result does not depend on hypotheses on the chemical nature of the substance or on its molecular structure. Even the special result for an ideal gas ($C_p - C_V = nR$) does not require - strictly speaking - hypotheses on the nature of the ideal gas at the atomic-molecular level. In CT, the ideal gas is simply the substance for which the law $pV=nRT$ (equation of state) holds and therefore the internal energy depends only on T (it does not depend on p or V considered separately, see also Exercise 2).

The example shows a feature of CT, which essentially distinguishes it from ST. *CT makes no reference to the structure of matter at the atomic-molecular level.* We can say that CT *does not know atoms or molecules*, but it is not incompatible with their existence.

It is clear that this is both an advantage and a disadvantage. It is an advantage because the results are valid regardless of the atomic-molecular structure of the substances under study. It is a disadvantage because it is obviously important to relate thermodynamic properties and structure, both to predict from structural information the values that a certain quantity assumes, and to obtain structural information from the value of experimentally measured quantities.

Statistical thermodynamics (ST), on the other hand, addresses the problem of **relating thermodynamics and structure** and of calculating thermodynamic properties for specific atomic or molecular structures. To continue with the previous example, the ST explains how the C_V of a mole of linear triatomic molecules is different from the C_V of a mole of diatomic molecules or non-linear triatomic molecules and predicts what values should be expected in the various cases. ST takes as its starting point the results of microscopic mechanics (for atoms, molecules and their interactions) and combines them following a treatment based on some statistical axioms. These statistical axioms appear more fundamental than the second law of CT. We can then say that the second law of CT can be derived from ST, in the sense that ST can give a statistical explanation or - if you prefer - a demonstration starting from its axioms.

The results of CT and ST are equally rigorous. Those of ST, when linked to a particular atomic-molecular structure, obviously have a validity that depends on the structure considered.

In both CT and ST it is generally useful to refer to specific models to better clarify some basic ideas. A particularly useful model is that of the ideal gas. However, it must be clear that this particular model is frequently used because its simplicity allows one to arrive more easily at the final relations and, therefore, to illustrate *in a simpler way* aspects that other models would require longer and more complicated derivations.

It is possible to develop CT and ST in completely independent ways, but on the contrary, it is very useful to mix the two treatments, as done here.

This first section discusses various topics common to the two formulations of thermodynamics and illustrates the principles or postulates that lie below each of them.

1.2 Basic Definitions

As with any scientific theory, to present thermodynamics it is first necessary to introduce its technical language. Instead of the deductive, rigorous, and abstract approach that is sometimes more difficult to understand, we here follow a constructive approach whereby as we proceed it becomes possible to better define and explain the meaning of terms previously introduced in a less precise way.

1.2.1 System and Surroundings

In the construction of the theory of thermodynamics, it was necessary to preliminarily distinguish between the thermodynamic **system** and its **surroundings**. These are, respectively, the portion of matter on which the investigation focuses and all the rest of the universe. The separation between a thermodynamic system and its surroundings can be only conceptual, but more typically is realized by walls through which exchanges of matter, or energy, or both can occur.

1.2.2 Equilibrium, Reversibility and Irreversibility

These are closely connected concepts that clarify each other. Let's start from the observation that our daily life is made up of **processes** or **transformations** that take place in a **specific direction** that we call **spontaneous**. We call them **irreversible processes**.

For example, if we heat a metal bar at one end, and then remove the heat source, the temperature of the bar will tend to become uniform. It does not happen, however, that a metal bar at a uniform temperature spontaneously cools at one end and becomes hot at the other.

An egg placed in a pot full of boiling water will become hard-boiled, but it does not happen that a hard-boiled egg spontaneously becomes raw again.

A short list of **simple** irreversible processes includes the kinetics of chemical reactions and phase transformations and transport processes: diffusion (transport of substance), thermal conductivity (transport of energy), electrical conductivity (transport of charge) and viscosity (transport of moment).

An irreversible process can be made to occur in the opposite direction to the spontaneous one, but this requires the *dissipation of energy*, and a way must be devised to carry out the transformation. In a sense, you can reverse the cooking of an egg. In a raw egg the proteins are folded on themselves by various intramolecular forces such as hydrogen bonds, van der Waals forces, coordination bonds, etc.... When the egg is hard boiled, the proteins unfold and form new intermolecular bonds. These new bonds can be broken, for example by grinding the hard-boiled egg, adding water and urea, and treating it in an ultracentrifuge. Then, the part of the sample closest to the wall of the tube spins faster than the part closest to the centre, and the differences in speed between different points create shear stresses, which cause the proteins to repeatedly stretch and contract, until they return to their original pre-boiled shapes.

It is possible to reverse the spontaneous flow of heat from a hot body to a cold one, but this requires a suitable device (a refrigeration machine) which in turn requires work to operate.

The situation in which *no irreversible transformations* occur is called thermodynamic **equilibrium**.

Thermodynamics deals with establishing **the equilibrium conditions**, that is, clarifying under which conditions irreversible transformations are not possible for a system.

There are some points to note.

Strictly speaking, a system can be in equilibrium only if it is **isolated**. In a non-isolated system, we must more properly speak of equilibrium of the system **with** the surroundings under certain conditions of exchange of energy or matter or both. The system and surroundings together are an isolated system.

A situation that does not change over time is called **stationary** (or **steady**) and is *not necessarily* an equilibrium situation: there are non-equilibrium stationary **states**. For example, a case that occurs quite frequently in biochemical processes is that of a series of chemical reactions in which the concentrations of reagents and products are constant over time thanks to the coupling with processes that supply reagents from outside the system and remove products to the outside.

Moreover, it is not uncommon to find systems in conditions such that one or more chemical reactions and/or phase transitions show a spontaneous direction of evolution but occur at such low speeds that they cannot be measured on the time scale of a laboratory experiment. We then speak of **frozen** (or **quenched**) **equilibria**. This is the case - for example - of many geochemical processes.

Thermodynamic equilibrium takes various forms that will be gradually presented later. For now, we can mention:

- **Thermal Equilibrium**: when in the entire system, the temperature is uniform, that is, it does not depend on the position within the system,
- **Mechanical Equilibrium**: when the pressure is uniform,
- **Compositional Equilibrium**: when each phase (Section 1.2.6) has a uniform composition.

The irreversible heat transfer processes that lead to thermal equilibrium are thermal conduction, convection and radiation. The irreversible processes that homogenize the composition of a phase are the various forms of mechanical mixing for liquids and gases and diffusion.

Unlike irreversible processes, a **reversible transformation** is an *ideal* transformation between equilibrium states. It can be thought of as the limiting case of a *quasi-static process*, that is, the sequence of *a very large number* of transformations that all keep the system *very close* to equilibrium states.

For example, we could think of heating a pot full of cold water by throwing in, grain after grain, red-hot sand. In this way, the temperature of water will increase with each addition by a very small amount, and the overall process will pass through a succession of states very close to equilibrium.

Another example is a pure liquid that evaporates if in the surroundings the partial pressure of its vapour (p') is lower than a T-dependent value p_0 called vapour pressure. If we subtract vapour as the liquid evaporates keeping $p'=p_0-\delta p$, where δp is a positive quantity as small as desired, the process occurs through a sequence of quasi-static processes.

It can be remembered that - precisely because of the interest focused on equilibrium - it has been proposed several times without success to change the name of this discipline from thermodynamics to thermostatics. There is also, but it is not considered here, the **thermodynamics of irreversible processes**, which deals with systems out of equilibrium and their transformations using methods that are extensions of those of thermodynamics.

1.2.3 Content and Limits of Thermodynamics

For simplicity and clarity, the fundamental results of CT are typically obtained - at least initially - under relatively restricted conditions. The contradiction with the generality of the problems that CT and ST deal with is however only apparent: the treatment, in fact, can be extended - gradually and in an easy and direct way - to include progressively broader conditions, even if all possible extensions are not considered here.

Perhaps the most important aspect in this regard concerns the forces that act from outside on the system under consideration. Normally, only the effect of **hydrostatic pressure** exerted by a homogeneous fluid is treated. This is typically the Earth atmosphere, and the particularity is due to the simple fact that it is an almost unavoidable condition in applications. The inclusion of **electric** or **magnetic** fields, static or slowly variable in time, does not pose particular problems and is solved with the simple addition of the appropriate contributions, as briefly illustrated for example in Section 3.7 for the electric field and in Section 6.2.6 for the magnetic field.

More complicated is the case of the **gravitational field**, which introduces significant changes even in the formulation of some general statements about equilibrium (*e.g.* for the simple cases of Section 1.2.2). In particular, it causes a non-uniformity of the thermodynamic equilibrium functions.

At equilibrium, for example, the density of a column of gas in the Earth gravitational field is not uniform but depends on the height above sea level. Systems in the gravitational field are ignored in this textbook.

Non-trivial complications also arise when the material properties of the system under consideration are **not isotropic**. The thermodynamics of anisotropic bodies constitutes a specialist topic that is also completely ignored here.

Another case concerns the linear dimensions of the systems studied and is strictly connected both to the energy that is localized at the interface between mono-phasic regions and to the concept of phase itself. Postponing the more detailed discussion to point 1.2.6, it is anticipated here that the simplest and most usual approach of thermodynamics is valid for massive phases. The problems of surfaces, colloids and nano-systems also are a specialized subject. For some *very simple* considerations on the chemistry of surfaces and nano-systems see Sections 1.2.6, 3.3.4.2 and 6.2.4.

1.2.4 Thermodynamic State and State Variables

A thermodynamic system is made up of a large number of particles, molecules for example. Even a small amount of matter, such as a millimole, contains more than 10^{20} particles. If we wanted to specify the **mechanical state** of such a system, we would have to specify the initial conditions for each particle. It is indeed necessary to specify for each particle the position (three spatial coordinates), the moment (three other variables) and, for each non-monatomic particle, other quantities relative to the internal degrees of freedom. The number of independent microscopic variables, for even a small portion of matter, easily approaches Avogadro's constant.

The point is further clarified in relation to the postulates of ST (see Section 1.4).

On the contrary, to define the thermodynamic state of a system it is necessary to specify many **fewer macroscopic parameters** (usually: three, four or a little more) which emerge as spatial and temporal averages of the mechanical coordinates of each single particle.

The possibility of averaging over time and space is because the measurements that are generally made on a thermodynamic system are coarse-grained with respect to the molecular dimensions and slow on the atomic time scale. The derivation of the results of thermodynamics from the equations of motion of individual particles is - as already mentioned - the aim of ST. Here it is only worth noting that the meaning of some of these averages can be understood quite intuitively: for example, volume is in a certain sense the spatial average of the positions of the particles constituting the system.

Some of these macroscopic parameters are used in thermodynamics to uniquely specify the state of a system, and for this reason they are called **state variables**. State variables define a thermodynamic space in which it is possible to define thermodynamic functions. These quantities are called **state functions**, because they *depend only on the thermodynamic state* of the system, in the sense that they *do not depend on the path* that was followed to bring the system to that state.

The distinction between state variables (*i.e.*, independent variables) and state functions (which depend on the state variables) is not fixed a priori. It is generally possible to choose in various ways which are the independent variables, and which are the dependent variables, but there are some constraints. One of the constraints concerns the choice between *extensive* and *intensive* variables, as specified below. Other constraints are summarised in the *phase rule* (Section 3.3.2).

In thermodynamics it is fundamental to distinguish between **extensive** and **intensive** quantities.

- Extensive quantities depend – broadly speaking – on *how big* the system we are studying is or, more precisely, on the amounts of the different substances of the system.
- Intensive quantities do not depend on the amounts of the various substances.

Extensive quantities include, for example, the various forms of energy (U, H, G, A), the volume (V), the total mass of a system, the masses of the substances that constitute the system, that is the *components*, the amount n_i of each component i, the total amount of all components $n = \sum_{i=1}^{m} n_i$.

Intensive quantities include, for example, temperature, pressure, concentrations, which are in turn measured in various ways. More generally include the integral molar quantities and partial molar quantities which will be discussed in Section 3.2.1.

Extensive quantities are additive, in the sense that if a system at equilibrium is made up of several parts, the value relative to the entire system is the sum of the values relative to the constituent parts (see also Section 1.2.6). Furthermore, if a generic transformation can be broken down into several partial transformations, the total variation of an extensive quantity is given by the sum of the variations due to the partial transformations. Intensive variables are not additive. With the examples above: the volume and amount of substance of a system are the sums of the volumes and amounts of substance of the parts; when the system is in thermal equilibrium, the temperature is the same in the various parts. Mathematically, intensive quantities are **homogeneous functions of degree zero** and extensive quantities are **homogeneous functions of degree one** (see Section 8.1.2). To fully specify the state of a system, **at least one extensive quantity is needed**. However, if only the relationships between intensive variables are of interest, it is not necessary to know any extensive variable.

1.2.5 Thermodynamic Systems

The walls that ideally or concretely delimit a system and their permeability to substances and/or energy flows establish the fundamental classification of thermodynamic systems. If the walls are impermeable to flows of substances, the system is said to be **closed**. In the opposite case (absence of walls, or walls permeable to substances) the system is said to be **open**. If the walls, in addition to being impermeable to flows of substance, do not allow energy exchanges, the system is said to be **isolated**. Obviously, if a system exchanges matter with its surroundings, it automatically also exchanges energy.

CT and ST deal with thermodynamic equilibrium by clearly distinguishing between systems placed in different external conditions. Among all the possible cases, it is worth mentioning some that correspond to frequent experimental arrangements and to wide applications, or because they are also important in ST.

Isolated System: the quantities N_i (number of molecules) or n_i (molar amount) of the different substances, V (volume) and U (total energy) are fixed. An isolated system does not exchange matter (the N_i or n_i are fixed) or energy (either in the form of heat or in the form of work) with the surroundings. The system does not exchange work with the surroundings because mechanical work (pV) is the only type of work to be considered, since there are no electrical, magnetic or other interactions, and the exchange with the surroundings is zero since the volume is fixed. The first law and the condition of constant total energy tell us that no energy is exchanged, not even in the form of heat.

Closed Isothermal-Isochoric System: N_i (number of molecules) or n_i (molar amount) of the different substances, V (volume) and T (temperature) are fixed. A system of this type does not exchange matter with the surroundings (it is closed) and exchanges energy with the surroundings only in the form of heat (not in the form of work, because V is fixed).

Closed Isothermal-Isobaric System: N_i (number of molecules) or n_i (molar amount) of the different substances, P (pressure) and T (temperature) are fixed. A system of this type does not exchange matter with the surroundings (it is closed) and exchanges energy with the surroundings both in the form of heat and in the form of expansion work.

This is considerably more important than the previous system because in an experiment it is usually much easier to keep the pressure, rather than volume, constant.

Open Isothermal-Isochoric System: the quantities V (extensive volume) and T (temperature) are fixed, as well as one or more quantities of a type, introduced later, which goes under the name of *chemical potential* (a chemical potential - μ_i - for each substance i). It exchanges matter with the surroundings but does not exchange energy in the form of work (V is constant). The chemical potential (specified in Eq. (2.10)) is essentially the (intensive) contribution of a substance to the energetic quantities of the entire system and, as the name suggests, represents a sort of exchange energy level of a substance. In a transformation, each substance is transferred from regions (internal or external to the system) to other regions (internal or external to the system) where its chemical potential is lower.

Isothermal-Isobaric Open System: the quantities μ_i (chemical potential of the different substances), p (intensive pressure) and T (temperature) are fixed. This system exchanges matter with the surroundings and exchanges energy both in the form of heat and in the form of expansion work.

1.2.6 Phases and Heterogeneity

The concept of thermodynamic **phase** emerges as a form of generalization of the observation of natural phenomenology. In CT and ST, it is required to describe thermodynamic equilibrium when it corresponds to the subdivision of a system into *regions of space*, so that at the boundaries some *intensive thermodynamic properties show discontinuities*. The regions separated by these boundaries are named *domains* and the set of all domains with the same intensive properties is named *phase*. In other words, a phase is *not* required to occupy *a connected region of space*: it can certainly consist of *disjoint domains*, such as the grains of a precipitate.

For example, a glass of liquid water plus five ice rocks is a two-phase system.

In CT, phase is essentially a concept of empirical origin, as happens for the laws of CT themselves. For example, the observation of gas/liquid and liquid/solid transitions in pure substances leads to assigning different thermodynamic properties to the different *states of aggregation*. On this basis, it is inferred that the optimal value of the relevant *global* thermodynamic potential can correspond not only to the presence of only a solid, or only a liquid, or only a gas, but also to the simultaneous presence of regions filled by one or another phase, *i.e.* the *coexistence* of two or more phases (see Sections 3.3.3, 3.3.4.3, 3.3.6). In ST, however, it is not necessary to preliminarily assume the existence of different phases, which can instead be derived by carrying out the calculations for appropriate statistical models. Point 6.4 shows an example that corresponds to the case of Section 3.3.4.3.

A system is said to be **homogeneous** if at equilibrium there is only one phase. The term does not imply that the properties are *uniform*, as already said in Section 1.2.3 when discussing the effects of the gravitational field. If there is more than one phase, the system is said to be **heterogeneous**. Sometimes, for better clarity, the terms **mono-phasic** and **poly- (bi-, tri-, ...) phasic** are used alternatively. In particular conditions, a system made of *a pure substance* can show at equilibrium the coexistence of two or *three* phases in different states of aggregation, *e.g.* solid, liquid or gaseous. The properties that are discontinuous at the boundary between different phases can be the structure (crystalline or not) as well as physical properties such as the density. A simple example is the melting point of a substance where a solid and liquid phase coexist. In many substances it is possible to find the coexistence of two or even three *distinct crystalline* phases.

The equilibrium of two phases of different aggregation states is even more easily achieved with a system of two or more substances. An example is a solution in which crystallization or melting of the solvent may occur. Depending on T, on the composition, and marginally on p, the system has or may have only one liquid phase, or it may show the coexistence of a liquid phase and a solid phase, the latter consisting exclusively or predominantly of the solvent. A liquid solution can be also in equilibrium with a certain amount of solute in a solid precipitate.

As regards the number of possible phases in the different states of aggregation, **only one gaseous phase** is considered. In the case of isotropic liquids and systems consisting of at least two substances, liquid phases with different compositions but similar structure possibly coexist. Finally, the coexistence of crystalline phases differing in structure and symmetry as well as in composition is also possible.

It is worth remembering here that there are phases, known as **liquid crystals**, which have characteristics somewhat intermediate between those of the liquid state and those of the crystalline state, because they have an ordered anisotropic structure but low viscosity. Under the effect of gravity, they therefore assume the shape of the container. There is also the somewhat opposite situation in which a phase lacking crystalline order behaves phenomenologically as a solid due to its very high viscosity: this is then called **amorphous**. An amorphous is called **glass** when produced by rapid solidification of a liquid. A liquid crystal can be thermodynamically stable. A glass is instead *metastable*. For the clarification of thermodynamic stability, instability and metastability, see Section 3.3.5.

The value of an extensive state function of the entire system is the sum of the same extensive functions of the coexisting phases, consistently with the principles of CT (see Section 2.1 and Eq. (2.1)) and with elementary ST considerations (see Section 5.1.1). For more details on the thermodynamic functions of different phases, see Section 3.3.6.

As already said in Section 1.2.3, when there are domains of nanometric dimensions in one or more phases, **interface** (or **interphase**) **effects** occur.

Interface effects occur both between domains of different phases, such as a gas bubble in a liquid matrix, and between domains of the same phase, such as at the grain boundaries of a polycrystalline sample of a pure solid. Focusing on the interactions between structural units (molecules, ions, etc.), we expect that the structural units belonging to two adjacent domains can interact significantly only if placed within a suitable boundary zone, whose depth depends on the type of interaction. When the domains are sufficiently large in the direction orthogonal to the interface, it is reasonable to ignore the contribution due to the interaction between structural units of different domains, compared to the contribution due to the interactions between structural units belonging to the same domain. Going into detail and considering specifically the energy, the interfacial contribution to the extensive energy grows monotonically and roughly in a proportional way with the extent of the separation area between the domains, while the bulk contribution is proportional to the total volume or mass, or to the volumes or masses of the individual phases, if they are different. Comparing the two trends, it emerges that the surface contributions are negligible when the linear dimensions of the domains are largely greater than a threshold value and experiments and calculations then show that the threshold is typically in the order of tens or hundreds of nm.

When interface or *interphase effects can be neglected*, the thermodynamic (intensive) functions for each phase of the system are independent of the linear dimensions of the domains of the phase. On the contrary, when the linear dimensions of the domains occupied by one or more phases are in the order of hundreds or tens of nm or less, **the surface terms influence the values of the state functions** both of the whole system and of the various phases. The values are different from the values that the system would have if it were made only by bulk phases. This is, in a nutshell, an important explanation of the particular properties of **nano-systems**. For the sake of completeness, it should be said that the properties of nano-systems also have an intrinsic explanation due to quantum effects characteristic of every system that occupies a spatially limited region. This is the so-called *quantum confinement* effect. For example, the quantum mechanical model of the *particle in a box* produces a quantization of the energy levels inversely proportional to the square of the linear dimensions of the "box".

An extremely simplified example of how the addition of an interface effect can modify the thermodynamic equilibrium between different phases is discussed in Section 3.3.4.2.

1.3 The Laws of Classical Thermodynamics

CT was developed during the 19th and first half of the 20th centuries. During this time, it was made clear that the theory could be derived from four fundamental postulates, which are essentially empirical in nature. These postulates are called **the laws of thermodynamics**.

They are not numbered from one to four, but from zero to three for historical reasons.

Here the statements and a brief discussion of the laws are presented. The derivation of the fundamental results of chemical thermodynamics from the four laws will be addressed later (see Section 2).

1.3.1 Zeroth Law

The zeroth law (Sommerfeld and Fowler, 1935)[3] essentially postulates the existence of a state variable called **temperature** and establishes that thermal equilibrium is an equivalence relation.

By equivalence relation on a set, we mean the *partition* of the set into subsets, so that each element of the set belongs to one and only one subset. In this case this is the set of systems that are in thermal equilibrium with themselves, and therefore on these systems a temperature can be defined.

Although this partition may appear arbitrary, it allows us to label each subset with a real number, which is the temperature of the systems that are in thermal equilibrium with each other. It also allows us to identify a reference system that we can use as a *thermometer*, for defining an *empirical scale*, that is, an *operational* definition, of temperature. Once this empirical scale is established, the second law allows us to define the thermodynamic temperature (Eq. 2.8) and finally verify whether a certain empirical scale adequately approximates the thermodynamic temperature.

> If a body A and a body B are both in thermal equilibrium with a third body C, then they are also in thermal equilibrium with each other and their **temperature** is equal to that of C.

The zeroth law of thermodynamics was formulated after the other three. However, it can be shown that: (i) it cannot be derived from the other three, and (ii) it is of a more fundamental nature. Hence the numbering (the ordinals first, second and third were already occupied).

Nowadays, it is preferred to define temperature in terms of the average kinetic energy of a set of freely moving molecules, even if the average kinetic energy is a concept that belongs more properly to the kinetic theory. The thermodynamic definition remains valid in any case and does not require any assumptions about how matter is microscopically constituted (for example, we could speak of the temperature of a gas of photons, or of a colour temperature).

It can be shown that the thermodynamic temperature coincides with the ideal gas temperature: $T = pV/(nR)$.

1.3.2 First Law

The first law of thermodynamics (Hess, Joule, Mayer, Clausius and Rankine 1840–1850) consists of two parts. The first part states that

> There is an extensive state function called **internal energy**.

The second part states that

> In an isolated system the internal energy is constant.

Usually, the internal energy of thermodynamics is denoted by U to distinguish it from many other forms of energy.

The first law of thermodynamics states the conservation of energy. The law of conservation of energy can be derived from first principles. For example, in a dynamical system governed by the laws of classical mechanics, conservation of energy follows from the symmetry of the laws of motion with respect to an arbitrary (continuous) translation in time (Noether theorem). In a macroscopic system consisting of classical particles, therefore, conservation of energy is a direct consequence of the symmetry of the dynamical laws. Once again, however, we require that classical thermodynamics be valid independently of how the system is constituted from the microscopic point of view. Hence the need to introduce conservation of energy as a postulate.

1.3.3 Second Law

The second law of thermodynamics (Clausius 1850–1865) is also divided into two parts.

The first part states that

> There exists an extensive, **monotonically increasing** state function of the internal energy, called **entropy**.

Usually, entropy is denoted by S. The second part states that:

> In an isolated system entropy can only increase.

We have seen previously that in an isolated system the equilibrium state is defined as the state in which all irreversible transformations have been completed. In a system at equilibrium the state variables will have reached a constant value, dictated precisely by the equilibrium conditions. For example, if a chemical reaction takes place in the system, the equilibrium concentrations are constant. In this case, entropy, being a state function, cannot vary:

$$\Delta S = 0 \text{ at equilibrium} \tag{1.1}$$

The operator Δ is regularly used to indicate the variation of a function as a consequence of a transformation, that is, the difference "after minus before".

If the system is prepared in out of equilibrium conditions and is then isolated from its surroundings, then irreversible transformations can take place in the system that bring it to equilibrium. For example, if we still consider a system in which a reaction is taking place, then the reaction will occur spontaneously in the direction that brings the system towards equilibrium, and the concentrations of reactants and products will change from their initial values to the equilibrium values. In this case, the second law states that the entropy of the system can only increase:

$$\Delta S > 0 \text{ out of equilibrium} \tag{1.2}$$

We can summarize the two equations:

$$\Delta S \geq 0 \tag{1.3}$$

Equation (1.3) is called the **Clausius inequality** and establishes that the direction of a spontaneous transformation in an isolated system is that of increasing entropy. In a fundamental article of 1865, Clausius summarizes the first and second laws of thermodynamics with a pair of sentences that later became famous:

Die Energie der Welt ist konstant. (The energy of the universe is constant).

Die Entropie der Welt strebt einem Maximum zu. (The entropy of the universe tends to a maximum).

1.3.4 Third Law

The content of the third law (Nernst, 1907)[4] is of a more technical nature:

For a reversible transformation, the entropy change tends to zero as the temperature tends to be zero.

That is:

$$\lim_{T\to 0} \Delta S = 0 \tag{1.4}$$

This law has two fundamental implications.

1. For $T = 0$, **entropy has a constant value**, which can be set **equal to zero for a perfect pure crystalline substance and is independent of any other external parameter.**
2. It is not possible to reach zero temperature in a finite number of steps.

The lowest temperature ever reached (as of 2021) is 38 pK, *i.e.* 38×10^{-12} K (kelvin).

1.4 The Postulates of Statistical Thermodynamics

1.4.1 Microstates, Macrostates, and Thermal Quantities

The conceptual starting point of statistical thermodynamics (ST) is the observation, already mentioned, that the mechanical description is enormously more detailed than that provided by CT. We call the situation specified at the mechanical level a **microstate** and the situation specified at the thermodynamic level a **macrostate**. Let's consider the example of a simple macroscopic system, the ideal gas with N equal and mono-atomic molecules. To specify the microstate, in classical mechanics it is necessary to specify the initial conditions for $3N$ coordinates and $3N$ moments or velocities. In quantum mechanics it is necessary to specify $3N$ quantum numbers relating to the three translations of each molecule, not to mention the quantum numbers possibly necessary to specify the electronic state. A macrostate is instead specified by three independent variables. For example, a closed isothermal-isobaric system is specified by three of the variables V, p, T, and n which are linked to each other by $pV = nRT$ while an isolated system is specified by n, V, E. Let us consider the latter case.

- The quantity n determines the number of molecules (N) of the microstate: $N = nL$, where L is the Avogadro's constant.
- The mechanical quantity V determines the possible translational energies of a molecule. If we use the quantum mechanical model of the particle in a cubic 3D box: $\epsilon_i = \hbar^2 \cdot \pi^2 \cdot \left(n_{i,x}^2 + n_{i,y}^2 + n_{i,z}^2\right) / \left(2m \cdot V^{2/3}\right)$, where the integers n_i are now quantum numbers.
- In the ideal gas model, there are no interactions between molecules and therefore the total energy of the entire system of N molecules is the sum of the energies of all the molecules: $E = \sum_{i=1}^{N} \epsilon_i$.

The microstates are all those for which the energy of a single molecule is given by the formula reported above and the constraint on the total energy E is respected. The total energy does not change if, for example, the three quantum numbers n_i of the same molecule i are permuted. However, other changes of quantum numbers are also possible: for example, limiting ourselves to a single triplet, the exchanges between triplets such as 333 and 511 do not change the energy, thanks to the particular relationship between the energy and quantum numbers. More generally, energy is conserved whenever the same quantity $\Delta\varepsilon$ is added to one or more molecules that is subtracted from other molecules. Moreover, in the counting of microstates, the indistinguishability of the molecules must also be respected, a topic that will be specified later in Section 5.2.

The extension to the case of the ideal **poly-atomic** gas and again for an isolated system is straightforward: it is enough to add to the energy of a single molecule the vibrational and rotational contributions. Two microstates can now also differ by a "transfer" of energy from one type to another, *e.g.* from a vibration to a rotation.

The complete description of a macroscopic system at the level of the motions of the various molecules or atoms is prevented by the *incomplete knowledge of the initial conditions* of motion. Regardless of the possibility of solving the equations of motion, the thermodynamic variables alone do not allow us to know which microstate we are dealing with among the many microstates compatible with the external conditions of the macrostate.

The considerations made for the ideal gas example are valid in general and should make it clear that, except for extremely rare cases:

A macrostate corresponds to a **huge number** of microstates.

This relationship between micro and macrostates can be illustrated with an example: suppose we have 100 coins, and that the macrostate of this system is specified by the number of coins that show heads, while to specify the microstate it is necessary to specify for each coin whether it shows heads or tails. To determine the number of microstates, we need to know how many ways there are to choose a certain number of heads on 100 coins. This is the number of combinations and is given by the binomial coefficient. Therefore, the macrostate for which 50 coins show heads and 50 show tails, is given by $\binom{100}{50} = \frac{100!}{50!50!}$ which amounts to approximately 10^{29} microstates. If there are 1,000 coins, the number of microstates associated with the macrostate in which 500 coins show heads, and 500 show tails is aroximately 2.7×10^{299}. It is evident by this example that as the number of coins becomes larger, the number of microstates grows very quickly to huge numbers. The example is also illustrative of the exceptions mentioned above. In fact, the macrostate in which all the coins show heads corresponds *to one microstate only.*

ST essentially provides a method for **averaging** over the microstates of a macrostate:

$$\langle M\rangle \overset{\text{def}}{=} \sum_j P_j \cdot M_j, \text{ where} \tag{1.5}$$

- M_j is a particular value of the quantity M obtained for microstate (j) from the microscopic mechanical treatment,
- $\langle M\rangle$ is its mean value, and
- P_j is the statistical probability of a microstate (j) of the macrostate.

Then, $\langle M\rangle$ is the thermodynamic counterpart of the quantity M_j defined at the microstate level. It is also the quantity that in CT is simply denoted with M: the specification of mean value with angle brackets is here useful for highlighting the statistical meaning but is generally omitted. Equation (1.5) is valid for **non-thermal quantities**, that is, the quantities proper to the theory on which the statistical treatment is based: mechanics (classical or quantum) and any other physical theories such as electromagnetism. For a **thermal quantity**, one cannot write an analogue of Eq. (1.5), because the M_j are not defined in microscopic mechanical treatment. Then, the goals of ST are:

a. for non-thermal quantities: to tell how the average of Eq. (1.5) is calculated, *i.e.* what are the P_j, and
b. to provide a statistical *definition* of thermal quantities.

According to this approach, thermal quantities, first of all temperature and entropy, are meaningful only in the context of a thermodynamic treatment, that is, only for a macroscopic system made of a large number of atoms, molecules, etc. in thermal equilibrium. Here, *large* number means a number of the order of the Avogadro constant, for example. So, temperature is – essentially – a statistical quantity and it makes no sense to speak of the temperature of *one* electron, *one* atom, *one* molecule. In both CT and ST, temperature is an *essentially positive quantity.*

An average is conceptually derived from a probability distribution, and the ST can also give an evaluation of the *deviations* around the averages. In summary, the results are as follows.

- The *standard deviations* of *extensive quantities* depend on the size of the system (number of molecules, for example).
- For macroscopic systems, a *relative* standard deviation, that is the standard deviation divided by the mean value is generally negligible in comparison with experimental data, *i.e.* it is largely lower than the errors of an experimental measurement. For example, in a closed isothermal system, the relative standard deviation, $\Delta E/E$, is of the order of $\sqrt{N}/N = 1/\sqrt{N}$: for a system of 10^{18} molecules (about 1 μmol) it is therefore of the order of 10^{-9}, *i.e.* 10^{-7} %.
- ST deals little with statistical dispersions around the mean values of the various quantities precisely because dispersions are *usually* negligible.
- However, there are particular cases in which the standard deviations take values comparable with the values of the quantity. For example, this occurs at a critical point (see Section 3.3.6).

1.4.2 The Gibbs Ensemble

Formalizing the considerations made so far and following the historical approach introduced by J. W. Gibbs, the statistical treatment is built on the idea of a

> Gibbs ensemble.

that is, a mental collection of

> M replicas of the same thermodynamic system (macrostate).

Each replica corresponds to one of the microstates compatible with the macrostate and *includes the external conditions* imposed on the system. Also, each microstate appears in the ensemble with a frequency that depends – in a way that is yet to be determined – on the external conditions. The meaning of the Gibbs ensemble is clarified by the

> First postulate of the ST: the *average value* of a mechanical variable in the given thermodynamic system is *equal* to the *average over the ensemble* of the same variable at the limit $M \to \infty$ (M is the size of the ensemble).

1.4.3 Gibbs Ensembles and Thermodynamic Potentials

ST, like CT, deals with the equilibrium of systems placed in different external conditions. The basic treatment of ST derives probabilities (Eq. 1.5) for various important cases, which are partly the same as those seen in Section 1.2.5: *isolated system, closed isothermal-isochoric system* and *open isothermal-isochoric system.* Compared to the list of Section 1.2.5, this one lacks the cases in which the fixed state variable is the *pressure.* The reason is that it is particularly difficult *in a calculation* to keep the pressure constant (it would be necessary to keep constant the moment exchanged by all the molecules in their collisions with the walls). It is of course possible to consider other Gibbs ensembles, but those listed above are the most used cases in ST.

For this reason, specific names have been given to the corresponding Gibbs ensembles:

> Isolated system ⇒ micro-canonical ensemble
> Closed isothermal-isochoric system ⇒ canonical ensemble
> Open isothermal-isochoric system ⇒ grand-canonical ensemble.

Moreover, if the system is not made by one substance, as tacitly assumed so far, but by several substances, then N must be replaced by a set of N_i and μ must be replaced by a set of μ_i.

To consider interactions with the external surroundings in addition to the pV expansion or compression work, other extensive variables may be needed in addition to V. In the presence of external electric fields, for example, the electric charge should be specified. For a more detailed discussion of the various forms of work in thermodynamics, see Section 2.1.

1.4.4 Equal a Priori Probability and Ergodic Hypothesis

The second postulate of the ST deals with the distribution of microstates in the Gibbs ensemble and the probabilities to be assigned to the various microstates in the average of Eq. (1.5). The probabilities depend on the type of external conditions (isolated system, closed and isothermal system, open system, ...) and **the second postulate is specific for the isolated system**, *i.e.* for the **micro-canonical ensemble**.

The ergodic hypothesis is certainly well known:

> An **isolated** system spends its time **equally** among **all** its possible microstates.

The ergodic hypothesis can be directly derived from the most useful form of the postulate:

> In a micro-canonical ensemble, *i.e.* the ensemble of an isolated system, and in the limit as M approaches infinity ($M \to \infty$, M=size of the ensemble), the systems are distributed with **equal probability over all the microstates** corresponding to the macrostate.

It should be emphasized again that

> The second postulate is valid only for **isolated** systems.

2 CLASSICAL THERMODYNAMICS

2.1 Energy, Heat, and Work

The presentation of classical thermodynamics (CT) starts with the first law, the thermodynamic function *energy* and the relationships between *energy, heat,* and *work*.

We shall think of the internal energy of a system as *the sum of all the forms of energy of the system, which are taken as independent of the reference frame.* The statement is equivalent to say that the potential and kinetic energies of the centre of mass of the system *do not* contribute to the internal energy. Considering a system at rest at the sea level and considering the same system in the pressurized and thermostated cabin of a plane flying at 800 km/h at 10,000 m of altitude, the internal energy U does change. Obviously, this statement is valid for the most common cases and should be reconsidered if we are willing to study, for example, the distribution of molecular velocities in a gas column 10 km high above sea level.

As said (1.1) the different forms of energy here referred to are defined in other branches of physics, for which thermodynamics acts as a *superstructure*. The first law (Section 1.3.2) defines internal energy as an extensive quantity (Section 1.2.4). For a system consisting of two parts, the total internal energy of the system is the sum of the internal energies of the two parts.

$$U = U_1 + U_2 \tag{2.1}$$

We can further observe, generalizing our experience, that the internal energy of a system **can be changed** in two different ways. Take as an example a pot of water. We are going to heat the water. An obvious way is to turn on the stove under the pot. We have in any case to consider an alternative and less obvious way. We can attach some strings and pulleys to a paddle wheel inside the water: weights attached to the strings fall, the paddle wheel starts to turn, and the result is that water warms up. This method is known as Joule's paddle wheel, and posed the historical basis for the first law, together with the observation by Rumford of the heat produced during cannon boring. This is a fundamental difference: in the former way, when the stove is turned on, *we cannot observe any ordered motion in the surroundings.* This way is called heat (Q). In the latter way (Joule's paddle wheel) *we can indeed observe an ordered motion in the surroundings (weights fall vertically).* This way is called **work** (W).

The difference between heat and work *concerns the surroundings*: as far as the system is concerned, it makes no difference at all if we change the internal energy by adding (or subtracting) heat or work, as internal energy is a state function. Any given variation of internal energy ΔU can be realized by any combination of exchanges of heat and work. In other words, while internal energy is a state function, work and heat are not (they are **path functions**). This fact is expressed by:

$$\Delta U = Q + W \tag{2.2}$$

For a system surrounded by thermally insulating (*adiabatic*) walls, there are no heat exchanges and $\Delta U = W$. Note that the Δ operator is not required for Q and W because these are exchanged quantities by definition. The differential form of Eq. (2.2) is:

$$\mathrm{d}U = đQ + đW \tag{2.3}$$

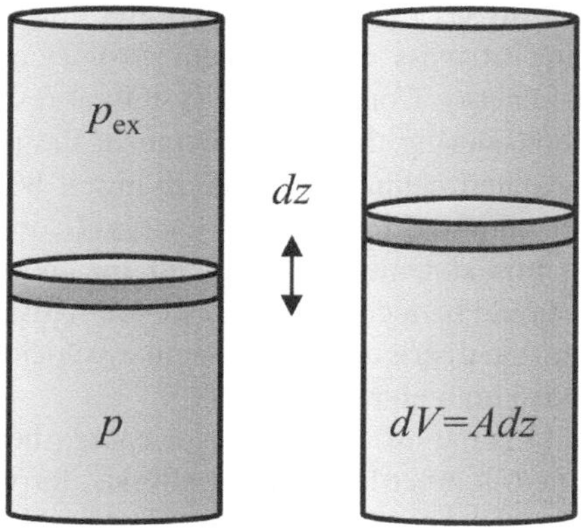

FIGURE 3.1 Work done by a gas expansion: calling p_{ex} the external pressure, the force exerted by the gas onto the piston is $|F| = p_{ex}A$, being A the area of the piston. If the piston is moved by dz, the work is: $đW = -p_{ex}Adz = -p_{ex}dV$.

We use two distinct notations to make clear the difference between the differential dU of the state function U (which is called *exact* differential) and the *differential forms* $đQ$ e $đW$, which are often called *inexact differentials or imperfect differentials.* Indeed, the different notation reflects the fact that heat and work are *not* state functions. As an example, a possible differential form for $đW$ is $đW=-pdV$, as we shall see in the following. It is not possible to define a function X of which the differential dX is $-pdV$. For instance, if we define $X=-pV$, then $dX=-pdV-Vdp$) (Figure 3.1).

For reasons that will be made explicit in the following, in thermodynamics it is convenient to make a distinction between mechanical work (expansion or compression) and *all other possible forms* of work. Considering now that $đW$ in Eq. (2.3) contains only expansion work, and the transformation is reversible *i.e.*, the internal gas pressure – which is a property of the system - is always equal to the external pressure:

$$dU = đQ - pdV \quad \text{(see Figure 3.1)} \tag{2.4}$$

Equation (2.4) assumes as *positive* all the exchanges that *increase* the internal energy of the system. This is the preferred convention in physics and chemistry, where attention is focused on the system. In engineering, attention is instead focused on how much work can be obtained for an internal transformation of an engine, so that the other convention is preferred. It is also preferred in electrochemistry (see Section 3.7).

In an adiabatic transformation, the exchanged heat is null, $đQ = 0$, and $dU = -pdV$. For a reversible and isothermal (= constant temperature) transformation of an ideal gas, for which $pV=nRT$, the expansion work is:

$$W = \int_{V_1}^{V_2} -pdV = -nRT\int_{V_1}^{V_2}\frac{dV}{V} = -nRT\ln\frac{V_2}{V_1}$$

Two notes.

1. In the previous equations, work is expressed as the product of an intensive variable (p) and a variation of an extensive variable (V). This is of general validity: variations of extensive functions of state are always expressed as products of intensive variables and variations of extensive variables (as in $-pdV$ as above), or as products of extensive variables and variations of intensive variables (for example, as in $dG_p=-SdT$, see Eq. (3.37))

Work	đW	Comments
Mechanical (volume expansion)	$-pdV$	p=pressure (intensive), V= volume (extensive)
Mechanical (surface expansion)	$-\gamma d\sigma$	γ =surface tension (intensive), σ =surface area (extensive)
Electric	ϕdQ or $Qd\phi$	ϕ=electric potential (intensive), Q=electric charge (extensive)
Magnetic	HdM	H= magnetic field (intensive), M=magnetization (extensive[5])

2. Work can always be written as the product of two vectors: mechanical work, for instance, is the dot product of a force and a displacement.

2.2 General Equilibrium Conditions

The second law concerns only the equilibrium of an *isolated system.* In this case, equilibrium corresponds to the *maximum* of entropy in the space of thermodynamic state variables. We now need to find – starting from the laws of thermodynamics – the relations between thermodynamic quantities that define the equilibrium conditions in the **general case**. According to the second law, entropy is a function of internal energy, and the first law guarantees that internal energy is constant for an isolated system. In an isolated system V is constant and all the amounts of substance of the different components of the system that we shall write as $\{n_i\}$ are also constant. Therefore, for the *equilibrium state in an isolated system* we must have:

$$\begin{cases} dS_U = 0 \\ d^2S_U < 0 \end{cases} \tag{2.5}$$

2.2.1 The Energy Function

Let us consider in some detail the internal energy U as a function of state. It is an extensive function and therefore it depends on the $\{n_i\}$, and on volume as well (see Eq. (2.4)). As entropy is a monotonously increasing function of U, we can invert the relation and obtain U as a function of S. Therefore:

$$\begin{cases} U = U(S,V,\{n_i\}) \\ S = S(U,V,\{n_i\}) \end{cases} \tag{2.6}$$

Equation (2.6) are a form of **fundamental** thermodynamic relation.

Taking the differential form of the first of Eq. (2.6) we have:

$$dU=\left(\frac{\partial U}{\partial S}\right)_{V,n_j} dS+\left(\frac{\partial U}{\partial V}\right)_{S,n_j} dV+\sum_{i=1}^{N}\left(\frac{\partial U}{\partial n_i}\right)_{V,S,n_k\neq n_i} dn_i \tag{2.7}$$

Here, n is the number of components of the system and the derivatives $\left(\frac{\partial U}{\partial S}\right)_{V,n_j}$ and $\left(\frac{\partial U}{\partial V}\right)_{S,n_j}$ are made keeping constant the amounts of *all components* (all n_j), while each derivative $\left(\frac{\partial U}{\partial n_i}\right)_{V,S,n_j\neq n_i}$ is evaluated keeping constant the amounts of all components other than i.

We now assign a physical meaning to each of the partial derivatives in Eq. (2.7). To this aim, it is convenient to consider a system which is isolated as a whole, but is divided into two subsystems, that in turn are separated by a wall. Then, we analyse the three different cases where only a single exchange occurs between the two subsystems.

Let us start by considering that the wall allows only heat exchanges (**diathermic or diabatic wall**). Keeping constant volume and all the amounts of substance, we have:

$$dU_{V,n_j}=\left(\frac{\partial U}{\partial S}\right)_{V,n_j} dS$$

As the whole system is isolated, its internal energy is constant. Therefore:

$$dU_{V,n_j}=dU(1)_{V,n_j}+dU(2)_{V,n_j}$$
$$=\left(\frac{\partial U(1)}{\partial S}\right)_{V,n_j} dS(1)+\left(\frac{\partial U(2)}{\partial S}\right)_{V,n_j} dS(2)=0$$

where (1) and (2) refer to subsystem 1 and 2, respectively. Given that we are considering a system which, as a whole, is in equilibrium, the second law ensures that entropy cannot change, and therefore $dS=dS(1)+dS(2)=0$, or $dS(1)=-dS(2)$. This last equation is an example of **detailed balance**: at *equilibrium, each process is equilibrated by the reverse process*. This condition is assured by the uniqueness of thermodynamic equilibrium, and by the condition that equilibrium is unchanged by time reversal. Substituting, we have:

$$\left(\frac{\partial U(1)}{\partial S}\right)_{V,n_j}=\left(\frac{\partial U(2)}{\partial S}\right)_{V,n_j}$$

Therefore, for the equilibrium state, the partial derivative of internal energy with respect to entropy, at constant volume and amount of substance, is uniform in the system. As the wall allows only heat exchanges, the equilibrium state is a state of thermal equilibrium. This derivative, therefore, has the property that the zeroth law assigns to temperature (see Section 1.3.1), and we are thus allowed to identify the derivative with temperature. In addition, we are taking the derivative of a state function and using the first and second laws: this is a *thermodynamic definition* of temperature, and not the simple empiric quantity referred to by the zeroth law.

The first result is therefore:

$$\left(\frac{\partial U}{\partial S}\right)_{V,n_j}\overset{\text{def}}{=}T \tag{2.8}$$

Let us now consider the same system and assume that it is globally isolated, and that thermal equilibrium is already attained. If the two subsystems are now separated by an insulating but **freely moving wall**, we have to consider possible variations in volume:

$$dU_{S,n_j}=dU(1)_{S,n_j}+dU(2)_{S,n_j}$$
$$=\left(\frac{\partial U(1)}{\partial V}\right)_{S,n_j} dV(1)+\left(\frac{\partial U(2)}{\partial V}\right)_{S,n_j} dV(2)=0$$

As the total volume is constant, $dV=dV(1)+dV(2)=0$. Therefore, mechanical equilibrium is attained when the partial derivative of the internal energy with respect to volume is uniform within the system. This derivative has the physical meaning of **pressure**.

Comparison with Eq. (2.4) gives:

$$\left(\frac{\partial U}{\partial V}\right)_{S,n_j}\overset{\text{def}}{=}-p \tag{2.9}$$

Finally, we shall consider that now thermal and mechanical equilibrium are already attained, and the wall is adiabatic, fixed, but **permeable** to the flow of the i-component.

$$dU_{S,V,n_{j\neq i}}=dU(1)_{S,V,n_{j\neq i}}+dU(2)_{S,V,n_{j\neq i}}$$

$$=\left(\frac{\partial U(1)}{\partial n_i}\right)_{S,V,n_{j\neq i}} dn_i(1)+\left(\frac{\partial U(2)}{\partial n_i}\right)_{S,V,n_{j\neq i}} dn_i(2)=0$$

The total amount of substance of the i-th component cannot change, as the system is isolated and then, *a fortiori*, closed. Then: $dn_i(1)+dn_i(2)=0$. The equilibrium condition is now that the partial derivative of internal energy with respect to the amount of substance of the generic i-th component is uniform in the system. Each partial derivative of U with respect to the amount of a substance (n_i) is known as the **chemical potential** (μ_i):

$$\left(\frac{\partial U}{\partial n_i}\right)_{S,V,n_{j\neq i}}\overset{\text{def}}{=}\mu_i \tag{2.10}$$

Summarizing, Eqs. (2.8), (2.9) and (2.10) are the *thermodynamic definitions of* **temperature, pressure,** and **chemical potential.** Taking these definitions into account, the fundamental Eq. (2.7) is rewritten as:

$$dU = TdS - pdV + \sum_{1}^{N} \mu_i dn_i \tag{2.11}$$

2.2.2 The Entropy Function

For a *closed system*, Eq. (2.11) can be written as: $dU = TdS - pdV$. Considering now a reversible transformation, Eq. (2.4) gives $dU + pdV = đQ_{rev}$. Therefore:

$$dS = \frac{đQ_{rev}}{T} \quad \text{or} \quad \Delta S = \int_{T_1}^{T_2} \frac{đQ_{rev}}{T} \tag{2.12}$$

Equation (2.12) *allows to determine entropy variations* from calorimetric measurements.

It is worth noting that dividing by T the reversibly exchanged heat gives an exact differential of a function of state (S). We say that $1/T$ is an *integrating factor of heat.*

The absolute values of entropy are known once a proper reference state is fixed. Hence the need of introducing the third law, stating that the entropy of a pure and perfect crystalline solid is zero for $T \to 0$ (see Section 1.3.4).

For an irreversible transformation: $dS > \frac{đQ_{rev}}{T}$. So, generally:

$$dS \geq \frac{đQ_{rev}}{T} \tag{2.13}$$

Equation (2.13) is a form of **Clausius inequality**. In an isolated system, the exchanged heat is zero, and therefore: $dS \geq 0$ or $\Delta S \geq 0$. These last inequalities are equivalent to Eqs. (1.1)–(1.3), and they give a synopsis of the second law.

2.2.3 The Fundamental Equations

The first Eq. (2.6) states that internal energy is a function of entropy, volume, and all amounts of substance. Internal energy is an extensive function, and therefore the Euler theorem on homogenous functions allows us to write a different form for the fundamental equation of thermodynamics.

$$\begin{aligned} U &= S \cdot \left(\frac{\partial U}{\partial S}\right)_{V,n_i} + V \cdot \left(\frac{\partial U}{\partial V}\right)_{S,n_i} + \sum_i n_i \cdot \left(\frac{\partial U}{\partial n_i}\right)_{S,V,n_{j \neq i}} \\ &= TS - pV + \sum_i \mu_i n_i \end{aligned} \tag{2.14}$$

Then, pressure and temperature can be expressed as the derivatives of internal energy. Internal energy, in turn, is a function of entropy, volume and amounts of substance. Therefore, it is possible to express pressure and temperature as functions of entropy, volume and amounts of substance:

$$\begin{cases} p = p(S,V,n_i) \\ T = T(S,V,n_i) \end{cases}$$

For convenience, here and in the following we indicate the set of all the amounts of substance by writing only the generic quantity n_i.

2.2.4 The State Equation

Eliminating entropy from the two equations above, we can obtain a function relating to each other pressure, temperature, volume and amounts of substance:

$$\Phi(V,T,p,n_i) = 0 \tag{2.15}$$

Equation (2.15) is an **equation of state of the system**. A well-known example is the ideal gas law: $pV - nRT = 0$.

2.3 Thermodynamic Potentials

The fundamental equations (Eqs. 2.6 and 2.14) express the functional dependence of internal energy U on entropy, volume and amounts of substance, and the dependence of entropy S on internal energy, volume and amounts of substance. These equations, therefore, contain all the information needed to specify the thermodynamic state of a system. The direct use of these equations is, however, hampered by two main difficulties:

a. In most cases, using *volume* as an independent variable is challenging, as *experimentally controlling the volume* is not an easy task: it is much simpler to control *pressure.*
b. For an isolated system, the sign of entropy variations indicates the direction of spontaneous evolution of a system, and the maximum of entropy corresponds to the equilibrium conditions. This is, in general, *not valid* for *systems* that are *not isolated.* Then, we should consider the variation of entropy of **both the system and its surroundings.** However, this is not what we generally want. What we are interested in is finding the equilibrium conditions **considering only the properties of the system.**

Our task is therefore to find **new functions of state** capable of overcoming the limitations of Eqs. (2.6) and (2.14). The task is accomplished using the Legendre transformation (see Section 8.1.3 for technical details). In the following (Sections 2.3.1 and 2.3.2) we will derive the new functions of state, while in Section 2.3.4 we will find the new equilibrium conditions expressed by these functions. We will show that the new functions of state are **thermodynamic potentials**, in the sense that *thermodynamic equilibrium corresponds to the minimum of these functions under proper external conditions.*

2.3.1 Enthalpy

We are now going to derive from the internal energy U a new function of state, where the dependence on volume is replaced by the dependence on pressure.

Neglecting the negative sign, pressure is the partial derivative of internal energy with respect to volume, so we can use the Legendre transformation (Eq. 8.2):

$$f(x,y) \leftarrow U(V,S,\{n_i\}); \quad \left(\frac{\partial f}{\partial x}\right)_y = u \leftarrow \left(\frac{\partial U}{\partial V}\right)_{S,n_i} = -p$$

$$\text{and} \quad g(x,z) = f(x,y) - ux \leftarrow U + pV$$

The result is a function of state $\boldsymbol{H}$, called **enthalpy** and defined by:

$$H \stackrel{\text{def}}{=} U + pV \tag{2.16}$$

By considering the differential $\mathrm{d}H = \mathrm{d}U + p\mathrm{d}V + V\mathrm{d}p$ and substituting $\mathrm{d}U = T\mathrm{d}S - p\mathrm{d}V + \sum_1^N \mu_i \mathrm{d}n_i$ we have $\mathrm{d}H = T\mathrm{d}S + V\mathrm{d}p + \sum_1^N \mu_i \mathrm{d}n_i$. Therefore: $H = H(S, p, n_i)$.

2.3.2 Helmholtz and Gibbs Energies

Using the same procedure, we can eliminate the dependence on entropy from internal energy and enthalpy. In this way, two new functions of state are obtained:

$$A \stackrel{\text{def}}{=} U - TS \tag{2.17}$$

$$G \stackrel{\text{def}}{=} H - TS \tag{2.18}$$

A and G are called **Helmholtz energy** and **Gibbs energy**, respectively. Historically, these functions were also known as *Helmholtz free energy* and *Gibbs free energy*, or, in the latter case, simply *free energy*. This terminology is now rejected by IUPAC.

The differentials are:

$$dA = -p\mathrm{d}V - S\mathrm{d}T + \sum_1^N \mu_i \mathrm{d}n_i \tag{2.19}$$

$$\mathrm{d}G = V\mathrm{d}p - S\mathrm{d}T + \sum_1^N \mu_i \mathrm{d}n_i \tag{2.20}$$

Therefore: $A = A(V, T, n_i)$ and $G = G(p, T, n_i)$.

Helmholtz and Gibbs energies have an **important direct physical meaning.** Indeed, the *reversible isothermal* work produced by a system is (see Eq. (2.17)):

$$đW = dU - T\mathrm{d}S = \mathrm{d}A \tag{2.21}$$

Reversible work is also the maximum amount of work we can *obtain* from a system. Therefore:

> The variation of Helmholtz energy is the maximum amount of work that can be obtained from a system reversibly and in isothermal conditions.

This explains the A symbol for Helmholtz energy: it is the initial letter of the German word *Arbeit* (work). In a similar way, considering the **useful work** W_u, *i.e.*, the amount of *work other than the expansion work against an external pressure*:

$$\mathrm{d}G = V\mathrm{d}p - S\mathrm{d}T + \sum_1^N \mu_i \mathrm{d}n_i$$

$$đW_u = đw - (-p\mathrm{d}V) = \mathrm{d}U - T\mathrm{d}S + p\mathrm{d}V = \mathrm{d}H - T\mathrm{d}S = \mathrm{d}G \tag{2.22}$$

> The variation of Gibbs energy is the **maximum useful work** that can be obtained from a system reversibly and in isothermal conditions.

When the system does an amount of work on its surroundings (it can be mechanical expansion work pV, or another form of work) A *decreases*: $A_{\text{final}} < A_{\text{initial}}$. The reversible work $W = A_{\text{final}} - A_{\text{initial}}$ is negative, in agreement with the already said convention that considers as *negative* the energy amounts that are *lost* by the system. When speaking of maximum work, we then mean that the absolute value of $W\left(=|W|\right) = |A_{\text{final}} - A_{\text{initial}}|$ is the greatest amount of work that can be *obtained* from the system. The actual work can be less, even zero, due to irreversibility losses. Considering the special case of mechanical work pV, then, in reversible, isothermal, and isobaric conditions $0 > W = -p\Delta V$, meaning that $\Delta V > 0$ *i.e.*, we are dealing with *expansion* work against p. The same line of thinking also applies if we change the signs. When the surroundings do some work on the system, the thermodynamic functions of the system *increase*: $A_{\text{final}} > A_{\text{initial}}$ and the reversible work $W = A_{\text{final}} - A_{\text{initial}}$ is positive, in agreement with the convention that considers as *positive* the energy amounts that are *gained* by the system. When reversible conditions apply, $W = A_{\text{final}} - A_{\text{initial}}$ reaches its maximum. W can be lower or even null because of irreversibility. Again, considering the particular case of mechanical work pV, and considering again reversible, isothermal, and isobaric conditions $-p\Delta V > 0$, meaning that $\Delta V < 0$ *i.e.*, we are dealing with *compression* work. The same considerations apply to G with the difference that now we are dealing with useful $W_u = G_{\text{final}} - G_{\text{initial}}$.

An important warning: a different convention is used in **electrochemistry**. For an electrochemical cell, we take as positive the amounts of energy that are *produced by the system*: $W_{\text{ele}} = -(G_{\text{final}} > G_{\text{initial}})$. This is better discussed in Section 3.7.

2.3.3 Maxwell Relations

In summary, for the differentials of energy and derived quantities we have the following expressions:

$$\mathrm{d}U = \left(\frac{\partial U}{\partial S}\right)_{V,n_i} \mathrm{d}S + \left(\frac{\partial U}{\partial V}\right)_{S,n_i} \mathrm{d}V + \sum_i \left(\frac{\partial U}{\partial n_i}\right)_{S,V,n_{j\neq i}} \mathrm{d}n_i \tag{2.23}$$

$$\mathrm{d}H = \left(\frac{\partial H}{\partial S}\right)_{p,n_i} \mathrm{d}S + \left(\frac{\partial H}{\partial p}\right)_{S,n_i} \mathrm{d}p + \sum_i \left(\frac{\partial H}{\partial n_i}\right)_{S,p,n_{j\neq i}} \mathrm{d}n_i \tag{2.24}$$

$$\mathrm{d}A = \left(\frac{\partial A}{\partial T}\right)_{V,n_i} \mathrm{d}T + \left(\frac{\partial A}{\partial V}\right)_{T,n_i} \mathrm{d}V + \sum_i \left(\frac{\partial A}{\partial n_i}\right)_{T,V,n_{j\neq i}} \mathrm{d}n_i \tag{2.25}$$

$$\mathrm{d}G = \left(\frac{\partial G}{\partial T}\right)_{p,n_i} \mathrm{d}T + \left(\frac{\partial G}{\partial p}\right)_{T,n_i} \mathrm{d}p + \sum_i \left(\frac{\partial G}{\partial n_i}\right)_{T,p,n_{j\neq i}} \mathrm{d}n_i \tag{2.26}$$

So, the definitions of pressure, volume, temperature, entropy and chemical potential are:

$$p = -\left(\frac{\partial U}{\partial V}\right)_{S,n_i} = -\left(\frac{\partial A}{\partial V}\right)_{T,n_i} \tag{2.27}$$

$$V = \left(\frac{\partial H}{\partial p}\right)_{S,n_i} = \left(\frac{\partial G}{\partial p}\right)_{T,n_i} \tag{2.28}$$

$$T = \left(\frac{\partial U}{\partial S}\right)_{V,n_i} = \left(\frac{\partial H}{\partial S}\right)_{p,n_i} \tag{2.29}$$

$$S = -\left(\frac{\partial A}{\partial T}\right)_{V,n_i} = -\left(\frac{\partial G}{\partial T}\right)_{p,n_i} \tag{2.30}$$

$$\mu_i = \left(\frac{\partial U}{\partial n_i}\right)_{S,V,n_{j\neq i}} = \left(\frac{\partial H}{\partial n_i}\right)_{S,p,n_{j\neq i}} = \left(\frac{\partial A}{\partial n_i}\right)_{T,V,n_{j\neq i}} = \left(\frac{\partial G}{\partial n_i}\right)_{T,p,n_{j\neq i}} \tag{2.31}$$

The previous equations Eqs. (2.23)–(2.31) allow us to derive useful relationships between thermodynamic quantities. For example, if we rewrite Eq. (2.23) *keeping constant the amounts of the various substances* and considering the definitions of Eq. (2.27) and (2.29), we obtain

$$\mathrm{d}U = T\mathrm{d}S - p\mathrm{d}V \tag{2.32}$$

Now considering that dU is an exact differential and therefore that the two mixed second order partial derivatives of U must be equal, we obtain:

$$\left(\frac{\partial T}{\partial V}\right)_S = -\left(\frac{\partial p}{\partial S}\right)_V \tag{2.33}$$

This is known as a **Maxwell** relation. Similarly, we can get that:

$$\left(\frac{\partial T}{\partial p}\right)_S = \left(\frac{\partial V}{\partial S}\right)_p \tag{2.34}$$

$$\left(\frac{\partial S}{\partial V}\right)_T = \left(\frac{\partial p}{\partial T}\right)_V \tag{2.35}$$

$$-\left(\frac{\partial S}{\partial p}\right)_T = \left(\frac{\partial V}{\partial T}\right)_p \tag{2.36}$$

In Eqs. (2.33)–(2.36) the condition that the quantities of substance are constant is omitted for simplicity of notation. Maxwell relations (Eqs. 2.33–2.36) allow us to directly obtain many important properties of the state functions and in particular their dependence on the state variables variously chosen. For example, for the dependence of U on the volume we have:

$$\begin{aligned}\left(\frac{\partial U}{\partial V}\right)_T &= \left(\frac{\partial U}{\partial S}\right)_V \left(\frac{\partial S}{\partial V}\right)_T + \left(\frac{\partial U}{\partial V}\right)_S \\ &= T\left(\frac{\partial S}{\partial V}\right)_T - p = T\left(\frac{\partial p}{\partial T}\right)_V - p\end{aligned} \tag{2.37}$$

In the particular case of an ideal gas: $p = \frac{nRT}{V}$ and $\left(\frac{\partial p}{\partial T}\right)_V = \frac{nR}{V}$, and therefore $T\left(\frac{\partial p}{\partial T}\right)_V - p = T\frac{nR}{V} - p = 0.$

At constant temperature, the internal energy of an ideal gas does not depend on the volume. Similarly, we can show that it does not depend on the pressure either (see Exercise 2.2). It is therefore shown that the internal energy of an ideal gas *depends only on the temperature.*

2.3.4 Equilibrium at Constant *p* or *V*

Holding the volume constant, Eq. (2.4) becomes:

$$\mathrm{d}U = đQ_v \tag{2.38}$$

In other words, *at constant volume, the heat exchanged by the system is equal to the variation of* a state function, the **internal energy *U*.**

At constant pressure, from the definition of enthalpy Eq. (2.16) we have that:

$$\mathrm{d}H = \mathrm{d}U + p\mathrm{d}V = đQ_p \tag{2.39}$$

So, even *at constant pressure the heat exchanged by the system is equal to the variation of a state function*, which is now **the enthalpy *H*.**

Let us now consider a system at constant pressure, which is in contact with its surroundings by diathermal walls. Let us also consider the complex system+surroundings as globally isolated. It is assumed that the surroundings of a system are large enough so that the heat transferred from the system to the surroundings (or vice versa) does not change the temperature of the surroundings. We therefore have at equilibrium:

$$dS_{tot} = dS_{syst} + dS_{amb} = dS_{syst} - \frac{dH_{syst}}{T} = \frac{-dG_{syst}}{T} \quad (2.40)$$

where we have considered that at equilibrium the temperature of the system and the surroundings are equal, that the heat exchanged by the system at constant pressure is equal to the change in enthalpy, and finally that when an amount of heat $đQ_p$ goes from the surroundings to the system, then an amount of heat $-đQ_p$ goes from the system to the surroundings. But at equilibrium Eq. (2.5) written for the global entropy (of system+surroundings) must be valid. Therefore, *for the system* the following applies:

$$\begin{cases} dG_{p,T} = 0 \\ d^2G_{p,T} > 0 \end{cases} \quad (2.41)$$

Equation (2.41) expresses the equilibrium conditions at constant pressure and temperature: the Gibbs energy must be minimal. Furthermore, we see that Eq. (2.40), rewritten as $dG_{syst} = dH_{syst} - TdS_{syst}$, allows us to focus our **attention only on variations of state functions of the system**.

As regards the **Helmholtz energy**, it can be similarly derived (just replace $dH_{syst} \to dU_{syst}$ and $dG \to dA$ in Eq. (2.40)) that **at constant temperature and volume** the equilibrium conditions are:

$$\begin{cases} dA_{V,T} = 0 \\ d^2A_{V,T} > 0 \end{cases} \quad (2.42)$$

In applications, we usually limit ourselves to considering the extremum condition; for a discussion on the meaning of the second derivatives, *i.e.* the second parts of Eqs. (2.41) and (2.42), see point 3.3.5.

3 EQUILIBRIA IN CHEMICAL SYSTEMS

After having derived the fundamental relations of thermodynamic equilibrium, we now discuss the various particular equilibria, or at least those of greatest interest to chemistry, namely chemical equilibrium, phase equilibrium and their combination. For this purpose, it is evidently necessary to study in more detail the dependence of the thermodynamic functions on the state variables involved in the various situations.

The order of presentation of the various topics is as follows. First, the dependence on T and p of the thermodynamic functions of pure substances in single-phase systems is discussed. Then, single-phase systems containing several non-interacting substances, reactions in homogeneous phases, and phase transitions are discussed. A further step regards heterogeneous equilibria, *i.e.* systems in which both chemical reactions and phase transformations occur. Some transformations that extend the notion of chemical reaction in various ways are also discussed.

3.1 A Pure Substance in a Single Phase

3.1.1 Temperature Dependence of the Thermodynamic Functions

The starting point is given by Eqs. (2.38) and (2.39), from which we know that, at constant volume, the heat exchanged by the system is equal to the variation in internal energy, while at constant pressure it is equal to the variation in enthalpy:

$$\Delta U = Q_V \quad \text{and} \quad \Delta U = \Delta H = Q_p \quad (3.1)$$

3.1.1.1 Heat Capacities

The heat capacity is defined as the limit for $\Delta T \to 0$ of $Q/\Delta T$. Q is the heat absorbed by a certain amount of substance and ΔT is the corresponding temperature increase:

$$C = \lim_{\Delta T \to 0} \frac{Q}{\Delta T} \quad (3.2)$$

From Eq. (3.1) it follows that it is necessary to distinguish between heat capacities at constant volume or at constant pressure:

$$C_V = \left(\frac{\partial U}{\partial T}\right)_V \quad \text{and} \quad C_p = \left(\frac{\partial H}{\partial T}\right)_p \quad (3.3)$$

C_V and C_p are **extensive** quantities and are always positive. In fact, it is sufficient to consider that the internal energy is a monotonically increasing function of the temperature. Therefore, in Eqs. (3.2) and (3.3) the exchanged heat is always positive if we speak of a T increase: $Q>0$ if $\Delta T>0$.

C_p is always greater than C_V and this fact has an intuitive interpretation. At constant volume, the expansion work is zero, so all the heat exchanged by the system is translated into temperature variations. At constant pressure, instead, part of the heat exchanged by the system can be translated into expansion work, and therefore it is to be expected that the temperature variations in these conditions are smaller. The determination of the heat capacities (experimentally or with ST) allows to obtain the temperature dependence of U and H:

$$\Delta U \stackrel{\text{def}}{=} U(T_2) - U(T_1) = \int_{T_1}^{T_2} C_V dT$$

$$\Delta H \stackrel{\text{def}}{=} H(T_2) - H(T_1) = \int_{T_1}^{T_2} C_p dT \quad (3.4)$$

Here, we are always talking about a single-phase system. If the pure substance shows a phase transition, the heat involved at constant pressure and temperature is $\Delta_{trs}H$ (cfr. Section 3.3.1). For example, in the case of fusion we will have $\Delta_{fus}H$. The experimental determination of heat exchanged in various situations (*i.e.* the determination of heat capacities, heats of phase transition, reaction, solution) is an important specialist sector of chemical thermodynamics: **calorimetry**. The determination of heat capacities, in particular, is usually performed at constant pressure on substances in the condensed phases, while on substances in the gaseous phase it is preferably performed at constant volume. However, it is not difficult to obtain one from the other, since the two quantities are related by a very important general thermodynamic relation:

$$C_P - C_V = \frac{\alpha^2 VT}{\beta} \tag{3.5}$$

Consider the Clausius inequality (Eq. 2.13). For a reversible transformation $dS = \frac{\delta Q_{rev}}{T}$. We need to express how entropy varies with temperature and volume variations:

$$dS = \left(\frac{\partial S}{\partial T}\right)_V dT + \left(\frac{\partial S}{\partial V}\right)_T dV \quad \text{Then: } C_V = T\left(\frac{\partial S}{\partial T}\right)_V \tag{3.6}$$

Analogously: $C_p = T\left(\frac{\partial S}{\partial T}\right)_p$.

Then, we can express the changes in volume following changes in temperature and pressure as:

$$dV = \left(\frac{\partial V}{\partial T}\right)_p dT + \left(\frac{\partial V}{\partial p}\right)_T dp$$

by substitution in Eq. (3.6):

$$dS = \left(\frac{\partial S}{\partial T}\right)_V dT + \left(\frac{\partial S}{\partial V}\right)_T \left(\frac{\partial V}{\partial T}\right)_P dT + \left(\frac{\partial V}{\partial T}\right)_T \left(\frac{\partial V}{\partial p}\right)_T dp$$

$$= \left[\left(\frac{\partial S}{\partial T}\right)_V + \left(\frac{\partial S}{\partial V}\right)_T \left(\frac{\partial V}{\partial T}\right)_p\right] dT + \left(\frac{\partial S}{\partial V}\right)_T \left(\frac{\partial V}{\partial p}\right)_T dp$$

so that: $\left(\frac{\partial S}{\partial T}\right)_p = \left[\left(\frac{\partial S}{\partial T}\right)_V + \left(\frac{\partial S}{\partial V}\right)_T \left(\frac{\partial V}{\partial T}\right)_p\right]$

and finally:

$$C_p - C_V = T\left[\left(\frac{\partial S}{\partial T}\right)_p - \left(\frac{\partial S}{\partial T}\right)_V\right] = T\left(\frac{\partial S}{\partial V}\right)_T \left(\frac{\partial V}{\partial T}\right)_p \tag{3.7}$$

Introducing the coefficient of thermal expansion $\alpha \overset{\text{def}}{=} \frac{1}{V}\left(\frac{\partial V}{\partial T}\right)_p$ and Maxwell relation Eq. (2.35), Eq. (3.7) becomes: $C_p - C_V = \alpha VT\left(\frac{\partial p}{\partial T}\right)_V$

Finally, introducing the isothermal compressibility $\beta \overset{\text{def}}{=} -\frac{1}{V}\left(\frac{\partial V}{\partial p}\right)_T$ and using Eq. (8.1)

$$\left[\left(\frac{\partial V}{\partial T}\right)_p \left(\frac{\partial p}{\partial V}\right)_T \left(\frac{\partial T}{\partial p}\right)_V = -1\right]: \left(\frac{\partial p}{\partial T}\right)_V = \frac{-\left(\frac{\partial V}{\partial T}\right)_p}{\left(\frac{\partial V}{\partial p}\right)_T} = \frac{\alpha}{\beta}$$

so that Eq. (3.7) becomes: $C_p - C_V = \frac{\alpha^2 VT}{\beta}$ (Eq. 3.5). For *one mole of perfect gas*, it is easy to show that $C_p - C_V = R$.

3.1.1.2 Dependence of Entropy on Temperature

The entropy changes at constant volume and pressure are:

$$\Delta S_V \overset{\text{def}}{=} S(V,T_2) - S(V,T_1) = \int_{T_1}^{T_2} C_V \, d\ln T \text{ at constant volume} \tag{3.8}$$

$$\Delta S_p \overset{\text{def}}{=} S(p,T_2) - S(p,T_1) = \int_{T_1}^{T_2} C_p \, d\ln T \text{ at constant pressure} \tag{3.9}$$

For a transformation at constant U: $đQ = -\delta W = p dV$. In a perfect gas, the internal energy depends only on the temperature, so the transformation is also isothermal. Using the equation of state of the ideal gas we have that:

$$\Delta S = \int_{V_1}^{V_2} nR \, d\ln V = nR \ln\frac{V_2}{V_1} \tag{3.10}$$

As for Eq. (3.4), here we are talking about a single-phase system. If the pure substance shows a phase transition at temperature T_{trs}, the heat involved at constant pressure and temperature is $\Delta_{trs}H$ (see Section 3.3.1). For example, for fusion we will have $\Delta_{fus}H$ (cfr. Section 3.3.1) and an additional contribution to entropy:

$$\Delta S = \frac{\Delta_{fus}H}{T_{fus}} \tag{3.11}$$

3.1.1.3 Dependence of Gibbs and Helmholtz Energies on Temperature

For convenience, let's rewrite Eq. (2.20) for a closed system with fixed composition:

$$dG = \left(\frac{\partial G}{\partial T}\right)_p dT + \left(\frac{\partial G}{\partial p}\right)_T dp; \quad \left(\frac{\partial G}{\partial T}\right)_p = -S; \quad \left(\frac{\partial G}{\partial p}\right)_T = V \tag{3.12}$$

Then:

$$\left(\frac{\partial G}{\partial T}\right)_p = -S = \frac{G-H}{T} \tag{3.13}$$

This relationship can be used to determine the temperature dependence of G/T.

$$\left(\frac{\partial G/T}{\partial T}\right)_p = \frac{1}{T}\left(\frac{\partial G}{\partial T}\right)_p - \frac{G}{T^2} = \frac{1}{T}\left(\frac{G-H}{T} - \frac{G}{T}\right) = -\frac{H}{T^2} \tag{3.14}$$

This is an example of a **Gibbs-Helmholtz equation**. Similarly, there is a Gibbs-Helmholtz equation for A:

$$\left(\frac{\partial A/T}{\partial T}\right)_p = -\frac{U}{T^2} \tag{3.15}$$

For a pure substance we have in fact (Eq. 2.19): $dA = \left(\frac{\partial A}{\partial T}\right)_V dT + \left(\frac{\partial G}{\partial V}\right)_T dT$ and $\left(\frac{\partial A}{\partial T}\right)_V = -S = \frac{A-U}{T}$.

As already seen for U, H, S, if the pure substance shows a phase transition, the variations must take this into account.

3.1.2 Pressure and Volume Dependence

From Eq. (3.12) we have, always for a pure substance:

$$\left(\frac{\partial G}{\partial p}\right)_T = V \tag{3.16}$$

In an ideal gas, $V = nRT/p$ and therefore by integration we obtain:

$$G = G^{\ominus} + nRT \ln \frac{p}{p^{\ominus}} \tag{3.17}$$

where the superscript $\ominus$ denotes a particular state called the **standard state** which, for convenience, is specified further on Sections 3.2.3 and 3.2.6.

Similarly (see Eq. (2.19)):

$$\left(\frac{\partial A}{\partial V}\right)_T = -p \tag{3.18}$$

3.2 Mixtures

To discuss in general the dependence of thermodynamic properties on the amounts $\{n_i\}$ of the various substances, it is convenient to begin by examining the case in which the system is made up of *several substances*, but *only one phase* and there are *no chemical reactions. Phase equilibria* are discussed in Section 3.3 while the problems of *chemical reactions* are discussed in Section 3.4. The amounts of substance are almost always measured in *moles*.

In the technical language of thermodynamics, the various substances that make up a system are referred to as **components**. For further clarifications on the term *component*, see Section 3.4.3.

A single-phase system with multiple components (regardless of whether the phase is gaseous, liquid, or solid) is called a **mixture**. The word **solution**, according to the IUPAC recommendation, is reserved for cases in which a component is treated differently from the others. Usually, but not necessarily, this is the majority component and is called **solvent**, while the others (see Section 3.2.6) are called **solutes**. The solution is then said to be **diluted** if the sum of the amounts of substance of the solutes is "small" compared to the amount of substance of the solvent. The term **mixture** was also used, with serious ambiguity, to indicate a poly-phasic system, sometimes specified as a *heterogeneous mixture*. This use must be rejected.

The fundamental quantity is evidently the chemical potential defined in Eq. (2.31): for the most common applications in chemistry the most useful definition is the last one, which allows us to treat a system at given p and T:

$$\mu_i = \left(\frac{\partial G}{\partial n_i}\right)_{T,p,n_{j\neq i}} \tag{3.19}$$

3.2.1 Integral Molar and Partial Molar Properties

The thermodynamic treatment of mixtures *does not* use the usual state variables, *i.e.* T, p (or V) and the amounts of the different substances $\{n_i\}$ which we assume to be k in number. Instead, a transformation is preliminarily carried out from the k extensive variables $\{n_i\}$ to the *extensive* variable $n = \sum_{i=1}^{k} n_i$ and the *intensive* quantities defined as $x_i = n_i/n$ and called **mole fractions**. From k independent quantities $\{n_i\}$ we switch to an equal number of independent quantities. With k components, in fact, there are **only $k-1$ independent mole fractions** because any one of them (usually the first is chosen) is given by $x_1 = 1 - \sum_{i=2}^{k} x_i$. This change of variables makes simpler the distinction between the **extensive** aspect now quantified by a single variable n and the **intensive** aspect quantified by the x_i, which measure the composition of the mixture. The advantage of the change of variables $[\{n_i\} \rightarrow n, \{x_i\}]$ is evident if we consider that, to characterize the properties (internal energy U, enthalpy H, entropy S, Helmholtz energy A, …) of a mixture (at constant T, p or T, V) the dependence on n is then a simple multiplicative factor: by increasing the total quantity without changing the composition, an extensive property (U, H, S, A, …) increases proportionally, while the corresponding intensive quantity (U/n, H/n, S/n, A/n, …) remains unchanged.

In addition to the mole fraction, other quantities are commonly used for compositions. These are the partial pressure for gases and, for condensed phases, the **molar** concentration and the **molal** concentration, as well as technical units such as percentage by weight and by volume. The *molar* concentration or **molarity** of component i is defined as the amount of substance in moles per volume of mixture (n_i/V), is measured in mol/L and is indicated by the letter M. The *molal* concentration or **molality** of component i in a solution is defined as the amount of substance in moles per kg of solvent (n_i/m_{solvent}) and is measured in mol/kg. Mole fractions are the simplest and most coherent choice for deriving useful relationships. If you prefer to express the results using other intensive compositions, it is not difficult to derive the appropriate relationships, but it must be considered that in some cases the conversion factors depend on other external variables. For example, a molarity - unlike a molality or a mole fraction - changes with temperature.

The quantities (E/n, H/n, S/n, A/n, …) are called **integral molar properties**. For these, a widely used convention, not followed here, is to use the same symbol as the extensive quantity, but in **lowercase**: $U/n=u$, $H/n=h$, $G/n=g$, $S/n=s$, $A/n=a$, etc. Another convention, which is that adopted by IUPAC and in this textbook, is to write the same symbol as the extensive with an m subscript. For example, the integral molar volume $v=V/n$ would be written as V_{m}; the integral molar volume of an A species would be written as $V_{\mathrm{m,A}}$.

For mixtures, it is necessary to distinguish an **integral molar** property from the corresponding k **partial molar** properties which are the *derivatives* of the same extensive property with respect to the amount of each component. In these partial derivatives, all other state variables including the amounts of *other* substances are held constant. For example, the partial molar volume of the second component in a ternary solution (using the state variables p, T in addition to the amounts of substance) is given by:

$$\left(\frac{\partial V}{\partial n_2}\right)_{n_1,n_3,p,T}$$

With the exception of the chemical potential, for which the use of the symbol μ_i is now consolidated, for partial molar quantities the convention is followed here of using the same symbol as the extensive quantity, but with the index of the component as a subscript. For example, the partial molar volume of component 2 is written $\bar{V}_2$.

In the case of single-component systems, obviously, the partial molar quantities coincide with the corresponding integral molar quantities. Equation (3.19) defines *the chemical potential* as the **partial molar Gibbs energy** so that, at constant pressure and temperature:

$$\mathrm{d}G_{T,p}=\sum_i \mu_i \mathrm{d}n_i \qquad (3.20)$$

Since G is extensive, it is a homogeneous function of degree 1 in the amounts of substance. Euler theorem must therefore hold (see Section 8.1.2):

$$G=\sum_i \mu_i n_i \text{ so that } G_{\mathrm{m}}\overset{\mathrm{def}}{=}\frac{G}{n}=\sum_{i=1}^{k}\mu_i x_i \qquad (3.21)$$

These relations can be extended to all extensive quantities and related integral and partial molar properties. From Eq. (3.21), we have: $\mathrm{d}G_{T,p}=\sum_i \mu_i \mathrm{d}n_i+\sum_i n_i \mathrm{d}\mu_i$. A comparison with Eq. (3.20) immediately shows that:

$$\sum_i n_i \mathrm{d}\mu_i=0 \quad \text{this is a } \textbf{Gibbs – Duhem } \boldsymbol{equation} \qquad (3.22)$$

These relations can be extended to all extensive quantities and related integral and partial molar properties. For example, for volume:

$$\sum_i n_i \mathrm{d}\bar{V}_i=0 \qquad (3.23)$$

where the $\bar{V}_i$ are partial molar volumes. The Gibbs-Duhem equations show that - for any extensive state function - the corresponding *partial molar properties do not vary independently* of each other and *quantify their mutual dependence.*

3.2.2 Graphical Representation

The relationships between the partial molar properties and the corresponding integral molar property have a *graphical representation* that is particularly effective and, when applied to the Gibbs energy or the Helmholtz energy, becomes a useful tool for discussing phase equilibria, as will be done extensively in Section 3.3.4. For generality, let us consider a generic extensive quantity denoted by F and its molar integral and partial properties denoted by F_{m} and $\bar{F}_i$. For a two-component mixture:

$$F_{\mathrm{m}}\overset{\mathrm{def}}{=}\frac{F}{n}=\bar{F}_1(1-x)+\bar{F}_2x=\bar{F}_1+\left(\bar{F}_2-\bar{F}_1\right)x$$

where $x=x_2$ and $(1-x)=x_1$, from which (using the only independent mole fraction x):

$$\left(\frac{\partial F_{\mathrm{m}}}{\partial x}\right)=\left(\bar{F}_2-\bar{F}_1\right) \quad \text{so that} \quad F_{\mathrm{m}}(x)=\bar{F}_1+\left(\frac{\partial F_{\mathrm{m}}}{\partial x}\right)x$$

By reversing the relationships, $\bar{F}_1=F_{\mathrm{m}}(x)-\left(\frac{\partial F_{\mathrm{m}}}{\partial x}\right)x$ and $\bar{F}_2=F_{\mathrm{m}}(x)+\left(\frac{\partial F_{\mathrm{m}}}{\partial x}\right)(1-x)$ or $\bar{F}_2=\bar{F}_1+\left(\frac{\partial F_{\mathrm{m}}}{\partial x}\right)$. Using the two linearly *dependent* mole fractions we have: $F_{\mathrm{m}}=\sum_1^2 \bar{F}_i x_i$ and $\bar{F}_i=F_{\mathrm{m}}(x_i)+\left(\frac{\partial F_{\mathrm{m}}}{\partial x_i}\right)(1-x_i)$. Figure 3.2 shows that for each x in the interval $0\geq x\geq 1$, the tangent to the curve F_{m} (x) at a specific point x graphically identifies the contribution ($\bar{F}_1$) of component 1 to the molar value of the property with the intercept of the tangent on the vertical axis placed at $x=0$ (*i.e.* $x_1=1$). Similarly, the contribution ($\bar{F}_2$) of component 2 to the molar value of the property is given by the intercept on the vertical axis placed at $x_2=x=1$.

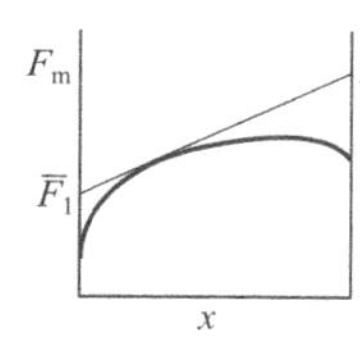

FIGURE 3.2 Integral and partial molar quantities.

If, for example, the quantity F_m in Figure 3.2 were the integral molar volume V_m of a liquid mixture, the meaning would be that - at the composition corresponding to the point of tangency in the drawing - both components have partial molar volumes greater than the molar volumes of the corresponding pure compounds in the liquid phase (at the same T, p), an indication that the mixture has a less compact structure than the pure compounds. The figure makes it clear that:

a. both these contributions, $\bar{F}_1$ and $\bar{F}_2$, generally change with the composition of the mixture (x),
b. they are different from (*i.e.*: not necessarily equal to) the values that the same component shows in the absence of the other and
c. the changes of $\bar{F}_1$ and $\bar{F}_2$ with x are constrained to each other and their ratio is given by the Gibbs-Duhem Eqs. (3.22) and (3.23).

The generalization to k components is straightforward. Again, using the x_i with $i \neq 1$ as independent variables):

$$F_m \overset{\text{def}}{=} \frac{F}{n} = \sum_{i=1}^{k} \bar{F}_i x_i,$$

$$\bar{F}_i = \bar{F}_1 + \left(\frac{\partial F_m}{\partial x_i}\right)_{x_{j\neq i}} \quad (\text{for } i \neq 1) \text{ and}$$

$$\bar{F}_1 = F_m(x) - \sum_{i=2}^{k} \left(\frac{\partial F_m}{\partial x_i}\right)_{x_{j\neq i}} x_i$$

With three components, the tangent plane to the curve F_m (x_2, x_3) at a specific point (x_2, x_3) intercepts the vertical axes corresponding to the pure ones (the $x_1=1$, axis *i.e.* $x_2=0$, $x_3=0$, the $x_2=1$, axis, *i.e.* $x_1=0$, $x_3=0$ and finally the $x_3=1$ axis, *i.e.* $x_1=0$, $x_2=0$) at the values corresponding to the partial molar quantities $\bar{F}_1$, $\bar{F}_2$, $\bar{F}_3$ (respectively). For $k>3$ components, the $(m-1)$-dimensional hyperplane intercepts the vertical axes corresponding to the pure ones at the values corresponding to the partial molar quantities $\bar{F}_1$, $\bar{F}_2$, ..., $\bar{F}_k$ (respectively).

3.2.3 Ideal Gas Mixture

For both methodological and historical reasons, the treatment of the thermodynamics of mixtures starts from a simple and important model called **ideal mixture**, which is easily introduced with the gaseous mixture of two substances that exactly follow the behaviour of the *ideal gas*. It is important to note that the two qualifiers "ideal" have different meanings but, for a mixture of gases, the ideal behaviour of the mixture can be derived from the ideal behaviour of each component gas: a mixture of ideal gases is an ideal mixture. The opposite is not true, as is obvious considering the fact that it is possible (see Section 3.2.4) to speak of *ideal liquid* and *solid* mixtures. Let's start by looking at how the chemical potential of each substance depends on pressure:

$$\left(\frac{\partial \mu_i}{\partial p}\right)_T = \frac{\partial}{\partial p}\left(\frac{\partial G}{\partial n_i}\right)_{T,p,n_{j\neq i}} = \frac{\partial}{\partial n_i}\left(\frac{\partial G}{\partial p}\right)_{T,n_{j\neq i}} = \bar{V}_i. \text{ Then:} \tag{3.24}$$

$$d\mu_i = \bar{V}_i dp \tag{3.25}$$

For a mixture of gases (not necessarily an ideal mixture) the **partial pressure** of each component is *defined* as the product of its molar fraction x_i by the total pressure p:

$$p_i = p \cdot x_i \tag{3.26}$$

In an *ideal gas*, in particular, $V=nRT/p$ and the partial molar volume is given by

$$\bar{V}_i = \left(\frac{\partial(nRT/p)}{\partial n_i}\right)_{T,\,p,\,n_{j\neq i}} = RT\frac{\partial}{\partial n_i}(n_i/p_i) \text{ and}$$

therefore, is equal to $\dfrac{RT}{p_i}$.

Inserting in Eq. (3.25) and integrating we get:

$$\mu_i = \mu_i^{\ominus} + RT\ln\frac{p_i}{p^{\ominus}} \tag{3.27}$$

The values $\mu_i^{\ominus}$ and $p^{\ominus}$ denote the chemical potential and the pressure *of a particular state* called the **standard state**. These values do not depend on the composition and not even on the pressure p, since they refer to a particular value $p^{\ominus}$ of pressure, but can depend - and usually do depend - on T. For convenience, the standard state is discussed further on Section 3.2.6.

Rewriting Eq. (3.27) still using the definition $p_i = px_i$ we have:

$$\mu_i = \mu_i^{\ominus} + RT\ln\frac{p_i}{p^{\ominus}} = \mu_i^{\ominus} + RT\ln\frac{p}{p^{\ominus}} + RT\ln(x_i) \tag{3.28}$$

From Eq. (3.21), for a mixture of two gases:

$$G = \sum_i \mu_i n_i = n_1\left[\mu_1^{\ominus} + RT\ln\frac{p}{p^{\ominus}}\right] + n_2\left[\mu_2^{\ominus} + RT\ln\frac{p}{p^{\ominus}}\right]$$

$$+n_1\left[RT\ln(x_1)\right] + n_2\left[RT\ln(x_2)\right] \tag{3.29}$$

where the terms that do not depend on the mole fractions (the first two) are separated from the terms that depend on them.

In the derivation we used the definition (Eq. 3.26), which is also valid for a non-ideal gas, and the property of ideal gases that the partial molar volume is $\bar{V}_i = V/n_i = RT/p_i$, which gives Eq. (3.27). This proves that the ideal behaviour of a mixture of gases (as specified in the following) derives from the ideal behaviour of each component. Moving from the variables (n_1, n_2) to (n, x):

$$G = G_m \cdot n = n \cdot \sum_i \mu_i x_i$$

$$G_m = \mu_1^0 + x\left(\mu_2^0 - \mu_1^0\right) + RT\left[(1-x)\ln(1-x) + x\ln(x)\right]$$

$$\text{using: } \mu_i^0 \stackrel{\text{def}}{=} \mu_i^\ominus + RT\ln\frac{p}{p^\ominus} \tag{3.30}$$

The point to note is that the Gibbs energy (G_m) of one global mole of a binary mixture of composition x, is given by the sum of:

a. a linear term in the variable x (the standard potentials $\mu_i^\ominus$ and $p/p^\ominus$ do not depend on the compositions and the same is true for the potentials μ_i^0 of the pure components)
b. plus a term that depends on x as $F_m(x) = (1-x)\ln(1-x) + x\ln(x)$.

This is what defines an ideal mixture.

The function $F_m(x) = (1-x)\ln(1-x) + x\ln(x) \to 0$ both for $x \to 1$ and $x \to 0$.

The corresponding expression for *enthalpy* can be directly obtained from the definition of Gibbs energy (Eq. 2.18) and from the dependence of G on T (Eq. 3.13): $H = G + TS = G - T\cdot\left(\frac{\partial G}{\partial T}\right)_{p,n_i}$. G contains $RT\left[(1-x)\ln(1-x) + x\ln(x)\right]$, while $S = -\left(\frac{\partial G}{\partial T}\right)_{p,n_i}$ contains $-R\left[(1-x)\ln(1-x) + x\ln(x)\right]$: the two logarithmic terms cancel out and only the linear dependence on x remains for enthalpy.

By performing the calculations for the various extensive properties (U, V, S, A, ...) we find that some quantities (S, A, ...) behave like Gibbs energy (Eq. 3.34), that is, they contain the expression $x_1\ln x_1 + x_2\ln x_2$, while the others (U, V, ...) behave like enthalpy. To be precise, the extensive quantities that in their definition contain entropy behave in the first way, and those that do not contain entropy (*i.e.* those that we called *non-thermal quantities* at the beginning of 1.1) behave in the second way.

Finally, considering the cases with k components, we still obtain the same results for any thermodynamic integral molar property (F_m) and

a. the linear term is $\sum_{i=1}^{k} \overline{F}_i^0 x_i$ where the values with the superscript 0 are for the pure components,
b. the logarithmic term (when required) is $\sum_{i=1}^{k} x_i \cdot \ln(x_i)$ and, as already said, $x_1 = 1 - \sum_{i=2}^{k} x_i$.

The logarithmic term (b) is equal to 0 for each pure compound, for the same reason already seen above.

In ST, the behaviour of an ideal gas mixture can be derived (see Section 6.3) from the assumption that the molecules of the different components do not interact with each other. The condition that must be satisfied in ST is that each molecule is an independent subsystem, a requirement that is satisfied by the definition of an ideal gas in CT.

3.2.4 Ideal Mixtures in General

The ideal mixture model, although derived for ideal *gas* mixtures, can be applied also to mixtures in *liquid* or *solid* phases. At the atomic-molecular level, the model requires that the mixtures *do not show preferential interactions* between *equal* particles with respect to those between *different* particles, and that the particles have a *completely random distribution*. Sometimes the term *random mixture* is used to indicate an ideal mixture. In summary, for a *binary* ideal mixture the following relations hold:

$$G/n = G_m = (1-x)\mu_1^0 + x\mu_2^0 + RT\left[(1-x)\ln(1-x) + x\ln(x)\right]$$

$$S/n = S_m = (1-x)\overline{S}_1^0 + x\overline{S}_2^0 - R\left[(1-x)\ln(1-x) + x\ln(x)\right] \stackrel{\text{def}}{=} S_m^{id}$$

$$H/n = H_m = (1-x)\overline{H}_1^0 + x\overline{H}_2^0 \stackrel{\text{def}}{=} H_m^{id}$$

$$U/n = U_m = (1-x)\overline{U}_1^0 + x\overline{U}_2^0 \stackrel{\text{def}}{=} U_m^{id}$$

$$V/n = V_m = (1-x)\overline{V}_1^0 + x\overline{V}_2^0 \stackrel{\text{def}}{=} V_m^{id}$$

where the superscript 0 again denotes the values of the pure components. For other thermodynamic functions, the same considerations seen for the ideal gas mixture (Section 3.2.3) apply. These expressions can be directly generalized to the case of k components, for example for the Gibbs energy:

$$G/n = G_m^{id} = \sum_{i=1}^{k} x_i \mu_1^0 + RT\sum_{i=1}^{k} x_i \ln x_i \quad \left(\text{with } x_1 = 1 - \sum_{j=2}^{k} x_j\right)$$

Obviously, no real system shows exactly the trends predicted by the model of the ideal solution, but some systems consisting of chemically *very similar* structural units (molecules, atoms, ions, ...) show a trend that approaches the model with good or excellent approximation and the *agreement* with experiments *is better the more the structural units are chemically similar.*

In the case of gases, as already seen, the model is justified by the *ideal gas* hypothesis that *all interactions* between molecules *can be neglected*, regardless of whether the molecules are the same or different. Instead, the less stringent approximation that the interactions between *equal* molecules can be assumed equal to the interactions between *different* molecules can be applied also to liquids or solids, in which the structural units certainly interact. As an example, Section 6.4 reports a simple ST model for an ideal

solid mixture. Obviously, the model is more acceptable the smaller the interactions, that is - generally speaking - in the order gas - liquid - solid.

For solutions, the accuracy of the model is greater the **more diluted the solution is**, that is, the greater the amount of solvent compared to the total amount of solutes. The traditional and well-known treatments of the colligative properties of liquid solutions, for example, are based on the ideal solution model. These describe methods such as **cryoscopy** (measurement of freezing point depression), **ebulliometry** (measurement of boiling point elevation), and **osmometry** (measurement of osmotic pressure) as a function of the solute concentration.

In general, the ideal mixture model is very important because

a. it is the simplest mixture behaviour,
b. in some cases, it is an adequate representation of the real behaviour,
c. it is often a useful first approximation and
d. generally, it is the basis of more refined ST models or more accurate fits of experimental measurements.

However, *the most important point of interest in the model* is the expression of the chemical potential as proportional to the logarithm of a concentration measure (C):

$$\mu_i = \mu_i^* + RT\ln(C_i) \tag{3.31}$$

The term with the asterisk here contains, in addition to the value of the chemical potential for the standard state (as specified in Section 3.2.6) and the pressure dependence (Eq. 3.28), also the constants relating to the conversion factors for the concentration measured in units different from molar fractions, such as molarity, molality or others.

By *expressing chemical potentials as proportional to the logarithm of a concentration* (except for an additive term independent of concentrations) *one can easily obtain the basic thermodynamic treatment of the equilibrium of chemical reactions*, including the *equilibrium constants* (see Section 3.4.1).

3.2.5 Non-ideal Mixtures, Thermodynamic Activity and Fugacity

The thermodynamic treatment (both in CT and ST) of non-ideal mixtures essentially focuses on the contributions to the thermodynamic functions *in addition to the ideal mixture* term, *i.e.* in addition to the linear term and possibly to $\sum_{i=1}^{k} x_i \cdot \ln(x_i)$. This additional term is referred to as the **excess** contribution. The integral molar Gibbs energy of a two-component system is then:

$$G_\mathrm{m} = x\mu_1^0 + (1-x)\mu_2^0 + RT\left[(1-x)\ln(1-x) + x\ln(x)\right] + G_\mathrm{m}^\mathrm{E}$$

The excess contribution G_m^E is constrained to 0 for mole fraction=1 of each component. For a binary mixture, it is frequently written as the series expansion:

$$G_\mathrm{m}^\mathrm{E} = x\cdot(1-x)\cdot\left[A + B\cdot(2x-1) + C\cdot(2x-1)^2 + \cdots\right]$$

where A, B, C, ... are coefficients evaluated with ST models or from experimental data and, in general, depend on T and p. The factor $x\cdot(1-x)$ guarantees that the excess term vanishes for $x=0$ and 1, while the part in square brackets is essentially a power series in the interval $0\geq x\geq 1$. The case $B = C = \cdots = 0$ is called **regular mixture** and an ST derivation is reported in Section 6.4.

Due to the importance of the ideal solution model and the simplicity of the expression of the chemical potentials, especially in the subsequent treatment of chemical reactions, it is preferable to maintain the form of Eq. (3.31) also for mixtures in which a significant deviation from the ideal behaviour occurs. This is achieved by introducing a quantity called **activity** (a_i) and thought of as a "modified" concentration by means of a coefficient γ_i:

$$a_i \overset{\mathrm{def}}{=} c_i \cdot \gamma_{c,i} \tag{3.32}$$

so that the chemical potential can be written as:

$$\mu_i = \mu_i^* + RT\ln(a_i) \quad \text{and} \quad RT\ln(\gamma_{c,i}) = \left(\frac{\partial G_\mathrm{m}^\mathrm{E}}{\partial x_i}\right)_{p,T,x_{j\neq i}} \tag{3.33}$$

where μ_i^* is independent of the composition.

Equations (3.32) and (3.33) apply also when a partial pressure is used instead of a concentration or a mole fraction: instead of activity we then speak of **fugacity** (and *fugacity coefficient*).

3.2.6 The Standard State

The standard state of a substance is defined according to a number of conventions.

First, we need to define the standard pressure, as the pressure of 1 bar (10^5 Pa). Temperature is not included in the definition of standard state and must therefore be indicated separately.

The standard state of a substance is defined as the most stable state for that pure substance at the given standard pressure and at the indicated temperature. For example, for water at 25 °C, the standard state is pure liquid water, for carbon at 25 °C pure solid graphite, etc.

By convention, the enthalpies and Gibbs energies of the elements in their standard state are set equal to zero. In addition, for ions in aqueous solution, the standard enthalpies and Gibbs energies of formation of H^+_{aq} are also set to zero.

This convention allows the determination of all thermodynamic quantities.

The experimental determination of the standard state values of thermodynamic functions does not need that the standard state be experimentally accessible. The thermodynamic properties in the standard state are in any case well defined, at least by extrapolation. For example, we can evaluate the enthalpy of formation of water vapour at 25°C and 1 bar even if in these conditions water vapour does not exist.

Some further clarifications are necessary:

- The standard state of a **gas** is defined by the standard pressure of one bar, but *with the additional constraint that the gas behaves as an ideal gas.* At a pressure of one bar all real gases deviate significantly from ideal behaviour. This means that experimental data at one bar pressure must be corrected for the non-ideal behaviour of the gas with an equation of state.
- For a liquid **solution**, two distinct conventions are used for the *solute* and the *solvent*. For the solvent, the standard state is that of the pure solvent at the given temperature and standard pressure. For the solute, the standard state is defined by a concentration (molal or molar) = 1, but with the additional condition that the *solute behaves as if it were at infinite dilution*. The reason for this odd convention lies in the fact that in this way the chemical potential of a species in solution can be written in a form analogous to Eq. (3.31).

3.3 Phase Equilibrium

As already said in Section 1.2.6, the phase equilibrium concerns the partitioning of one or more substances between two or more phases, that is regions or *domains*, not necessarily connected spatially to each other, where the thermodynamic quantities do not show discontinuities. The essential aspects of phase equilibrium can be introduced by considering two-phase systems, for example a pure liquid in equilibrium with its vapour, or a liquid mixture in the presence of a solid precipitate. For each of the two phases in equilibrium (hereafter denoted with α and β) the domains can be considered as a (single) *open* system in equilibrium with their surroundings, with the constraint of mass conservation: if a given amount $\mathrm{d}n_i^\alpha$ of component i goes from phase α to phase β, then an amount $\mathrm{d}n_i^\beta = -\mathrm{d}n_i^\alpha$ goes from phase β to phase α. Therefore:

$$\mathrm{d}G = \sum_i \left(\mu_i^\alpha - \mu_i^\beta\right)\mathrm{d}n_i^\alpha = 0 \tag{3.34}$$

This equation must be satisfied for any value of $\mathrm{d}n_i^\alpha$. Then, the equilibrium condition (the minimum of G/n) becomes the equality of chemical potentials:

$$\mu_i^\alpha = \mu_i^\beta \tag{3.35}$$

Equation (3.35) must obviously be valid independently for each component of the system and can be directly generalized to a multiphase system. The equilibrium condition is therefore that

> At constant pressure and temperature, the chemical potentials of each component are equal in all phases.

3.3.1 Phase Equilibria for a Pure Component

We rewrite the equilibrium condition Eq. (3.35) as

$$\mathrm{d}\mu_i^\alpha = \mathrm{d}\mu_i^\beta \tag{3.36}$$

For a system consisting of a pure component:

$$\mathrm{d}\mu = \mathrm{d}G_\mathrm{m} = V_\mathrm{m}\mathrm{d}p - S_\mathrm{m}\mathrm{d}T \tag{3.37}$$

where G_m, V_m, and S_m denote the molar Gibbs energy, molar volume, and molar entropy, respectively. Equation (3.36) then becomes:

$$V_\mathrm{m}^\alpha \mathrm{d}p - S_\mathrm{m}^\alpha \mathrm{d}T = V_\mathrm{m}^\beta \mathrm{d}p - S_\mathrm{m}^\beta \mathrm{d}T \quad \text{and} \tag{3.38}$$

$$\frac{\mathrm{d}p}{\mathrm{d}T} = \frac{\Delta_\mathrm{trs}S_\mathrm{m}}{\Delta_\mathrm{trs}V_\mathrm{m}} = \frac{\Delta_\mathrm{trs}H_\mathrm{m}}{T\Delta_\mathrm{trs}V_\mathrm{m}} \tag{3.39}$$

Here, $\Delta_\mathrm{trs}S_\mathrm{m}$, $\Delta_\mathrm{trs}V_\mathrm{m}$ and $\Delta_\mathrm{trs}H_\mathrm{m}$ indicate the molar entropy, volume and enthalpy *of the phase transition*, *i.e.* the difference in molar values between one phase and another: $\Delta_\mathrm{trs}S_\mathrm{m} = S_\mathrm{m}^\beta - S_\mathrm{m}^\alpha$, $\Delta_\mathrm{trs}V_\mathrm{m} = V_\mathrm{m}^\beta - V_\mathrm{m}^\alpha$, $\Delta_\mathrm{trs}H_\mathrm{m} = H_\mathrm{m}^\beta - H_\mathrm{m}^\alpha$. Similarly to what we will see for Δ_r in Eq. (3.51), the operator Δ_trs corresponds dimensionally to the ratio between two extensive quantities (an energy, volume, entropy, …, divided by an amount of substance) and therefore defines an *intensive* property. Equation (3.39) is an example of a **Clapeyron equation** and expresses the slope of the $p(T)$ function.

In the case of equilibria between the gas phase and a condensed phase, Eq. (3.39) can be *simplified* by assuming that the volume of the condensed phase is negligible compared to the volume of the vapour and that the vapour behaves as an ideal gas. We then obtain:

$$\frac{\mathrm{d}p}{\mathrm{d}T} = \frac{p\Delta_\mathrm{trs}H_\mathrm{m}}{RT^2} \tag{3.40}$$

This relationship, once integrated by assuming, in addition to the other assumptions, that the transition enthalpy does not depend on the temperature, gives the well-known **Clausius-Clapeyron equation**:

$$\ln p = \frac{-\Delta_\mathrm{trs}H_\mathrm{m}}{RT} + c \quad \text{where } c \text{ is a constant of integration} \tag{3.41}$$

3.3.2 Phase Equilibria in Systems with More Than One Component: The Phase Rule

Let us now consider the equilibrium of a system with P coexisting phases and C **independent components**. For example, a solution of hexane in benzene is a two-component system. A solution of NaCl in water is also a two-component system, even if the solution phase contains water, Na^+ ions, and Cl^- ions, because the amounts of Na^+ and Cl^- cannot be varied independently without altering the electro-neutrality of the solution. Specifying the *correct* number of *independent* components is important when there are chemical reactions (see Section 3.4.3) and other stoichiometric constraints (see Section 3.6).

To discuss phase equilibrium, we must first specify **two external variables**. If we discuss equilibrium using Gibbs energy, these are temperature and pressure. Evidently, this would not change if we considered T and V as external variables and then used the Helmholtz energy as the thermodynamic potential.

Next, we need to specify a number of *compositions* for each phase at equilibrium. The compositions are not all independent of each other. For example, if we use mole fractions as compositions, their sum must be 1. In general, the number of compositional variables is $C-1$ for each phase. So, we will have a total of $P(C-1)$ compositions in addition to the two external physical variables.

However, relations Eq. (3.35) constrain the number of **independent thermodynamic variables**, that is, the number of thermodynamic variables needed to specify the state of the system. A simple example is the one already seen of a pure substance in the liquid state in equilibrium with its vapour. The coexistence of the two phases occurs at a given T (if we fix the p from the outside) or at a given p (if we fix the T) and therefore we do not have two independent variables but only one. This number of independent thermodynamic variables is defined as the number of macroscopic degrees of freedom or **thermodynamic variance** of the system and is denoted by F, from freedom. Let's count the variance. Equation (3.35) holds for each component and for each of them establish $P-1$ relations between the chemical potentials in P phases. It follows that we will have $C(P-1)$ constraints. The thermodynamic variance is therefore:

$$F = 2 + P(C-1) - C(P-1) = C - P + 2. \text{ This is} \tag{3.42}$$

Gibbs phase rule: the thermodynamic variance is $C-P+2$.

It must be remembered that *the components must be independent.*

The phase rule is a relation between *intensive variables* and therefore *does not depend on the total amount of the substances involved nor does it say anything about it.* For example, the same considerations on equilibrium apply both to a laboratory cryoscopy experiment and to the melting of the Antarctic glaciers.

3.3.3 Phase Diagrams

A phase diagram is a *graphical representation* of the relationships between two or, more rarely, three *intensive state variables* of a system that may be split into several phases. The intensive variables generally considered are T, p and the appropriate number of compositions, one less than the number of independent components. Without describing in detail the various cases, here we discuss some general considerations.

When there is *only* **one** (independent) **component**, the phase diagram is two-dimensional as expected by the phase rule (Section 3.3.1). Essentially, it describes the coexistence conditions between pairs of phases that can be in thermodynamic equilibrium. For the simplest case of a substance with only one solid phase, there are three equilibria between solid and gas, solid and liquid and liquid and gas and there are three $p(T)$ monotonous curves that represent as a function of T the pressure p at which two phases coexist or by the three inverse relations $T(p)$ that express as a function of p the temperature T at which two phases coexist. These equilibria are then called **univariant**. The 2D regions bounded by the univariant curves describe the states in which the system is stable as a single phase and **two** state variables can be **modified independently** without changing the number and nature of the coexisting phases. These are **bivariant** regions. To consider a simple case, at $T = 300$ K (about 25 °C) and $p = 10^5$ Pa a sample of pure water is in the liquid state and remains liquid even by independently modifying both variables *by small quantities*, for example, by bringing the T to 310 K or the p to $1.2 \cdot 10^5$ Pa, or with both variations. If, however, without modifying the T, the p is brought to the value of 10^2 Pa, the new equilibrium is that the sample is made up of only the gas phase (vapour). In the interval between the two p, there is a particular value at which the two phases coexist. This is the vapour pressure of water at 300 K. At 300 K and a pressure p equal to the vapour pressure of water, the equilibrium is said to be **univariant** because **only one** state variable can be **modified** without changing the number and nature of the coexisting phases. For instance, at 300 K and a pressure p slightly above the vapour pressure, water is liquid, at the same T and p slightly below the vapour pressure, there is only water vapour. Finally, the singular (or **invariant**) **point** where the three curves cross and the three univariant regions meet each other represents **the coexistence equilibrium of three phases** and is called **the triple point**. When a substance has *more than one solid phase*, there are other triple points. This is, for example, the well-known case of water.

With **two-component systems**, two-dimensional plots are generally used: $p(x)$ at constant T or $T(x)$ at constant p, where x is a measure of composition, such as the mole fraction of the second component. A three-dimensional plot is also used, but less commonly.

With **three components** the 2D space of the compositional variables (typically: molar fractions $\{x_i\}$) is frequently represented by an equilateral because this graphically expresses the constraint between the three molar fractions. Usual representations are then the projections onto the compositional triangle of the equilibrium surfaces $T(\{x_i\})$ between pairs of phases at constant p or the sections at constant T and p of the relations between the compositions of phases in equilibrium with each other.

In all cases, *a phase diagram* is a consequence of the relationships between the *Gibbs energies of the phases* that *can be present* in the system and their dependencies on T, p, and the necessary compositions. The knowledge of the Gibbs energy functions of all phases therefore allows the phase diagram to be calculated.

To discuss phase diagrams, the graphical representation of the integral molar properties at constant T and p (introduced in Section 3.2.2) is particularly useful and must now be extended to multiphase systems. Each phase is characterized by its own extensive (let us say F) or integral molar ($F_m = F/n$) property, and we must discuss how to find the values of the F or F_m of a *heterogeneous* system, that is when *two or more phases coexist*, and then to find the corresponding equilibrium conditions.

To simplify the discussion, it is convenient to first examine a system with *only two components* (1 and 2) and *only two* phases (α and β) since the extension to more components and/or more phases is trivial. If n_1^α, n_2^α, n_1^β and n_2^β are the amounts of components 1 and 2 in phases α and β, the total amount of substances (component 1 + component 2) in both phases is $n = n_1^\alpha + n_2^\alpha + n_1^\beta + n_2^\beta$. Also, $n^\alpha = n_1^\alpha + n_2^\alpha$ and $n^\beta = n_1^\beta + n_2^\beta$ are the amount of both substances (1 and 2) in one phase, while $n_1 = n_1^\alpha + n_1^\beta$ and $n_2 = n_2^\alpha + n_2^\beta$ are the amounts of a substance in both phases. Using $\bar{x}$ to indicate the mole fraction of the second component (that of the first component is $1-\bar{x}$), $\bar{x} = n_2/n$, while $x^\alpha = n_2^\alpha/n^\alpha$ is the mole fraction of component 2 in the α phase and $x^\beta = n_2^\beta/n^\beta$ is the mole fraction of component 2 in the β phase.

For the conservation of the amount of component 2, we have $n\bar{x} = n^\alpha x^\alpha + n^\beta x^\beta \Rightarrow n^\alpha \bar{x} - n^\alpha x^\alpha = n^\beta x^\beta - n^\beta \bar{x}$. This finally gives $n^\alpha(\bar{x} - x^\alpha) = n^\beta(x^\beta - \bar{x})$. Therefore, in a system of global composition $\bar{x}$ with two coexisting phases, the amounts (n^α and n^β) of the two phases are constrained by the compositions of the two phases (x^α e x^β) through the

Lever rule: $\dfrac{n^\alpha}{n^\beta} = \dfrac{x^\beta - \bar{x}}{\bar{x} - x^\alpha}$ or :

$$n^\alpha = n\frac{x^\beta - \bar{x}}{x^\beta - x^\alpha} \quad \text{and} \quad n^\beta = n\frac{\bar{x} - x^\alpha}{x^\beta - x^\alpha} \quad \text{with} \quad n = n^\alpha + n^\beta \tag{3.43}$$

These relations merely express the **conservation of the amounts of substance**.

Then, let us consider a system with global amounts n_1 and n_2 of the two substances and two phases (α and β) with compositions x^α and x^β. For an extensive property F and its integral molar property F_m, the additivity of the thermodynamic properties gives:

$$F_m(\bar{x}) = \frac{F(\bar{x})}{n} \quad \text{and}$$
$$F_m(\bar{x}) = \frac{x^\beta - \bar{x}}{x^\beta - x^\alpha} \cdot F_m^\alpha(x^\alpha) + \frac{\bar{x} - x^\alpha}{x^\beta - x^\alpha} F_m^\beta(x^\beta) \tag{3.44}$$

The relation can also be written $F_m(\bar{x}) = F_m^\alpha(x^\alpha) + \dfrac{F_m^\beta(x^\beta) - F_m^\alpha(x^\alpha)}{x^\beta - x^\alpha}(\bar{x} - x^\alpha)$ and we see that in a (F_m, x) plot it graphically represents the segment from the point (F_m^α, x^α) to the point (F_m^β, x^β) and includes the points with abscissa $\bar{x}$ and ordinate $F_m(\bar{x})$. Once the equilibrium compositions of the coexisting phases are known, Eqs. (3.43) and (3.44) specify the value of an integral molar property F_m of a two-phase system as *linear combination* of the properties of the pure phases. This holds also for the integral molar Gibbs energy.

However, we are here instead interested in *finding the equilibrium conditions* using the characteristic of G_m being the thermodynamic potential at given T and p. For this purpose, Eqs. (3.43) and (3.44) are **necessary but not sufficient conditions**, as well shown by the graphical representation of Figure 3.3, where the integral molar Gibbs energies of both phases are plotted as a function of composition at constant T and p. Indeed, let us consider a system that has the global composition $\bar{x}$ of the dashed vertical line and is made up of quantities to be determined of *both* phases. Different cases are possible. The system can be represented, for example, by the segments through points c, or d in Figure 3.3. The compositions are the abscissas of the points where the segments touch the curves G_m^α and G_m^β and the amount of the phases n^α and n^β are then given by Eq. (3.43). However, the system can contain other mixtures of the same two phases or even phase α alone, or phase β alone, with representative point a or b. Now, the indeterminacy disappears if we also require that there is thermodynamic equilibrium. To that purpose it is sufficient to evaluate which of the various possibilities (points a, b, c, d in Figure 3.3, or any other possibility) corresponds to the **lowest Gibbs energy**. The graph makes it clear that *the equilibrium corresponds to case d*. The compositions (x^α and x^β) for case d are the equilibrium compositions, the amounts of the two phases obtained with the lever rule (Eq. 3.43) are then equilibrium amounts, and the molar **Gibbs** energy of the two-phase system is given by Eq. (3.44) using these x^α and x^β and considering $G_m(\bar{x})$ instead of $F_m(\bar{x})$. Geometrically, the *minimum of the Gibbs molar integral energy* (G_m) is given by

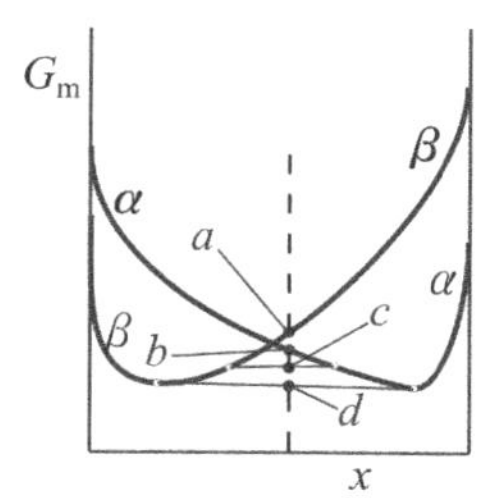

FIGURE 3.3 A biphasic binary system. The integral molar Gibbs energies of the two phases are plotted against the molar fraction of the second component (x) at constant T and p. The dashed vertical line marks the composition of the whole system. The states represented by a, b, c, d are all in agreement with Eqs. (3.43) and (3.44), but only d represents equilibrium.

the ordinate of the point (d) which is the intersection of the **common tangent** *to the two* G_m *curves of the phases* with the vertical line corresponding to the global composition ($\bar{x}$) of the system. The common tangent condition (or **double tangent**, as it is also called) is the geometric counterpart of the analytical condition of equality of the chemical potentials (Eq. 3.35) in the two phases for each of the two components, since the chemical potential is given graphically by the intercept of the tangents to the curves $G_m(x)$ on the axes $x=0$ and $x=1$ (see Section 3.2.2).

As said, the extension of the graphical method to *more than two coexisting phases* is trivial: it is enough to add other integral molar Gibbs energy curves and discuss each couple of phases (see Section 3.3.4.1 for a three-phase example). For the extension to *more than two components*, the graphic method evidently requires a greater number of coordinates. With three components, for example, we must speak of a *tangent plane* which is common to the surfaces $G_m(x_2, x_3)$ at constant T, p of two or three different phases.

3.3.4 Examples

As said, the graphic approach is fully equivalent to the analytical treatment (Eq. 3.35) but generally it is a more effective tool. Some examples can clarify the simplicity and usefulness of this method of discussing phase equilibrium.

3.3.4.1 Eutectic Equilibrium

In a **two-component system**, the phase rule (Eq. 3.42, $F=C-P+2=4-P$) predicts that (i) a monophasic system, $P=1$, is tri-variant, $F=3$, (ii) a biphasic system is bivariant, (iii) a triphasic system is univariant and finally that (iv) the coexistence of four phases occurs only at an invariant point, that is, a particular set of values of p, T and global composition.

A remarkable simplification of the graphical representation occurs when only condensed phases need to be considered. The Gibbs energy of condensed phases indeed depends much less on the pressure than that of the gas phase, since the molar volumes of condensed phases are *usually* orders of magnitude smaller. Then, the pressure can be kept constant, the graphical representations require one less coordinate, and a three-variant equilibrium corresponds to a region of a two-dimensional (T, x) graph. Under the p=constant constraint, a bivariant equilibrium corresponds to a curve $T=f(x)$ or $x=f(T)$ and an univariant equilibrium appears as a singular (invariant) point. It occurs at T and x values that weakly depend on the fixed p and correspond to the coexistence of three phases: $F=C-P+1$, where +1 is added instead of +2 because p is fixed and $F=2-3+1=0$.

Among various possible situations, an interesting case is the **eutectic equilibrium**.[6] At a particular temperature, called eutectic temperature T_E, a **liquid mixture** (L) **is in equilibrium with two solid mixtures** (α and β) that for simplicity we assume are different. The equilibrium at the assigned p and at T_E is represented by the left part of Figure 3.4. The system shows: (i) a range of global compositions ($0 \le x \le x_a$) where it is stable as a monophasic system (only phase α), (ii) a biphasic range ($x_a < x < x_b$) with coexistence of phases α (of composition x_a) and L (of composition x_b), (iii) the invariant three-phase eutectic point at $x=x_b$, (iv) another biphasic range ($x_b < x < x_c$) with coexistence of phases L (of composition x_b) and β (with composition x_c) and finally (v) a monophasic range (only phase β, $x_c \le x \le 1$).

By changing T but without changing p, the equilibrium changes because the G_m curves shift. At $T<T_E$, the L curve is always *above* the common tangent of the two solid phases, the liquid phase is always out of equilibrium and there is equilibrium only between the solid phases. At $T>T_E$, the L curve cuts the common tangent of the two solid phases. In addition to the previous single-phase (α and β) fields at the extremes of the compositional range, two solid/liquid equilibria then appear (α/L at low compositions and L/β at high compositions, right part of Figure 3.4) and these biphasic fields enclose the monophasic field of the liquid.

3.3.4.2 Interfacial Energy and Gibbs-Thomson Effect

Another result that the graphical method based on the $G_m(x)$ curves directly explains is the effect of interfacial energy on the solubility of precipitates with particular small grain sizes.

Let us consider a liquid mixture α and a solid mixture β, but the example also applies to other cases, for example if α is another solid mixture. If the overall composition $\bar{x}$ is slightly higher than x_a, from what we have seen so far, the equilibrium corresponds to a majority phase α with

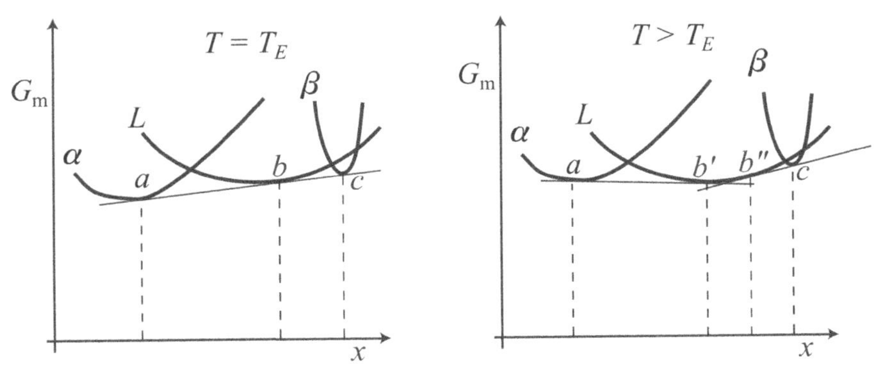

FIGURE 3.4 Eutectic equilibrium. Left: if the overall composition is x_b, at eutectic temperature (T_E) three phases, solid α, solid β and liquid (L) coexist. Right: at $T>T_E$ the system is monophasic (liquid) if the overall composition is between $x_{b'}$ and $x_{b''}$.

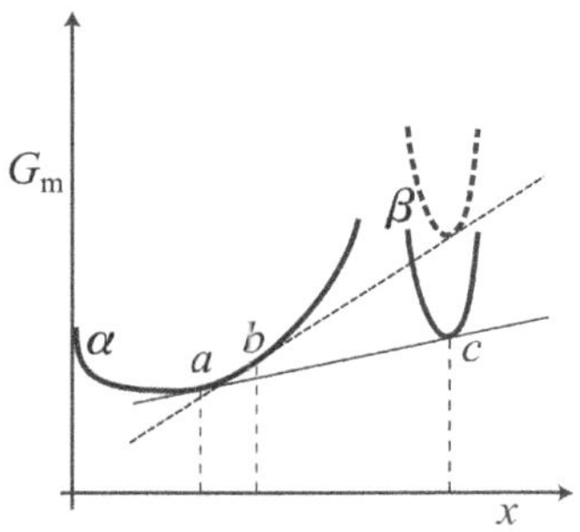

FIGURE 3.5 Solubility of small precipitates. If the global composition is $x_a \leq \bar{x} < x_b$, the precipitates of phase (β) dissolve when the contribution of the interfacial energy is large (dotted curve of Gibbs energy).

composition x_a, and a much smaller amount of phase β. The molar ratio between phases α and β is $\frac{n^\alpha}{n^\beta} = \frac{x_c - \bar{x}}{\bar{x} - x_a} \gg 1$.

Let us now suppose that the precipitate is made up of grains so small that the contribution of the interfacial energy to G_m is no longer negligible. Then we must consider for the β phase a new $G_\mathrm{m}(x)$, the dotted curve in the Figure 3.5, that is shifted upwards by an amount equal to the contribution of the interfacial effects (see Section 1.2.6), that we assume as independent of x for the sake of simplicity. The common tangent touches G_m^α at a point with abscissa $x_b > x_a$. We see that - if we consider a global composition $\bar{x}$ slightly higher than x_a, but$<x_b$ - the equilibrium between the two phases now corresponds (dotted line) to the presence of only the α phase. This is called the **solubility of small precipitates** effect and is also known as *Gibbs-Thomson effect.*

Experimentally, the situation can be achieved starting from a system with nanometric-sized precipitates at lower temperatures: rising T then dissolves the precipitates. Under these conditions, samples with the same composition but with a lower amount of surface energy due to a smaller number of larger precipitates remain biphasic. The effect also plays an important role in the kinetics of phase transformations because it represents an energetic barrier to the formation of precipitates, which can grow only if they exceed some critical dimensions.

3.3.4.3 Demixing

Finally, it is interesting to discuss the problem of *demixing* using the graphical method of $G_\mathrm{m}(x)$ curves, for example a two-component liquid mixture where a *phase separation* into two liquid mixtures occurs.

Thermodynamically, the phenomenon is the consequence of a particular shape of the $G_\mathrm{m}(x)$ curve that is concave downwards in an appropriate range of compositions. The shape of $G_\mathrm{m}(x)$ is in turn the consequence of a particular type of interaction between the components in the liquid phase, in which *attractive interactions between equal molecules* (A-A and B-B interactions) *predominate over interactions between different molecules* (A-B interactions). This situation is specifically discussed with a ST

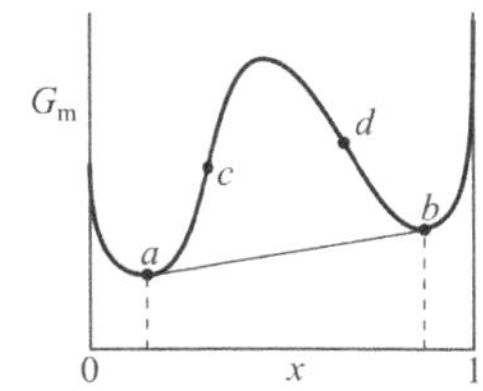

FIGURE 3.6 Phase separation of a liquid mixture into two mixtures.

model in Section 6.4. The effect is essentially enthalpic and, at sufficiently low temperatures, becomes predominant over the entropic contribution. An equilibrium is then possible between *distinct parts* of the *same* $G_\mathrm{m}(x)$ *curve*, with the formation of two liquid mixtures of composition x_a and x_b, if the overall composition lies between these values. The mixtures with compositions $x \leq x_a$ and $x \geq x_b$ do not have a fundamentally different structure, which is in any case that of the liquid, and are different only in composition (Figure 3.6).

The whole discussion is also valid for the demixing of a *crystalline* phase. For example, the treatment of eutectic deposition (Section 3.3.4.1) would also be valid if the two *curves* of the solid phases α and β were instead two *parts* of the same curve $G_\mathrm{m}(x)$.

3.3.5 Stability, Instability and Metastability

We now have the basis to better clarify the meaning of **thermodynamic stability** and related concepts. As a premise, it must be specified that the actual possibility of realizing the equilibrium states predicted and described by thermodynamics, the time necessary to reach them and the time law according to which they are reached are topics that thermodynamics does **not** deal with. This is the field of kinetics of chemical reactions and phase transformations.

There are – as is well known – many processes that require times that are orders of magnitude longer than a typical laboratory experiment.

The fundamental equations for discussing the stability of a thermodynamic state are Eq. (2.41) or Eq. (2.42) when the external variables held fixed are (T, p) or (T, V) respectively:

$$\begin{cases} \mathrm{d}G_{p,T} = 0 \\ \mathrm{d}^2G_{p,T} > 0 \end{cases} \tag{2.41}$$

$$\begin{cases} \mathrm{d}A_{V,T} = 0 \\ \mathrm{d}^2A_{V,T} > 0 \end{cases} \tag{2.42}$$

Briefly speaking, there is **stability** or **metastability** when both conditions hold. **Stability** is an absolute minimum of the thermodynamic potential, **metastability** is a secondary minimum at a value of the thermodynamic potential higher than stability. In other words, we speak of **metastability** in cases that would be correctly recognized as thermodynamic stability if we could ignore the formation of a chemical species or an additional phase that is thermodynamically

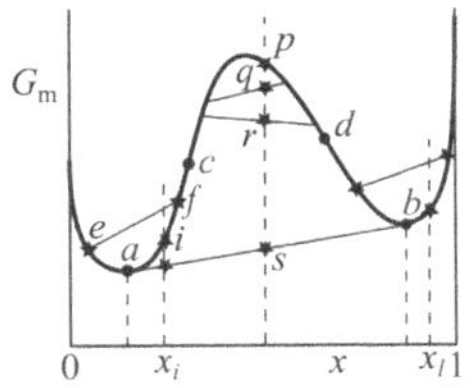

FIGURE 3.7 Stability, metastability and instability in a binary system undergoing demixing.

predicted but does not form for kinetic reasons. **Instability** is when the second constraint is not fulfilled and, therefore, it is not a minimum of the thermodynamic potential.

A clear distinction must first be made between **stability** and **metastability** on the one hand and **instability** on the other and it is convenient to refer to the example of demixing (Section 3.3.4.3). The whole composition range ($0 \le x \le 1$) can be split into three intervals by the compositions (c and d in Figure 3.7) corresponding to the **inflections**, that is, to the zeros of the second derivative: $\left(\frac{\partial^2 G_m}{\partial x^2}\right) = 0$. These are called **spinodal** points.

Let us first discuss the systems with compositions *inside* the $x_c < x < x_d$ interval, for example x_p. Now, the two-phase system is always characterized by a global G_m lower than that of the single-phase system, as shown by the representative points q, r, s, where the last one corresponds to the equilibrium. This is a case of **instability,** and the sign of the second derivative $\partial^2 G_m / \partial x^2$ agrees with the general rule of thermodynamic equilibrium (Eq. 2.41).

The situation is completely different for compositions (x') outside the interval ($x_c < x < x_d$) specified by the spinodal points. A monophasic system with composition $x' \le x_c$, always has a G_m smaller than the G_m of a system consisting of appropriate quantities of two mixtures having one composition $> x'$ and the other $< x'$ *but both in the same interval* ($0 \le x' \le x_c$). For example, we see from Figure 3.7 that for composition x_i, the molar Gibbs energy of a monophasic mixture $G_m(x_i)$ is smaller than the linear combination according to the lever rule of $G_m(x_e)$ and $G_m(x_f)$ however we select x_e and x_f provided that $0 \le x_e \le x_i \le x_f \le x_c$. The same applies to the other interval, $x' \ge x_d$, as shown in Figure 3.7 for a composition x_l with $x_d \le x_l \le 1$. However, in these two intervals a monophasic system does not necessarily correspond to thermodynamic equilibrium, that is to the minimum of the thermodynamic potential. We see indeed that $G_m(x_i)$ is higher than the linear combination (Eq. 3.44) of $G_m(x_a)$ and $G_m(x_b)$ and the same applies to any x' in the interval $x_a < x' < x_c$ as well as to any x' in the interval $x_d < x' < x_b$. In these two intervals thermodynamic equilibrium corresponds to a biphasic system, and the monophasic system is said to be **metastable**.

In conclusion, summarizing the discussions on equilibrium and stability/metastability/instability, the system:

- for $0 \le x \le x_a$ a monophasic system is stable,
- for $x_a < x \le x_c$ a biphasic system is stable with compositions x_a and x_b and a monophasic system is metastable,
- for $x_c < x < x_d$ a biphasic system is stable with compositions x_a and x_b and a monophasic system is intrinsically unstable,
- for $x_d \le x < x_b$ is a biphasic system is stable with compositions x_a and x_b and a monophasic system is metastable,
- for $x_b \le x \le 1$ a monophasic system is stable.

The example of demixing also shows the **meaning** of the **positivity condition** (Eq. 2.41) of the second derivative. Let us consider a *compositional fluctuation* δx such that a domain of the phase adopts a composition $(x' - \delta x)$ slightly lower than the mean value (x') and a nearby domain a composition $(x' + \delta x)$. The Gibbs energy G_m of this inhomogeneous mixture is the linear combination according to the lever rule (Eq. 3.43) of the G_m values at $(x' - \delta x)$ and $(x' + \delta x)$. By looking at any point of the curve $G_m(x)$ where the second derivative is negative, for instance point p, we see that $G_m(x_p)$ is higher than the G_m of the inhomogeneous mixture. Considering instead the points such as e, a, h, f, l where the second derivative is positive, we see that the G_m of the monophasic system is lower than the G_m of the inhomogeneous mixture.

Instability means *lack of stability with respect to fluctuations of every small size*, which must be expected as an inevitable consequence of thermal disorder. Sometimes, it is specified by saying that it is an **intrinsic instability**. It is also a **local property**: it does *not* depend on whether there are more stable compositions different for quantities that are not small (see Section 3.3.4.2, for example).

In **metastability** ranges, on the contrary, a system is **stable with respect to thermal fluctuations**. Metastability is found when the external conditions of a system are changed and a new phase is stable, either alone or in addition to the previous phases, but the **kinetics of the processes leading to the new phase** is so slow that experimentally it does not occur.

This can occur even in a pure substance. Examples are the undercooling of the liquid due to the slowness of the crystallization kinetics once brought below the melting point and the liquid/vapour transition discussed below (see Section 3.3.6 and Figure 3.8).

Another example is the two-component system of Figure 3.3, if initially prepared under conditions where only one phase (α or β) is stable. Once brought to the situation of the figure, equilibrium (point d) corresponds to a biphasic mixture ($\alpha + \beta$). If the kinetics of this phase transformation is slow, the **frozen equilibrium** actually achieved is represented by point a or point b.

A third example is the Gibbs-Thomson effect (Section 3.3.4.2).

In these examples, to reach a thermodynamic stability corresponding to a poly-phasic system, it is necessary to somehow provide a **nucleus** on which the second phase can grow.

We incidentally note that metastability plays an important role in many geochemical problems, where the stable or metastable nature has long been debated and sometimes solved only by computational means or by thermodynamic

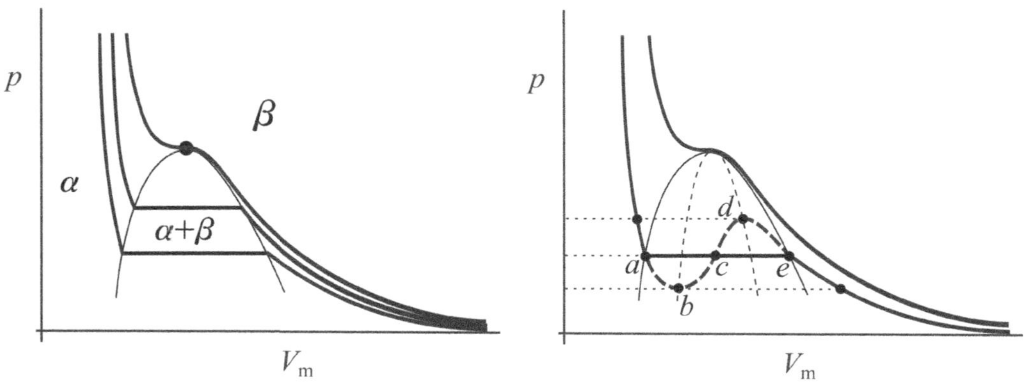

FIGURE 3.8 Stability, metastability and critical behaviour in the liquid/vapour (α/β) equilibrium of a pure substance. Left: experimental behaviour; right: Van der Waals equation. The tick lines are isotherms and the dashed part marks non-equilibrium states; the bell-shaped thin lines enclose the stability field of the biphasic system.

cycle calculations, because the direct experimental determination has been found impossible.

3.3.6 Critical Points

A critical point marks the boundary between the range of a thermodynamic variable where *instability occurs* and another range where *instability does not occur.* The intrinsic instability and the critical points are described by models of thermodynamic potentials with ranges where the second derivative is negative (Eq. 2.41). The example of demixing is useful also in this aspect. Recalling the previous graphs of $G_m(x)$ at constant T and p, we understand that what is needed is a **Gibbs energy model** of the mixture with **downwards concavity**.

Following the regular solutions model mentioned in Section 3.2.5, we write:

$$G_m^E = A \cdot x \cdot (1-x)$$

$$G_m = (1-x)\mu_1^0 + x\mu_2^0 + RT\left[(1-x)\ln(1-x) + x\ln(x)\right] + A \cdot x \cdot (1-x)$$

Assuming for simplicity that the A parameter is independent of T:

$$\left(\frac{\partial^2 G_m}{\partial x^2}\right) = -2A + RT/\left[x(1-x)\right]$$

since the derivative of $(1-x)\cdot\ln(1-x)+x\cdot\ln\ x$ is $-\ln(1-x)-1+x/x+\ln\ x$ and $1/(1-x)+1/x=1/(x(1-x))$. The second derivative can be <0 only if $A>0$ and its minimum is at $x=0.5$, where it is equal to $4\ RT-2A$. At $T<A/2R$ the derivative is <0 in a wide compositional interval around $x=0.5$. On the contrary, at $T>A2/R$, the derivative is always positive in the whole composition interval. $T_c=2A/R$ is called the *critical* unmixing *temperature* because it separates the two intervals of T in which there can or cannot be unmixing. The *critical composition* is $x_c=0.5$. The critical point is identified by T_c and x_c.

Another well-known type of critical point concerns **the equilibrium between liquid** (α) **and vapour** (β) **of a pure substance**. It is usually discussed (Figure 3.8) by the (p, V_m) graph, which reports several isotherms $p=p(V_m)$ where V_m is the molar volume. The coexistence of the two phases is marked by a horizontal isothermal-isobaric segment that extends from the value of the liquid on the left to the value of the vapour on the right. For increasing temperatures, the two values of molar volume at the same p show a convergent trend (the thin lines in Figure 3.8) that finally merge in a single point. This point is marked by a dot on the left part of Figure 3.8 and identifies the critical pressure p_c and the critical molar volume $V_{m,c}$, while the T of the isotherm is the critical temperature T_c. At $T>T_c$ the isotherms show a monotonously decreasing trend, describing a gas that cannot be liquified only by increasing pressure and getting closer to ideal behaviour with increasing temperature.

The analogy of the liquid/vapour critical behaviour with demixing is better clarified using a *single equation of state* for the gaseous state and the liquid state. Historically, the first example has been the Van der Waals (VdW) equation (see Exercise 3.4). At a given T, the VdW equation is of the third degree in the molar volume. For each (T, p) couple, that is for each p along an isotherm, the equation gives three solutions which are either (i) one real and two complex numbers, or (ii) three real numbers. Only real solutions matter. Case (ii) is clearly shown on the right part of Figure 3.8 by the lower thick curve which is partly continuous and partly dashed. For this VdW isotherm there is *only one solution* $V_m=V_m\ (p)$ when $p>p_d$ or $p<p_b$. In the range ($p_b<p<p_d$) there are three distinct v roots at each p. These are for example the abscissae of points a, c and e. At $p=p_b$ and $p=p_d$ two solutions merge, so that there are only two *distinct* solutions.

Now, the point to remark is that a phase transition is marked by a *discontinuity* of some state functions while the VdW equation describes the *continuity* of liquid and vapour. The phase transition arises from the stability conditions given by thermodynamics, and *the thermodynamic constraints must be applied to the VdW equation* to find if and where the transition occurs. The T and p values where liquid and vapour coexist are characterized by two molar volumes V_m(liquid) < V_m(vapour). Therefore, the equilibrium is described (see the right part of Figure 3.8) by a *horizontal segment* drawn between two isobaric points along a VdW isotherm. The fact that the molar volumes of the coexisting phases must appear at the same p along an isotherm also derives from the condition of mechanical equilibrium.

Since the two phases differ in molar volume, it is easier to discuss the equilibrium between liquid and vapour using as thermodynamic potential the Helmholtz energy $A = U - TS$ and its integral molar quantity A/n. Considering that the partial derivative of the molar Helmholtz energy with respect to V_m is the pressure with changed sign (see Eq. (2.27)), the *slope* with changed sign of an isotherm $p = p(V_m)$ is the second partial derivative of the thermodynamic potential (molar Helmholtz energy) with respect to V_m:

$$\left(-\frac{\partial p}{\partial V_m}\right)_T = \left(\frac{\partial^2 A_m}{\partial V_m{}^2}\right)_T$$

and therefore directly expresses the thermodynamic constraint (Eq. 2.42). Conversely, the integral of $-p(V_m)$ along an isotherm is the difference between initial and final values of the thermodynamic potential $A_m = A/n$:

$$\int_1^2 (-p)\, d(V_m)_T = A_{m,2} - A_{m,1} \qquad (3.45)$$

Along the dashed VdW isotherm of the right part of Figure 3.8, points b and d define an interval in which the slope of $p(V_m)$ is positive and therefore the second derivative of the thermodynamic potential A_m is negative. This is where there is **intrinsic instability**: points b and d correspond to the spinodal points of the example of demixing (see Section 3.3.5). The middle solution of the VdW equation at this T (point c) is inside the instability range and is meaningless for what concerns equilibrium. The horizontal line of the liquid - vapour equilibrium connects the left solution a and right solution e: $V_{m,a}$ is the molar volume of the liquid while $V_{m,e}$ is the molar volume of the vapour.

Finally, the common **ordinate** of a and e, which is the *equilibrium vapour pressure* at the selected isotherm, is determined by the constraint that the area of the rectangle between the horizontal segment $a-e$ and the V_m axis *must be equal to the integral of the curve* $p = p(V_m)$ between a and e because both areas are equal to the difference between $A_{m,liquid}$ and $A_{m,vapour}$ (see Eq. (3.45)). In an equivalent way, we can say that the $a-e$ segment must be at such a pressure that it encloses with the two branches of the dashed curve *two equal areas*.

Along the lower VdW isotherm of the right part of Figure 3.8, therefore, the branches to the left of a ($V_m < V_{m,a}$) or to the right of e ($V_m > V_{m,e}$) correspond to **monophasic stability** of the liquid and vapour phases, respectively. Along the dashed branch (*abcde*), the intervals $V_{m,a} < V_m < V_{m,b}$ and $V_{m,d} < V_m < V_{m,e}$ correspond to *biphasic stability* and *monophasic metastability*: the isothermal-isobaric paths $a \rightarrow b$ and $e \rightarrow d$ can be experimentally realised, at least partially, starting from the liquid or from the vapour. In conclusion, the *continuous* parts of the lower VdW isotherm of the right part of Figure 3.8 and *the horizontal segment* describe *thermodynamic equilibrium*, while the *dashed* part describes *out-of-equilibrium states*.

The same considerations apply to other isotherms. By increasing T, the three solutions of the VdW equation get closer, and a temperature is reached where the equation shows three *equal* real solutions at a point that is also a horizontal inflection. This identifies the critical temperature T_c, and the coordinates of the horizontal inflection point are the *critical* molar *volume* and the *critical pressure*. The VdW curve $p = p(V_m)$ at T_c is shown by the upper thick curve in the right part of Figure 3.8. The bell-shaped curves drawn as thin continuous or dashed lines connect the points a and e, or b and d, along the various isotherms.

Above T_c all isotherms show only one real solution $V_m = V_m(p)$ and a monotonously decreasing trend. $T > T_c$ is the range of supercritical fluid and does not need to be discussed for clarifying liquid-vapour equilibrium, metastability or critical behaviour. Temperature, therefore, is the thermodynamic variable that controls the presence/absence of intrinsic instability and T_c is the value that marks the boundary between the two behaviours.

Several correspondences with the case of the unmixing of a two-component system are well evident. There are (i) the stability, metastability and instability ranges of the state variables, (ii) the stable bivariant monophasic ranges, (iii) the stable monovariant biphasic ranges defined by the equality condition of the appropriate thermodynamic potential of the coexisting phases, (iv) the metastable monophasic ranges within the biphasic ranges, due to the slow formation of the second phase, (v) the positive values of the second derivative of the thermodynamic potential for both the stability and metastability ranges, and (vi) a critical point between stability and instability and T as the thermodynamic variable that controls the boundary.

The two examples, demixing and one-component liquid/vapour equilibrium, show that a *single* continuous thermodynamic potential (G/n or A/n) can represent *two* phases. This may appear to be of little significance in the case of demixing, where the two phases differ only in composition, but it is much more noteworthy in the liquid/vapour case, where they are commonly considered as essentially different phases.

3.4 Chemical Reactions

Let us now look at the case of a **single-phase system** in which a chemical reaction occurs:

$$aA + bB + \cdots \rightleftharpoons cC + dD + \cdots \qquad (3.46)$$

where A, B, C, D *are substances* and a, b, c, d the pertinent stoichiometric coefficients.

The equilibrium composition can be calculated by determining the Gibbs energy of the mixture and imposing the minimum conditions already seen:

$$\begin{cases} dG_{T,p} = 0 \\ d^2G_{T,p} > 0 \end{cases} \tag{2.41}$$

3.4.1 The Equilibrium Constant

If the reaction (Eq. 3.46) is not at equilibrium, the quantities of the different substances may change, and the infinitesimal transformation corresponds to:

$$dG_{T,p} = \sum_i \mu_i dn_i < 0 \tag{3.47}$$

To use Eq. (3.47) or the first of Eq. (2.41) we must consider the fact that the various dn_i are not independent of each other because they are linked by the stoichiometry of the reaction. It is convenient to rewrite a generic chemical equation (Eq. 3.45) as:

$$\sum_i \nu_i J_i = 0 \tag{3.48}$$

where the J_i are the *chemical species* (A, B, ..., C, D, ...) *and the various* $\nu_i = \nu(J_i)$ *are the stoichiometric coefficients* (a, b, ..., c, d, ...) *with sign*, >0 for the products and <0 for the reactants, also known as **stoichiometric numbers**. For example, for the reaction $CO + \frac{1}{2}O_2 \leftrightarrows CO_2$ we have: $\nu(CO) = -1$, $\nu(CO_2) = 1$, $\nu(O_2) = -1/2$. Note that if a reactive mixture is prepared with initial quantities $n_0(CO)$, $n_0(CO_2)$, $n_0(O_2)$ of the three reagents and the reaction is allowed to proceed, for example with the formation of an additional quantity x of carbon dioxide, then:

$$n(CO) = n_0(CO) - x; \quad n(CO_2) = n_0(CO_2) + x;$$

$$n(O_2) = n_0(O_2) - x/2$$

Starting from the initial (out-of-equilibrium) situation specified by $[n_0(CO), n_0(CO_2), n_0(O_2)]$:

a. only one variable (x) is sufficient to describe the progression of the reaction towards equilibrium and
b. once equilibrium is reached at a particular value of x, the equilibrium is specified by only two amounts of substance n_i. It is said that there are **two independent components**, corresponding to the three total components minus one constraint, the stoichiometry of the chemical reaction.

Note, on the other hand, that there is ample freedom, but also ambiguity, in writing a reaction: all coefficients (as defined in Eq. (3.48)) can be multiplied by the same common factor, which may be negative. For example, instead of writing $CO + \frac{1}{2}O_2 \leftrightarrows CO_2$ with $\nu(CO) = -1$, $\nu(CO_2) = 1$, $\nu(O_2) = -1/2$ you can write the reaction backwards and then it would be $\nu(CO) = 1$, $\nu(CO_2) = -1$, $\nu(O_2) = 1/2$ or you could write $2CO + O_2 \leftrightarrows 2CO_2$ and then $\nu(CO) = -2$, $\nu(CO_2) = 2$, $\nu(O_2) = -1$. Therefore, as a measure of the progress of the reaction from an initial out-of-equilibrium situation, it is not convenient to use the change of one or the other of the n_i, because we would use different quantities depending on the chosen reagent or product. Instead, it is convenient to define **the extent of reaction** as:

$$d\xi = \frac{1}{\nu_i} dn_i \tag{3.49}$$

Still in the previous example, $d\xi = -dn(CO) = dn(CO_2) = -dn(O_2)/2$. ξ is measured in the same units as n, and therefore has units of moles.

A dimensionless extent of reaction can be generally defined in the range between 0 (reactants only) and 1 (products only). For the example of reaction $CO + \frac{1}{2}O_2 \leftrightarrows CO_2$ and assuming that *the limiting reactant is* CO (*i.e.* that there is enough O_2 to convert all the CO to CO_2), we can define a dimensionless extent of reaction by normalizing to the situation corresponding to the total consumption of CO (ξ_{max}, and $\xi' \overset{def}{=} \xi/\xi_{max}$): then, the situation $\xi' = 0$ corresponds to $[n(CO) = n_0(CO), n(CO_2) = n_0(CO_2), n(O_2) = n_0(O_2)]$, the situation $\xi' = 1$ corresponds to $[n(CO) = 0, n(CO_2) = n_0(CO_2) + n_0(CO), n(O_2) = n_0(O_2) - 2n_0(CO)]$.

The example also clarifies that the definitions for which $\xi' = 0$ is the initial state and $\xi' = 1$ the (hypothetical) final state of the reactive process, with the condition that the equilibrium corresponds to a given ξ' within the interval 0–1, are not univocal, because in any case they are insufficient to specify the state of the system. For example, it is legitimate to exchange the definitions for the initial state with the final state or to use a definition such that the equilibrium corresponds to a negative advancement. These considerations are also important for the kinetics of chemical reactions: often a reaction rate is defined by relations such as $d[A]/dt$, where [A] is the molar concentration of a reactant or a product, while the correct definition is $d\xi/dt$.

From Eqs. (3.49) and (3.47):

$$dG_{p,T} = \sum_i \mu_i \nu_i d\xi \tag{3.50}$$

The Gibbs energy of reaction is the derivative of the Gibbs energy with respect to the advancement:

$$\Delta_r G = \sum_i \mu_i \nu_i = \left(\frac{\partial G}{\partial \xi}\right)_{p,T} \tag{3.51}$$

Note. The operator Δ_r corresponds dimensionally to the ratio between two extensive quantities (*e.g.*, for Eq. (3.51) it is an energy divided by a quantity of substance) and therefore defines an intensive quantity, even if - for uniformity with the consolidated notation - it is written $\Delta_r G$, *not* $\Delta_r G_m$ as one would expect. Similar specifications apply to the variations of internal energy, enthalpy, volume, Helmholtz energy, …. The corresponding quantities relating to **phase transitions** are defined analogously (see Eq. (3.39)).

At equilibrium $dG_{p,T}=0$ so that:

$$\left(\frac{\partial G}{\partial \xi}\right)_{p,T} = \sum_i \nu_i \mu_i = \Delta_r G = 0 \qquad (3.52)$$

We now use Eq. (3.33), write the chemical potentials in terms of activity:

$$\begin{aligned}\Delta_r G = \left(\frac{\partial G}{\partial \xi}\right)_{p,T} &= \sum_i \nu_i \mu_i = \sum_i \nu_i \mu_i^{\ominus} + RT\sum_i \nu_i \ln a_i \\ &= \Delta_r G^{\ominus} + RT \ln \prod_i a_i^{\nu_i}\end{aligned} \qquad (3.53)$$

and define the **reaction quotient** as:

$$Q = \prod_i a_i^{\nu_i} \qquad (3.54)$$

As the reaction proceeds, the concentrations and activities generally change, and the reaction quotient also changes. At equilibrium:

$$-RT \ln Q_{\text{equilibrium}} = -RT \ln \left(\prod_i a_i^{\nu_i}\right)_{\text{equilibrium}} = \Delta_r G^{\ominus} \qquad (3.55)$$

and therefore it is constant, that is, it does not depend on the concentrations.

At equilibrium, the reaction quotient does not depend on the concentrations and is called the **equilibrium constant (*K*)**.

$$K = \left(\prod_i a_i^{\nu_i}\right)_{\text{equilibrium}} \qquad (3.56)$$

We therefore have that the equilibrium condition (Eq. 3.52) becomes:

$$\Delta_r G^{\ominus} = -RT \ln K \qquad (3.57)$$

Remark 3.1

By direct extension, the whole discussion is also valid when among m substances we can write not one but $r>1$ chemical reaction: we will then have r advancements and r equilibrium constants, as clarified further on (3.9).

Remark 3.2

Like the reaction quotient, the equilibrium constant is also dimensionless, as is obvious from the fact that they are products and/or ratios of dimensionless quantities (the activities, Eq. (3.54)).

3.4.1.1 Practical Constants

The quantities (partial pressures, mole fractions, molar or molal concentrations, or other …) we are interested to determine when solving the equilibrium equations are usually grouped in quotients that correspond to the equilibrium constants and are called *practical* (equilibrium) constants. Similarly, we group the activity or fugacity *coefficients* in corresponding quotients. These practical equilibrium constants, of course, are not - strictly speaking - constant and have dimensions and units that depend on the stoichiometry of the reaction. For example, for the usual reaction $CO + \frac{1}{2}O_2 \leftrightarrows CO_2$, using f for the fugacities and γ for the fugacity coefficients, Eq. (3.56) is written as:

$$K = \frac{f(CO_2)}{f(CO)\cdot\sqrt{[f(O_2)]}} = K_p \cdot K_\gamma \text{ where}$$

$$K_p \stackrel{\text{def}}{=} \frac{p(CO_2)}{p(CO)\cdot\sqrt{[p(O_2)]}} \quad \text{and} \quad K_\gamma \stackrel{\text{def}}{=} \frac{\gamma(CO_2)}{\gamma(CO)\cdot\sqrt{[\gamma(O_2)]}}$$

K_p has the dimensions $p^{-1/2}$ and, for example, is measured in $Pa^{-1/2}$. K_γ has the inverse dimensions ($p^{1/2}$). In all these relations, the only quantity that is a real equilibrium constant is K, which depends only on T as state variable. Instead, the quotient of the fugacity coefficients (K_γ) and the so-called practical constants (K_p and K_x) vary with the partial pressures or mole fractions. If the fugacity coefficients can be assumed to be all reasonably close to 1, then it is allowed to discuss the equilibrium using K_p alone. Anyway, the quotient of the fugacity coefficients generally varies more slowly than K_p with the amounts of reactants and products in the reaction and therefore it may be reasonable to treat it as invariant as a first approximation. If necessary, the variations in the fugacity coefficients are considered as a final correction.

Remembering that - by definition and even in case of non-ideal systems - the partial pressure of a component (p_i) is the product of the mole fraction (x_i) by the total pressure p, we also have:

$$K_p = K_x \cdot p^{-1/2} \quad \text{where} \quad K_x \stackrel{\text{def}}{=} \frac{x(CO_2)}{x(CO)\cdot\sqrt{[x(O_2)]}}$$

In the case of reactions in a homogeneous liquid phase, the compositional quantities of choice in discussing chemical equilibria are the molar concentrations, indicated by convention by square brackets. For example, for the reaction between an alcohol (ROH) and a carboxylic acid (R′COOH) in aqueous solution we write:

$ROH + R'COOH \leftrightarrows R'COOR + H_2O$ and

$$K = K_c \cdot K_\gamma = \frac{a(\mathrm{R'COOR}) \cdot a(\mathrm{H_2O})}{a(\mathrm{ROH}) \cdot a(\mathrm{R'COOH})}$$

where $K_c = \dfrac{[\mathrm{R'COOR}][\mathrm{H_2O}]}{[\mathrm{ROH}][\mathrm{R'COOH}]}$ and

$$K_\gamma = \frac{\gamma(\mathrm{R'COOR}) \cdot \gamma(\mathrm{H_2O})}{\gamma(\mathrm{ROH}) \cdot \gamma(\mathrm{R'COOH})}$$

Surely, the procedures for solving an equilibrium in an exact or approximate way are well known and therefore a quick example may suffice. Ignoring the problems related to fugacity coefficients, let's consider the reaction $CO + \frac{1}{2} O_2 \leftrightarrows CO_2$ and an equimolar starting mixture of CO and CO_2, *e.g.* both at a partial pressure of 1 atm. Let's also suppose that at the temperature at which we work K is equal to 10^{10}. Then, the reaction produces some amount of oxygen, but it is easy to note that the partial pressure of oxygen must be very small due to the large value of K and therefore we do not make a significant error by assuming that at equilibrium the ratio $p(CO_2)/p(CO)$ does not deviate from the initial value (= 1). Then $p(O_2) \approx 10^{-5}$.

When it is not possible to resort to this further approximation, using p' to denote the initial values of the pressures and still assuming ideal behaviour of the gas, one can write:

$$K = \frac{[p'(\mathrm{CO_2}) - x]}{[p'(\mathrm{CO}) + x]\sqrt{x/2}}$$

and solve the resulting quadratic equation, with one root to be discarded as it yields a negative amount of substance.

3.4.2 *T* and *p* Dependence of the Equilibrium Constant

Equation (3.57) shows that the *equilibrium constant does not depend on pressure.* In fact, the standard state of a chemical species is defined at a given pressure (the standard pressure) and therefore the change in standard Gibbs energy $\Delta_r G^\ominus$ and K are **independent of pressure**.

This *does not mean* that the *equilibrium is independent of pressure*, meaning by equilibrium the values in equilibrium conditions of the extensive or intensive quantities (concentrations, partial pressures, molar fractions, moles, etc.) of the various substances of the reaction. For a reaction between gaseous components, for example, it is well known that an increase in total pressure leads to an increase in the quantities of products (or reagents) if the sum of the stoichiometric coefficients of the products (or reagents) in the reaction is lower than the sum of the stoichiometric coefficients of the reagents (or - respectively - of the products).

The **temperature dependence** is easily obtained by differentiating Eq. (3.57), rewritten as:

$$\frac{\Delta_r G^\ominus}{RT} = -\ln K, \quad \text{and therefore} \quad -\frac{1}{R}\frac{\mathrm{d}(\Delta_r G^\ominus / T)}{\mathrm{d}T} = \frac{\mathrm{d}\ln K}{\mathrm{d}T} \tag{3.57}$$

which can be simplified using the Gibbs-Helmholtz equation (Eq. 3.14):

$$\frac{\mathrm{d}\ln K}{\mathrm{d}T} = \frac{\Delta_r H^\ominus}{RT^2} \tag{3.58}$$

where $\Delta_r H^\ominus$ is the standard reaction enthalpy. Equation (3.58) is called **van't Hoff equation** and shows that

For reactions that - under standard conditions - are **exothermic** ($\Delta_r H^\ominus < 0$) the equilibrium constant decreases as the temperature increases and for reactions that - under standard conditions - are **endothermic** ($\Delta_r H^\ominus > 0$) the equilibrium constant increases as the temperature increases.

3.4.3 Independent Components

If a system contains m components, between which a reaction can occur, the state of the system is specified by specifying only $m-1$ amounts of substance: we say that there are only $m-1$ **independent components**. In the usual example of the reaction $CO + \frac{1}{2} O_2 \leftrightarrows CO_2$, assuming the ideality of the mixture and writing Eq. (3.56) as a function of the partial pressures only:

$$K = \frac{p(\mathrm{CO_2})}{p(\mathrm{CO}) \cdot \sqrt{[p(\mathrm{O_2})]}}$$

We can therefore see that the p_i and the amounts $n_i = \frac{p_i \cdot V}{RT}$ of the various substances (CO, O_2 and CO_2) are linked to each other by K: only two of them are freely variable.

Generalising, in a system in which chemical reactions occur, the number of *independent* components is obtained by subtracting the number of chemical reactions from the total number of substances, because each reaction is a constraint between the amounts of substance. With many reactions, however, it is necessary to pay attention to the fact that the **reactions must be linearly independent**, that is, it is necessary to check that it is not possible to write even one of them in the form of a linear combination of the others. The number of independent components is therefore given by $m-r$, where r is the number of **independent reactions**.

It should also be noted that adding association or dissociation equilibria between the components do not change the situation. For example, for a reaction in aqueous phase, one could hypothesise in addition to the simple H_2O species the presence of dimeric, trimeric, etc. species of the solvent. There would then be other components, but also other formation reactions, for example:

$$2H_2O \leftrightarrows H_4O_2$$

and the count of independent components does not change. The example also shows a more general aspect: the choice of the species and reactions to consider is essentially *practical.* If there are measurements or computational results that for a given species allow us to identify and quantify polymeric forms (*e.g.*: H_4O_2, etc.) alongside the monomeric form, or other associations, then it makes sense to add these to the components and the related formation reactions to the constraints. If this is not possible, or is not of interest, only one component is considered, which implicitly includes any polymeric forms alongside the monomer. It is also clear that in the last case the hypothesis of ideal behaviour of the mixture might be less reliable.

3.5 Heterogeneous Equilibrium

To extend the previous discussions and include cases in which there are both phase equilibria and chemical reactions, we once again resort to the general principle of equality of chemical potentials for each component in the different phases:

$$\mu_i^\alpha = \mu_i^\beta \tag{3.35}$$

combined with the relationship between chemical potentials of the different substances of a reaction:

$$\left(\frac{\partial G}{\partial \xi}\right)_{p,T} = \sum_i \nu_i \mu_i = \Delta_r G = 0 \tag{3.52}$$

As already seen (see Section 3.4.3), the **number of independent components** of a system is given by the total number of substances that appear in the system minus one for each (independent) chemical reaction that can be written between those substances and for each additional constraint. A simple example is the thermal decomposition of calcium carbonate: $CaCO_3 \leftrightarrows CaO + CO_2$. There are three phases (carbonate, oxide and gas) and three substances but only two independent components (three substances and a chemical reaction): the variance is $C-P+2=2-3+2=1$. A given T corresponds to an equilibrium partial pressure of CO_2. If the partial pressure of CO_2 in the system is lower (or higher) than the equilibrium value, the reaction proceeds to the right (or to the left) until the value reaches the equilibrium value. For more precise determinations, one must consider the fugacity rather than the partial pressure (Section 3.2.5).

3.6 Reactions of Charged Species

In gas-phase reactions, reactants and products are well-defined and only weakly interacting molecules, ideally not interacting at all, not considering the reactions. Considering condensed-phase reactions means moving to a situation in which interactions with other components become significant and less clearly defined species are possibly involved. In this regard, it is sufficient to compare the usual reaction between CO and O_2 in the gaseous phase with some of the reactions in the liquid phase already considered (the esterification of a carboxylic acid, the dimeric forms of a solvent, …) to understand how each of the apparently simple formulas, such as CH_3COOH, CH_3CH_2OH, $CH_3COOCH_2CH_3$, … and their associations with a solvent, can in reality represent a multiplicity of atomic arrangements, even very complicated ones.

As a matter of fact, *thermodynamics leaves ample freedom* in the choice of what can be considered as a species taking part in a chemical reaction and can therefore be treated with the formalism of $\Delta_r G$ and the equilibrium constant.

The first extension is that of reactions between charged species (ions).

3.6.1 Ionic Reactions

While a particularly relevant case of reactions involving ions is that of acid/base equilibria, for example in aqueous solution:

Self-ionization of the solvent:

$$2H_2O \leftrightarrows OH^- + H_3O^+, \quad \text{or}$$

Dissociation of a weak acid:

$$CH_3COOH + H_2O \leftrightarrows CH_3COO^- + H_3O^+$$

it is also possible to write reactions that involve only charged species, for example, in aqueous solution:

$$Fe^{2+} + Ce^{4+} \leftrightarrows Fe^{3+} + Ce^{3+}$$

Of course, such a reaction can be written as a normal reaction between neutral species, at the cost of inserting as reactants and products appropriate amounts of other ions of opposite sign. However, since these other ions do not actually take part in the reaction, it is preferable to neglect them in the discussion. In these cases, it is correct to write an equilibrium constant:

$$K = \frac{[Fe^{3+}][Ce^{3+}]}{[Fe^{2+}][Ce^{4+}]}$$

as seen for reactions between neutral species.

However, the extension of the normal treatment of chemical reactions to charged species **must be justified** because the basic definition of the chemical potential:

$$\mu_i = \left(\frac{\partial G}{\partial n_i}\right)_{T,p,n_{j\neq i}} \tag{3.19}$$

cannot be applied. In fact, one can vary the n_i of a neutral species *holding constant all other amounts of substance*, but one cannot do so for a charged species, because it breaks the condition of electro-neutrality.

The problem is solved by splitting the chemical potentials of neutral species (for which the definition of Eq. (3.19) holds) according to appropriate dissociation reactions. The quantities thus obtained for charged species can be used **operationally** in the same way as chemical potentials (see *e.g.* Eq. (3.47)) and are therefore also referred to as chemical potentials. With this extended definition, for example, the chemical potentials of the ions in an aqueous solution of Na^+ and CO_3^{2-} ions are derived from the chemical potential of sodium carbonate according to the reaction $2Na^+ + CO_3^{2-} \leftrightarrows Na_2CO_3$ which translates to $2\mu(Na^+) + \mu(CO_3^{2-}) = \mu(Na_2CO_3)$. Experimentally, the chemical potentials on the left cannot be determined independently, that on the right can, but it is necessary to establish how it should be distributed among the chemical potentials of the ions. The essential point is that, once the chemical potential of a single ion is fixed in some way, the chemical potentials of all ions can be determined with appropriate sequences of reactions, not mentioning technical difficulties. For example, having fixed $\mu(H^+)$ and using the reaction $H^+ + Cl^- \leftrightarrows HCl, \mu(Cl^-)$ is determined by measuring the chemical potential of HCl. Then, considering the reactions of formation of various chlorides and knowing their chemical potentials, the chemical potentials of the corresponding cations are determined, etc....

The decomposition of the chemical potential of a neutral species into the sum of the chemical potentials of charged species therefore requires an appropriate **convention**, which essentially fixes the chemical potential of a particular ion (H^+ or, better, H_3O^+) under certain conditions.

Summarizing, in spite of the impossibility of using the general definition (Eq. 3.19), the chemical potentials of charged species can be defined using their formation reactions from neutral species and an appropriate convention. The quantities so defined share the properties of the *normal* chemical potential (defined by Eq. (3.19)) and can be operationally used in the same way, but for two aspects. One point is that **their values are conventional**, the other is that, when counting the independent components $m-r$ (Section 3.4.3), the number of constraints r is no longer given by the number of independent reactions: in the case of reactions between charged species, the **constraint** relating to **the balance of the charges** must be added.

For example, in aqueous solution and under the approximation of replacing activities with concentrations, the equilibrium constant of the reaction $Fe^{2+} + Ce^{4+} \leftrightarrows Fe^{3+} + Ce^{3+}$ is a constraint between the amounts of the ions, but another constraint is that the charges appearing on the right must equal those on the left: $2 \cdot [Fe^{2+}] + 4 \cdot [Ce^{4+}] = 3 \cdot [Fe^{3+}] + 3 \cdot [Ce^{3+}]$. In conclusion, there are two constraints and two independent components.

Similarly, the dissociation of a weak acid (denoted by HA) in water is described by the two equilibria:

$$K_a = \frac{[H_3O^+][A^-]}{[HA]} \quad \text{and} \quad K_w = [H_3O^+][OH^-]$$

and there is the charge balance: $[H_3O^+] = [A^-] + [OH^-]$. The components are H_2O, HA, A^-, H_3O^+ and OH^-. In total there are three constraints (two reactions and the charge balance) and five components of which two (5–3) are independent. From the point of view of the phase rule, the variance of a system with two independent components in a single phase is $2-1+2=3$. For example, one can freely change (within finite limits) the temperature, the pressure and the overall acid concentration $c = [HA] + [A^-]$.

In general, the charge balance condition is $\sum_i c_i z_i = 0$, where z_i is the charge, c_i is the concentration, and the summation is over all charged species. With acid/base reactions, to obtain the various concentrations at given values of c, K_a and K_w, one usually starts by writing the charge balance:

$$[H_3O^+] - [OH^-] - [A^-] = 0$$

From $K_a = \frac{[H_3O^+][A^-]}{c - [A^-]}$ we get $[A^-] = \frac{cK_a}{K_a + [H_3O^+]}$

and from $K_w = [H_3O^+][OH^-]$ we get $[OH^-] = \frac{K_w}{[H_3O^+]}$.

Then: $$[H_3O^+] - \frac{K_w}{[H_3O^+]} - \frac{cK_a}{K_a + [H_3O^+]} = 0$$ and:

$$[H_3O^+]^3 + K_a[H_3O^+]^2 - [H_3O^+]\{cK_a - K_w\} - K_w K_a = 0 .$$

It is necessary to solve a third-degree equation in the concentration of H_3O^+, from which the other concentrations can then be obtained. If $[H_3O^+] >> [OH^-]$ (*e.g.*, if $[H_3O^+] > 10^{-5}$) then the equilibrium K_w can be completely neglected and the equation to be solved becomes a second-degree equation:

$$[H_3O^+]^2 + K_a[H_3O^+] - \{cK_a\} = 0$$

and, in the further hypothesis of poor dissociation (($[H_3O^+] << c$) it produces the well-known approximate formula:

$$[H_3O^+] = \sqrt{cK_a}$$

3.6.2 The Chemical Potential of Electrons

It may look strange to deal with relations such as

$$2\mu(H^+) + 2\mu(e^-) = \mu(H_2) \quad \text{and}$$

$$2\mu(H_2O) = 4\mu(e^-) + 4\mu(H^+) + \mu(O_2)$$

for the chemical reactions

$$H_2 \leftrightarrows 2H^+ + 2e^- \quad \text{and} \quad O_2 + 4H^+ + 4e^- \leftrightarrows 2H_2O \quad (3.59)$$

where *electrons* are treated in the same way as conventional chemical species such as atoms, ions, or molecules. However, there is no *thermodynamic* reason for *not* applying to electrons the treatment outlined in the previous points. Furthermore, these reactions (Eq. 3.59) are particularly important because they are, let us say, the building blocks of a very important class of (neutral) chemical reactions, the **oxidation-reduction (redox) reactions**, and are therefore known as **redox half reactions**.

The two half reactions of Eq. (3.59) are indeed combined in the redox reaction:

$$2H_2 + O_2 \rightleftharpoons 2H_2O \tag{3.60}$$

where the four electrons exchanged between hydrogen and oxygen do not explicitly appear. Our main interest in splitting a redox reaction into half reactions is that *the interaction of an electric field with the charged species of the half reactions can be used to control the whole redox reaction.* In particular, a spontaneously occurring redox reaction can be used as an electric energy producer, or else an external electric energy source can be used to drive a redox reaction contrary to its spontaneous way for obtaining a chemical product. The wide area of chemistry dealing with these and related processes is **electrochemistry** and is not treated here extensively: the following point 3.7 simply outlines the basic thermodynamic aspects of the electrochemical equilibrium and the fundamental role played there by the chemical potential of the electron.

Actually, *the chemical potential of the electron requires some comments* with respect to the other charged species. In short, the reason is that electrons follow the Fermi-Dirac (FD) statistics instead of the Maxwell-Boltzmann (MB) statistics. Let us summarize here the main features of the Fermi-Dirac (FD) statistical distribution $s_{FD}(\varepsilon)$ of the **occupancies of single electron states** as a function of their energy ε, as widely discussed in Sections 5.2.2.1, 5.3, and 8.6. The important difference from the other distribution laws (Bose-Einstein, BE, and MB) are (i) that the FD never exceeds 1, (ii) that the chemical potential of an electron $\mu(e)$ corresponds to the inflection point of the distribution and to half-occupancy ($s_{FD}(\varepsilon) = 1/2$, see Section 5.2.2.4), and finally (iii) that the FD goes from a value "very close" to 1 to a value very close to 0 in an energy range of the order of kT (or RT on a molar scale) around $\mu(e)$ (see Section 8.6).

In most atomic and molecular systems, the electronic states are characterized by discrete energy levels that are widely spaced from each other, so that the promotion of an electron from a normally filled state to a normally empty state requires an energy that is typically two orders of magnitude, or more, higher than kT at typical temperatures. Therefore, the states at energy below $\mu(e)$ are fully occupied while those above $\mu(e)$ are completely empty.

In condensed matter, the electronic energies are grouped in energy bands, and the bands are separated by gaps that are almost of the same order of magnitude as a typical spacing between the discrete levels of atoms or molecules, while the energy levels within a band are much more closely spaced and can be well approximated by a continuous energy distribution.

Now, in **insulators**, $\mu(e)$ falls within a *band gap* so that the result does not change significantly with respect to atoms or molecules. On the contrary, in a **metallic system** $\mu(e)$ falls within a band or an envelope of bands and a kT-wide energy interval around $\mu(e)$ contains plenty of empty states and filled states available for exchanging electrons. This is achieved through thermal promotion or doping also in **semiconductor**s, where $\mu(e)$ falls within a *band gap* but the gap is smaller than in insulators. In the last two kinds of materials, an applied electric *potential* rises or lowers in energy all the band structure, including $\mu(e)$, while an applied electric *field* gives rise to electronic conductivity. Concerning the macroscopic level:

- in an electronic conductor, any injected charge spreads out throughout the whole body.
- The chemical potential of the electrons can be finely tuned using an externally applied electric potential.
- If the electronic conductor is placed in contact with another phase containing chemical species that can undergo oxidation or reduction, the reaction possibly injects or removes "large" amounts of electrons without "large" changes in the chemical potential of the electrons and
- the thermodynamic equilibrium reached after a short kinetic transient corresponds to the onset of a net electric charge in the electronic conductor and an electric potential difference between the phases.

3.7 Electrochemical Equilibrium

An **electrochemical cell** is a device that allows to carry out **in a particular way** a **redox reaction** by spatially separating the half-reactions. Oxidation and reduction occur at the interfaces between electron-conducting phases (electrodes) and ion-conducting phases (electrolytes).[7]

Using two or more electrodes, a different spatial localization of the two half-reactions can be obtained and the separation of charges thus achieved corresponds to a potential difference (ΔE) between the electrode on which the oxidation occurs (anode) and the one where the reduction occurs (cathode). The overall redox reaction of an electrochemical cell is a reaction with neutral reactants and products and (at least in principle) can also be carried out outside an electrochemical cell. For example:

$$2H_2 + O_2 \leftrightarrows 2H_2O \tag{3.60}$$

It is worth noting the difference between

a. the thermal reaction carried out in the gas phase at high temperature and characterized by a radical chain reaction mechanism and typically explosive kinetics and
b. the same reaction carried out in an electrochemical cell at room temperature, mechanistically characterized by the charge transfers between electrodes and solution and easily controlled with chemical and electrical parameters.

Despite the significant differences in mechanism, mode of use and yield in energy conversion, from the point of view of thermodynamics it is the same reaction and the thermodynamic parameters ($\Delta_r G$, $\Delta_r H$, $\Delta_r S$, …) obtained from the experimental or computational study of one process can be applied to the other. Incidentally, electrochemical measurements are a widespread and precise method for the experimental determination of the thermodynamic parameters of chemical reactions.

The two half-reactions (Eq. 3.59) can also be written differently while keeping the overall reaction unchanged. For example, in a basic aqueous medium it is more appropriate to write Eq. (3.59a) as $H_2(\text{gas}) + 2OH^- \leftrightarrows 2H_2O + 2e^-$ and Eq. (3.59b) as $O_2(\text{gas}) + 2H_2O + 4e^- \leftrightarrows 4OH^-$.

From a strictly thermodynamic point of view, the redox reactions that occur in an electrochemical cell are an important example of both a **heterogeneous equilibrium** and a **thermodynamic system** in which **it is necessary to consider a work term in addition to** the almost inevitable ***pV* work**.

Generally, in an electrochemical device the overall chemical reaction is out of *chemical* equilibrium and $\Delta_r G \neq 0$. When a reagent of a half-reaction is placed in contact with an electrode, an exchange of electrons initially occurs between the electrode and the other phases involved and an accumulation of electric charge occurs on the electrodes due to the electrons exchanged in the respective half-reactions. If the electrodes are electrically insulated from each other, after an initial transient an **electrochemical equilibrium is established** in which (i) no more net exchanges of electrons occur, (ii) no electric current flows in the cell, (iii) the two half reactions and the overall redox reaction stop, and (iv) the chemical imbalance ($\Delta_r G < 0$, as we are supposing that a reaction does occur) is equalized by the electrical work corresponding to the potential difference generated in the initial transient.

As seen before (Eq. 2.22), the variation of Gibbs energy gives the maximum net amount of reversible work that can be obtained. Here, the $\Delta_r G$ is that of the global reaction: $\Delta_r G = \sum_i \nu_i \mu_i$, while the work involved is the **electric work**, given by the product of the potential difference between anode and cathode (ΔE_{eq}) multiplied by the charge exchanged in the reaction nF.[8] Therefore:

$$\Delta E_{eq} = -\Delta_r G / nF$$

Using Eqs. (3.53) and (3.54):

$$\Delta_r G = \Delta_r G^{\ominus} + RT \ln Q; \quad Q = \prod_i a_i^{\nu_i} \quad \text{we have:}$$

$$\Delta E_{eq} = -\frac{\Delta_r G^{\ominus}}{nF} - \frac{RT}{nF} \ln Q = \Delta E_{eq}^{\ominus} - \frac{RT}{nF} \ln Q; \; \Delta E_{eq}^{\ominus} \overset{\text{def}}{=} -\frac{\Delta_r G^{\ominus}}{nF} \tag{3.61}$$

This is perhaps the best-known form of the famous **Nernst equation**. The negative sign in Eq. (3.61) is conventional and expresses the fact that we want to treat as positive the work that can be obtained from a chemical reaction taking place in its spontaneous direction ($\Delta_r G < 0$). This choice is justified by considerations of convenience in applied science but is contrary to the more general convention (see Section 1.3.2), which instead considers the work **done on** the system to be positive and the work **done by** the system to be negative.

Although outside the scope of thermodynamics, it is appropriate to briefly mention the behaviour of a cell out of electrochemical equilibrium. If an electric potential difference $\Delta E > \Delta E_{eq} = -\Delta_r G / nF$ is applied in opposition to the electrodes, the reaction can proceed in the opposite direction to the spontaneous one. The electrochemical device acts as a **substance producer** using the energy supplied by the external source: a typical device is an *electrolysis cell*.

If, instead, the electrodes of an electrochemical cell are connected to a device capable of using electrical energy (a resistor or an electric motor, for example), the cell reaction can take place spontaneously with the production of heat (due to the Joule effect) and/or work at the expense of the $\Delta_r G$. The device acts as an **energy producer** and is usually called a *battery* or a *pile*.

In both cases, the electrochemical cell is out of **electrochemical equilibrium**. Electric currents flow in both cases, with opposite signs, depending on the actual value of ΔE, which in turn is different from the value at electrochemical equilibrium. The difference is negative ($\Delta E < \Delta E_{eq}$) in a battery, positive ($\Delta E > \Delta E_{eq}$) in an electrolysis cell. The actual value of ΔE is also dependent on various dissipative processes, primarily those internal to the cell, *i.e.* the kinetics of the charge and substance transport processes at the interphases and in the electrolyte.

A third very important case from the application point of view is **corrosion**, in which the energy is entirely dissipated within the cell itself. The situation can be trivially realized when the cell is short-circuited, or when the anode and cathode are two areas of the same material.

The half-reactions at the two electrodes are always **kinetically coupled** (in series): the rate of both is that of the slowest process and is the rate of the overall redox reaction. Note also that the measurement of the electric current is a direct measure of the rate of an (electro)chemical reaction.

While electrochemical equilibrium implies zero current, zero current **does not imply** electrochemical equilibrium. A null current may be due to kinetic hindrance for even just one half-reaction. The reversibility of electrochemical equilibrium can be experimentally verified by exploring a range of values of the externally applied electric potential around the electrochemical equilibrium value $\Delta E = -\Delta_r G / nF$.

As regards the efficiency of the conversion of chemical energy into electrical energy, it should be noted that the energy supplied by an electrochemical generator is given by $\Delta_r G$. For a heat engine, on the other hand, it is necessary to consider the $\Delta_r H$, which is typically, but not necessarily, greater. On the other hand, the heat engine is limited by the Carnot efficiency, which does not control the electrochemical conversion. In applications, obviously, much more than the purely thermodynamic efficiency, what counts are the dissipative effects of the processes, *i.e.* the efficiency in terms of *power* delivered, rather than energy.

Finally, note (Eq. 3.34) that with electrochemical measurements it is possible to determine the thermodynamic activities (or fugacities) of species in solution or in the gaseous phase. A well-known example is a pH-metre, which measures the activity of the H_3O^+ ion.

3.8 Quasi-Chemical Equilibria

The list of structural arrangements that can be treated as reactants or products of a chemical reaction can be extended well beyond the case of reactions between ions (3.6) or of redox half-reactions (3.7). However, to show further extensions, it is convenient to proceed by examples. The approach is always to write a reaction, assign a chemical potential to its reactants and products, make explicit the equilibrium constant and finally "solve" the equilibrium in a similar way to what is done, for example, for acid-base equilibria in solution or for gas phase equilibria.

3.8.1 Adsorption

A first interesting case concerns solid/gas adsorption which can written:

$$\mathrm{A} + \mathrm{S} \leftrightarrows \mathrm{AS}$$

where A indicates a molecule in the gas phase, while AS indicates the molecule adsorbed on a site S. Using K to denote the equilibrium constant of this reaction and assuming ideal behaviour both in the gaseous phase and on the adsorption surface, the equilibrium constant is:

$$K = \frac{[\mathrm{AS}]}{p \cdot [\mathrm{S}]}$$

where for short, p indicates the ratio between the partial pressure of the gaseous species A and the standard pressure. Finally, we arrive at the well-known **Langmuir equation** which expresses the dependence of the coverage ϑ of the adsorbent surface on the partial pressure of the species being adsorbed:

$$\vartheta \stackrel{\mathrm{def}}{=} \frac{[\mathrm{AS}]}{[\mathrm{AS}]+[\mathrm{S}]} = \frac{K \cdot p}{K \cdot p + 1} \tag{3.62}$$

Indeed, from the expression for K, we have $K{\cdot}p = [\mathrm{AS}]/[\mathrm{S}]$. The fraction of sites occupied by an adsorbed molecule is 1 minus the fraction of free sites, and then: $[\mathrm{AS}]/([\mathrm{S}]+[\mathrm{AS}]) = 1 - [\mathrm{S}]/([\mathrm{S}]+[\mathrm{AS}]) = 1 - 1/(K{\cdot}p + 1)$ from which Eq. (3.62) follows.

The comparison with the corresponding derivation in ST (see Section 6.2.4), besides validating the treatment of CT, clarifies what ideal behaviour means for the "species" S and AS.

3.8.2 Pure Semiconductors

In extending the approach based on chemical equations, several important cases concern problems of *solid-state chemistry*. Let us start with the equilibrium of formation of an **electron/hole pair in a** (undoped) **semiconductor**. This involves the thermal promotion of an electron from the valence band to the conduction band.

Briefly, the valence band (VB) is the band of electronic states just below the chemical potential of the electrons. At 0 K it is completely substitute with filled. The conduction band (CB) is the band immediately above in energy. At 0 K it is completely empty.

Starting from the reference situation, indicated by the symbol ∅ and corresponding to a completely filled VB and a completely empty CB, the formation of the electron/hole pair is described by the equation (significantly called **quasi-chemical**):

$$\varnothing \leftrightarrows e' + h^{\bullet}$$

$$K_i = [e'][h^{\bullet}] \tag{3.63}$$

where e′ is an electron in the CB and $h^{\bullet}$ is a hole in the VB.

The introduction of the pseudo-particle called **hole** is essentially based on a result of the theory of solid-state electronic bands relating to the behaviour in an electric field of the electrons remaining in the valence band. While a completely filled band corresponds to the electrical behaviour of an insulator, an empty state in an otherwise filled band behaves like a particle with a positive charge. This (pseudo)particle is called a hole and is denoted by $h^{\bullet}$, while e′ denotes an electron promoted to the VB.

Instead of the usual − and + signs, negative and positive charges are now indicated by the symbols ′ and, respectively, a usage that is discussed in detail below. Note that $h^{\bullet}$ and e′ do **not** refer to **particles** formed or released by the reaction but denote particular *states of the entire electronic system of the bands of the solid*, which globally neither loses nor acquires electrons. For comparison, let us remember the case of electrode half-reactions (Eqs. 3.59 and 3.60) which instead correspond to the transfer of an electron from one phase to another, electrode and electrolyte.

3.8.3 Doped Semiconductors

The quasi-chemical approach can be extended to *doped* semiconductors. Let us start with a semiconductor, *e.g.* silicon, where n-type doping is achieved by dissolving aliovalent atoms (*e.g.* As) that enter the regular Si sites of the crystal structure. The electronic states in the valence band of undoped silicon are completely filled by a total amount of four electrons/Si atom. The additional electron of As can enter localized states around the dopant or states in the conduction band. These two cases are denoted as

$\mathrm{As}_{\mathrm{Si}}^{x}$ (the additional electron of As is localized near the dopant) and

$\mathrm{As}_{\mathrm{Si}}^{\bullet}$ (the additional electron in CB)

Note the need for a notation to indicate (i) the atomic species, (ii) where it is placed and (iii) its charge. So, $\mathrm{As}_{\mathrm{Si}}^{\bullet}$ means that an As^{+} ion is placed on a site that, according to the regular crystal structure of silicon is occupied by a Si atom and its formal charge with respect to the regular structure of silicon is +1, because a cation with 1+ charge replaces a neutral atom (Si) and its four electrons of the outermost shell enter the VB. The superscript dot (·) instead of (+) specifies that a relative charge is implied instead of an absolute charge. The analogous case of a relative negative charge is denoted by ('). This notation was already used for electrons and holes in Eq. (3.63). Finally, when the charge is not different from the reference structure, the x superscript is used for better clarity (*e.g.* $\mathrm{As}_{\mathrm{Si}}^{x}$).

The species $\mathrm{As}_{\mathrm{Si}}^{x}$ and $\mathrm{As}_{\mathrm{Si}}^{\bullet}$ are examples of **solid-state point defects**, situations that *can* reach thermodynamic equilibrium and can be treated according to the chemical reaction formalism:

$$\mathrm{As}_{\mathrm{Si}}^{x} \leftrightarrows \mathrm{As}_{\mathrm{Si}}^{\bullet} + e' \quad K_{\mathrm{d}} = \frac{[\mathrm{As}_{\mathrm{Si}}^{\bullet}\mathrm{As}_{\mathrm{Si}}][e']}{[\mathrm{As}_{\mathrm{Si}}^{x}]} \tag{3.64}$$

In the doping equilibrium of semiconductors, Eq. (3.63) plays a role that is in many ways analogous to the role played in acid-base equilibria in aqueous solution by the ionic product of water $K_w=[H_3O^+]\cdot[OH^-]$: it is an ever-present equilibrium that combines with those of the reactions due to the addition of various solutes.

Combining Eq. (3.64) with Eq. (3.63) gives the electronic concentration $[e']$ corresponding to a given overall dopant concentration $c=[\mathrm{As}_{\mathrm{Si}}^{x}]+[\mathrm{As}_{\mathrm{Si}}^{\bullet}]$. The situation is formally analogous to the case of the dissociation of an acid in aqueous solution, where the role of Eq. (3.64) is played by the dissociation reaction of the acid. Performing the calculations in parallel: the reactions are:

$$\varnothing \leftrightarrows e' + h^{\bullet} \qquad 2\mathrm{H_2O} \rightleftharpoons \mathrm{H_3O^+} + \mathrm{OH^-}$$

$$\mathrm{As}_{\mathrm{Si}}^{x} \leftrightarrows \mathrm{As}_{\mathrm{Si}}^{\bullet} + e' \qquad \mathrm{HA} + \mathrm{H_2O} \rightleftharpoons \mathrm{A^-} + \mathrm{H_3O^+}$$

and the equilibrium constants are:

$$K_{\mathrm{i}} = [e'][h^{\bullet}] \qquad K_{\mathrm{w}} = [\mathrm{H_3O^+}][\mathrm{OH^-}]$$

$$K_{\mathrm{d}} = \frac{[\mathrm{As}_{\mathrm{Si}}^{\bullet}][e']}{[\mathrm{As}_{\mathrm{Si}}^{x}]} \qquad K_{\mathrm{a}} = \frac{[\mathrm{H_3O^+}][\mathrm{A^-}]}{[\mathrm{HA}]}$$

Furthermore, the global concentrations of the solute are:

$$c = [\mathrm{As}_{\mathrm{Si}}^{x}] + [\mathrm{As}_{\mathrm{Si}}^{\bullet}] \qquad c = [\mathrm{HA}] + [\mathrm{A^-}]$$

These equilibria are solved in analogous ways. Following the usual approach, three cases are discussed. The first solution concerns the case in which no dopant (or, respectively, acid) is added or the added amount is negligible. Then:

$$[e'] = [h^{\bullet}] = \sqrt{K_{\mathrm{i}}} \qquad [\mathrm{H_3O^+}] = [\mathrm{OH^-}] = \sqrt{K_{\mathrm{w}}}$$

A second case concerns the case where the acid is weak (not completely dissociated) or the dopant is only partially ionized. This requires solving a third-degree equation and a well-known *approximation* (see Section 3.6.1) is:

$$[e'] = \sqrt{c \cdot K_{\mathrm{d}}} \qquad [\mathrm{H_3O^+}] = \sqrt{c \cdot K_{\mathrm{a}}}$$

The third case concerns the case where the dopant is completely ionized, or the acid is strong (completely dissociated) and the solution is:

$$[e'] = c \qquad [\mathrm{H_3O^+}] = c$$

Note that the solutions follow one another in the order 1 → 2 → 3 as the value of K_a (or K_d) increases or else as T increases. The quasi-chemical formalism allows us to obtain the concentrations of electronic charge carriers (electrons in CB and/or holes in VB) in the various cases (without or with doping) much more quickly and easily than the direct treatment based on the Fermi-Dirac statistical distribution. The comparison with this treatment also shows that the result is valid when it is possible to apply the **Maxwell-Boltzmann distribution** to the "chemical species" e' and $h^{\bullet}$, that is, under the condition that they have energies higher - many times kT - than the chemical potential of the electrons (which is also that of the holes). The consideration is of general validity.

3.8.4 Point Defect Equations

In addition to **substitutional** defects (such as $\mathrm{As}_{\mathrm{Si}}^{x}$ and $\mathrm{As}_{\mathrm{Si}}^{\bullet}$), *i.e.* cases in which an atom or ion is replaced on its regular site by another atom or ion, typical solid-state point defects include:

- **interstitial defects** in which an atom or ion enters a position that is empty in the reference structure and
- **vacancies** which are the lack of an atom or ion in a site where it is required by the reference structure.

As an example, showing various aspects of the notation and thermodynamic treatment of point defects, one can consider the case of doping zirconia with yttria[9]:

$$\mathrm{Y_2O_3} \leftrightarrows 2\mathrm{Y}_{\mathrm{Zr}}' + 3\mathrm{O}_{\mathrm{O}}^{x} + \mathrm{V}_{\mathrm{O}}^{2\bullet} \tag{3.65}$$

In Eq. (3.65), on the left there is pure yttria, the species on the right are in the zirconia phase. The reaction describes the dissolution of some amount of the solute yttria in the zirconia solvent. In the zirconia phase, a Y^{3+} ion enters substitutionally (only) the site that in the unit cell of the reference structure is occupied by a Zr^{4+} ion. This substitutional defect has a charge one unit lower than the regular one and in the notational convention is indicated with a superscript<′>as a relative charge −1. Furthermore, two primitive unit cells are involved to accommodate the two substitutional ions corresponding to the Y_2O_3 dopant. The three O^{2-} ions of yttria enter regularly on the sites that in the reference structure belong to the oxide ions. As there are two equivalent O sites per primitive unit cell, two unit cells are involved, and a vacant O site is left. This defect is a vacancy and is denoted by the letter V (for vacancy) with two positive charges (symbol (•)) as a result of the lack of the two negative charges of the oxide ion with respect to the reference situation.

To conclude with the notation, in the case of **interstitial** defects (indicated by placing an "i" as a subscript) the true charge is equal to the effective charge, since the reference charge is 0. Finally, to avoid confusion in the case of reactions involving vanadium, Va can be used instead of V to denote a vacancy.

The extension of the formalism of chemical reaction to progressively more complex cases beyond the reactions between molecules in the gas phase also translates into new constraints on the balance of stoichiometric coefficients. In addition to the balance of atomic species, we must consider the charge balance for electrochemical reactions and reactions between ions (see Sections 3.6 and 3.7) and now the balance of the sites of the reference structure. This constraint does not require that the sites of the various types appear in equal numbers on the right and left of the stoichiometric equation. Instead, for all the sites that appear in the stoichiometric equation the differences between left and right must correspond to the same integer multiple of the content of a unit cell. For example, in Eq. (3.65) on the left there are no sites of the zirconia structure while on the right there are 2Zr sites and 4O sites, and these sites correspond to two cells because a primitive cell contains 1Zr site and 2O sites.

Note the need to preliminarily specify **a reference crystal structure**.

3.9 Thermodynamic Degrees of Freedom

In the phase rule (Eq. 3.43) $V=C-P+2$, C indicates the number of independent components of a system, that is, the total number of components minus the number of constraints. The constraints are (i) the chemical reactions (each one is a constraint, provided that it is a reaction independent of the others), (ii) the charge balances (for reactions between charged species) and (iii) the site balances (for defect reactions in solids).

To give an example related to point defects in solids, one can consider the sub-stoichiometry of oxygen in a simple oxide by formation of oxide ion vacancies with injections of electrons into the CB and evolution of oxygen:

$$O_O^{x} \leftrightarrows V_O^{2\bullet} + 2e' + \frac{1}{2}O_2(\text{gas})$$

The detailed count is as follows. Total components=4 (anion on its regular site, anion vacancy, electron and oxygen gas), constraints=3 (reaction, charge balance, anion site balance), independent components=4−3=1, phases=2 (oxide and gas), $F=1-2+2=1$. The equilibrium of the reaction can be achieved at various temperatures, and, at a given T, the partial pressure of oxygen, the non-stoichiometry of oxygen and the concentration of electrons in the conduction band assume precise values.

If we also considered the cation, which should then appear unchanged on both sides of the reaction, we would have five total components and four constraints because the balance of the cationic sites would be added. The variance would still be 1.

Finally, it can also be specified that the **number 2 in the phase rule** (Eq. 3.43, $V=C-P+2$) is valid when the only external force field is that due to pressure. In the presence of other external fields, the variance increases because, in addition to the pV work, it is necessary to consider other kinds of work and other variables are required to specify the thermodynamic state of a system, as occurs for example in electrochemical or magnetic systems.

4 STATISTICAL THERMODYNAMICS

As seen in Section 1.4 when discussing the general characteristics of ST, each kind of thermodynamic system, that is each set of external variables that are held constant, corresponds to a specific Gibbs ensemble. There is also a specific *thermodynamic potential*, *i.e.* a property that at constant external conditions indicates with its minimum the thermodynamic equilibrium of the system. The corresponding thermodynamic potentials are those obtained from CT. Considering only the cases of greatest interest these are:

Closed isolated system (micro-canonical ensemble)	$\Rightarrow -S$
Closed isothermal system (canonical ensemble)	$\Rightarrow A = U - TS$
Open system (grand-canonical ensemble)	$\Rightarrow pV$

4.1 Partition Functions

ST essentially gives the probability of each microstate of a macrostate.

What is first obtained is a statistical weight (g_i) which is then transformed into a probability P by dividing by the sum of the weights of all microstates within the specific Gibbs ensemble:

$$P_i = \frac{g_i}{\sum_i g_i} \tag{4.1}$$

Taking the micro-canonical ensemble as an example, all the microstates have the given energy E, and the quantum mechanical degeneracy of the energy level (eigenvalue of the Hamiltonian of the macroscopic system) is Ω. As stated by the second postulate of the ST, all microstates have the same statistical weight $g_i = 1$. Each microstate therefore has probability $= 1/\Omega$.

For other Gibbs ensembles, the microstates have different (= not necessarily equal) statistical weights.

The sum $\sum_i g_i$ (Eq. 4.1) plays a much more important role than a simple normalization factor for probabilities. In fact, it is related in a simple way to the pertinent thermodynamic potential and therefore allows us to directly obtain all thermodynamic quantities, without applying Eq. (1.5). So, it is given a specific name: **partition function**. There is a *micro-canonical, canonical, grand-canonical,* ... partition function.

In the previous example of the micro-canonical ensemble, the partition function is Ω and the thermodynamic potential is $-S$ with $S = k\ln\Omega$ (k is the Boltzmann constant). From the thermodynamic potential we obtain the other quantities; for example, we obtain $\frac{1}{T} = \left(\frac{\partial S}{\partial E}\right)_{N,V}$ and $p = T \cdot \left(\frac{\partial S}{\partial V}\right)_{N,E}$. See also Exercise 3.6.

4.1.1 Most Important Ensembles

Except for the micro-canonical ensemble, it is not easy to derive for the various ensembles the expression of the partition function and its relation to the thermodynamic potential: we must necessarily limit ourselves here to reporting the results. In 8.2 there is a simplified derivation for the canonical ensemble. For a complete derivation, for other ensembles and for further information, one should refer to a specialised text in statistical mechanics or ST.

4.1.1.1 Isolated System (Micro-Canonical Ensemble)

Here, the partition function is the quantum mechanical degeneracy $\Omega(N, E, V)$ of the energy level corresponding to the fixed energy E. Its relationship with the thermodynamic potential of the isolated system, *i.e.* the entropy S, is the famous **Boltzmann relation**: $S = k \ln \Omega$.

4.1.1.2 Isothermal Closed System (Canonical Ensemble)

The result (see Section 8.2) is well known: the statistical weight of each microstate (i) is given by the Boltzmann exponential [$\exp(-E_i/kT)$] and therefore the canonical partition function is:

$$Q(N,V,T) = \sum_i e^{-E_i/kT} \tag{4.2}$$

This summation can be written in two ways. It can be left it as it is, but then it must remember that i indicates a **state** and that each Boltzmann exponential can appear many times due to degeneracies. Instead, it can be a **sum** over the **energy levels** and then:

$$Q = \sum_i \omega_i \cdot e^{-E_i/kT} \tag{4.3}$$

where i now specifies an **energy level** and ω_i is its quantum mechanical degeneracy. Equation (4.2) is most frequently reported for simplicity. It may appear ambiguous and in fact it is evaluated through Eq. (4.3). From CT we know that the thermodynamic potential at given T, V, N is the Helmholtz energy defined as $A = U - TS$. Its relation to the partition function $Q(N, V, T)$ is (see Section 8.2):

$$A = -kT \ln Q \tag{4.4}$$

4.1.1.3 Open System (Grand-Canonical Ensemble)

The grand-canonical partition function is:

$$\Xi = \sum_N \left[e^{N\mu/kT} \cdot \sum_i e^{-\frac{\varepsilon_i}{kT}} \right] \tag{4.5}$$

Here, the external summation is over the number of molecules, that is considering $N = 0, 1, 2, 3, \ldots$ molecules. For each N, $e^{N\mu/kT}$ multiplies the inner summation at given T and μ values.

Warning: The expression for Ξ is more complicated than it might appear: the ε_i are the energies of a system of N molecules in a volume V and therefore depend on N and V.

Even for Eq. (4.5) there are two versions, summing over states or levels.

4.1.2 From the Partition Functions to the State Variables

The most important relations are reported in the following Table 3.1.

The quantities assigned and therefore to be treated as exact and the quantities obtainable from the formulas and therefore to be considered as statistical averages change depending on the ensemble. For example, in the canonical system, N is an exact quantity from which the average value of the chemical potential is calculated; in the grand canonical system, however, the chemical potential is the exact quantity, from which the average value of N is obtained: in the canonical system: $\langle \mu \rangle = f(N,V,T)$, in the grand-canonical system: $\langle N \rangle = g(\mu,V,T)$.

The important result is that, regardless of the difficulty of the calculations, using one ensemble or another does not matter: the results are the same (see Sections 6.2.4, 6.2.6 and 5.3 for examples).

TABLE 3.1
Statistical Function for the Gibbs Ensembles

Ensemble	Canonical	Micro Canonical	Grand Canonical
Independent variables	N, V, T	N, V, U	μ, V, T
Partition function	$Q(N,V,T)=\sum_i e^{-E_i/kT}$	Ω (QM degeneracy of the state with given N, V, U values)	$\Xi(m, V, T)=\sum_N\left[e^{N\mu/kT}\cdot\sum_i e^{\frac{-\varepsilon_i}{kT}}\right]$
Thermodynamic potential	$A=-kT\ln Q$	$S=k\ln\Omega$	$pV=kT\ln\Xi$
U (internal energy)	$kT^2\left(\frac{\partial\ln Q}{\partial T}\right)_{N,V}$	Assigned	$kT\mu\left(\frac{\partial\ln\Xi}{\partial\mu}\right)_{V,T}+kT^2\left(\frac{\partial\ln\Xi}{\partial T}\right)_{V,\mu}$
S (entropy)	$kT\left(\frac{\partial\ln Q}{\partial T}\right)_{N,V}+k\ln Q$	$S=k\ln\Omega$	$kT\left(\frac{\partial\ln\Xi}{\partial T}\right)_{V,\mu}+k\ln\Xi$
p (pressure)	$kT\left(\frac{\partial\ln Q}{\partial V}\right)_{N,T}$	$kT\left(\frac{\partial\ln\Omega}{\partial V}\right)_{N,U}$	$kT\left(\frac{\partial\ln\Xi}{\partial V}\right)_{N,\mu}$
N (number of molecules)	Assigned	Assigned	$kT\left(\frac{\partial\ln\Xi}{\partial\mu}\right)_{V,T}$
μ (chemical potential)	$-kT\left(\frac{\partial\ln Q}{\partial N_i}\right)_{V,T,N_{j\neq i}}$	$-kT\left(\frac{\partial\ln\Omega}{\partial N_i}\right)_{V,T,N_{j\neq i}}$	Assigned

4.1.2.1 Notes

A. One may ask why we discuss several cases (statistical system/external variables/Gibbs ensemble/thermodynamic potential/partition function). Wouldn't it be enough to discuss just one case? The fact is that the formulas in Section 4.1.2 need to be developed for a particular atomic-molecular model to derive the dependence on the microscopic parameters (molecular masses, moments of inertia, strength constants of chemical bonds, …) and on the state variables. For each model, it may be easy to derive the partition function using Eqs. (4.2)–(4.5) of an ensemble, but difficult or impossible using another ensemble. *Almost* all the *analytical* applications of ST (see point 6 below) are obtained using the partition function of the *canonical ensemble.*

Generally, in ST the canonical ensemble is used.

An important exception is given by the Fermi-Dirac and Bose-Einstein distribution functions of a system of independent, equal and indistinguishable microscopic particles: to derive these distribution functions the grand canonical ensemble is used instead, because the grand canonical partition function makes possible to consider the indistinguishability.

B. Why, in addition to the canonical ensemble (N, V, T), or instead of it, is a system (N, p, T) not considered? This, in fact, would seem to be the most obvious choice, since the Gibbs energy G (N, p, T) is by far the most widely used thermodynamic potential in chemistry, since experiments are preferably conducted at constant (atmospheric) pressure. The reason is, essentially, that it is difficult to impose the constraint of constant pressure in the treatment of microscopic mechanics: it would be necessary to keep constant in the calculations the global quantity of motion exchanged on the walls of the container. Instead, doing the calculations at constant volume is much simpler and more direct, as can be seen for example from the quantum mechanical formulas for the particle in the box (see Section 6.2.2).

C. The results for the various ensembles show (as is evident by looking at Eq. (4.3)) that in ST it is sufficient to know the energy levels and their degeneracies, it is not necessary to specify how the states of a degenerate level are chosen.

4.2 Analytic Approach and Computational Approach

Historically, the development of ST can be divided into two different steps. The approach was first focused on finding analytical solutions (Eq. 1.5); later, the computational approach was developed. Once again, the point to note is that the quantities M_j (Eq. 1.5) or the quantities E_j (*e.g.*: Eq. (4.3)) concern macroscopic systems, *i.e.* with a number of particles (atoms, ions, molecules, …) of the order of the Avogadro constant, for which - evidently - an analytic solution of the equations of motion is *almost* impossible to access directly.

In the simplest **analytic approach**, the problem is made tractable essentially by dividing a complex system into *independent subsystems* and then combining the solutions for these simpler systems in the solution for the entire system. The meaning of independent subsystems is that the particles of one subsystem do not interact with the particles of another subsystem. Then, if the subsystems are simpler than the whole system by many orders of magnitude, they can be treated analytically. Finally, the solutions for the subsystems can be combined to solve the problem for the entire system thanks to the fact that - unless otherwise specified later - *the canonical partition function of the complex system is the product of the canonical partition functions of the subsystems* (see Section 5.1).

The most obvious example is once again the ideal gas. The molecules are independent, *i.e.* do not interact. Then, a problem of N molecules is solved by combining N times the possible solutions for the problem concerning a single molecule. The energy of the entire system is obtained as the sum of the energies of the individual molecules. Besides the ideal gas case (Section 6.2.1), we will see as examples the Langmuir model for adsorption (Section 6.2.4), the Einstein model for lattice vibrations (Section 6.2.5) and paramagnetism (Section 6.2.6).

An important special case is the **solid** (*i.e. crystalline*) **state**, where the treatment at the mechanical level is made possible by the translational symmetry. Consider for example the problem of vibrations of a solid with N unit cells related by translational symmetry. Each cell contains n atoms so that there are $3nN$ vibrational degrees of freedom.[10] Translational symmetry allows us to classify and separate these vibrations into N symmetry species and for each symmetry species we face an independent mechanical problem with only $3n$ degrees of freedom and the solutions are obtained as functions of a specific parameter dependent on the symmetry species.

The next step of the analytical approaches has been focussed on devising methods to treat interacting particles. **Mean field theories** are an example. This approach introduces a potential energy field that reproduces, in an evidently approximate way, the effects of the interactions of a particle with the others. A well-known application is the Weiss field for ferromagnetism, in which the interaction between adjacent magnetic moments is simulated by introducing a fictitious external magnetic field. The **quasi-chemical approach** is based on the idea of treating a system of interacting particles following the approach used for chemical reactions. Several **lattice theories** have also been developed where equal or different molecules (or units of polymer chains) are placed on the sites of a lattice and the interactions between the nearest neighbours and then possibly between more distant sites are modelled in various ways, considering different coordinations, different shapes and different types of lattice. A simple example of this approach is discussed in Section 6.4.

The extraordinary development of a great variety of **computational methods** has greatly surpassed the previous analytical treatments both in terms of the vastness of tractable problems and accuracy of results. In these computational methods, a reduced version of the macroscopic system is approached at the level of microscopic mechanics, including all the interactions that are of interest. Part 7 of this chapter very concisely summarises the main computational methods used by ST.

5 THE ANALYTIC APPROACH TO ST

5.1 Independent Sub-Systems

If a thermodynamic system is made up of several parts (subsystems) and the atoms, ions, molecules, … of a subsystem *do not interact with those of another subsystem, i.e.* if the subsystems are *independent,*
then the canonical *partition function of the global system* is the product of the partition functions of the subsystems.

Let us look at the case of only two subsystems, A and B, and for simplicity we consider them different: the case of equal subsystems is discussed later. We introduce two quantities: $Q_A = \sum_i e^{\frac{-\varepsilon_{A_i}}{kT}}$ where A_i indicates the state i of subsystem A

and similarly, $Q_B = \sum_i e^{\frac{-\varepsilon_{B_j}}{kT}}$ where B_j indicates the state j of subsystem B. The states of the global system are specified by the two indices ij and hence: $= \sum_{ij} e^{\frac{-\varepsilon_{ij}}{kT}}$ where the double summation is over all values of and j. We do not care whether the particles of A interact with each other or not, and similarly for the particles of B, but we assume that there are no interactions between one or more particles of A and one or more particles of B. This is what we mean by saying that the two subsystems are independent. Then, the energy of state ij will be given by $\varepsilon_{ij} = \varepsilon_{A_i} + \varepsilon_{B_j}$. So:

$$Q = \sum_{ij} e^{\frac{-(\varepsilon_{A_i}+\varepsilon_{B_j})}{kT}} = \sum_i \left(\sum_j e^{\frac{-(\varepsilon_{A_i}+\varepsilon_{B_j})}{kT}} \right)$$ but, for each j:

$$\left(\sum_j e^{\frac{-(\varepsilon_{A_i}+\varepsilon_{B_j})}{kT}} \right) = e^{\frac{-\varepsilon_{A_i}}{kT}} \cdot \left(\sum_j e^{\frac{-\varepsilon_{B_j}}{kT}} \right) = e^{\frac{-\varepsilon_{A_i}}{kT}} \cdot Q_B$$ and then

$$Q = \sum_i \left(\sum_j e^{\frac{-(\varepsilon_{A_i}+\varepsilon_{B_j})}{kT}} \right) = \left(\sum_i e^{\frac{-\varepsilon_{A_i}}{kT}} \right) \cdot Q_B = Q_A \cdot Q_B \quad (5.1)$$

To generalise to more subsystems, one can treat three independent subsystems A, B, C as made of two subsystems (A+B) and C, etc.…. Then:

For a system of many *independent* subsystems $(A + B + \cdots + Z)$ the partition function is the product of the partition functions of the subsystems:

$$Q = Q_A \cdot Q_B \cdot \ldots \cdot Q_Z \tag{5.2}$$

For example, a mixture of two ideal gases (see Section 6.3) can be considered as consisting of two disjoint parts, the subsystem of molecules A and the subsystem of molecules B. Then: $Q=Q_A+Q_B$. In fact, these are ideal gases, the molecules do not interact with each other and therefore certainly a molecule A does not interact with a molecule B. The two subsystems are therefore independent.

Some direct consequences of the independence of the subsystems are hereafter discussed, while most applications are shown later.

5.1.1 Additivity of Thermodynamic Functions

If $A=-kT\ln Q$ and $Q=Q_AQ_B$, then $A_{A+B}=-kT\ln(Q_AQ_B)=-kT(\ln Q_A+\ln Q_B)=A_A+A_B$, that is, **the Helmholtz energy of the global system is the sum of the Helmholtz energies** $A_A=-kT\ln Q_A$ and $A_B=-kT\ln Q_B$ that can be calculated for each subsystem ignoring the other.

It is easily seen that the same property applies to all extensive thermodynamic functions.

When the degrees of freedom are separable, the additivity of the extensive thermodynamic functions follows as a direct consequence.

Another important consequence concerns the phase equilibrium.

If interphase effects can be neglected, the extensive quantities of a polyphasic system are the sum of the extensive quantities of the individual phases.

5.2 Distinguishable and Indistinguishable Sub-Systems

With *many independent* sub-systems (i) we should have:

$$Q = \prod_i q_i \tag{5.3}$$

However, this is *generally* true *only if the subsystems are different* from each other. Otherwise, Eq. (5.3) *cannot be arbitrarily extended down to the microscopic level* as, for example, in the case of equal molecules. In other words, for N independent (non-interacting) and *identical* particles, we *may think* to write:

$$Q = q^N\text{, \textbf{but this result is wrong}} \tag{5.4}$$

when applied to microscopic particles, such as electrons, nuclei, atoms, ions, molecules, because in this case the subsystems are *indistinguishable.*

Indistinguishability of microscopic particles is a direct consequence of the Pauli principle, and, as such, it should be a fully understood concept. However, the problem may appear counterintuitive in a first approach to ST, so that it is reasonable to spend a few words. It must be clear that the problem concerns *independent, identical, and microscopic* subsystems. Let us consider a system which is split into two subsystems, let us say A and B. As the two subsystems are independent, from Eq. 5.1 we have:

$$Q = \sum_{kl} e^{\frac{-(\varepsilon_{Ak}+\varepsilon_{Bl})}{kT}} = Q_A \cdot Q_B$$

If A and B are microscopic and identical systems (*i.e.*, two identical molecules, two electrons, …) the indexes (k, l) refer now to the same set of *single particle states*, and therefore apply to both A and B. The double summation contains:

- both the Boltzmann exponential where A is in state i and B is in state j, that is the state for which, for the whole system, $k=i$, $l=j$, let us say state $|ij\rangle$,
- and the exponential where A is in the j state and B is in the i state, *i.e.*, $|ji\rangle$.

Pauli principle predicates that the states for a system of microscopic particles *must* have a precise behaviour with respect to the exchange of any two particles. These states are requested to be eigenfunctions of all permutation operators. For the case of only two particles, the states $|ij\rangle$ and $|ji\rangle$ are not allowed states: they must be **symmetrized** (case of **integer spin particles**) or **anti-symmetrized** (for **half-integer spin particles**, such as **electrons**). In particular, given the couple of states ($|ij\rangle$ and $|ji\rangle$), anti-symmetrization gives only the state $1/\sqrt{2}\,(|ij\rangle-|ji\rangle)$, while symmetrization gives only the state $1/\sqrt{2}\,(|ij\rangle+|ji\rangle)$. *In any case, from the two states that are seemingly possible for the two particles we are left with only one state because of indistinguishability.* For the **antisymmetric case** only, we must also reject (**exclusion principle**) all states such as $|ii\rangle$ or $|jj\rangle$, *i.e.*, the states in which two (or more) particles occupy the *same single particle state.*

Let us take as an example a system with two electrons and an energy level with six spin-orbitals, the $2p^2$ atomic configuration that we have already considered when discussing antisymmetry and multi-electron systems in QM. If antisymmetry is not considered, the number of states for the bi-electronic system is $6^2=36$. Antisymmetry removes the six states where both electrons are in the same spin-orbital and pairwise combines the other states. In conclusion, from the original 36 states, we obtain 15 antisymmetric or 21 symmetric linear combinations.

In addition to photons, nuclei, electrons, …, Pauli principle applies also to composite particles. The symmetry/antisymmetry that we need to consider is given by the global spin. This is in turn given by the vector quantum summation of the constituent spins. While the vector addition of spins can give *more than one value*, the results always all belong to the same (integer/half-integer) case. For instance, adding an integer spin to a half-integer spin always gives a half-integer spin ($0+1/2 \rightarrow 1/2$, $1+1/2 \rightarrow 1/2$ or $3/2$, …);

adding two half integer spins always gives an integer spin ($1/2+1/2 \rightarrow 0$ or 1, $3/2+1/2 \rightarrow 1$ or 2, ...); adding two integer spins always gives an integer spin.

Summarizing, symmetrization and anti-symmetrization are at the basis of indistinguishability of microscopic particles and give rise to **a lower number of states compared to what is seemingly possible**. The expression q^N contains spurious terms, corresponding to states that cannot exist as they are not allowed by symmetry or antisymmetry. There are two methods for finding the right partition function.

- The first method simply introduces a corrective factor f: in Eq. (5.4) $Q=q^N$ becomes $Q=f \cdot q^N$. This is an **approximation**, which is valid only under conditions that *must* be verified *a posteriori*, but is straightforward, and easy to use in applications.
- The second method is exact and has general validity but is more complex both in its derivation and in its applications. Also, there are two different results for integer spin particles (**Bose Einstein** statistic, BE), and for half-integer spin particles (**Fermi-Dirac** statistic, FD).

A preliminary warning: as already said, the problem, *in principle*, occurs only when we are treating **independent** particles. For **interacting particles**, the interactions *should be* explicitly included in handling the problem and finding the states and energies. For interacting electrons, for example, one should derive from QM the over-all electronic states for the atom or molecule. An example are the electronic states of atoms derived by the LS coupling. However, we are here concerned with systems with many electrons, approached using single electron states (spin-orbitals, SO) and looking for the occupancies of the SOs as a function of temperature.

5.2.1 The Approximate Approach

In a system of independent, identical, and indistinguishable particles, a very easy method exists for evaluating the partition function **if** the **number** of states available for each particle (the *single particle states*) **is** vastly **greater than the number of particles**. It is an approximation, but its accuracy is often excellent. We start by noting that, among *all the* ***overall*** *states* (let us say microstates) *that we are considering when writing* $Q=q^N$,

A. in some microstates, the *single particle states* are **all** different from each other,
B. in other microstates, the *single particle states* are **at least partly the same**.

Saying that the *number* of states available for each particle is vastly greater than the number of particles means saying that the number of A microstates is vastly greater than the number of B microstates. It is therefore reasonable to neglect the B microstates and limit our attention to the A microstates, where all the single particle states are different. This is the very meaning of the approximation.

So, we are now considering *N different* single-particle states, hereafter referred to as 1, 2, ..., *N*. Writing $Q=q^N$ accounts for **both** the microstate in which particle A is in state 1, particle B is in state 2, ..., particle Z is in state *N*, **and** *all* the microstates obtained from the same (1, 2, ..., *N*) single-particle states, with *all possible* permutations of the particles (A, ..., Z) between the states. Overall, these micro-states are $N!$[11] On the contrary, taking symmetrization (or anti-symmetrization) into account, we must consider **just one microstate**, *i.e.*, the one for which: *one* particle is in state 1, *one* particle is in state 2, ..., *one* particle is in state *N*. That is, we are considering that *N* different states are occupied, and there is no reason to *specify which particle is in which state*. In this approximation, $Q=q^N$ incorrectly counts $N!$ microstates, all at the same energy, when just one microstate is allowed by anti-symmetrization (or symmetrization): the corrective factor is $1/N!$ Therefore:

Approximately, the canonical partition function of a system of *N* independent, equal, and indistinguishable particles is:

$$Q = \frac{q^N}{N!} \tag{5.5}$$

The corresponding Helmholtz energy is:

$$\begin{aligned} A &= -kT \ln Q = -kT\left(N \ln q - \ln N!\right) \quad \text{instead of} \\ A &= -kT \ln Q = -kTN \ln q \end{aligned} \tag{5.6}$$

When this **approximation** is used, **we must verify its applicability**, that is we must verify that the number of single particle states is largely greater than the number of particles. For example, we refer to the cases of translational degrees of freedom of an ideal gas (see Section 6.2.2) or free electrons in a metallic solid (see Section 5.3).

5.2.2 Quantum Statistics

We are now looking for the *exact* expressions for the statistical distributions of *independent, identical, and indistinguishable* particles. As seen above (see Section 5.2), the difficulty is due to Pauli principle: we must count just once all permutations of the particles over single particle states. In this respect, the canonical ensemble does not work. On the contrary, useful results can be obtained within the framework of the *grand canonical ensemble*. The grand canonical partition function is:

$$\Xi = \sum_N \left(e^{N\mu/kT} \cdot \sum_i e^{-\epsilon_{Ni}/kT} \right) \tag{5.7}$$

Each term of the *external* summation is on a given number of particles $N=0, 1, 2, \ldots$ while each term of the *internal* summation is on a system of *N* particles at a given *T*. A micro-state is therefore specified by *N* and the *i* index.

For better clarity, we write here some terms in the external summation:

$$\Xi = \sum_N \left(e^{N\mu/kT} \cdot \sum_i e^{-\epsilon_{Ni}/kT} \right) =$$

$$\begin{array}{ll} 1+ & \text{no particles } \left(e^0 = 1\right) \\ +e^{\mu/kT} \cdot \sum_i e^{-\epsilon_{1i}/kT} + & \text{one particle} \\ +e^{2\mu/kT} \cdot \sum_{i,j} e^{-\epsilon_{2ij}/kT} + & \text{two particles} \\ + \cdots & \ldots \end{array}$$

With $e^{\mu/kT} \overset{\text{def}}{=} \lambda$, we obtain a more compact form:

$$\Xi = 1 + \lambda \cdot \sum_i e^{-\epsilon_{1i}/kT} + \lambda^2 \cdot \sum_{i,j} e^{-\epsilon_{2ij}/kT} + \cdots$$

We are here considering that each state for a system with two, three, … N (independent and *identical*) particles is the product of 2, 3, …, N, single particle states, and its energy is given by the sum of the single particle energies. In detail, each state of a two particles system has an energy given by the sum of two single particle energies: $\varepsilon_{2ij} = \varepsilon_i + \varepsilon_j$; each state of a three particles system has an energy given by the sum of three single particle energies: $\varepsilon_{3ijk} = \varepsilon_i + \varepsilon_i + \varepsilon_k$, etc.

Therefore:

$$\Xi = \sum_N \left(e^{N\mu/kT} \cdot \sum_i e^{-\epsilon_{Ni}/kT} \right) =$$

$$\begin{array}{ll} 1+ & \text{no particles} \\ +\lambda \cdot \sum_i e^{-\epsilon_i/kT} + & \text{one particle} \\ +\lambda^2 \cdot \sum_{i,j} e^{-(\epsilon_i+\epsilon_j)/kT} + & \text{two particles} \\ +\lambda^3 \cdot \sum_{i,j,k} e^{-(\epsilon_i+\epsilon_j+\epsilon_k)/kT} + & \text{three particles} \\ + \cdots & \end{array}$$

Indistinguishability implies that state *ij* should be identified with state *ji*, state *ijk* with states *ikj, jik, jki, kij, kji*, and so on for 4, 5, … indexes, *i.e., each of the N! permutations of N indexes* must appear **only once**. This is easily done: in the summations we must simply *let the second index start from* $j = i$, the third index from $k = j$, etc.

$$\Xi = \sum_N \left(e^{N\mu/kT} \cdot \sum_i e^{-\epsilon_i/kT} \right) =$$

$$\begin{array}{ll} 1+ & \text{no particles} \\ +\lambda \cdot \sum_i e^{-\epsilon_i/kT} + & \text{one particle} \\ +\lambda^2 \cdot \sum_{i,j\geq i} e^{-(\epsilon_i+\epsilon_j)/kT} + & \text{two particles} \\ +\lambda^3 \cdot \sum_{i,j\geq i,k\geq j} e^{-(\epsilon_i+\epsilon_j+\epsilon_k)/kT} + & \text{three particles} \\ + \cdots & \end{array} \tag{5.8}$$

It is easily seen that the $k \geq j \geq i$ constraint fulfils the requirement that the *ijk* index triplet is considered just once in the summation, and that, for that triplet, the energy has the correct value. Meanwhile, we are not making a distinction between symmetric systems (where equal indexes are allowed, and therefore the relations $k \geq j \geq i$ hold) and anti-symmetric systems (where equal indexes are not allowed, and we must use the strict inequalities $k > j > i$). The choice will be easily done in the following.

Equation (5.8) is rather intricate but can be simplified by defining an additional quantity:

$$\begin{aligned} \xi_i &= 1 + \lambda \cdot e^{-\varepsilon_i/kT} + \lambda^2 \cdot e^{-2\varepsilon_i/kT} + \lambda^3 \cdot e^{-3\varepsilon_i/kT} \\ &\quad + \lambda^4 \cdot e^{-4\varepsilon_i/kT} + \cdots \quad (\text{NB}: \lambda \overset{\text{def}}{=} e^{\mu/kT}) \end{aligned} \tag{5.9}$$

that we can call the *partition function of a single particle state*. Indeed, combining Eq. (5.8) with Eq. (5.9) and using the Multinomial Theorem of algebra, we can obtain:

$$\Xi = \prod_i \xi_i \tag{5.10}$$

The Multinomial Theorem is an advanced topic and is rarely taught in non-specialized courses. An assessment of Eq. (5.10) in simple words is given in Section 8.3, by comparing Eqs. (5.8), (5.9) and (5.10). Just to *remember* the result, we can note that Eq. (5.10) is somehow analogous to $Q = \prod_i q_i$, which holds for the *canonical* partition function of a system of *independent and distinguishable* particles (whether identical or different): the product is now on *states* (*that are different* and therefore *distinguishable*). Then, the results summarized in Section 4.1.2 for the grand canonical ensemble allow to calculate from the specified quantities (μ and T), the statistical *average number of particles* (in all the states) $\langle N \rangle$:

$$\langle N \rangle = kT \cdot \sum_i \left(\frac{\partial \ln \xi_i}{\partial \mu} \right)_{V,T}$$

and finally, the *occupancy* (s_i) *of a single particle state i*, that is the average number of particles in the i state:

$$s_i = kT \cdot \left(\frac{\partial \ln \xi_i}{\partial \mu} \right)_{V,T} \quad \text{and} \quad \langle N \rangle = \sum_i s_i$$

Moreover, $\lambda \cdot e^{-\varepsilon_i/kT} \stackrel{\text{def}}{=} e^{-(\varepsilon_i-\mu)/kT}$ using again: $e^{\mu/kT} \stackrel{\text{def}}{=} \lambda$.

5.2.2.1 Half-Integer Spin Particles: Fermions

For anti-symmetric particles, we must use the strict inequalities ($k>j>i$) in Eq. (5.8). However, we can directly get the result by exploiting the fact that two particles are not allowed to occupy the same state (i): a state can be either empty or occupied by only one particle. Therefore, the partition function of a single state (Eq. 5.9) is obtained by truncating the series to its second term, which indeed means: "no more than one particle in the state."

$$\xi_i = 1 + \lambda \cdot e^{-\varepsilon_i/kT} = 1 + e^{(\mu-\varepsilon_i)/kT}$$

$$s_i = kT \cdot \left(\frac{\partial \ln \xi_i}{\partial \mu}\right)_{V,T} = kT \cdot \frac{e^{(\mu-\varepsilon_i)/kT}}{1+e^{(\mu-\varepsilon_i)/kT}} \left(\frac{\partial\left(\dfrac{\mu-\varepsilon_i}{kT}\right)}{\partial \mu}\right)_{V,T}$$

$$= kT \cdot \frac{e^{(\mu-\varepsilon_i)/kT}}{1+e^{(\mu-\varepsilon_i)/kT}} \frac{1}{kT} = \frac{1}{e^{(\varepsilon_i-\mu)/kT}+1}$$

We finally obtain:

$$\text{Fermi-Dirac Statistics: } s_i = \frac{1}{e^{(\varepsilon_i-\mu)/kT}+1} \tag{5.11}$$

5.2.2.2 Integer Spin Particles: Bosons

When two or more particles are allowed to occupy the same state (i), we can easily evaluate Eq. (5.9), by noting that it is a geometric series.

Using $a \stackrel{\text{def}}{=} \lambda \cdot e^{-\varepsilon_i/kT} = e^{(\mu-\varepsilon_i)/kT}$, the sum is $\sum_l a^l = \frac{1}{1-a} = \frac{1}{1-e^{-(\varepsilon_i-\mu)/kT}}$. Indeed:

$$\xi_i = 1 + \lambda \cdot e^{-\varepsilon_i/kT} + \lambda^2 \cdot e^{-2\varepsilon_i/kT} + \lambda^3 \cdot e^{-3\varepsilon_i/kT} + \lambda^4 \cdot e^{-4\varepsilon_i/kT} + \cdots$$

$$= \sum_l \lambda^l \cdot e^{-l \cdot \varepsilon_i/kT} = \sum_l a^l$$

$$s_i = \left(\frac{\partial \ln \xi_i}{\partial \mu}\right)_{V,T} = \frac{1}{\xi_i}\left(\frac{\partial \xi_i}{\partial \mu}\right)_{V,T}$$

Then,

$$= (1-a)\left(\frac{\partial \xi_i}{\partial a}\right)_{V,T}\left(\frac{\partial a}{\partial \lambda}\right)_{V,T}\left(\frac{\partial \lambda}{\partial \mu}\right)_{V,T}$$

$$= \frac{(1-a)}{(1-a)^2} e^{-\varepsilon_i/kT} \cdot \frac{e^{\mu/kT}}{kT} = \frac{a}{(1-a)kT}$$

$$= \frac{1}{kT} \frac{1}{\left(\dfrac{1}{a}-1\right)} = \frac{1}{kT} \frac{1}{e^{-(\varepsilon_i-\mu)/kT}-1}$$

This gives the

$$\textbf{Bose - Einstein Statistics: } s_i = \frac{1}{e^{(\varepsilon_i-\mu)/kT}-1} \tag{5.12}$$

Convergence of the series requires $|a|<1$ and therefore $e^{(\mu-\varepsilon_i)/kT} < 1, (\varepsilon_i - \mu) > 0$, so that $\varepsilon_i > \mu$ for any i.

5.2.2.3 Maxwell-Boltzmann Statistics

The occupancy of a single particle state can also be derived from the approximated equation $Q=q^N/N!$ From $\ln Q = N \ln q - \ln(N!) \approx N \ln q - [N \ln(N) - N]$:

$$\mu = -kT\left(\frac{\partial \ln Q}{\partial N}\right)_{T,V} = -kT \ln \frac{q}{N}$$

that is $\frac{N}{q} = e^{\mu/kT}$ and, using $q = \sum_i e^{-\varepsilon_i/kT}$,

$$N = \sum_i s_i = \sum_i e^{(-\varepsilon_i+\mu)/kT}$$

Finally, (see also Section 8.5) the occupancy of the k state is:

$$\text{Maxwell-Boltzmann Statistics: } s_k = \frac{1}{e^{(\varepsilon_k-\mu)/kT}} \tag{5.13}$$

5.2.2.4 The Comparison

The three statistical laws have very similar *analytical* forms: the difference is only in +1, −1 or 0 in the denominator. Obviously, all distributions are strictly decreasing with increasing ε_k. However, the predictions are very different at low energies, as better shown by plotting the occupancy (s_k) of a single particle state k as a function of the energy (ε_k) (Figure 3.9). The FD is peculiar for it displays an inflection point (ordinate = 1/2, abscissa = 0, *i.e.*, $\varepsilon=\mu$). A characteristic of BE is that the chemical potential is always lower than the energy of any single particle state. In the limit of high energy, the distributions merge to each other. In this case, it is not important which function we are going to use. As a matter of fact, MB is preferred, because it is the simplest for the applications. Finally, note that the energy always appears divided by kT, which is a kind of scale factor of the energy at the atomic-molecular level (see Section 5.4.1).

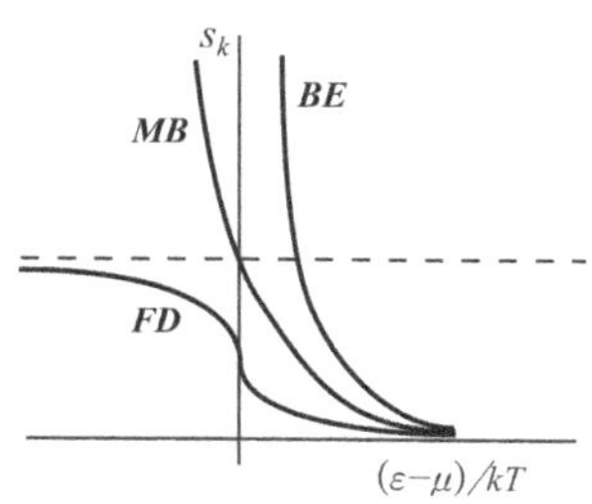

FIGURE 3.9 Mean (statistical) occupancy of single electron states (s_k) according to the FD (Eq. 5.11), MB (Eq. 5.13) and BE (Eq. 5.12) statistics. The dashed horizontal line marks the asymptotic value (= 1) of the FD at low energy.

5.3 Occupancy of One-Electron States

Although it is derived in the grand-canonical ensemble scheme, *i.e.* knowing the chemical potential μ, the FD distribution

$$s_i = \frac{1}{e^{(\varepsilon_i-\mu)/kT}+1} \tag{5.14}$$

is frequently used also for obtaining the chemical potential of the electrons knowing their number: this is another example of the thermodynamic equivalence of ensembles (see Section 4.1.2). At absolute zero, the FD distribution is a step function. Its discontinuity translates the exclusion principle in its elementary form into a statement for full shells: "up to a given energy the states are all completely occupied, from that energy onwards they are all completely empty". The chemical potential is also called "Fermi energy": the two terms are often used as alternatives, but it is better to distinguish them. The Fermi energy is the chemical potential at absolute zero, it is a characteristic of the energetic structure of the electron system and does not depend on the temperature. The chemical potential can instead vary with temperature, as discussed in Section 8.6.

An important feature of the FD distribution is that it is symmetric with respect to the inflection point: two energy levels ε_A and ε_B at the same distance from the inflection but on opposite sides $[\varepsilon_A-\mu=-(\varepsilon_B-\mu)]$, are occupied in a complementary way: $s_A=1-s_B$. The energy range in which the FD is significantly different from 1 or 0 is of the order of kT. Exercise 3.7 shows, for example, that the statistical distribution deviates from the value 1 (at low energies) or from the value 0 (at high energies) by an amount $<10^{-4}$ when $|\varepsilon-\mu|=10\,kT$. To discuss the consequences of this feature of FD it is useful to distinguish between two cases, depending on whether the energies of the states (of the spin orbitals) are discrete or continuously distributed.

The case of **discrete energy levels** for electrons is typical for *atomic and molecular systems*. In **closed-shell configurations**, μ lies between two orbitals denoted as HOO (highest occupied orbital) and LUO (lowest unoccupied orbital). These two levels are generally completely full (HOO) and completely empty (LUO) since their separation in energy is usually much greater than kT. Indeed, a HOO-LUO separation is typically some eV, while $kT\sim0.03$ eV at room temperature. The occupancy of the HOO is very close to 1 and the occupancy of the LUO is very close to 0: the differences from 1 or 0, respectively, are of the order of 10^{-20} or less. In **open-shell configurations**, the occupancy of the last level is equal to the number of electrons that remain to be accommodated after having completely occupied the shells at lower energies. The energy difference from any lower (or higher) level is anyway much greater than kT (as seen before) and therefore also in this case the occupancies do not change appreciably by increasing T. This is why, for AO and MO, the *FD distribution is usually not explicitly considered*, it is treated as a step function and the effect of temperature on orbital occupancy is completely ignored.

On the contrary, in **condensed matter**, the energy levels are grouped into **bands** that, in the range of a few eV, contain a number of states which is of the order of 10^{23}. Therefore, the energy levels can be treated with a **continuous distribution**. Then, there is possibly an energy range of the order of *a few kT* in which the spin orbitals have occupations significantly different from 1 or 0.

A. *If the chemical potential (μ) falls well inside a band* (this is the case of a **metal**), then no matter how low the temperature, a kT-wide energy interval contains a number of states that is several powers of ten.
B. *If the chemical potential falls in an energy gap between bands*, but the *gap "is not much larger" than kT*, the tails of the FD distribution still extend to adjacent bands. This is the case of **semiconductors**, which occurs – approximately – when the gap is at most 100 kT.
C. *If the chemical potential falls in an energy gap between bands* and the *gap "is much larger" than* kT the tails of the FD distribution have such low values that the nearby bands can be considered completely full or completely empty, as in the case of AOs and MOs. This is the case of **insulators**.

5.4 High-Temperature Limit and *Classical* Results

5.4.1 Energy and Temperature

In the previous discussions, the partition functions are given by expressions such as:

$$\sum_i \omega_i \cdot e^{\frac{-E_i}{kT}} \tag{4.3}$$

where *energies always appear divided by temperature*. Reversing the point of view, since energy spectra do not change with T (or change very little and or indirectly, for example through the effect of thermal expansion), temperature is the scale with which energy is evaluated. At **low** T, kT corresponds to a narrow energy range: the separation between the levels is large compared to kT and the levels to be considered are only those with the lowest energy and the quantization of the energy plays an important role. By **increasing** T, kT covers wider energy ranges, the separation in energy is less marked and the levels to be considered increase in number. **At "high" temperature**, kT covers a very large energy range, the levels included in a kT interval have grown enormously in number and the separation in energy between the successive levels has become barely appreciable.

In other words, what matters is the order of magnitude of the *ratio* between a measure of the separation in energy δE between successive levels and the so-called "thermal quantum" kT. "High" temperature means $T \gg \delta E/k$, "low" temperature means $T \ll \delta E/k$.

When kT is very large the energy is better treated as a **continuous variable**. Essentially, the degeneracy ω_i is replaced with its continuous version $\rho(E)$ called *density of states*. The density of states $\rho(E)$ is the quantity that, integrated between E_1 and E_2, counts the states in that energy range. It is derived from the detailed analysis of the pertinent quantum mechanical model (see Sections 6.2.2 and 6.2.3 for two examples). Then, the canonical partition function:

$$q=\sum_i \omega_i \cdot e^{\frac{-E_i}{kT}} \quad \text{must be replaced by its continuous expression:} \tag{4.3}$$

$$q=\int_0^\infty \rho(\varepsilon)\cdot e^{\frac{-\varepsilon}{kT}}\cdot d\varepsilon \tag{5.15}$$

When it is then necessary to statistically treat another quantity (say X), the dependence (X_i) on the state index is replaced by the functional dependence on the continuous variable ε: $X=X(\varepsilon)$. The summary is:

Discrete Version	Continuous Version
Level i: $\omega_i \cdot e^{\frac{-\epsilon_i}{kT}}$	Given ε: $\rho(\varepsilon)\cdot e^{-\frac{\epsilon}{kT}}$
$X=\sum_i X_i \cdot \omega_i \cdot \frac{-\epsilon_i}{kT}$	$X=\int_0^\infty X(\epsilon)\cdot\rho(\epsilon)\cdot e^{-\frac{\epsilon}{kT}}\cdot d\epsilon$

The approximation of ST in which the energy is treated as continuous is known as the **high-temperature limit**.

5.4.2 Classical Results

Since in CM the energy assumes continuously distributed values, it is sometimes reasonable to use the CM results in ST for the same mechanical model, *i.e.* to use the Hamiltonian function $H(q, p)$. This goal can be pursued following two different approaches.

A first approach essentially uses *classical statistical thermodynamics*, that is the version of ST historically developed before the discovery of QM and therefore entirely based on CM. Otherwise, you can use the relationship:

$$Q_{class}=\frac{1}{h^n}\cdot\int e^{-H(q,p)}dqdp \tag{5.16}$$

where n is the number of degrees of freedom of the system, h is Planck constant, q and p denote the **set** of coordinates $\{q_1, \ldots, q_n\}$ and moments $\{p_1, \ldots, p_n\}$, $H(q, p)$ is the Hamiltonian function of CM, and the integral is over all coordinates and all moments. Equation (5.16) is not a formula of the classical version of the ST (as is clear from the presence of the Planck constant) but is a way to quickly obtain its result. For the derivation of Eq. (5.16) we refer to specialistic texts. The results obtained by following these paths coincide and are known as the **classical limit of ST**: the expressions *high-T limit, classical limit* and *classical statistical thermodynamics* are equivalent. Equation (5.16) is also used when only a part of the system can be treated according to CM, if it is a separable part (see Section 5.1) and the total Q is the product of a classical Q and a quantum Q.

5.4.3 Classical Approximation and Quantum Statistics

An aspect closely related to the classical limit or high temperature limit, concerns the possibility of using MB statistics instead of FD or BE when dealing with systems consisting of equal and indistinguishable microscopic particles. An example in this regard is the case of translational degrees of freedom (see Section 6.2.2).

In general, the use of the various statistics is based on temperature.

At sufficiently high temperatures, the MB applies and the high temperature limit can be used.

It is necessary to verify that the conditions for the MB (Section 5.2.1) are satisfied. If they are, we write $Q=q^N/N!$ and for q we can use the discrete or continuous formulation (see Section 6.2.2).

At sufficiently **low temperatures** one can assume that a system is in its ground state: the partition function is equal to the degeneracy of the ground state and is **independent of temperature.**

At **intermediate temperatures** quantum statistics (FD or BE) must be applied.

6 SOME ANALYTIC RESULTS

6.1 Energy Equipartition

Let us consider a Hamiltonian function H, generalized coordinates[12] $q=\{q_1, \ldots, q_N\}$, generalized moments $p=\{p_1, \ldots, p_N\}$ and a partition function derived as:

$$Q_{class}=\frac{1}{h^N}\cdot\int e^{-H(p,q)}dqdp \tag{5.16}$$

N is here the number of degrees of freedom of the system.

The case of interest is the one in which three conditions on $H(q, p)$ hold:

- The total energy of a system can be separated into two or more parts and each depends on *disjoint sets* of variables (coordinates and/or moments),
- one (or more) of these parts depends on a single coordinate (or a single moment) and
- for this disjoint part, the dependence on the coordinate (or moment) is quadratic.

In formulas, the first two conditions mean that:

$$H\left(q_1,\ldots,q_{j-1},q_j,q_{j+1},\ldots,q_N,p_1,\ldots,p_{j-1},p,p_{j+1},\ldots,p_N\right)$$

$$=H_A\left(q_1,\ldots,q_{j-1},q_{j+1},\ldots,q_N,p_1,\ldots,p_{j-1},p_j,p_{j+1},\ldots,p_N\right)+H_B\left(q_j\right)$$

or

$$= H_A(q_1,\ldots,q_{j-1},q_j,q_{j+1},\ldots,q_N,p_1,\ldots,p_{j-1},p_{j+1},\ldots,p_N)$$
$$+H_B(p_j)$$

The third condition means: $H_B(q_j)=\alpha\cdot q_j^2$ or, respectively: $H_B(q_j)=\beta\cdot p_j^2$

If these conditions hold, **then** the average contribution of the disjoint part to the total energy of the system is ½*kT*.

It is sufficient to prove $H_B(q_j)=\alpha\cdot q_j^2$: the other case is proved analogously. Due to the independence of the part in q_j, Eq. (5.13) becomes:

$$Q=\left(\frac{1}{h^{1/2}}\int e^{-\alpha\cdot q_j^2/kT}dq_j\right)\cdot$$
$$\left(\frac{1}{h^{N-1/2}}\int e^{\frac{-E_A}{kT}}dq_1\cdots dq_{j-1}dq_{j+1}\cdots dq_N dp_1\cdots dp_N\right)$$
$$=Q_B\cdot Q_A$$

Then: $$E=kT^2\left(\frac{\partial\ln Q_B}{\partial T}\right)+kT^2\left(\frac{\partial\ln Q_A}{\partial T}\right)=E_B+E_A$$

and: $Q_B=\frac{1}{\sqrt{h}}\sqrt{\frac{\pi kT}{\alpha}}$ where the integral is evaluated using: $\int e^{-\gamma\cdot x^2}dx=\sqrt{\frac{\pi}{\gamma}}$

Finally: $$E_B=kT^2\left(\frac{\partial\ln Q_B}{\partial T}\right)=kT^2\left(\frac{\partial\ln\sqrt{T}}{\partial T}\right)=\frac{1}{2}kT$$

6.2 Independent and Equal Sub-Systems

There are several important ST results that hold for *microscopic*, independent, and equal sub-systems.

6.2.1 The Ideal Gas

The fundamental formula: $Q=q^N/N!$ is the basis of the ideal gas ST, a topic that has played a very important role in the development of ST in general and in its applications to chemistry. To evaluate the q of a single molecule, one can then go further and analyse its various degrees of freedom (DOF). If they are separated from each other, or can be treated with good approximation as separated as is usually assumed, then the partition function of the single molecule can be written as the product of the partition functions of the various DOFs:

q(molecule)=*q*(translations) × *q*(rotations) × *q*(vibrations) × ... (6.1)

The degrees of freedom to be considered[13] are all these.

- The *translation* of the molecule as a dimensionless body of mass *m* (3 translational degrees of freedom, see Section 6.2.2).
- The *rotation* of the molecule as a rigid body: there are 2 or 3 degrees of freedom for linear or non-linear molecules, respectively.
- The *vibrations*, *i.e.* variations in bond lengths and angles, etc.: for a molecule of *n* atoms, the vibrational degrees of freedom are $3n-5$ or $3n-6$, respectively for linear or non-linear molecules.
- The *electronic states.*
- The states relating to the internal structure of the *nuclei* and their angular moments.

As general considerations:

- In the ideal gas model, the 3 translational degrees of freedom (DOFs) are exactly separated from all the other DOFs, and we therefore speak of *external* DOFs - the translations - and *internal* DOFs - the others; the translations are also separated from each other.
- The separation between rotations and vibrations is approximate.
- The separation of the vibrations from each other is also approximate.
- The separation of the electronic DOFs from the other DOFs is based on the Born-Oppenheimer approximation.
- Due to the wide intervals between the energy levels of the DOFs pertinent to the internal structure of the nuclei, only the nuclear ground state is populated, so that the contribution of these DOFs to the thermodynamic functions is usually neglected in Chemistry.
- When the ideal gas is in an external electric or magnetic field, a degeneracy removal occurs for the energy levels pertinent to the DOFs of the electrons and the orientation of nuclear moments: this is important also in spectroscopy (Stark and Zeeman effects, EPR, NMR).

6.2.2 The Translational Partition Function of an Ideal Gas Molecule

The determination of the *translational q* of a molecule is an important example of the application of the continuous treatment of Eq. (5.16) for discrete but very dense energy levels. From QM, using the particle-in-a-box model we get that the energy levels related to the translations of a molecule of mass *m* are:

$$\epsilon_{i,j,k}=\frac{\hbar^2\cdot\pi^2}{2m\cdot V^{2/3}}\left(i^2+j^2+k^2\right)=\frac{h^2}{8m\cdot V^{2/3}}\left(i^2+j^2+k^2\right)$$

where *i, j, k*=1,1,1; 2,1,1; 1,2,1; ...; 5,1,1; 1,5,1; 1,1,5; 3,3,3; ... are the quantum numbers of the model.

To obtain the density of states $\rho(\varepsilon)$, we first look for the cumulative distribution $W(\varepsilon)$. $W(\varepsilon)$ essentially counts how many states there are at ε less or equal to a given value and can be calculated thanks to the simple trick shown (in its simplified 2D version) by Figure 3.10. Each point

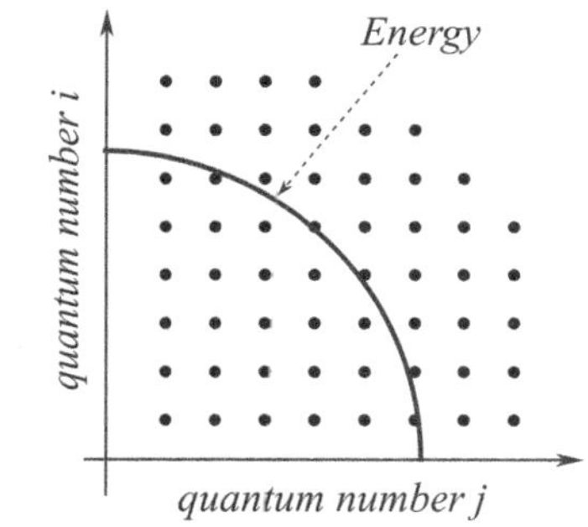

FIGURE 3.10 Simplified (2D) space of the quantum numbers of the particle in a box QM model. In the 3D model, a state corresponds to three quantum numbers instead of 2.

represents a state, *i.e.* a set of three quantum numbers and occupies a *unit volume* in the 3D space of quantum numbers. The distance of a point from the origin is R, where $R^2 = \left(i^2 + j^2 + k^2\right) = 8 \cdot m \cdot \varepsilon \cdot V^{2/3} / h^2$ (see Exercise 3.8). Since the quantum numbers are positive integers ≥ 1, only one octant of the sphere is needed and the number of points inside this volume gives the state counting. Therefore:

$$W(\varepsilon) = \frac{1}{8} \cdot \frac{4 \cdot \pi \cdot R^3}{3} = \frac{\pi}{6} \cdot \left(\frac{8 \cdot m \cdot \varepsilon \cdot V^{\frac{2}{3}}}{h^2} \right)^{\frac{3}{2}} \quad \text{and}$$

$$= \frac{4\pi}{3} \cdot \frac{(2m)^{3/2} \cdot \varepsilon^{3/2}}{h^3} \cdot V$$

$$\rho(\varepsilon) = \frac{\partial W(\varepsilon)}{\partial E} = \frac{4\pi}{3} \cdot \frac{(2m)^{3/2}}{h^3} \cdot V \cdot \frac{3}{2} \cdot \sqrt{\varepsilon}$$
$$= \frac{2\pi \cdot (2m)^{3/2}}{h^3} \cdot V \cdot \sqrt{\varepsilon} \qquad (6.2)$$

According to Eq. (5.15), $q = \sum_i \omega_i \cdot e^{\frac{-\varepsilon_i}{kT}}$ must now be written:

$$q = \int_C^{\infty} \rho(\varepsilon) \cdot e^{\frac{-\varepsilon}{kT}} \cdot d\varepsilon \qquad (6.3)$$

so that, for *one* ideal gas molecule and using $\int_0^{\infty} \sqrt{u} \cdot e^{-u} \cdot du = \frac{\sqrt{\pi}}{2}$:

$$q = \frac{V(2 \cdot \pi \cdot m \cdot k \cdot T)^{3/2}}{h^3} = \frac{V}{\Lambda^3} \quad \text{where:} \quad \frac{1}{\Lambda} \overset{\text{def}}{=} \frac{\sqrt{2 \cdot \pi \cdot m \cdot k \cdot T}}{h} \qquad (6.4)$$

From the molecular partition function (q) many quantities can then be obtained:

- the **probability density** that a single molecule has kinetic energy ε:

$$f(\varepsilon) = \frac{\rho(E) \cdot e^{-\varepsilon/kT}}{q}$$
$$= 2\pi \left(\frac{2m}{h^2} \right)^{3/2} \cdot V \cdot \varepsilon^{1/2} \cdot e^{-\varepsilon/kT} \cdot \left(\frac{h^2}{2\pi mkT} \right)^{3/2}$$
$$= \frac{2\pi \cdot \varepsilon^{1/2} \cdot e^{-\varepsilon/kT}}{(\pi kT)^{3/2}}$$

- the **probability density** of **the number of molecules** (out of the total N molecules) with **kinetic energy ε**:

$$N(\varepsilon) = N \cdot f(\varepsilon)$$

- the corresponding functions for the **speed** (v) (see Figure 3.11), which is the **Maxwell-Boltzmann distribution**:

$$f(v) = f(\varepsilon)\frac{d\varepsilon}{dv} = 4\pi \left(\frac{m}{2\pi kT} \right)^{3/2} v^2 e^{-mv^2/2kT} \quad \text{and}$$

$$N(v) = 4\pi N \left(\frac{m}{2\pi kT} \right)^{3/2} v^2 e^{-mv^2/2kT}$$

- some **statistical parameters** of the **speed distribution:**

  the most probable speed [the v for which $f(v)$ is maximum]: $v_p = \sqrt{\frac{2kT}{m}}$

  the average speed: $\langle v \rangle = \sqrt{\frac{8kT}{\pi m}}$

  the mean square velocity: $v_{rms} = \sqrt{\langle v^2 \rangle} = \sqrt{\frac{3kT}{m}}$

- the **thermodynamic** functions: $Q = q^N / N!$

$$\ln Q = N \ln q - N \ln N + N = N \ln(qe / N); \quad N = N \ln e \qquad (6.5)$$

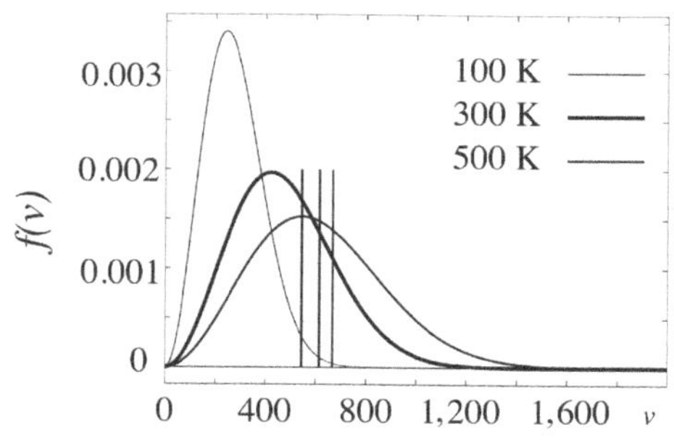

FIGURE 3.11 Probability density distribution, $f(v)$, $m^{-1} \cdot s$, for the speed v, $m \cdot s^{-1}$, of nitrogen molecules at three different temperatures. The vertical lines mark $v_p = 545$, $\langle v \rangle = 615$ and $v_{rms} = 667$ m/s at 500 K.

$$A(N,V,T) = -kT\ln Q = -NkT\ln\left[\left(\frac{2\pi mkT}{h^2}\right)^{3/2}\cdot V\cdot\frac{e}{N}\right] \tag{6.6}$$

$$p = kT\left(\frac{\partial \ln Q}{\partial V}\right)_{T,N} = \frac{NkT}{V} \text{ this is the } \textbf{ideal gas law}$$

$$U = kT^2\left(\frac{\partial \ln Q}{\partial T}\right)_{V,N} = NkT^2\frac{\mathrm{d}\ln T^{3/2}}{\mathrm{d}T} = \frac{3}{2}NkT$$

$$C_V = \frac{3}{2}Nk$$

$$S = \frac{E-A}{T} = Nk\left\{\frac{3}{2} + \ln\left[\left(\frac{2\pi mkT}{h^2}\right)^{3/2}\cdot V\cdot\frac{e}{N}\right]\right\}$$

$$= Nk\ln\left[\left(\frac{2\pi mkT}{h^2}\right)^{3/2}\cdot V\cdot\frac{e^{5/2}}{N}\right]\left(\begin{array}{c}\text{Sackur-Tetrode}\\ \text{equation}\end{array}\right)$$

$$\mu = -kT\ln\left[\left(\frac{2\pi mkT}{h^2}\right)^{3/2}\cdot\frac{V}{N}\right] - kT\ln e + NkT\frac{1}{N}$$

$$= -kT\ln\left[\left(\frac{2\pi mkT}{h^2}\right)^{3/2}\cdot\frac{kT}{p}\right] = \mu^{\ominus} + kT\ln p \tag{6.7}$$

using $\mu^{\ominus} \stackrel{\text{def}}{=} -kT\ln\Lambda^3$; $\Lambda \stackrel{\text{def}}{=} \left(\frac{2\pi mkT}{h^2}\right)^{1/2}$ and Stirling approximation (see Eq. (6.11)) for Eq. (6.5).

It can be seen that the values for U and C_V are those of the equipartition of energy.

Can the MB approximation be applied? The question is: is it fair to say that the number of states available for a molecule is much greater than the number of molecules so that the relationship $Q=q^N/N!$ can be applied? The condition is therefore: $\Phi(\varepsilon) >> N$ when ε is of the order of kT. Substituting (and up to factors of the order of unity): $\Phi(kT) \sim V/\Lambda^3$. So, the condition becomes: $\Lambda^3 N/V << 1$. We check using the density of the liquid at the boiling point: this gives us at least two orders of magnitude as a 'safety margin'. As shown in Table 3.2, with the only (possible) exceptions of He and H_2, the condition is *largely satisfied*.

It is interesting to verify the same relationship in a different case, that of **free electrons in a metallic solid**. The relationships are exactly the same, but now we have m=mass of the electron, T=300 K, V/N=10 cm³/mol. We obtain: $\Lambda^3 N/V \sim 5{,}000$. The result shows that it is not allowed to apply the classical approximation to free electrons in a solid. It is not allowed either to use the relation $Q=q^N/N!$ or to use the Maxwell-Boltzmann statistics (one must use the Fermi-Dirac one, since we are dealing with electrons). It is also worth recalling here the comment on the chemical potential of the electron in classical thermodynamics, see Section 3.6.2.

TABLE 3.2
Check of the MB approximation

Substance	T/K	$\Lambda^3 N/V$
He	4.2	1.5
H_2	20.4	0.44
Ne	27.2	0.015
A	87.4	0.00054

6.2.3 Vibrations and Rotations of an Ideal Gas Molecule

For **rotations**, in the limit of high temperatures the result is (see Exercises 3.10 and 3.11):

$$q = \frac{\sqrt{\pi}}{\sigma}\sqrt{\frac{(kT)^3}{ABC}}$$

Here A, B and C are *rotational constants*, and each corresponds to the reciprocal of a moment of inertia multiplied by $\hbar^2/2$, while σ is the so called *rotational symmetry number* and corresponds to the number of indistinguishable orientations for a given molecule (for example, $\sigma=2$ for a homo-nuclear diatomic molecule, and $\sigma=1$ for a hetero-nuclear diatomic molecule, $\sigma=12$ for methane).

As for **vibrations**, from QM we have that - in the harmonic oscillator model - the levels are $\epsilon_n = \left(n+\frac{1}{2}\right)\cdot\hbar\cdot\omega$ and they are all non-degenerate. Then:

$$q = \sum_{n=0} e^{-\left(n+\frac{1}{2}\right)\frac{\hbar\omega}{kT}} = e^{\frac{-1}{2}\frac{\hbar\omega}{kT}}\cdot\sum_{n=0}^{\infty} e^{-n\frac{\hbar\omega}{kT}}$$

In the sum we recognize a geometric series $1+a+a^2+a^3+\cdots = 1/(1-a)$ with $a \stackrel{\text{def}}{=} e^{\frac{-\hbar\cdot\omega}{k\cdot T}}$. Therefore:

$$q = e^{\frac{-\hbar\omega}{2kT}}\left(1-\frac{\hbar\omega}{kT}\right)^{-1} \tag{6.8}$$

6.2.4 Langmuir Adsorption Isotherm

In this model, the molecules **adsorbed** on specific points of the solid/gas *separation surface* called *adsorption sites* are treated as **independent** of each other, in the sense that a molecule adsorbed on a site *does not influence* the adsorption of another molecule on another site, in particular on an adjacent site. The statistical treatment of the system of N molecules adsorbed on M sites ($N \le M$) then leads to:

$$Q(M,N,T) = \binom{M}{N}q^N;\quad \binom{M}{N} = \frac{M!}{N!\cdot(M-N)!} \tag{6.9}$$

The binomial coefficient considers the distinct ways in which N equal and indistinguishable molecules can arrange themselves on M equal and distinguishable sites, while the single-molecule partition function is:

$$q(T) = q_x q_y q_z e^{-U_0/kT} \tag{6.10}$$

Here $U_0<0$ is the energy of the bottom of the attractive potential of the adsorption site with respect to a zero corresponding to the free molecule in the gas phase. The three contributions q_x, q_y, and q_z of the vibrational DOFs around the site (see Eq. (6.8)) replace the three translational DOFs of the free molecule.

The M sites are distinguishable because they are parts of a macroscopic structure (the surface) and therefore it is possible to say which site we are talking about. The canonical partition function Q of Eq. (6.9) provides the thermodynamic functions for defined values of N, M and T, but this is not what is usually needed. In fact, in the study of adsorption equilibrium, we are *not* interested in the *closed system* of the adsorbed molecules but in the *open system* of the *adsorbed molecules in equilibrium with the gas phase* at partial pressure p and the same temperature. The variables T and p fix the chemical potential μ of the molecules in the gas phase, then the chemical potential of the adsorbed molecules and finally the quantity of adsorbed substance N. Strictly speaking, we should treat adsorption with the grand canonical partition function (fixed M, μ, T) to obtain the average number $\langle N \rangle$ of adsorbed molecules. More simply, what is done instead is that:

a. the canonical partition function (Eq. 6.9) is used to obtain the statistical mean value $\langle \mu \rangle$ of the chemical potential at a given N:

$$Q(M,N,T)=\binom{M}{N}q^N \Rightarrow \mu=-k\cdot T\cdot\left(\frac{\partial \ln Q}{\partial N}\right)_{T,M} \quad \text{that is:}$$

$$\mu=-k\cdot T\cdot N\cdot \ln q-k\cdot T\cdot\frac{\partial}{\partial N}\{\ln(M!)-\ln(N!)-\ln[M-N]!\}$$

The logarithms of factorials are approximated by Stirling's formula:

$$\ln(N!) \cong N\cdot\ln(N)-N \quad \text{and then:}$$

$$\ln(M!)-\ln(N!)-\ln[M-N]!$$
$$= M\cdot\ln(M)-M-N\cdot\ln(N)$$
$$+N-(M-N)\cdot\ln(M-N)+M-N$$
$$=-\ln(N)-1+\ln(M-N)-(M-N)\frac{-1}{M-N}=\ln\frac{M-N}{N} \tag{6.11}$$

If we now define the fractional occupancy of the adsorption sites:

$$\frac{N}{M-N}=\left(\frac{M-N}{N}\right)^{-1}=\left(\frac{1}{\vartheta}-1\right)^{-1}$$
$$=\frac{\vartheta}{1-\vartheta}, \text{ we obtain } \mu=R\cdot T\cdot\ln\left(\frac{\vartheta}{q\cdot(1-\vartheta)}\right)$$

and then

$$\frac{N}{M-N}=\left(\frac{M-N}{N}\right)^{-1}=\left(\frac{1}{\vartheta}-1\right)^{-1}$$
$$=\frac{\vartheta}{1-\vartheta}, \text{ we obtain } \mu=R\cdot T\cdot\ln\left(\frac{\vartheta}{q\cdot(1-\vartheta)}\right)$$

b. This average value is equated to the chemical potential of the molecules in the gas phase:

$$\mu=R\cdot T\cdot\ln\left(\frac{\vartheta}{q\cdot(1-\vartheta)}\right)=\mu^{\ominus}+R\cdot T\cdot\ln p$$

where p is a shorthand for the ratio of the pressure to the standard state pressure, denoted by the symbol $\ominus$.

The derivation is a typical example of thermodynamic equivalence of Gibbs ensembles (see Section 4.1.2). Note that no distinction is made between exact values (N in the canonical case, μ in the grand-canonical case) and statistical means ($\langle N \rangle$ in the grand-canonical case, μ in the canonical case) and that the relations between variables are freely replaced by the inverse relations, *e.g.* from $\mu(N)$ and $\mu(p)$ we obtain $N(\mu)$ and then $N(p)$.

Finally, we arrive at the relation between the fractional occupancy of the sites and the equilibrium pressure p of the gas phase:

$$\vartheta=\frac{\chi\cdot p}{1+\chi\cdot p} \quad \text{which is known as the **Langmuir isotherm**} \tag{6.12}$$

The $\chi \overset{\text{def}}{=} q(T)\cdot e^{\mu^{\ominus}/RT}$ contains the energy U_0 (Eq. 6.10) and is essentially an equilibrium constant, as is clearly evident from the derivation in CT (Eq. 3.62).

The Langmuir isotherm is also often used to describe adsorption from liquid solution, with obvious modifications: the concentration of the solution, for example, replaces the partial pressure of the gas phase.

It is interesting to derive the entropy of the system (adsorbent+adsorbate). From the fundamental relation between entropy and the canonical partition function (see Section 4.1.2):

$$S=kT\left(\frac{\partial \ln Q}{\partial T}\right)_{N,V}+k\ln Q \text{ and } Q(M,N,T)=\frac{M!}{N!(M-N)!}q^N$$

and considering that the combinatorial term in Q is independent of T, we obtain that the entropy can be decomposed into:

$$S = S_{\text{config}} + S_{\text{vibrat}}$$

$$\text{where } S_{\text{config}} = k \cdot \ln \frac{M!}{N!(M-N)!} \quad \text{and}$$

$$S_{\text{vibrat}} = Nk\left(\ln q + T\frac{\mathrm{d}\ln q}{\mathrm{d}T}\right)$$

The surprising aspect of the result is that, while the vibrational part correctly depends on T and, in particular $\rightarrow 0$ for $T \rightarrow 0$, the configurational part of the entropy of the system is independent of T and therefore the entire entropy does not $\rightarrow 0$ for $T \rightarrow 0$ as one would expect from the third law. The explanation lies in the simple fact that the system (adsorbed molecules+adsorption lattice) is not a pure substance, as the third law would require, but is a binary system (molecules+sites).

6.2.5 The Monoatomic Solid in the Einstein Model

A monoatomic crystalline solid, *i.e.* with only one atom per unit cell, made up of N atoms has 3 translational degrees of motion and 3 rotational degrees of motion (due to the motions of the entire solid as a macroscopic object) and therefore $3N-6$ vibrational degrees of motion, which approximates to $3N$ neglecting translations and rotations ($3N >> 6$), The atoms of the solid are considered **independent** and **equal**:

$$Q(N,T) = q^N$$

The $N!$ factor is missing because now the N atoms must be treated as distinguishable, as one can label them by specifying their site in the crystal. Obvious additional hypotheses are that the solid has no defects (such as vacancies or interstitials) and is not a mixture.

The atomic partition function q is the product of the exponential of $-U_0/kT$ (where U_0 is the interaction energy of an atom with all the others) multiplied by three vibrational partition functions, each of them being the same expression as the vibrational partition function of a gaseous molecule (Eq. 6.8).

A historical curiosity: the model was developed by Einstein as a **teaching example** to show in a simple way how the quantization of energy at the atomic-molecular level can lead to a particular trend with the temperature of the thermodynamic functions, in particular to a trend that disagrees with the principle of equipartition of energy. For example, according to this model, the specific heat of a monatomic solid vanishes for $T \longrightarrow 0$, in agreement with the experimental data but in disagreement with the constant value ($3R$) predicted by equipartition (see Exercise 3.9).

6.2.6 Non-interacting Magnetic Dipoles: Paramagnetism

In this model, M equal magnetic dipoles are placed in specific crystallographic sites of a crystalline solid and can be variously oriented due to the interaction of each dipole with an external constant magnetic field $\vec{H}$. The interaction between the dipoles is neglected: the dipoles are treated as *independent*. For simplicity, we consider the case in which the dipoles can only be oriented "up" and "down", that is, with a direction parallel or antiparallel with respect to the external field. The situation is then partly similar to that of the Langmuir adsorption model, where we had the states "full site" and "empty site".

Indicating with $\vec{\mu}$ the magnetic moment of a single dipole, for a "down" dipole (energetically favoured situation) we define:

$$\alpha = -(\vec{\mu} \cdot \vec{H}) / kT = -(\mu \cdot H)/kT$$

where μ and H are the projections of the vectors along the same line.

Let us consider a system of M dipoles, where $N \leq M$ are in the "down" state and $M-N$ are in the "up" state. The total contribution of the "down" dipoles is $-N\alpha < 0$. The total contribution of the "up" dipoles is $(M-N)$ α. Globally the energy of the system of M dipoles is $(M-2N)\alpha = (M-2N)$ $(\mu \cdot H)/kT$ and the canonical partition function is:

$$Q(M,N,T) = \binom{M}{N} q^M \cdot \mathrm{e}^{(M-2N)\alpha} \tag{6.13}$$

Here q denotes the partition function of a single dipole (e.g. atom or ion) *in the absence of the field*.

The difference with respect to the Langmuir model is that N is here a quantity determined by the value of the externally applied magnetic field (H) and it is necessary to express the dependence of N on H. For this purpose we consider all the possible values of N and their energies. We then arrive at:

$$Q(M,H,T) = q^M \cdot \mathrm{e}^{M(\mu \cdot H)/kT} \cdot \sum_{N=0}^{N=M} \binom{M}{N} \cdot \mathrm{e}^{-(2N)(\mu \cdot H)/kT}$$

Then:

$$Q(M,H,T) = q^M \cdot \mathrm{e}^{M(\mu \cdot H)/kT} \cdot \sum_{N=0}^{N=M} \frac{M!}{N! \cdot (M-N)!} \cdot \mathrm{e}^{-2N(\mu \cdot H)/kT}$$

Writing $y \overset{\text{def}}{=} \mathrm{e}^{-2(\mu \cdot H)/kT}$ the M-th power of the binomial $(1+y)$ is recognized in the sum, so that:

$$Q(M,H,T) = q^M \cdot \mathrm{e}^{M(\mu \cdot H)/kT} \cdot \left(1 + \mathrm{e}^{-2(\mu \cdot H)/kT}\right)^M$$

The expression can be made more concise by using the definitions of hyperbolic functions, in particular:

$$\sinh x = \frac{e^x - e^{-x}}{2} \qquad \cosh x = \frac{e^x + e^{-x}}{2} \qquad \text{and}$$

$$\tanh x \overset{\text{def}}{=} \frac{\sinh x}{\cosh x} \overset{\text{def}}{=} \frac{e^x - e^{-x}}{e^x + e^{-x}} = \frac{e^{2x} - 1}{e^{2x} + 1}$$

So:

$$Q(M,H,T) = \left[q \cdot \left(e^{\alpha} + e^{-\alpha}\right)\right]^M = \left[2q \cdot \cosh\left((\mu \cdot H)/kT\right)\right]^M$$

Finally introducing the magnetic work in analogy with the pV work, the intensive variable corresponding to the pressure is the external magnetic field H, while the extensive variable corresponding to V is the global dipole moment $(M-2N)\mu$. From $A=-kT\ln Q$ we obtain:

$$(M-2N)\mu = kT \cdot \left(\frac{\partial \ln Q}{\partial H}\right)_{M,T} = M \cdot \mu \cdot \tanh\frac{\mu \cdot H}{kT} \tag{6.14}$$

Typically, however, one instead prefers to introduce another quantity, which is intensive, usually is called **magnetization**, is denoted by m and is obtained by dividing by its maximum value ($M\mu$):

$$\frac{(M-2N)\mu}{M \cdot \mu} = 1 - 2\frac{N}{M} = \tanh\frac{\mu \cdot H}{kT} \tag{6.15}$$

Less frequently, the name *magnetization* is assigned to the extensive quantity $(M-2N)\mu$ (Eq. 6.14).

6.3 An Experimental Macroscopic Validation

The problem of distinguishability/indistinguishability of electrons in an atomic, molecular or crystalline system is a very familiar topic, as well as its consequence, namely the exclusion principle. It does not seem equally obvious that the problem of distinguishability/indistinguishability must also be posed for molecules. It may therefore be useful to verify more directly, *i.e.* to show that it is *necessary* to bring indistinguishability into play to explain also some important experimental results at the macroscopic level. In other words, if one were to develop a treatment of Statistical Thermodynamics in which equal molecules are treated as distinguishable, then one would obtain results that disagree with the experiments.

Since the statistical consequence of indistinguishability is the term $N!$ in some thermodynamic functions, we are particularly interested in seeing whether this term produces effects that have an experimental outcome: in this regard, a particularly significant case is that of an *ideal mixture*. In an ideal gas, by definition, the molecules do not interact with each other, regardless of whether they are equal or different: it is therefore useful to examine the particular case of an ideal mixture of two ideal gases (A and B) in amounts N_A and N_B.

The results of our interest concern the trend of a molar integral quantity as a function of the molar fraction x in isothermal-isobaric conditions (since these are ideal gases, they are also isochoric conditions, *i.e.* the volume is constant). The derivations are in Section 8.4.1, and we see that the trends corresponding to the definition of ideal solution according to CT (Section 3.2.3) are obtained. In fact, for quantities such as (E/n, p, H/n, C_p/n, …), the result is that in the usual isothermal-isochoric graphs that represent the molar integral quantity as a function of the molar fraction x, the trend is a straight line passing through the molar values of the pure gases for $x=0$ and for $x=1$. For the thermal quantities (S, A, …), however, we have an additional contribution with a non-linear trend. The difference with respect to the straight line passing through the molar values of the pure gases at $x=0$ and $x=1$ is:

$$R \cdot T \cdot \{(1-x) \cdot \ln(1-x) + x \cdot \ln(x)\} \quad \text{for Helmholtz energy } (A/n) \tag{6.16}$$

$$-R \cdot \{(1-x) \cdot \ln(1-x) + x \cdot \ln(x)\} \quad \text{for entropy } (S/n) \tag{6.17}$$

NB. It can be proved (cfr.3.2.3) that these expressions $\rightarrow 0$ both as $x \rightarrow 0$ and $x \rightarrow 1$.

The result is therefore an important experimental proof of the indistinguishability at the molecular level: indistinguishability explains the entropy of ideal mixtures.

6.4 Non-Ideal Systems: A Simple Model

As said in Section 4.2, many different *analytical* methods have been developed in ST to deal with systems of interacting particles. This topic goes well beyond our scope and is therefore left to specialised books. We report here only a brief example, selected for its simplicity and easy derivation and because it allows to illustrate how the interactions between molecules can lead to a phase separation (see Section 1.2.6).

Lattice theories are a wide class of analytical methods of ST for interacting particles and are essentially based on a lattice of equal sites hosting equal or different structural units, such as atoms, free molecules, units of polymer chains, electric or magnetic dipoles, vacancies, etc. Equal structural units are indistinguishable, while the sites are distinguishable, since the lattice allows one to identify a site.

The lattices are different in dimensionality and geometry (for example: 1D, triangular, square or hexagonal for the 2D case or simple cubic, fcc or bcc for the 3D case) and are essentially characterized by the coordination number c, that is, the number of *nearest neighbours* of each site. Some ST treatments already seen are in fact examples of lattice theories: these are Langmuir adsorption (see Section 6.2.4), the Einstein solid (see Section 6.2.5) and paramagnetism (see Section 6.2.6). They are all *ideal* cases in the sense that the structural units on different sites were *independent*, *i.e.* non-interacting.

Unlike these cases, *interactions* between first neighbours, second neighbours, etc. are now introduced. To illustrate the method and the results, it is sufficient to consider a system with only two structural units (denoted as A and B) having N_A and N_B particles and M equal sites (with the constraint $M=N_A+N_B$). Furthermore, it is convenient to limit us to considering only *interactions between first neighbours* and therefore only three interaction energies w_{AA}, w_{AB} and w_{BB} are needed.

With this simplification, *each configuration* of the system (which is specified by saying which unit - A or B - occupies each site) contributes to the partition function Q simply with:

$$q_A{}^{N_A} \cdot q_B{}^{N_B} \cdot \left(e^{-w_{AA}/kT}\right)^{N_{AA}} \cdot \left(e^{-w_{BB}/kT}\right)^{N_{BB}} \cdot \left(e^{-w_{AB}/kT}\right)^{N_{AB}}$$

where N_{AA}, N_{BB} and N_{AB} are the numbers of pairs of nearest neighbours AA, BB and AB (respectively) in the configuration considered. This must be multiplied by a degeneracy, that is, the number of configurations that, having the same N_{AA}, N_{BB} and N_{AB}, by the previous hypothesis have the same energy.

In the 2D and 3D cases, for this apparently simple model, no expression of the partition function is known that is both *exact* and *complete*, *i.e.* valid for all possible values of the specific parameters of the model (dimensionality of the lattice, coordination number, ratios between the interaction energies). The difficulty lies in evaluating the possible values of N_{AA}, N_{BB} and N_{AB} for given values of N_A and N_B.

It is then necessary to resort to various approximate methods. The simplest method (known as **BW** by the names of *Bragg* and *Williams*), considers only the interactions between nearest neighbours, as said, and calculates *approximately* both (a) the energies of the various configurations, and (b) their degeneracies. Both approximations derive from of the fundamental assumption that the units *A* and *B* are completely distributed at random, an assumption that would clearly be acceptable only as a limit for $T \to \infty$.

Concerning the first approximation (a) of the BW model, the energy of a configuration is evaluated using the *mean values* of N_{AA}, N_{BB} and N_{AB} instead of the different possible values and, in particular, in using $W = N_{AA} \cdot w_{AA} + N_{AB} \cdot w_{AB} + N_{BB} \cdot w_{BB}$ for the interaction between different units. Consequently, given that N_A and N_B are fixed, that is, that the composition of the binary system is fixed, the possible configurations all have the same energy and contribute to the partition function with the same quantity:

$$q_A^{N_A}\left(e^{-w_{AA}/kT}\right)^{N_{AA}} \cdot q_B^{N_B}\left(e^{-w_{BB}/kT}\right)^{N_{BB}}$$

Then, the second approximation (b) of the BW model uses the same configurational (combinatorial) factor of the ideal case, *i.e.* the situation in which there are no interactions between the structural units:

$$\frac{(N_A+N_B)!}{N_A!\,N_B!}$$

which, as already seen, is the total number of configurations of N_A and N_B structural units on $M=N_A+N_B$ sites.

These approximations make possible to calculate the average values N_{AA}, N_{BB} and N_{AB} of the various pairs once the number (c) of nearest neighbours of a site has been specified. An A unit has on average cN_A/M units as its nearest neighbours, so the average number of AA pairs as nearest neighbours is $N_{AA} = (cN_A/M)(N_A/2) = c(N_A)^2/2M$ where the factor 2 avoids counting the same AA pair twice. Similarly, the average number of BB pairs as nearest neighbours is $N_{BB} = (cN_B/M)(N_B/2) = c(N_B)^2/2M$. An A unit has on average cN_B/M B units as nearest neighbours and the average number of AB pairs as nearest neighbours is given by $N_{AB} = (cN_B N_A/M)$ where now the two has been omitted to distinguish the AB pair from the BA pair.

so that the partition function becomes:

$$Q(N_A, N_B, T) = \frac{(N_A+N_B)!}{N_A!\,N_B!}[X+Y+Z] \quad \text{where}$$

$$X = \left(q_A e^{-c \cdot w_{AA}/2kT}\right)^{N_A}$$

$$Y = \left(q_B e^{-c \cdot w_{BB}/2kT}\right)^{N_B}$$

$$Z = \left(e^{w/2kT}\right)^{-cN_A N_B/(N_A+N_B)} \tag{6.18}$$

Equation (6.18) can be directly applied to the treatment of a solid solution of interacting structural units (*e.g.* atoms). From the basic relations (Section 4.1.2: $A = -k \cdot T \ln Q$ and $S = k \cdot T\left(\frac{\partial \ln Q}{\partial T}\right)_{N,V} + k \cdot \ln Q$), we derive that $A(N_A, N_B, T)$ is given by the sum of four parts:

$$1: -N_A\left(kT \cdot \ln q_A - c \cdot w_{AA}/2\right) \quad 2: -N_B\left(kT \cdot \ln q_B - c \cdot w_{BB}/2\right)$$

$$3: -kT \ln\frac{(N_A+N_B)!}{N_A!\,N_B!} \quad \text{and} \quad 4: \frac{cN_A N_B}{2(N_A+N_B)} w$$

For **pure** A we derive[14]:

$$\frac{A}{kT} = -N_A \ln\left(q_A \cdot e^{-c \cdot w_{AA}/2kT}\right) \quad \mu_A^{\ominus} = -k \cdot T \cdot \ln\left(q_A \cdot e^{-c \cdot w_{AA}/2kT}\right)$$

So, the four parts give the contributions of component A (see 1), component B (2), ideal mixture (3) and excess enthalpy (4) to the Helmholtz energy of the mixture. In particular, there is a T-independent excess enthalpy (see Section 3.2.5):

$$A^{ex}(N_A, N_B) = \frac{-cN_AN_B}{2(N_A + N_B)} w \text{ or, in molar units:}$$

$$A_m^{ex}(x) = -x \cdot (1-x) \cdot \frac{c \cdot w}{2} \tag{6.19}$$

which is equal to the empirical model of the **regular solutions** already seen in Section 3.2.5.

Then, the model is able to predict a **phase separation** of the mixture, as seen in Section 3.3.4.3, if the excess term is positive *i.e.* if $w = w_{AA} + w_{BB} - 2w_{AB} < 0$ and when the attractive AA and BB interactions (w_{AA} and $w_{BB} < 0$) prevail over the AB interactions, which are less attractive or even repulsive: $w_{AA} + w_{BB} > 2w_{AB}$.

Below the critical temperature $T_c = cw/4k$, a solution with overall composition close to $x = 0.5$ is not thermodynamically stable and splits into two solutions - with compositions $x' < 0.5$ and $x'' > 0.5$ - which, by lowering the T, progressively move away from the value 0.5.

Indeed, the simple BW model can predict also (i) a **gas/liquid phase separation** in a pure substance when applied to a system of only N_A molecules on $M > N_A$ sites or (ii) the **ferromagnetism** when applied to a system of N_A "up" magnetic dipoles and N_B "down" magnetic dipoles on $M = N_A + N_B$ sites.

7 SHORT NOTES ON THE COMPUTATIONAL APPROACH

Computational chemistry is so broad, varied and pervasive throughout chemistry that it is not easy to illustrate the aspects strictly pertaining to statistical thermodynamics: here we limit ourselves to very schematically outlining some ideas and procedures thanks to which, within the computational approach to a chemical problem, results concerning thermodynamic equilibrium can be obtained.

7.1 Molecular Dynamics (MD)

The methods that go under the name of Molecular Dynamics (MD) have found applications in chemically very different problems: simple molecules, polymeric structures, biological systems, crystalline solids or amorphous condensed matter.

At least in its original version, MD is based on the direct solution of the equations of motion of a suitable set of microscopic particles, *i.e.* Newton's equations:

$$F_i = m_i \cdot a_i \tag{7.1}$$

which are integrated numerically.

7.1.1 Outline

Very briefly:

A. the initial conditions are set, that is, both the number N and the masses m_j of all the particles ($j = 1, \ldots, N$), and the initial values of the coordinates $x_1, \ldots, x_{3N}$ and of the corresponding derivatives with respect to time, that is, the velocities $v_1, \ldots, v_{3N}$,
B. an appropriate time step δt is chosen
C. and a procedure is iterated N_t times: for each time step t_j
   1. from the values of the coordinates the procedure evaluates the potential energy $V(x_1, \ldots, x_{3N})$, and
   2. for each coordinate ($i = 1, \ldots, 3N$):
      a. evaluates the force F_i [$F_i = \partial V / \partial x_i$],
      b. from this the acceleration $a_i = F_i/m_j$ in which $j = j(i)$ is the particle of the coordinate i;
      c. from the acceleration (by numerical integration) obtains the variations of velocity δv_i and coordinate δx_i
      d. and then updates the previous values of time: $t_{j-1} + \delta t \rightarrow t_j$,
      e. of the velocities: $v_i + \delta v_i \rightarrow v_i$,
      f. and of the coordinates $x_i + \delta x_i \rightarrow x_i$,
      g. and evaluates all the quantities of interest, in particular the kinetic energy:

$$K = \sum_{j=1}^{N} \frac{1}{2} m_j v_j^2 \tag{7.2}$$

In Eq. (7.2) v_j^2 is the sum of the squares of the three components of the velocity of the j-th particle.

7.1.2 Potential Energy

The functional form of the potential energy evidently plays a fundamental role in the MD method and very different expressions are found in relation to the different types of systems.

As a matter of fact, a very wide range of problems have been shown to be easily tractable by resorting to a potential energy expressed as a sum of *atom-atom potentials* (that is pairwise potentials).

In the case of **crystalline solids**, we speak more specifically of *lattice dynamics* (LD). A category of crystalline solids to which LD has been widely applied is that of ionic solids: the potential energy is expressed as the sum of the Coulomb term, a short-range attractive term (negative and proportional to r^{-6}, where r indicates the interatomic distance) and a repulsive term (positive and proportional to e^{-kr}) known as the *Buckingham* potential.

In applications to **molecular systems** and in particular to problems in organic chemistry and biochemistry, atom-atom potentials are typically expressed as a function of quantities such as bond lengths, bond angles, dihedral angles, etc..... Typical expressions are:

- for the stretching of the bond distance (r_{ij}) between atom i and atom j: $V_{ij} = k_{ij}\left(r_{ij} - r_{0,ij}\right)^2$ where $r_{0,ij}$ is the equilibrium value of r_{ij}.

- For the bending of the angle φ_{ijl} between the i–j bond and the j–l bond: $V_{ijl} = k_{ijl}\left(\varphi_{ijl} - \varphi_{0,ijl}\right)^2$ where $\varphi_{0,ijl}$ is the equilibrium value of the angle.
- For the twisting of the dihedral angle ψ_{ijlm} between the i–j and l–m bonds of four atoms bonded in the sequence $i\,j\,l\,m$: $V_{ijlm} = k_{ijlm}\left[1-\cos\left(n\cdot\psi_{ijlm}\right)\right]$ in the case where there are n equivalent configurations.
- Long-range interactions are also modelled, for example Coulomb interactions, when net charges are involved, or Van der Waals interactions using a 6/12 potential (*i.e.* attractive proportional to r^{-6} and repulsive proportional to r^{-12}) between atoms that are spatially close even if they are not directly connected by covalent bonds.

The parameters of the various expressions of potential energy are obtained through a *parameterization* step, which includes the best fit of a large set of data (experimental and/or computational) relating to various structural, spectroscopic, functional, thermal properties and to many different substances within a given category.

Finally, it has been recognized that in various applications of MD, in particular to metallic and semiconductor solids, the expression of the potential energy as a sum of atom-atom potentials is not reliable and it is necessary to resort to quantum determinations to correctly evaluate the contribution of the electrons in a band. Among the various methods of this type, the method developed by *Car* and *Parrinello* is particularly well known.

7.1.3 Statistics

Conceptually, numerical integration of the equations of motion is a deterministic procedure that essentially computes the trajectory of the system in and phase space (coordinates and moments) starting from a specific initial situation and based on the knowledge of the potential energy of the system. The trajectory becomes a sampling of microstates when interpreted according to the ergodic hypothesis, *i.e.* as a list of states in which the system is at successive (discrete) times. Then, for a certain quantity A, which we can compute for each (n) of the N_t time steps, we can evaluate the statistical mean from:

$$\langle A\rangle = \frac{1}{n}\sum_{n=1}^{N_t} A_n \tag{7.3}$$

The meaning of an MD calculation is therefore that the average over the microstates, sampled following their temporal evolution according to (Eq. 7.3) is representative of the average over the Gibbs ensemble. Among the various quantities that are of interest to determine, the potential energy is necessarily calculated at each integration step because it is needed for the MD algorithm, while the kinetic energy is calculated according to Eq. (7.2) and the total energy from the sum of the two, while the temperature is obtained from the equipartition:

$$T_n = \frac{2}{3\cdot N\cdot k_B}K_n \tag{7.4}$$

The procedure described so far is, therefore, a **micro-canonical** procedure in which the system is isolated, the total energy is fixed (at least in principle, as specified further on) and the T can vary from step to step if the system reaches a configuration with different potential energy.

This is not a obliged choice, however, and if desired, statistical averages can be calculated for a canonical ensemble. One can then resort to a thermostat, that is, embed the system in a sufficiently large container of inert molecules with which it can exchange energy. In this case, what remains (at least approximately) constant is the kinetic energy of the system alone, while the potential energy and the total energy vary in parallel step by step.

In reality, the MD procedure is not strictly deterministic and contains elements of randomness. In addition to the fact that several runs can be made for the same macrostate by choosing different initial conditions, numerical algorithms with rounding and truncation and discrete time sampling introduce errors, which behave as statistical noise and are detectable in a single MD run by the fact that the total energy (in the micro-canonical case) or the temperature (in the canonical case) do not remain exactly constant. As a consequence, an MD algorithm must include corrections, *i.e.* a restoration procedure of the total energy or kinetic energy to the initial value after a number of computational steps.

Going into more detail, consider a particular quantity A, for example a bond angle or a bond length. During an MD run starting from a particular set of initial conditions, A shows variations to be interpreted as statistical fluctuations around the mean value (Eq. 7.3). Actually, over the fluctuations with a zero mean value it is often clearly observed a monotonic trend, well represented by an inverse exponential law as a function of time. The reason is that the value of A in the initial configuration of the run can be more or less distant from the statistical mean value of the macrostate, *i.e.* from the value corresponding to thermodynamic equilibrium. The implementation of the method therefore requires that each MD run follows an initial phase called *equilibration*, in which, to speak metaphorically, the system must be given the time necessary to reach thermodynamic equilibrium. Obviously, the data significant for the statistics (Eq. 7.3) are sampled only in the computational steps following equilibration.

7.2 The Monte Carlo (MC) Approach

MC is the common name of several computational methods based on the generation of random numbers. MC methods are found in very different computational fields, which may even have nothing to do with ST. A curious and frequently recalled example is the calculation of π as the ratio $4C/Q$ between the area of a circle $C=\pi r^2$ and that of the circumscribed square $Q=4r^2$, sampling pairs of random numbers generated with uniform probability density in the interval between −1 and +1. For each pair (x, y), if $x^2+y^2<1$ the point is within the circle C and a counter c is incremented. After n steps, the ratio c/n provides an estimate of $\pi/4$.

Another example is the search for the minimum of a function of several (n) real variables done by generating many successive n-tuples of random numbers x_i and computing the values of the function at the points $(x_1, x_2, \ldots, x_n)$. Obviously not very performant compared to local methods that explore the neighbourhood of a starting point somehow known or hypothesized, the MC method can be useful for exploring large regions of the space of variables and provides a reasonable starting point when one does not have better information on the behaviour of the function.

In the MC method of ST, random number generation is used to sample the space of states accessible to a system. *As in MD*, the mean value of a quantity is obtained by averaging the values (A_i) obtained in the various (n_{MC}) samplings:

$$A = \frac{1}{n_{MC}} \sum_{n=1}^{n_{MC}} A_n \tag{7.5}$$

Unlike MD, the sequence of sampled states is independent of the temporal evolution of the system and is generated randomly by an algorithm according to a pre-assigned probability law. The crucial problem of the MC method of ST is precisely the sampling.

It is useless to uniformly sample a configuration space or a phase space, *i.e.*, by generating random numbers with uniform distribution for coordinates and moments, because that is not what is needed. Consider for example an isolated system and therefore a micro-canonical ensemble. The sampling must be done in such a way as to have the same probability of selecting one state or another from a population of states that *all have the same pre-assigned value of energy*. There is no simple and direct method for doing this.

The problem is solved by first resorting to a sampling method that is easily implemented even if it is characterized by a probability distribution that is not adequate for the problem of interest. Then, this sampling is combined with a mechanism for rejecting some of the states thus produced. The aim to be achieved is that the composite probability (sampling probability×rejection probability) corresponds to the desired probability.

7.2.1 Metropolis Algorithm

In the specific case of closed isothermal – isochoric systems (N, V, T), *i.e.* in the case of the canonical ensemble, the MC method is known as the Metropolis method from the name of one of the researchers who introduced it.[15] Operationally, an algorithm is first created that generates a sequence of possible states ($\ldots, i, j, k, l, \ldots$), *i.e.* an algorithm that starting from a possible state (i) generates a "move" towards another possible state (j).

Then, given the states i and j relative to a "move", the Metropolis algorithm provides a mechanism (a) to accept the final state (and then this is effectively "promoted" to microstate j of the sequence and a new "move" is made, this time starting from state j), or (b) to reject the final state which is not used and then a different move is attempted (always starting from state i) and on the result of this new attempt it is decided again whether to accept or reject, and so on …

A straightforward way to justify the Metropolis algorithm is to invoke the *principle of detailed balance* (or *microscopic reversibility*), according to which the probability of transition between one microstate (i) and another (j) must be the same in both directions:

$$p_{i\to j} = p_{j\to i} \tag{7.6}$$

These are *global* probabilities and can be expressed as the product of three *partial* probabilities:

$$p_{i\to j} = \pi_i \cdot T_{i\to j} \cdot A_{i\to j}$$

where π_i is the probability that the system is in microstate i; $T_{i\to j}$ is the probability that the move from microstate i to microstate j is attempted, *i.e.* if the system is in microstate i, microstate j is considered as the next state; $A_{i\to j}$ is the probability that the attempted move is accepted.

A significant aspect is that only the two states between which the transition possibly occurs are considered. For example, no states prior to i appear: this is a process *without memory*.

As for the probabilities T, since it is an algorithm that we can design at will, it is possible and convenient to consider the symmetric case $T_{j\to i} = T_{i\to i}$. Then, we say that the sequence of states ($\ldots, i, j, k, l, \ldots$) is a *Markov chain* and the principle of the detailed balance simplifies to:

$$\begin{aligned} p_{i\to j} &= \pi_i \cdot T_{i\to j} \cdot A_{i\to j} = p_{j\to i} \\ &= \pi_j \cdot T_{j\to i} \cdot A_{j\to i} \Rightarrow \pi_i \cdot A_{i\to j} = \pi_j \cdot A_{j\to i} \end{aligned} \tag{7.7}$$

With the canonical ensemble, the two microstates i and j have probabilities of occurring:

$$\frac{\pi_i}{\pi_j} = \frac{e^{-E_i/kT}}{e^{-E_j/kT}} \tag{7.8}$$

Then, let us understand what form the acceptance probability $A_{i\to j}$ takes under the constraints of Eqs. (7.6) and (7.8). The result is quite simple:

$$A_{i\to j} = \min\left(1, e^{-(E_j-E_i)/kT}\right) \tag{7.9}$$

meaning that you have to choose the smaller value between 1 and $e^{-(E_j-E_i)/kT}$.

Equation (7.9) is straightforward to verify. Just insert it into the detailed balance of Eq. (7.7) and check whether the probability ratio of the canonical ensemble (Eq. 7.8) is obtained:

$$\frac{A_{i\to j}}{A_{j\to i}} = \frac{e^{-E_j/kT}}{e^{-E_i/kT}} = \frac{\pi_j}{\pi_i} \tag{7.10}$$

Indeed, in the case of a "downward" transition $E_i > E_j$: $-(E_j - E_i)/kT > 0$, $A_{i\to j} = e^{-(E_j-E_i)/kT} > 1$, $A_{j\to i} = 1$ and $A_{i\to j}/A_{j\to i} = e^{-(E_j-E_i)/kT}$. In the case of an "upward" transition $E_i < E_j$, $-(E_j - E_i)/kT < 0$, $A_{i\to j} = 1$, $A_{j\to 1} = e^{-(E_i-E_j)/kT} < 1$ and $A_{i\to j}/A_{j\to i} = e^{-(E_j-E_i)/kT}$. In both cases, Eq. (7.10) is satisfied.

The random number generation comes into play to simulate a rejection probability in Eq. (7.9). Given the two states (i and j), we then calculate the two energies (E_i and E_j) and compare them

- If $E_i > E_j$, *i.e.* $e^{-(E_j-E_i)/kT} > 1$, then the "move" is accepted (Eq. 7.9 gives 1).
- If $E_i < E_j$, *i.e.* $e^{-(E_j-E_i)/kT} < 1$, then a random number (r) between 0 and 1 is generated with *uniform probability*.
  - If $e^{-(E_j-E_i)/kT} > r$ the "move" is accepted.
  - If $e^{-(E_j-E_i)/kT} < r$ the "move" is rejected.

A critical point of the method is how to generate a new state (j) from the previous state (i) of the sequence. In designing the algorithm, attention must be placed first of all on the capability to reach all regions of the configuration space and only later on computational simplicity.

To conclude, by fixing the initial values of the coordinates and using the "move" algorithm instead of Newton's law used in MD, the various states are generated that are accepted or rejected from step to step according to the Metropolis algorithm. For each accepted state, the quantities of interest are calculated, and the averages are calculated at the end of the simulation.

7.2.2 An MD/MC Comparison

The two methods are *not* (broadly speaking) to be considered *in competition* but rather as complementary.

- Other things being equal, MD typically requires computation times several orders of magnitude longer.
- On the other hand, with the increase in available computing power, many problems previously addressable only via MC have progressively entered the field of MD applications.
- Always very indicatively, MD follows the evolution of a system for *short* times, and its results may *not* correspond to a *true equilibrium* but rather to a *partial* equilibrium. In these cases, the MC technique allows sampling a much wider configuration space and therefore better simulations over longer time scales.
- MD is intrinsically able to describe the time evolution of the system and therefore is the technique of choice for studying the **dynamics** of a system. An example concerns the study of *diffusion coefficients*.
- Although not strictly necessary, MD typically describes the system according to the laws of motion of CM. Instead, MC methods can be used (and are widely used) also to describe interactions that cannot be treated within the formalism of CM, as in the case of spin systems.

8 REMARKS

We report here some clarifications and insights, as well as some derivations of particular length or complexity.

8.1 Some Properties of Thermodynamic Functions

8.1.1 Partial Derivatives

We here consider for simplicity a function of two variables $f(x, y)$ since the extension to three, four, …, independent real variables is trivial. If one of the two variables (e.g.: y) is kept constant, then the function can be considered as a function of the other one only (x) and the partial derivatives can therefore be defined as:

$$\left(\frac{\partial f(x,y)}{\partial x}\right)_y \overset{\text{def}}{=} \lim_{h\to 0} \frac{f(x+h,y)-f(x,y)}{h} \quad \text{and}$$

$$\left(\frac{\partial f(x,y)}{\partial y}\right)_x \overset{\text{def}}{=} \lim_{h\to 0} \frac{f(x,y+h)-f(x,y)}{h}$$

For example, for $p = \frac{nRT}{V}$ which is a function of T, V we have $\left(\frac{\partial p}{\partial T}\right)_V = \frac{nR}{V}$ and $\left(\frac{\partial p}{\partial V}\right)_T = \frac{-nRT}{V^2}$.

Partial derivatives can in turn be derived, to obtain second or higher order derivatives:

$$\frac{\partial^2 f}{\partial x^2} = \left[\frac{\partial}{\partial x}\left(\frac{\partial f}{\partial x}\right)_y\right]_y \qquad \frac{\partial^2 f}{\partial x \partial y} = \left[\frac{\partial}{\partial x}\left(\frac{\partial f}{\partial y}\right)_x\right]_y$$

If the mixed derivatives $\frac{\partial^2 f}{\partial x \partial y}$ and $\frac{\partial^2 f}{\partial y \partial x}$ are continuous, then the order of derivation is commutative:

$$\frac{\partial^2 f}{\partial x \partial y} = \frac{\partial^2 f}{\partial y \partial x}$$

As an example, with $p = \frac{nRT}{V}$, $\left(\frac{\partial^2 p}{\partial T \partial V}\right) = -\frac{nR}{V^2}$ and $\left(\frac{\partial^2 p}{\partial V \partial T}\right) = -\frac{nR}{V^2}$.

If the same variable is kept constant, the products of partial derivatives can be treated like the products of ordinary derivatives; for $f(t(x), z)$ we have: $\left(\frac{\partial f}{\partial t}\right)_z \left(\frac{\partial t}{\partial x}\right)_z = \left(\frac{\partial f}{\partial x}\right)_z$

The total differential is defined as:

$$df = \left(\frac{\partial f}{\partial x}\right)_y dx + \left(\frac{\partial f}{\partial y}\right)_x dy$$

For f=constant:

$$df = \left(\frac{\partial f}{\partial x}\right)_y dx + \left(\frac{\partial f}{\partial y}\right)_x dy = 0$$

from which we obtain the Euler chain rule:

$$\frac{\left(\frac{\partial f}{\partial x}\right)_y}{\left(\frac{\partial f}{\partial y}\right)_x} = -\left(\frac{\partial y}{\partial x}\right)_f \qquad (8.1)$$

or: $$\left(\frac{\partial f}{\partial x}\right)_y \left(\frac{\partial y}{\partial f}\right)_x \left(\frac{\partial x}{\partial y}\right)_f = -1$$

8.1.2 Homogeneous Functions

A function $f(x_1, x_2, x_3, \ldots)$ is homogenous of degree k if multiplication of the variables by a scalar λ gives the function multiplied by λ to the kth power:

$$f(\lambda x_1, \lambda x_2, \lambda x_3, \ldots) = \lambda^k f(x_1, x_2, x_3, \ldots)$$

In thermodynamics, homogeneous functions of degree 1 in the amount of substance n_i are called **extensive** functions, while homogeneous functions of degree zero are called **intensive**.

The ratio between two homogeneous functions of degree 1 is a homogeneous function of degree 0. For homogeneous functions, the Euler theorem holds:

$$f(x_1, x_2, x_3, \ldots) = \sum_i \left(\frac{\partial f}{\partial x_i}\right) x_i$$

For example, for the chemical potentials, for which the corresponding function of degree 1 is $G(n_1, n_2, \ldots)$:

$$G = \sum_i \mu_i n_i$$

8.1.3 Legendre Transformation

The Legendre transformation converts a function f of two (or more) independent variables, say: $f(x, y)$ into a g function of new variables (i.e.: u, y) provided that:

- only one variable is involved in the transformation, and
- the independent variable u that substitutes x is

$$u = \left(\frac{\partial f}{\partial x}\right)_y$$

For example, starting from the function f in the two variables (x, y) with differential:

$$df = \left(\frac{\partial f}{\partial x}\right)_y dx + \left(\frac{\partial f}{\partial y}\right)_x dy = u dx + v dy$$

we look for a new function g in the variables u and y. This is defined as:

$$g(u, y) \stackrel{\text{def}}{=} f(x, y) - ux \qquad (8.2)$$

Indeed, by differentiation:

$$dg = df - u dx - x du = u dx + v dy - u dx - x du = -x du + v dy$$

From which:

$$\left(\frac{\partial g(u, y)}{\partial u}\right)_y = -x, \quad \left(\frac{\partial g(u, y)}{\partial y}\right)_u = v$$

As can be seen, the dependence on the remaining variable (or variables, in the case of functions of more than two variables) is not changed.

The mnemonic rule is that the new function is obtained from the old one by subtracting the product of the incoming and outgoing pair of variables. The transformation can be applied many times.

8.2 The Partition Function of the Canonical Ensemble

We here show a somewhat *simplified* derivation of the expression of the partition function for the canonical ensemble, and its relationship with the corresponding thermodynamic potential ($A = U - TS$), to illustrate the logic of some basic arguments of ST, as well as to highlight the difficulties and the mathematical knowledge required.

8.2.1 The Super-System

The canonical Gibbs ensemble is made up of a very large number M of replicas (which possibly tends to infinity) of the same macroscopic thermodynamic system, which is made up of N molecules in the volume V and at the temperature T. The last requirement means that the ensemble is more precisely made up of *M replicas of the system including the pertinent thermostatic bath.*

For the whole ensemble, however, one can think of using a *single* thermostatic bath large enough to enclose all the systems, which must then be delimited by rigid walls (to keep V constant) and impermeable to molecules (to keep N constant), provided however that these *walls* are also *diathermal.* After the time needed to bring the entire ensemble of M systems to the same temperature T, the thermostat is removed and replaced by a container made up of rigid, *adiabatic* walls, impermeable to molecules.

In this way,

- **the ensemble itself is the thermostatic bath** for each of the constituent systems, and
- the **whole ensemble is an isolated system**.

Notes.

- V and N are the same for all the systems of the ensemble.
- The E_j are continuous with V but discontinuous with the number of molecules N.
- If j refers to the **quantum state** (microstate) of a **system**, the energy E_j are given by QM, and the *spectrum* of the eigenvalues of energy is the same for all the systems. All the systems are equal, but the state of each of the systems is not the same; each of the systems *can be* in a different state, but it can happen that two or more systems are in the same state j.
- The actual calculation of E_j is impossible, as the number of molecules in the systems is way too large. This "practical" difficulty is not relevant for the validity of the derivation, as we can assume to have the E_j that are needed.
- We can expect that degeneracy is large. Using a single index j and listing the states in order of non-decreasing energies $[E_1 \leq E_2 \leq \cdots \leq E_{j-1} \leq E_j \leq E_{j+1} \leq \cdots]$, if a level is m times degenerate, we will have for example that the states $k+1, \ldots, k+m$ have the same energy. It must be clear that only energies and degeneracy count in the statistical treatment: it is not important *how* the states (the wave functions) are chosen within an m-times degenerate level, it only matters that there are m states for the *energy level* $E_{k+1} = E_{k+2} = \cdots = E_{k+m}$.
- T has no mechanical meaning, but it is anyway fixed *without ambiguity*, albeit *implicitly*, through the boundary condition imposed by the thermostatic bath surrounding all the systems of the ensemble.

A *microstate of the ensemble* is specified when the microstate of *each* system is specified because the systems are macroscopic, and thus distinguishable. A given *microstate of the ensemble* can be specified by assigning microstate l to the first system, microstate n to the second system, microstate l to the third system also, … However, it is more convenient to count (for a given microstate of the ensemble) how many systems are in the same microstate. Thus, n_1 is the number of systems in the state with energy $E=E_1, \ldots, n_r$ is the number of systems in the state with energy $E=E_r, \ldots$, and so on. The numbers: $\{n_1, \ldots, n_r, \ldots\}$ form a ***distribution***, hereafter denoted $\boldsymbol{n}$. A distribution, $\boldsymbol{n}=\{n_1, \ldots, n_r, \ldots\}$ corresponds to many microstates of the canonical ensemble, as – for example – there is more than one choice to assign $E=E_1$ to n_1 systems among the M systems, and so on.

This simplification of the treatment by *grouping* the microstates of the entire set is justified by the obvious fact that the various microstates of the set corresponding to the same distribution provide the same contribution to the average of the mechanical quantities: in other words, what matters is *how many* systems are in a certain microstate, not *which ones*.

8.2.2 Analytical Workout

The **whole canonical ensemble is a system** (we can call it a super-system) which is **isolated.** Therefore, **we can apply the second postulate** to this super-system: its volume is MV, the number of particles is MN, the energy$=E_{tot}$, which is fixed, as the ensemble is isolated.

Considering now a distribution $(\boldsymbol{n})$, this groups $\Gamma(\boldsymbol{n})$ microstates of the *super-system*:

$$\Gamma(\boldsymbol{n}) = \frac{(n_1 + n_2 + \cdots + n_j + \cdots)!}{n_1! \cdot n_2! \cdots n_j! \cdots} = \frac{M!}{\prod_j (n_j!)} \tag{8.3}$$

Equation (8.3) arises since the number of microstates of the super-system is given by the number of permutations of the systems on the j state of a single system divided by $n_{1!}$, $n_{2!}$, etc.…. Indeed, a *distribution* groups all the microstates having n_1 systems in microstate 1, n_2 systems in microstate 2, etc.…, etc.…. In other words, there are $M!$ ways to arrange M different objects, but there are no M different objects, but only n_1 indistinguishable objects in the 1 state, n_2 indistinguishable objects in the second state, and so on.

The probability of $(\boldsymbol{n})$ is:

$$\frac{\Gamma(\boldsymbol{n})}{\sum_n \Gamma(\boldsymbol{n})} \tag{8.4}$$

where we applied the **postulate of a priori equal probability of microstates of the super-system**. Finally, the probability P_i that any system is in the i state is:

$$P_i = \frac{1}{M} \frac{\sum_n \Gamma(\boldsymbol{n}) \cdot \boldsymbol{n}_i}{\sum_n \Gamma(\boldsymbol{n})}$$

In fact, we are not interested in knowing which system is in state i, but how many systems are in that state, *i.e.* n_i. The sum then groups all the systems that are in the same microstate of single system; the probability of a distribution (Eq. 8.4) is multiplied by the ratio (n_i/M) between the number of systems (in *that* distribution) that are in the i state and the number of systems.

At the end, the mean value of a mechanical quantity F for the (canonical) ensemble is:

$$\langle F \rangle = \sum_i P_i \cdot F_i$$

where $\langle \ldots \rangle$ is the mean on the Gibbs ensemble.

Calculating P_i is difficult, but it may be eased by explicitly considering the limit $M \to \infty$.

In fact, because of 'large numbers', the most probable distribution (denoted as $\boldsymbol{n}^* = \{n_1^*, n_2^*, \ldots n_j^*, \ldots\}$) completely dominates the calculation of the average and we are allowed to write:

$$P \cong \frac{n_i^*}{M}$$

where n_j^* is the value of n_j in $\boldsymbol{n}^*$.

Note: it is possible to tackle the problem with a more rigorous method. This is called *steepest descent* and requires more difficult mathematics. In the end, the two methods provide the same results. So, the derivation based on the limit $M \to \infty$ and also known as the *maximum term method*, is not strictly necessary and is used here only to simplify the presentation.

In this method, the problem is solved by finding the distribution $\boldsymbol{n}^*$ which gives a maximum value for Γ. This is a **constrained maximum** problem, as the relations

$$\sum_i n_i = M \quad \text{and} \quad \sum_i n_i \cdot E_i = E_{\text{tot}}$$

constrain the numbers $\{n_1, n_2, \ldots, n_j, \ldots\}$, that therefore are not *independent* variables. The problem is solved with the **Lagrange multiplier method:**

$$\frac{\partial}{\partial n_i}\left[\ln\Gamma(n_1, n_2, \ldots, n_i, \ldots) + \alpha \cdot \sum_i n_i + \beta \cdot \sum_i n_i \cdot E_i\right] = 0;$$

$$n_i = 1, 2, \ldots$$

Where α and β are the Lagrange multipliers.

we also used the fact that the *logarithm* of Γ is a monotonic function of Γ, so that we can look for the maximum of its logarithm to find the maximum of Γ.

From $\Gamma(n) = \dfrac{(n_1 + n_2 + \cdots + n_j + \cdots)!}{n_1! \cdot n_2! \cdot \ldots n_j! \cdot \ldots} = \dfrac{M!}{\prod_j (n_j!)}$ with Stirling's approximation, we have:

$$\ln(\Gamma(n)) = \ln(M!) - \sum_j \ln(n_j!)$$

$$= [M \cdot \ln(M) - M] - \sum_j [n_j \cdot \ln(n_j) - n_j]$$

$$= M \cdot \ln(M) - \sum_j n_j \cdot \ln(n_j)$$

Moreover, $M = \sum_j n_j$. Therefore:

$$\frac{\partial[M \cdot \ln(M)]}{\partial n_i} = M\frac{\partial[\ln(M)]}{\partial n_i} + \ln(M)\frac{\partial[M]}{\partial n_i}$$

$$= M \cdot \frac{1}{M} + \ln M = \ln M$$

and

$$\frac{\partial\left[\sum_j n_j \cdot \ln(n_j)\right]}{\partial n_i} = \frac{\partial n_i \cdot \ln(n_i)}{\partial n_i} = n_i \cdot \frac{\partial \ln(n_i)}{\partial n_i} + \ln(n_i) \cdot \frac{\partial n_i}{\partial n_i}$$

$$= 1 + \ln(n_j) \cong \ln(n_j)$$

Recall that n_i^* are the number of systems in the microstate i that maximise Γ. We then need to solve the equation $\ln M - \ln(n_i^*) - \alpha - \beta \cdot E_i = 0$, which gives:

$$n_i^* = M \cdot e^{-\alpha} \cdot e^{-\beta \cdot E_i} \tag{8.5}$$

Furthermore, as $P_i \cong \dfrac{n_i^*}{M}$,

$$P_i = \frac{n_i^*}{M} = e^{-\alpha} \cdot e^{-\beta \cdot E_i} = \frac{e^{-\beta \cdot E_i}}{Q}, \text{ where we used the definition:} \tag{8.6}$$

$$Q \overset{\text{def}}{=} \sum_i e^{-\beta \cdot E_i} = e^{\alpha}; \quad \alpha = \ln Q \tag{8.7}$$

that is needed for P_i to be a probability. Now we need (i) to determine β and (ii) to establish a 'bridge' with CT to define (in a consistent way) the non-mechanical quantities (which up to this point 'do not exist').

8.2.3 Boltzmann Constant

The internal energy U of CT can be defined as $\langle E \rangle = \sum_i E_i P_i$, and then:

$$E = \sum_i E_i \cdot dP_i + \sum_i P_i \cdot dE_i \quad \text{and} \quad E_i = \frac{-1}{\beta}(\ln P_i - \ln Q),$$

which follows from Equation (8.6)
As the number of molecules is fixed, the energy is a function of V only, and therefore:

$$dE_i = \left(\frac{\partial E_i}{\partial V}\right)_N dV \tag{8.8}$$

substituting:

$$dU \overset{\text{def}}{=} dE = \frac{-1}{\beta}\sum_i (\ln P_i - \ln Q) \cdot dP_i + \sum_i P_i \cdot \left(\frac{\partial E_i}{\partial V}\right)_N dV.$$

As $\sum_i P_i = 1, \sum_i dP_i = 0,$ and $\sum_i (\ln Q) \cdot dP_i = \ln Q \cdot \sum_i dP_i = 0,$ which gives:

$$\sum_i (\ln P_i - \ln Q) \cdot dP_i = \sum_i (\ln P_i) \cdot dP_i \quad \text{and}$$

$$dE = \frac{-1}{\beta}\sum_i (\ln P_i)\cdot dP_i + \sum_i P_i \cdot \left(\frac{\partial E_i}{\partial V}\right)_N dV$$

The derivative of energy with respect to volume is the (negative) pressure: $\left(\frac{\partial E_i}{\partial V}\right)_N = -p_i$.

Therefore:

$$dU = dE = \frac{-1}{\beta}\sum_i (\ln P_i)\cdot dP_i + \left[\sum_i P_i \cdot \left(\frac{\partial E_i}{\partial V}\right)_N\right]\cdot dV$$

$$= \frac{-1}{\beta}\sum_i (\ln P_i)\cdot dP_i - p\cdot dV$$

Using $dU = TdS - pdV$, we have:

$$T\cdot dS = \frac{-1}{\beta}\sum_i (\ln P_i)\cdot dP_i \quad \text{or:} \tag{8.9}$$

$$dS = \frac{-1}{\beta\cdot T}\sum_i (\ln P_i)\cdot dP_i \tag{8.10}$$

For a generic function $G \overset{\text{def}}{=} -\sum_i (P_i\cdot \ln P_i)$, the differential is:

$$dG = -\sum_i \frac{d(P_i\cdot \ln P_i)}{dP_i}\cdot dP_i = -\sum_i (\ln P_i + 1)\cdot dP_i$$

$$= -\sum_i (\ln P_i)\cdot dP_{ii} - \sum_i dP_i = -\sum_i (\ln P_i)\cdot dP_i \text{ as } \sum_i dP_i = 0$$

Then, Eq. (8.10) can be rewritten as

$$dS = \frac{1}{\beta\cdot T} dG \tag{8.11}$$

Entropy is a state function, and therefore dS is an exact differential: this implies that $1/\beta T$ is a function of G which we write as $r(G)$. Thus, $r(G)\cdot dG$ is the differential of a new function $f(G)$, which differs from S by an additive constant c:

$$S = f(G) + c \leftarrow dS = df(G) \leftarrow dS = r(G)\cdot dG$$

c is a constant of integration, and, as such, does not depend on the variables of G, that is on T, β, nor V (it is independent of the state of the system): it has nothing to do with the value of entropy at absolute zero, with the third law, etc. For this reason, **we can set c to zero** (we only have one relation between the differentials of S and G: we have to make a choice, and we make the simplest one).

The function f can be determined by the fact that entropy is extensive, and therefore additive: if we merge two systems A and B in thermal equilibrium in a system AB, then in CT we have:

$$S_{AB} = S_A + S_B$$

Which means that in ST we should have $G_{AB} = G_A + G_B$

$$\text{Or: } S_{AB} = f(G_{AB}) = f(G_A + G_B) = S_A + S_B$$
$$= f(G_A) + f(G_B)$$

The function f is linear: $f(x+y) = f(x) + f(y)$, and $f(kx) = k\cdot x$. Therefore $S = k\cdot G$.

From Eq. (8.11) we then have:

$$1/\beta T = k \rightarrow \beta = 1/kT$$

In this derivation we did not make any assumption on *the couple of systems* A and B. If we changed A, with fixed B, we would get the same k, which is therefore a **universal constant**. To determine its value, the properties of any system for which the ST calculations (which include k) can be considered sufficiently accurate are compared with experimental data. The simplest calculation is done with the ideal gas, giving:

$$k \approx 1.38\times 10^{-23}\,\text{m}^2\text{kg/s}^2/\text{K}\,(\text{J/K})$$

This is the **Boltzmann constant** (the ratio between the gas constant and Avogadro constant).

8.2.4 Partition Function and Helmholtz Energy

We then have:

$$S = S(N,V,T) = -k\sum_i (P_i\cdot \ln P_i)$$

$$= -k\sum_i P_i\cdot\left(\frac{-E_i}{kT} - \ln Q\right) = \frac{1}{T}\sum_i P_i\cdot E_i - k\cdot \ln Q \sum_i P_i$$

and $S = \frac{\langle E\rangle}{T} - k\cdot \ln Q$ where $\langle E\rangle$ is equal to the CT function U.

The Helmholtz energy is $A \overset{\text{def}}{=} U - T\cdot S$, and therefore

$$A(N,V,T) = -k\cdot T\cdot \ln(Q); \quad Q \overset{\text{def}}{=} \sum_i e^{\frac{-E_i}{kT}}$$

where Q is the canonical partition function, and $A(N,V,T)$ is the thermodynamic potential in the canonical ensemble. In Q, the dependence on N and V is contained in the energies of the microstates E_i.

8.3 Quantum Statistics and the Grand-Canonical Partition Function

For the calculation of the partition function of a system of *independent, equal and indistinguishable particles* (see Section 5.2.2) we need to show that:

$$\Xi = 1 + \lambda \cdot \sum_i \mathrm{e}^{-\epsilon_i/kT} + \lambda^2 \cdot \sum_{i,j \geq i} \mathrm{e}^{-(\epsilon_i+\epsilon_j)/kT} + \lambda^3 \cdot \sum_{i,j\geq i,k\geq j} \mathrm{e}^{-(\epsilon_i+\epsilon_j+\epsilon_k)/kT} + \cdots \tag{8.12}$$

is equal to:

$$\prod_i \xi_i \tag{8.13}$$

where $\xi_i \stackrel{\text{def}}{=} 1 + \lambda \cdot \mathrm{e}^{-\varepsilon_i/kT} + \lambda^2 \cdot \mathrm{e}^{-2\varepsilon_i/kT} + \cdots + \lambda^n \cdot \mathrm{e}^{-n\varepsilon_i/kT} + \cdots$ and thus:

$$\Xi = \prod_i \xi_i \tag{8.14}$$

This result is a consequence of the multinomial theorem. To *verify* the equality let us define:

$$g_i \stackrel{\text{def}}{=} \mathrm{e}^{-\epsilon_i/kT}$$

and inspect how these g_i appear in the coefficients of the same λ in both equations.

From Eq. (8.12):

λ^0: 1	Only one term = 1.
λ^1: $\sum_i g_i$	The sum of all g_i terms.
λ^2: $\sum_i \left[g_i \sum_{j\geq i} g_j \right]$	The sum of all **products** of **two** g, both **pure** products [*i.e.* $g_i\, g_i$], and **mixed** products [*i.e.* $g_i\, g_j$].
λ^3: $\sum_i \left[g_i \sum_{j\geq i} \left(g_j \sum_{k\geq j} g_k \right) \right]$	The sum of all triple **products**, with both **pure** products [*i.e.* $g_i\, g_i\, g_i$], and **mixed** products [*i.e.* $g_i\, g_j\, g_k$, that is $(g_i)^2 g_j$].
…	and so on

Concerning the sum in Eq. (8.12), each λ^n multiplies a sum of terms: a term is the product of n factors g_i and the factors contained in a product can all be distinct but can also repeat themselves in part or totally, given that the inequalities $j \geq i$, $k \geq j$, etc. hold, instead of the *strict* inequalities $j > i$, $k > j$, etc.…. Each product appears just once in the sum, it contains no more than n *distinct* g_i, each g_i appears to the power r_i, and $\sum r_i = n$ (if a g_k is not contained in the product, we write $r_k = 0$).

Going to Eq. (8.13), we have different products of $(\lambda g_i)^{r_i}$ monomials of the ξ_i polynomials. In these products each monomial comes from the ξ_i polynomial and multiplies other monomials coming from other polynomials $\xi_{j\neq i}$: one (and only one) monomial from each $\xi_{j\neq i}$. The final expression contains all monomials from all polynomials, including monomials such as $(\lambda g_i)^0 =$ land, for a given i, each exponent r_i (including $r_i = 0$) is present, as it is included in the ξ_i polynomial. A given power λ^n multiplies $\prod_i (\lambda g_i)^{r_i}$ and $r_1 + r_2 + \cdots + r_i + r_j + r_k + \cdots = n$. Each $(\lambda g_i)^{r_i}$ monomial in ξ_i is multiplied by the monomials of the $\xi_{j\neq i}$ with all the combinations of exponents $r_{j\neq i}$ ($r_1, r_2, \ldots, r_i, r_j, r_k, \ldots$) *without repetitions*.

Now, it can be seen that this is exactly what is said for Eq. (8.12): therefore Eqs. (8.12) and (8.13) are coincident, and Eq. (8.14) is verified.

Concerning the grand canonical partition function, as reported in Eqs. (5.7), (8.14), (8.12) and (8.13) confirm that all the products of the n single particle state appear in Ξ, and also that *each of the different products appears in only one permutation of single particles states.*

8.4 The Ideal Mixture in ST

8.4.1 Mixture of Ideal Gases

We will examine here the case of two different gases for the sake of simplicity:

$$Q = Q_A \cdot Q_B = \frac{(q_A)^{N_A}}{(N_A)!} \cdot \frac{(q_B)^{N_B}}{(N_B)!} \quad \text{and}$$

$$\ln Q = N_A \ln(q_A) - \ln(N_A!) + N_B \ln(q_B) - \ln(N_B!)$$

Considering the internal energy as an example of **non-thermal quantity**:

$$U = kT^2 \left(\frac{\partial \ln Q}{\partial T} \right), \quad \text{that is}$$

$$U = kT^2 \left(N_A \frac{\partial \ln q_A}{\partial T} + N_B \frac{\partial \ln q_B}{\partial T} \right) = N_A \cdot \varepsilon_A + N_B \cdot \varepsilon_B$$

$$= (N_A + N_B) \left[\frac{N_A}{(N_A + N_B)} \cdot \varepsilon_A + \frac{N_B}{(N_A + N_B)} \cdot \varepsilon_B \right]$$

ε_A and ε_B are specific to the two different molecules and depend on temperature only.

Defining $N = (N_A + N_B)$ and $x = N_B/(N_A + N_B)$ we have:

$$U = N_A \cdot \varepsilon_A + N_B \cdot \varepsilon_B = N \cdot [(1-x) \cdot \varepsilon_A + x \cdot \varepsilon_B]$$

The molar internal energy is U/n with $n = N/L$, where L is the Avogadro constant. Therefore:

$$\frac{U}{n} = \frac{L}{N_A + N_B}(N_A \cdot \varepsilon_A + N_B \cdot \varepsilon_B) = L \cdot [(1-x) \cdot \varepsilon_A + x \cdot \varepsilon_B]$$

For the pure gases we can define the molar internal energies as: $u_A^0 \stackrel{\text{def}}{=} L \cdot \varepsilon_A$ and $u_B^0 \stackrel{\text{def}}{=} L \cdot \varepsilon_B$. Therefore:

$$\frac{U}{n} = \frac{L}{N}(N_A \cdot \varepsilon_A + N_B \cdot \varepsilon_B) = (1-x) \cdot u_A^0 + x \cdot u_B^0$$

The plot of the intensive variable (U/n) as a function of x, **is a straight line** from $U/n = u_A^0$ for $x=0$ to $U/n = u_B^0$ for $x=1$. *This is the expected trend for an ideal mixture.* This result does not contain the $N!$ corrections for indistinguishability: in this case, indistinguishability is irrelevant, and the same result would have been obtained considering it or not.

As an example of a **thermal quantity,** we will consider Helmholtz energy: $A = -kT(\ln Q)$:

$$\begin{aligned} A &= -kT(\ln Q) \\ &= -kT[N_A \cdot \ln(q_A) + N_B \cdot \ln(q_B) - \ln(N_A!) - \ln(N_B!)] \end{aligned}$$

and for a mole of mixture, as $N_A + N_B = L$

$$A/n = \frac{-L \cdot kT}{N_A + N_B}\begin{bmatrix} N_A \cdot \ln(q_A) + N_B \cdot \ln(q_B) \\ -\ln(N_A!) - \ln(N_B!) \end{bmatrix} \text{or, as } R \stackrel{\text{def}}{=} k \cdot L:$$

$$A/n = -RT\begin{bmatrix} (1-x) \cdot \ln(q_A) \\ +x \cdot \ln(q_B) \end{bmatrix} + kT\begin{bmatrix} \ln(N_A!) \\ +\ln(N_B!) \end{bmatrix}$$

For a mole of the pure components:

$$A_m^0(\text{A}) = -RT \cdot \ln(q_A) + kT \cdot \ln(L!) \quad \text{and}$$
$$A_m^0(\text{B}) = -RT \cdot \ln(q_B) + kT \cdot \ln(L!)$$

The *linear trend* the integral molar function is given by:

$$\begin{aligned} &(1-x) \cdot A_m^0(\text{A}) + x \cdot A_m^0(\text{B}) \\ &= -RT[(1-x) \cdot \ln(q_A) + x \cdot \ln(q_B)] + kT \cdot \ln(L!) \end{aligned}$$

Comparing now this equation with the A/n derived above we are left with:

$$\begin{aligned} &\frac{A}{n} - \{(1-x) \cdot A_m^0(A) + x \cdot A_m^0(B)\} \\ &= kT[\ln(N_A!) + \ln(N_B!) - \ln(L!)] \end{aligned} \tag{8.15}$$

Using Stirling's approximation: $\ln(N!) \cong N \cdot \ln(N) - N$. then, adding and subtracting $N_A \cdot \ln(N_B)$

$$\begin{aligned} &N_A \cdot \ln(N_A) + N_A \cdot \ln(N_B) - N_A \cdot \ln(N_B) \\ &+N_B \cdot \ln(N_B) - L \cdot \ln(L) \\ &= N_A \cdot \ln(N_A / N_B) + L \cdot \ln(N_B) - L \cdot \ln(\text{L}) \\ &= N_A \cdot \ln((1-x)/x) + L \cdot \ln(x) \\ &= N_A \cdot \ln(1-x) + N_B \cdot \ln(x) = L \cdot [(1-x) \cdot \ln(1-x) + x \cdot \ln(x)] \end{aligned}$$

At the end:

$$\begin{aligned} &\frac{A}{n} - \{(1-x) \cdot \overline{A}_A^0 + x \cdot \overline{A}_B^0\} \\ &= R \cdot T \cdot \{(1-x) \cdot \ln(1-x) + x \cdot \ln(x)\} \end{aligned}$$

where the configurational term $(1-x) \cdot \ln(1-x) + x \cdot \ln(x)$ is present.

For the entropy, using: $S = -\left(\frac{\partial A}{\partial T}\right)_{V,N}$ the configurational term is:

$$-R \cdot \{(1-x) \cdot \ln(1-x) + x \cdot \ln(x)\}$$

In summary (refer to Section 4.1.2), we have that U, p, H, that are the derivatives of state variables V, T do not contain the logarithm of a factorial of a number of molecules and therefore do not contain the configurational term. On the contrary, *thermal quantities such as A, S, μ, …,* containing Q or the derivatives with respect to a number of molecules, contain the configurational term for the properties of the mixture.

8.5 Maxwell-Boltzmann Statistics

It is instructive to derive, for distinguishable and indistinguishable particles, the state occupancy. In the table below, $\overline{C}_k / N \stackrel{\text{def}}{=} s_k$ indicates - for indistinguishable particles - the fraction of particles in the k state, that is the **occupancy of *k***. For distinguishable particles, p_k is the probability that a *given particle* (for example, particle 1) is in the k state and therefore the occupancy of the k state by that particle.

Distinguishable Particles

For system of independent, equal and distinguishable particles, we have:

$$Q = q^N$$

Indistinguishable Particles

For system of independent, equal and indistinguishable particles: **with the Maxwell-Boltzmann approximation**, we have:

$$Q = \frac{q^N}{N!}$$

The quantum state $|j\rangle$ of the whole system is the product of the quantum single particle states $|k\rangle, |l\rangle, |m\rangle, |n\rangle, \ldots$

Distinguishable: The products corresponding to different permutations of the indices $\{k, l, m, n, \ldots\}$ **are distinct states.**

Indistinguishable: All indexes are different (necessary for writing $Q=q^N/N!$). All products corresponding to different permutations of the indices $\{k, l, m, n, \ldots\}$ **count as a single state.**

In **both cases**, the probability to find the **system** in the j quantum state is:

$$P_j(N,V,T) = \frac{e^{-E_j/kT}}{Q}$$

N.B.: the formula *looks* the same for the two cases, but do not forget that the Qs are different, since *many more addends* appear in Q(distinguishable) than in Q(indistinguishable).

Let us now evaluate $\bar{C}_k$ = average number of particles that are in the **particular quantum single particle state** k.

Distinguishable: It is the product of the probability (p_k) that a particular particle (e.g. 1) is in the k state by the total number of particles (N).

$$p_k = \frac{\sum_{l,m,n,\ldots} e^{-(\epsilon_k+\epsilon_l+\epsilon_m+\epsilon_n+\ldots)/kT}}{Q} = \frac{e^{-\epsilon_k/kT} \cdot \sum_{l,m,n,\ldots} e^{-(\epsilon_l+\epsilon_m+\epsilon_n+\ldots)/kT}}{Q}$$

p_k is the sum of the Boltzmann weights of all the possible values of $l, m, n, \ldots$, by fixing to k the particle index (in our case, the first index):

The Σ is q^{N-1}: $p_k = e^{-\epsilon_k/kT}\frac{q^{N-1}}{q^N} = \frac{e^{-\epsilon_k/kT}}{q}$ and then: $\bar{C}_k = N\frac{e^{-\epsilon_k/kT}}{q}$

Indistinguishable: It is the probability that *any* particle is in k state.

This probability is the sum of the Boltzmann weights for all the possible values of $l, m, n, \ldots$, by fixing k to 1, **but dividing by $(N-1)!$**, because otherwise we would mistakenly include permutations of the values $l, m, n, \ldots$ on the remaining $(N-1)$ molecules.

Thus:

$$\bar{C}_k = \frac{1}{(N-1)!}\frac{\sum_{l,m,n,\ldots} e^{-(\epsilon_k+\epsilon_l+\epsilon_m+\epsilon_n+\ldots)/kT}}{Q} = \frac{e^{-\epsilon_k/kT} \cdot \sum_{l,m,n,\ldots} e^{-(\epsilon_l+\epsilon_m+\epsilon_n+\ldots)/kT}}{Q\cdot(N-1)!}$$

$$= \frac{e^{-\epsilon_k/kT}}{Q\cdot(N-1)!}\cdot q^{N-1} = \frac{e^{-\epsilon_k/kT}}{q^N\cdot(N-1)!}\cdot N!\cdot q^{N-1}$$

that is: $\bar{C}_k = \frac{e^{-\epsilon_k/kT}}{q}\cdot N$

N.B. p_k for distinguishable particles and $s_k = \bar{C}_k / N$ for indistinguishable particles are equal; on the contrary Q(distinguishable) and Q(indistinguishable) are different. Also:

- in the case of *indistinguishable* particles, the formulas are only *approximately* valid (*i.e.* only if the number of states accessible to a particle is much higher than the number of particles, so that the case of two particles in the same k state is statistically negligible);
- in the case of *distinguishable* particles, they are exact.

Considering the case of **indistinguishable particles only**, we conclude the derivation of the MB occupation function by assuming that the partition function can be written as $Q=q^N/N!$

Then: $\ln Q = N \ln q - \ln(N!) \approx N \ln q - [N \ln(N) - N]$,

$$\mu = -kT\left(\frac{\partial \ln Q}{\partial N}\right)_{T,V} = -kT\ln\frac{q}{N} \text{ and}$$

$$\left(\frac{\partial \ln Q}{\partial N}\right)_{T,V} = \left[\frac{\partial(N\cdot\ln q)}{\partial N}\right]_{T,V} - \left[\frac{\partial(\ln N!)}{\partial N}\right]_{T,V}$$

$$= \ln(q) - N\left[\frac{\partial(\ln N)}{\partial N}\right]_{T,V} - \ln(N)\left[\frac{\partial(\ln N)}{\partial N}\right]_{T,V}$$

$$+\left[\frac{\partial(\ln N)}{\partial N}\right]_{T,V} \quad \text{so that}$$

$$\mu = -kT\left[\ln(q) - 1 - \ln(N) + \frac{1}{N}\right]$$

$$\approx kT\left[\ln(N) - \ln(q)\right] = kT\cdot\ln\frac{N}{q}$$

Writing $\frac{N}{q} = \lambda \overset{\text{def}}{=} e^{\mu/kT}$, $N=\lambda q$, the occupancy s_k of the k state *of a single indistinguishable particle*, as a function of energy and chemical potential is:

$$s_k = \lambda q\frac{e^{-\epsilon_k/kT}}{q} = \frac{1}{e^{(\epsilon_k-\mu)/kT}}$$

8.6 Temperature Dependence of the Chemical Potential of the Electrons

The chemical potential of the electron is defined as the energy corresponding to $s_{FD}=1/2$ and to the infection point of the FD distribution, and is constrained to the number of electrons (n) by:

$$n=\sum_i\left(\omega_i\cdot s_{FD}(\varepsilon_i)\right)\text{ for discrete energy, or}$$

$$n=\int_0^\infty\left[\rho(\varepsilon)\cdot s_{FD}(\varepsilon)\right]\mathrm{d}\varepsilon\text{ for continuous energy}$$

In both cases: $s_{FD}(\varepsilon)=\dfrac{1}{\mathrm{e}^{(\varepsilon-\mu)/kT}+1}$

These two relations implicitly define the chemical potential μ at the temperature T, and hence the possible dependence of μ on T. In fact, μ is not necessarily constant as T changes and a counterexample best illustrates this fact. Let us consider the case with only two energy levels $\varepsilon_A \neq \varepsilon_B$, and a number of electrons such that at $T=0$ the *lower level is fully occupied* ($s_A=1$) and *the upper level is empty* ($s_B=0$). Increasing T, s_A decreases and s_B increases.

If the two levels have the same degeneracy, at $T=0$ the chemical potential is at the centre of the interval $\varepsilon_A-\varepsilon_B$ and remains there even when the temperature increase, because each electron promoted from A to B produces in s_A and s_B variations equal in magnitude but of opposite sign and the FD is symmetric with respect to μ (see Sections 5.2.2.4 and 5.3).

Let us instead consider the case where **the two levels** have **different degeneracies** and the degeneracy of ε_B is x times the degeneracy of ε_A.

For the sake of simplicity, we can assume that $\varepsilon_A=0$ and define $\lambda=\mathrm{e}^{-\mu/kT}$, $r=\varepsilon_B/kT$ and $\sigma=\mathrm{e}^r$. The occupancies are then $s_A=1/(1+\lambda)$ and $s_B=1/(1+\lambda\sigma)$. For the conservation of the number of electrons, "number of empty states in level A"="number of occupied states in level B" and therefore: $(1-s_A)=xs_B$, that is: $1-1/(1+\lambda)=\lambda/(1+\lambda)=x/(1+\lambda\sigma)$, from which $\lambda+\lambda^2\sigma-x\lambda=x$ or $\lambda^2\sigma+\lambda(1-x)-x=0$.

The solutions are: $\lambda=\dfrac{(x-1)\pm\sqrt{(1-x)^2+4x\sigma}}{2\sigma}$ but the exponential of a real number is positive definite, $\sqrt{(1-x)^2+4x\sigma}>\lfloor x-1\rfloor$, and then only the solution with the + sign is physical. Then: $\lambda=\dfrac{(x-1)+\sqrt{(1-x)^2+4x\sigma}}{2\sigma}$.

In the present case, $x\neq1$ (A and B have different degeneracies) and we can consider the trend at "low" temperature only, when

$$r=\frac{\varepsilon_B}{kT}>>1,\quad \sigma=e^r>>1,$$

$$\lambda=\frac{(x-1)+\sqrt{(1-x)^2+4x\sigma}}{2\sigma}\approx\sqrt{\frac{x}{\sigma}}\quad\text{and}\quad\mu\approx\frac{\varepsilon_B}{2}-kT$$

Thus $\mu<\varepsilon_B/2$ for $x>1$ and $\mu>\varepsilon_B/2$ for $x<1$, and for $x=1$: $\mu=\dfrac{\varepsilon_B}{2}$.

It is also easy to see that, in the case of a **continuous distribution**, the dependence on T of the chemical potential is governed by the derivative of $\rho(\varepsilon)$ at the Fermi level and is a direct consequence of the symmetry of the FD distribution on the right and left of its inflection point: $s_{FD}(\mu-\varepsilon)=1-s_{FD}(\mu+\varepsilon)$. By increasing T, the population of the states with $\varepsilon<\mu$ decreases and that of the states with $\varepsilon<\mu$ increases.

Let us assume, for example, that the derivative $\left(\dfrac{\partial\rho(\varepsilon)}{\partial\varepsilon}\right)_{\varepsilon=\mu}$ is positive - that is $\rho(\varepsilon)$ is increasing - at $\varepsilon=\mu$. Without a change of the chemical potential there would be a larger number of states filled at $\varepsilon>\mu$ than the number of states emptied at $\varepsilon<\mu$. Therefore, the conservation in the number of electrons requires an *increase* of μ.

It is worth noting that the temperature dependence of μ is generally weak and is often neglected; however, it is essential for some applications, for example for the evaluation of thermoelectric power.

9 Exercises

Exercise 3.1

Use conservation of energy to determine how many Hiroshima-like nuclear explosions are needed to raise the temperature of the atmosphere by 2 degrees (energy of one Hiroshima-like nuclear explosion: 15 kT (kiloton); 1 kT=4.184 TJ; the specific heat capacity of air at constant pressure (C_p) is approximately 1.005 kJ/kg/K; mass of the atmosphere: 5.1480×10^{18}kg.

Exercise 3.2

Prove that the internal energy of an ideal gas does not depend on pressure [hint: start with enthalpy and write the derivative $\left(\dfrac{\partial H}{\partial p}\right)_T$; then use the proper Maxwell relation (Section 2.3.3), and the definition of enthalpy to obtain the derivative $\left(\dfrac{\partial U}{\partial p}\right)_T$.

Exercise 3.3

Thermodynamic data are determined experimentally with some uncertainty. Suppose that the standard enthalpy of formation of a given substance is obtained with an uncertainty ε_H. What are the corresponding uncertainties related to the equilibrium constant of the formation reaction of the substance (a) 300 K and (b) 1,000 K?

Exercise 3.4

Prove that the heat transferred in a system after an isothermal change in volume can be expressed as $Q=T\left(\dfrac{\partial p}{\partial T}\right)_V\mathrm{d}V$. Calculate this expression for the van der Waals gas for which: $p=\dfrac{RT}{V-b}-\dfrac{a}{V^2}$.

Exercise 3.5

Show explicitly that in a multicomponent system the configurational entropy of mixing $\Delta_{\text{mix}}S = -R\sum_i x_i \ln x_i$ is maximum when the mixture is equimolar.

Exercise 3.6

Use the fundamental equation $dU = TdS - pdV$ to prove that:

$$\frac{1}{T} = \left(\frac{\partial S}{\partial U}\right)_{N,V} \quad \text{and} \quad p = T\cdot\left(\frac{\partial S}{\partial V}\right)_{N,U}$$

Exercise 3.7

Calculate the values of the FD (Fermi-Dirac) distribution function for the following values of $(\varepsilon - \mu)/kT$: 1, 2, 3, 4, 5, 7, 10, 20, 50.

Exercise 3.8

Rearrange the equation

$$\epsilon_{i,j,k} = \frac{h^2}{8m\cdot V^{2/3}}\left(i^2 + j^2 + k^2\right) \text{ to show that}$$

$$\left(\frac{1}{h}V^{1/3}\sqrt{8m\varepsilon}\right)^{-2}\left(i^2 + j^2 + k^2\right) = 1$$

This is the equation of a sphere. Find the radius of the sphere in the space defined by the quantum numbers i, j and k.

Exercise 3.9

Use equipartition to show that for a solid, $C_V = 3R$.

Exercise 3.10

For a linear rotor, the energy levels are $\varepsilon_J = hcBJ(J+1)$, where B is called the rotational constant and J is the rotational quantum number with J=0, 1, 2, 3, ... Each level has degeneracy $2J$+1. Write the canonical molecular partition function for the rotor. In the high temperature limit, the energy difference between two levels is much smaller than kT, the levels can be treated as continuous, and the sum in the partition function can be written as an integral. Show then that the partition function is $q = kT/(hcB)$.

Exercise 3.11

For a symmetric rotor, the energy levels can be written as

$$\varepsilon_{J,K,M_j} = hcBJ(J+1) + hc(A-B)K^2$$

where A and B are the rotational constants, J, K, M_j are integers with $-\infty < |K| < \infty$, $J = |K|, |K|+1, |K|+2, \ldots$, and $-J < M_J < +J$. Use the same approximation of Exercise 3.10 to show that the partition function is $q = \left(\frac{kT}{hc}\right)^{3/2}\left(\frac{\pi}{AB^2}\right)^{1/2}$.

Extend this result to the asymmetric rotor, which has three different rotational constants A, B and C (see Section 6.2.3).

Exercise 3.12

An electron in a magnetic field $\vec{B}$ has two energy levels, depending on the orientation of spin with respect to the field. These levels are labelled with a quantum number $m_s = \pm\frac{1}{2}$. The energies of the two levels can be written as $\varepsilon_{-1/2} = 0$ and $\varepsilon_{1/2} = g_e\mu_e|\vec{B}|$. Write the canonical spin partition function.

NOTES

1 Stanley W. Angrist, and Loren G. Helper, *Order and Chaos – Laws of Energy and Entropy* (p. 215). New York: Basic Books, 1967.

2 The derivation follows with simplifications and comments the seminal work by T. L. Hill. *Statistical Thermodynamics*. Reading & London: Addison Wesley Publishing Company Inc, 1960.

3 Sommerfeld A, Thermodynamics and Statistical Mechanics, p. 1, vol. 5 of Lectures on Theoretical Physics, 1951/1955, edited by F. Bopp, J. Meixner, translated by J. Kestin, Academic Press, New York.

4 Nernst W, *Experimental and Theoretical Applications of Thermodynamics to Chemistry*. New York: Charles Scribner's Sons, 1907.

5 In most cases, magnetization is rather defined as an intensive quantity i.e., per unit volume; we can then obtain the corresponding extensive quantity by multiplying by volume (see Section 6.2.6).

6 It is called eutectic (= well melting) because a system that has the precise eutectic composition (x_b) behaves phenomenologically like a pure substance and shows a melting point (T_E) instead of a range of T of solid/liquid coexistence, as typically occurs with mixtures.

7 Electrodes are not necessarily solid metals. Liquid metals (Hg, for instance), semiconductors and other electronic conductors are used. Electrolytes are not necessarily aqueous solutions. Other liquid solvents and solid electrolytes are widely used.

8 n is the stoichiometric coefficient of the electrons in the two half-reactions, while the Faraday F is the charge of 1 mol of electrons: $F = Le$, L is the Avogadro constant, and e is the absolute value of the electron charge.

9 These are the common names for zirconium oxide ZrO_2 and yttrium oxide Y_2O_3.

10 Actually, the vibrational degrees of freedom are $3nN - 6$, but this does not matter, since N is of the order of $10^{20} - 10^{24}$.

11 This is easily found: for the first particle the possible choices are N, for the second particle they are $(N - 1)$ .... At the end, we have $N\cdot(N-1)\cdot(N-2)\cdot\ldots\cdot3\cdot2\cdot1 = N!$ choices.

12 The reader is warned that q has been frequently used to denote a partition function: this is the standard practice, and we adopted it because there is no risk of confusion.

13 See also the Appendices (1.3).

14 Taking $U_0 = cw_{AA}$, this corresponds to the Einstein model of a monoatomic solid (see Section 6.2.5).

15 N. Metropolis, A. W. Rosenbluth, M. N. Rosenbluth, A. H. Teller, and E. Teller, Equation of State Calculations by Fast Computing Machines, *J. Chem. Phys.* **21**, 1087–1092, 1953.

4 Symmetry in Chemistry

Symmetry concerns both the microscopic and macroscopic words and plays a fundamental role at the most abstract level: the preface of a well-known treatise on symmetry,[1] for instance, starts saying that "Symmetry can be seen as the most basic and important concept in physics". Concerning Physical Chemistry, symmetry applies to problems treated either with Classical Mechanics or Quantum Mechanics, about structure as well as spectroscopic, electric, magnetic, and optical properties, and reactivity.

Symmetry shows several contradictory aspects. While being a most abstract and theoretical tool, it gives results that are of fundamental importance also for the experimentalists, who use symmetry widely. The basic aspects of symmetry appear quite simple and obvious, but a deeper approach implies a non-commutative algebra, which is surely far from familiar. Also, symmetry does not appear explicitly in the most elementary introductions to Classical Mechanics (CM) and Quantum Mechanics (QM) but is indeed extremely powerful in solving many mechanical problems. Finally, symmetry solves some problems in a very direct way, but after changing a problem only slightly a much more intricate procedure and more advanced theoretical tools are required.

Considering all the above features, symmetry is here discussed in somehow particular ways. An important point is that - before entering topics such as matrix representations, characters, and orthogonality theorems – this chapter follows a simpler and more direct approach that is valid only for commutative groups. Besides being obviously useful as an introduction to the most general and complete part, this partial approach is indeed so effective in showing how powerful group theoretical methods are, that a shorter presentation may easily end at that point.

A second point is related to the typical focus, in Physical Chemistry textbook, on the use of character tables to find symmetry orbitals and spectroscopic selection rules. As a matter of fact, the pertinent procedures are necessarily disclosed as black box recipes: aim of this chapter is then to provide a sound and comprehensive understanding and explain the underlying basic aspects. However, this is not a chapter on Group Theory, a branch of Algebra: in deriving the results we use as far as possible plain words and a simple language, as required for readers that generally have scarce familiarity with the mathematical style. The line of presentation is almost complete (we preferred to skip Schur's lemma and the derivation of the great orthogonality theorem) but not all results are rigorously derived: some are presented with an informal argumentation, that however can be upgraded to a rigorous form without much difficulty. Along the same line of presentation, we do not care about repeating an argument when it can be useful for better clarity, we sometimes disclose in advance topics that can be fully proved only later, and in front of possible sources of difficulty we proceed by explaining and exemplifying most frequently and giving many (perhaps not always necessary) details.

On the other hand, we do *not* address here topics that are reasonably well known to the reader and widely treated in basic textbooks, such as the procedure for identifying a point group.

Finally, molecular and solid-state symmetries are considered in a parallel way, assuming that a sharp distinction is more detrimental than useful when the aim is to go beyond technicalities.

1 INTRODUCTION AND EXAMPLES

1.1 Symmetry Groups

1.1.1 Symmetry Operations

To introduce symmetry, it is expedient to proceed by examples about molecular symmetries focusing attention on some simple and well-known molecules: water, carbon dioxide, ammonia, butadiene, and benzene.

We see that there are *geometric entities*, points, lines and planes called *symmetry elements*, with at least one associated *operation* of symmetry. In the water molecule, for example, we recognize a *reflection plane* (σ' in Figure 4.1) that perpendicularly cuts the molecular plane along a line through the O and the midpoint of the two H atoms. Similarly, in the carbon dioxide molecule we recognize an *inversion centre* on the C atom. In the ammonia molecule we recognize a *ternary axis* and *two rotations* of one-third and two-third of a turn, and the second one is equivalently seen as a rotation of one-third of a turn in the opposite direction.

Generalizing the examples, symmetry operations essentially relate two or more points of ordinary three-dimensional (3D) space that are in *the same physical situation*. They:

- transform one *vector* of ordinary 3D space ($\vec{r}$) into another *vector* ($\vec{r}'$),
- are *linear* operations,

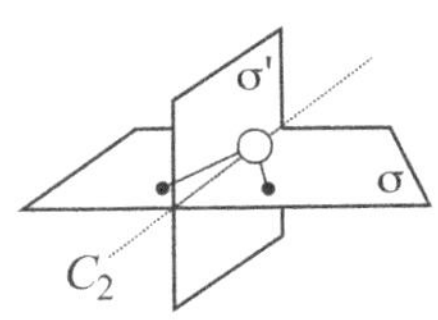

FIGURE 4.1 Symmetry elements of the water molecule. The open circle is the oxygen atom; the filled circles are the hydrogen atoms.

DOI: 10.1201/9781003644422-4

- preserve the lengths and angles of the vectors (technically we say that they are *unitary*) and
- do not change the physics of the problem: the transformation $\vec{r} \to \vec{r}'$ leaves the Hamiltonian of the problem *unchanged*.

The **linearity** of a symmetry transformation can be easily verified by comparing its effect on any two vectors and on a linear combination of them. In words: the transformation of the linear combination is the linear combination, with the same coefficients, of the results of the transformations of the original vectors.

The property of **unitarity** will be discussed in the following (see Section 5.1.4 and the example in Section 6.1.1.2). Here it is sufficient to say that a symmetry operation does not alter (or, better, *must* not alter, otherwise it is not considered a symmetry operation) the lengths of vectors and the angles between vectors.

The **invariance** can be verified *analytically* by considering the effect of a symmetry transformation on the Hamiltonian function of CM or the Hamiltonian operator of QM. Consider, for example, the one-dimensional harmonic oscillator and the symmetry operation that transforms x into $-x$. In CM the total energy is $H = \frac{1}{2}m\dot{x}^2 + \frac{1}{2}kx^2$. By substitution, we obtain the *transformed* $H = \frac{1}{2}m(-\dot{x})^2 + \frac{1}{2}k(-x)^2$. As the derivative of the opposite is the opposite of the derivative, we have, $\frac{1}{2}m(-\dot{x})^2 + \frac{1}{2}k(-x)^2 = \frac{1}{2}m\dot{x}^2 + \frac{1}{2}kx^2$ and we see that the transformation leaves $\hat{H}$ unchanged. In a similar way, the invariance is proved for the corresponding QM operator:

$$\frac{1}{2m}\frac{\partial^2}{\partial(-x)^2} + \frac{1}{2}k(-x)^2 = \frac{1}{2m}\frac{\partial^2}{\partial(x)^2} + \frac{1}{2}k(x)^2$$

Often, invariance is hastily recognized by observing a geometric figure that represents the problem. For example, it is clear that an equilateral triangle is invariant under rotation of one-third (or two-third) of a turn around a perpendicular axis passing through its centre. The same can be said for the equilibrium configuration of the ammonia molecule. Of course, great care must be taken not to introduce spurious features. Following the example, in the equilateral triangle there are symmetries that do not exist in the ammonia molecule, such as binary rotations around axes lying in the plane of the triangle.

It can be said that a system is symmetric (*i.e.* that a transformation is a symmetry operation of the system) by performing the following **thought experiment**. The system is first shown to an observer who then goes away. The transformation is applied to the system, the observer is called back and asked to indicate whether the operation has been done or not. If the observer cannot notice any change, then the transformation is a symmetry operation.

The graphical representation of symmetry poses a problem that is solved here with the use of shading. To represent a clockwise rotation by one-third of a turn of a system with the symmetry of a triangle we can for example use the left scheme of Figure 4.2. However, we would have the problem of making clear that the transformation has actually been done. We can therefore use the artifice of the right part of Figure 4.2. It must then be clear that the shading is *our addition*, it is an artefact that has the sole purpose of explaining what is being done, but the symmetry of the system is that represented by the triangle *without shading*.

Symmetry operations of *different nature* can coexist. In the water molecule, for example (Figure 4.1), we recognize that also the plane containing the three atoms is a reflection plane and that the intersection of the two reflection planes is a binary rotation axis. In the carbon dioxide molecule, in addition to the inversion, there is the reflection across a plane passing through the C atom and perpendicular to the molecular axis. In the ammonia molecule we recognize, in addition to the two rotations around the ternary axis, three reflections across the three *vertical* planes (*i.e.* perpendicular to the plane of the three hydrogens) each containing an N–H bond. Even an infinite number of symmetry elements and symmetry operations can coexist. For the water and ammonia molecules, the operations are finite in number, but in the carbon dioxide molecule, in addition to those already said, we have:

- rotations of 1/2 of a turn, of 1/3 and 2/3 of a turn, of 1/4, 2/4=1/2 and 3/4 of a turn, …, of $1/n$, $2/n$, … and $(n-1)/n$ of a turn, … (any integer n) around the O-C-O axis and
- reflections across any of the infinite planes containing the O-C-O axis.

Somewhat different is the **translational symmetry** typical of solid-state matter (see Sections 3 and 6.6). While rotation, reflection and inversion operations leave at least one point of ordinary 3D space unaltered, translations do not have this feature. This complicates the association of translations with other symmetry operations. We can also

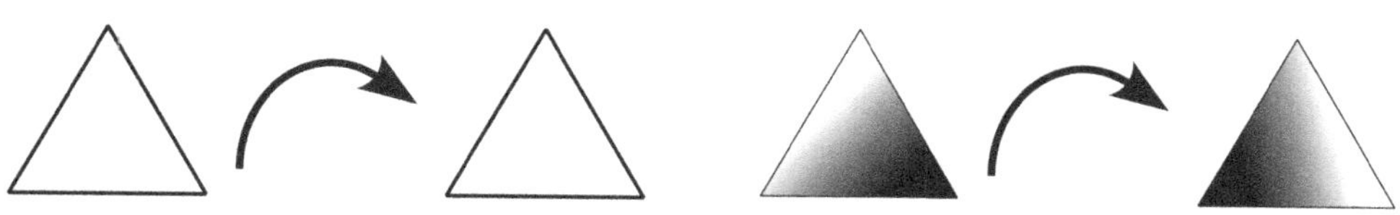

FIGURE 4.2 Left part: the clockwise rotation of a triangle by one-third of a turn. Right part: the transformation is made apparent by the shadow.

mention here the symmetry elements known as **screw axes** and **glides** (see Section 6.6.3).

Various **extensions** must be considered. A first extension is when the symmetry operations do not act on vectors of the 3D ordinary space, but are applied to **other vector spaces**, even of higher dimensionality than three and, in particular, to the *configuration spaces* of molecular systems and to the *wave function spaces* of QM. A simple example is the set of H and N atomic orbitals (AOs) that combine to form the molecular orbitals (MOs) of the NH_3 molecule according to the LCAO method. Each ternary rotation operation transforms, for example, the 1s AO of an H into the 1s AO of another H.

Symmetry operations can operate on mathematical objects of greater complexity than vectors, such as *tensors*. These are quantities characterized by two indices and used to express linear relations between vector quantities. Examples are the *inertia tensor* used in Section 1.3.1 of the Appendices to describe molecular rotations, and the *polarizability tensor* used in Eq. (2.21) and discussed in Section 6.5 of Chapter 1. The application of symmetry to tensors is not discussed here.

There are also symmetry operations that interest chemistry but *do not derive from geometric entities* of ordinary 3D space. In a system of n indistinguishable particles, it makes sense to consider as a symmetry operation also the exchange of two particles, that is:

$$\hat{P}_{kl}\psi(\tau_1,\tau_2,\ldots,\tau_k,\ldots,\tau_l,\ldots,\tau_n) = \psi(\tau_1,\tau_2,\ldots,\tau_l,\ldots,\tau_k,\ldots,\tau_n)$$

where τ_k denotes the set of spatial and spin coordinates of the k-th particle. This is also a coordinate transformation of the entire space of configurations of the system; it is a linear transformation and preserves orthogonality and normalization of the wave functions. If there are n particles, the configuration space is a $4n$-dimensional space.

Another symmetry of some importance in chemistry is the *time reversal symmetry* (see Section 2.1.3). Finally, there is a strange symmetry of the hydrogen atom (see Section 1.3.3).

1.1.2 Relations between Operations

Saying that symmetry operations can coexist, *does not mean* that they can be put together at will. The copresence of two or more symmetry operations is not possible *in general* and there is a **compatibility** problem. Even in the simplest cases with operations such as inversion, reflections or rotations there are constraints on nature and reciprocal position and orientation of the symmetry elements. For example, a threefold and a fivefold axis can appear together only if they are aligned to each other and therefore correspond to a 15-fold axis. A ternary axis and a quaternary axis that are *not* aligned *can* coexist, but the axes must be oriented in a particular way with respect to each other and there is not only one ternary axis and one quaternary axis but four ternary axes and three quaternary axes.

We now try to clarify discursively the relations between the different symmetry operations: for convenience, let us call a **symmetry group** a set of *compatible* symmetry *operations*, such as those of a molecule or a crystal structure. Here, the word **group** has a precise and very important *technical meaning*. This will be specified later (see Section 1.1.3) and then used regularly and intensively: for now, we speak of a group only to make clear that we are not dealing with a simple *set* of symmetry operations *chosen at random* but with operations that are *compatible with each other*.

1.1.2.1 Symmetry Operations Can Be Combined

The first point to note is that the symmetry operations of a group can be *combined* with each other. By *combination* of symmetry operations, we mean their *execution in sequence*. For example, we can combine a *clockwise* rotation of one-third of a turn with itself, executing it twice in a row. From Figure 4.3 we understand that the result of the combination is a *counterclockwise* rotation of one-third of a turn. This result (like the subsequent results) can be rigorously proved analytically.

Let's look at another example, again taken from the symmetry group of the ammonia molecule. As can be seen on Figure 4.4, the combination of a rotation with a reflection in a vertical plane is a different symmetry operation: it is still a reflection but is made with respect to another vertical plane.

1.1.2.2 How Combinations Are Denoted

Different notations are used quite regularly to denote the combination of symmetry operations.

- The most common is a **multiplicative notation**. Given two operations, A and B, their combination is simply written $C=AB$ or with a multiplication sign in between: $A{\cdot}B$ or $A \times B$. It is understood that the operation A is *applied* first, and it is *written* first according to the usual direction of writing from left to right. The operation $D=BA$ instead means to first apply operation B. With obvious extension, a repeated operation is then indicated with the power: AA becomes, A^2, etc. … Using the

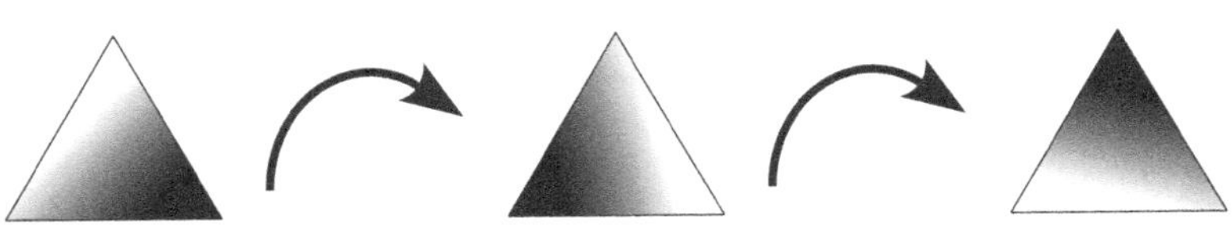

FIGURE 4.3 A symmetry operation performed twice is (generally) another symmetry operation.

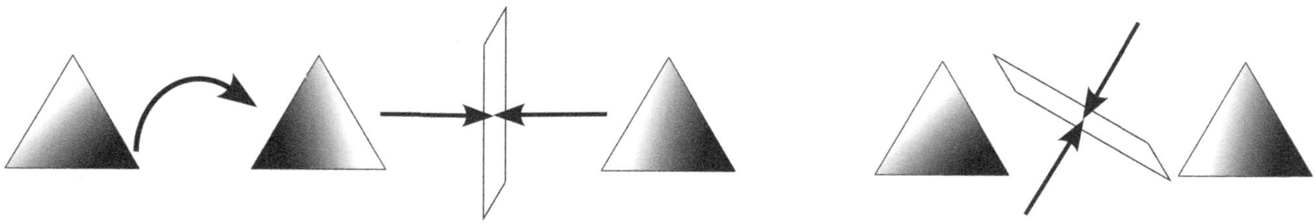

FIGURE 4.4 A clockwise ternary rotation followed by a reflection (left part) is a reflection with respect to a different mirror plane (right part).

multiplicative notation and indicating the symmetry operations with C_3, C_{-3} (the counterclockwise rotation) and with σ_v, $\sigma_v^{'}$ and $\sigma_v^{''}$ the three reflections in the vertical planes, the combinations of Figures 4.3 and 4.4 are written: $C_3 \cdot C_3 = (C_3)^2 = C_{-3}$ and $C_3 \cdot \sigma_v = \sigma_v^{'}$.

- In applications concerning QM, however, it is quite natural to follow the **operator notation**. If the transformation A produces the function g starting from the function f, we write: $\hat{A}f = g$. When doing two or more transformations in sequence, the operation to be performed first is now written on the right instead of left. The combination written in multiplicative notation as $C = AB$ (that is: first apply A, then B) must then be written $\hat{C}f = \hat{B}\left(\hat{A}f\right) = \hat{B}\hat{A}f$.
- Finally, there is also an **additive** notation, which is especially used for translation symmetries (see Sections 3 and 6.6).

1.1.2.3 Closure Property

The previous examples show that the list of the symmetry operations of a system includes not only the operations that are first recognised but also all the results of their combinations in pairs, then their combinations with the results, those between the results, and so on.... Now, thinking at the symmetry groups with a finite number of operations it becomes obvious that these combinations do not always produce new operations. At most, they do nothing more than complete the set of operations of the group if, for some reason, the set we started from was incomplete. This procedure is indeed a convenient tool to complete or verify the set of operations of a group. Another useful method is to consider that *a symmetry operation also acts on the other operations* and replicates them. For example, for the symmetry group of the ammonia molecule, one can think that a reflection (σ_v) on a vertical plane is replicated by the ternary rotations and produces the other two reflections ($\sigma_v^{'}$ and $\sigma_v^{''}$), about planes that are rotated by 120° and 240° (or −120°) with respect to the plane of σ_v.

With obvious modifications, these considerations also apply to *infinite groups*, that is, groups with an infinite number of elements and operations. In the case of the carbon dioxide molecule, for example, we write C_n to refer globally to the clockwise rotations of $1/n$ of a turn. The symmetry group must surely include all the successive rotations, that is 2, ..., $n-1$ times C_n, and we write $(C_n)^2$, ..., $(C_n)^{n-1}$. Also, rotating twice by a $1/2n$ of a turn is equivalent to rotating only once by $1/n$ of a turn, that is $(C_{2n})^2 = C_n$, and so on...

Reversing the argument, if we start from a complete list of all the symmetry operations of a group, by performing any combination of these elements we do *not* find something new. This is the first and fundamental property of algebraic groups, and symmetry groups in particular: **closure**. By combining elements of the group, we *do not exit the group.*

In a symmetry group, the combination of two symmetry operations is always equal to a symmetry operation of the group itself.

1.1.2.4 Identity

Let us now discover other properties of a symmetry group through examples. When applying three times the clockwise ternary rotation (see Figure 4.5) we go back to the top. Three equal ternary rotations produce no transformation at all.

Similarly, we go back to the top by applying twice a binary rotation, a reflection or an inversion or applying four times a quaternary rotation, etc.... For logical consistency, that is, to keep valid in general the property that *a combination of symmetry operations is a symmetry operation of the group*, it is necessary to include also the transformation... that does not transform. We call it **identity**.

Among the operations of a symmetry group there is always the identity.

FIGURE 4.5 With three ternary rotations we go back to the top.

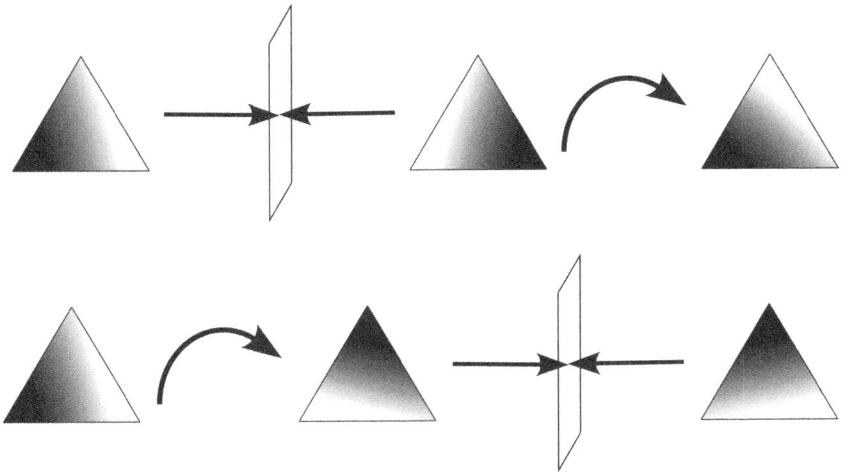

FIGURE 4.6 C_3 and σ_v do not commute

The identity of a group (E) plays a role similar to the role that the number 1 plays in the multiplication of numbers: the combination of the identity with any symmetry operation of the group always reproduces that symmetry operation, just as the number 1, multiplied by any number, gives that number as a result. The example in the previous figure is therefore written: $C_3C_3C_3=(C_3)^3=E$.

1.1.2.5 The Inverse

In multiplication of numbers, the existence of unity is associated with the possibility of defining the inverse of a number. The analogy holds: there is the inverse of a symmetry operation and is the operation that, when applied after the given operation, produces the identity as result. The examples show that in a group there is the inverse of any element. Considering for example the group of symmetry operations of the ammonia molecule, the two ternary rotation operations, clockwise and anticlockwise, are the inverse of each other: $C_3C_{-3}=E$. With an obvious extension of the multiplicative notation, we write: $C_{-3}=(C_3)^{-1}$ and $C_3=(C_{-3})^{-1}$. In additive notation, the inverse operation is obviously written as the opposite using the negative sign.

Examples show that if A is the inverse of B, then B is the inverse of A: if you go back to the top by first applying B and then A, then you go back to the top also by first applying A and then B.

> For every symmetry operation there **always exists** the inverse symmetry operation and is **unique.**

It *can* happen that a symmetry operation is the inverse of itself: for example, $(\sigma_v)^{-1}=\sigma_v$ and the same is true for a binary rotation and an inversion.

1.1.2.6 Commutation

As seen, the combination of symmetry operations, and in general the combination of elements of a group, shows some analogies with the multiplication of numbers. Among the analogies, however, the property of *commutativity is not necessarily included.*

Figure 4.6, for example, shows that first applying a clockwise ternary rotation and then a reflection or applying the same two operations but in the other order produces different results. We therefore write $C_3\sigma_v \neq \sigma_v C_3$. Commutativity is however *compatible* (*i.e.*: it *can* coexist) with the other features of a group, it simply *does not have to* coexist (see Section 1.1.3.5 for more details).

1.1.3 The Algebraic Structure

Having seen *discursively* the most important points, we can go to the abstract level. In algebra, a group G is defined as:

- a collection of objects (a **set**) {A, B, …} with
- an **internal composition law** [often called *multiplication* and denoted as such] that to *each ordered pair of elements* of the set (A, B) associates one and only one element C of G and therefore we write: $AB=C$.

In principle, one should distinguish between the *set* {A, B, …} of elements and the *group*, which is the set *and* the internal composition law. The distinction is not useless pedantry: it is possible, in fact, to find more than one composition law that makes a given set a group. However, it is usually difficult to confuse a group with its underlying set (its *support*), and therefore it is rarely necessary to pay attention to the detail. Note the abstraction: no reference is made to the nature of the objects that are the *support* of the group or to the internal composition law.

The set and the internal composition law form a group if the following axioms hold.

1.1.3.1 First Axiom: The Law Is Associative[2]

$\forall A,B,C \in G : A(BC)=(AB)C$ and it's simply written ABC

In words: the result **does not depend** on whether:

- we proceed according to $A(BC)$, that is, we first combine B with C (in this order, that is, we perform B and then C) forming the product $S=BC$ and then we combine A with S (in this order), or
- we proceed according to $(AB)C$, that is, we first combine A with B (in this order) forming the product $R=AB$ and then we combine the result R with C (in this order).

When using operator notation, we write on the right the operator that acts first. Then, applying to a set of coordinates $\{q\}$ what in multiplicative notation is the product ABC, means first applying operator $\hat{A}$, then operator $\hat{B}$, and finally operator $\hat{C}$, obtaining $q' = \hat{C}\left(\hat{B}\left(\hat{A}q\right)\right)$. If we indicate with R the product AB (which means applying A first and then B) then $q' = \hat{C}\left(\hat{B}\left(\hat{A}q\right)\right)$ has the meaning of applying R first and then C: $\hat{C}\left(\hat{R}q\right)$. Associativity then means that $\hat{C}\left(\hat{B}\left(\hat{A}q\right)\right) = \hat{C}\left(\hat{R}q\right) = \hat{S}\left(\hat{A}q\right)$, where $\hat{S} = \hat{B}\hat{C}$, that is the application of $\hat{B}$ followed by the application of $\hat{C}$. Then, this axiom establishes that both the multiplicative notation ABC and the operator notation $q' = \hat{A}\hat{B}\hat{C}q = \hat{C}\left(\hat{B}\left(\hat{A}q\right)\right)$ are *unambiguous*. It is important to remember that ABC is not necessarily equivalent to ACB or to one of the other four permutations of the symbols: associativity should not be confused with commutativity (Section 1.1.2.6).

1.1.3.2 Second Axiom: Existence of a Unit Element or Identity

There exists an element (usually denoted by E) such that:

$$\forall R \in G: \quad RE = ER = R$$

1.1.3.3 Third Axiom: Existence of the Inverse for Every Element of the Group

"$\forall R \in G$ there exists an element (T) such that $RT=TR=E$, and we write $T=R^{-1}$.

1.1.3.4 Direct Consequences

Actually, the axioms are more simply (ii) that there exists an element E such that "$\forall R \in G$: $ER=R$ and (iii) that "$\forall R \in G$ there exists an element (T) such that $TR=E$. Then it is proved that E is unique and that (for a given R): $RE=ER$, $RT=TR$ and T is unique.

1.1.3.5 Commutative or Abelian Groups

As seen with a counterexample (Figure 4.6), the group structure does not require **commutativity**:

> There are both commutative and non-commutative groups and there are both commutative *symmetry* groups, and non-commutative *symmetry* groups.

For this reason, the definition of the internal composition law requires that it acts on *ordered pairs*. If, instead, *all the elements of a group commute with each other*, we are dealing with an **additional property**. A commutative group is also called **Abelian** from the name of the Norwegian mathematician Niels Abel who in the 19th century gave an essential contribution to the foundation of modern algebra. Using the notation in the form of operators, we say that $\hat{A}\left(\hat{B}(q)\right)$ is in general different from $\hat{B}\left(\hat{A}(q)\right)$. If $\hat{A}\left(\hat{B}(q)\right) = \hat{B}\left(\hat{A}(q)\right)$, the two operators **commute**. Note: even in a non-commutative group there are pairs of elements that do commute. For example, in any group, the identity commutes with any other element of the group and every element commutes with its inverse.

The lack of commutativity (AB not necessarily $=BA$) implies that one must distinguish between a multiplication on the right and a multiplication on the left. For this reason – in general – one must be careful in the algebraic manipulation of symbols. For example, from $AC=BA$, one cannot derive $C=B$, a result that holds only for commutative groups. In general, instead, $C=A^{-1}BA$. Commutativity – when it occurs – means that the multiplication on the right is always equal to the multiplication on the left.

1.1.3.6 Order of a Group and an Element

A group with g elements is said to have **order g**. For a finite group, the closure property has the obvious consequence that each element multiplied by itself a sufficient number of times (m), necessarily produces as a result the identity: m is called the **order of the element**.

1.1.3.7 Cyclic Groups

These are the groups consisting only of the powers R, R^2, R^3, …, $R^g=E$ of a single element (g is the order of the group).

1.1.4 Point Groups

The **finite** point groups[3] are the simplest kind of symmetry groups and are described in detail in most textbooks with the procedure to identify the symmetry group of a molecule once its geometry is known. A notable feature is that *each operation leaves at least one point of the ordinary 3D space unchanged.*

As for the symbols of operations and groups, two notations are in use.

1.1.4.1 Schoenflies Notation

Here is the list of the symbols for the symmetry **operations**.

- E: **identity.**
- C_n: **rotation** of $1/n$ of a turn.[4] We call *principal axis* the axis with the greatest n when in a group there are more rotation axes with different n, or the only rotation axis of the group.

- σ: **reflection** in a plane. A *horizontal* reflection (σ_h) is across a plane perpendicular to the principal axis, a *vertical* reflection (σ_v) is across a plane containing the principal axis (or if there is no rotation axis, and then the group is C_s). If there are C_2 axes perpendicular to a principal axis C_n ($n>2$), the vertical reflection planes halfway between two of these C_2 axes are instead called *diagonal or dihedral planes* (σ_d) and the notation σ_v denotes only the vertical planes containing the C_2 axes.
- i: **inversion** with respect to a centre, *i.e.* the transformation from (x, y, z) to $(-x, -y, -z)$.
- S_n: **improper rotation** of $1/n$ of a turn (n even). This is a C_n rotation followed by a σ_h reflection or, in an equivalent way, σ_h followed by C_n, and is also called **roto-reflection**. Warning: an S_n is a symmetry operation *by itself* and does not necessarily imply that C_n and σ_h are elements of the group.

The finite point **groups** are denoted with the following symbols.

- C_1 is the trivial group with only one operation: identity.
- C_n: the only symmetry element is the axis of order n. There are n rotations and $(C_n)^n=E$.
- C_i, C_s and C_2 are three groups with only one operation other than identity (respectively i, σ and C_2). C_i and C_s are called *non-axial groups.*
- C_{nv}: there is the axis of order n and n vertical planes. These are all σ_v if n is odd or half are σ_v and half are σ_d if n is even.
- C_{nh}: there is the axis of order n and the horizontal reflection. Consequently, there is also the improper rotation axis (S_n) and, when n is even, also the inversion centre i.
- D_n: there is the axis of order n and n perpendicular C_2 axes.
- D_{nh}: there is the axis of order n, n perpendicular C_2 axes, and the horizontal reflection. Depending on n (even or odd) there are consequently other elements such as vertical planes, improper axes and inversion centre.
- D_{nd}: there are also vertical planes of type σ_d.
- S_m: there are the improper rotations S_m, $m=2n$; group S_2 is preferably denoted as C_i.
- The tetrahedral groups, T, T_h and T_d, are characterized by ternary axes and binary axes that are not perpendicular to the ternaries.
- The octahedral groups O and O_h: compared to the tetrahedral groups, they also contain quaternary axes.
- The icosahedral groups: one of them is indicated with I, another with I_h, the others are included among the groups with rotation axes of order 5.

1.1.4.2 Hermann-Maugin (H-M) Notation

While the above Schoenflies notation is the most used for molecular symmetries, the international notation (also called Hermann-Maugin notation) is the most used for solid-state symmetries (see also Section 6.6) and in particular for the space groups (see next point). Considering for now only the point groups, *operations* and *elements* are **partly the same as for the Schoenflies notation** but with **different symbols**: $i \to \bar{1}$; $C_n \to n$ (the number); $\sigma \to m$ (the letter), while an n axis orthogonal to a reflection is denoted as n/m or $\frac{n}{m}$. However, instead of the roto-reflection S_n, the **roto-inversion** axis is used: it is denoted by $\bar{n}$ and corresponds to a rotation followed by an *inversion* (instead of the reflection in the horizontal plane). Warning: similarly to what seen in Schoenflies notation for S_n, the inversion and the rotation taken individually are not necessarily group operations, $\bar{n}$ only indicates their combination. Some care is then needed. For example, the operation $\bar{1}$ (Hermann-Maugin) corresponds to S_2 (Schoenflies), which is also i (Schoenflies); the operation $\bar{2}$ (H-M) corresponds to σ (Schoenflies); $\bar{3}$ (H-M) corresponds to $(S_6)^{-1}$ (Schoenflies); $\bar{4}$ (H-M) corresponds to $(S_4)^{-1}$ (Schoenflies); $\bar{5}$ (H-M) corresponds to $(S_{10})^{-1}$ (Schoenflies); $\bar{6}$ (H-M) corresponds to $(S_3)^{-1}$ (Schoenflies).

The notation for groups is different. Here is a short sample list:

$$C_s \to m;\quad C_i \to \bar{1};\quad C_2 \to 2;\quad C_{2v} \to mm2;$$

$$C_{2h} \to 2/m;\quad D_2 \to 222;\quad D_{2h} \to mmm;$$

$$C_{2d} \to \bar{4}2m;\quad T_d \to \bar{4}3m;\quad T_h \to m3;$$

$$O \to 43;\quad O_h \to m3m$$

Furthermore, there is sometimes a *double notation*, with full and short symbols: for example, the *mmm* group is denoted as $\frac{2}{m}\frac{2}{m}\frac{2}{m}$ in full form.

1.1.5 Other Symmetry Groups

Besides the above groups, other symmetry groups are of interest to chemistry and, for various reasons, are better considered separately.

A. The point group $C_{\infty v}$ of **linear molecules lacking the inversion and the reflection** across the *horizontal* plane perpendicular to the molecular axis. There are all rotations C_n ($n=2, 3, \ldots$) about the molecular axis and all reflections across any *vertical* plane σ_v. Example are: CO, HCN, COS, …
B. The point group $D_{\infty v}$ of **linear molecules with inversion and reflection**. In addition to the elements of $C_{\infty v}$, there is also the inversion i, the reflection in the horizontal plane σ_h, the binary rotations C_2' about any axis perpendicular to the molecular axis and all the improper rotations

S_n ($n = 2, 3, \ldots$) about the same axis as C_n. Examples are O_2, CO_2, HCCH, …
C. The point group of rotations of a sphere, denoted as SO(3) and also known as **full rotation group**, which is important for the atomic structure. Note however, that no molecule belongs to this group.
D. The **translation groups** involved in the crystalline state of matter (see Section 3) and described (in the 3D case) by the 14 Bravais lattices (Section 6.6).
E. The **space groups** (Section 6.6.4) that combine the translation and point symmetries.
F. The **permutation** groups of 2, 3, … equal particles, denoted as $S(2)$, $S(3)$, …

The groups of cases A, B, C are also point groups but are anomalous because they have an *infinite number* of symmetry elements (see Section 6.4) and are *continuous*, while all the others are discrete groups. The category is also known as *Lie groups*. The translation groups (D), while having an infinite number of symmetry operations, are in some respects simpler than the cases A, B, C due to their nature of commutative groups (see Section 3). Unlike all the previous groups, whose elements are geometric operations in ordinary 3D space, the permutation groups (case F) are made up of operations of a completely different nature and therefore their combination with geometric groups has particular features.

1.1.6 The Cayley Table

The **multiplication table**, also known as *Cayley table*, is a very useful tool for understanding, finding or verifying the properties of a group. It lists all the combinations between the elements of a group by inserting in each cell the product of the element in the row header by the element in the column header, performed in this order.

Let us first consider the groups that have only one operation in addition to identity: these are the operations of inversion, reflection, binary rotation as well as the exchanges between two equal particles. The groups are denoted as C_i, C_s and C_2 (Schoenflies) and $S(2)$. The symmetry operations are combined as shown in Table 4.1, which is obviously the same for the three groups because there is no other possibility:

TABLE 4.1
Multiplication Table of Groups C_i, C_s, C_2, and $S(2)$

↓×→	E	X
E	E	X
X	X	E

E is the identity; X is the reflection (group C_s), the inversion (group C_i), the binary rotation (group C_2) or else the exchange (1,2) → (2,1) in group $S(2)$.

TABLE 4.2
Multiplication Table of Group C_{2v}

↓×→	E	C_2	σ	σ'
E	E	C_2	σ	σ'
C_2	C_2	E	σ'	σ
σ	σ	σ'	E	C_2
σ'	σ'	σ	C_2	E

TABLE 4.3
Multiplication Table of C_{3v}

↓×→	E	C_3	C_{-3}	$\sigma^{(1)}$	$\sigma^{(2)}$	$\sigma^{(3)}$
E	E	C_3	C_{-3}	$\sigma^{(1)}$	$\sigma^{(2)}$	$\sigma^{(3)}$
C_3	C_3	C_{-3}	E	$\sigma^{(3)}$	$\sigma^{(1)}$	$\sigma^{(2)}$
C_{-3}	C_{-3}	E	C_3	$\sigma^{(2)}$	$\sigma^{(3)}$	$\sigma^{(1)}$
$\sigma^{(1)}$	$\sigma^{(1)}$	$\sigma^{(2)}$	$\sigma^{(3)}$	E	C_3	C_{-3}
$\sigma^{(2)}$	$\sigma^{(2)}$	$\sigma^{(3)}$	$\sigma^{(1)}$	C_{-3}	E	C_3
$\sigma^{(3)}$	$\sigma^{(3)}$	$\sigma^{(1)}$	$\sigma^{(2)}$	C_3	C_{-3}	E

Let us then examine the symmetry group C_{2v} (Schoenflies) to which, for example, the water molecule belongs. There are three symmetry operations plus identity: the reflection in the plane of the molecule σ, the binary rotation C_2, and the reflection σ' in the plane perpendicular to the plane of the molecule. The symmetry operations are combined as shown in Table 4.2.

The symmetry group of the ammonia molecule, C_{3v} in Schoenflies notation, is a slightly more complicated example (see Table 4.3). The group contains: the identity (E), two ternary rotations C_3 and C_{-3} of 1/3 and 2/3, (or −1/3) of a turn, and three “vertical” planes $\sigma^{(1)}$, $\sigma^{(2)}$ and $\sigma^{(3)}$.

Let's look at the features of multiplication tables using this last group as an example. The table agrees with the order of the factors, as indicated in the top left box: the product $C_3\sigma^{(1)}$ (that is, the application of C_3 followed by the application of $\sigma^{(1)}$) is at the intersection between the second row (heading C_3) and the fourth column (heading $\sigma^{(1)}$) and gives $\sigma^{(3)}$. The product in the other order instead gives $\sigma^{(2)}$ which is on the opposite side to the main diagonal of the table (the one drawn from the top left to the bottom right). The table shows that,

> in a Cayley table each row contains all the elements of the group in a different order (*i.e.* a different permutation).

and

> each column contains all the elements of the group in a different permutation.

As derived in Section 7.1, *these properties are consequences of the group axioms.*

> The identity is the header of the row that is equal to the row of column headers.

In Table 4.3, the column header contains the elements of the group in the order E, C_3, C_{-3}, $\sigma^{(1)}$, $\sigma^{(2)}$, and $\sigma^{(3)}$: this is the same order as the first row and therefore the identity is the first element (E) of the list. Wanting to prove the property, we can say that if X is the header of the row that contains the elements in the same order as the column header, then for any element (Y) of the group it is true that $XY=Y$, from which $XYY^{-1}=YY^{-1}=E$ and therefore $X=E$. In a completely equivalent way, the identity is obtained from the header of the column that is equal the row header. The following statements on the properties of the multiplication table are easily proved in a similar manner.

> The **inverse** of an element is in the cell placed in the column identified by the given element and in the row containing the identity.

For example, to find the inverse of the rotation C_3, we look at the column with the header C_3. In this column, E is in the third row. The inverse of C_3 is C_{-3}, because this is the header of the third row.

> If the rows are arranged in the same order as the columns a commutative group is recognized by the *symmetry* of the table with respect to *the main diagonal.*

Then, Table 4.3 clearly shows that the group C_{3v} is *not commutative*. Despite the advantage of making commutativity evident, sometimes (see Section 5.1.3.7) it is useful to rewrite the multiplication tables using a different order for the columns, *i.e.* using for the columns the order of the inverses of the elements placed at the headers of the rows: in this case the diagonal of the table always contains the identity.

As an example of a symmetry group quite richer in symmetry operations (they are 24), we can consider that of the benzene molecule with its D_{6h} group. The symmetry operations are:

a. the identity,
b. two senary rotations, clockwise and counterclockwise: C_6 and $C_{-6}=(C_6)^{-1}$,
c. two ternary rotations, clockwise and anticlockwise: C_3 and $C_{-3}=(C_3)^{-1}$, which also correspond to $(C_6)^2$ and $(C_{-6})^2=(C_6)^{-2}$,
d. a vertical binary rotation (see b and c): $C_2=(C_6)^3$,
e. three binary rotations around horizontal axes containing two opposite C–H bonds: $C_2'^{(1)}$, $C_2'^{(2)}$, $C_2'^{(3)}$,
f. three other binary rotations around horizontal axes halfway between a C–H bond and the adjacent one: $C_2''^{(1)}$, $C_2''^{(2)}$, $C_2''^{(3)}$,
g. the reflection in the horizontal plane: σ_h,
h. the inversion with respect to the centre of the molecule: i,
i. two vertical ternary roto-inversions clockwise and anticlockwise: S_3 and $S_{-3}=(S_3)^{-1}$,
j. two vertical senary roto-inversions clockwise and anticlockwise S_6 and $S_{-6}=(S_6)^{-1}$,
k. three reflections in vertical planes halfway between two C–H bonds: $\sigma_d^{(1)}$, $\sigma_d^{(2)}$, $\sigma_d^{(3)}$,
l. three reflections in "vertical" planes containing two opposite C–H bonds: $\sigma_v^{(1)}$, $\sigma_v^{(2)}$, $\sigma_v^{(3)}$.

NB: the sets (e) and (f) must be considered separately: the reasons for the distinct classification can be clarified in Section 4.1.3. The same applies to operations (k) and (l). For ease of writing, Table 4.4 reports only half of the multiplication table, that is, all the rows but only half of the columns. The group is not commutative, as can be seen by a detailed comparison (see Table 4.4).

Note how intricate and difficult to read the multiplication table of this symmetry group is, and this is not among the most complicated point groups. It can be anticipated from now that there are internal structures to the groups and relations between groups that simplify the treatment of the most intricate groups. An example of the possibilities of simplifying the calculations of molecular orbitals is given in Section 6.2.1.6 for the D_{6h} group.

1.1.7 Abstract Groups

The multiplication table fully describes the relations between the elements of a group *regardless of the meaning of the elements* (symmetry operations or others) and *the internal composition operation* (sequential execution or otherwise) and therefore defines an *abstract* group. The algebraic study of an abstract group is the study of all groups that share the multiplication table.

There is only one abstract group of order 2 (see Table 4.1) as is obvious, since there are only two permutations (AB and BA) of a list of two symbols A, B. There is only one abstract group of order 3, two of order 4, one of order 5 and two of order 6 (see Exercises 4.1–4.4) in spite of the fact that there are $3!=6$, $4!=24$ and $5!=120$ permutations of 3, 4, 5 and 6 symbols, because the algebraic properties (closure, associativity, existence and uniqueness of a neutral element, existence and uniqueness of the inverse of each element) strongly limit the number of groups.

Table 4.5 well explains that the same abstract group represents groups consisting of mathematical objects of different nature. Table 4.3 shows that the abstract group corresponds to the *point* group C_{3v}. Now, let's consider *permutation* groups. In addition to the simple case of the group $S(2)$ (Table 4.1), let's look at the group $S(3)$ of permutations of three objects (which we call a, b, c). There are six permutations of three objects and we represent each of them with a two-row scheme, writing in the lower row how the objects in the upper row are re-ordered by the permutation:

$$A=\begin{pmatrix} a & b & c \\ a & b & c \end{pmatrix}; B=\begin{pmatrix} a & b & c \\ b & c & a \end{pmatrix};$$

$$C=\begin{pmatrix} a & b & c \\ c & a & b \end{pmatrix}; D=\begin{pmatrix} a & b & c \\ a & c & b \end{pmatrix};$$

TABLE 4.4
Left Half of the Multiplication Table of Group D_{6h}

↓×→	E	C_6	C_{-6}	C_3	C_{-3}	C_2	$C_2'^{(1)}$	$C_2'^{(2)}$	$C_2'^{(3)}$	$C_2''^{(1)}$	$C_2''^{(2)}$	$C_2''^{(3)}$
E	E	C_6	C_{-6}	C_3	C_{-3}	C_2	$C_2'^{(1)}$	$C_2'^{(2)}$	$C_2'^{(3)}$	$C_2''^{(1)}$	$C_2''^{(2)}$	$C_2''^{(3)}$
C_6	C_6	C_3	E	C_2	C_{-6}	C_{-3}	$C_2''^{(2)}$	$C_2''^{(3)}$	$C_2''^{(1)}$	$C_2'^{(2)}$	$C_2'^{(3)}$	$C_2'^{(1)}$
C_{-6}	C_{-6}	E	C_{-3}	C_6	C_2	C_3	$C_2''^{(3)}$	$C_2''^{(1)}$	$C_2''^{(2)}$	$C_2'^{(3)}$	$C_2'^{(1)}$	$C_2'^{(2)}$
C_3	C_3	C_2	C_6	C_{-3}	E	C_{-6}	$C_2'^{(3)}$	$C_2'^{(1)}$	$C_2'^{(2)}$	$C_2''^{(3)}$	$C_2''^{(1)}$	$C_2''^{(2)}$
C_{-3}	C_{-3}	C_{-6}	C_2	E	C_3	C_6	$C_2'^{(2)}$	$C_2'^{(3)}$	$C_2'^{(1)}$	$C_2''^{(2)}$	$C_2''^{(3)}$	$C_2''^{(1)}$
C_2	C_2	C_{-3}	C_3	C_{-6}	C_6	E	$C_2''^{(1)}$	$C_2''^{(2)}$	$C_2''^{(3)}$	$C_2'^{(1)}$	$C_2'^{(2)}$	$C_2'^{(3)}$
$C_2'^{(1)}$	$C_2'^{(1)}$	$C_2''^{(3)}$	$C_2''^{(2)}$	$C_2'^{(2)}$	$C_2'^{(3)}$	$C_2''^{(1)}$	E	C_3	C_{-3}	C_2	C_{-6}	C_6
$C_2'^{(2)}$	$C_2'^{(2)}$	$C_2''^{(1)}$	$C_2''^{(3)}$	$C_2'^{(3)}$	$C_2'^{(1)}$	$C_2''^{(2)}$	C_{-3}	E	C_3	C_6	C_2	C_{-6}
$C_2'^{(3)}$	$C_2'^{(3)}$	$C_2''^{(2)}$	$C_2''^{(1)}$	$C_2'^{(1)}$	$C_2'^{(2)}$	$C_2''^{(3)}$	C_3	C_{-3}	E	C_{-6}	C_6	C_2
$C_2''^{(1)}$	$C_2''^{(1)}$	$C_2'^{(3)}$	$C_2'^{(2)}$	$C_2''^{(2)}$	$C_2''^{(3)}$	$C_2'^{(1)}$	C_2	C_{-6}	C_6	E	C_3	C_{-3}
$C_2''^{(2)}$	$C_2''^{(2)}$	$C_2'^{(1)}$	$C_2'^{(3)}$	$C_2''^{(3)}$	$C_2''^{(1)}$	$C_2'^{(2)}$	C_6	C_2	C_{-6}	C_{-3}	E	C_3
$C_2''^{(3)}$	$C_2''^{(3)}$	$C_2'^{(2)}$	$C_2'^{(1)}$	$C_2''^{(1)}$	$C_2''^{(2)}$	$C_2'^{(3)}$	C_{-6}	C_6	C_2	C_3	C_{-3}	E
i	i	S_{-3}	S_3	S_{-6}	S_6	σ_h	$\sigma_d^{(1)}$	$\sigma_d^{(2)}$	$\sigma_d^{(3)}$	$\sigma_v^{(1)}$	$\sigma_v^{(2)}$	$\sigma_v^{(3)}$
S_3	S_3	i	S_6	S_{-3}	σ_h	S_{-6}	$\sigma_v^{(3)}$	$\sigma_v^{(1)}$	$\sigma_v^{(2)}$	$\sigma_d^{(3)}$	$\sigma_d^{(1)}$	$\sigma_d^{(2)}$
S_{-3}	S_{-3}	S_{-6}	i	σ_h	S_3	S_6	$\sigma_v^{(2)}$	$\sigma_v^{(3)}$	$\sigma_v^{(1)}$	$\sigma_d^{(2)}$	$\sigma_d^{(3)}$	$\sigma_d^{(1)}$
S_6	S_6	S_3	σ_h	i	S_{-6}	S_{-3}	$\sigma_d^{(2)}$	$\sigma_d^{(3)}$	$\sigma_d^{(1)}$	$\sigma_v^{(3)}$	$\sigma_v^{(1)}$	$\sigma_v^{(2)}$
S_{-6}	S_{-6}	σ_h	S_{-3}	S_6	i	S_3	$\sigma_d^{(3)}$	$\sigma_d^{(1)}$	$\sigma_d^{(2)}$	$\sigma_v^{(2)}$	$\sigma_v^{(3)}$	$\sigma_v^{(1)}$
σ_h	σ_h	S_6	S_{-6}	S_3	S_{-3}	i	$\sigma_v^{(1)}$	$\sigma_v^{(2)}$	$\sigma_v^{(3)}$	$\sigma_d^{(1)}$	$\sigma_d^{(2)}$	$\sigma_d^{(3)}$
$\sigma_d^{(1)}$	$\sigma_d^{(1)}$	$\sigma_d^{(1)}$	$\sigma_v^{(3)}$	$\sigma_v^{(2)}$	$\sigma_d^{(2)}$	$\sigma_d^{(3)}$	i	S_{-6}	S_6	σ_h	S_3	S_{-3}
$\sigma_d^{(2)}$	$\sigma_d^{(2)}$	$\sigma_v^{(1)}$	$\sigma_v^{(3)}$	$\sigma_d^{(3)}$	$\sigma_d^{(1)}$	$\sigma_v^{(2)}$	S_6	i	S_{-6}	S_{-3}	σ_h	S_3
$\sigma_d^{(3)}$	$\sigma_d^{(3)}$	$\sigma_v^{(2)}$	$\sigma_v^{(1)}$	$\sigma_d^{(1)}$	$\sigma_d^{(2)}$	$\sigma_v^{(3)}$	S_{-6}	S_6	i	S_3	S_{-3}	σ_h
$\sigma_v^{(1)}$	$\sigma_v^{(1)}$	$\sigma_d^{(3)}$	$\sigma_d^{(2)}$	$\sigma_v^{(2)}$	$\sigma_v^{(3)}$	$\sigma_v^{(1)}$	σ_h	S_3	S_{-3}	i	S_{-6}	S_6
$\sigma_v^{(2)}$	$\sigma_v^{(2)}$	$\sigma_d^{(1)}$	$\sigma_d^{(3)}$	$\sigma_v^{(3)}$	$\sigma_v^{(1)}$	$\sigma_v^{(2)}$	S_{-3}	σ_h	S_3	S_6	i	S_{-6}
$\sigma_v^{(3)}$	$\sigma_v^{(3)}$	$\sigma_d^{(2)}$	$\sigma_d^{(1)}$	$\sigma_v^{(1)}$	$\sigma_v^{(2)}$	$\sigma_v^{(3)}$	S_3	S_{-3}	σ_h	S_{-6}	S_6	i

The group is not commutative as shown, *e.g.* by the products of senary rotations with primed binary rotations.

TABLE 4.5
The Abstract Group of Order 6, Isomorphous to C_{3v} and $S(3)$

↓×→	A	B	C	D	F	G
A	A	B	C	D	F	G
B	B	C	A	G	D	F
C	C	A	B	F	G	D
D	D	F	G	A	B	C
F	F	G	D	C	A	B
G	G	D	F	B	C	A

$$F = \begin{pmatrix} a & b & c \\ b & a & c \end{pmatrix}; G = \begin{pmatrix} a & b & c \\ c & b & a \end{pmatrix};$$

where $A = E$.

The combination of two permutations is their execution in sequence. So, for instance,

$$BD = \begin{pmatrix} a & b & c \\ b & c & a \end{pmatrix}\begin{pmatrix} a & b & c \\ a & c & b \end{pmatrix} = \begin{pmatrix} a & b & c \\ c & b & a \end{pmatrix}$$

and it is easy to see that $BD = G$, while

$$BF = \begin{pmatrix} a & b & c \\ a & c & b \end{pmatrix} = D$$ and so on. We say that groups

and $S(3)$ are **isomorphic**.

1.2 Symmetry and Wave Functions

We now pay attention to the equations that in CM and QM describe the motion of nuclei, electrons and molecules, to their solutions (the electronic wave functions and the normal modes of vibration, in particular), to the selection rules for spectroscopic transitions, to the functional properties, …. For all these problems we analyse examples of the features of a symmetric system.

It is often said that symmetry is the **cause** of some particular results. This can be misleading. Obviously, it is the constituent particles and their interactions that determine both the symmetry of a system and their properties. Therefore, symmetry can be used (i) to obtain information about the solutions of the equations of motion and about various properties that result from them, and (ii) to simplify the solution procedure, sometimes even to obtain the solution.

1.2.1 Some Elementary Effects of Symmetry

To begin with, let us consider some results that can be obtained by considering only the symmetry group.

1.2.1.1 The Polarity of Molecules

A molecule has a permanent electric dipole moment if the centre of the distribution of positive charges does not coincide with that of the negative charges. The knowledge of symmetry allows us to draw some general conclusions in an elementary way.

The presence of an *inversion centre* is clearly incompatible with a permanent electric dipole moment, since the centres of the positive and negative charge distributions should then be both located on the inversion centre. However, the reverse statement is not true: a zero value of the permanent electric dipole moment is not in fact a *generally* valid indication of the presence of an inversion centre, as demonstrated by the example of tetrahedral or triangular molecules (for example CH_4 or BF_3).

When there is only *one axis of rotational symmetry*, the permanent electric dipole moment, if it is not zero, must be aligned along the axis (*e.g.* water and ammonia molecules) and is certainly zero if there are symmetry operations that, generically speaking, "flip up and down" (considering the axis oriented vertically), such as a reflection plane σ_h or binary axes perpendicular to a principal axis. In conclusion, a molecule exhibits a permanent electric dipole moment only if its symmetry belongs to the groups C_s, C_n, C_{nv} and $C_{\infty v}$.

1.2.1.2 Chirality

The *simple* definition of a chiral system is the *non-superposition* with its mirror image. In more technical terms and skipping the derivation based on the analysis of the interaction with electromagnetic radiation, we accept the fact that a molecule is optically inactive (better: cannot be optically active) if its symmetry group contains at *least* one improper rotation S_n.

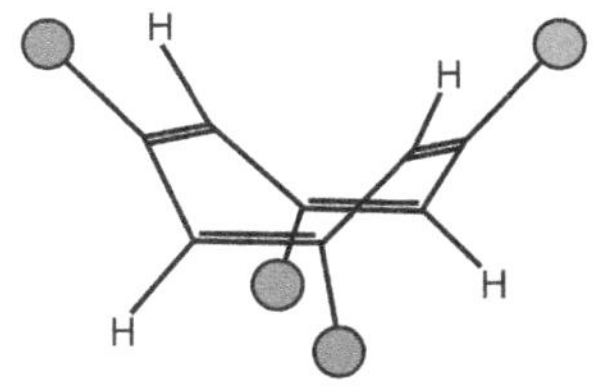

FIGURE 4.7 Molecules lacking both inversion and reflection symmetry but optically inactive are not very common. An example pertinent to the S_4 group is 1,3,5,7-tetrachlorocycloocta 1,3,5,7-tetraene (gray circles are Cl atoms).

Reflection is the same as an improper rotation of order 1: $S_1=\sigma$. *Inversion* is a binary improper rotation: $S_2=i$. Then, the well-known rule follows according to which all groups containing the inversion operation surely describe optically inactive molecules. However, it is not true that all molecules without an inversion centre and without a reflection plane are optically active: optical inactivity can derive – for example – from an S_4 operation (see Figure 4.7).

1.2.2 Symmetry Species

It would seem obvious that, in a system characterized by some symmetry, the solutions of the equations of motion *necessarily* exhibit the symmetric behaviour of the system itself. The statement is somehow correct, but not in the most trivial sense: things are more complicated, and are better introduced with examples, starting from the cases in which the energy levels are *not degenerate*, because – as we will see – these cases are considerably simpler than the others.

1.2.2.1 Examples without Energy Degeneracy

Let's look at a very simple QM problem: the harmonic oscillator, where there is only one symmetry operation in addition to the identity, that is the transformation from x to $-x$ (*parity*). From the solution of the Schrödinger equation, we know that all energy levels are non-degenerate. Figure 4.8 shows the potential energy with eigenvalues and eigenfunctions of the Hamiltonian for the first four levels. It is immediately clear that, while the system (its kinetic and potential energy) is invariant under the effect of the transformation of coordinates from x to $-x$, this is not true for the eigenfunctions, which instead show *two distinct behaviours*:

a. eigenfunctions with $n=0, 2, 4, 6, \dots$ are invariant: $f(-x)=f(x)$,
b. eigenfunctions with $n=1, 3, 5, 7, \dots$ change sign: $f(-x)=-f(x)$.

It would seem natural to say that the solutions of the first type are symmetric and those of the second type are not, but this is not the case. Also the solutions of the second type must be considered symmetric, because they too have a behaviour predictable from symmetry. Once given the value of the function at x, we know its value at $-x$. Essentially, what changes from the solutions (a) to the solutions (b) is only the eigenvalue of the symmetry operator.

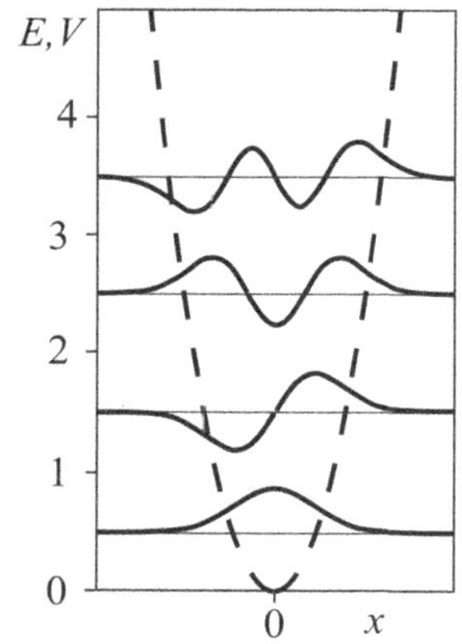

FIGURE 4.8 Potential energy, energy eigenvalues and eigenfunctions for the 1D harmonic oscillator model. The horizontal lines indicate the eigenvalues ($n+½$, in units of $h\nu$), the dashed parabola is the potential energy of the system, the solid lines are the spatial (real) parts of the eigenfunctions of *Hamiltonia operator* for $n = 0, 1, 2, 3$ in order from bottom. The plots of eigenfunctions are vertically shifted so that each zero is at the level of the eigenvalue of energy.

The eigenfunctions with eigenvalue=1 ($n = 0, 2, 4, 6, \ldots$) are called *even*; those with eigenvalue=−1 are called *odd* ($n=1, 3, 5, 7, \ldots$). We will see later that +1 *and* −1 *are all possible eigenvalues* for the parity transformation.

Let us now look at a very similar symmetry operation, the *inversion* in 3D space, *i.e.* the transformation $(x, y, z) \rightarrow (-x, -y, -z)$. We find two behaviours analogous to those of the harmonic oscillator by observing the example of the hydrogen molecule and two if its MOs expressed as LCAOs of the 1s AOs. There are (Figure 4.9) only two independent linear combinations: the bonding MO (top) and the anti-bonding MO (bottom), which have different energies and have eigenvalues +1 and −1 for the inversion, respectively.

The next situation to consider is the behaviour with respect to *more* symmetry operations. We can take the example of the C_{2v} group (see Table 4.2) which, as shown by Figure 4.10, has three non-trivial symmetry operations, namely the reflection (σ) in the plane of the molecule, the

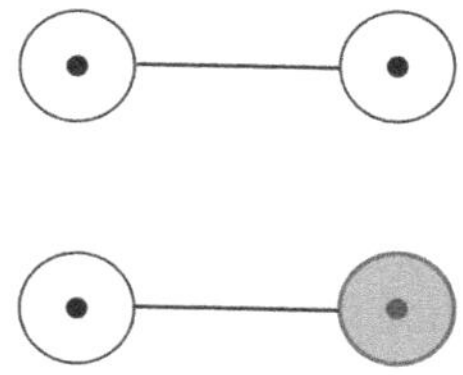

FIGURE 4.9 Two MOs of H_2 molecule and inversion symmetry. White and grey circles denote positive and negative lobes of the MOs.

reflection (σ') in the plane that in the figure appears vertical and the binary rotation (C_2) around the axis placed at the intersection of the two planes.

Let us consider two different applications for two different molecules. The first one concerns the butadiene molecule in its (improperly said) *cis* conformation and the π-type MO orbitals. It is a simple problem because only one type of AO has to be examined: the $2p_z$ for each carbon atom. The results have been obtained with the simple Hückel-LCAO method (see Section 2.2.1). The eigenvalues of the symmetry operations globally summarise how the MOs behave, and the grey levels in the boxes highlight that the four MOs follow only two behaviours.

The second application concerns the water molecule and its normal modes of vibration (for the derivation, see Section 2.2.3.1). The solutions are now characterized by two different *sets* of eigenvalues, which are all different from the sets found for the MOs of butadiene.

Overall, we have identified four distinct sets of eigenvalues. Let's summarize.

> The solutions of the equations of motion "respond" in different ways to the complex of symmetry operations of the system. These different possible responses have the technical name of **symmetry species**.

A warning: on the left and right of Figure 4.10, the Cartesian axes are oriented differently, because the usual choice for the π MOs is to combine 2p AOs oriented along the z-axis, while z is the usual choice for the direction of the principal axis of symmetry groups (see Tables 4.6 and 4.21).

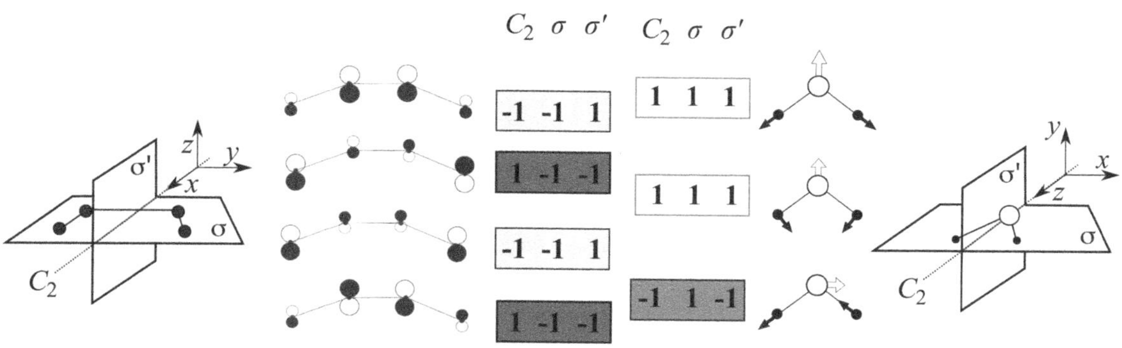

FIGURE 4.10 Two applications concerning the C_{2v} group. On the left, the π MOs of "cis" butadiene, on the right the normal modes of vibration of water. The central parts report the behaviours with respect to the three non-trivial symmetry operations, and the grey levels highlight four different species of symmetry (see text and Table 4.8 for the symbols of the symmetry species).

We derive below (Table 4.8) the complete list of eigenvalues of the various symmetry species for the C_{2v} group, and we see that there are no other symmetry species besides the four species found with the two examples just discussed. Furthermore:

> There is a particular species of symmetry, called *totally symmetric* which has eigenvalue=1 for every symmetry operation. In *all symmetry groups* there is the totally symmetric species.

1.2.2.2 Examples with Energy Degeneracy

Let us now go to the case where the eigenfunctions of the Hamiltonian exhibit degeneracy and consider for example the MO π of benzene obtained with the same method used for butadiene.

Compared to butadiene we have a simplification, because the AOs are all equivalent to each other by symmetry, but also a complication, due to the symmetry group which is much richer in operations (see Table 4.4). The result of the direct calculation is that the six MO π belong to four energy levels, with energy increasing from left to right (Figure 4.11). The first and the last levels are non-degenerate, each of the two intermediate levels is twice degenerate. It is easy to see that the non-degenerate MO on the left has an eigenvalue=+1 for some operations (senary and ternary rotations, reflections on vertical planes, …) and an eigenvalue=–1 for the others (reflection on the molecular plane, binary rotations with axis in the molecular plane, …). The last MO also has an eigenvalue of +1 for some symmetry operations and –1 for the others, although they are not the same as for the first MO. With time and patience, we can verify that what is said is valid for all 24 symmetry operations. Then, for these two MOs, the situation is *analogous to what we have just seen with the examples for non-degenerate cases.*

The novelty regards the two degenerate levels of the two central parts of Figure 4.11. In fact, *three* distinct behaviours can be noted (see Figure 4.12) looking at the left MO of the second level (considering the other MO or an MO of the third level leads to the same conclusions).

a. The MO is an eigenfunction with eigenvalue=1 for some symmetry operations (inversion or reflection on the dotted plane in Figure 4.12, part a).
b. For other operations (binary rotation around the dashed axis in Figure 4.12, part b) the MO is an eigenfunction with eigenvalue=–1.
c. Finally, for operations such ternary or senary rotations, (part c-1 and c-2 of Figure 4.12, respectively) the MO **is not an eigenfunction** of the symmetry operation.

In cases (c) a symmetry operation transforms an MO into **another** MO (c-1) or into a **linear combination of MOs** (c-2). When we say that the symmetry operation *transforms a function into another function*, we mean that the two functions are linearly independent of each other.

> If the symmetry operations linearly combine two, three, … independent degenerate solutions, the symmetry species is said to be **two, three, … -dimensional**.

Note that the functions schematized in Figures 4.11 and 4.12 for the two degenerate levels are linearly independent but are *not orthogonal*. In the treatment of symmetry, the use of orthogonal and normalized functions is always preferable and sometimes necessary. The point is illustrated in

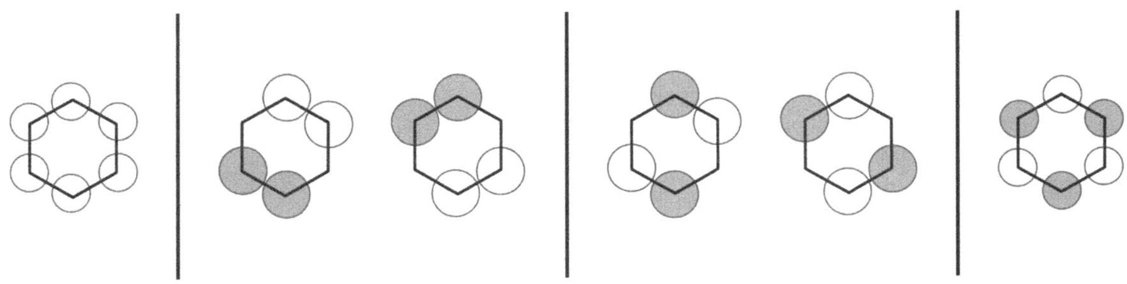

FIGURE 4.11 Benzene, π - type MOs. The vertical bars separate the four energy levels. The $2p_z$ AOs have lobes of different signs indicated by white or grey: the figure shows the lobes *above* the molecular plane, those below the plane have the opposite sign. See also Figure 4.32.

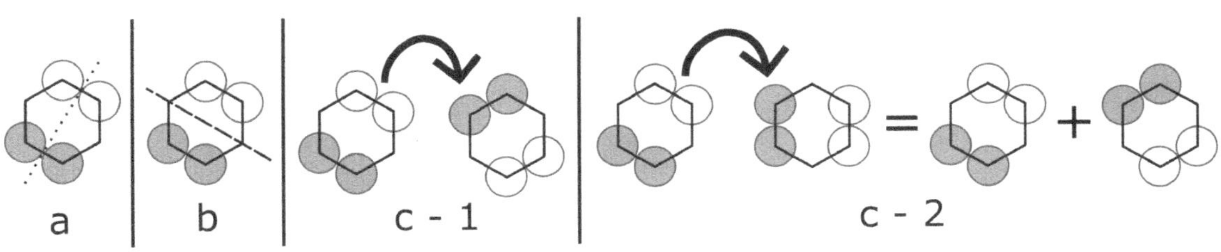

FIGURE 4.12 Effects of some symmetry operations on an eigenfunction of a degenerate level of the Hamiltonian (see text).

particular in Section 6.2.1.6, where Figure 4.32 reports a couple of orthonormal eigenfunctions for both degenerate levels of the Hückel Hamiltonian of benzene.

1.2.3 SAFs, SALCs and Wavefunctions

Let us now analyse in a discursive way the examples discussed so far, to better understand what symmetry can or cannot say. In the example of the MOs of the homo-nuclear diatomic molecule (Figure 4.9) it is evident that, if we limit ourselves to a basis of two 1s AOs only, the symmetry gives us directly the MOs and the QM calculations are used only for obtaining the energy values. A similar result (with $2p_z$ AOs) is valid for the π MOs of benzene (see Figures 4.11 and 4.12). The fundamental characteristic of the two examples is that the AOs to be combined are *all equivalent to each other by symmetry* (same atoms in symmetry equivalent positions and same type of AO).

In the case of the π MOs of butadiene (see Figure 4.10), it is evident that the relative contributions of two carbon atoms placed in positions that are equivalent by symmetry (carbons 1 and 4, or 2 and 3) is consistent with the symmetry. This is not true for the relative contributions of atoms that are not equivalent by symmetry, carbon 1 (or 4) on one side and 2 (or 3) on the other. Even more evident is the case of the normal modes of the water molecule (right of Figure 4.10) in which the atoms (O and H) are also *chemically* different: it is obvious that the symmetry determines how the motions of the two Hs combine with each other, but it is even more obvious that for the combination of the motions of the Hs and the O, other parameters must be taken into account, for example the masses of the atoms. Similarly, it is evident that symmetry cannot directly give MOs when combining AOs of the same atom but of different types, *e.g.* 1s with 2s, or with 2p.

It is important to emphasize that these results do not depend on the details of the LCAO method used, *i.e.* on the Hückel approximations. For benzene π MOs, for example, in the Hückel calculation the resonance integrals and the overlap integrals between non-adjacent AOs are neglected. *With these assumptions*, the energies of the four levels are $\alpha-2|\beta|, \alpha-|\beta|, \alpha+|\beta|, \alpha+2|\beta|$, where $\alpha<0$ is the Coulomb integral of a carbon AO and $\beta<0$ is the resonance integral between the AOs of two adjacent C atoms. If, for example, we also considered the resonance integrals between second neighbouring carbons, or even beyond, the energies would be different but the coefficients of the AOs within the MOs of the various symmetries would remain the same, being an expression of the symmetry. Furthermore, if we used an LCAO method that allows mixing with other AOs of the C atoms, for example, the np_z AOs ($n=3, 4, \ldots$) we would see that the relative coefficients between the C-$2p_z$ (and between the $3p_z$) would be the same as in the previous example.

In general, what symmetry does is

a. recognize whether some assigned functions behave in a consistent way with symmetry, that is, whether they belong to a symmetry species of the group (if so, they are called **SAF** - *symmetry adapted functions*) and
b. derive SAF starting from other functions that are not already adapted to symmetry, using appropriate linear combinations (and then we speak of **SALCs** - *symmetry adapted linear combinations*).

As we will see systematically later (Sections 2.2 and 6.2), *group theory provides the tools to recognize and classify SAFs and to construct SALCs.*

Then, the examples show that the SAFs *can* themselves be solutions of the equations of motion, a result that occurs only in the cases (very rare or due to a drastically simplified treatment) in which the functions to combine (the AOs or the displacements of the various atoms) are all equivalent to each other by symmetry (*i.e.* they are all transformed into one another by symmetry operations). In general, however, the SAFs must be combined with other SAFs to provide the MOs or the normal modes of vibration. In this regard, the fundamental result from group theory (see Section 6.2 and the proof in Section 7.4) is that:

> SAFs of different symmetry species are orthogonal to each other.

and that (see Sections 6.1 and 6.2):

> each solution of the equations of motion belongs to a specific symmetry species and is obtained by combining only *SAFs of that species.*

1.3 Symmetry and Degeneracy

Let us now try to understand why the eigenfunctions of the Hamiltonian can also be eigenfunctions of one or more symmetry operators, and – when they are –why they have certain eigenvalues of the symmetry operators and not others and finally why in some cases they are eigenfunctions of *all the symmetry operators of the group,* but in other cases they are not, in the sense that they are eigenfunctions of *only some symmetry operators.*

The essential aspect is the **commutation** of the symmetry operations between themselves and with the Hamiltonian operator. To understand the reasons and consequences we start from a result already seen at point 3.3.2 of the chapter pertinent to the basic aspects of QM (Chapter 1):

> **Every symmetry operator commutes with the Hamiltonian operator.**

Going from a single symmetry operation to a **group** of symmetry operations, we are faced with two completely different situations that require different treatments with different complexity. For example, we can compare the group D_{6h} (see the π MOs of benzene, Figures 4.11 and 4.12, with two degenerate levels) with the group C_{2v} seen with the example of the MOs π of the butadiene molecule (Figure 4.10) without degenerate levels. The essential aspect is that C_{2v} (Table 4.1) is commutative, but D_{6h} (Table 4.4) is not.

1.3.1 Commutative Groups

The simplest case is that of commutative symmetry groups (for example, C_{2v}) in which the Hamiltonian commutes with each symmetry operator as seen above, while the symmetry operators all commute with each other because the group is commutative. Based on the rules of QM concerning the commutation of operators, it is therefore possible to find a set of eigenfunctions of the Hamiltonian that are also eigenfunctions of each symmetry operator, as seen for the examples of Figure 4.10. Given an eigenfunction of the Hamiltonian, all the symmetry operations transform it essentially into itself, up to a numerical constant which is the eigenvalue of the symmetry operation. It is usually said that *there cannot be degeneracy*. However, what should be said more precisely is that - due to commutativity - each symmetry operation transforms a certain eigenfunction in a way that is completely independent of the others and leaves it essentially unchanged. Then, *there is no reason derived from symmetry why linearly independent eigenfunctions should have the same energy*. Therefore, the result is that

> for **commutative** groups one expects that there is **no degeneracy** of the Hamiltonian.

In other words: *degeneracy is not forbidden*, but since it is not due to symmetry, it is expected to occur by chance and rarely. When it occurs, it is then called *accidental degeneracy*. The following point 1.3.3 reports some important clarifications on accidental degeneracy.

1.3.2 Non-commutative Groups

Let us now go to the case where there are two or more symmetry operators that *do not commute* with each other, as for example happens for the D_{6h} group (see Table 4.4). Let us consider a specific eigenfunction of the Hamiltonian (φ) that is also an eigenfunction of a symmetry operator $\left(\hat{R}\right)$: $\hat{H}\varphi = E_\varphi \cdot \varphi$; $\hat{R}\varphi = r_\varphi \cdot \varphi$. Let us analyse the effect of the other symmetry operators ($\hat{S}$) on φ. We face two possibilities:

1. For *all* operators $\hat{S}$, it happens that φ is also an eigenfunction of $\hat{S}$. This is what is found for the first and last energy levels in Figure 4.11.
2. For some operator ($\hat{S}$) it happens, instead, that φ is *not* an eigenfunction. This is what is found for the second and third energy levels of Figure 4.11 and shown in detail by Figure 4.12c. The meaning is that, by applying this operator to φ, we obtain an *intrinsically different* function, in the sense that it is not a simple multiple of the starting function: technically, we say that it is *linearly independent* of φ. However, also $\chi = \hat{S}\varphi$ must be an eigenfunction of the Hamiltonian with the same eigenvalue because also $\hat{S}$ commutes with the Hamiltonian. In fact: $\hat{H}\chi = \hat{H}\left(\hat{S}\varphi\right) = \hat{S}\left(\hat{H}\varphi\right)$ so that $\hat{H}\chi = \hat{S}\left(E_\varphi \cdot \varphi\right) = E_\varphi \cdot \hat{S}\varphi = E_\varphi \cdot \chi$. So, we say that we have **degeneracy due to symmetry**.

Case (1) makes it clear that *degeneracy is not the rule*: it is certainly possible to have non-degenerate eigenvalues of the Hamiltonian even with non-commutative symmetries. Indeed, this is a relatively frequent case, but not an exclusive one.

If, instead, we find an eigenfunction χ which is linearly *independent* of φ, the procedure for case (2) can be repeated to verify whether a third linearly independent function possibly exists, and so on.

In conclusion, an n-times degenerate level of the Hamiltonian defines (provided it is not an accidental degeneracy) a subspace of eigenfunctions of dimension n, and we say that the symmetry species of these eigenfunctions is n-dimensional.

By acting with the different symmetry operators on a single eigenfunction of that space, we obtain a whole set of n linearly independent eigenfunctions relative to the assigned level (see Figure 4.12). This basis is not uniquely specified but must be selected and the choice determines which basis function is (or not) an eigenfunction of a given symmetry operator.

Group theory also allows us to know the *possible degeneracies*. In general, for molecular symmetries, the following are possible: (a) only degeneracy 1, (b) degeneracies 1 and 2 (this is a frequent case), (c) degeneracies 1, 2 and 3, for tetrahedral or octahedral molecules, (d) icosahedral molecules, which can also have degeneracies 4 and 5 and finally (e) linear molecules and systems with spherical symmetry (atoms), for which degeneracies 1, 3, 5, 7, … (all odd numbers) are possible.

The difference between commutative and non-commutative groups also concerns the *number of symmetry species* in the group. Anticipating once again the result (derived in Section 5.4.4.1).

- Commutative groups have as many symmetry species as there are elements in the group; each symmetry species is 1D.
- Non-commutative groups also have symmetry species with dimensions>1 and the number of elements of the group is *greater* than a number of symmetry species and equal to the *sum of the squares of the dimensions* of the symmetry species (see Section 5.4.3 and Eq. 5.22).

1.3.3 Accidental Degeneracy

In front of this simple situation, there is an important warning: the degeneracy found in solving a problem is sometimes greater than that predicted by symmetry. For example, sometimes degeneracy is found even with commutative groups. As we have seen, the situation is not forbidden, because nothing *requires* that eigenfunctions of different symmetry species have different eigenvalues. We then say that there is *accidental degeneracy*, in the sense that it is not due to symmetry. In fact, however, true *accidental* degeneracy is very rare: more frequently one is faced with a degeneracy that is actually due to a form of *additional symmetry*. There are various reasons for this additional symmetry:

a. (trivially) it has been intentionally neglected to simplify a treatment, or
b. it has not been detected because it has a completely different nature with respect to other symmetries and can be difficult to recognize (we then speak of a *hidden symmetry*), or
c. it is the consequence of a specific approximate calculation method that somehow adds a symmetry that in reality is not present in the system.

An example of hidden symmetry (case b) that is often ignored is the so-called *time reversal* symmetry as discussed in detail in Section 2.1.3. There is another curious and important example, which we will not go into here because we do not have a sufficiently simple treatment. This is the symmetry that characterizes a central field of forces where the dependence on the radial coordinate is of the form $1/r$: it is the symmetry underlying the degeneracy between AOs with different values of l but equal values of n of one-electron atoms (hydrogen-like).

As for case (c), a curious example is the calculation with the HMO method of the electronic structure of Buckminsterfullerene. Buckminsterfullerene, the C_{60} fullerene, has an alternating structure of rings of five and six C atoms closed to form a spheroid that impressively resembles a football ball. The symmetry group is icosahedral (I) and the possible degeneracies are 1, 3, 4 and 5. An HMO calculation set up in analogy with that for conjugated organic molecules with the usual assumptions (same Coulomb integral for all Cs, same resonance integral between adjacent Cs) gives energy levels with degeneracies 1, 3, 4 and 5 but for a bonding level with an unexpected degeneracy=9. Upon re-examination of the calculation, however, it is recognized that the π interactions must be classified into two categories, depending on whether they are between two atoms in a pentagon or a hexagon. Clearly, it is not correct to assign the same resonance integral to the two π interactions, since the C–C distances are different. If we repeat the calculations using a parameterization with two different values for the resonance integral, the 9-times degenerate level splits into two levels with degeneracies 4 and 5, in agreement with symmetry.

1.4 Symmetry Ascent and Symmetry Descent

A problem that arises very frequently is the comparison between two situations that are slightly different in symmetry, in the sense that one situation is described by a group with more elements (and therefore operations) of symmetry than the other. Typical cases are:

a. a molecule, or a crystal structure, which presents two forms, one more and the other less symmetric, for example: crystalline polymorphs of the same substance, or
b. addition or substitution reactions that lead to a less "symmetric" product, for example: the octahedral complexes MA_6 and MA_4B_2, or methane and chloromethane, or
c. a reduction of symmetry of the surroundings of a chemical species: a very well-known example is that of an ion in a crystal field; another example is a crystal in an axial mechanical deformation field.

The third case includes various forms of perturbation due to external fields, both static (*e.g.* Stark and Zeeman effects) and time-varying, and in particular the interaction with electromagnetic radiation.

An example of a relationship that, on the contrary, does not fall into this kind of problems, is an AL_5 complex, in which the L ligands can be in square pyramidal or trigonal bipyramidal coordination, with principal axes C_4 and C_3, respectively. From a physical point of view, it is worth remembering the famous

Curie Principle: For a system subjected to perturbation, the symmetry operations are those common to the unperturbed system and to the perturbation.

Generally speaking, we can say that systems with higher symmetry, that is with a greater number of symmetry operations, show greater degeneracy. When the description of a system is modified by introducing an additional energy contribution (a perturbation, in technical terms) characterized by lower symmetry, a reduction in the degeneracy of the levels *can* occur. In other words, it can happen that an unperturbed degenerate level is split into two or more sub-levels. The topic is discussed in detail in various points of Section 6.

Furthermore, groups that are particularly rich in symmetry operations are often difficult to deal with even with the tools of group theory. In these cases, then, one can obtain the solution of various problems by working with a simpler group instead of the theoretically correct but more complicated group (for an example, see Section 6.2.1.6).

1.5 Spectroscopic Selection Rules

Before concluding this introductory part, it is worth noting that the search, construction and use of functions that are suitable for the symmetry of a particular system do not only concern problems of molecular orbitals and normal modes of vibration or, more generally, the solutions of equations of motion in classical or quantum mechanics.

To study, for example, the interaction of electromagnetic radiation with an atomic, molecular or crystalline system, it is important to know the type of symmetry of some simple functions, such as x, y, z, or x^2, or y^2z, etc.... because x, y, z are the components of the electric dipole moment operator, which represents the main interaction between electromagnetic radiation and matter, while the higher degree functions are components of other interactions, such as magnetic dipole, electric quadrupole, magnetic quadrupole, ...

As a first example, let us consider a particularly simple case, the radiative electric dipole transition from one AO (i) to another (f) in a hydrogen atom, which has inversion symmetry ($x, y, z \rightarrow -x, -y, -z$). The probability is proportional to the following integral extended to the whole 3D space:

$$\iiint \psi_i(\vec{r}) \cdot \vec{r} \cdot \psi_f(\vec{r}) \cdot d\vec{r}$$

If the states i and f have the same eigenvalue (+1 or −1) for the inversion the product of the two functions does not change sign. Since the vector $\vec{r}$ has eigenvalue −1 for the inversion, the entire integrand has eigenvalue −1. In conclusion, the integral vanishes because it is given by the sum of two integrals (on opposite regions of the 3D space) equal in modulus but opposite in sign.

If, however, the states i, f have opposite behaviours with respect to the inversion (as, for example, for a transition from 1s to 2p), the integrand does not change sign and can be different from zero. This conclusion is not exclusive to atomic systems, and we can therefore formulate the

Laporte's rule: in systems with inversion symmetry ($\vec{r} \rightarrow -\vec{r}$) electric dipole transitions are allowed only between states of different parity.

For atomic systems, a more detailed analysis considering all symmetries, would eventually lead to the well-known selection rules $\Delta l = \pm 1$, which includes Laporte's rule, but is more restrictive.

As a second example, let's take again the C_{2v} group again and orient the Cartesian axes of Figure 4.10 so that the y-axis is aligned with C_2, σ is the xy-plane, and σ' is the yz-plane.

The three unit-vectors along x, y and z are SAF, because they are eigenfunctions of the symmetry operators with the eigenvalues indicated in Table 4.6. Making reference to Figure 4.10, we see that three of the four symmetry species are now found, while no vector in the 3D space belongs to the species with eigenvalues (1, −1, −1). We anticipate the fundamental result (see Table 4.8) that the C_{2v} group has only the four symmetry species reported in the examples so far discussed.

If the states (i, f) have the same symmetry, the products of the eigenvalues are always equal to 1 and therefore the product of i and f belongs to the totally symmetric species. The integrand then has the same eigenvalue as one of the functions x, y or z and is invariant. In particular, it does not vanish if the z component of the dipole moment is specifically considered. In conclusion, for the C_{2v} group, the electric dipole transition between states with the same symmetry is allowed.

TABLE 4.6
Eigenvalues of the Unit-Vectors of Ordinary 3D Space of Group C_{2v}

	C_2	$\sigma(xy)$	$\sigma'(yz)$
x	−1	1	−1
y	−1	−1	1
z	1	1	1
—	**1**	**−1**	**−1**

See also Table 4.21 for the conventional labels of the species.

Similar considerations can be made for other symmetries of states i and f and of the dipole components, as well as for other radiative transitions (magnetic dipole absorption, quadrupole absorption, Raman scattering, ...) which require studying functions other than x, y or z.

1.6 Summary

- The symmetry operations of a system are related to each other according to the algebraic structure of **group**, which is characterized by the properties of *closure* (the sequential execution of two symmetry operations is also a symmetry operation of the group), *associativity* and *existence* of a unique neutral element (*identity*) and of a unique *inverse* of each operation.
- The solutions of the equations of motion of a system, both in MC and in QM, have characteristics consistent with the symmetry group. This is a consequence of the commutation of symmetry operators with the Hamiltonian (see Section 1.3). In general, **several symmetry species are possible**, that is, various behaviours with respect to all symmetry operations.
- Whatever the symmetry group, among the symmetry species there is always the **totally symmetric species**, which shows invariance under the effect of *all* symmetry operations.
- The symmetry species of finite groups, that is, groups consisting of a finite number of symmetry operations, are finite in **number** (a result that will be proved later, Section 5.4.3).

The first and most important classification of symmetry species is based on their *dimension*. There are two cases.

- If an eigenfunction of the Hamiltonian is also an eigenfunction of *all* symmetry operations (each symmetry operation with its own eigenvalue), the symmetry species is **one-dimensional**.

- **Two-dimensional** (or **three-dimensional**, etc....) symmetry species **are also possible**. Then the eigenfunctions of the Hamiltonian are **not** eigenfunctions of **all symmetry operations** of the group, but only of some. As a result of some symmetry operation, an eigenfunction is transformed into the linear combination of two (or more) eigenfunctions of the same eigenvalue of the energy, which **therefore** turns out to be **degenerate** two (or more) times.

Commutation, degeneracy and dimensionality of symmetry species are strictly connected.

- Commutative groups have only one-dimensional (1D) symmetry species. Then:
  - different symmetry species have different sets of eigenvalues for all symmetry operations.
  - The eigenvalues are (all) numbers of modulus = 1.
  - There is no degeneracy of the energy levels, except for accidental degeneracies.
  - The number of symmetry species is equal to the order of the group.
- **Non-commutative groups,** in addition to one-dimensional symmetry species, also have two-dimensional (2D) or three-dimensional (3D) symmetry species, exceptionally of higher dimension. Then:
  - There is degeneracy due to symmetry.
  - The number of symmetry species is less than the order of the group.

2 THE SIMPLE RESULTS

Many chemical applications of group theory require theoretical results that are often presented without justification, because the derivations are difficult and very abstract. Aim of all this chapter is to show the foundation of these results and to discuss their relevance. Before introducing the pertinent theory, however, it is worth presenting a simpler approach, which can show the capabilities of symmetry-based methods, even if it *cannot* be applied to *all symmetry groups*. This simplified approach is based on the property that, as far as *commutative groups* are considered, the *eigenfunctions of the Hamiltonian* are also *eigenfunctions of all symmetry operators*. Then, each energy level is non-degenerate, each eigenfunction is characterized by a particular *set* of eigenvalues for the symmetry operators, and the set specifies the *symmetry species* of the eigenfunction and energy level. Moreover, obtaining *all possible eigenvalues of* a *symmetry operator* is not a difficult task: this will be done explicitly in Section 2.1.1. Finally, if the relation $C = AB$ holds for three symmetry operations, then, for *each symmetry species*, the eigenvalue associated to C cannot take on all its possible values, but only the one that is the product of the eigenvalues of A and B *for that species*. Using these considerations, we can obtain *all symmetry species* of a group and build a *complete table* of the eigenvalues of all symmetry operations for each species.

The complete table of the eigenvalues of the symmetry operators *of a commutative group* is usually named **character table**. The character table reports the eigenvalues of each symmetry operator (*i.e.*, of each column of the table) for each symmetry species (*i.e.*, for each row of the table). This is what we have seen with the previous examples. There is a character table also for *non-commutative* groups: the entries of the table then contain numbers that are known as characters and (generally) are *not* eigenvalues. Characters are introduced in section 5.3 and character tables are fully discussed in Section 5.5. The character table of a group contains all the information needed to build symmetry-adapted functions for each species of the group, both for commutative and non-commutative groups. We here just state the results (see Section 2.2 and Eq. (2.4)), because the derivation requires more powerfull theoretical tools (see Section 6.2).

2.1 Finding the Symmetry Species of Commutative Groups

A warning: for all groups that are relevant to chemistry, the character tables are reported in detail in technical publications and web sites (see endnote 4), and in several textbooks. Therefore, this part (2.1) could be skipped. However, the reasoning is particularly simple and easy to follow, and it is useful for a better understanding of the technicalities of group theory and for becoming acquainted with basic terminology and applications.

2.1.1 Eigenvalues of Symmetry Operations

The preliminary result here required is the set of all possible eigenvalues of a single symmetry operator. Let us start with a particular set of symmetries, the **self-inverse operations**, that is, operations that are equal to their inverse ($\hat{R}^{-1} = \hat{R}$). Then, repeating their application, we *get back to the top*. Examples are easily found. Inversion, reflection, rotation by 180°, coordinate exchange between two identical particles are all self-inverse operations.

Applying *twice* the transformation $\hat{R}$ on an eigenfunction of $\hat{R}$, we obtain the same function. Thus, the eigenvalue (r) of $\hat{R}$ is one of the two square roots of unity ($r = 1$ or $r = -1$).

$$\text{If } \hat{R}\psi(q,t) = r \cdot \psi(q,t) \quad \text{and} \quad \hat{R} = \hat{R}^{-1}\,(\text{self-inverse})$$
$$\text{then } r = \begin{cases} 1 \\ -1 \end{cases} \tag{2.1}$$

$$\text{In fact: } \hat{R}^2\psi(q,t) = \hat{R}\left[\hat{R}\psi(q,t)\right] = \hat{R}\left[r \cdot \psi(q,t)\right]$$
$$= r^2 \cdot \psi(q,t) = \psi(q,t)$$

Note: the eigenvalues of a symmetry operator are not necessarily real numbers, as a symmetry operator does not correspond to an observable, and, therefore, is not necessarily Hermitian. Here, the eigenvalues are real numbers because, also in the field of complex numbers, the square roots of unity are real numbers, and are only +1 and −1. Note also that the eigenvalues are =1 in modulus, in agreement with the fact that all the symmetry operations are unitary, as already seen in Section 1.1. and discussed in detail in Section 5.1.4.

The result is easily extended to operators that give the identity when applied not 2 but 3, 4, … n times. Examples are the rotation by 1/3, 1/4, …, $1/n$, … of a full turn. The possible eigenvalues are clearly the n-th roots of unity. We remind that *in a finite group* each point symmetry operation gives the identity if repeated an adequate number (n) of times: n is called the order of the element (see Section 1.1.3.6). As said (see Figure 1.1 of Chapter 1), in the complex field there are n distinct roots of unity that, in the Argand plane, are at the vertices of a regular polygon with n sides, centre in the origin, and radius = 1, starting from the real number 1. By enumeration:

$$n = 3 \Rightarrow \text{roots}: \exp(0) = 1, \quad \exp(2\pi/3), \quad \exp(4\pi/3);$$

$$n = 4 \Rightarrow \text{roots}: \exp(0) = 1, \quad \exp(2\pi/4) = i,$$

$$\exp(4\pi/4) = -1, \quad \exp(6\pi/4) = -i;$$

$$n = \cdots \Rightarrow \text{roots}: \exp(0) = 1, \quad \exp(2\pi/n),$$

$$\exp(4\pi/n), \ldots, \quad \exp[2(n-1)\pi/n].$$

These are the possible eigenvalues of the rotation of $1/n$ of full turn. The eigenvalues of multiple rotations [*i.e.*, $2/n$, …, $(n-1)/n$ of full turn] are the powers of these complex numbers. For a three-fold rotation, *e.g.*, the eigenvalues are $\exp(2\pi i/3)$ and $\exp(4\pi i/3) = \exp(-2\pi i/3)$ for C_3, $\exp(4\pi i/3) = \exp(-2\pi i/3)$ and $\exp(8\pi i/3) = \exp(2\pi i/3)$ for $(C_3)^2 = C_{-3}$, and finally only 1 for $(C_3)^3 = E$. If we call the complex number $\exp(2\pi i/3)$ with ε, and noting that $\varepsilon^2 = \varepsilon^*$, eigenvalues for C_3, $(C_3)^2$ and $(C_3)^3 = E$ are (ε, or ε^*, or 1), (ε^*, or ε, or 1), and 1 (always).

2.1.2 Groups with Real Eigenvalues

To understand how to find the symmetry species let us now consider the commutative groups and real eigenvalues; the case of complex eigenvalues will be considered later (see Section 2.1.4). The symbols identifying the symmetry species are explained in Section 5.5.1.

2.1.2.1 Groups C_i, C_s and C_2

Let us label A the identity, while B is the reflection (for group C_s) or inversion (for C_i), or twofold rotation (for C_2). The eigenvalue of identity is always +1. The possible eigenvalues for the other symmetry operation are +1 or −1, as it is self-inverse (see Eq. (2.1)). We therefore have only two possible pairs of eigenvalues (Table 4.7):

TABLE 4.7
Eigenvalues of Symmetry Operations for Point Groups C_i, C_s and C_2

	A	B
A_g	1	1
A_u	1	−1

The symbols of the symmetry species are those of group C_i; for group C_s the symbols are A' and A"; for group C_2 they are A and B.

TABLE 4.8
Eigenvalues and Symmetry Species for Group C_{2v}, see also Table 4.21

C_{2v}	E	C_2	σ	σ'
A_1	1	1	1	1
A_2	1	1	−1	−1
B_1	1	−1	1	−1
B_2	1	−1	−1	1

In Figure 4.10, the eigenvalues of the symmetry species are respectively marked with white, middle grey, light grey and dark grey background colours.

As already said without proof in Section 1.6, the number of symmetry species of a commutative group is equal to the number of symmetry operations and here we find just two **symmetry species**, in agreement with the fact that we have two symmetry operations The totally symmetric species is labelled A_g (or A' or A). The symmetry species A_u describes the eigenfunctions that change in sign under the inversion symmetry. These symmetry species agree with what we have already seen for the eigenfunctions of the Hamiltonian of the harmonic oscillator (see Section 2.1.1 and Figure 4.8): we have even and odd eigenfunctions.

2.1.2.2 Group C_{2v}

The non-trivial symmetry operations are two "vertical" reflections σ and σ' and the twofold rotation C_2 at their intersection (see Figure 4.1 and Table 4.2). Possible eigenvalues are +1 and −1 and the group is commutative. The symmetry species are obtained by considering the possible eigenvalues for the symmetry operations other than E, and the constraints due to their combinations. We have just four possibilities because for each reflection, σ and σ', the eigenvalue can be +1 or −1. Then, the eigenvalue for the rotation is specified by the product of these eigenvalues, and the eigenvalue of identity is always +1 (Table 4.8).

We now see that the π MOs of "cis" butadiene (Figure 4.10, left part) belong to the A_2 and B_2 species and the normal vibrational modes of water (Figure 4.10, right part) belong to the A_1 and B_1 species.

2.1.2.3 Groups D_2 and C_{2h}

The group D_2 contains three twofold rotations around three mutually orthogonal axes that are labelled as $C_2(x)$, $C_2(y)$ and $C_2(z)$. A molecule with this symmetry is biphenyl (the two benzene rings are *not* in the same plane). The group C_{2h} contains a twofold rotation, inversion, and a reflection with respect to a plane orthogonal to the rotation axis and therefore labelled as horizontal. A molecule with C_{2h} symmetry is the so called "trans" butadiene. D_2 and C_{2h} are isomorphous to C_{2v}. For the multiplication tables and symmetry species, see Exercise 4.5. The fact that three symmetry groups (C_{2v}, D_2 and C_{2h}) correspond to the same abstract group clearly shows that the character tables are due to the algebraic properties (relations between group elements) and are not *directly* due to geometry (nature or orientation of the symmetry elements).

2.1.2.4 Group D_{2h}

This group contains eight symmetry operations: three twofold rotations around the three Cartesian axes x, y, z, three reflections with respect to the xy, yz, zx planes, the inversion symmetry, and identity. This symmetry group is very common for molecules. Examples are ethylene, diborane B_2H_6, naphthalene, and anthracene. Table 4.9 can be easily obtained considering three different binary choices, so that we must fill in $2^3=8$ different rows using +1 or −1 for $C_2(x)$, $C_2(y)$ and $\sigma(xy)$. We also have +1 for the identity, while the eigenvalues of the other symmetry operations are derived from the relations $C_2(z)=C_2(y)\bullet C_2(x)$; $\sigma(yz)=\sigma(xy)\bullet C_2(x)$ and $\sigma(zx)=\sigma(xy)\bullet C_2(y)$.

2.1.3 Time Reversal Symmetry

With group D_{2h} we have completed the list of *commutative point groups* having *only real eigenvalues* for all symmetry operations. Following the previous considerations, we should say that we must have only non-degenerate energy levels also for commutative groups with *complex eigenvalues*. However, things are not so simple, because of a "hidden symmetry" (known as *time reversal symmetry*) which mimics an accidental degeneracy (see Section 1.3.3).

The *time reversal* symmetry is a widely discussed topic, also in science dissemination, and has several effects. Let us neglect the macroscopic effects, for example in thermodynamics, and consider here only the effects on microscopic mechanics (classical, but mainly quantum mechanics).

TABLE 4.9
Eigenvalues and Symmetry Species for Group D_{2h}

	E	$C_2(z)$	$C_2(y)$	$C_2(x)$	i	$\sigma(xy)$	$\sigma(yz)$	$\sigma(zx)$
A_g	1	1	1	1	1	1	1	1
B_{1g}	1	1	−1	−1	1	1	−1	−1
B_{2g}	1	1	1	−1	1	−1	1	−1
B_{3g}	1	−1	−1	1	1	−1	−1	1
A_u	1	1	1	1	−1	−1	−1	−1
B_{1u}	1	1	−1	−1	−1	−1	1	1
B_{2u}	1	−1	1	−1	−1	1	−1	1
B_{3u}	1	−1	−1	1	−1	1	1	−1

For discussing the time reversal transformation, we must distinguish two aspects. A first point is the evolution with time of the state of a system. In classical mechanics, time reversal (*i.e.* switching from t to $-t$) means that the trajectory is reversed, and the system goes backwards. A similar consideration holds in quantum mechanics for the time evolution of a stationary state $\psi(r,t)=e^{-iEt/h}\cdot\phi(r)$. However, in addition to this *global* meaning, we must also consider a *punctual* meaning, as time reversal involves *reversing the sign of velocities and moments* (both linear and angular moments) *including spins*. This second feature entangles that switching from t to $-t$, *must go with* the conjugation of the wave function, a nontrivial result that we leave to specialized textbooks for the derivation. A simple *justification* considers that the conjugate variable of time is energy. Thus, reversing time would reverse the energy scale and therefore would prevent the existence of a *ground state*. Passing to complex conjugates, $\exp(-iEt)) \rightarrow \{\exp[-iE(-t)]\}^*$, that is, keeping $\exp(-iEt))$ instead of passing from $\exp(-iEt))$ to $\exp[-iE(-t)]=\exp[iE(t)]$ solves the problem.

Now, let us write the time-dependent Schrödinger equation $\hat{H}\psi(r,t)=i\cdot\hbar\cdot\frac{\partial\psi(r,t)}{\partial t}$, and then apply the two transformations: conjugation ($\psi\rightarrow\psi^*$), and time reversal ($\psi(r,t)\rightarrow\psi(r,-t)$). By using *the additional constraint of real Hamiltonian*, we have that the function $\psi^*(r,-t)$ so obtained is a solution of the same time-dependent Schrödinger equation:

$$\hat{H}\psi^*(r,-t)=i\cdot\hbar\cdot\frac{\partial\psi^*(r,-t)}{\partial t} \tag{2.2}$$

This is the form of the time reversal symmetry as typically used in QM. Furthermore, if the Hamiltonian does not include time we obtain (see Exercise 4.6):

$$\hat{H}\phi^*(r)=E\cdot\phi^*(r) \tag{2.3}$$

This means that *the complex conjugate* of *an eigenfunction of a real Hamiltonian, is an eigenfunction as well and has the same eigenvalue.*

This constraint about the Hamiltonian does not apply to magnetism.

2.1.4 Groups with Complex Eigenvalues

The *commutative* point groups with *complex eigenvalues* are C_3, C_4, C_5, C_6, C_7, …, C_{3h}, C_{4h}, C_{5h}, C_{6h}, C_{7h}, …, and finally S_4, S_6, S_8, ….

Let us consider here the groups C_n, and in particular C_7. Possible eigenvalues for a rotation of one-seventh of full turn are $\exp(2\pi i/7)$, $\exp(4\pi i/7)$, $\exp(6\pi i/7)$, $\exp(8\pi i/7)$, $\exp(10\pi i/7)$, $\exp(12\pi i/7)$, and $\exp(14\pi i/7)=1$. Letting $\varepsilon=\exp(2\pi i/7)$, the possible eigenvalues are shown in (Table 4.10):

TABLE 4.10
Eigenvalues of the Symmetry Operators for Group C_7

	E	C_7	$(C_7)^2$	$(C_7)^3$	$(C_7)^4$	$(C_7)^5$	$(C_7)^6$
1	1	1	1	1	1	1	1
2	1	ε	ε^2	ε^3	ε^4	ε^5	ε^6
3	1	ε^2	ε^4	ε^6	ε	ε^3	ε^5
4	1	ε^3	ε^6	ε^2	ε^5	ε	ε^4
5	1	ε^4	ε	ε^5	ε^2	ε^6	ε^3
6	1	ε^5	ε^3	ε	ε^6	ε^4	ε^2
7	1	ε^6	ε^5	ε^4	ε^3	ε^2	ε

The first column is an index for identifying the species in Table 4.11.

TABLE 4.11
Table 4.10 Is Here Rewritten by Gathering the Species with Conjugate Eigenvalues

		E	C_7	$(C_7)^2$	$(C_7)^3$	$(C_7)^4$	$(C_7)^5$	$(C_7)^6$
1	A	1	1	1	1	1	1	1
2	E_1	1	ε	ε^2	ε^3	$(\varepsilon^3)^*$	$(\varepsilon^2)^*$	ε^*
7		1	ε^*	$(\varepsilon^2)^*$	$(\varepsilon^3)^*$	ε^3	ε^2	ε
3	E_2	1	ε^2	$(\varepsilon^3)^*$	ε^*	ε	ε^3	$(\varepsilon^2)^*$
6		1	$(\varepsilon^2)^*$	ε^3	ε	ε^*	$(\varepsilon^3)^*$	ε^2
4	E_3	1	ε^3	ε^*	ε^2	$(\varepsilon^2)^*$	ε	$(\varepsilon^3)^*$
5		1	$(\varepsilon^3)^*$	ε	$(\varepsilon^2)^*$	ε^2	ε^*	ε^3

Now, let us rewrite the table using the powers of ε and ε^* up to the third power, as $\varepsilon=\exp(2\pi i/7)$ is the complex conjugate of $\varepsilon^6=\exp(12\pi i/7)=\exp(-2\pi i/7)=\varepsilon^*$ and $(\varepsilon^n)^*=(\varepsilon^*)^n$. Let us also reorder all the lines, to highlight the fact that eigenvalues in lines 2, 3, 4 are the conjugates of those in lines 7, 6, 5, respectively (see Table 4.11). As seen in Section 2.1.3 the effect of time reversal symmetry is to *double the degeneracy of symmetry species having complex eigenvalues* because *the complex conjugate of an eigenfunction is also an eigenfunction and has the same eigenvalue* when the system is described by a ***real** Hamiltonian*:

$$\hat{H}\phi(r)=E\cdot\phi(r)\Rightarrow\hat{H}\phi^*(r)=E\cdot\phi^*(r).$$

This is the hidden symmetry

So, *for any couple of symmetry species with conjugate eigenvalues*, the time reversal symmetry assigns two 1D species to a single energy level. These *pairs of symmetry species with conjugate eigenvalues* are labelled as E_1, E_2, E_3, as it is usually done in character tables. The case $\varepsilon=1$ is the totally symmetric symmetry species. With respect to groups with only real eigenvalues (+1 or –1), the commutative groups having also complex eigenvalues *seemingly oblige* us *to use complex wave functions*. However, it is possible to use real wave functions, which are generally more convenient for discussing molecular symmetry, by using a *combined treatment* of the symmetry species of each couple with conjugate eigenvalues. Shortly, instead of using two complex conjugate coefficients ($\varepsilon=e^{i\alpha}=\cos\alpha+i\sin\alpha$ and $\varepsilon^*=e^{-i\alpha}=\cos\alpha-i\sin\alpha$), we use the sum and the difference multiplied by i (giving $\cos\alpha$ and $\sin\alpha$), that are two real numbers. The combined treatment emphasizes the degeneracy of the symmetry species with conjugate eigenvalues and therefore is preferred when counting states or dealing with the symmetry descent due to additional interactions. However, there is a feature that is anyway distinctive of commutative groups: the number of symmetry species is equal to the number of symmetry operations of the group. In non-commutative groups the number is lower: for group C_7 there are seven symmetry species *not* four as would be found counting for non-commutative groups (see Section 5.4.4.1).

However, it is *not necessary to follow the combined treatment*. It is surely possible to *ignore the time reversal symmetry*. For example, in forming the SALC, we can work using the complex eigenvalues *as if* the additional symmetry would not exist. Once the calculations have been completed, we can note that the energies of two species with conjugate eigenvalues are equal, and we take this into account in further applications. This is the preferred choice for a very important class of symmetry groups, those including **translational symmetry**, that is for the **space groups** of crystals. Reference is here made to Section 3.3.2 for the $\vec{k}\,/-\vec{k}$ degeneracy of electronic and vibrational wave functions in crystals.

2.2 Symmetry Adapted Functions (SAFs)

Let us now use symmetry for obtaining functions that belong to one or another symmetry species of a group. The goal is to form symmetry adapted linear combinations (SALCs) starting from functions or vectors that, generally, *are not* symmetry adapted, *e.g.* because they refer to a single atom in a molecule. The treatment is still restricted to commutative groups (the complete treatment is in Section 6.2). This fundamental application concerns the LCAO buildup of MOs, and the normal modes of vibration of a molecule as linear combinations of the motions of each atom, but also spectroscopic properties (absorption and Raman scattering, for instance), because the symmetry species specify the properties of the electronic or vibrational states and tell us which transitions possibly take place between which states.

For the MOs, the functions to combine are AOs, and the SALCs are linear combinations of AOs (i) of the same type *and* (ii) of *symmetry-equivalent* atoms. Each SALCs belongs to a symmetry species and gives one or more MOs that belong to the same species either by itself or combined with other SALCs of the same species but formed from AOs of a different type or other atoms.

For the normal modes of vibration, the vectors that are combined are the x, y, z components of the displacements of the atoms from the equilibrium positions. The SALC are linear combinations of symmetry-equivalent displacements. Each SALC belongs to a symmetry species and

(i) is a normal vibration mode of the whole molecule or (ii) more frequently, gives a normal mode when combined with SALCs of the same species but formed from vectors that point to a non-equivalent direction, or refer to a non-equivalent atom, or both.

The procedure starts by selecting some generalized vectors that we may call *prototype-functions*: a prototype-function is a particular AO of a particular atom, or a given Cartesian component of the displacement of a particular atom from its equilibrium position. Then, the different symmetry species (j) of the group are considered one after the other and the SALC of the j-th symmetry species formed from the ν-th prototype-function is given by:

$$\psi(j,\nu) = \sum_S a_{S,j} \cdot \hat{S}\varphi_\nu$$

or, in vector notation: (2.4)

$$\vec{v}_{j,\nu} = \sum_S a_{S,j} \cdot \hat{S}\vec{w}_\nu$$

In more detail:

- All local functions (AOs or displacement components) must be considered when selecting the prototype-functions but only one function in a set of symmetry equivalent functions is required: two functions are symmetry equivalent if a function is transformed into the other by a symmetry operation.
- Then, for each prototype function (φ_ν or $\vec{w}_\nu$) and for each symmetry species (j)
  - considering in sequence all the symmetry operations S,
  - one must add the prototype function as transformed by S ($\hat{S}\varphi_\nu$) and multiplied by the eigenvalue a_{Sj} of S in the j-th symmetry species.

Equation (2.4) is a particular case of a more general result proved later (see Section 6.2).

2.2.1 SALCs and the MO-LCAO Method

As an example, let us consider the problem of the SALCs for the π MOs of butadiene, that have already been shown (Figure 4.10) without explanation. The prototype-functions are the $2p_z$ AOs of the first and second C atom and the eigenvalues are in Table 4.8.

Figure 4.13 shows the effects of the symmetry on the $2p_z$ AO of C1, while Figure 4.14 applies the procedure of Eq. (2.4).

As Figure 4.14 shows, there is a *non-null result* only for species A_2 and B_2: for the other two species, the partial results cancel pairwise. Then, we say that the $2p_x$-C1 AO *is not a basis for the* A_1 *symmetry species* (nor for B_1). So, by combining two (symmetry equivalent) AOs ($2p_z$ of C1 and $2p_z$ of C4) we form two linearly independent combinations (any other combination of these AO is linearly dependent). Starting from a non-equivalent AO, for example the $2p_z$ of carbon 2, we obtain an independent but completely analogous result: from two symmetry equivalent AOs ($2p_z$ of carbon 2 and 3) we form 2 SALCs that belong to the same symmetry species (A_2 and B_2) as before.

Also, it is easily verified that, starting from the $2p_z$ AO of carbon 4 (C4), that is the AO that is symmetry-equivalent to the $2p_z$ of C1) gives the same results as starting from the $2p_z$ of C1, and so happens for the $2p_z$ AOs of C2 and C3.

A better understanding of the role played by symmetry is possibly obtained by a detailed calculation with MO-LCAO for this problem. Within the Hückel approximation, the simplest approach, and with the $2p_z$ AO basis:

$$\boldsymbol{H} = \begin{pmatrix} \alpha & \beta & 0 & 0 \\ \beta & \alpha & \beta & 0 \\ 0 & \beta & \alpha & \beta \\ 0 & 0 & \beta & \alpha \end{pmatrix} \quad (2.5)$$

where α is the energy of a $2p_z$-C AO, and β is the resonance integral of two $2p_z$ AO on neighbouring C atoms. Forming the SALC is a change of basis: the SALCs are obtained from the previous basis functions (the AOs) and the matrix of the transformation (see Eqs. (2.4) and (2.5)) is:

$$\boldsymbol{C} = \frac{1}{\sqrt{2}} \begin{pmatrix} 1 & 0 & 1 & 0 \\ 0 & 1 & 0 & 1 \\ 0 & 1 & 0 & -1 \\ 1 & 0 & -1 & 0 \end{pmatrix} \begin{pmatrix} 1/\sqrt{2} \text{ is for} \\ \text{normalization} \end{pmatrix} \quad (2.6)$$

In Eq. (2.6), each column of $\boldsymbol{C}$ gives a SALC, *i.e.*, the coefficients of the $2p_z$ AOs of C1, C2, C3 and C4 of the corresponding symmetry species, while the rows identify the AOs. The SALC belonging to the same symmetry species are placed side by side, in order to better show the final result (Eq. 2.7): the order is: (i) SALC B_2 formed from C1, (ii) SALC B_2 formed from C2, (iii) SALC A_2 formed from C1 and finally SALC A_2 formed from C2. The Hamiltonian matrix in the SALC basis (the *new basis*) is the similarity transformation of $\boldsymbol{H}$ (Eq. 2.5):

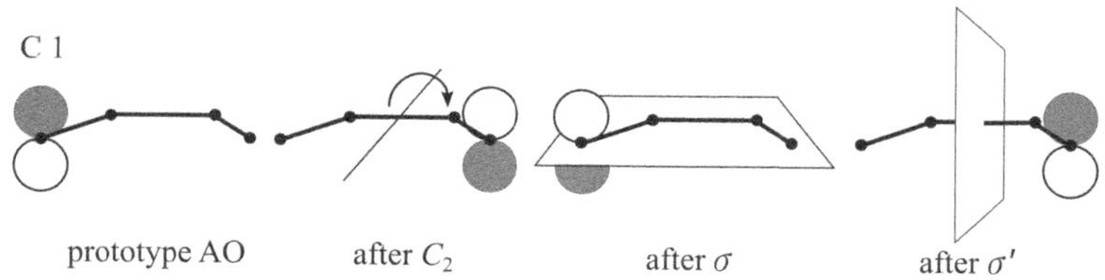

FIGURE 4.13 How the symmetry operations of C_{2v} transform the $2p_z$ AO of the first C atom (C1).

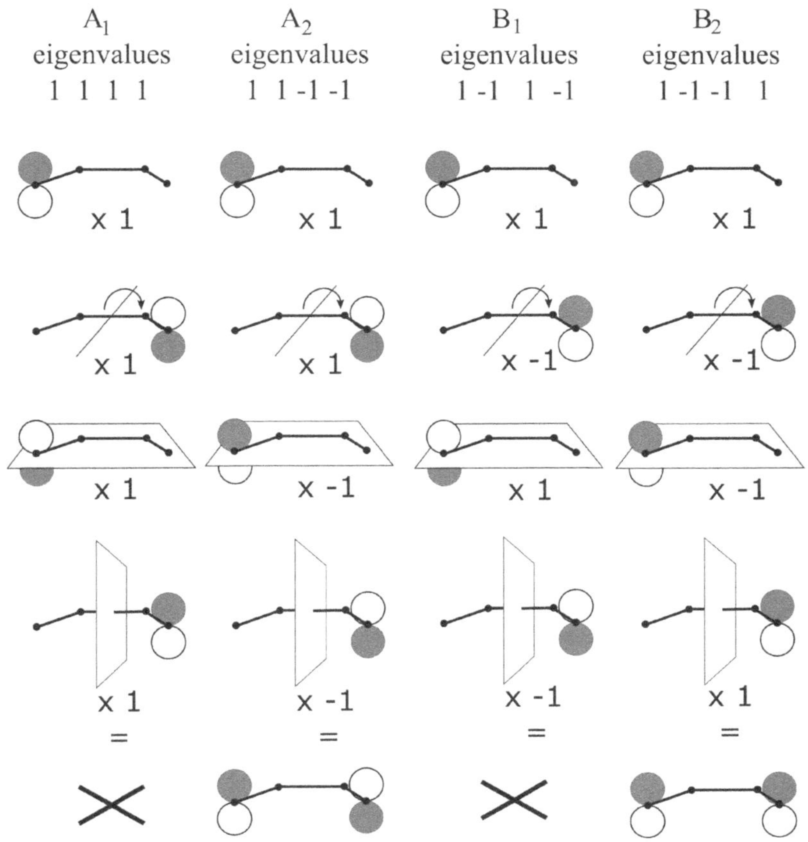

FIGURE 4.14 Use of Eq. (2.4) to form the SALCs for the four symmetry species from the $2p_z$-C1 prototype. Each column regards a symmetry species, each row regards a symmetry operation, the last row gives the sums along columns.

$$\boldsymbol{H}_{nb} = \boldsymbol{C}^{-1} \bullet \boldsymbol{H} \bullet \boldsymbol{C}.$$

$\boldsymbol{C}$ is orthogonal and therefore its inverse ($\boldsymbol{C}^{-1}$) is the transposed matrix ($\boldsymbol{C}^{T}$) so that $\boldsymbol{H}_{nb}$ is:

$$\frac{1}{\sqrt{2}}\begin{pmatrix} 1 & 0 & 1 & 0 \\ 0 & 1 & 0 & 1 \\ 0 & 1 & 0 & -1 \\ 1 & 0 & -1 & 0 \end{pmatrix}\begin{pmatrix} \alpha & \beta & 0 & 0 \\ \beta & \alpha & \beta & 0 \\ 0 & \beta & \alpha & \beta \\ 0 & 0 & \beta & \alpha \end{pmatrix}$$

$$\frac{1}{\sqrt{2}}\begin{pmatrix} 1 & 0 & 0 & 1 \\ 0 & 1 & 1 & 0 \\ 0 & 1 & 0 & -1 \\ 1 & 0 & -1 & 0 \end{pmatrix} \tag{2.7}$$

$$= \frac{1}{2}\begin{pmatrix} \alpha & \beta & 0 & 0 \\ \beta & \alpha+\beta & 0 & 0 \\ 0 & 0 & \alpha & \beta \\ 0 & 0 & \beta & \alpha-\beta \end{pmatrix}$$

So, in the SALC basis the Hamiltonian matrix is block-diagonal, with a block for each symmetry species. The elements of $\boldsymbol{H}_{nb}$ between *different symmetry species* are zero, as the SALC *are mutually orthogonal*: the result is due to symmetry. Its derivation and deeper explanations are reported in Section 6.1. For obtaining the MOs and the eigenvalues of $\boldsymbol{H}$, we **do not need** to perform the diagonalization of the whole $\boldsymbol{H}_{nb}$, but we can solve the two sub-matrixes separately.

From the top-left 2×2 matrix we have:

$$E = \frac{1}{2}\left[(2\alpha+\beta) \pm \sqrt{(2\alpha+\beta)^2 - 4(\alpha^2+\alpha\beta-\beta^2)}\right].$$

For the other sub-matrix:

$$E = \frac{1}{2}\left[(2\alpha-\beta) \pm \sqrt{5\beta^2}\right] = \alpha - \beta\frac{1 \pm \sqrt{5}}{2}$$

Thus, the eigenvalues are, in increasing energy order ($\beta < 0$):

$$(B_2)\ E = \alpha + \frac{1+\sqrt{5}}{2}\beta \simeq \alpha + 1.618\,\beta$$

$$(A_2)\ E = \alpha - \frac{1-\sqrt{5}}{2}\beta \simeq \alpha + 0.618\,\beta$$

$$(B_2)\ E = \alpha + \frac{1-\sqrt{5}}{2}\beta \simeq \alpha - 0.618\,\beta$$

$$(A_2)\ E = \alpha - \frac{1+\sqrt{5}}{2}\beta \simeq \alpha - 1.618\,\beta$$

Summarizing, for obtaining the MO we must combine SALC of the same symmetry species but coming from AOs that are NOT symmetry-equivalent to each other. For example, we combine the B_2 SALC coming from the $2p_z$-C1 with the B_2 SALC coming from the $2p_z$-C2. On the other side, symmetry has nothing to say about the relative weight of different SALCs in an MO.

2.2.2 SALCs and Vibration Modes

The second important application is about the vibrations of a polyatomic system and can be discussed within classical mechanics (CM), as illustrated in detail in part 1.3.2 of the Appendices. Briefly summarizing, two approximations are followed, the Born-Oppenheimer separation and the harmonic approximation. For a system of n point-like atoms:

$$\ddot{\boldsymbol{x}} = \boldsymbol{D}\cdot\boldsymbol{x} \tag{2.8}$$

The x_i are the Cartesian coordinates of the n atoms: $i=1$, 2, 3 for the first atom, 4, 5, 6 for the second, ..., $3n-2$, $3n-1$, $3n$ for the last and the matrix $\boldsymbol{D}=-\boldsymbol{M}^{-1}\boldsymbol{V}$ is known as the dynamic matrix of the system. $\boldsymbol{M}^{-1}$ is a diagonal matrix whose elements are the inverses of the masses: $\boldsymbol{M}^{-1} = diag\left(m_1^{-1}, m_1^{-1}, m_1^{-1}, m_2^{-1}, \ldots\right)$. $\boldsymbol{V}$ is a symmetric matrix: its elements $\boldsymbol{V}_{ij}=\boldsymbol{V}_{ji}$ are the second derivatives of the potential energy evaluated at the point corresponding to the equilibrium configuration and are known as *force constants*. For the system of second-order homogeneous ODEs (ordinary differential equations) of Eq. (2.8) the solution is in the form of a superposition of *independent* motions, which are linear combinations of the x's and are called *normal modes*:

$$\begin{pmatrix} Q_1 \\ \vdots \\ Q_{3s} \end{pmatrix} = \begin{pmatrix} c_{11} & \cdots & c_{1,3s} \\ \vdots & \ddots & \vdots \\ c_{3s,1} & \cdots & c_{3s,3s} \end{pmatrix} \cdot \begin{pmatrix} x_1 \\ \vdots \\ x_{3s} \end{pmatrix} \quad \text{that is } \boldsymbol{Q} = \boldsymbol{C}\cdot\boldsymbol{x} \tag{2.9}$$

The coefficients c_{ij} of the j-th column of **C** tell how much the coordinate x_i contributes to the normal mode Q_j. Each normal mode depends on time according to its own specific harmonic law independent from that of the other modes, except for degeneracies of the frequencies ω_i:

$$\ddot{Q}_i = -\omega_i^2 \cdot Q_i \Rightarrow Q_i = Q_{i,0}e^{i\omega_i t} \tag{2.10}$$

$$\begin{pmatrix} \ddot{Q}_1 \\ \vdots \\ \ddot{Q}_{3s} \end{pmatrix} = \begin{pmatrix} \omega_1^2 & \cdots & 0 \\ \vdots & \ddots & \vdots \\ 0 & \cdots & \omega_{3s}^2 \end{pmatrix} \cdot \begin{pmatrix} Q_1 \\ \vdots \\ Q_{3s} \end{pmatrix} \quad \text{that is } \ddot{\boldsymbol{Q}} = \boldsymbol{\Lambda}\cdot\boldsymbol{Q} \tag{2.11}$$

where $\boldsymbol{\Lambda} = \boldsymbol{diag}\left(\omega_1^2,\ldots,\omega_{3s}^2\right)$. The problem is then solved diagonalising $\boldsymbol{D}$:

$$\boldsymbol{\Lambda} = \boldsymbol{C}\cdot\boldsymbol{D}\cdot\boldsymbol{C}^{-1} \tag{2.12}$$

The squared frequencies ω^2 are the eigenvalues, and the columns of $\boldsymbol{C}$ are the eigenvectors.

2.2.3 Using SALCs

The above considerations about both the normal modes of vibration and the MO-LCAO method show us that symmetry simplifies the solution by transforming by similarity a matrix to be diagonalized ($\boldsymbol{D}$ or $\boldsymbol{H}$) into a *block diagonal matrix*, where a block concerns only one symmetry species. Apparently, the simplification lies in the fact that each block can be diagonalized separately from the others. In fact, what symmetry does is much more than making the solution of the equations of motion easier by transforming a matrix into block diagonal form.

Without having to solve the equations of motion, symmetry

a. **identifies** the symmetry species,
b. **combines** localized functions **into** SALCs, and
c. derives their properties (*e.g.* spectroscopic).

To explain this important result, let us use as an example the same molecule (water) and look at the applications to normal modes of vibration (Section 2.2.3.1), to MOs (Section 2.2.3.2), and spectroscopic transitions (Sections 2.3.1.1, 2.3.2.2, and 2.3.2.3). For these examples, let us orient the axes so that C_2 is along z, σ is the xz plane, and σ' is the yz plane as in the right part of Figure 4.10. The eigenvalues of the symmetry operations of the C_{2v} group are in Table 4.8 (see also Table 4.21).

2.2.3.1 Normal Vibration Modes

The normal modes of vibration of the water molecule have already been shown without derivation in Figure 4.10. They combine the *deviations* of the atoms from the respective equilibrium positions of the molecule that are hereafter denoted for simplicity as x, y, z. Tab. 4.6 and the right part of Figure 4.10 directly show that the deviations of the **O atom** (x_O, y_O and z_O) already belong, respectively, to the symmetry species B_1, B_2 and A_1. As regards the **two H** atoms, to use 1.4, we must first find how the deviation vectors are transformed (Table 4.12).

Then, we proceed to determine the SALCs based on the displacements of the Hs, as obtained by applying the

TABLE 4.12
Transformations of the Displacements *x, y, z* of the Two Hydrogen Atoms (1 and 2) of Water

C_{2v}	E	C_2	σ	σ'
x_1	x_1	$-x_2$	x_1	$-x_2$
y_1	y_1	y_2	y_1	y_2
z_1	z_1	z_2	z_1	z_2

relations of Eq. (2.4). Neglecting the normalization constants, the results are as follows.

From x_1 :
$A_1 : x_1 - x_2 + x_1 - x_2 \Rightarrow x_1 - x_2$ $A_2 : x_1 - x_2 - x_1 + x_2 \Rightarrow 0$
$B_1 : x_1 + x_2 + x_1 + x_2 \Rightarrow x_1 + x_2$ $B_2 : x_1 + x_2 - x_1 - x_2 \Rightarrow 0$

From y_1 :
$A_1 : y_1 - y_2 - y_1 + y_2 \Rightarrow 0$ $A_2 : y_1 - y_2 + y_1 - y_2 \Rightarrow y_1 - y_2$
$B_1 : y_1 + y_2 - y_1 - y_2 \Rightarrow 0$ $B_2 : y_1 + y_2 + y_1 + y_2 \Rightarrow y_1 + y_2$

From z_1 :
$A_1 : z_1 + z_2 + z_1 + z_2 \Rightarrow z_1 + z_2$ $A_2 : z_1 + z_2 - z_1 - z_2 \Rightarrow 0$
$B_1 : z_1 - z_2 + z_1 - z_2 \Rightarrow z_1 - z_2$ $B_2 : z_1 + z_2 - z_1 - z_2 \Rightarrow 0$

From each pair (x_1, x_2), (y_1, y_2), (z_1, z_2) we obtain two linear combinations (sum and difference) without combining coordinates in different directions. In total there are six SALCs belonging to all four symmetry species. Including the results for the O atom, the picture is therefore:

A_1	A_1	A_1	A_2	B_1	B_1	B_1	B_2	B_2
(x_1-x_2)	(z_1+z_2)	z_O	(y_1-y_2)	(x_1+x_2)	(z_1-z_2)	x_O	(y_1+y_2)	y_O

To assign these SALCs to the 9 degrees of freedom (dof), first of all we consider the translational and rotational dof: the vibrational dof are the remaining ones. The *translations* of the whole molecule along x, y, z belong to the same species as the O displacements, which must be combined with the SALCs of the same species in which the two H move in the *same* direction. Therefore:

Translation along x: species B_1: vectors (x_1+x_2), x_O.
Translation along y: species B_2: vectors (y_1+y_2), y_O.
Translation along z: species A_1: vectors (z_1+z_2), z_O.

The analysis of rotations is similar although a little more intricate. For the *rotation around the axis parallel to x*, the displacement y_O must be combined with the displacements (y_1+y_2) of the two hydrogens in the *other* direction: the species is B_2. As the translation along y seen before requires the combination of (y_1+y_2) and y_O displacements in the *same* direction, these two motions exhaust the contribution of the *two* SALCs B_2: two vectors, two independent motions.

For the *rotation around the axis parallel to y*, the (x_1+x_2) and (z_1-z_2) displacements of the hydrogens must be combined with x_O. With the translation along x seen before, this motion provides a second contribution of the three SALCs B_1: three vectors, so far two independent motions, a third linearly independent vector remains.

The *rotation around the axis parallel to z* requires the displacements (z_1-z_2) of the hydrogens of symmetry B_1 (the O atom does not contribute because it lies on the rotation axis). With the rotation around y, this leaves only one contribution of SALCs B_1.

We thus have the following picture. Species A_1: three SALCs, one is required by the translation along z. Two independent linear combinations of SALCs remain and therefore there must be two vibrations of this species. Species A_2: one SALC. The rotation around z belongs to this symmetry species and therefore there are no vibrations of this species. Species B_1: three SALCs. The translation along x and the rotation around y require two linear combinations: an independent linear combination is missing and therefore there must be a vibration of species B_1. Species B_2: the two SALCs are required by the translation along y and rotation around x.

In conclusion:

Two vibrations of species A_1 + one vibration of species B_1.

This analysis of symmetry species replaces the analytical solution of the equations of motion according to Section 2.2.2 (see Eq. (2.12)), which would require knowing of the force constant in matrix $\boldsymbol{V}$. Indeed, after the combination of the atomic displacements into SALCs and a reordering of the variables, the dynamic matrix $\boldsymbol{D}$ (see Eq. (2.8)) would take the block diagonal form $\boldsymbol{D}_{\mathrm{SALC}}$ of Eq. (2.13) with a 3×3 block (species A_1), a 1×1 block (species A_2), a 3×3 block (specie B_1), and a 2×2 block (species B_2):

$$\boldsymbol{D}_{\mathrm{SALC}} = \begin{pmatrix} * & * & * & 0 & 0 & 0 & 0 & 0 & 0 \\ * & * & * & 0 & 0 & 0 & 0 & 0 & 0 \\ * & * & * & 0 & 0 & 0 & 0 & 0 & 0 \\ 0 & 0 & 0 & * & 0 & 0 & 0 & 0 & 0 \\ 0 & 0 & 0 & 0 & * & * & * & 0 & 0 \\ 0 & 0 & 0 & 0 & * & * & * & 0 & 0 \\ 0 & 0 & 0 & 0 & * & * & * & 0 & 0 \\ 0 & 0 & 0 & 0 & 0 & 0 & 0 & * & * \\ 0 & 0 & 0 & 0 & 0 & 0 & 0 & * & * \end{pmatrix} \quad (2.13)$$

Here, the *first block* (3×3) provides a zero eigenvalue (corresponding to the translation along z) and two non-zero squared frequencies for two vibrations of symmetry A_1, which are the *symmetric stretching,* and the *bending* reported in Figure 4.10 and labelled 1, 1, 1. The *second block* (1×1) and the *fourth block* (2×2) have only zero eigenvalues, while the *third block* (3×3) provides two zero eigenvalues and the squared frequency of the *asymmetric stretching* of symmetry B_1 labelled 1, –1, 1, –1 in Figure 4.10.

2.2.3.2 *Molecular Orbitals of Water*

Let us now apply the procedure to MO problems. We begin by recognizing the symmetry species of the AOs of O and finding the SALCs based on the AOs of H atoms. For simplicity, it is convenient to consider only s-type or p-type AOs, both for O and H. Each AO of O belongs to a symmetry species, as shown in Figure 4.15. Figure 4.16 reports the SALCs based on the AOs of the two H.

In all cases Eq. (2.4) provides only two SALCs: the AOs of the two Hs must have the same weight because they are

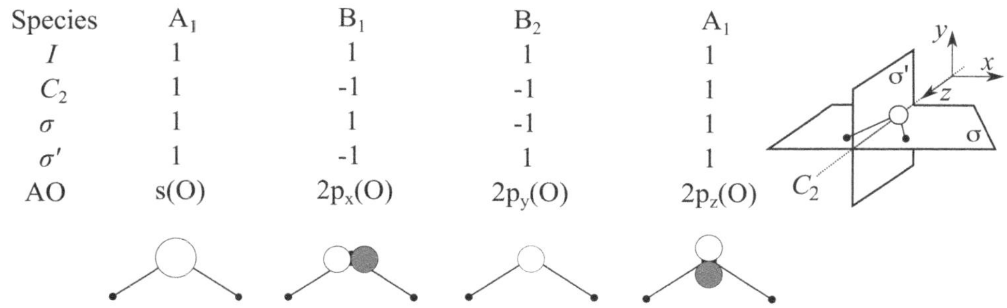

FIGURE 4.15 Symmetry species of oxygen AOs of water, group C_{2v}. The plane of the plot is the molecular plane (σ). The part of $2p_y$ AO below σ is negative.

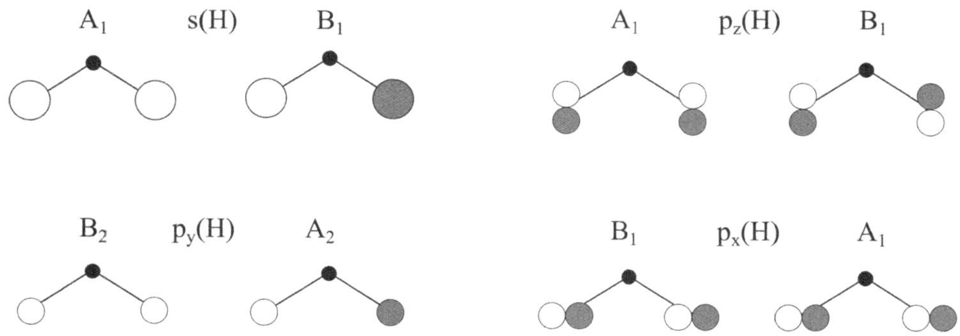

FIGURE 4.16 SALCs for the s(H) and p(H) AOs of water. The parts of the p_y AOs below the molecular plane (σ) have opposite signs. The character table of group C_{2v} is in Tables 4.8 and 4.21.

equivalent by symmetry and therefore the SALCs are their sum and difference.

By combining SAF of O and SALC of H belonging to the same symmetry species it becomes possible to obtain a global picture of the MOs. They can be classified as bonding, non-bonding or anti-bonding MOs and ordered in energy, based on considerations on the energy of the AOs and the greater or lesser stabilization or de-stabilization in bonding or anti-bonding combinations.

Considering initially a reduced set of AOs (the 1s, 2s and 2p AOs of O and only the 1s of the H) the MOs belong to the following symmetry species.

- Species A_1: 1s(O), 2s(O), $2p_z$(O), [1s(H1) + 1s(H2)].
- Species A_2: nothing.
- Species B_1: $2p_x$(O), [1s(H1) − 1s(H2)].
- Species B_2: $2p_y$(O).

From seven AOs we obtain seven SAFs: the MOs are now obtained by combining SAFs of the same species (see Figure 4.17). The usual practice is to label each MO using the symbol of the symmetry species in lowercase and preceded by the order number.

For the A_1 species, Figure 4.17 does not report the $1a_1$ MO (that is essentially the 1s(O) AO) because it has a much lower energy than the 2s and 2p, which are close in energy to the 1s(H) AOs. Then, the 2s(O) and $2p_z$(O) AOs and the SALC 1s(H1) + 1s(H2) combine in (i) $2a_1$ with an energy lower than that of all the constituent AOs, (ii) $4a_1$ at higher energy than all the constituent AOs and (iii) $3a_1$ at intermediate energy. In fact, the $2a_1$ MO has almost lone pair characteristics due to the high difference in energy between the combining AOs, while $3a_1$ and $4a_1$ are, respectively, strong bonding and antibonding in character.

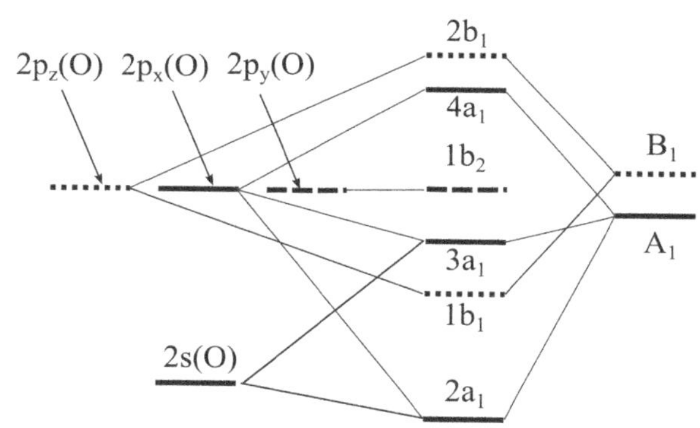

FIGURE 4.17 Energy diagram of the LCAO MOs of the water molecule (middle) with the O AOs (left) and the two SALCs due to the 1s AOs of the two Hs (right). Equal linestyle corresponds to the same symmetry species. The figure *does not show* the $1a_1$ MO which is at a much lower energy (see text).

For the B_1 species, the AO $2p_y$(O) does not combine into an MO (within the reduced base so far assumed).

There are two MOs for the B_2 species, $1b_2$ and $2b_2$, with bonding and antibonding character. Finally, there is no MO for species A_2.

In the fundamental configuration, MOs up to $1b_2$ included are doubly occupied.

Of course, one can refine this schematic picture by enlarging the basis set and including, for example, 2s and 2p AOs of the Hs as well as the 3s and 3p AOs of O. The new AOs of O do not add anything substantially new regarding symmetry, but their inclusion will obviously contribute to modifying the energies. The 2p of the hydrogens provide SALCs of species A_2 and B_2, with the result in particular that the $2p_y$-O AO (the $1b_2$ of Figure 4.17) now combines with the $2p_y$-H SALC of species B_2 to give two b_2 MOs: the additive combination has weakly bonding characteristics (π) and is therefore at a lower energy than the simple $2p_y$-O AO that in the picture with the smallest basis would be a nonbonding lone pair.

2.3 Spectroscopic Transitions

Spectroscopic transitions essentially require the evaluation of integrals such as:

$$\int_{\Omega} \psi_i^* \hat{V} \psi_f \, d\tau \qquad (2.14)$$

In Eq. (2.14), $\hat{V}$ is the time-independent part of the interaction operator of the system with the radiation, while the indices i and f denote the initial state and the final state. The time-dependent part of the interaction operator is factored out and gives the frequency selection rule.

Here we consider only the simplest problems. In particular, we assume the exact separation between the electronic and the other degrees of freedom, we discuss only vibrational and electronic spectroscopies, and we consider only the ground state as the initial state. Systems with an odd number of electrons or with hidden or accidental degeneracies, transitions from an excited state and open-shell electronic configurations, as well as all problems relating to non-commutative symmetries, are left to the more general approach of Section 6.

2.3.1 Electronic Spectroscopy

Under the hypothesis of considering only non-degenerate MOs, it is easy to see that a closed-shell ground state is *totally symmetric*. In fact, for any symmetry operation the eigenvalue of the product of two one-electron functions that differ only in the spin part is always = 1, being the square of the eigenvalues +1 or −1. Then, the totally symmetric species can be ignored:

The symmetry species of

$$\int_{\Omega} \psi_i^* \hat{V} \psi_f \, d\tau \text{ is the same as } \int_{\Omega} \hat{V} \psi_f \, d\tau \qquad (2.15)$$

In the final state there are two non-degenerate MOs, each occupied by a single electron: an initially filled MO φ_m and an initially empty MO φ_n, while the other MOs remain

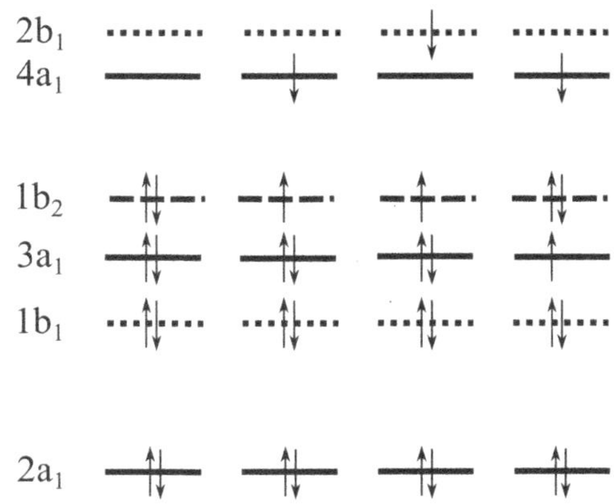

FIGURE 4.18 Electronic configurations of H_2O (not considering the $1a_1$MO): the left column shows the ground state; the other columns show the final electronic configurations of three possible transitions from the ground state (see text).

completely filled and their contribution can be ignored. Therefore, only the effects of φ_m and φ_n must be considered:

The symmetry species of

$$\int_{\Omega} \psi_i^* \hat{V} \psi_f \, d\tau \text{ is the same as } \int_{\Omega} \hat{V} \varphi_n \varphi_m \, d\tau \qquad (2.16)$$

2.3.1.1 *Electronic Transitions of Water*

With reference to Figures 4.17 and 4.18, let us discuss as an example three particular excited configurations: an electron is promoted:

i. from the highest occupied MO ($1b_2$) to the lowest empty MO ($4a_1$) or
ii. again from $1b_2$, but to a higher MO ($2b_1$) or
iii. from an intermediate level MO of symmetry A_1 ($3a_1$) again to $4a_1$.

The eigenvalue of a product of functions is the product of the eigenvalues. Let us see then what the products of the eigenvalues of these pairs of symmetry species are (Table 4.13).

TABLE 4.13
The Symmetry Species of the Products of the Singly Occupied MOs in Figure 4.18

Species	E	C_2	σ	σ'
a_1	1	1	1	1
b_2	1	−1	−1	1
⇒ the product is B_2:	1	−1	−1	1
b_1	1	−1	1	−1
b_2	1	−1	−1	1
⇒ the product is A_2:	1	1	−1	−1
a_1	1	1	1	1
a_1	1	1	1	1
⇒ the product is A_1:	1	1	1	1

TABLE 4.14
Eigenvalues and Symmetry Species of Some Electronic Transitions of the Water Molecule

Transition	Species and Eigenvalues	Eigenvalues of *x*, *y*, or *z*	Eigenvalues of the Product	Allowed (Yes/No)
$1b_1 \rightarrow 4a_1$	B_2: 1, −1, −1, 1	x (B_1): 1, −1, 1, −1	1, 1, −1, −1 ⇒ A_2	No
		y (B_2): 1, −1, −1, 1	1, 1, 1, 1 ⇒ A_1	**Yes**
		z (A_1): 1, 1, 1, 1	1, −1, 1, −1 ⇒ B_2	No
$1b_1 \rightarrow 4b_2$	A_2: 1, 1, −1, −1	x (B_1): 1, −1, 1, −1	1, −1, −1, 1 ⇒ B_2	No
		y (B_2): 1, −1, −1, 1	1, −1, 1, −1 ⇒ B_1	No
		z (A_1): 1, 1, 1, 1	1, 1, −1, −1 ⇒ A_2	No
$3a_1 \rightarrow 4a_1$	A_1: 1, 1, 1, 1	x (B_1): 1, −1, 1, −1	1, −1, 1, −1 ⇒ B_1	No
		y (B_2): 1, −1, −1, 1	1, −1, −1, 1 ⇒ B_2	No
		z (A_1): 1, 1, 1, 1	1, 1, 1, 1 ⇒ A_1	**Yes**

Finally, to know the symmetry of the integrals of Eq. (2.14), the symmetry of the interaction operator must be known. In absorption, the spatial part of the most important operator of the transition (called *electric dipole moment*) essentially is a vector of the ordinary 3D space, whose components have the same symmetry as the three functions x, y, z, already seen: x belongs to species B_1, y to species B_2 and z to species A_1. As shown by Table 4.14, two transitions (the first and the third one) are active in absorption, since they have a component of the dipole moment with a totally symmetric integrand.

Two warnings.

1. No mention has been made so far of the *spin* of electrons. It is obvious that the initial state has $S=0$, while there are actually four final states with different spin situations ($S=0$ with $M_S=0$; $S=1$ with $M_S=1, 0, -1$). Compared to the simple picture above, there are two consequences: (i) the electron-electron repulsion splits in energy the states with $S=1$ from those with $S=0$ and (ii) due to the rule $\Delta S=0$, only the transition to the singlet ($S=0$) is allowed.
2. All problems related to the spin-orbit interaction are neglected in this simplified discussion.

2.3.2 Vibrational Spectroscopy

There are four steps in the application of symmetry to vibrational spectroscopy, with the goal of identifying (a) the normal modes of vibration, (b) the energy levels and their degeneracies, (c) the symmetry species of the wave functions and (d) the selection rules for the transitions.

Point (a) essentially recognises how the motions of the different atoms of the molecule combine into SALCs, finds - among the different SALCs - those related to vibrations instead of translations and rotations, and finally combines one or more SALCs of the same species into a normal mode. This is what has already been seen in Section 2.2.2. However, compared to the long and detailed treatment of the example in Section 2.2.3.1, which is *valid only for commutative groups*, a more synthetic and general approach is available and will be discussed in Section 6.2.4.

As regards the energy levels (point b) one can resort, at least initially, to the *harmonic approximation*. Each normal mode (i) has *equally spaced* and non-degenerate *energy levels*:

$$E_i = \left(\frac{1}{2} + n_i\right)\hbar\omega_i \tag{2.17}$$

where n_i is the number of energy quanta (*phonons*) having angular frequency $\omega_i = 2\pi\nu_i$. We speak of phonons both in molecules and crystals. Differently from molecular phonons, crystal phonons are *travelling* excitations that propagate with wavevector $\vec{q}$ (see Section 3.4).

The energy levels for the entire molecule are then given by:

$$E(n_1, n_2, \ldots, n_s) = E_0 + \hbar\sum_{i=1}^{s} n_i\omega_i \tag{2.18}$$

where s is the number of *distinct* frequencies. Equation (2.18) holds regardless of whether the symmetry is described by a commutative or non-commutative group. If the *group is commutative*, s is the number of normal modes because all symmetry species of a commutative group are one-dimensional, as already seen. If the *group is non-commutative* (a case that will be discussed in Section 6.2.4) two or three (in some cases even more) normal modes of a species have the same frequency, there is degeneracy of the levels and most of the considerations that follow will have to be modified. For consistency with the terminology used in the technical literature, it is convenient to discuss separately several types of vibrational states of the whole molecule.

a. In the **ground state** of the molecule all n_i are=0, *i.e.* the quantum numbers are (0,0. …, 0). The ground state is non-degenerate (see also Section 6.2.5).
b. In the **fundamental levels**, one n_i is 1 and all the other n_i are 0. These are the states such as (1,0. …, 0), or (0, 1, …, 0), …. These are called *fundamental* because the *transitions* from the ground state to each of these levels directly gives the

frequencies of the vibrational modes of the molecule (the ω_i of Eq. (2.17)), but the name is a bit misleading because the adjective is generally used to indicate the minimum energy state of a quantized system, *i.e.* what we have called here ground state. With *commutative groups, a fundamental level is non-degenerate*. With *non-commutative groups, however, a fundamental level can be non-degenerate*, or *two* or *three* (or more) times degenerate, respectively with 1D, 2D or 3D, or … species.

c. Only one frequency is excited also in the **overtone levels**, but $n_i > 1$. These are the states such as (2,0. …, 0), or (0, 3, …, 0), …. With non-commutative groups, overtone levels are even more degenerate than the corresponding fundamental levels.
d. Finally, in the **combination levels**, normal modes with different characteristic frequencies are excited. These are the states such as (1, 1. …, 0), or (0, 3, …, 2).

The assumption of the harmonic approximation seemingly also solves the problem of the selection rules (point d). According to this approximation, for example, only transitions for which only one quantum number changes by 1 would be allowed in absorption. In reality, as explained in detail in Section 6.2.5.1, it is precisely on this point, even more than on the numerical accuracy of the energies, that the harmonic approximation falls seriously short. For this reason, it is necessary to discuss the vibrational spectroscopic transitions **only on the basis of the symmetry properties of the functions**, ignoring the more stringent, but often violated, selection rule relying on the harmonic approximation.

2.3.2.1 The Symmetry of the Vibrational Wave Functions

If there is more than one vibrational degree of freedom, *i.e.* the molecule is at least triatomic, the vibrational wave function of the whole molecule is the product of several functions, each dependent on a single normal coordinate, precisely because the normal coordinates are independent vibrational degrees of freedom. For clarity, let us call each of these factors a *single-mode wave function* and the product the *global* (or molecular) wave function.

Despite the importance of anharmonicity in specifying energy levels and wave functions, to discuss the *symmetry* of single-mode wave functions and hence also of the global wave function it is possible to refer to the *eigenfunctions of the harmonic oscillator model*, because the correct (anharmonic) eigenfunctions belong to the same symmetry species as the corresponding harmonic functions (see Section 6.2.5.1). From this premise, it follows, in particular, that the wave function of a normal mode is totally symmetric for n even, while for n odd it belongs to the symmetry species of the coordinate expressing the normal mode.

Up to the normalization constant, the wave function with $n=0$ of a normal mode is given by:

$$\varphi_0 = \exp\left[\frac{-m\omega}{2\hbar}\xi^2\right] \tag{2.19}$$

where ξ is the normal mode and $\omega = 2\pi\nu$ is the angular frequency of the mode. Any symmetry operation leaves this wave function unchanged.

The single-mode wave functions of excited levels are the product of the same exponential of Eq. (2.19) and a polynomial in ξ (called the Hermite polynomial) that contains only even powers (for even n) or only odd powers (for odd n) up to and including n. These polynomials are therefore totally symmetric (for even n) or belong to the same species as the coordinate ξ (for odd n). The wave function has the same behaviour, because the exponential is totally symmetric. For example, Figure 4.8 reports the first four eigenfunctions of the harmonic oscillator model for one vibrational degree of freedom and shows their symmetry $f(\xi)=f(-\xi)$ or $f(\xi)=-f(-\xi)$ for functions with even or odd quantum number. The following conclusions can now be derived (again, *only for commutative groups*).

- The *global* wave function of the vibrational ground state (the state without excited vibrations) is totally symmetric because it is the product of single-mode wave functions (as many as there are vibrational degrees of freedom), each of which is totally symmetric.
- The *global* wave function of a *fundamental* level belongs to the symmetry species of the excited normal mode.
- The *global* wave function of an *overtone* is either totally symmetric if the quantum number of the mode is even or belongs to the symmetry species of the excited normal mode if the quantum number of the mode is odd.
- The *global* wave function of a *combination* level belongs to the symmetry species corresponding to the eigenvalues of the product of the eigenvalues of the single mode functions.

2.3.2.2 Vibrational Absorption Transitions

Various transitions in vibrational absorption are possible. We here discuss only transitions from the ground state. Then, the integrals:

$$\int_\Omega \psi_i^* \hat{V} \psi_f \, d\tau \quad \text{and} \quad \int_\Omega \hat{V} \psi_f \, d\tau \tag{2.20}$$

have equal symmetry, as already seen for electronic transitions. For vibrational absorption spectroscopy, the electric dipole moment operator must be considered. For the water molecule, and for the fundamental transitions, the situation is summarized in Table 4.15.

TABLE 4.15
Eigenvalues and Symmetry Species for the Fundamental Transitions of the Water Molecule in Vibrational Absorption Spectroscopy

Excited Normal Mode	Species and Eigenvalues	Eigenvalues of *x*, *y*, or *z*	Eigenvalues of the Product	Allowed (Yes/No)
Symmetric stretching	A_1: 1, 1, 1, 1	x (B_2): 1, −1, 1, −1	1, −1, 1, −1 ⇒ B_2	No
		y (B_1): 1, −1, −1, 1	1, −1, −1, 1 ⇒ B_1	No
		z (A_1): 1, 1, 1, 1	1, 1, 1, 1 ⇒ A_1	**Yes**
Asymmetric stretching	B_2: 1, −1, 1, −1	x (B_2): 1, −1, 1, −1	1, 1, 1, 1 ⇒ A_1	**Yes**
		y (B_1): 1, −1, −1, 1	1, 1, −1, −1 ⇒ A_2	No
		z (A_1): 1, 1, 1, 1	1, −1, 1, −1 ⇒ B_2	No
Bending	A_1: 1, 1, 1, 1	x (B_2): 1, −1, 1, −1	1, −1, 1, −1 ⇒ B_2	No
		y (B_1): 1, −1, −1, 1	1, −1, −1, 1 ⇒ B_1	No
		z (A_1): 1, 1, 1, 1	1, 1, 1, 1 ⇒ A_1	**Yes**

It follows that all fundamental transitions are active in vibrational absorption spectroscopy, each with a specific polarization of radiation. Note that a radiation polarized along the direction perpendicular to the plane of the molecule is not active with any vibration. This agrees with the fact that there are no vibrations that cause any atom to exit the plane of the molecule.

For transitions that do not have the ground state as their initial state (the so-called **hot lines**) or involve the formation of more vibrational quanta (*overtone* or *combination lines*) see Section 6. Finally, Section 6.2.5.1 deals with the *limits of the harmonic approximation*.

2.3.2.3 Raman Vibrational Transitions

A brief introduction to Raman spectroscopy has been discussed in connection with time-dependent perturbations in the Chapter devoted to general aspects of quantum mechanics, and the dipole moment ($\vec{\mu}$) induced by the electric field ($\vec{E}$) of an em radiation was derived:

$$\vec{\mu} = \vec{\vec{\alpha}}.\vec{E} \tag{2.21}$$

where $\vec{\vec{\alpha}}$ is the polarization, a symmetric tensor

It was also seen that the dipole induced by a *monochromatic* incident radiation such as $\vec{E}_i \cdot \cos\omega t$ in a system characterized by a proper vibration of frequency ω_v is given by:

$$\begin{aligned}\vec{\mu} &= \vec{\vec{\alpha}}\,\vec{E} = \left(\overrightarrow{\overrightarrow{\alpha_0}} + \overrightarrow{\overrightarrow{\Delta\alpha}}\cos\omega_v t\right)\vec{E}_i\cos\omega t \\ &= \overrightarrow{\overrightarrow{\alpha_0}}\,\vec{E}_i\cos\omega t + \overrightarrow{\overrightarrow{\Delta\alpha}}\cos\omega_v t\cos\omega t\end{aligned} \tag{2.22}$$

that is:

$$\vec{\mu} = \vec{\vec{\alpha}}\,\vec{E}\cos\omega t + \frac{\overrightarrow{\overrightarrow{\Delta\alpha}}}{2}\vec{E}_i\left[\cos(\omega-\omega_v)t + \cos(\omega+\omega_v)t\right] \tag{2.23}$$

Equation (2.23) shows three scattered radiations, called *Rayleigh scattering* ($\omega_s = \omega$), *Stokes scattering* ($\omega_s = \omega + \omega_v$) and *anti-Stokes scattering* ($\omega_s = \omega - \omega_v$). A Stokes line is the transition from the ground state to the state in which only one normal mode is excited, while an anti-Stokes line is the inverse transition from the state in which only one normal mode is excited to the ground state. Ignoring the Rayleigh line (which adds to the absorption transition), for Stokes and anti-Stokes scattering

$$\text{The symmetry of } \int_\Omega \psi_i^* \hat{V} \varphi_f \, d\tau \text{ is the same as } \int_\Omega \hat{V} \varphi_j \, d\tau \tag{2.24}$$

where φ_j is the wave function of the fundamental level of normal mode j, which is both the final state of a Stokes line and the initial state of an anti-Stokes line.

As regards symmetry, the information of interest is that $\overrightarrow{\overrightarrow{\Delta\alpha}}$ is a symmetric tensor ($\Delta\alpha_{xy} = \Delta\alpha_{yx}$, …) and behaves like the quadratic functions x^2, y^2, z^2, xy, yz, zx. The symmetry behaviour of these functions can be obtained in the same way as for the linear functions x, y, z but is available thanks to specific tables, as will be seen in Section 5.5 (see Table 4.21). What is obtained for the C_{2v} group is that the quadratic functions belong to all symmetry species: there is no symmetry species to which at least one quadratic function does not belong.

Consequently, in the case of the water molecule, all the normal modes are symmetry-allowed in Raman spectroscopy. However, for the asymmetric stretching, the polarizability tensor changes only slightly, giving rise to a faint transition. For this reason, this transition is usually quoted as weakly forbidden.

3 TRANSLATIONAL SYMMETRY

Translational symmetry (see Section 3.1.1) is traditionally placed at the basis of all treatments of the *solid state*, more properly meaning the *crystalline state of matter*. The historical paradigm of crystallography, which based long-range order and discrete diffraction on three-dimensional translational symmetry, has been challenged by the works of various authors on incommensurable structures and composite crystals and by the fundamental work of Shechtman on quasicrystals, which earned him the Nobel Prize in Chemistry in 2011. Discrete diffraction occurs, in fact, even in the *absence* of 3D translational symmetry or has characteristics that cannot be *explained* by it alone: this is what happens in the different classes of *aperiodic* crystals. Now, it is recognized that **3D translational symmetry is not a necessary condition for the crystalline state of matter** and since 1992 the *International Union of Crystallography* has established that "by *crystal* we mean any solid having an essentially discrete diffraction pattern".

This clarification, however, does not diminish the extreme importance that 3D translational symmetry continues to play both in the description and determination of crystal structures (which in most cases are 3D periodic), in the explanation of the properties of solids and in the interpretation of the related experiments. It should also be kept in mind that many results of solid-state theory – first and foremost Bloch theorem, see Section 3.2 – require translational symmetry and, therefore, *do not* apply to aperiodic crystals.

In addition to 3D translational symmetry, 2D and 1D translations also are important in chemistry: consider, for example, adsorption phenomena on crystalline surfaces. Here, however, they are only marginally considered as an introduction to the 3D case for the treatment of some problems.

A brief presentation of the many and different consequences of translational symmetry would require discussing topics that are more properly pertinent to crystallography and solid-state science. Therefore, only brief references are made here and, essentially, we limit ourselves to highlighting the role that 3D translational symmetry plays in making solvable problems characterized by a very large number (of the order of 10^{20} or more) of atoms, ions or molecules.

Furthermore, a crystal structure *usually* presents other symmetries in addition to translational symmetry: for example, there may be reflections, rotations (proper or improper) and inversions, as for the point groups of molecular symmetry. The problems related to the **combination** with the translational symmetry operations in the algebraic group structure, and all the applicative implications are addressed later (Section 6.6). Instead, considering only translational symmetry, which is commutative (see Section 3.2), we remain in the same type of problems as in Chapter 2 and we can discuss the results in a similar way. Also, by ignoring non-translational symmetries various specific aspects of the solid state are made more explicit.

Ignoring non-translational symmetry does not affect the validity of the treatment: it is possible to have crystal structures in which there is *only* translational symmetry. On the other hand, it is quite obvious that non-translational symmetry operations, when present, play an important role both in the theoretical treatment and in the applications in crystallography and solid-state science.

3.1 The Crystal Structure

The description of the structure, *i.e.* the arrangement in space of the constituent elements (atoms, ions, or polyatomic groups) of a crystalline solid with translational symmetry, is based on the **ideal crystal model**, which possesses the translational symmetry described in Section 3.1.1 and is unlimited and free of defects.

It is indeed important to recall that a **real** crystal always presents **defects** of various types (point, line, etc.), which are due to entropic reasons (in the case of equilibrium defects) or kinetic reasons (for example, defects formed in a high-temperature synthesis and then *frozen* by the sluggishness of the processes required to reach a different equilibrium) or both. Even the external surface of a single crystal has to be considered a defect, as well as the boundaries separating the grains of a monophasic polycrystalline sample.

3.1.1 The Translation Lattice

Translational symmetry is based on **lattices**, that are (infinite) sets of vectors $\vec{l}$ (in the 3D space) being *linear combinations with integer* coefficients of a triplet of **basis vectors** $(\vec{a}, \vec{b}, \vec{c})$:

$$\vec{l} = m\vec{a} + n\,\vec{b} + o\,\vec{c}$$
$$\left(m,\ n,\ o \text{ are three negative, positive or null integers}\right) \quad (3.1)$$

We require that the basis vectors are *non-coplanar, i.e. linearly independent*. When the origin of the ordinary 3D space is fixed, the $\vec{l}$ vectors of Eq. (3.1) specify **lattice points** that are typically used as a graphical representation of the lattice. *Lattice* is commonly but improperly used to denote a crystal structure or part of a crystal structure. However, for the sake of clarity, it is preferred to consider a lattice as a simple set of vectors or immaterial points, without reference to atoms or ions or other material objects (molecules or molecular fragments) that may be found there.

> Translational symmetry consists in the translational invariance for any $\vec{l}$ vector, *i.e.* for the coordinate transformation $\vec{r} \to \vec{r} + \vec{l}$.

Equation (3.1) is the underlying set of **the translation group**: the identity is the vector $\vec{0} = 0\vec{a} + 0\vec{b} + 0\,\vec{c}$, while the inverse operation of the $\vec{l} = m\vec{a} + n\,\vec{b} + o\,\vec{c}$ translation is the $-\vec{l} = -m\vec{a} - n\vec{b} - o\vec{c}$ translation. Anticipating a topic that will be addressed in Section 6.6.2, it is useful to clarify that there exist (in the 3D case) 14 *types* of lattices that are essentially distinguished by **their compatibility with other** (non-translational) **symmetry operations** and are known as **Bravais lattices**.

We also stress here a fundamental difference with respect to the point symmetry operations: *no points in the ordinary 3D space are left unmoved by any of the translational symmetry operations* (with the obvious exception of the identity).

3.1.2 The Unit Cell

The non-coplanar triplet ($\vec{a}$, $\vec{b}$, $\vec{c}$) in Eq. (3.1) defines a **parallelepiped,** which is the ***unit cell*** of the translation lattice and of the crystal structure: by repeating the unit cell by translation, the ordinary 3D space is completely filled. The lattice points are placed on the vertices of the parallelepiped cells. Each unit cell contains one lattice point, because each lattice point is shared between eight adjacent cells and a parallelepiped has eight vertices. The unit cell is defined by ***six*** scalars: the three cell sides ($a = |\vec{a}|$, $b = |\vec{b}|$, $c = |\vec{c}|$) and three angles (α=angle between $\vec{b}$ and $\vec{c}$, β=angle between $\vec{c}$ and $\vec{a}$, γ=angle between $\vec{a}$ and $\vec{b}$). The three components of each basis vector add to *nine* scalars, and the remaining three degrees of freedom describe the orientation of the $\vec{a}$, $\vec{b}$, $\vec{c}$ triplet with respect to an external reference frame.

It is apparent that **the same lattice can be described by multiple basis triplets,** and therefore by more than one unit cell: Figure 4.19 shows for a 2D lattice two different unit cells (displayed by different shadows) and it should clarify that other triplets of basis vectors can be chosen and therefore other unit cells, always of parallelepiped shape.

Crystallography specifies the *conventional* criteria for choosing the **cell** and therefore the **standard triplet**. In short: the choice is for the shortest possible edges and the angles closest to (but ≥) 90°, apart from cases with rotational symmetries of order 3 or 6. In these cases, two basis vectors perpendicular to the C_3 or C_6 axis are chosen forming an angle of 120°, the third basis vector is along the C_3 or C_6 axis.

Another important crystallographic convention concerns the *graphical* representation of parallelepiped cells, where the **origin** is placed at the **top** left vertex (instead of the bottom left, as is used in other scientific areas).

The parallelepiped is not the *only* possible shape for a *unit cell.* For some applications it is useful to use a unit cell (called **Wigner-Seitz** cell) which usually has a less regular shape, contains **one lattice point in its centre** and no other lattice points inside the cell or on its external surface. Figure 4.20 shows (in 2D) the construction of this unit cell as the region of space included within the envelope of the planes that bisect the vectors joining a lattice point at the centre of the cell with its nearest points. The Wigner-Seitz cell *has no* ambiguity of choice (Figure 4.19) and has the *particular property of containing the entire region of ordinary 3D space that is closer to a particular lattice point than to any other lattice point.*

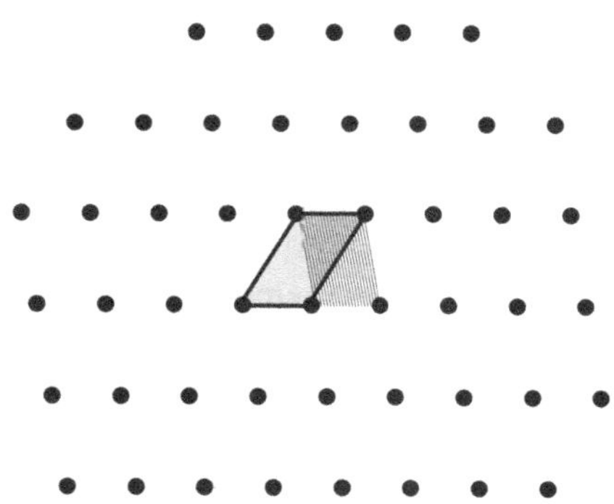

FIGURE 4.19 Two unit-cells of the same lattice. The standard cell is that with the market contour.

For convenience, **multiple** parallelepiped cells are also frequently used. These *do not* contain a single lattice point but 2, 3, or 4 lattice points and are called:

- *centred*: "A face centred" means that there is a lattice point at the centre of the face defined by the vectors $\vec{b}$ and $\vec{c}$, and, in the same way "*B*" or "*C* face centred" cells,
- *body centred,* denoted by *I* - the capital letter "i", which has a lattice point (in the centre) not shared with other cells, or
- *face* (all) *centred* (*F*), and a
- triple *hexagonal* cell (*H*).

Multiple cells are used with some types of lattices to better highlight their symmetry features and facilitate some processing. By choosing a multiple cell, however, you lose the property that - using as basis vectors $\vec{a}$, $\vec{b}$, $\vec{c}$ defining its sides – all the lattice points are described by Eq. (3.1), $\vec{l} = m\vec{a} + n\vec{b} + o\vec{c}$ where *m*, *n*, *o* are all integers (negative, positive or zero): we need to add points with fractional *m*, *n*, *o*. For further clarifications on multiple cells, see Section 6.6.2.

Non-multiple cells (both those in the shape of a parallelepiped and the Wigner-Seitz cell) are then more precisely called **primitive cells**. All primitive cells share the same volume, which is 1/2, 1/3 or 1/4 of possible double (*A*, *B*, *C* centred or I), triple (*H*) or quadruple (*F*) cells. Double, triple or quadruple cells contain two, three or four lattice points, respectively.

3.1.3 Atomic Positions

As said, the structure of a 3D periodic crystalline solid is described by a unit cell that is repeated by translational symmetry to fill the whole crystal. Its content is specified by specifying the coordinates of the positions (sites) occupied by the various atoms. The positions are conventionally reported as fractions (*x*, *y*, *z*) of the sides of the parallelepiped unit cell (*a*, *b*, *c*): they are therefore *parallel*

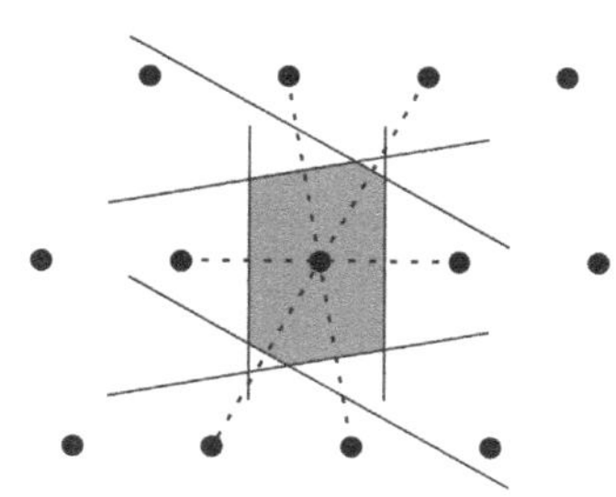

FIGURE 4.20 The Wigner-Seitz cell (2D version).

coordinates. Considering the atomic motions, the positions are to be considered average values.

3.1.4 Reciprocal Vectors

For the following topics it is useful to introduce, alongside the $\vec{a}$, $\vec{b}$, $\vec{c}$ basis triplet (Eq. 3.1), a complementary triplet defined as:

$$\vec{a}^* \stackrel{\text{def}}{=} \frac{\vec{b} \wedge \vec{c}}{\vec{a} \cdot \vec{b} \wedge \vec{c}}; \quad \vec{b}^* \stackrel{\text{def}}{=} \frac{\vec{c} \wedge \vec{a}}{\vec{b} \cdot \vec{c} \wedge \vec{a}}; \quad \vec{c}^* \stackrel{\text{def}}{=} \frac{\vec{a} \wedge \vec{b}}{\vec{c} \cdot \vec{a} \wedge \vec{b}} \tag{3.2}$$

In Eq. (3.2), $\vec{a} \wedge \vec{b}$, $\vec{b} \wedge \vec{c}$ and $\vec{c} \wedge \vec{a}$ are **vector products**. A vector product $\vec{a} \wedge \vec{b}$ is a vector having: (i) modulus given by the product of the moduli of the factors multiplied by the sine of the included angle: $|\vec{a} \wedge \vec{b}| = a \cdot b \cdot \sin(\gamma)$, which is the area of the parallelogram defined by $\vec{a}$ and $\vec{b}$ and (ii) direction perpendicular to the plane defined by $\vec{a}$ and $\vec{b}$, and given by the *right-hand rule*: if $\vec{a}$ and $\vec{b}$ point as the forefinger and middle finger of a righthand, respectively, $\vec{a} \wedge \vec{b}$ points as the thumb. In Eq. (3.2), the *triple products* at the denominator [$\vec{a} \cdot \vec{b} \wedge \vec{c} \stackrel{\text{def}}{=} \vec{a} \cdot (\vec{b} \wedge \vec{c}) \stackrel{\text{def}}{=} \vec{b} \cdot (\vec{c} \wedge \vec{a}) \stackrel{\text{def}}{=} \vec{c} \cdot (\vec{a} \wedge \vec{b})$] are scalars and correspond to the *volume of the unit cell* with sides $\vec{a}$, $\vec{b}$, $\vec{c}$. Indeed, considering the scalar product of $\vec{a} \wedge \vec{b}$ and $\vec{c}$, we have the area of the parallelogram defined by $\vec{a}, \vec{b}$ (= modulus of vector $\vec{a} \wedge \vec{b}$) multiplied by the projection of $\vec{c}$ on the perpendicular to the *ab* plane, that is the altitude of the parallelepiped with $\vec{c}$ as the third side. Definitions (Eq. 3.2) contain $\vec{a}$, $\vec{b}$, $\vec{c}$ in *cyclic permutations*, and can be expressed as:

$$\vec{x}^* \stackrel{\text{def}}{=} \frac{\vec{y} \wedge \vec{z}}{\vec{x} \cdot \vec{y} \wedge \vec{z}}$$

where $\vec{x} \cdot \vec{y} \wedge \vec{z}$ is $\vec{a} \cdot \vec{b} \wedge \vec{c}$, $\vec{b} \cdot \vec{c} \wedge \vec{a}$, or $\vec{c} \cdot \vec{a} \wedge \vec{b}$ (the other three permutations change the sign). Lattice vectors $\vec{l} = m\vec{a} + n\vec{b} + o\vec{c}$, including the $\vec{a}$, $\vec{b}$, $\vec{c}$ basis vectors, are measured as length, while the vectors in Eq. (3.2) are measured as [length]$^{-1}$, which explains the name "reciprocal triplet". The reciprocal triplet (as well as the subsequent notion of reciprocal lattice, see Section 3.2.2) is introduced to express in a compact way many important relations, first those concerning (a) the functions that have the periodicity of the translation lattice (see Section 3.2.2.1) and (b) elastic *scattering* of radiation by crystals. The fundamental property that is used for such applications is that scalar products of a basis vector by a vector of the corresponding reciprocal triplet are 1 (if the vectors "have the same label") or 0 (if the vectors "have different labels"):

$$\begin{aligned} &\vec{a}\cdot\vec{a}^* = \vec{b}\cdot\vec{b}^* = \vec{c}\cdot\vec{c}^* = 1 \quad \text{and} \\ &\vec{a}\cdot\vec{b}^* = \vec{b}\cdot\vec{a}^* = \vec{b}\cdot\vec{c}^* = \vec{c}\cdot\vec{b}^* = \vec{c}\cdot\vec{a}^* = \vec{a}\cdot\vec{c}^* = 0 \end{aligned} \tag{3.3}$$

The first three equalities are trivial (by calculation, we obtain a fraction with the denominator equal to the numerator). For the others, let's see $\vec{a}\cdot\vec{b}^*$. The $\vec{b}^*$ vector **is** perpendicular to the plane containing $\vec{a}$ and $\vec{c}$; thus, is perpendicular to $\vec{a}$ and $\vec{c}$; therefore: $\vec{a}\cdot\vec{b}^* = 0 = \vec{c}\cdot\vec{b}^*$. The rest of the equalities are demonstrated analogously starting from $\vec{c}^*$ and $\vec{a}^*$. These scalar products, evidently, are dimensionless.

3.2 Bloch Theorem

The fundamental *algebraic* property of translational symmetry is that the operations described by the vector $\vec{l}$ (Eq. 3.1) are commutative:

> The symmetry group of lattice translations is *commutative* and has (only) one-dimensional symmetry species.

In words: the order in which we add *m* times $\vec{a}$, *n* times $\vec{b}$ and *o* times $\vec{c}$ is irrelevant: the resulting translation $\vec{l}_1 + \vec{l}_2$ is $(m_1 + m_2)\vec{a} + (n_1 + n_2)\vec{b} + (o_1 + o_2)\vec{c}$. This has a fundamental consequence.

> The effect of a translational symmetry operation is simply to multiply an eigenfunction of the Hamiltonian by a number which depends on the symmetry species.

The situation is the same as already seen for commutative *point* groups. If, given the symmetry species, we indicate with ϵ_a, ϵ_b, ϵ_a the eigenvalues of the translations corresponding to the basis vectors ($\vec{a}$, $\vec{b}$, $\vec{c}$), then, the eigenvalue of $\vec{l} = m\vec{a} + n\vec{b} + o\vec{c}$, is $(\epsilon_a)^m \cdot (\epsilon_b)^n \cdot (\epsilon_c)^o$. For convenience (and for homogeneity with the additive notation - see Eq. (3.1)- of the translational symmetries), it is preferred to introduce three **real** numbers ρ, σ, τ, called phases, that are fractions of a full angle. In this way the eigenvalues, which are complex numbers of modulus = 1, are expressed in exponential notation, as seen in Section 2.1 for the commutative groups of molecular symmetries (see also Section 5.1.4). In other words, it is preferred to write:

$$\epsilon_a = e^{2\pi i\rho}, \quad \epsilon_b = e^{2\pi i\sigma}, \quad \epsilon_a = e^{2\pi i\tau} \tag{3.4}$$

where the phases ρ, σ, τ are characteristic of the symmetry species. Thus, the eigenvalue for the $\vec{l} = m\vec{a} + n\vec{b} + o\vec{c}$ translation is:

$$e^{2\pi i(m\rho + n\sigma + o\tau)} = \text{eigenvalue of } \vec{l} = m\vec{a} + n\vec{b} + o\vec{c} \tag{3.5}$$

and the phase $m\rho + n\sigma + o\tau$ for the translation $\vec{l}$ is written as a linear combination of (ρ, σ, τ) with *the same coefficients as in the translation vector.*

Let's consider a linear combination of the reciprocal triplet constructed using as coefficients the phases (ρ, σ, τ) of a given symmetry species:

$$\vec{k} = 2\pi\left(\rho\vec{a}^* + \sigma\vec{b}^* + \tau\vec{c}^*\right) \tag{3.6}$$

From Eq. (3.3) we obtain that the imaginary exponential of the scalar product of a vector $\vec{k}$ and a direct vector $\vec{l}$.

$$\vec{k}\cdot\vec{l} = 2\pi(m\rho+n\sigma+o\tau) \quad \text{and:} \quad e^{i\vec{k}\cdot\vec{l}} = e^{2\pi i(m\rho+n\sigma+o\tau)} \tag{3.7}$$

gives the eigenvalue (Eq. 3.5) of translation $\vec{l}$ in the given symmetry species which can now be synthetically identified by $\vec{k}$ (Eq. 3.6). We can therefore conclude by stating the

> **Bloch's Theorem**: The eigenvalue of the translation $\vec{l}$ is $e^{i\vec{k}\cdot\vec{l}}$ where the vector $\vec{k}$ is characteristic of the symmetry species.

It follows that:

> A $\vec{k}$ vector identifies a symmetry species.

3.2.1 Equivalent Statements of the Theorem

Bloch's theorem has an importance that is difficult to underestimate and therefore it is useful to discuss its meaning from several points of view.

The first meaning – just derived – is that, if $\psi(\vec{r})$ is an eigenfunction of a Hamiltonian that is characterized by translational invariance and $\hat{T}_{\vec{l}}$ is a translation operator, then:

$$\hat{T}_{\vec{l}}\left(\psi(\vec{r})\right) \overset{\text{def}}{=} \psi\left(\vec{r}+\vec{l}\right) = e^{i\vec{k}\cdot\vec{l}}\cdot\psi(\vec{r}) \tag{3.8}$$

where $\vec{k}$ is specific of $\psi(\vec{r})$ and therefore it is specified as a subscript:

$$\hat{T}_{\vec{l}}\left(\psi_{\vec{k}}(\vec{r})\right) \overset{\text{def}}{=} \psi_{\vec{k}}\left(\vec{r}+\vec{l}\right) = e^{i\vec{k}\cdot\vec{l}}\cdot\psi_{\vec{k}}(\vec{r}) \tag{3.9}$$

A second way of looking at Bloch's theorem can be obtained by considering:

$$\psi_{\vec{k}}(\vec{r}) = e^{i\vec{k}\cdot\vec{r}}\cdot u_{\vec{k}}(\vec{r}) \text{ where } u_{\vec{k}}(\vec{r}) \text{ is a function to be determined} \tag{3.10}$$

Note: the decomposition of Eq. (3.10) is *always* allowed, as $e^{i\vec{k}\cdot\vec{r}} \neq 0$ for any $\vec{k}$ or $\vec{r}$. Warning: $e^{i\vec{k}\cdot\vec{r}}$ *is not* a Bloch's factor! The exponential contains a generic 3D vector ($\vec{r}$), *and not* a lattice vector $\vec{l}$. By elaborating the formula and using the Bloch's theorem (Eq. 3.8), we derive that $u_{\vec{k}}(\vec{r})$ **is a periodic function**, that is, a function that has the periodicity of the translation lattice:

$$u_{\vec{k}}(\vec{r}) = u_{\vec{k}}\left(\vec{r}+\vec{l}\right) \tag{3.11}$$

Indeed $\hat{T}_{\vec{l}}\left(\psi_{\vec{k}}(\vec{r})\right) \overset{\text{def}}{=} \psi_{\vec{k}}\left(\vec{r}+\vec{l}\right) = e^{i\vec{k}\cdot(\vec{r}+\vec{l})}\cdot u_{\vec{k}}\left(\vec{r}+\vec{l}\right)$, and $\hat{T}_{\vec{l}}\left(\psi(\vec{r})\right) \overset{\text{def}}{=} e^{i\vec{k}\cdot\vec{l}}\cdot\psi(\vec{r})$

That is $\hat{T}_{\vec{l}}\left(\psi(\vec{r})\right) \overset{\text{def}}{=} e^{i\vec{k}\cdot\vec{l}}\cdot e^{i\vec{k}\cdot\vec{r}}\cdot u_{\vec{k}}(\vec{r})$, from which $u_{\vec{k}}\left(\vec{r}+\vec{l}\right) = u_{\vec{k}}(\vec{r})$.

The solution of a Schrödinger equation for a periodic potential is then the product of:

- an oscillating function (a wave) $e^{i\vec{k}\cdot\vec{r}} = \cos\left(\vec{k}\cdot\vec{r}\right) + i\sin\left(\vec{k}\cdot\vec{r}\right)$ with wavevector $\vec{k}$, times
- **a periodic function** $u_{\vec{k}}(\vec{r})$, *having the periodicity of the translation lattice.*

The third form of Bloch's theorem, finally, concerns specifically SALCs, as will be discussed in Section 3.3 and specifies the coefficients, which are given by the Bloch factor $e^{i\vec{k}\cdot\vec{l}}$, as this is the eigenvalue for the translation $\vec{l}$ for the species $\vec{k}$.

3.2.2 The Reciprocal Lattice

Each symmetry species is identified by a $\vec{k}$. Different species have different $\vec{k}$ but the correspondence *is not* biunivocal. A symmetry species corresponds to *an infinite number of distinct* $\vec{k}$. To explain, we start from the observation that, with the reciprocal triplet ($\vec{a}^*$, $\vec{b}^*$, $\vec{c}^*$, Eq. (3.2)) we can construct a whole **reciprocal lattice** (**RL**) defined as:

$$\left(\vec{a}^*, \vec{b}^*, \vec{c}^*\right) \Rightarrow \vec{g}_{rst} = 2\pi\left(r\vec{a}^* + s\vec{b}^* + t\vec{c}^*\right) \quad (r,s,t \text{ positive, negative or null integers}) \tag{3.12}$$

Notes. (i) When appropriate for the sake of clarity, the "original" lattice constructed from the ($\vec{a}$, $\vec{b}$, $\vec{c}$) triplet and formed by the $\vec{l}$ vectors of Eq. (3.1) is then called *direct lattice* (DL). (ii) The 2π *factor in* Eq. (3.12) is only inserted to make formulas that use the RL simpler - see for example Eqs. (3.13) and (3.14) - and could be omitted. In many cases (particularly in crystallography), in fact, the choice is to omit this factor, thus creating a complete parallelism with Eq. (3.1) but at the cost of a more complicated notation in subsequent relations, in which 2π must then appear explicitly. In other cases, 2π is inserted even earlier, that is, in the definition of reciprocal triplet, and Eq. (3.2) is modified consequently.

From the definition and basic property of the reciprocal triplet (Eqs. 3.2 and 3.3), we can derive a fundamental **property of the reciprocal lattice**, *i.e.* that

> The scalar product of *any* RL vector by *any* DL vector

$$\forall\vec{g} = 2\pi\left(r\vec{a}^* + s\vec{b}^* + t\vec{c}^*\right) \quad \text{and} \quad \forall\vec{l} = m\vec{a} + n\vec{b} + o\vec{c} \tag{3.13}$$

> is an integer number of turns: $\vec{g}\cdot\vec{l} = 2\pi(rm+sn+to) = 2\pi \cdot$ integer, so that the **imaginary exponential** of the product **is equal to 1**:

$$e^{i\vec{g}\cdot\vec{l}} = e^{2\pi i(rm+sn+to)} = 1 \quad (3.14)$$

Returning now to the correspondence between symmetry species and reciprocal space vectors, let us examine two vectors, $\vec{k}$ and $\vec{k}'$, differing by a *reciprocal lattice vector* $\vec{g}$: $\vec{k}' = \vec{k} + \vec{g}$. We will then have that the exponential of $\vec{k}'$ is $e^{i\vec{k}\cdot\vec{l}} \cdot e^{i\vec{g}\cdot\vec{l}} = e^{i\vec{k}\cdot\vec{l}}$.

The Bloch's vector $\vec{k}$ of a symmetry species is defined up to a RL vector.

To specify a symmetry species, it is sufficient to specify *one point within a* (primitive) *unit cell of the reciprocal lattice.* A point in the equivalent position of *another unit cell of the RL* identifies *the same symmetry species*.

The vector space containing the $\vec{k}$ of Bloch's theorem and the RL vectors is called **reciprocal space**. The relationship, in the reciprocal space, between the $\vec{k}$ vectors of Bloch's theorem and the RL vectors is the same as the relation between $\vec{r}$ and $\vec{l}$ vectors in ordinary space (Eq. 3.1).

As the unit cell of the RL, it is usually selected the cell that is constructed by performing in the RL the same operations that allow to obtain the Wigner-Seitz cell of the DL (see Section 3.1.2). This cell is primitive, is centred on the origin of the reciprocal space and contains the $\vec{k}$ vectors closest to the origin, *i.e.* of the smallest possible modulus. Therefore, it is the most convenient cell for the applications: for historical reasons it is known as of the **first Brillouin zone**.

3.2.2.1 Periodic Functions

The functions that have the periodicity described by a translational lattice can be written as a **Fourier series**, that is, as a series of sine and cosine functions (+ a constant) or, equivalently, as a series of complex exponentials. The argument of the functions of the series is the scalar product $\vec{g}\cdot\vec{r}$ ($\vec{g}$ = reciprocal lattice vector) and the series is over all the vectors of the RL:

$\sum_{\vec{g}} A_{\vec{g}} \cdot e^{i\vec{g}\cdot\vec{r}}$ is a function with the periodicity of the DL

Indeed: $f(\vec{r}+\vec{l}) = \sum_{\vec{g}} A_{\vec{g}} \cdot e^{i\vec{g}\cdot(\vec{r}+\vec{l})} = \sum_{\vec{g}} A_{\vec{g}} \cdot e^{i\vec{g}\cdot\vec{r}} \cdot e^{i\vec{g}\cdot\vec{l}} =$
$\sum_{\vec{g}} A_{\vec{g}} \cdot e^{i\vec{g}\cdot\vec{r}} = f(\vec{r})$, as $e^{i\vec{g}\cdot\vec{l}} = 1$ (Eq. 3.14).

A periodic function is identified by the set of the ($A_{\vec{g}}$) coefficients, each relating to a $\vec{g}$. A trivial consequence is that

If a Bloch's vector $\vec{k}$ is equal to a RL vector, it identifies the **total symmetric species**.

This is obviously valid when $\vec{k} = 0$, that is the standard choice.

3.2.3 Counting the Symmetry Species

To arrive at the results of interest here, we need to overcome the difficulties caused by having to deal with an *infinite number of symmetry operations*.

An important difference must be noted with respect to other symmetry groups, for example those of linear molecules, which are also characterized by an infinite number of symmetry operations. Dealing with an infinite number of symmetry operations *here* implies dealing with an *infinite number of unit cells*, that is, an "*infinitely large* system" in terms of extension in space and in terms of the number of constituent atoms, ions or molecules and degrees of freedom. For comparison, consider the infinite rotation operations present in a *single* molecule of carbon dioxide.

A *real* crystalline solid is made up of a "very large" number of unit cells, for example: $10^{23} \div 10^{24}$, if we refer to a *single crystal* containing an amount of substance of the order of a mole. It is evident that the difficulty cannot be avoided by simply modifying the definition of translation lattice (Eq. 3.1) by introducing appropriate limits for the m, n, o numbers for example, between -10^8 and $+10^8$. Following this reasoning, the closure property of the translational symmetry group would fail.

On the other hand, the structural elements (atoms, ions, molecules) that are close to the surface of a real crystalline solid are in a bonding situation that is certainly different from those well inside the crystal. Indeed, various **surface phenomena** occur, primarily adsorption and catalysis. It is appropriate to keep the *bulk* properties of a solid separate from the surface properties: they are in fact clearly different from the conceptual point of view and are somehow operationally distinguishable from the experimental point of view, for example by varying the ratio between volume and surface area of a sample or the penetration depth of a characterization technique.

To deal with a crystalline solid, a theoretical model is introduced in which the external surface simply does not exist, and the translational symmetry is rigorously respected even with a finite number of unit cells thanks to special boundary conditions. For clarity, it is convenient to first present the model in its 1D version, in which there is only one basis vector (a), the lattice is defined by $l = na$ and translational invariance is expressed by $r \rightarrow r + l$. We now have a single basis reciprocal vector $a^* = 1/a$, and symmetry species are identified by the scalar $k = 2\pi\alpha a^*$, where α specifies the symmetry species, and the eigenvalues of translation $l = na$ are $e^{i(kl)} = e^{2\pi i(n\alpha)}$.

To bring together translational invariance and the absence of an external surface with a finite, albeit "very large", number of unit cells N, we resort to **cyclic boundary conditions**, that is, we carry out the calculations "as if", after having moved to a very large distance $L = Na$, we were to return to the origin of the lattice. Whatever the symmetry species and therefore the parameter α, the eigenvalue $e^{2\pi i(N\alpha)}$ must then be $= 1$, because the cell at the origin and

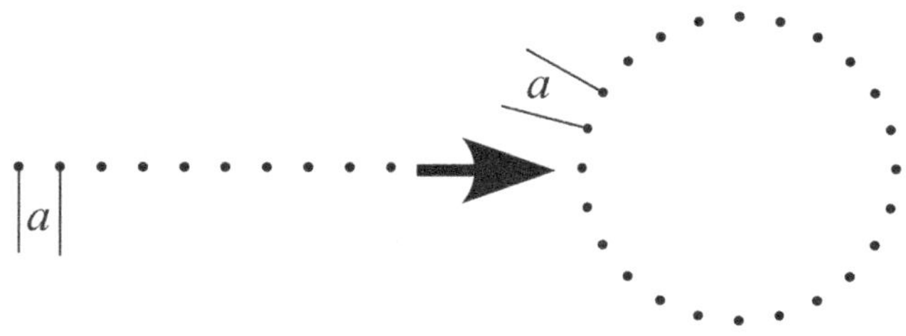

FIGURE 4.21 Born-von Karman boundary condition for a 1D lattice.

the cell translated by Na are not only equivalent by symmetry but are now *the same cell.*

The cyclic boundary conditions are known as **Born–von Karman** (B-vK) conditions and can be visualized as a folding of the 1D crystal into a second dimension (or into the third dimension, for a 2D crystal), as illustrated by Figure 4.21.

The *physical meaning* of the model is that a *finite portion of a crystal* is *immersed* in other identical portions of crystal and therefore the *surface effects are cancelled*, and the translational symmetry is not broken. The essence of the model, therefore, is that:

$$e^{2\pi i(N\alpha)} = 1 \Rightarrow N\alpha = j; \quad (j \text{ is an integer}) \Rightarrow \alpha = \frac{j}{N} \tag{3.15}$$

For a given symmetry species identified by j, or by $k = 2\pi j/(Na)$ the following applies.

- We need to consider only N distinct translations: $l = 0a, a, \ldots, (N-1)a$, because the translations: $l = Na, (N+1)a, \ldots$ do nothing more than reproduce the same crystal.
- The eigenvalues of distinct translations are: $e^0 = 1$, $e^{2\pi i(j/N)}, e^{2\pi i(2j/N)}, \ldots, e^{2\pi i((N-1)j/N)}$.
- Finally, using $\varepsilon \overset{\text{def}}{=} e^{2\pi i(j/N)}$, the eigenvalues are $1 = \varepsilon^0, \varepsilon, \varepsilon^2, \ldots, \varepsilon^{N-1}$, with $\varepsilon^N = 1$.

Each symmetry species has its own j. This integer can take all the distinct values between 0 and $N-1$ and each j corresponds, for the fundamental translation a, to eigenvalues that are the N distinct N-th roots of unity:

$$\varepsilon_j \overset{\text{def}}{=} e^{2\pi i\left(\frac{j}{N}\right)}; \quad j = 0, \ldots, N-1$$

However, the interval $0 \div N-1$ is usually substituted by the equivalent interval that is symmetric with respect to 0 and goes from $-N/2$ to $+N/2$, neglecting the distinction between even and odd N, since N is "very large". This choice corresponds to the first Brillouin zone.

Considering a value outside this interval means moving an integer number of $2\pi/a$ steps in k, that is, moving by a vector of the reciprocal lattice, exiting the first Brillouin zone, reaching an equivalent position in another cell of the RL and therefore finding the eigenvalue already considered.

If the 1D solid is made up of N unit cells, it also has N symmetry species, because it has N distinct values of k, in agreement with the result already found that – in a commutative group – *the number of symmetry species is equal to the number of symmetry operations.*

Now, the extension to 3D is trivial. The $\vec{k} = 2\pi\left(\rho\vec{a}^* + \sigma\vec{b}^* + \tau\vec{c}^*\right)$ vector (Eq. 3.6) becomes:

$$\vec{k} = 2\pi\left(\frac{p}{N_a}\vec{a}^* + \frac{q}{N_b}\vec{b}^* + \frac{r}{N_c}\vec{c}^*\right) \tag{3.16}$$

Here N_a, N_b, N_c are of the order of $10^7 \div 10^8$ for a macroscopic crystal, and $N = N_a N_b N_c$ is the total number of unit cells in the crystal. The integers p, q, and r vary in the intervals $-N_a/2 \div N_a/2$, $-N_b/2 \div N_b/2$, and $-N_c/2 \div N_c/2$. The total amount of non-equivalent $\vec{k}$ vectors is $N = N_a N_b N_c$, and each represents a different symmetry species. If we want to refer specifically to a mole of substance and to an isotropic crystal, we put $N_a = N_b = N_c = L^{1/3}$ for a solid with L *unit cells* (L = Avogadro constant).

As already mentioned, the $\vec{k}$ vectors are in the **first Brillouin zone** and are distributed discretely and with constant spacing in each of the three directions of the RL basis vectors, as indicated by Eq. (3.17) and (very) schematically illustrated by Figure 4.22 for the 2D case. Neglecting the discrete nature, one can speak of uniform density $= N/V^*$, where V^* is the volume of a unit cell of the reciprocal lattice which is, up to the factor $(2\pi)^3$, the inverse of the volume (V_c) of the unit cell in the DL:

$$V^* = 8\pi^3 \cdot \vec{a}^* \cdot \vec{b}^* \wedge \vec{c}^* = 8\pi^3\left(\vec{a} \cdot \vec{b} \wedge \vec{c}\right)^{-1} = 8\pi^3 / V_c \tag{3.17}$$

3.2.4 The Density of States

In a solid, the separation in energy between successive electronic or vibrational levels is much lower than that occurring in molecules in the gaseous phase. To give a rough indication, in an energy interval of the order of magnitude

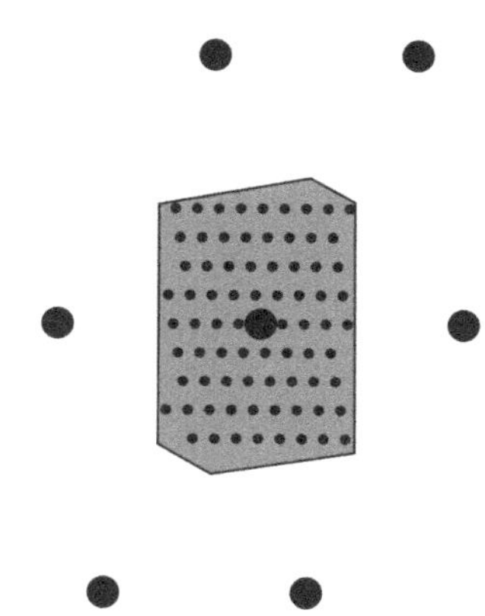

FIGURE 4.22 A Wigner-Seitz cell (grey area) of the RL (large dots). The smaller dots only *suggest* the vectors of Bloch's theorem, because these vectors are actually much more finely spaced (*e.g.* 10^8 along a row in one cell).

of τηε typical HOMO-LUMO separation in a molecule, the crystalline orbitals present a number of energy levels of the order of N (N=number of unit cells of which the macroscopic solid is made). The difference in energy between a level and the next is therefore smaller by a factor of $N \sim 1020$.

For this reason, we speak of **energy bands**, a fundamental characteristic of the electronic or vibrational energy levels of a solid. The point deserves some clarification. The formation of bands of very close levels *is not* the effect of translational symmetry, but of the extension over macroscopic distances of the interactions between adjacent atoms. As a counterproof, the metallic-like electrical conductivity associated with partially occupied bands of mono-electronic levels, also occurs in *liquid metals*. The role of symmetry is essentially to **allow the calculation** of the band structure and to provide a relatively easy treatment to explain the experimental phenomenology.

The energy separation is also much smaller than the "thermal quantum" kT and therefore it is reasonable - and usually convenient - to treat the energy levels "as if" they were continuously distributed, thus introducing a density of states $\rho(E)$, such that the number of states in the energy range $E_1 \div E_2$ is given by the integral of $\rho(E)\,dE$ between E_2 and E_1. We here just give the definition:

$$\rho(\bar{E}) = \frac{V_c}{8\pi^3} \oiint_{E=\bar{E}} \frac{dS_k}{\nabla_k\left[E(k)\right]} \tag{3.18}$$

We need a continuous function $E(\vec{k})$ and its gradient in the $\vec{k}$ space: $\nabla_{\vec{k}}\left[E(\vec{k})\right]$, which is sometimes written as $\partial E(\vec{k})/\partial\vec{k}$. Then, for the energy ($\bar{E}$) at which one wants to determine the density of states, one must calculate the integral of the inverse of the gradient on the level surface at energy=$\bar{E}$. In non-rigorous terms, this integral is the volume of the film having infinitesimal width in energy. Considering that the volume of the Brillouin zone is $8\pi^3/V_c$ (see Eq. (3.17)) and contains N states, one switches from a volume to a number of states.

3.3 Crystal Orbitals as LCAO

Using the same treatment already used for the MOs of molecules with commutative symmetry (see Section 2.2), we can write the SALC of symmetry $\vec{k}$ and based on the prototype $\varphi(\vec{r})$ AO as:

$$\psi_{\vec{k}}(\vec{r}) = \sum_{\vec{l}} e^{i\vec{k}\cdot\vec{l}}\varphi_{\vec{l}}(\vec{r}) \tag{3.19}$$

where $\varphi_{\vec{l}}(\vec{r})$ is the corresponding AO of the corresponding atom in unit cell $\vec{l}$. The $e^{i\vec{k}\cdot\vec{l}}$ is the eigenvalue of the $\vec{l}$ translation in the $\vec{k}$ species and the summation is over all unit cells in the crystal, that is over all the translational symmetries. There are as many equations like Eq. (3.19) as are the different values of $\vec{k}$, in agreement with the B-vK boundary conditions (see Eq. (3.15)), considering equivalent two $\vec{k}$ vectors that differ by a RL vector (see Section 3.2.2). As already seen for the MOs, each *crystal orbital* (hereafter: CO) belongs to a specific symmetry species, here identified by $\vec{k}$ and is obtained by combining SALCs that belong to the same $\vec{k}$. If $\vec{k}$ is treated as a *continuous variable*, the entire Hamiltonian matrix $\boldsymbol{H}$ has dimensions given by the **number of linearly independent prototype AOs**, but each element of $\boldsymbol{H}$ is a function of the variable $\vec{k}$ and contains the kinetic energy and the interactions of an electron with all nuclei:

$$\left\langle \varphi_{rs\vec{l}}(\vec{r})\middle|\hat{H}\middle|\varphi_{r's'\vec{l}'}(\vec{r})\right\rangle = \left\langle rs\vec{l}\middle|\hat{H}\middle|r's'\vec{l}'\right\rangle \tag{3.20}$$

where $\vec{l}$ and $\vec{l}'$ indicate two unit-cells, s and s' indicate two atoms in their cells, and r and r' indicate two AOs of the pertinent atoms. Because of translational invariance, these integrals *do not* depend on individual cells ($\vec{l}$ and $\vec{l}'$) but only on the difference $\vec{L} \overset{\text{def}}{=} \vec{l}' - \vec{l}$ and we write $\left\langle rs\middle|\hat{H}\middle|r's'\vec{L}\right\rangle \overset{\text{def}}{=} \left\langle rs\vec{0}\middle|\hat{H}\middle|r's'\vec{L}\right\rangle$. Equation (3.19) combines AOs multiplied by a Bloch's factor:

$$\left\langle e^{i\vec{k}\cdot\vec{l}}\cdot\varphi_{rs\vec{l}}(\vec{r})\middle|\hat{H}\middle|e^{i\vec{k}\cdot\vec{l}'}\cdot\varphi_{r's'\vec{l}'}(\vec{r})\right\rangle$$
$$= \cdots = e^{i\vec{k}\cdot\vec{L}}\cdot\left\langle rs\middle|\hat{H}\middle|r's'\vec{L}\right\rangle \tag{3.21}$$

Finally, considering the double summation: $\sum_{\vec{l}}\sum_{\vec{l}'} e^{i\vec{k}\cdot(\vec{l}'-\vec{l})}\cdot\left\langle rs\vec{l}\middle|\hat{H}\middle|r's'\vec{l}'\right\rangle$, and putting $\vec{L} = \vec{l}' - \vec{l}$, we have:

$$\sum_{\vec{l}}\sum_{\vec{L}} e^{i\vec{k}\cdot\vec{L}}\cdot\left\langle rs\vec{l}\middle|\hat{H}\middle|r's'(\vec{L}+\vec{l})\right\rangle$$

The first summation merely reproduces N times the integral of Eq. (3.21) so that:

$$\sum_{\vec{l}}\sum_{\vec{l}'} e^{i\vec{k}\cdot(\vec{l}'-\vec{l})}\cdot\left\langle rs\vec{l}\middle|\hat{H}\middle|r's'\vec{l}'\right\rangle$$
$$= N\sum_{\vec{L}} e^{i\vec{k}\cdot\vec{L}}\cdot\left\langle rs\middle|\hat{H}\middle|r's'\vec{L}\right\rangle \tag{3.22}$$

In a completely analogous way, for the overlap integrals:

$$\sum_{\vec{l}}\sum_{\vec{l}'} e^{i\vec{k}\cdot(\vec{l}'-\vec{l})}\cdot\left\langle rs\vec{l}\middle|r's'\vec{l}'\right\rangle$$
$$= N\cdot\sum_{\vec{L}} e^{i\vec{k}\cdot\vec{L}}\cdot\left\langle rs\middle|r's'\vec{L}\right\rangle \tag{3.23}$$

3.3.1 Simple Examples

To show method and results, it is convenient to use the simple LCAO-HMO method (HMO: Hückel Molecular Orbital) as already done for butadiene (see Section 2.2.1) and benzene (see Section 1.2.2.2):

- The $\langle rs|\hat{H}|rs\vec{0}\rangle = \alpha$ are *coulomb integrals* and depend on the AO (r) and on the chemical nature of s,
- The $\langle rs|\hat{H}|r's'\vec{L}\rangle \overset{\text{def}}{=} \langle rs\vec{0}|\hat{H}|r's'\vec{L}\rangle = \beta(s \neq s')$ are *resonance integrals*, depend on r, r', s, s', $\vec{L}$ and are different from zero if and only if the s and $s'\vec{L}$ atoms are adjacent,
- The normalization integrals are = 1, the *overlap* integrals between different AO are = 0.

For simplicity, in the examples we restrict ourselves to a 1D lattice of lattice constant a. Then $L = 0, a, -a, 2a, -2a, 3a, -3a, \ldots$.

3.3.1.1 Only One AO Per Unit Cell

Let us start with the simplest case: the unit cell contains only one atom, with only one AO. Equation (3.19) becomes:

$$\psi_k(r) = \sum_l e^{ik\cdot l}\varphi_l(r) \tag{3.24}$$

and describes the CO with k symmetry. The SALC is itself a CO, as seen for similar cases of MOs built on AOs that are all *symmetry equivalent* to each other. The energy is calculated as expectation value using Eqs. (3.22) and (3.23):

$$E(k) = \frac{\langle k|\hat{H}|k\rangle}{\langle k|k\rangle} = \frac{N\cdot\sum_L e^{ik\cdot L}\cdot\langle 0|\hat{H}|L\rangle}{N\cdot\sum_L e^{ik\cdot L}\cdot\langle 0|L\rangle}$$

$$= \sum_L e^{ik\cdot L}\cdot\langle 0|\hat{H}|L\rangle / \langle 0|L\rangle \tag{3.25}$$

Adding at the numerator and dividing by the denominator we get:

$$E(k) = \alpha + \beta\cdot\left(e^{ik\cdot a} + e^{-ik\cdot a}\right) = \alpha + 2\cdot\beta\cdot\cos(k\cdot a) \tag{3.26}$$

Equation (3.26) explicitly shows that the solution to the CO problem is obtained as a function of the variable k. From N AOs, we write N independent linear combinations (SALCs) using Bloch's factors with N different values of k.

The CO energy (Eq. 3.26) varies between $\alpha+2\beta$ and $\alpha-2\beta$: the **band width** (4β) is directly given by the interaction between adjacent AOs (β). Each energy level is two-fold degenerate, except for $k=0$. This is a general feature, called $\vec{k}\,/-\vec{k}$ **degeneracy**: in this respect, in addition to what already seen in Section 2.3, see also Sections 3.3.2 and 6.6.6.

3.3.1.2 More Than One AO Per Unit Cell

We can then consider a 1D solid with two atoms per unit cell (atom A in the origin and atom B in the middle of the cell) and a single s AO, which we label as the atom. The Hamiltonian matrix elements for equal or different AO are

$$\alpha_A = \langle A|\hat{H}|A\rangle, \alpha_B = \langle B|\hat{H}|B\rangle \text{ and } \beta = \langle B|\hat{H}|A\rangle = \langle A|\hat{H}|B\rangle.$$

Using normalized AO, the normalized SALC are:

$$\psi_{Ak}(r) = \left(1/\sqrt{N}\right)\sum_l e^{ik\cdot l}\varphi_{Al}(r) \quad \text{and}$$

$$\psi_{Bk}(r) = \left(1/\sqrt{N}\right)\sum_l e^{ik\cdot l}\varphi_{Bl}(r)$$

and can be written as $|Ak\rangle$ and $|Bk\rangle$ in Dirac notation. In the SALC basis, the Hamiltonian is:

$$\boldsymbol{H} = \begin{pmatrix} \langle Ak|\hat{H}|Ak\rangle & \langle Ak|\hat{H}|Bk\rangle \\ \langle Bk|\hat{H}|Ak\rangle & \langle Bk|\hat{H}|Bk\rangle \end{pmatrix} \tag{3.27}$$

With the same methods as in Eq. (3.22):

$$\boldsymbol{H} = \begin{pmatrix} \alpha_A & \beta\left(1+e^{-ika}\right) \\ \beta\left(1+e^{+ika}\right) & \alpha_B \end{pmatrix} \tag{3.28}$$

so that

$$E = \frac{1}{2}\left\{\alpha_A + \alpha_B \pm \sqrt{(\alpha_A-\alpha_B)^2 - 8\beta^2[1+\cos(ka)]}\right\}$$

For the graphical representation (see Figure 4.23), it is worth noting that:

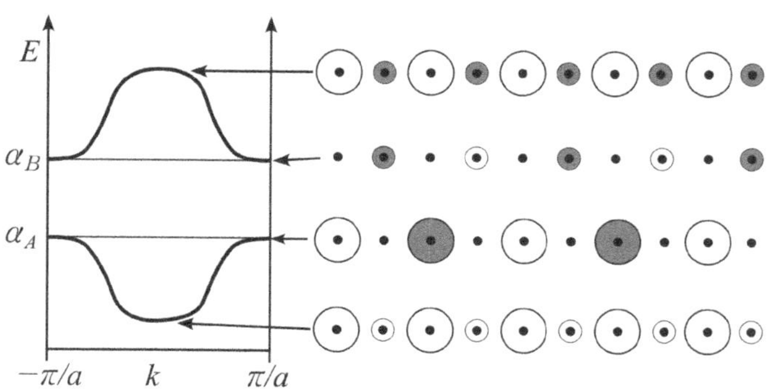

FIGURE 4.23 Electron energy relations $E(k)$ for the 1D case with two atoms per unit cell (A, B) and only one AO per atom. The drawings show the COs for k at the origin and at the edge of the Brillouin zone. Large and small circles are the A and B AOs. The lattice constant is a.

For $|k|=0,\ E=\frac{1}{2}\left\{\alpha_A+\alpha_B\pm\sqrt{(\alpha_A-\alpha_B)^2-16\cdot\beta^2}\right\}$

For $ka=\pi,\ E=\frac{1}{2}\left\{\alpha_A+\alpha_B\pm\sqrt{(\alpha_A-\alpha_B)^2}\right\}=\begin{cases}E=\alpha_A\\E=\alpha_B\end{cases}$

3.3.2 Dispersion Relations for Electrons

Much of the considerations in these examples can be generalized, provided we overcome some limitations.

- The approximations of the HMO method are independent of symmetry considerations and are pure computational details used for the sole purpose of obtaining illustrative results in a simple way. Obvious extensions would be, for example, the introduction of *overlap* integrals different from zero and resonance integrals between non-nearest neighbouring atoms.
- Moving from a 1D to a 2D or 3D model has several consequences: the bandwidth increases (in the HMO approach it becomes 8β or 12β), anisotropy effects may be encountered with noncubic crystalline systems, and the graphical representation becomes more complicated. However, even in these cases, the treatment of translational symmetry does not change substantially.

We can now see how translational symmetry simplifies the construction of CO as LCAO. Consider, in fact, a basis of m AO per unit cell. For a macroscopic crystal of N primitive cells (N of the order of 10^{20}–10^{24}), the Hamiltonian matrix to be diagonalised would have dimensions $mN\times mN$ and the direct solution would require a drastic reduction of N by several orders of magnitude. Thanks to translational symmetry and Bloch's theorem, the original matrix is transformed into an $m\times m$ matrix whose elements are functions of $\vec{k}$ (see Eq. 3.27 and Figure 4.23 for the simple 1D case with m=2). This sometimes gives the analytical solution (see Eqs. (3.27) and (3.29)). In any case, the numerical solution is available sampling on selected values of $\vec{k}$, and gives both the eigenvalues (the energy) and the eigenvectors, *i.e.* the coefficients of the various AOs within a CO). The results $E(\vec{k})$ are known as **dispersion relations**. It is found that the bandwidth is of the same order of magnitude as the HOMO-LUMO separation between MOs, as indicated by the examples above, where a small multiple of β is obtained. In this energy range, a solid has a number of CO of the order of N.

Knowing the total number of electrons in the system, one can then calculate the total electronic energy, charge densities and bond indices. Finally, by examining the different crystal structures that are possible for a given composition, it is possible to compare the contributions of the electronic energy to the relative stability of different crystal structures.

As seen in Section 2.1.3, when the Hamiltonian operator is real, time-reversal symmetry essentially implies that if a function is a solution of the time-independent Schrödinger equation, its conjugate is also a solution of the same equation with the *same eigenvalue*:

$$\hat{H}\psi(\vec{r})=E\cdot\psi(\vec{r})\Rightarrow\hat{H}\psi^*(\vec{r})=E\cdot\psi^*(\vec{r})$$

For Bloch's theorem:

$$\hat{T}_{\vec{l}}\left(\psi_{\vec{k}}(\vec{r})\right)\overset{\text{def}}{=}\psi_{\vec{k}}\left(\vec{r}+\vec{l}\right)=e^{i\vec{k}\cdot\vec{l}}\cdot\psi_{\vec{k}}(\vec{r}) \tag{3.9}$$

The last equality for the conjugates is: $\psi^*_{\vec{k}}\left(\vec{r}+\vec{l}\right)=e^{-i\vec{k}\cdot\vec{l}}\cdot\psi^*_{\vec{k}}(\vec{r})$.

Therefore, for the conjugate $\psi^*_{\vec{k}}(\vec{r})$, the eigenvalue of the translation operator is $-\vec{k}$ and the eigenvalue for the Hamiltonian in the same $E(\vec{k})$. This has two consequences for the dispersion relations that apply both to electrons and phonons.

The dispersion relations have

1. inversion symmetry in the reciprocal space: $E(\vec{k})=E(-\vec{k})$ and
2. zero derivative at the zone border.

The latter statement is proved by considering a vector $\vec{k}=\vec{g}/2$ lying at a zone boundary. Two $\vec{k}$ close to the zone boundary can be written as: $\vec{k}=\vec{g}/2\pm\delta\vec{k}$. For *time reversal*: $E(\vec{g}/2+\delta\vec{k})=E(-\vec{g}/2-\delta\vec{k})$; periodicity of $E(\vec{k})$ implies that $E(-\vec{g}/2-\delta\vec{k}+\vec{g})=E(\vec{g}/2-\delta\vec{k})=E(\vec{g}/2+\delta\vec{k})$.

Other important consequences of *time-reversal symmetry* (not discussed here) concern all the problems involving **spin**, in particular the spin-orbit interaction.

3.4 "Lattice" Vibrations

The treatment of vibrations[5] of a periodic crystalline solid is similar to that already seen for molecules and based on the **Born-Oppenheimer** (BO) **separation** and the **harmonic approximation**. When necessary, the BO approximation can be overcome by treating electron-phonon interactions (phonons are the quanta of crystal vibrations).

In a crystal with N primitive unit cells and s atoms per primitive cell there are $3Ns$ atomic degrees of freedom. Using Cartesian coordinates for the deviations u of atoms from their equilibrium positions and writing the potential energy with only harmonic terms, *i.e.* with only the coefficients relating to the second derivatives, we have:

$$\begin{aligned}V&\cong\frac{1}{2}\sum_{\vec{l},\alpha,j,\vec{m},\beta,k}u_{\vec{l},\alpha,j}\cdot u_{\vec{m},\beta,k}\cdot\frac{\partial^2 V}{\partial u_{\vec{l},\alpha,j}\cdot\partial u_{\vec{m},\beta,k}}\\&=\frac{1}{2}\cdot\boldsymbol{u}^{\mathrm{T}}\cdot\boldsymbol{V}\cdot\boldsymbol{u},\ \text{where}\end{aligned} \tag{3.29}$$

- The $\vec{l}$, $\vec{m}$ vectors refer to two (equal or different) unit cells,
- The α and β indexes refer to two (equal or different) atoms in the $\vec{l}$ and $\vec{m}$ unit cells, respectively, while:
- The j and k indexes refer to two (equal or different) cartesian x, y, z components.

We put $V_{\vec{m},\beta,k}^{\vec{l},\alpha,j} \overset{\text{def}}{=} \dfrac{\partial^2 V}{\partial u_{\vec{l},\alpha,j} \cdot \partial u_{\vec{m},\beta,k}}$. These **force constants** are expected to be particularly strong for first neighbour interactions and then decay rapidly with distance. The matrix $\boldsymbol{V}$ is symmetric because the order of derivation is irrelevant.

The motion equations are:

$$\ddot{u} = \boldsymbol{D} \cdot \boldsymbol{u} = -\boldsymbol{M}^{-1} \cdot \boldsymbol{V} \cdot \boldsymbol{u}, \text{ where}$$
$$\boldsymbol{D} = -\boldsymbol{M}^{-1} \cdot \boldsymbol{V} \text{ is the dynamical matrix, and}$$
$$(\boldsymbol{M}^{-1})_{ij} = \delta_{ij} / m_k \ (i, j = 3k\text{-}2, 3k\text{-}1, 3k, k = 1, \ldots, Ns) \quad (3.30)$$

contains the reciprocal of the masses.

Translational invariance implies that the $\boldsymbol{V}$ *does not* depend on $\vec{l}$ and $\vec{m}$, but *only on* $\vec{l} - \vec{m}$:

$$V_{\vec{m},\beta,k}^{\vec{l},\alpha,j} = V_{\vec{0},\beta,k}^{\vec{l}-\vec{m},\alpha,j}, \text{ where } \vec{0} \text{ is the cell in the origin} \quad (3.31)$$

Symmetry also transforms the $\vec{u}$ coordinates into coordinates *adapted to translational symmetry*, according to Bloch's theorem. In detail, we consider that the Cartesian i components of the $\vec{u}$ displacements for each α atom in the $\vec{l}$ and $\vec{l}'$ unit cells are *equivalent* for translational symmetry. Thus, all coordinates that differ only by the $\vec{l}$ index are combined in SALC of $\vec{q}$ symmetry according to Bloch's theorem:

$$U_{i,\alpha}(\vec{q}) = \sum_{l} e^{i \cdot \vec{q} \cdot \vec{l}} \cdot u_{i,\alpha,\vec{l}} \quad \text{and}$$

$$U_{i,\alpha}(\vec{q},t) = e^{i \cdot \omega(\vec{q}) \cdot t} \sum_{\vec{l}} e^{i \cdot \vec{q} \cdot \vec{l}} \cdot u_{i,\alpha,\vec{l}} \quad (3.32)$$

Following the current use, the Bloch's vector is here indicated with $\vec{q}$ and not with $\vec{k}$, which is used for electronic states. The $\vec{u}$ coordinates are written as functions of the $\vec{U}$ by inverting Eq. (3.32), then their time derivatives are obtained, and finally we obtain the kinetic and potential energy as a function of the transformed variables. In the end, as already seen for molecules, we obtain a block-diagonal dynamical matrix $\boldsymbol{D}$ with N blocks (one for each $\vec{q}$ allowed by the B-vK conditions), each block of size $3s$. As seen for CO (see Section 3.3.1), diagonalization can be done by considering $\vec{q}$ as a continuous variable. Then, the dynamical matrix is $3s \times 3s$ and its elements are *function* of $\vec{q}$, and we obtain $3s$ functions $\omega = \omega(\vec{q})$) called *phonon dispersion relations*.

As already seen (Section 2.2.2), the solution is in the form of independent harmonic motions

$$Q_i = Q_{i,0} e^{i\omega_i t} \quad (2.11)$$

which can be found by diagonalising the dynamical matrix and then solving the eigenvalue equation:

$$= \boldsymbol{C} \cdot \boldsymbol{D} \cdot \boldsymbol{C}^{-1} \quad (2.12)$$

The diagonal $\boldsymbol{\Lambda}$ matrix contains the squared (ω^2) frequencies of normal vibrational modes, and the c_{ij} coefficients of column j of $\boldsymbol{C}$ are the linear contributions of u_i to the normal mode j:

$$\begin{pmatrix} Q_1 \\ \vdots \\ Q_{3s} \end{pmatrix} = \begin{pmatrix} c_{11} & \cdots & c_{1,3s} \\ \vdots & \ddots & \vdots \\ c_{3s,1} & \cdots & c_{3s,3s} \end{pmatrix} \cdot \begin{pmatrix} u_1 \\ \vdots \\ u_{3s} \end{pmatrix} \text{ that is: } \boldsymbol{Q} = \boldsymbol{C} \cdot \boldsymbol{u} \quad (2.9)$$

These equations are identical to what we have already seen for molecular vibrations, apart from the enormously larger number of atoms and therefore degrees of freedom involved.

3.4.1 Some Simple Models

As for the problem of COs, the examples given here are deliberately simple, because the aim is to illustrate the essential aspects of the problem, not to provide accurate results. For this purpose, two (further) approximations are made: (a) only the interactions between nearest neighbouring atoms are considered, and (b) we restrict to the 1D case for both the positions of the structural elements and the vibrational atomic coordinates.

3.4.1.1 1D Model with One Atom Per Unit Cell

With the above approximations, the potential energy is:

$$V = \frac{k}{2} \sum_{l} (u_l - u_{l+a})^2$$
$$= \frac{k}{2} \left[\left(\cdots + u_{l-a}^2 - 2u_{l-a}u_l + u_l^2 \right) + \left(u_l^2 - 2u_l u_{l+a} + u_{l+a}^2 \right) + \cdots \right]$$

The kinetic energy is: $T = \frac{m}{2} \sum_l (\dot{u}_l)^2$. We write Bloch's linear combinations of cartesian coordinates:

$$U_q = \sum_l e^{iql} \cdot u_l \qquad \Rightarrow u_l = \sum_q e^{-iql} \cdot U_q;$$
$$\Rightarrow \dot{u}_l = \sum_q e^{-iql} \cdot \dot{U}_q \qquad \Rightarrow \dot{U}_q = \sum_l e^{iql} \cdot \dot{u}_l$$

The (Newton) motion equation for the atom in the l cell is:

$$m\ddot{u}_l = \frac{-\partial V}{\partial u_l} = -k(-u_{l-a} + 2u_l - u_{l+a})$$

Without need of specifying j and α (that are unique), this gives the contribution to the motion of the atom in cell

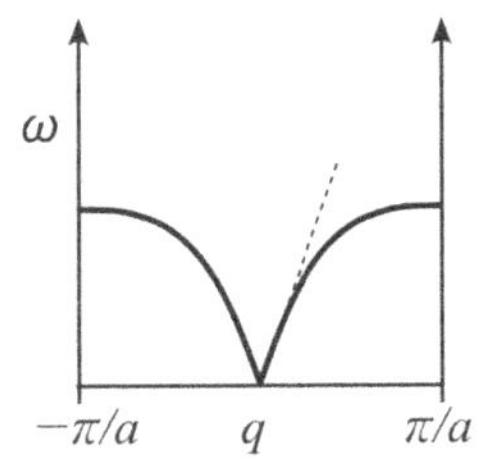

FIGURE 4.24 Phonon dispersion relation for the 1D model with one atom per primitive cell, Eq. (3.33). The dotted line is the tangent at the origin, and its slope indicates the speed of sound. Limiting values for ω are $\omega(0)=0$ and $\omega(\pi/a)=\sqrt{2k/m}$.

l (u_l) to the vibrational mode with q Bloch's vector and frequency $\omega(q)$:

$$u_l(q,t)\mathrm{e}^{i\cdot\omega(q)\cdot t}\cdot\mathrm{e}^{i\cdot q\cdot l}\cdot u_l^0$$

The complex amplitude u_l^0 includes the actual (real) amplitude of the vibration and its phase. Substituting: (Figure 4.24):

$$\begin{aligned} m\ddot{u}_l &= m\left(-\left(\omega(q)\right)^2\right)\cdot u_l = k\left(-2u_{l-a}+2u_l-2u_{l+a}\right) \\ &= k\left(2-\mathrm{e}^{-i\cdot q\cdot a}-\mathrm{e}^{i\cdot q\cdot a}\right) \text{ that is: } \left(\omega(q)\right)^2 \\ &= \frac{2k}{m}\left(1-\cos(q\cdot a)\right) \end{aligned} \tag{3.33}$$

(Figure 4.24 reports the result.

3.4.1.2 1D Model with Two Atoms Per Unit Cell

The two atoms are indicated as 1 and 2.

As shown in Figure 4.25, the potential energy for nearest neighbour interaction only is:

$$\frac{k}{2}\begin{bmatrix} \cdots+\left(u_{1,l-a}-u_{2,l-a}\right)^2+\left(u_{2,l-a}-u_{1,l}\right)^2 \\ +\left(u_{1,l}-u_{2,l}\right)^2+\left(u_{2,l}-u_{1,l+a}\right)^2+\cdots \end{bmatrix}$$

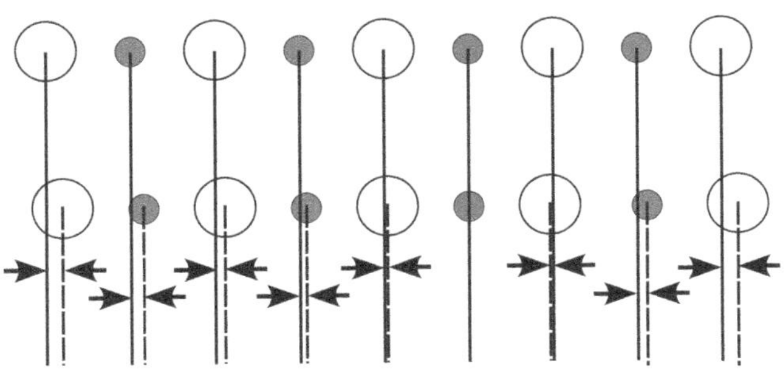

FIGURE 4.25 1D Vibrational model with two atoms per unit cell. Atom 1: large white circles, atom 2: small grey circles. The unit cells are at $l-a$, l, $l+a$, $l+2a$. Upper part: equilibrium atomic positions, lower part: displacements from equilibrium at a given time.

Thus: $$\frac{-\partial V}{\partial u_{1,l}}=-k\left(-u_{2,l-a}+2u_{1,l}-u_{2,l}\right) \text{ and}$$

$$\frac{-\partial V}{\partial u_{2,l}}=-k\left(-u_{1,l}+2u_{2,l}-u_{1,l+a}\right)$$

The motion equations for the displacements of 1 and 2 from their equilibrium positions are:

$$m_1\ddot{u}_{1,l}=-k\left(-u_{2,l-a}+2u_{1,l}-u_{2,l}\right) \quad \text{and}$$

$$m_2\ddot{u}_{2,l}=-k\left(-u_{1,l}+2u_{2,l}-u_{1,l+a}\right)$$

For each atom and each cell, we write a contribution to a SALC:

$$u_{1,l}(q,t)\mathrm{e}^{i\cdot\omega(q)\cdot t}\cdot\mathrm{e}^{i\cdot q\cdot l}\cdot u_1^0 \quad \text{and} \quad u_{2,l}(q,t)\mathrm{e}^{i\cdot\omega(q)\cdot t}\cdot\mathrm{e}^{i\cdot q\cdot l}\cdot u_2^0$$

By substitution:

$$-\omega^2 m_1\cdot\mathrm{e}^{i\cdot q\cdot l}\cdot u_1^0=-k\left(\mathrm{e}^{i\cdot q\cdot(l-a)}\cdot u_2^0-2\cdot\mathrm{e}^{i\cdot q\cdot l}\cdot u_1^0+\mathrm{e}^{i\cdot q\cdot l}\cdot u_2^0\right)$$

$$-\omega^2 m_2\cdot\mathrm{e}^{i\cdot q\cdot l}\cdot u_2^0=-k\left(\mathrm{e}^{i\cdot q\cdot l}\cdot u_1^0-2\mathrm{e}^{i\cdot q\cdot l}\cdot u_2^0+\mathrm{e}^{i\cdot q\cdot(l+q)}\cdot u_1^0\right)$$

$$\begin{pmatrix} -\omega^2 m_1\cdot u_1^0 \\ -\omega^2 m_2\cdot u_2^0 \end{pmatrix}=\begin{pmatrix} -2k & k\left(1+\mathrm{e}^{-i\cdot q\cdot a}\right) \\ k\left(1+\mathrm{e}^{i\cdot q\cdot a}\right) & -2k \end{pmatrix}\begin{pmatrix} u_1^0 \\ u_2^0 \end{pmatrix} \tag{3.34}$$

From Eq. (3.34), $\omega_1=\sqrt{\frac{2k}{m_1}}$, $\omega_2=\sqrt{\frac{2k}{m_2}}$ and $\omega_3=\sqrt{\frac{2k(m_1+m_2)}{m_1\cdot m_2}}$ (see Figure 4.26).

3.4.2 Phonon Dispersions

Generalizing the examples and comparing with the problem of crystal symmetry orbitals (3.3), the problem of vibrations of a periodic crystal translates into the calculation - for each $\vec{q}$ within the first Brillouin zone - of the frequencies of normal vibrations as square roots of the eigenvalues of a dynamical matrix of dimension n, where n is the number of vibrational degrees of freedom **of a primitive unit cell**, *i.e.*, $n=3m$ if we denote by m the number of atoms contained

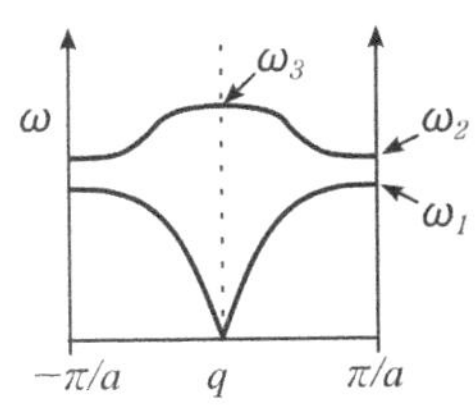

FIGURE 4.26 Phonon dispersion for the 1D model with two atoms/primitive cell.

in the unit cell. Non-translational symmetries can further simplify the partition of the dynamical matrix.

Reversing the point of view, we have n phonon dispersion relations, *i.e.*, $n\omega(\vec{q})$ functions defined in the whole first Brillouin zone. Three of these relations or branches are called **acoustic** and the sense is that $\omega(\vec{q}) \rightarrow 0$ as $\vec{q} \rightarrow 0$ (*i.e.*, the origin); the others are called **optical**.[6] The name is because they are the active vibrations in **vibrational** spectroscopy in the case of a diatomic ionic solid, *e.g.*, NaCl. For both acoustic and optical branches, a distinction is then made between **longitudinal** vibrations (in which atoms move parallel to $\vec{q}$) and transverse **vibrations** (in which atoms move in two directions perpendicular to $\vec{q}$).

To sum up: one longitudinal acoustic branch, two transverse acoustic branches, $(n-1)$ transverse optical branches $2(n-1)$ longitudinal optical branches. Structures with only one atom per primitive cell thus have only acoustic branches.

3.5 Conclusions

Translational symmetry provides the most striking example of the power of symmetry-based methods. In fact, for a system consisting of nN degrees of freedom (N=number of unit cells of the solid, n=degrees of freedom per unit cell, typically of the order of unity or tens), translational symmetry allows one to obtain n distinct solutions as functions of a parameter (the Bloch's vector) that takes N discrete values. This can make tractable problems otherwise addressable only by a reduction of N by many orders of magnitude. By comparison, the use of molecular symmetry has a vastly smaller impact, of the order "only" of a number comparable with n, to the point that in simple problems the direct solution of MO problems is easier than the use of symmetry.

4 SUBSTRUCTURES AND SUPERSTRUCTURES OF GROUPS

What has been described so far is a sufficiently broad picture of the basic features and applications of the algebraic theory of symmetry to relatively simple problems and - above all - *limited to commutative symmetry groups*. For the extension to non-commutative groups, more in-depth theory is required, in particular on the possibility of representing the elements of a symmetry group by means of matrices, on the properties of these matrix representations and on the properties of related parameters called characters. These tools not only generalise the approach but also achieve greater conciseness and practicality than what is already seen for the commutative case.

Before tackling the more technical part, it is worth introducing some internal structures of algebraic groups and some operations that combine smaller groups into a larger group or break down large groups into smaller groups.

4.1 Inner Structures

4.1.1 Generator Set

Let us consider the elements C_3 and $\sigma^{(1)}$ of group C_{3v} (Table 4.3). We see that using only these elements as factors we can obtain all the elements of the group, in fact:

$$E = (C_3)^3 = (\sigma^{(1)})^2, \quad C_{-3} = (C_3)^2,$$
$$\sigma^{(2)} = \sigma^{(1)}C_3, \quad \sigma^{(3)} = C_3\sigma^{(1)}$$

Clearly, there are other possibilities, for example we could use $\sigma^{(2)}$ [or $\sigma^{(3)}$] instead of $\sigma^{(1)}$: $E=(\sigma^{(2)})^2$, $\sigma^{(1)}=C_3\sigma^{(2)}$ and $\sigma^{(3)}=\sigma^{(2)}C_3$. The choice of generators can change, but not their number. The example justifies the definition:

> A subset P of a group G is called a generator system of G if every element of G is the product of a finite number of factors and every factor is an element of P.

A given factor can obviously appear more than once and is then written in power form. A system of generators identifies a group: significantly, the international classification (Hermann-Mauguin) of point groups (Section 1.1.4.2) and space groups (Section 6.6.4) is essentially the list of generators (abbreviated, when there is no ambiguity, and in a given order).

4.1.2 Subgroups

Still using as an example group C_{3v}, let's consider the set $\{E, C_3, C_{-3}\}$. We see that this is a group in itself. It is group C_3, and we say that C_3 is a *subgroup* of group C_{3v}. It is also the cyclic group generated by C_3. Other subgroups of C_{3v} are $\{I, \sigma^{(1)}\}$, $\{I, \sigma^{(2)}\}$, $\{I, \sigma^{(3)}\}$.

> In a group (G), any subset that in itself is a group is called a **subgroup** of G.

Referring to Section 4.1.4 for the proof, we anticipate that

> The order h of every subgroup is a divisor of the order g of the group.

For logical consistency of the definitions, every group of g elements also has subgroups of order 1 (the identity alone) and of order g (the entire group), which are called *trivial subgroups*. So, C_{3v} has subgroups of order 1 (just one, the identity), order 2 (the three above), order 3 (one, see above) and order 6 (the entire group).

4.1.3 Conjugate Elements and Classes

We can now ask whether there is a particular *algebraic* relation between elements of the same type (in a *geometric* sense) and whether they can somehow be treated jointly. For example: in the group C_{3v}, it is obvious to ask whether it makes sense to group the two ternary rotations or the three reflections. In fact, treating jointly symmetry operations of the same type (*i.e.* n-ary rotations with n-ary rotations, reflections with reflections, ...) greatly simplifies the development of the theory. It is convenient to introduce the topic step by step. An algebraic definition of the notion of *elements of the same type* is based on the property of **conjugation**. Let us consider three elements of a group, A, B, R for which: $B=RAR^{-1}$. We say that:

> A and B are *conjugate* to each other, or B is conjugate to A with respect to R, or B is the *similarity transform* of A by R.

The next step is that:

> Keeping the element A fixed and considering the various elements of the group as R, we find the set of all the elements conjugated to A. This set is called the **class of elements conjugated** to A.

From Table 4.16 we see that in the usual group C_{3v}, $\{E\}$ is the class of elements conjugated to E, $\{C_3, C_{-3}\}$ is the class of elements conjugated to C_3 and C_{-3}, and $\{\sigma^{1)}, \sigma^{(2)}, \sigma^{(3)}\}$ is the class of elements conjugated to $\sigma^{(1)}$, $\sigma^{(2)}$ and $\sigma^{(3)}$. Note that the *set* of elements conjugated to C_3 is equal to the set of elements conjugated to C_{-3} because, when comparing sets, the order of the elements is not relevant. Then, in group C_{3v} there are three classes: $\{I\}$, $\{C_3, C_{-3}\}$, e $\{\sigma^{(1)}, \sigma^{(2)}, \sigma^{(3)}\}$.

In the example, a class groups symmetry operations of the same type, namely the two rotations in a class and the three reflections in another class. However, this is not always true: a class is a *stronger relation*. For example, the symmetry group of the benzene molecule (D_{6h}, see Section 1.1.6 and Table 4.4) includes *seven* reflections that belong to three different classes, respectively of one (σ_h), three (σ_v) and three (σ_d) reflections (see Table 4.31). Anticipating two later results, the classes are important because:

a. the elements of a class share an important parameter, called **character** (see Section 5.3) and
b. the **number of classes** equals the **number of symmetry species** (see Section 5.4.4.1).

TABLE 4.16
The First Column Contains the Elements Conjugated to E, the Second Column Contains the Elements Conjugated to C_3, etc.

R	RER^{-1}	RC_3R^{-1}	$RC_{-3}R^{-1}$	$R\sigma^{(1)}R^{-1}$	$R\sigma^{(2)}R^{-1}$	$R\sigma^{(3)}R^{-1}$
E	E	C_3	C_{-3}	$\sigma^{(1)}$	$\sigma^{(2)}$	$\sigma^{(3)}$
C_3	E	C_3	C_{-3}	$\sigma^{(2)}$	$\sigma^{(3)}$	$\sigma^{(1)}$
C_{-3}	E	C_{-3}	C_{-3}	$\sigma^{(3)}$	$\sigma^{(1)}$	$\sigma^{(2)}$
$\sigma^{(1)}$	E	C_{-3}	C_3	$\sigma^{(1)}$	$\sigma^{(3)}$	$\sigma^{(2)}$
$\sigma^{(2)}$	E	C_{-3}	C_3	$\sigma^{(3)}$	$\sigma^{(2)}$	$\sigma^{(1)}$
$\sigma^{(3)}$	E	C_{-3}	C_3	$\sigma^{(2)}$	$\sigma^{(1)}$	$\sigma^{(2)}$

The row headers indicate which R we are using to form the conjugate.

When discussing group properties (Section 1.1.2.6), we saw that - when a group is not commutative - we must distinguish between right and left multiplication and that the relation $AB=BX$ does not necessarily give $X=A$. We now note that conjugation $X=B^{-1}AB$ can be seen as an extension of equality and therefore we sometimes write $X \sim A$.

Furthermore,

> The classes are a **partition** of the underlying set of a group.

The meaning is that the classes as a whole

- contain all the elements of the group (no element is outside a class) and
- there are no elements shared by two classes.

To derive this fundamental property, it is first necessary to verify that being conjugated has the following properties: (i) it is reflexive (every element of a group is conjugated to itself), (ii) it is symmetric (if A is conjugated to B, then B is conjugated to A), (iii) it is transitive (if A is conjugated to B and B is conjugated to C, then A and C are also conjugated to each other). Then, we refer to the mathematical set theory according to which a relation between elements of a set that has the properties 1, 2, 3 is called an *equivalence relation* and defines a *partition* of the set, in the sense that it divides the set into non-empty subsets in such a way that every element is included in exactly one subset. Other properties of classes are left as an exercise (number 7).

Finally, classes allow us to distinguish between two fundamentally different types of subgroups, normal and non-normal subgroups.

> A subgroup A of a group G is said to be **normal**, or **invariant** or **self-conjugate** if it contains all the classes conjugated to its elements.

This means that if we take an element of a normal subgroup and apply to it the conjugation for a given element of the group, the result is an element of the subgroup itself: this applies to every element of the subgroup for every element of the group.

A group that has no normal subgroups is said to be *simple* (see also Exercise 4.8).

4.1.4 Cosets

Another important internal structure of groups is called *coset* (from *co*mplement and *set*) and has mainly theoretical interest because it provides the tools to derive many results concerning representations and characters in Section 5. The structure is defined starting from the subgroup structure. A left coset of subgroup A of group G is the set of all the elements of G that are obtained by choosing an element (X) of the entire group G and multiplying from the left by this element all the elements of subgroup A: to indicate this set we write XA. Similarly, the *right coset* is defined as AX. Obviously, the coset possibly changes when X is changed.

Some properties of the coset are derived in Exercises 4.9–4.12.

Like classes, the cosets of a given subgroup are a *partition* of the group. Unlike classes, they all have the same number of elements (see Exercise 4.13). This proves (as anticipated in Section 4.1.2) that

> The order of a subgroup is a divisor of the order of the group.

4.2 Group Products and Product Group

A group can be expanded to larger groups. This plays an important role in the abstract study of the relations between symmetry groups and in applications. On the other side, breaking down a complicated group into smaller groups is a useful tool for simplifying calculations: an example concerning the AOs of benzene is reported in Section 6.2.1.6.

The topic is difficult, includes different cases and is outlined here very concisely.

The expansion of a group G ({A, B, ..., E, ...}, *) is *not* a simple *addition* of one or more elements to the underlying set of G. Due to the need to comply with the algebraic group structure, the set H of additional elements {X, Y, ...} must include the inverses (X^{-1}, ...) of each new element and all the combination (XY, ...) of the *new* elements among themselves. Moreover, the enlarged group (P) necessarily includes all the combinations (AX, BX, AY, BY, ...) of an element of G with an element of H. Now, let us include in H also the identity, so that also the elements {A, B, ..., E, ...} of P are obtained by combining the elements of G with the identity in H. Therefore, we can say that *all the elements of P are given by the combination of an element of G with an element of H*. Finally, this proves that when adding symmetry elements to a group G we actually **combine two groups *G* and *H* that are subgroups** of their combination ***P***.

There are cases where *some elements of group H other than identity* are already present in group G. An example discussed later in Section 6.3.2 is that of group C_{2v}, when it is "enlarged" to C_{4v} by adding a C_4 rotation, which means adding also $(C_4)^2$ and $(C_4)^3$. As $(C_4)^2=C_2$ is already present in C_{2v}, the groups C_{2v} and $C_4=\{C_4, (C_4)^2, (C_4)^3, E\}$ that are combined have *non-trivial intersection*. However, a frequent and more interesting case is when G and H have *only trivial intersection*, meaning that G and H share only the identity. P is then called the **group product** and is written $P = G \otimes H$. We can have a **direct product** or a **semi-direct product** when more or less restrictive conditions hold. In particular, the direct product requires **commutativity** for all products $AX=XA$. When a group product *expands* a group, it is referred to as an **external** product, while the corresponding **internal** product deals with the breakdown of a group into simpler groups. Some consequences of these relations between smaller and larger groups are briefly discussed in Section 6.3.

5 REPRESENTATIONS AND CHARACTERS

5.1 Matrix Representations

Matrix representations[7] are a basic topic in group theory. On the one hand, representing symmetries with matrices is an operation that arises naturally from many applications. On the other hand, representations allow us to obtain many important theoretical results.

Considering the applications to QM, the matrix representations of symmetries derive, first of all, from the fact that the wave functions of QM are a linear space and the operators are linear. The notation is typically a matrix notation, as well shown by the nomenclature (*e.g.* the matrix element of an operator), and the transformations of the wavefunctions due to a symmetry operation are essentially matrix multiplications. Also, operations on symmetries are generally much more conveniently dealt with matrix algebra than in a purely geometrical manner, not to mention symmetries of a non-geometric nature.

To **represent** a set of symmetry operations, it would seem simpler to use a set of numbers (integer, or real, or possibly complex) and to combine them according to the usual arithmetic rules. This is, in essence, what has been done with the commutative groups in Section 2. However, this is not always possible: as said, the product and the sum of two numbers are commutative operations, while many symmetry groups are non-commutative. Matrix multiplication is generally non-commutative. For the sake of completeness, it can be anticipated that a form of representation with numbers can be realised also for non-commutative groups, albeit with some limitations and with a specific technicality: this is the character system and will be introduced in Section 5.3.

The basic idea behind the matrix representation of a group is that:

if

> for a group G of order g one can find g square matrices
>
> $$\{\boldsymbol{A},\boldsymbol{B},\ldots,\boldsymbol{R},\boldsymbol{S},\boldsymbol{T},\ldots\}$$
>
> and associate them with the g elements (A, B, ..., R, S, T, ...) of G:
>
> $$[\boldsymbol{A}\to A,\boldsymbol{B}\to B,\ldots,\boldsymbol{R}\to R,\boldsymbol{S}\to S,\boldsymbol{T}\to T,\ldots]$$
>
> in such a way that from $T=RS$, (in this order) one can derive $\boldsymbol{T}=\boldsymbol{RS}$ (in this order)

then

> one says that the matrices {$\boldsymbol{A}$, $\boldsymbol{B}$, ..., $\boldsymbol{R}$, $\boldsymbol{S}$, $\boldsymbol{T}$, ...} are a **representation** of the group.

It is important to point out immediately that, in the idea of representation,

It is **not** required that the matrices $\boldsymbol{A}, \boldsymbol{B}, \ldots, \boldsymbol{R}, \boldsymbol{S}, \boldsymbol{T}$ are **all different from each other**.

If the matrices are *all different from each other*, then the representation is an **isomorphism**, a one-to-one correspondence. If, instead, a matrix corresponds to more than one symmetry operation, we have a **homomorphism**. We can therefore better understand the previous statements: a representation with numbers, rather than with square matrices, is always possible (the numbers can always be seen as 1×1 matrices), but it does not exhaust the cases of interest. Any representation with numbers of a non-commutative group is a *homomorphic* representation (*non-isomorphic*). In particular, the **totally symmetric** representation for which each element of a group is represented by the number 1 is a homomorphism that holds for all groups. A *group of symmetry operations* and *its isomorphic representations* belong to the same abstract group and share the multiplication table. By definition:

The **dimension of a representation** is the dimension of its square matrices.

As shown in detail later, a group has infinitely many matrix representations of different sizes, some isomorphic, some not. Some representations, although different from each other, somehow convey the same information and are said to be **equivalent**, an important aspect that will be discussed in Section 5.1.4.4.

5.1.1 Matrix Operations

We systematically use **boldface** *italics* type for matrices, row and column matrices in lowercase, square matrices in uppercase. When required, the superscript T is used to indicate transposition $(\boldsymbol{A}^{\mathrm{T}})_{ij}=\boldsymbol{A}_{ji}$, while the superscript † indicates the Hermitian-conjugate, that is the transpose of the conjugated matrix or – it is equal – the conjugate of the transposed matrix: $(\boldsymbol{A}^{\dagger})_{ij}=(\boldsymbol{A}_{ji})^{*}=(\boldsymbol{A}^{*})_{ji}$. If we want to clarify that matrix $\boldsymbol{A}$ represents the symmetry operation A in the j-th representation, we write $\boldsymbol{A}=\boldsymbol{D}_j(A)$.

The operations that must be performed on the matrices of group representations are, first of all, the usual ones: matrix addition and the row×column product. These are described in better detail in the Appendix dedicated to matrix calculus. Briefly: it is possible to *add* two matrices $\boldsymbol{A}$ and $\boldsymbol{B}$ if and only if they have the same dimensions, that is, they are both $(m \times n)$ matrices with m rows and n columns. The result $\boldsymbol{C}=\boldsymbol{A}+\boldsymbol{B}$ is the $(m \times n)$ matrix such that $\boldsymbol{C}_{ij}=\boldsymbol{A}_{ij}+\boldsymbol{B}_{ij}$ for any i $(1 \le i \le m)$ and any j $(1 \le j \le n)$.

It is possible to *multiply* two matrices $\boldsymbol{A}$ and $\boldsymbol{B}$ if and only if the second dimension of the first matrix is equal to the first dimension of the second matrix. If $\boldsymbol{A}$ is an $(m \times n)$ matrix and $\boldsymbol{B}$ is an $(n \times p)$ matrix, then the product $\boldsymbol{C}=\boldsymbol{AB}$ is the $(m \times p)$ matrix such that $\boldsymbol{C}_{ij} = \sum_k \boldsymbol{A}_{ik} \bullet \boldsymbol{B}_{kj}$ whit $(1 \le i \le m)$, $(1 \le j \le p)$ and $(1 \le k \le n)$.

To these, two operations must now be added.

5.1.1.1 Direct Sum of Matrices

The first additional operation is called **direct sum**, operates *on square matrices*, not necessarily of equal size, and produces *a square matrix* that has size equal to the sum of the sizes of the addend matrices. Given an $(n \times n)$ matrix $\boldsymbol{A}$ and an $(m \times m)$ matrix $\boldsymbol{B}$, their direct sum is the square $(n+m) \times (n+m)$ matrix obtained by placing the elements of $\boldsymbol{A}$ in the upper left block and matrix $\boldsymbol{B}$ in the lower right block:

$$\boldsymbol{A} \oplus \boldsymbol{B} = \begin{pmatrix} a_{11} & \cdots & a_{n1} & 0 & \cdots & 0 \\ \vdots & \ddots & \vdots & \vdots & \ddots & \vdots \\ a_{1n} & \cdots & a_{nn} & 0 & \cdots & 0 \\ 0 & \cdots & 0 & b_{11} & \cdots & b_{n1} \\ \vdots & \ddots & \vdots & \vdots & \ddots & \vdots \\ 0 & \cdots & 0 & b_{1n} & \cdots & b_{nn} \end{pmatrix}$$

$$\text{also written as } \boldsymbol{A} \oplus \boldsymbol{B} = \begin{pmatrix} \boldsymbol{A} & \boldsymbol{0} \\ \boldsymbol{0}^{\mathrm{T}} & \boldsymbol{B} \end{pmatrix}$$

where $\boldsymbol{0}$ and $\boldsymbol{0}^{\mathrm{T}}$ are matrices ($n \times m$ and $m \times n$, respectively) with all elements $=0$.

It is evident that the direct sum of two matrices is not commutative: $\boldsymbol{A} \oplus \boldsymbol{B} \neq \boldsymbol{B} \oplus \boldsymbol{A}$. The direct sum of two matrices is a so-called **block-diagonal** matrix. There are two important properties of sum and multiplication of *block-diagonal matrices with the same block structure*. Let us consider for instance two such matrices ($\boldsymbol{A}$ and $\boldsymbol{B}$) and let us write $\boldsymbol{A} = \boldsymbol{A}_1 \oplus \boldsymbol{A}_2$ and $\boldsymbol{B} = \boldsymbol{B}_1 \oplus \boldsymbol{B}_2$, where $\boldsymbol{A}_1$ and $\boldsymbol{B}_1$ are square matrices of the same dimension, and so is for $\boldsymbol{A}_2$ and $\boldsymbol{B}_2$. It is trivial to see that, when two such matrices are added, the result is obtained by summing the corresponding blocks along the diagonals:

$$\boldsymbol{A}+\boldsymbol{B} = (\boldsymbol{A}_1 \oplus \boldsymbol{A}_2)+(\boldsymbol{B}_1 \oplus \boldsymbol{B}_2) = (\boldsymbol{A}_1 + \boldsymbol{B}_1) \oplus (\boldsymbol{A}_2 + \boldsymbol{B}_2)$$

It is somewhat longer but not difficult to prove that a similar property also holds for the product:

$$\boldsymbol{A} \cdot \boldsymbol{B} = (\boldsymbol{A}_1 \oplus \boldsymbol{A}_2) \cdot (\boldsymbol{B}_1 \oplus \boldsymbol{B}_2) = (\boldsymbol{A}_1 \cdot \boldsymbol{B}_1) \oplus (\boldsymbol{A}_2 \cdot \boldsymbol{B}_2).$$

We also speak of **direct sum of representations** when, for each element of the group, the related matrices of the two representations are combined with the direct sum to form a third representation.

5.1.1.2 Direct Product of Matrices

The second additional operation is called **direct product** or *Kronecker product*: if the matrix $\boldsymbol{C}$ is the direct product of the matrices $\boldsymbol{A}$ and $\boldsymbol{B}$ (in this order), we write $\boldsymbol{C}=\boldsymbol{A} \otimes \boldsymbol{B}$. The operation produces a generally larger matrix $\boldsymbol{C}$ in which each element of $\boldsymbol{A}$ corresponds to a block of elements containing $\boldsymbol{B}$ multiplied by that element of $\boldsymbol{A}$. More precisely, if $\boldsymbol{A}$ is an $n \times m$ matrix and $\boldsymbol{B}$ is a $p \times q$ matrix, the direct product can be schematized as follows:

$$A \otimes B = \begin{pmatrix} a_{11}B & \cdots & a_{1m}B \\ \vdots & \ddots & \vdots \\ a_{n1}B & \cdots & a_{nm}B \end{pmatrix}$$ where $a_{jk}B$ is a $p \times q$ block given by B multiplied by a_{jk}

In detail:

For example, if A is a 2×3 matrix and B is a 4×5 matrix, then C is an 8×15 matrix.

The direct matrix product is often denoted (or even defined) with a *double index* system:

$$\text{if } A \otimes B = C\text{: } C_{ij,kl} = A_{ij}B_{kl} \tag{5.1}$$

in the sense that the matrix element $C_{ij,kl}$ is in position kl within the block (of equal shape to matrix B) that occupies the partition ij of matrix C. Do not confuse the double-index notation ij with the multiplication of indices ($i \cdot j$). The double-index notation is useful in proofs. Section 3.3.2 of the Appendices reports the detailed relations between the two systems of indices.

Differently from the row×column multiplication, there are *no constraints on the dimensions of the matrices* that can be combined in this way. An important relationship between the direct product and the more usual row×column product is as follows. If two matrices (A and C) can be multiplied together with the row×column multiplication, and if the same is true for two other matrices (B and D), then

$$(A \otimes B)(C \otimes D) = (AC) \otimes (BD)$$

Another important property concerns the **trace**, that is, the sum of the elements along the main diagonal of a square matrix (see Section 5.3.1): if A and B are square matrices, **the trace of the direct product is the product of the traces**: $tr(A \otimes B) = tr(A)\, tr(B) = tr(B)\, tr(A)$. There is commutativity of the traces, while the direct product itself is not necessarily commutative: $A \otimes B$ is not necessarily $= B \otimes A$.

Analogously to the direct sum, we speak of the **direct product of representations** when for each element of the group we perform the direct product of the matrices of the two representations.

5.1.2 Matrices and Linear Operators

A fundamental way to obtain matrices representing a group is by considering how symmetry operations transform vectors. The following point 5.1.3 shows many examples in this regard. In general, they can be either *vectors in the strict sense*, belonging to the ordinary 3D space or to a higher-dimensional configuration space and then are labelled as $\vec{r}$, $\vec{s}$, …, but can be also elements of function spaces and then the Greek letters φ, χ, ψ, …, are in common use. An essential aspect of the linear space structure is that any element of the space can be written as a linear combination of a set of **basis vectors** (*e.g.*: $\vec{e_1}$, $\vec{e_2}$, $\vec{e_3}$, …, or φ_1, φ_2, …):

$$\vec{r} = r_1 \cdot \vec{e_1} + r_2 \cdot \vec{e_2} + r_3 \cdot \vec{e_3} \quad \text{or} \quad \varphi_r = r_1 \cdot \varphi_1 + r_2 \cdot \varphi_2 + r_3 \cdot \varphi_3 + \cdots$$

and a vector ($\vec{r}$) is represented in matrix notation using a column matrix such as $r = \begin{pmatrix} r_1 \\ r_2 \\ r_3 \end{pmatrix}$ that gathers its **components** with respect to the **basis**, while the basis vectors are written as row matrices.

The above expressions then become:

$$\vec{r} = \begin{pmatrix} \vec{e_1} & \vec{e_2} & \vec{e_3} \end{pmatrix} \cdot \begin{pmatrix} r_1 \\ r_2 \\ r_3 \end{pmatrix} = \begin{pmatrix} \vec{e_1} & \vec{e_2} & \vec{e_3} \end{pmatrix} \cdot r = e \cdot r$$

or

$$\varphi_r = \begin{pmatrix} \varphi_1 & \varphi_2 & \cdots \end{pmatrix} \cdot \begin{pmatrix} r_1 \\ r_2 \\ \vdots \end{pmatrix} = \begin{pmatrix} \varphi_1 & \varphi_2 & \cdots \end{pmatrix} \cdot r = \varphi \cdot r$$

Note that writing the components as a column matrix and the basis as a row matrix is a *conventional* choice that must be followed strictly (see Section 3.2 on the Appendices).

Square matrices transform (i) a column matrix into another column matrix by multiplying the column from the left ($s = A \cdot r$) or (ii) transform a row matrix into another row matrix by multiplying the row from the right ($e' = e \cdot A$) or else (iii) transform a square matrix into another square matrix.

Case (i) concerns the representation of symmetry operators. The **linear transformation** A of a vector $\vec{r} = e \cdot r$ into a vector $\vec{s} = e \cdot s$, that is: $\vec{s} = A\vec{r}$, in matrix notation becomes:

$$s = A \cdot r \Leftrightarrow \begin{pmatrix} s_1 \\ s_2 \\ s_3 \end{pmatrix} = \begin{pmatrix} a_{11} & a_{12} & a_{13} \\ a_{21} & a_{22} & a_{23} \\ a_{31} & a_{32} & a_{33} \end{pmatrix} \begin{pmatrix} r_1 \\ r_2 \\ r_3 \end{pmatrix} \tag{5.2}$$

where the matrix A depends on A, but also on the basis.

Case (ii) concerns a **change of basis**. As described in detail in Section 5.1.3.3, the basis vectors are transformed according to the symmetry operation A:

$$\begin{pmatrix} \vec{e_1'} & \vec{e_2'} & \vec{e_3'} \end{pmatrix} = \hat{A} \begin{pmatrix} \vec{e_1} & \vec{e_2} & \vec{e_3} \end{pmatrix}.$$

Then: $e' = \begin{pmatrix} \vec{e_1'} & \vec{e_2'} & \vec{e_3'} \end{pmatrix} = \begin{pmatrix} \vec{e_1} & \vec{e_2} & \vec{e_3} \end{pmatrix} \cdot A$,

that is

$$e' = e \cdot A \tag{5.3}$$

Case (iii) is related to case (B) as a change of basis also transforms the **matrix representation of linear operators.** The significant relation is:

$$A'' = T^{-1} \cdot A \cdot T$$

where A and A'' are the matrices that represent a symmetry operation (A) in the two bases and T is the matrix of the change of basis. More details on this point are in Section 5.1.3.3.

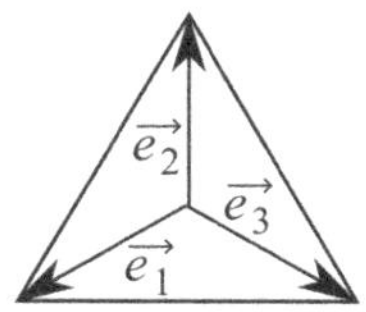

FIGURE 4.27 A possible basis of 3D vectors for a system with C_{3v} symmetry. The origin of the basis vectors is along the ternary axis, which is perpendicular to the plane of the drawing. The versors $\vec{e_1}, \vec{e_2}, \vec{e_3}$ point *downwards* and are at right angles with each other.

5.1.3 Examples of Representations

For simplicity, all following examples consider the symmetry operations of group C_{3v}, because its characteristics ease the treatment of various topics. In fact, C_{3v} is a non-commutative group and has a number of operations small enough to make the discussion and analysis of the results quite simple, without being too small to be trivial.

5.1.3.1 A First Representation in the Ordinary 3D Space

Let us consider vectors in ordinary 3D space and a set of orthonormal vectors oriented in the way shown by Figure 4.27.

With this basis, the matrices of the various transformations of the C_{3v} group are:

$$\boldsymbol{D}(E)=\begin{pmatrix}1&0&0\\0&1&0\\0&0&1\end{pmatrix};\quad \boldsymbol{D}(C_3)=\begin{pmatrix}0&0&1\\1&0&0\\0&1&0\end{pmatrix};$$
$$\boldsymbol{D}(C_{-3})=\begin{pmatrix}0&1&0\\0&0&1\\1&0&0\end{pmatrix};\quad \boldsymbol{D}\left(\sigma^{(1)}\right)=\begin{pmatrix}1&0&0\\0&0&1\\0&1&0\end{pmatrix};$$
$$\boldsymbol{D}\left(\sigma^{(2)}\right)=\begin{pmatrix}0&0&1\\0&1&0\\1&0&0\end{pmatrix};\quad \boldsymbol{D}\left(\sigma^{(3)}\right)=\begin{pmatrix}0&1&0\\1&0&0\\0&0&1\end{pmatrix} \tag{5.4}$$

The fact that these matrices are a matrix representation of group C_{3v} means, for example, that $\boldsymbol{D}(C_3)\,\boldsymbol{D}(\sigma^{(1)})=\boldsymbol{D}(\sigma^{(3)})$ (as can be verified directly) because $C_3\sigma^{(1)}=\sigma^{(3)}$.

5.1.3.2 Another Representation of the Same Group

The particularly simple form of the matrices of Eq. (5.4) is due to the particular choice of the basis. Let us now consider the same problem using a different basis. The new basis (marked with double quotation marks) consists of three orthogonal versors as follows:

- $\vec{e_1''}$ (not drawn in Figure 4.28) *is perpendicular to the plane of the triangle and points upwards* (towards the observer), while
- the other versors lie in the plane of the triangle as shown by Figure 4.28.

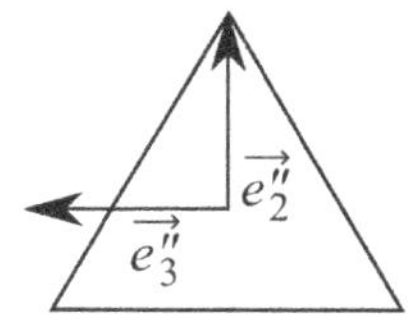

FIGURE 4.28 Another basis for the same problem as Figure 4.27. The first versor ($\vec{e_1''}$, not drawn) is perpendicular to the triangle and points towards the observer.

With this new basis, the representative matrices are:

All these matrices, both in the new (Figure 4.28) and old basis (Figure 4.27) are orthogonal (see Eq. (5.14)). Further comments on this transformation are reported in Section 5.1.5.1.

5.1.3.3 Changing the Basis

The matrices of the last example can be obtained by considering the change from one basis $\begin{pmatrix}\vec{e_1} & \vec{e_2} & \vec{e_3}\end{pmatrix}$ to the other $\begin{pmatrix}\vec{e_1''} & \vec{e_2''} & \vec{e_3''}\end{pmatrix}$:

$$\vec{e_1''}=-\left(\vec{e_1}+\vec{e_2}+\vec{e_3}\right)/\sqrt{3};\quad \vec{e_2''}=\left(2\vec{e_2}-\vec{e_1}-\vec{e_3}\right)/\sqrt{6};$$
$$\vec{e_3''}=\left(\vec{e_1}-\vec{e_3}\right)/\sqrt{2} \tag{5.6}$$

where the $\sqrt{3}$, $\sqrt{2}$, $\sqrt{6}$ coefficients are due to normalization. In matrix notation, Eq. (5.6) is:

$$\begin{pmatrix}\vec{e_1''} & \vec{e_2''} & \vec{e_3''}\end{pmatrix}=\begin{pmatrix}\vec{e_1} & \vec{e_2} & \vec{e_3}\end{pmatrix}\cdot\boldsymbol{T}, \tag{5.7}$$

where

$$\boldsymbol{T}=\begin{pmatrix}-\frac{1}{\sqrt{3}} & -\frac{1}{\sqrt{6}} & \frac{1}{\sqrt{2}}\\ -\frac{1}{\sqrt{3}} & \frac{2}{\sqrt{6}} & 0\\ -\frac{1}{\sqrt{3}} & -\frac{1}{\sqrt{6}} & -\frac{1}{\sqrt{2}}\end{pmatrix}=\frac{1}{\sqrt{6}}\begin{pmatrix}-\sqrt{2} & -1 & \sqrt{3}\\ -\sqrt{2} & 2 & 0\\ -\sqrt{2} & -1 & -\sqrt{3}\end{pmatrix} \tag{5.8}$$

The components ($\boldsymbol{r}$ and $\boldsymbol{r}''$) of a vector $\vec{r}$ are then transformed according to:

$$\vec{r}=\begin{pmatrix}\vec{e_1} & \vec{e_2} & \vec{e_3}\end{pmatrix}\cdot\boldsymbol{r}=\begin{pmatrix}\vec{e_1''} & \vec{e_2''} & \vec{e_3''}\end{pmatrix}$$

$$\boldsymbol{r}''=\left[\begin{pmatrix}\vec{e_1} & \vec{e_2} & \vec{e_3}\end{pmatrix}\cdot\boldsymbol{T}\right]\cdot\boldsymbol{r}'':\ \boldsymbol{r}=\boldsymbol{T}\cdot\boldsymbol{r}''\quad\text{or}\quad \boldsymbol{r}''=\boldsymbol{T}^{-1}\cdot\boldsymbol{r} \tag{5.9}$$

Now, the representative matrix of a symmetry operation $\boldsymbol{R}=\boldsymbol{D}(R)$ transforms a vector $\vec{r}$ into $\vec{s}$ according to $\boldsymbol{Rr}=\boldsymbol{s}$.

Changing the basis, $\boldsymbol{R}$, $\boldsymbol{s}$, and $\boldsymbol{r}$ also change, but the relation remains: $\boldsymbol{R}''\cdot\boldsymbol{r}''=\boldsymbol{s}''$. Then:

$$\vec{r}=\begin{pmatrix}\vec{e_1} & \vec{e_2} & \vec{e_3}\end{pmatrix}\cdot\boldsymbol{r}=\begin{pmatrix}\vec{e_1} & \vec{e_2} & \vec{e_3}\end{pmatrix}\cdot\boldsymbol{T}\cdot\boldsymbol{r}''$$

that is $\boldsymbol{r}=\boldsymbol{T}\cdot\boldsymbol{r}''$ as already seen and

$\vec{s}=\begin{pmatrix}\vec{e_1} & \vec{e_2} & \vec{e_3}\end{pmatrix}\cdot\boldsymbol{s}=\begin{pmatrix}\vec{e_1} & \vec{e_2} & \vec{e_3}\end{pmatrix}\cdot\boldsymbol{T}\cdot\boldsymbol{s}''$ that is $\boldsymbol{s}=\boldsymbol{T}\cdot\boldsymbol{s}''$.

This means that: $\boldsymbol{R}''\cdot\boldsymbol{r}''=\boldsymbol{s}''=\boldsymbol{T}^{-1}\cdot\boldsymbol{s}=\boldsymbol{T}^{-1}\cdot\boldsymbol{R}\cdot\boldsymbol{r}=\boldsymbol{T}^{-1}\cdot\boldsymbol{R}\cdot\boldsymbol{T}\cdot\boldsymbol{r}''$, and finally

$$\boldsymbol{R}''=\boldsymbol{T}^{-1}\cdot\boldsymbol{R}\cdot\boldsymbol{T} \tag{5.10}$$

In summary:

1. $\boldsymbol{T}$ is used to describe the transformation of the components of a given vector (Eq. 5.9),
2. $\boldsymbol{T}^{-1}$ is used to describe the transformation of the basis (Eq. 5.7) and
3. both $\boldsymbol{T}$ and $\boldsymbol{T}^{-1}$ are used to describe the transformation of the operators (Eq. 5.10).

5.1.3.4 Similarity Transformations and Equivalence of Representations

The transformation of Eq. (5.10) is called *similarity transformation*. By applying the same similarity transformation to all the matrices $\boldsymbol{D}_\alpha(R)$ we obtain the representation $\boldsymbol{D}_\beta(R)$. Let us consider three elements of the group that are related by $AB=C$, and therefore $\boldsymbol{D}_\alpha(A)\cdot\boldsymbol{D}_\alpha(B)=\boldsymbol{D}_\alpha(C)$. With the transformation $\boldsymbol{T}$ we have $\boldsymbol{D}_\beta(A)=\boldsymbol{T}^{-1}\cdot\boldsymbol{D}_\alpha(A)\cdot\boldsymbol{T}$, $\boldsymbol{D}_\beta(B)=\boldsymbol{T}^{-1}\cdot\boldsymbol{D}_\alpha(B)\cdot\boldsymbol{T}$ and $\boldsymbol{D}_\beta(C)=\boldsymbol{T}^{-1}\cdot\boldsymbol{D}_\alpha(C)\cdot\mathrm{T}$. Then, $\boldsymbol{D}_\beta(A)\cdot\boldsymbol{D}_\beta(B)=\boldsymbol{T}^{-1}\cdot\boldsymbol{D}_\alpha(A)\cdot\boldsymbol{T}\cdot\boldsymbol{T}^{-1}\cdot\boldsymbol{D}_\alpha(B)\cdot\boldsymbol{T}=\boldsymbol{T}^{-1}\cdot\boldsymbol{D}_\alpha(A)\cdot\boldsymbol{D}_\alpha(B)\cdot\boldsymbol{T}=\boldsymbol{T}^{-1}\cdot\boldsymbol{D}_\alpha(C)\cdot\boldsymbol{T}$ which proves that $\boldsymbol{D}_\beta(C)=\boldsymbol{D}_\beta(A)\cdot\boldsymbol{D}_\beta(B)$.

Two representations $\boldsymbol{D}_\alpha(R)$ and $\boldsymbol{D}_\beta(R)$ related by a similarity transformation have obviously the same dimension. Even if they are apparently different, the two representations derive from a simple change of basis and therefore *do not* carry *essentially different information*. For this reason, they are called *equivalent* representations, and we write:

$$\boldsymbol{D}_\alpha(R)\sim\boldsymbol{D}_\beta(R)$$

For the sake of clarity, it is best to avoid saying that they are the *same* representation.

5.1.3.5 Representations with Function Spaces

Let us now consider a space of atomic orbitals (AOs) that we want to use to express the molecular orbitals (MOs), for example, of the ammonia molecule. For simplicity, let us begin by considering a basis consisting of only the 1s AOs of the H atoms (we indicate them with the letters *a*, *b*, *c*) plus the 2s AO of the N atom. We immediately see that the rotation operation around the ternary axis transforms the AOs in the following way:

$$C_3\text{: } a\rightarrow b;\ b\rightarrow c;\ c\rightarrow a;\ \mathrm{N}\rightarrow\mathrm{N}.$$

So, the transformation is:

$$\begin{pmatrix} a & b & c & \mathrm{N}\end{pmatrix}\begin{pmatrix}0&0&1&0\\1&0&0&0\\0&1&0&0\\0&0&0&1\end{pmatrix}\begin{pmatrix}a\\b\\c\\\mathrm{N}\end{pmatrix} \tag{5.11}$$

In a similar way we obtain the matrices for the other symmetry operations:

The procedure can be extended as desired, for example by considering an 8-D space with a basis consisting of the AO 1s of the H and the AO 1s, 2s and 2p of the N, or a larger basis, ….

5.1.3.6 The Representation of Equivalence of Atomic Sites (EAS)

Thanks to the spherical symmetry of *s*-type AOs, the transformations of the previous example (Table 4.17) can also be identified with the representation of equivalence of atomic sites ($\boldsymbol{D}_{\text{eas}}$). This representation is very important for various applications, as we will see in Section 6. The $\boldsymbol{D}_{\text{eas}}$ of a particular problem is typically obtained by considering the position vectors of the various atoms in ordinary 3D space or by formally treating a row of symbols denoting the atoms. For example, for NH_3 and denoting the three equivalent H by *a*, *b*, *c*, the positions $\begin{pmatrix}\vec{r_a} & \vec{r_b} & \vec{r_c} & \vec{r_\mathrm{N}}\end{pmatrix}$ and the symbols ($a\ b\ c$ N) are transformed into $\begin{pmatrix}\vec{r_b} & \vec{r_c} & \vec{r_a} & \vec{r_\mathrm{N}}\end{pmatrix}$ and ($b\ c\ a$ N) by C_3 in the following way:

$$\begin{pmatrix}\vec{r_a} & \vec{r_b} & \vec{r_c} & \vec{r_\mathrm{N}}\end{pmatrix}\begin{pmatrix}0&0&1&0\\1&0&0&0\\0&1&0&0\\0&0&0&1\end{pmatrix}=\begin{pmatrix}\vec{r_b} & \vec{r_c} & \vec{r_a} & \vec{r_\mathrm{N}}\end{pmatrix}$$

$$\begin{pmatrix} a & b & c & \mathrm{N}\end{pmatrix}\begin{pmatrix}0&0&1&0\\1&0&0&0\\0&1&0&0\\0&0&0&1\end{pmatrix}=\begin{pmatrix} b & c & a & \mathrm{N}\end{pmatrix}$$

5.1.3.7 The Regular Representation

Among the possible matrix representations of a group, in the development of the theory an important role is played by a representation that is essentially the transformation into a set of matrices of the multiplication table. The construction

TABLE 4.17
4D Representation Based on the s-Type AOs of H and N

E	C_3	C_{-3}
$\begin{pmatrix}1&0&0&0\\0&1&0&0\\0&0&1&0\\0&0&0&1\end{pmatrix}$	$\begin{pmatrix}0&0&1&0\\1&0&0&0\\0&1&0&0\\0&0&0&1\end{pmatrix}$	$\begin{pmatrix}0&1&0&0\\0&0&1&0\\1&0&0&0\\0&0&0&1\end{pmatrix}$
$\sigma^{(1)}$	$\sigma^{(2)}$	$\sigma^{(3)}$
$\begin{pmatrix}1&0&0&0\\0&0&1&0\\0&1&0&0\\0&0&0&1\end{pmatrix}$	$\begin{pmatrix}0&0&1&0\\0&1&0&0\\1&0&0&0\\0&0&0&1\end{pmatrix}$	$\begin{pmatrix}0&1&0&0\\1&0&0&0\\0&0&1&0\\0&0&0&1\end{pmatrix}$

TABLE 4.18
Rearrangement of the Cayley Table of Group C_{3v} from Table 4.3

↓×→	E	C_3	C_{-3}	$\sigma^{(1)}$	$\sigma^{(2)}$	$\sigma^{(3)}$
E	E	C_3	C_{-3}	$\sigma^{(1)}$	$\sigma^{(2)}$	$\sigma^{(3)}$
$(C_3)^{-1} = C_{-3}$	C_{-3}	E	C_3	$\sigma^{(2)}$	$\sigma^{(3)}$	$\sigma^{(1)}$
$(C_{-3})^{-1} = C_3$	C_3	C_{-3}	E	$\sigma^{(3)}$	$\sigma^{(1)}$	$\sigma^{(2)}$
$(\sigma^{(1)})^{-1} = \sigma^{(1)}$	$\sigma^{(1)}$	$\sigma^{(2)}$	$\sigma^{(3)}$	E	C_3	C_{-3}
$(\sigma^{(2)})^{-1} = \sigma^{(2)}$	$\sigma^{(2)}$	$\sigma^{(3)}$	$\sigma^{(1)}$	C_{-3}	E	C_3
$(\sigma^{(3)})^{-1} = \sigma^{(3)}$	$\sigma^{(3)}$	$\sigma^{(1)}$	$\sigma^{(2)}$	C_3	C_{-3}	E

of this representation, which is called **regular**, is explained by the example for the case of our usual group C_{3v}.

The first operation is to permute the rows of the multiplication table, so that the identity is on the main diagonal. To achieve this arrangement, first a certain order is chosen - starting from the identity - for the elements of the group that identify the columns, then the order of the respective inverse elements is followed for the rows (see Table 4.18).

Finally, in the matrix representing a given element of the group, the number 1 is inserted where the element appears in the reorganized table, or the number 0 where the given element does not appear. Consequently, *each row* (each column) of the regular representation of any element of the group *has only one element* = 1 *and all the others* = 0. For example, here are the matrices of the identity and C_3 operation in the regular representation:

$$\boldsymbol{D}_{\text{reg}}(E) = \mathbf{1}_6;$$

$$\boldsymbol{D}_{\text{reg}}(C_3) = \begin{pmatrix} 0 & 1 & 0 & 0 & 0 & 0 \\ 0 & 0 & 1 & 0 & 0 & 0 \\ 1 & 0 & 0 & 0 & 0 & 0 \\ 0 & 0 & 0 & 0 & 1 & 0 \\ 0 & 0 & 0 & 0 & 0 & 1 \\ 0 & 0 & 0 & 1 & 0 & 0 \end{pmatrix} \quad (5.12)$$

The regular representation has a size equal to the order of the group. The proof that the matrices thus defined are effectively a representation, *i.e.* that the matrices are multiplied like the elements of the group is given in Section 8.2.

In addition to the multiplication table of the group, the regular representation can be originated by a transformation of coordinates, if each of them is transformed (only) into another coordinate by the different operations of the symmetry group, similarly to what we have seen for other representations. A curious example are the 12 bond angles of cyclopropane.

5.1.3.8 A 12-D Representation

Let us now consider a different vector space, the configuration space of the ammonia molecule. A generic configuration of the entire molecule (deformed or not) will be expressed by a point in a 12-D space (three coordinates for each atom). To describe the deviations from the equilibrium positions, let us use four Cartesian triplets as basis vectors.

As Figure 4.29 shows, the triplets are:

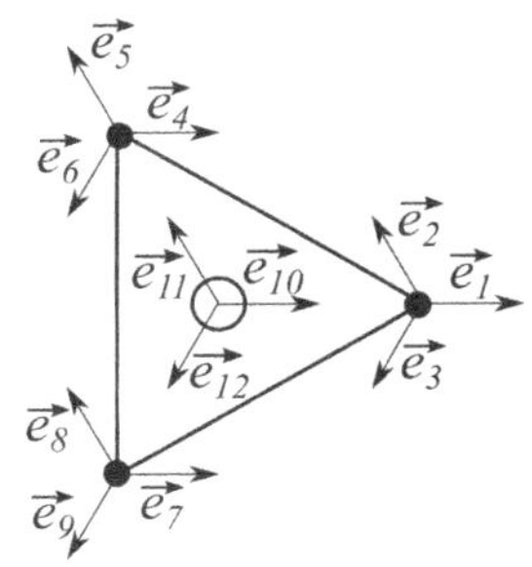

FIGURE 4.29 Basis vectors for the configuration of NH_3. N is the large open circle; the Hs are the small, filled circles.

- centred on the various atoms [*i.e.* the triplet $(\vec{e_1}, \vec{e_2}, \vec{e_3})$ centred on the first H, $(\vec{e_4}, \vec{e_5}, \vec{e_6})$ centred on the second H, $(\vec{e_7}, \vec{e_8}, \vec{e_9})$ centered on the third H, and finally the triplet $(\vec{e_{10}}, \vec{e_{11}}, \vec{e_{12}})$ centered on the N] and
- parallel to the three basis vectors of the example in Section 5.1.3.2.

As usual, the transformation matrix of a symmetry operation is obtained by considering the effect of the operation on the basis vectors. For example, C_3 transforms the basis in this way:

$$\vec{e_1} \to \vec{e_5},\quad \vec{e_2} \to \vec{e_6},\quad \vec{e_3} \to \vec{e_4},\quad \vec{e_4} \to \vec{e_8},$$

$$\vec{e_5} \to \vec{e_9},\quad \vec{e_6} \to \vec{e_7},\quad \vec{e_7} \to \vec{e_2},\quad \vec{e_8} \to \vec{e_3},$$

$$\vec{e_9} \to \vec{e_1},\quad \vec{e_{10}} \to \vec{e_{11}},\quad \vec{e_{11}} \to \vec{e_{12}},\quad \vec{e_{12}} \to \vec{e_{10}}$$

Then, the transformation matrix $\boldsymbol{D}_{12\text{D}}(C_3)$, for example, is:

$$\boldsymbol{D}_{12\text{D}}(C_3) = \begin{pmatrix} 0 & 0 & 0 & 0 & 0 & 0 & 0 & 0 & 1 & 0 & 0 & 0 \\ 0 & 0 & 0 & 0 & 0 & 0 & 1 & 0 & 0 & 0 & 0 & 0 \\ 0 & 0 & 0 & 0 & 0 & 0 & 0 & 1 & 0 & 0 & 0 & 0 \\ 0 & 0 & 1 & 0 & 0 & 0 & 0 & 0 & 0 & 0 & 0 & 0 \\ 1 & 0 & 0 & 0 & 0 & 0 & 0 & 0 & 0 & 0 & 0 & 0 \\ 0 & 1 & 0 & 0 & 0 & 0 & 0 & 0 & 0 & 0 & 0 & 0 \\ 0 & 0 & 0 & 0 & 0 & 1 & 0 & 0 & 0 & 0 & 0 & 0 \\ 0 & 0 & 0 & 1 & 0 & 0 & 0 & 0 & 0 & 0 & 0 & 0 \\ 0 & 0 & 0 & 0 & 1 & 0 & 0 & 0 & 0 & 0 & 0 & 0 \\ 0 & 0 & 0 & 0 & 0 & 0 & 0 & 0 & 0 & 0 & 0 & 0 \\ 0 & 0 & 0 & 0 & 0 & 0 & 0 & 0 & 0 & 1 & 0 & 1 \\ 0 & 0 & 0 & 0 & 0 & 0 & 0 & 0 & 0 & 0 & 1 & 0 \end{pmatrix} \quad (5.13)$$

The matrices of C_{-3}, $\sigma^{(1)}$, $\sigma^{(2)}$ and $\sigma^{(3)}$ are obtained in the same way, while the identity is obviously represented by $\mathbf{1}_{12}$. The complete list is in Section 5.1.5.2.

5.1.4 Unitarity Transformations

A symmetry operation (A) **does not modify the moduli of vectors and the angles between them**; in more technical terms it means that it *preserves scalar products*. On linear spaces, whether 3D vectors ($\vec{r}$, $\vec{s}$, …), or functions (φ, χ, ψ, …), the scalar product is a real or complex number (respectively) is defined[8]:

$$(\vec{t},\vec{v})\overset{\text{def}}{=}|\vec{t}|\cdot|\vec{v}|\cdot\cos\left(\widehat{\vec{t},\vec{v}}\right) \text{ and is: } (\vec{t},\vec{v})=(\vec{v},\vec{t}) \text{ or}$$

$$\langle t\mid v\rangle\overset{\text{def}}{=}\int_\Omega \varphi_t^*\varphi_v \mathrm{d}\tau \quad \text{and is:} \quad \langle t\mid v\rangle^*=\langle v\mid t\rangle\overset{\text{def}}{=}\int_\Omega \varphi_v^*\varphi_t \mathrm{d}\tau$$

and is linear with respect to the second element (v) and anti-linear with respect to the first element (t), so that the matrix notation is:

$$\langle\chi_t|\chi_v\rangle=\boldsymbol{t}^\dagger\cdot\boldsymbol{S}\cdot\boldsymbol{v} \quad \text{or} \quad (\vec{t},\vec{v})=\boldsymbol{t}^T\cdot\boldsymbol{S}\cdot\boldsymbol{v}$$

where the $S_{ij}=\langle\varphi_i|\varphi_j\rangle$ or $(\vec{e_i},\vec{e_j})$ are the *scalar products of the basis vectors.*

If we consider the symmetry operation A and $\vec{w}=A\vec{t}$, or $\varphi_w=\hat{A}\varphi_t$ and respectively $\vec{z}=A\vec{v}$, or $\varphi_z=\hat{A}\varphi_z$, in matrix notation the scalar products of the *transformed* vectors are:

$$(\vec{w},\vec{z})=(\boldsymbol{A}\cdot\boldsymbol{t})^\mathrm{T}\cdot\boldsymbol{S}\cdot(\boldsymbol{A}\cdot\boldsymbol{v})=\boldsymbol{t}^\mathrm{T}\cdot\boldsymbol{A}^\mathrm{T}\cdot\boldsymbol{S}\cdot\boldsymbol{A}\cdot\boldsymbol{v} \text{ or}$$

$$\langle w|z\rangle=(\boldsymbol{A}\cdot\boldsymbol{t})^\dagger\cdot\boldsymbol{S}\cdot(\boldsymbol{A}\cdot\boldsymbol{v})=\boldsymbol{t}^\dagger\cdot\boldsymbol{A}^\dagger\cdot\boldsymbol{S}\cdot\boldsymbol{A}\cdot\boldsymbol{v}$$

The requirement for *preserving scalar products* is:

$$\boldsymbol{S}\to\boldsymbol{S}'=\boldsymbol{A}^\dagger\cdot\boldsymbol{S}\cdot\boldsymbol{A}=\boldsymbol{S} \quad \text{or} \quad \boldsymbol{S}\to\boldsymbol{S}'=\boldsymbol{A}^\mathrm{T}\cdot\boldsymbol{S}\cdot\boldsymbol{A}=\boldsymbol{S}$$

Now, **if the basis vectors are an orthonormal set**, $\boldsymbol{S}=\boldsymbol{1}=\boldsymbol{A}^\dagger\cdot\boldsymbol{A}$ and

$$\boldsymbol{A}^{-1}=\boldsymbol{A}^\dagger$$

A matrix ($\boldsymbol{U}$) is said to be unitary if it is invertible and its inverse ($\boldsymbol{U}^{-1}$) equals the adjoint matrix $\boldsymbol{U}^\dagger$: $\boldsymbol{U}^{-1}=\boldsymbol{U}^\dagger$: therefore, the $\boldsymbol{A}$ matrices are all unitary. A *transformation* represented by a *unitary matrix* is said to be *unitary.*

> The matrices of any representation of a group of symmetry operations must be unitary.

A point to stress is that, when the matrices $\boldsymbol{R}$ of a representation of a symmetry group are obtained from the transformations of a set of basis functions, these functions **must be orthogonal to each other and normalized**, otherwise the matrices of the transformations so obtained are **not** unitary. Section 6.1.1.2 shows a transformation that does not correspond to a unitary matrix because of the choice of a non-orthonormal basis.

In the real case, we speak of **orthogonal** (or orthonormal) matrices: the inverse ($\boldsymbol{O}^{-1}$) of an orthogonal matrix ($\boldsymbol{O}$) equals the transpose ($\boldsymbol{O}^{-1}=\boldsymbol{O}^\mathrm{T}$). It is easy to verify that a matrix $\boldsymbol{A}$ is orthonormal (or unitary): multiplying it by its transpose $\boldsymbol{A}^\mathrm{T}$ (or its adjoint $\boldsymbol{A}^\dagger$) we obtain the unit matrix of the same order. For real matrices, in particular, (a) the sum on the rows (columns) of the products of the coefficients of two different columns (rows) is equal to zero and (b) the sum on the rows (columns) of the squares of the coefficients of any column (row) is equal to one; for complex matrices slightly different properties hold because it is necessary to multiply by the conjugates. For example, the matrix:

$$\frac{1}{\sqrt{6}}\begin{pmatrix}\sqrt{2} & 2 & 0\\ \sqrt{2} & -1 & \sqrt{3}\\ \sqrt{2} & -1 & -\sqrt{3}\end{pmatrix} \quad \text{is orthonormal}$$

$$\text{Indeed:}\ \frac{1}{\sqrt{6}}\begin{pmatrix}\sqrt{2} & 2 & 0\\ \sqrt{2} & -1 & \sqrt{3}\\ \sqrt{2} & -1 & -\sqrt{3}\end{pmatrix}\cdot \tag{5.14}$$

$$\frac{1}{\sqrt{6}}\begin{pmatrix}\sqrt{2} & \sqrt{2} & \sqrt{2}\\ 2 & -1 & -1\\ 0 & \sqrt{3} & -\sqrt{3}\end{pmatrix}=\begin{pmatrix}1 & 0 & 0\\ 0 & 1 & 0\\ 0 & 0 & 1\end{pmatrix}$$

Also, let's consider for example the second and third columns. The sum of the products on the rows is $\frac{1}{6}\left[2\cdot 0-1\cdot\sqrt{3}-1\cdot\left(-\sqrt{3}\right)\right]=0$, while, considering for example the first column, the sum of the squares on the rows: $\left(\frac{1}{\sqrt{6}}\right)^2(2+2+2)=1$

There is an important exception to unitary symmetry transformations: the *time reversal symmetry* (see Section 2.1.3) is **anti-unitary** and is described by an anti-unitary matrix. For an anti-unitary matrix ($\boldsymbol{A}$) $\boldsymbol{A}\bullet\boldsymbol{A}^\dagger=\boldsymbol{-1}$ (instead of $\boldsymbol{A}\bullet\boldsymbol{A}^\dagger=\boldsymbol{1}$). In functional notation, for an anti-unitary operator $\hat{A}$ the following applies: $\langle\hat{A}\varphi|\hat{A}\psi\rangle=-\langle\varphi|\psi\rangle$ instead of $\langle\hat{A}\varphi|\hat{A}\psi\rangle=\langle\varphi|\psi\rangle$.

5.1.5 Sum and Product of Representations

5.1.5.1 Direct Sum of Representations

Let us consider the 3D representation of the second example (Section 5.1.3.2). We see that each matrix in the 3D representation is block-diagonal partitioned with a 1×1 block at the top left and a 2×2 block at the bottom right. Also, it is not difficult to verify that the two sets of 1×1 and 2×2 matrices (Table 4.19) are two matrix representations of the group because they multiply as required. So, we write: $\boldsymbol{D}_{3\mathrm{D}}=\boldsymbol{D}_{1\mathrm{D}}\oplus\boldsymbol{D}_{2\mathrm{D}}$.

The example shows the property that:

> The direct sum of two representations is a representation of the group and has dimension equal to the sum of the dimensions of the summed representations.

The *proof* is based on a property (see Section 5.1.1.1) of the direct sum of matrices: if a and b are two

TABLE 4.19
The Two Representations That Give the 3D Representation of Eq. (5.5) by Direct Sum (see Also Section 5.1.3.2)

	E	C_3	C_{-3}	$\sigma^{(1)}$	$\sigma^{(2)}$	$\sigma^{(3)}$
D_{1D}	1	1	1	1	1	1
D_{2D}	$\begin{pmatrix} 1 & 0 \\ 0 & 1 \end{pmatrix}$	$\begin{pmatrix} \frac{-1}{2} & \frac{-\sqrt{3}}{2} \\ \frac{\sqrt{3}}{2} & \frac{-1}{2} \end{pmatrix}$	$\begin{pmatrix} \frac{-1}{2} & \frac{\sqrt{3}}{2} \\ \frac{-\sqrt{3}}{2} & \frac{-1}{2} \end{pmatrix}$	$\begin{pmatrix} \frac{-1}{2} & \frac{-\sqrt{3}}{2} \\ \frac{-\sqrt{3}}{2} & \frac{1}{2} \end{pmatrix}$	$\begin{pmatrix} 1 & 0 \\ 0 & -1 \end{pmatrix}$	$\begin{pmatrix} \frac{-1}{2} & \frac{\sqrt{3}}{2} \\ \frac{\sqrt{3}}{2} & \frac{1}{2} \end{pmatrix}$

representations and two symmetry operations A and B are represented by $\boldsymbol{D}_a(A)$ and $\boldsymbol{D}_a(B)$ in the a representation and by $\mathbf{D}_b(A)$ and $\mathbf{D}_b(B)$ in the b representation, then the matrices $\boldsymbol{D}_c(A) = \boldsymbol{D}_a(A) \oplus \boldsymbol{D}_b(A)$ and $\boldsymbol{D}_c(B) = \boldsymbol{D}_a(B) \oplus \boldsymbol{D}_b(B)$ represent A and B in the representation c that is the sum of the a and b representations, because $\boldsymbol{D}_c(A) \cdot \boldsymbol{D}_c(B) = [\boldsymbol{D}_a(A) \oplus \boldsymbol{D}_b(A)] \cdot [\boldsymbol{D}_a(B) \oplus \boldsymbol{D}_b(B)] = [\boldsymbol{D}_a(A) \cdot \boldsymbol{D}_a(B)] \oplus [\boldsymbol{D}_b(A) \cdot \boldsymbol{D}_b(B)]$.

5.1.5.2 Direct Product of Representations

Let us now consider the 12D representation of the example in Section 5.1.3.8 and, in particular, compare the $\mathbf{D}_{12D}(C_3)$ matrix (see Eq. (5.13)) with the matrix of the same operation in the 3D representation of the example of Section 5.1.3.1 (see Eq. (5.4)):

$$\boldsymbol{D}_{3D}(C_3) = \begin{pmatrix} 0 & 0 & 1 \\ 1 & 0 & 0 \\ 0 & 1 & 0 \end{pmatrix}$$

First, we note that the $\boldsymbol{D}_{12D}(C_3)$ matrix can be partitioned into 16 3×3 blocks and that four of these blocks contain replicas of the above $\boldsymbol{D}_{3D}(C_3)$ matrix, while the other blocks are null (3×3) matrices $\mathbf{0}_3$ (*i.e.* matrices in which all elements are=0):

$$\boldsymbol{D}_{12D}(C_3) = \begin{pmatrix} \mathbf{0}_3 & \mathbf{0}_3 & \boldsymbol{D}_{3D}(C_3) & \mathbf{0}_3 \\ \boldsymbol{D}_{3D}(C_3) & \mathbf{0}_3 & \mathbf{0}_3 & \mathbf{0}_3 \\ \mathbf{0}_3 & \boldsymbol{D}_{3D}(C_3) & \mathbf{0}_3 & \mathbf{0}_3 \\ \mathbf{0}_3 & \mathbf{0}_3 & \mathbf{0}_3 & \boldsymbol{D}_{3D}(C_3) \end{pmatrix} \quad (5.15)$$

Second, the four replicas of $\mathbf{D}_{3D}(C_3)$ are placed in the blocks that correspond to 1 in the matrix:

$$\boldsymbol{D}' = \begin{pmatrix} 0 & 0 & 1 & 0 \\ 1 & 0 & 0 & 0 \\ 0 & 1 & 0 & 0 \\ 0 & 0 & 0 & 1 \end{pmatrix}$$

while the elements=0 of $\boldsymbol{D}'$ denote null (3×3) matrices $\mathbf{0}_3$ in the block partition of $\boldsymbol{D}_{12D}(C_3)$.

Third, we note that $\boldsymbol{D}' = \boldsymbol{D}_{4D}(C_3)$ (see Eq. (5.4)).

Finally, we analyse the full set of $\boldsymbol{D}_{12D}$ matrices (see below), and we note that for each symmetry operation R the three considerations above still hold. We see that each $\boldsymbol{D}_{12D}$ matrix is partitioned as described by its correspondent $\boldsymbol{D}_{4D}$ matrix of the example in Section 5.1.3.6:

$$\boldsymbol{D}_{12D}(E) = \mathbf{1}_{12D}$$

$$\boldsymbol{D}_{12D}(C_{-3}) = \begin{pmatrix} \mathbf{0}_3 & \boldsymbol{D}_{3D}(C_{-3}) & \mathbf{0}_3 & \mathbf{0}_3 \\ \mathbf{0}_3 & \mathbf{0}_3 & \boldsymbol{D}_{3D}(C_{-3}) & \mathbf{0}_3 \\ \boldsymbol{D}_{3D}(C_{-3}) & \mathbf{0}_3 & \mathbf{0}_3 & \mathbf{0}_3 \\ \mathbf{0}_3 & \mathbf{0}_3 & \mathbf{0}_3 & \boldsymbol{D}_{3D}(C_{-3}) \end{pmatrix}$$

$$\boldsymbol{D}_{12D}\left(\sigma^{(1)}\right) = \begin{pmatrix} \boldsymbol{D}_{3D}\left(\sigma^{(1)}\right) & \mathbf{0}_3 & \mathbf{0}_3 & \mathbf{0}_3 \\ \mathbf{0}_3 & \mathbf{0}_3 & \boldsymbol{D}_{3D}\left(\sigma^{(1)}\right) & \mathbf{0}_3 \\ \mathbf{0}_3 & \boldsymbol{D}_{3D}\left(\sigma^{(1)}\right) & \mathbf{0}_3 & \mathbf{0}_3 \\ \mathbf{0}_3 & \mathbf{0}_3 & \mathbf{0}_3 & \boldsymbol{D}_{3D}\left(\sigma^{(1)}\right) \end{pmatrix}$$

$$\boldsymbol{D}_{12D}\left(\sigma^{(2)}\right) = \begin{pmatrix} \mathbf{0}_3 & \mathbf{0}_3 & \boldsymbol{D}_{3D}\left(\sigma^{(2)}\right) & \mathbf{0}_3 \\ \mathbf{0}_3 & \boldsymbol{D}_{3D}\left(\sigma^{(2)}\right) & \mathbf{0}_3 & \mathbf{0}_3 \\ \boldsymbol{D}_{3D}\left(\sigma^{(2)}\right) & \mathbf{0}_3 & \mathbf{0}_3 & \mathbf{0}_3 \\ \mathbf{0}_3 & \mathbf{0}_3 & \mathbf{0}_3 & \boldsymbol{D}_{3D}\left(\sigma^{(2)}\right) \end{pmatrix}$$

$$\boldsymbol{D}_{12D}\left(\sigma^{(3)}\right) = \begin{pmatrix} \mathbf{0}_3 & \boldsymbol{D}_{3D}\left(\sigma^{(3)}\right) & \mathbf{0}_3 & \mathbf{0}_3 \\ \boldsymbol{D}_{3D}\left(\sigma^{(3)}\right) & \mathbf{0}_3 & \mathbf{0}_3 & \mathbf{0}_3 \\ \mathbf{0}_3 & \mathbf{0}_3 & \boldsymbol{D}_{3D}\left(\sigma^{(3)}\right) & \mathbf{0}_3 \\ \mathbf{0}_3 & \mathbf{0}_3 & \mathbf{0}_3 & \boldsymbol{D}_{3D}\left(\sigma^{(3)}\right) \end{pmatrix}$$

Each matrix of $\boldsymbol{D}_{12\mathrm{D}}$ corresponds to a direct product: $\boldsymbol{D}_{12\mathrm{D}}(R) = \boldsymbol{D}_{4\mathrm{D}}(R) \otimes \boldsymbol{D}_{3\mathrm{D}}(R)$.
The example shows the general property that

> The direct products of the matrices that belong to two representations are a representation of the group.

In other words: if we have two matrix representations (1 and 2) of a group and for each element R of the group we write the direct matrix product $\boldsymbol{D}_1(R) \otimes \boldsymbol{D}_2(R)$, the matrices thus obtained are also a representation of the group and we say that it is the *direct product of the two representations*. For a proof, see Exercise 4.14. For other notes on the direct product of representations, see Sections 5.2 and 5.3.2. For the physical meaning, see Section 6.1.2.

5.2 Reducible and Irreducible Representations

Considering all the examples, it should now be evident that there are no limits to the representations that can be found for a given group. In addition to the cases shown by the examples, in which different vector spaces have been considered, there are in fact *purely algebraic procedures* for making new representations at will starting from already known representations. They are

- the similarity transformation (see Section 5.1.3.3),
- the direct sum (see Sections 5.1.5.1 and 5.1.1.1) and
- the direct product (see Sections 5.1.5.2 and 5.1.1.2).

In the case of similarity transformations, a new representation has the same size as the original representation and is equivalent to it, because the transformation is a simple change of basis. The direct sum of two or more representations instead generates a representation that has *larger dimensions* than the original representations and is *not* equivalent to them. The same is true for the direct product, except when at least one representation to be multiplied is 1D.

In applications, however, it is much more important to **decompose** *a high-dimensional representation* **into smaller** representations, *i.e.*:

- in particular, to derive the representations with the **smallest dimensions** that are contained, in a way that is not always evident, in a representation with higher dimensions, and
- in general, to derive **all** the **distinct** (*i.e.* non-equivalent) **simple** representations of a symmetry group.

What exactly is meant by *simple* is made clear hereafter by the description of a procedure, which is called **reduction**. The reduction begins by attempting to subject a representation to a change of basis (a similarity transformation) to obtain an equivalent representation that is the direct sum of two or more smaller representations. *If this can be done*, the original **representation** is said to be **reducible** (hereafter written **RR**). An example is given by what discussed in Section 5.1.5.1, where a 3D representation is transformed into the direct sum of a 1D representation and a 2D representation using a different basis: the change of basis is a similarity transformation (see Section 5.1.3.3) and the original 3D representation is equivalent to the direct sum: $\boldsymbol{D}_{3\mathrm{D}} \sim \boldsymbol{D}_{1\mathrm{D}} \oplus \boldsymbol{D}_{2\mathrm{D}}$.

The representations produced can in turn be examined for the possibility of a further reduction, and so on, until only **irreducible representations** (**IR** or **irreps**) are obtained. A 1D representation is obviously irreducible, but a group can also have higher-dimensional IRs. For example, if we continued with the example in Section 5.1.5.1, we would find that the 2D representation thus obtained is irreducible. So, what is required is:

- in particular, to know if a high-dimensional representation is reducible and, if it is, to decompose it into the direct sum of the IRs that are contained in it and,
- in general, to derive all the distinct IRs of a given symmetry group,

with the essential specification that we do not consider *distinct* two IRs that are linked by a similarity transformation or, in other words, are equivalent (*distinct* means *non-equivalent*).

Also, looking back at the examples of Section 2, and noting that an eigenvalue can be seen as a 1D matrix and that the product of eigenvalues corresponds to the product of the matrices of a representation, we now understand that:

> The distinct irreducible representations (IRs) correspond to the *symmetry species* of the group and the two terms (IR and symmetry species) are synonymous.

Group theory provides the algebraic tools needed to rigorously and systematically perform the reduction of a reducible representation and to find all the distinct IRs of a group and their characteristics. Let us anticipate here some fundamental results. First of all (it will be proved in Section 5.4.4):

> The reduction of an RR into its constituent IRs is unique.

As for the **number** of IRs, (see Section 5.4.3):

> A finite group has a finite number of distinct IRs.

More precisely (it will be derived in Section 5.4.4.1):

> The number of distinct IRs is equal to the number of classes.

About their dimensions (see again Section 5.4.3):

The sum of the squares of the dimensions of all distinct IRs is equal to the number of elements of the group.

A consequence, already discussed (see Section 1.3.1), is that:

For commutative groups, there are only 1D IRs.

A commutative group has as many IRs as there are elements in the group, while a non-commutative group has a smaller number. For example, the group C_{3v} has six elements and three classes and three IRs, one of dimension 2 and two of dimension 1: $2^2+1+1=6$.

Finally. it must be paid attention to the fact that:

The direct product of two IRs is *not necessarily* an IR,

as trivially shown by the product of two 2D IRs of a group having only 1D and 2D IRs. For example, the product of two 2D representations (both labelled E, see Table 20) of group C_{3v}: $\boldsymbol{D}_E \otimes \boldsymbol{D}_E$ is a 4D representation that is reducible.

5.3 Characters

Reducing a representation seems to be an operation that is not easily achieved, because it would require searching, among the infinite equivalent representations, for one that is in block-diagonal form (see Section 5.1.1.1), without even knowing in advance that it exists, that is, without knowing that the reduction is possible. The problem can actually be solved thanks to a fundamental parameter (a *list* of numbers called a **character system**) that *is the same for all representations that are equivalent* to each other.

The character system is an extraordinarily powerful tool for using symmetry and establishing fundamental relationships both between representations of a given group and between representations of different groups. This is because characters are transformed in a particularly simple way by the effect of operations that generate new representations. In fact:

- similarity leaves the character sistem unchanged,
- the characters of the direct sum of representations are the sums of the characters,
- the characters of the direct product of representations are the products of the characters.

Due to its properties (derived later), the character system avoids having to proceed by trial and error and provides procedures and operational formulas allowing to recognize two equivalent representations, and to know whether a representation is reducible or not. If the representation is reducible, the character system also tells us in which IRs it can be reduced and how many times each IR is contained in the RR.

In general, the character system indicates how many distinct IRs there are in a given group and what dimensions they have.

5.3.1 Trace and Character

Let's see step by step the definition and properties of the characters of a representation.

The **trace** of a square matrix ($\boldsymbol{D}$) is the sum of its elements along the main diagonal. For an $n\times n$ matrix:

$$tr(\boldsymbol{D}) \overset{\text{def}}{=} \sum_{i=1}^{n} \boldsymbol{D}_{ii}.$$

Examples: the trace of $\boldsymbol{D}(E)=\begin{pmatrix}1&0&0\\0&1&0\\0&0&1\end{pmatrix}$ is 3; the trace of $\boldsymbol{D}(C_3)=\begin{pmatrix}0&0&1\\1&0&0\\0&1&0\end{pmatrix}$.

The character χ of element R of a group $G=\{R, S, T, \dots\}$ in the **representation** $\alpha = \{\boldsymbol{D}_\alpha(R), \boldsymbol{D}_\alpha(S), \boldsymbol{D}_\alpha(T), \dots\}$ is the trace of its matrix $\boldsymbol{D}_\alpha(R)$: $\chi_\alpha(R)=tr[\boldsymbol{D}_\alpha(R)]$.

Finally,

The **character system** of the group $G=\{R, S, T, \dots\}$ in the representation $\alpha = \{\boldsymbol{D}_\alpha(R), \boldsymbol{D}_\alpha(S), \boldsymbol{D}_\alpha(T), \dots\}$ is the set of characters of the group: $\{\chi_\alpha(R), \chi_\alpha(S), \chi_\alpha(T), \dots\}$.

For example, the character system of the 3D RR of group C_{3v} shown in Section 5.1.3.1 is $\chi(E)=3$, $\chi(C_3)=\chi(C_{-3})=0$, $\chi(\sigma^{(1)})=\chi(\sigma^{(2)})=\chi(\sigma^{(3)})=1$.

5.3.2 Properties of Characters

The fundamental property of the character system of a representation is that:

The characters are invariant under similarity transformations.

The proof is simple, since it derives directly from the definition. The trace of an $n\times n$ matrix $\boldsymbol{R}'$ is $\sum_{i=1}^{n}(R')_{ii}$. If $\boldsymbol{R}'=\boldsymbol{T}^{-1}\bullet\boldsymbol{R}\bullet\boldsymbol{T}$, then $(R')_{ii}=\sum_{k=1}^{n}(T^{-1})_{ik}\cdot(RT)_{ki}$. The (RT) product is $(RT)_{ki}=\left(\sum_{j=1}^{n}R_{kj}T_{ji}\right)$ and therefore: $(R')_{ii}=\sum_{k=1}^{n}(T^{-1})_{ik}\cdot\left(\sum_{j=1}^{n}R_{kj}T_{ji}\right)$. After changing the order of the summations, the trace becomes: $\sum_{j=1}^{n}\sum_{k=1}^{n}R_{kj}\cdot\sum_{i=1}^{n}(T^{-1})_{ik}\cdot T_{ji}$.

The innermost summation $\sum_{i=1}^{n}(T^{-1})_{ik}\cdot T_{ji}=\sum_{i=1}^{n}T_{ji}\cdot(T^{-1})_{ik}$ is the element jk of $\boldsymbol{T}\cdot\boldsymbol{T}^{-1}=\delta_{kj}$.

Therefore: $\sum_{j=1}^{n}\sum_{k=1}^{n} R_{kj}\cdot\delta_{kj} = \sum_{j=1}^{n} R_{jj}$, which is the trace of $\boldsymbol{R}$.

Some properties of the character system are derived directly.

> In any representation, the characters of operations belonging to the *same class* are equal.

In fact, in any representation, the matrices of two group elements belonging to the same class are related by a similarity transformation, which does not change the trace of a matrix.

Furthermore, the trace of a block-diagonal matrix is the sum of the traces of the constituent blocks. If a representation is a direct sum (see Section 5.1.5.1) of representations, then all matrices are block-diagonal in the same way, and the consideration holds for the entire character system:

> The characters of the direct sum of representations are the sums of the respective characters.

The character system is therefore the tool to verify that a representation is the direct sum of other representations (better: that it is equivalent to their direct sum) without the need to find the equivalent representation in block-diagonal form. For example, the character systems of the representations of examples of Sections 5.1.3.2 and 5.1.5.1 are:

- 3D representation: $\chi(I)=3$, $\chi(C_3)=0$, $\chi(C_{-3})=0$, $\chi(\sigma^{(1)})=1$, $\chi(\sigma^{(2)})=1$, $\chi(\sigma^{(3)})=1$,
- 2D representation: $\chi(I)=2$, $\chi(C_3)=-1$, $\chi(C_{-3})=-1$, $\chi(\sigma^{(1)})=0$, $\chi(\sigma^{(2)})=0$, $\chi(\sigma^{(3)})=0$,
- 1D representation: $\chi(I)=1$, $\chi(C_3)=1$, $\chi(C_{-3})=1$, $\chi(\sigma^{(1)})=1$, $\chi(\sigma^{(2)})=1$, $\chi(\sigma^{(3)})=1$,

and it is immediately verified that the 3D representation is equivalent to the direct sum of the 2D and 1D, as already shown in detail. For the product of two representations (Section 5.1.5.2):

> The characters of the direct product of two representations are the products of their respective characters,

Indeed, the trace of a matrix of the product representation is the sum of the diagonal elements of the first factor (*i.e.* its character) multiplied by the sum of the diagonal elements of the second factor. For example, the character systems of the representations shown in Sections 5.1.3.8, 5.1.3.5 and 5.1.3.1 are:

- representation 12D: $\chi(I)=12$, $\chi(C_3)=0$, $\chi(C_{-3})=0$, $\chi(\sigma^{(1)})=2$, $\chi(\sigma^{(2)})=2$, $\chi(\sigma^{(3)})=2$,
- representation 4D: $\chi(I)=4$, $\chi(C_3)=1$, $\chi(C_{-3})=1$, $\chi(\sigma^{(1)})=2$, $\chi(\sigma^{(2)})=2$, $\chi(\sigma^{(3)})=2$,
- representation 3D: $\chi(I)=3$, $\chi(C_3)=0$, $\chi(C_{-3})=0$, $\chi(\sigma^{(1)})=1$, $\chi(\sigma^{(2)})=1$, $\chi(\sigma^{(3)})=1$.

Finally, anticipating a result that can only be demonstrated later (Section 5.4.3):

> A necessary and sufficient condition for the equivalence of two representations is the equality of the character system.

The condition *is necessary* because, if the representations are equivalent, each matrix of a representation is obtained by similarity from the corresponding matrix of the other representation, and the similarity transformation does not change the trace. Proving that the condition is also sufficient will be done in Section 5.4.3 by resorting to theoretical results obtained later.

For example, the character system of the 3D representation of Sections 5.1.3.2 is: $\chi(I)=3$, $\chi(C_3)=0$, $\chi(C_{-3})=0$, $\chi(\sigma^{(1)})=1$, $\chi(\sigma^{(2)})=1$, $\chi(\sigma^{(3)})=1$, *i.e.* the same as the representation of Section 5.1.3.1: the two representations are indeed related by a change of basis *i.e.* by similarity.

5.3.3 Characters for Some Representations

It would seem obvious that you need the matrix representation to have the character system. In reality, for some representations, which are also of considerable importance in applications, this is not necessary. In this regard, it is also useful to remember that only one operation per class must be considered since all operations of the class have the same character, and that the identity always has a character equal to the dimension of the representation.

For the **representation of equivalence of atomic sites** (eas, see Section 5.1.3.6), one obtains the character of a symmetry operation by counting how many atomic sites are left unchanged by the operation. The explanation is simple: if a certain atom is left unchanged, it appears as 1 along the diagonal of the representative matrix and thus contributes 1 to the character; if it is not left unchanged, there is 0 along the diagonal of the matrix and thus it does not contribute to the character.

In addition to the eas representation, one can similarly obtain the **equivalence representations** of **particular sets of atoms**. For example, for the ammonia molecule, the equivalence representation of the H sites has the characters (3, 0, 1) because the identity leaves all the hydrogens in place, the ternary rotations move them all, and the vertical reflections keep one H fixed and exchange the other two. The representation of the N site is totally symmetric, because no symmetry operation moves the N. The equivalence representation of *all* atomic sites has then characters (4, 1, 2), in agreement with the fact that it is the *direct sum* of the other two.

The **representation of the AOs** is almost analogous: the AOs that are equivalent by symmetry are counted: for each of them, if it is left unchanged, it contributes 1 to the character; if its sign changes, it contributes –1 to the character, otherwise it does not contribute to the character.

Also the characters of the **regular representation** can be written directly without processing matrices. In fact, considering how the regular representation is defined (see Section 5.1.3.7), the characters are = 0 for all elements of the group, except the identity, which has character = g. In fact, only E has non-zero matrix elements along the main diagonal, and these are all = 1.

5.4 Orthogonality Theorems

Many important results of group theory can be derived comparing **two IRs of the same group** by globally examining all the symmetry operations of the group. They are essentially orthogonality theorems, which in the simplest formulation provide a null result (= orthogonality) when two different (or better, non-equivalent) IRs are involved. The compared quantities are matrix elements or characters.

5.4.1 GOT: Great Orthogonality Theorem

The theorem is also called the **orthogonality theorem of the IRs**.

$$\sum_R \left[\boldsymbol{D}_\alpha(R)\right]_{ij}^* \left[\boldsymbol{D}_\beta(R)_{kl}\right] = \left(\frac{g}{g_\alpha}\right)\delta_{\alpha\beta}\delta_{ik}\delta_{jl} \quad (5.16)$$

On the left side, the summation is to be performed on all the elements of the group, R indicates a generic element of the group, α and β are two (equal or different) IRs of the group, $\boldsymbol{D}_\alpha(R)$ is the matrix that represents R in the α IR, $\boldsymbol{D}_\alpha(R)_{kl}$ is the element ij of this matrix, and similarly for the β IR, while the asterisk as usual indicates the complex conjugate.

On the right-hand side, g and g_α are respectively the order of the group (the number of symmetry operations) and the dimension of the α IR. As usual, the symbols $\delta_{\alpha\beta}$, ... are equal to 1 or 0 if the two indices are equal or different.

Basically, if we are dealing with the same IR ($\alpha=\beta$) *and* the same matrix element ($i=k$ and $j=l$), the expression on the right is g / g_α. Otherwise, it is zero.

For example, let us consider element 1,1 of the two IRs (1D and 2D, respectively) of group C_{3v} shown in Section 5.1.5.1. On the left: $1\cdot1+1\cdot(-1/2)+1\cdot(-1/2)+1\times1+1\cdot(-1/2)+1\cdot(-1/2)=0$, and this agrees with the right part, which is zero because these are two different IRs. Let us now consider twice the same 2D IR and the 2,1 elements of the matrices. On the left: $0+3/4+3/4+0+3/4+3/4=12/4=3$. On the right, each $\delta=1$ and what remains is $6/2=3$.

The **proof** of the GOT is particularly long and technical, requiring the proof of two lemmas without applicative relevance, and is not reported here. However, it is important to know that the proof explicitly requires the matrix representations to be unitary: *anti-unitary transformations* and in particular *time reversal symmetry* fall outside the validity of the GOT. We will not go into the details of this problem. As previously done (Section 2.1.4), in the following we will proceed as if time reversal symmetry did not exist and simply treat its effects as accidental degeneracies.

5.4.2 SOT: Small Orthogonality Theorem

$$\sum_R \left[\chi_\alpha(R)\right]^* \left[\chi_\beta(R)\right] = g\delta_{\alpha\beta} \quad (5.17)$$

The SOT is also called **orthogonality theorem of the characters** and is proved directly by the GOT by considering the diagonal elements of the matrices of two IRs: $\sum_R \left[\boldsymbol{D}_\alpha(R)\right]_{ii}^* \left[\boldsymbol{D}_\beta(R)\right]_{ll} = \left(\frac{g}{g_\alpha}\right)\delta_{\alpha\beta}\delta_{il}\delta_{il}$. We add over the diagonal elements to obtain the traces. To the left of the equality sign then remains the expression $\sum_R \left[\chi_\alpha(R)\right]^* \left[\chi_\beta(R)\right]$. On the right, the double summation over $\delta_{il}\delta_{il}$ resolves to a simple summation δ_{ii} which is equal to the number of elements of the matrix, $\boldsymbol{D}_\alpha(R)$ *i.e.* g_α, thus obtaining $g\delta_{\alpha\beta}$.

As with the GOT, the non-zero result on the right is obtained not only when the two representations are equal, but more generally when they are *equivalent*, which means that they have the same character system.

5.4.3 Direct Consequences of GOT and SOT

The orthogonality theorems consider the elements of the matrices of the IRs (the GOT) or the characters of the IRs (the SOT) as components of abstract vectors of a generally complex g-dimensional vector space.[9] The vectors are:

$$\boldsymbol{v}_s = \begin{pmatrix} \boldsymbol{D}_\alpha(R_1)_{ij} \\ \boldsymbol{D}_\alpha(R_2)_{ij} \\ \cdots \end{pmatrix} \text{ for the GOT or}$$

$$\boldsymbol{v}_\alpha = \begin{pmatrix} \chi_\alpha(R_1) \\ \chi_\alpha(R_2) \\ \cdots \end{pmatrix} \text{ for the SOT} \quad (5.18)$$

s globally denotes $\{\alpha, i, j\}$, g is the order of the group and R_1, R_2, ..., R_g are its elements.

In other words, each vector has one component for each symmetry operation (R). In the case of the SOT, two vectors $\boldsymbol{v}_\beta$ and $\boldsymbol{v}_\gamma$ that belong to non-equivalent IRs are orthogonal, in the case of the GOT two orthogonal vectors belong to non-equivalent IRs or concern different matrix elements or both. The corresponding normalised vectors are $\boldsymbol{v}_\alpha$ and $\boldsymbol{v}_s$:

$$\boldsymbol{v}_s = \frac{\sqrt{g_\alpha}}{\sqrt{g}} \begin{pmatrix} \boldsymbol{D}_\alpha(R_1)_{ij} \\ \boldsymbol{D}_\alpha(R_2)_{ij} \\ \cdots \end{pmatrix} \quad (5.19)$$

$$\text{and } \boldsymbol{v}_\alpha = \frac{1}{\sqrt{g}} \begin{pmatrix} \chi_\alpha(R_1) \\ \chi_\alpha(R_2) \\ \cdots \end{pmatrix} \quad (5.20)$$

Exercise 4.15 reports some examples of the usual group C_{3v}.

We can now derive some important results from the properties of these abstract vectors. Since a g-dimensional space there cannot have more than g vectors pairwise orthogonal to each other, we immediately find that

> A finite group has a finite number of IRs.

Considering the GOT, the number m of different orthogonal vectors is given by the sum of the squares of the dimensions (l_i) of the distinct IRs:

$$m = \sum_i l_i^2 \leq g \tag{5.21}$$

In Section 7.2.2 is proved tha the equality holds:

> The sum of the squares of the dimensions of all the IRs is equal to the order of the group:

$$m = \sum_i l_i^2 = g \tag{5.22}$$

From these, it is also proven that

> The number of IRs of a group is $\leq g$ (the order of the group).

It is finally proven that

> The equality of the character system is a sufficient (as well as necessary) condition for the equivalence of two representations.

as already said without proof in Section 5.3.2. We can say that the condition is also sufficient [*i.e.*: (the character systems are equal) ⇒ (the two representations are equivalent)] because, if the character systems of two representations are equal, the respective vectors (see Eq. (5.20)) in the g-dimensional space cannot be orthogonal, according to the SOT. Therefore $\delta_{\alpha\beta}=1$ and the two representations *must* be *equivalent*. Note once again the meaning of $\delta_{\alpha\beta}=1$ regarding the orthogonality theorems.

5.4.4 Some Relations for RRs

After having analysed the theorems for the IRs, let us now consider the meaning of the *equations* which correspond to the theorems but one or both representations are **reducible**.

If β is an RR, it means that there exists an RR (let us say γ) that is equivalent to β and all its matrices are in block-diagonal form *in the same way*. In any matrix of γ there are $a_{1=}$ 0, 1, 2, ... blocks equal or equivalent to the matrices of the first IR of the group, a_2 blocks equal or equivalent to the matrices of the second IR of the group, and so on. For each R we will therefore have $\chi_\beta(R) = \sum_i a_i\chi_i(R)$, that is, a linear combination of the basis vectors $\vec{v}_i$ (i denotes an IR) of Eq. (5.20) with coefficients a_i. Since the basis vectors are orthonormal, the scalar product of two linear combinations such as $a_i\chi_i(R)$ is simply given by the sum of the products of the homologous components (possibly with the conjugate for the first factor of each product, in the case of complex values).

Let us now consider one of the RRs, say the j-th. Its R-th component in the character system of the β RR is $a_i\chi_j(R)$. The *expression* $\sum_R [\chi_j(R)]^*[\chi_\beta(R)]$ is *not* the STO because β is an RR, so that we cannot say that the expression is equal to $g\delta_{j\beta}$. However, let us evaluate it. Changing the order of the summations, $\sum_R [\chi_j(R)]^*\left[\sum_i a_i\chi_i(R)\right] = \sum_i a_i \sum_R [\chi_j(R)]^*[\chi_i(R)]$ and we apply the SOT to the inner summation, where only IRs appear. The result is $g\delta_{ij}$. So we end up with $\sum_i a_i \cdot g \cdot \delta_{ij} = a_j \cdot g$ and:

> an IR (α) is contained in an RR (β) a number of times a_α given by:

$$a_\alpha = \frac{1}{g}\sum_R \left\{[\chi_\alpha(R)]^*[\chi_\beta(R)]\right\} \tag{5.23}$$

The meaning of Eq. (5.23) is that the characters of an RR are linear combinations of the orthonormal vectors of Eq. (5.20), and a particular component is given by the scalar product of the linear combination with the respective basis vector.

Let us use again as example the character systems of the representations in Sections 5.1.3.2 and 5.1.5.1:

- 3D representation: $\chi(I)=3$, $\chi(C_3)=0$, $\chi(C_{-3})=0$, $\chi(\sigma^{(1)})=1$, $\chi(\sigma^{(2)})=1$, $\chi(\sigma^{(3)})=1$,
- 2D representation: $\chi(I)=2$, $\chi(C_3)=-1$, $\chi(C_{-3})=-1$, $\chi(\sigma^{(1)})=0$, $\chi(\sigma^{(2)})=0$, $\chi(\sigma^{(3)})=0$,
- 1D representation: $\chi(I)=1$, $\chi(C_3)=1$, $\chi(C_{-3})=1$, $\chi(\sigma^{(1)})=1$, $\chi(\sigma^{(2)})=1$, $\chi(\sigma^{(3)})=1$,

The first one is an RR the others are IR: the 2D representation is contained in the 3D representation $(6+0+0+0+0+0)/6=1$ times, the 1D representation is also contained $(3+0+0+1+1+1)/6=1$ times.

Being constructive, the derivation of Eq. (5.23) also shows that:

> The reduction of a RR into the constituent IRs is **unique**.

Let us now consider the $\alpha=\beta$ case, that is the expression $\sum_R [\chi_\alpha(R)]^* [\chi_\alpha(R)]$. If α is an IR, this is the SOT and accordingly gives g. If it is an RR, $\chi_\alpha(R)$ is a linear combination of the characters of the various IRs contained in the RR: $\chi_\alpha(R) = \sum_i a_i \chi_i(R)$. Each $|\chi_\alpha(R)|^2$ in $\sum_R [\chi_\alpha(R)]^* [\chi_\alpha(R)]$, is in turn given by a sum of two types of contributions: (a) the square modulus of the product of the character of an IR multiplied by the coefficients of the linear combination: $|a_i \chi_i(R)|^2$, or (b) product of the characters of two different IRs (i and j) both multiplied by the coefficients: $(a_i \chi_i(R))^* (a_j \chi_j(R))$. The contributions (b) lead to a zero sum because of the SOT. Finally adding the contributions of type (a) we obtain g multiplied by the sum of the squares of the coefficients a_i. Then:

> The characters of a representation tell us whether it is reducible or irreducible. If it is an IR:
>
> $$\sum_R [\chi_\alpha(R)]^* [\chi_\alpha(R)] = g \tag{5.24}$$
>
> If it is an RR:
>
> $$\sum_R [\chi_\alpha(R)]^* [\chi_\alpha(R)] = g \sum_i |a_i|^2 > g \tag{5.25}$$
>
> and a_i tells us how many times the i-th IR is contained in the RR.

For example, using again the representations of Sections 5.1.3.2 and 5.1.5.1, $\sum_R [\chi_\alpha(R)]^* [\chi_\alpha(R)]$ is:

- 3D representation {3, 0, 0, 1, 1, 1}: 9+1+1+1=12,
- 2D representation {2, −1, −1, 0, 0, 0}: 4+1+1=6,
- 1D representation {1, 1, 1, 1, 1, 1}: 6.

So, in agreement with what was already been found, we see that only the 3D representation is an RR.

Note that Eq. (5.25) *is essentially used to know if a representation is reducible or not.* If it is an RR, each a_i *can* be obtained from Eq. (5.23) but is more easily obtained from Eq. (5.27).

5.4.4.1 Another Version of the SOT

All the formulas of the SOT and its consequences can then be written in *a different form* considering that all the elements of a group belonging to the same class have the same character in any representation, either reducible or not. Therefore, the elements of the same class can be grouped together. Using k to denote a class, N_k the number of its elements, $\chi_\alpha(k)$ and $\chi_\beta(k)$ the character in the α and β IRs of the elements of the k-th class, the SOT (Eq. 5.17) becomes:

$$\sum_{k=1}^{h} N_k [\chi_\alpha(k)]^* [\chi_\beta(k)] = g\delta_{\alpha\beta} \tag{5.26}$$

Generally, with respect to the previous version of the SOT (Eq. 5.17) the sum is over a smaller number of addends because the classes are h in number and $h<g$, if the group is not commutative and $h=g$, only if it is commutative.

This change can be usefully applied also to other results. For example, Eq. (5.23) can be rewritten:

> An IR (β) is contained in an RR (γ) a number of times given by:
>
> $$a_\beta = \frac{1}{g} \sum_{k=1}^{h} N_k [\chi_\gamma(k)]^* [\chi_\beta(k)] \tag{5.27}$$

Equation (5.26) also shows that the characters can be considered as vectors of a vector space that has dimension h. As for the original version of the SOT, there is a vector (orthogonal to the others) for each IR (e.g. β), but now each vector has h components (instead of g). Each component (k) is $\sqrt{N_k / g} \cdot \chi_\beta(k)$ and the vectors are:

$$w_\beta = \frac{1}{\sqrt{g}} \begin{pmatrix} \sqrt{N_1}\chi_\beta(1) \\ \sqrt{N_2}\chi_\beta(2) \\ \cdots \end{pmatrix}$$

where the indices 1, 2, ... denote the classes (5.28)

See also Exercise 4.16 on this topic.

This vector space has dimension h (= number of classes) and there are h independent and orthogonal basis vectors (see again footnote 10). As there is an independent vector for each IR, it is finally proven that

> The number of IRs is equal to the number of classes.

5.4.4.2 The Second Orthogonality Theorem of Characters

The SOT in the class version (Eq. 5.26) is an orthogonality between vectors (each relative to an IR) having components relative to each class. The fact that there is an equal number of IRs and classes makes it possible to write (the proof is in Section 7.3) a new orthogonality relation:

$$\sum_{i=1}^{h} N_l \cdot [\chi_i(l)]^* \cdot \chi_i(j) = g\delta_{jl} \quad (i \text{ is an IR, } l \text{ and } j \text{ are classes}) \tag{5.29}$$

Compared to the previous theorems, orthogonality is in a sense "reversed", because we consider vectors (each relative to a class) having one component for each IR:

$$v_j = \sqrt{\frac{N_j}{g}} \begin{pmatrix} \chi_1(j) \\ \chi_2(j) \\ \dots \end{pmatrix}$$

5.5 Character Tables

The relations shown so far can be used in various ways and with different difficulties from case to case to obtain the characters for all the IRs of a group. However, it is not necessary to delve into this purely specialist topic because:

For all the symmetry groups that interest chemistry the character systems of the IRs are available in tables.

5.5.1 The Table Content

In the tables, the various IRs are specified according to the following notations.

- A denotes 1D representations, generally when the character of the main rotation is +1 but it is also used for groups C_1 (the trivial group with only the identity), and for C_s and C_i even if the only operation different from the identity has character=−1. In these groups there are no pure rotation symmetry operations.
- B denotes 1D representations when the character of the main rotation is −1.
- E denotes 2D representations.
- T denotes 3D representations; then, less frequent,
- G denotes 4D representations 4D and
- H denotes 5D representations.

If the group contains the *inversion*, the subscripts g (from the German gerade=even) and u (from ungerade=odd) indicate that the character of the inversion operation is +1 or −1, respectively.

If the group contains a *horizontal* reflection plane, but does not contain an inversion centre, the single quotation mark "and double quotation mark" are used respectively to indicate that the character of the horizontal reflection operation is +1 or −1, respectively.

If further specification is needed, the subscripts 1 or 2 are used to indicate the character +1 or −1 of (a) a *horizontal* binary rotation, *i.e.* a rotation around an axis perpendicular to another axis of higher order, or (b) a vertical reflection, when the horizontal rotation is missing.

E_1, E_2, E_3, … are also used (see Section 2.1.4) for pairs of 1D IRs with eigenvalues that are conjugate to each other and which, as already seen, are degenerate in energy due to the time reversal symmetry (see Section 2.1.3).

In the character table of a symmetry group, each IR is reported on *a row,* and the columns report the characters of the various classes. Since there are as many IRs as there are classes, this part of the character tables is a square pattern, as shown by the grey background of part of the table below for the usual C_{3v} group.

	E	C_3, C_3	$\sigma^{(1)}$, $\sigma^{(2)}$, $\sigma^{(3)}$
A_1	1	1	1
A_2	1	1	−1
E	2	−1	0

It is preferred however to head the columns with $2C_3$ and $3\sigma_v$, to make immediately clear how many elements the classes have. As already seen in Section 2.3, additional columns typically report some functions which are important for treating interactions with e. m. radiation and for classifying AOs. These are the *linear* functions (x, y, z), the *quadratic* functions (x^2, y^2, z^2, xy, yz, zx and possibly their combinations, for example x^2+y^2 or x^2-y^2) and the *rotations* $R_x = y\dot{z} - \dot{y}z$, $R_y = z\dot{x} - \dot{z}x$ and $R_z = x\dot{y} - \dot{x}y$, where the dot superscript denotes the derivative with respect to time (Table 4.20).

Some tables also report higher order functions (cubic, etc., …) (Table 4.21).

The functions that combine linearly with each other as a result of group operations are gathered in brackets (see Table 4.20). An empty cell indicates that there are no suitable functions for the IR.

5.6 Operations on Representations

The properties of the matrices and their character systems make possible to operate on representations according to algebraic rules that are (surprisingly, at first sight) analogous to those relating to the sum and product of numbers. Without proof, let us explain the result with an example.

TABLE 4.20
The Usual Presentation of a Character Table, e.g. for Group C_{3v}

C_{3v}	E	$2C_3$	$3\sigma_v$			
A_1	1	1	1	z		x^2+y^2, z^2
A_2	1	1	−1		R_z	
E	2	−1	0	(x, y)	(R_x, R_y)	(x^2-y^2, xy) (zx, yz)

TABLE 4.21
The Extended Presentation of a Character Table, e.g. for Group C_{2v}

C_{2v}	E	C_2	$\sigma_v(xz)$	$\sigma_v(yz)$				
A_1	1	1	1	1	z		x^2, y^2, z^2	z^3, x^2z, y^2z
A_2	1	1	−1	−1		R_z	xy	xyz
B_1	1	−1	1	−1	x	R_y	zx	xy^2, xz^2, x^3
B_2	1	−1	−1	1	y	R_x	yz	x^2y, yz^2, y^3

Warning: in the *left* part of Figure 4.10 the x, y, z axes are labelled differently.

Let α, β, γ, κ, λ, ν, π, ρ indicate some representations of a specific group.

If $\alpha \sim \beta \otimes \gamma$, $\beta \sim \kappa \oplus \lambda$ and $\gamma \sim \nu \oplus \pi \oplus \rho$

then

$$\alpha \sim (\kappa \oplus \lambda) \otimes (\nu \oplus \pi \oplus \rho)$$

$$\sim \kappa \oplus (\nu \oplus \pi \oplus \rho) \oplus [\lambda \otimes (\nu \oplus \pi \oplus \rho)]$$

$$\sim (\kappa \otimes \nu) \oplus (\kappa \otimes \pi) \oplus (\kappa \oplus \rho) \oplus (\lambda \otimes \nu) \oplus (\lambda \otimes \pi) \oplus (\lambda \otimes \rho) \quad (5.30)$$

and finally, all the products in the brackets are representations, and therefore they can be expressed as direct sums of IRs and also α does. Furthermore, since the characters of a sum (or product) of representations are the sums (or products) of the relative characters, the analogous consideration also applies to the *characters* of the representations.

In these equations, the representation symbols (α, β, γ, κ, λ, ν, π, ρ) are added and multiplied as if they were numbers and we also find the distributive properties of the addition and multiplication of numbers.

Of course, one must not forget the important differences between operations on numbers and those on representations or on their characters. A first fundamental difference is that these operations are performed on lists of matrices or numbers (the characters). Another important difference is in the meaning of the "~" sign. For example, $\alpha \sim \beta \otimes \gamma$ does not mean that each matrix of the representation α is the direct product of the corresponding matrices of the representations β and γ, but it means that it is related to the direct product by a similarity transformation (which must be the same for all matrices of α), that is, it means that it is *equivalent* (not equal) to the direct product. The same holds for the direct sum.

The possibility of operating in this way on representations evidently derives from the properties of representations and characters but it still appears surprising if one considers that the direct matrix product is *not commutative* (Section 5.1.1.2). There are two essential points. The first is the commutativity of the respective traces. The second is that the characters uniquely identify the representations, if one requires simple equivalence instead of identity.

A frequently used application is when both the direct sum of several representations (*e.g.*: $\gamma \sim \nu \oplus \pi \oplus \rho$) and some of its addends (e.g.: π and ρ) are known. Then, by subtraction the remaining addend (ν) is obtained. This is the case of determining the RR of the vibrational degrees of freedom of a molecule knowing the RR of all the atomic motions and the RR of the rotational and translational degrees of freedom (see Sections 6.2.4.1 and 6.2.4.2).

5.7 Summary

In the first part of this Section 5, the **matrix representations** of the elements of a group have been introduced with several examples. From the representations, all the most important results of the theory are obtained and, in particular, the characters are derived, which are the main tool for using the theory of symmetry in its various applications.

If a representation of dimension > 1 can be transformed by similarity into the direct sum of two or more representations of lower dimension, it is said to be **reducible** (RR); otherwise, it is said to be **irreducible** (IR) (see Section 5.2). The reduction of an RR is unique (see Section 5.4.4).

The **IRs** of a group are the same as the **symmetry species** (Section 5.2) already seen for commutative groups and correspond to all the distinct ways in which the symmetry operations of a group transform the SAFs (symmetry-adapted functions).

The IRs of a finite group are **finite in number,** and this number is equal to the **number of classes** in the group (cf. Section 5.4.4.1) *if one does not distinguish* between *equivalent* IRs, *i.e.* related by a similarity transformation. Furthermore, the sum of the squares of the dimensions of distinct IRs is equal to the number of elements in the group (see Section 5.4.3 and Eq. (5.22)).

The **characters** of a representation are defined as sums of the elements along the main diagonal of the matrices of the representation, are invariant under similarity transformation (see Section 5.3.2) and uniquely identify a representation up to equivalences.

For the elements of the IR matrices and for the related characters, there are **orthogonality theorems** which allow us to obtain theoretical results which are also very important at the applicative level because they provide the tools

- to recognize two equivalent representations,
- to distinguish an IR from a RR (see Section 5.4.4),
- to know in which IRs a given RR is reducible and how many times each IR is contained in it (see again Section 5.4.4).

6 APPLICATIONS

Using character tables and other tools provided by representation theory and orthogonality theorems, one can now generalize the results and methods illustrated in Section 2 for commutative groups, tackle some pending issues and extend the discussion to more general considerations. For the reasons already said for representations (Section 5.1.3), most examples are made by using group C_{3v} and the NH_3 molecule.

6.1 Symmetry and QM

6.1.1 Eigenfunctions of H and Representations

Summarizing what already said in Section 1.3, the symmetry operations are linear transformations that leave the Hamiltonian of a system unchanged and therefore commute with H. These operations form a group. This is not difficult to prove: we need to check that they obey the group axioms, that is, that the closure property holds, that associativity holds, that an identity uniquely exists, and that the inverse of each symmetry operation exists unique.

The **basic relation** between the wavefunctions of a QM system and its symmetry is that a *symmetry operator*, acting on an eigenfunction (φ) of H, gives a function that is still an eigenfunction of H with the *same eigenvalue*: it can be (a) the starting φ multiplied by a number with modulus = 1, that is - operationally - the *same* function, or (b) a linearly *independent* function. The second case occurs only with non-commutative groups (see Section 1.3.2).

As discussed in Section 3.2 of Chapter 1, a fundamental feature of QM is that one can **arbitrarily** choose *l linearly independent eigenfunctions* of $H\left(\varphi_{n1},\ldots,\varphi_{nl}\right)$ as **basis functions** for an l-times degenerate level (n): any linear combination of them is still an eigenfunction of H with the same eigenvalue.

6.1.1.1 Degeneracy of Eigenfunctions and IRs

Let us summarise what is shown in Section 5.1.

1. Acting on *an* **orthonormal** *basis set* $\left(\varphi_{n1},\ldots,\varphi_{nl}\right)$ *of l linearly independent* **degenerate** *eigenfunctions*, **the symmetry operations generate an *l*-dimensional matrix representation (D_α)** of the symmetry group made of **unitary** matrices.
2. This is an irreducible representation (IR), excluding accidental degeneracies.
3. Written as a linear combination of the basis functions:

$$\chi_r = r_1\cdot\varphi_{n1}+\cdots+r_l\cdot\varphi_{nl} = \left(\varphi_{n1},\ldots,\varphi_{nl}\right)\cdot\begin{pmatrix} r_1 \\ \cdots \\ r_l \end{pmatrix}$$
$$= \left(\varphi_{n1},\ldots,\varphi_{nl}\right)\cdot \boldsymbol{r}$$

   any eigenfunction of the n-th *level* (χ_r) is transformed by a symmetry operation R into $\chi_s = \hat{R}\chi_r$ according to $\boldsymbol{s} = \boldsymbol{R}\,\boldsymbol{r}$, where $\boldsymbol{R} = \boldsymbol{D}_\alpha(R)$.
4. The same $\boldsymbol{R}$ matrix can also perform a basis change from $\left(\varphi_{n1},\ldots,\varphi_{nl}\right)$ to $\left(\psi_{n1},\ldots,\psi_{nl}\right)$:

$$\left(\psi_{n1},\ldots,\psi_{nl}\right) = \left(\varphi_{n1},\ldots,\varphi_{nl}\right)\bullet \boldsymbol{R}$$

5. This means that each symmetry operation (R) transforms *a function of the basis set* (say the kth function), according to the kth **column** of the corresponding $\boldsymbol{R} = \boldsymbol{D}_\alpha(R)$ matrix:

$$\psi_{nk} = \sum_j \varphi_{n,j}\cdot\left[\boldsymbol{D}_\alpha(R)\right]_{jk};\quad j,k = 1,\ldots,l \tag{6.1}$$

6. Starting from **any** eigenfunction of the Hamiltonian of a given level, *the symmetry operations of the group allow to obtain* **all** *the other basis eigenfunctions* of the level (Figure 4.29 shows an example).

These results trivially apply also to *non-degenerate levels* of the Hamiltonian when the symmetry operations simply multiply the eigenfunction by a number k, with $|k| = 1$.

As said (point 2 above), the representation is **irreducible**, *except for accidental degeneracies*. Basically, the reason is a direct extension of what we have already seen for commutative groups (see Section 1.3.1). If — by contradiction — it were a RR, it would be possible to find an equivalent representation that is the direct sum of two or more smaller IRs. Consequently, it would be possible to find a new basis of eigenfunctions such that some basis functions are transformed by symmetry only *among themselves* as indicated by one of the constituent IRs, while other basis functions transform only among themselves as indicated by *another* constituent IR, that is, without linearly combining functions from different IRs. Therefore, there *would not necessarily* be degeneracy between the eigenfunctions of the distinct subsets. Obviously, there *can be* degeneracy, which we therefore call *accidental* (cf. Section 1.3.3). We can therefore conclude that:

> Each eigenvalue of the Hamiltonian corresponds to (**only**) **one** IR of the symmetry group (excluding accidental degeneracies) The eigenfunctions of an eigenvalue of the Hamiltonian are the basis of an IR of its symmetry group. (excluding accidental degeneracies)

It is easy to verify that these are the same conclusions already shown for commutative groups, generalizing to IRs of dimensions > 1 the correspondence between a symmetry species and a set of eigenvalues of symmetry operations that holds for commutative groups and 1D representations. Finally, we have that, again excluding accidental degeneracies:

> The possible degeneracies for a system are the dimensions of the IRs of its symmetry group.

6.1.1.2 Unitarity

Along with arbitrariness, the choice of the basis functions for a degenerate level of the Hamiltonian raises the issue of **orthogonality** and normalization. Eigenfunctions belonging to *different* eigenvalues are always orthogonal: this is a standard result of QM. Instead, the orthogonality of eigenfunctions *of the same degenerate level is not automatic.* Obviously, the eigenfunctions of a basis set must be linearly independent and in the correct number, but they must **be made orthogonal** if they are not, and normalized, because the *matrices* $\boldsymbol{R}$ *of a representation are required to be unitary*: ($\boldsymbol{R}^\dagger = \boldsymbol{R}^{-1}$) (see Section 5.1.4).

Let us discuss the problem in some detail, using the example of the twofold degenerate second level of the π MOs of benzene and the simple HMO method (Figure 4.11). Indeed, this method combines only symmetry equivalent $2p_z$-C AOs and therefore gives MOs that are also the SALCs of these AOs. So, after solving the eigenvalue equation of the Hamiltonian matrix $\boldsymbol{H}$:

$$\Lambda = C \cdot H \cdot C^{-1}$$

each column of C is a SALCs of the $2p_z$ AOs of the various C atoms:

$$C = \begin{pmatrix} 1/\sqrt{6} & 1/2 & -1/2 & -1/2 & 1/2 & 1/\sqrt{6} \\ 1/\sqrt{6} & 1/2 & 0 & 1/2 & 0 & -1/\sqrt{6} \\ 1/\sqrt{6} & 0 & 1/2 & 0 & -1/2 & 1/\sqrt{6} \\ 1/\sqrt{6} & -1/2 & 1/2 & -1/2 & 1/2 & -1/\sqrt{6} \\ 1/\sqrt{6} & -1/2 & 0 & 0 & 0 & 1/\sqrt{6} \\ 1/\sqrt{6} & 0 & -1/2 & 1/2 & -1/2 & -1/\sqrt{6} \end{pmatrix} \quad (6.2)$$

The diagonalization routines generally also implement orthogonalization and normalization of the eigenfunctions. On the contrary, the result of Eq. (6.2), also shown in Figure 4.11, has been obtained by *expressly* using an HMO program that provides eigenfunctions that are not orthogonal but only linearly independent. As already seen (Section 1.2.2.2 and Figure 4.11), there are four eigenvalues with degeneracies 1, 2, 2, 1. This is also the order of the columns of Eq. (6.2). It is easily verified that each column of C is orthogonal to each column with a *different* eigenvalue, but the second and third columns are **not** orthogonal to each other and so are the fourth and fifth columns.

In the basis of the AOs, the clockwise ternary rotation C_3 is given by:

$$R = \begin{pmatrix} 0 & 0 & 0 & 0 & 1 & 0 \\ 0 & 0 & 0 & 0 & 0 & 1 \\ 1 & 0 & 0 & 0 & 0 & 0 \\ 0 & 1 & 0 & 0 & 0 & 0 \\ 0 & 0 & 1 & 0 & 0 & 0 \\ 0 & 0 & 0 & 1 & 0 & 0 \end{pmatrix} \quad (6.3)$$

This is easily found as the RR of the AOs is also the RR of the equivalence of C sites.

Let us now consider the linear combinations c_2 and c_3 corresponding to the second and third column, $\boldsymbol{c}_2$ and $\boldsymbol{c}_3$, of C, that is $c_2 = \boldsymbol{a}^T \cdot \boldsymbol{c}_2$ and $c_3 = \boldsymbol{a}^T \cdot \boldsymbol{c}_3$, where $\boldsymbol{a}^T$ is the row of the AOs of C atoms 1, …, 6. Figure 4.30 visually shows that these functions are **not** orthogonal to each other, as already said.

Figure 4.11 has already shown graphically that c_2 and c_3 are transformed into c_3 and $(-c_2 - c_3)$ by symmetry operation $R = C_3$. The same result is easily derived analytically (see Exercise 4.17). Incidentally, the example illustrates what already said at point 6 in Section 6.1.1.1: starting from a *single* basis function and by applying the different symmetry operators, one constructs the entire set of degenerate linearly independent basis functions.

Now, let us consider **only this degenerate level** using the SALCs c_2 and c_3 as the basis set. In this basis, c_2 and c_3 are respectively represented by $\boldsymbol{a} = \begin{pmatrix} 1 \\ 0 \end{pmatrix}$ and $\boldsymbol{b} = \begin{pmatrix} 0 \\ 1 \end{pmatrix}$. The transformation from (c_2, c_3) to $(c_3, -c_2 - c_3)$ is now represented by $\boldsymbol{R}' \cdot \boldsymbol{a} = \boldsymbol{b}$ and $\boldsymbol{R}' \cdot \boldsymbol{b} = -\boldsymbol{a} - \boldsymbol{b}$, where $\boldsymbol{R}'$ is the 2D matrix representation of *symmetry* C_3. Let us write $\boldsymbol{R}' = \begin{pmatrix} x & y \\ w & z \end{pmatrix}$. Then

$$\begin{pmatrix} x & y \\ w & z \end{pmatrix}\begin{pmatrix} 1 \\ 0 \end{pmatrix} = \begin{pmatrix} 1 \\ 0 \end{pmatrix} \text{ and } \begin{pmatrix} x & y \\ w & z \end{pmatrix}\begin{pmatrix} 0 \\ 1 \end{pmatrix} = \begin{pmatrix} -1 \\ -1 \end{pmatrix}$$

so that $\boldsymbol{R}' = \begin{pmatrix} 0 & -1 \\ 1 & -1 \end{pmatrix}$.

For a *unitary* matrix $\boldsymbol{A}$: $\boldsymbol{A}^\dagger \cdot \boldsymbol{A} = \boldsymbol{1}$ because $\boldsymbol{A}^\dagger = \boldsymbol{A}^{-1}$. Instead, $(\boldsymbol{R}')^T \cdot (\boldsymbol{R}')$ gives

$$\begin{pmatrix} 0 & 1 \\ -1 & -1 \end{pmatrix} \cdot \begin{pmatrix} 0 & -1 \\ 1 & -1 \end{pmatrix} = \begin{pmatrix} 1 & -1 \\ -1 & 2 \end{pmatrix}$$

which is not the unit matrix of second order $\boldsymbol{1}_2$ and therefore the **matrix representation** derived from the (c_2, c_3) basis **is not unitary.** We more appropriately say that **this set of matrices is not a representation.**

Let's now look for a representation with unitary matrices. The technique is discussed in more detail in Section 3.2.6.3 of the Appendix. Applying this procedure to the eigenfunctions of Figure 4.30, that is solving the eigenvalue equation of the Hückel Hamiltonian $\boldsymbol{H}$ with a diagonalization algorithm *that also implements the orthogonalization*, we obtain a new matrix of eigenvectors that we may call $\boldsymbol{D}$ and differs from $\boldsymbol{C}$ in the third and fifth columns:

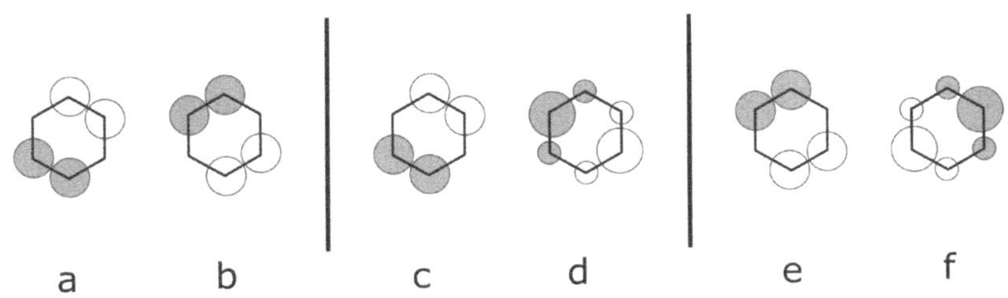

FIGURE 4.30 Orthogonalization of the second level of benzene π MOs: a and b are the non-orthogonal starting functions (c_1 and c_2 of Eq. (6.2)); c and d are the orthogonalized functions [d_1 and d_2 of Eq. (6.4)]; e and f are C_3d_1 and C_3d_2, respectively. Other MOs are shown in Figure 4.32. As said, these simple HMO MOs are the SALCs based on C-$2p_z$ AOs.

$$D = \begin{pmatrix} 1/\sqrt{6} & 1/2 & -1/\sqrt{12} & -1/2 & -1/\sqrt{12} & 1/\sqrt{6} \\ 1/\sqrt{6} & 1/2 & 1/\sqrt{12} & 1/2 & -1/\sqrt{12} & -1/\sqrt{6} \\ 1/\sqrt{6} & 0 & 2/\sqrt{12} & 0 & 2/\sqrt{12} & 1/\sqrt{6} \\ 1/\sqrt{6} & -1/2 & 1/\sqrt{12} & -1/2 & -1/\sqrt{12} & -1/\sqrt{6} \\ 1/\sqrt{6} & -1/2 & -1/\sqrt{12} & 1/2 & -1/\sqrt{12} & 1/\sqrt{6} \\ 1/\sqrt{6} & 0 & -2/\sqrt{12} & 0 & 2/\sqrt{12} & -1/\sqrt{6} \end{pmatrix} \quad (6.4)$$

A simple visual comparison (parts c and d of Figure 4.30) directly shows without calculations that the second and third columns of $\boldsymbol{D}$ are orthogonal to each other. In fact, considering the reflection in the vertical plane passing through the third and sixth C atoms, we see that the function c is reflected with a change of sign, while function d is reflected without changing sign.

Now, let us call d_1, d_2 these orthogonal functions, use them as the basis set of the second level of Hückel Hamiltonian and apply symmetry C_3. Graphically, C_3d_1 is shown in part e of Figure 4.30, while C_3d_2 is shown in part f. Analytically:

$$C_3d_1 = -\frac{1}{2}\cdot d_1 + \frac{\sqrt{3}}{2}\cdot d_2 \quad \text{and} \quad C_3d_2 = -\frac{\sqrt{3}}{2}\cdot d_1 - \frac{1}{2}\cdot d_2$$

Thus, the matrix for C_3 is: $\begin{pmatrix} \frac{-1}{2} & \frac{-\sqrt{3}}{2} \\ \frac{\sqrt{3}}{2} & \frac{-1}{2} \end{pmatrix}$ and is unitary:

$$\begin{pmatrix} \frac{-1}{2} & \frac{-\sqrt{3}}{2} \\ \frac{\sqrt{3}}{2} & \frac{-1}{2} \end{pmatrix} \cdot \begin{pmatrix} \frac{-1}{2} & \frac{\sqrt{3}}{2} \\ \frac{-\sqrt{3}}{2} & \frac{-1}{2} \end{pmatrix} = \begin{pmatrix} 1 & 0 \\ 0 & 1 \end{pmatrix}.$$

To summarize what the example shows, **a representation is unitary** by definition (see Section 5.1.4), or, if you prefer, it ***must*** be unitary, otherwise it is not considered a representation. Therefore:

- When the matrices of a representation are obtained from the eigenfunctions of the Hamiltonian, it is necessary to preliminary select a set of orthogonal and normalized eigenfunctions to guarantee the unitarity of the representation.
- When - on the contrary - we already have the *unitary* matrices of a representation and we use them to construct SAFs, then the so obtained SAFs are automatically orthogonal. Usually, however, they are not yet normalized, because the procedures for constructing SALCs (discussed in Section 6.2) do not have this additional property.

Each *orthonormal* eigenfunction of an l-fold degenerate level of the Hamiltonian is also referred to as a **basis partner**, and the designation is also applied to the **columns** of the matrices of the representation.

Finally, generalizing and including in the definition the case of 1D representations (their eigenfunctions and their characters), we have that:

> Every partner of every level of the Hamiltonian is orthogonal to any other partner of the same or different level.

Note however, that the partners within the same IR (more than 1D) are chosen *arbitrarily*, while the IRs reflect the algebraic properties of the symmetry group.

These statements are derived in Section 7.4. Remind, however, that the **choice** of the *partners* within a level is **arbitrary**. On the contrary, there is no arbitrariness in the choice of the IRs (see below).

6.1.2 Applications of Direct Sums and Direct Products

It is now appropriate to discuss the meaning in the applications of the two operations of direct sum and direct product, that were introduced as pure matrix calculus (Sections 5.1.1.1 and 5.1.1.2) or as operations on the representations (Sections 5.1.5.1 and 5.1.5.2).

Basically, the **direct sum** is the inverse of the procedure of reducing an RR into its constituent IRs. A representation is reducible because *different vectors of the same nature* exhibit different symmetry characteristics or have components that behave according to different IRs. Examples are (a) the motions of atoms in a molecule and (b) the AOs used to build SALCs that are in turn combined with each other to obtain MOs with the LCAO method.

The **direct product of matrices** of the *same* group is essentially used to provide the matrix representation of **products of functions**, with each other or with other quantities. The pertinent result is here that:

> A product of functions belongs to the representation which is the direct product of the IR of the factors.

For the proof, we use the result already seen (Section 5.1.5.2) that the direct product of the corresponding matrices $\boldsymbol{D}_\alpha(R)\otimes \boldsymbol{D}_\beta(R)$ of two representations $\boldsymbol{D}_\alpha$ and $\boldsymbol{D}_\beta$ of the same group is a representation $\boldsymbol{D}_{\alpha\otimes\beta}(R)=\boldsymbol{D}_\alpha(R)\otimes \boldsymbol{D}_\beta(R)$ of the group. The result can be extended to more representations. For simplicity, we can then consider the product of only two functions, the first of which belongs to an m-dimensional IR $\boldsymbol{D}_m(R)$, while the second belongs to an n-dimensional IR $\boldsymbol{D}_n(R)$. Correspondingly, we have an orthonormal basis of m functions ($\chi_1, \ldots, \chi_m$) and an orthonormal basis of n functions ($\xi_1, \ldots, \xi_n$). There are $m \times n = p$ functions, in which all the possible products of a function of the first set by one of the second set appear. For a given symmetry operation (R) there will be a p-dimensional matrix $D_p(R)$ and the various products of basis functions are transformed according to:

$$(\chi_1\xi_1,\ldots,\chi_1\xi_n,\ldots,\chi_m\xi_1,\ldots,m\xi_n)\cdot D_p(R)$$

We must verify that $D_p(R)$ is $D_p(R)=D_m(R)\otimes D_n(R)$. Considering the transformation of the generic product kl we obtain in fact:

$$\hat{R}(\chi_k\cdot\xi_l)=\sum_k\sum_l \chi_k\cdot[D_m(R)]_{ki}\cdot\xi_l\cdot[D_n(R)]_{lj} \text{ and then}$$

$$\hat{R}(\chi_k\cdot\xi_l)=\sum_{kl}(\chi_k\cdot\xi_l)\cdot[D_{m\otimes n}(R)]_{kl,ij} \tag{6.5}$$

In Eq. (6.5), the double index notation is used (see Section 5.1.4) for the direct product between matrices.

The result is important when the exact or approximate separation between degrees of freedom allows the Hamiltonian operator (see Section 3.5 of the Chapter 1) to be written as the sum of operators related to separate degrees of freedom or group of degrees of freedom (i):

$$\hat{H}=\sum_i^n \hat{H}_i \tag{6.6}$$

Due to the linearity of QM, the total wave function is then written as the product

$$\psi_j=\prod_i^n \chi_{i,\alpha} \tag{6.7}$$

of wavefunctions $\chi_{i,\alpha}$ relative to the different partial Hamiltonian i (α is an index for the eigenvalue, possibly degenerate, chosen for the ith partial Hamiltonian):

$$\hat{H}_i\chi_{i,\alpha}=E_{i,\alpha}\cdot\chi_{i,\alpha} \tag{6.8}$$

while the eigenvalue for the global H is the sum of the eigenvalues:

$$\hat{H}\psi_j=E_j\psi_j;\quad E_j=\sum_i^n E_{i,\alpha} \tag{6.9}$$

Equation (6.9) is directly verified by substitution.

A **first application** of the direct product of representations concerns the approximate treatment of a *polyelectronic problem* as a sum of monoelectronic problems and therefore the expression of the polyelectronic function as a Hartree product of monoelectronic functions.

A **second application** is to treat vibrations as combinations of normal modes and then the global vibrational function of the entire molecule as a product of single-mode vibrational functions.

A **third case** is the evaluation of integrals, such as those relating to the interaction between a quantum system and electromagnetic radiation or to the addition of a perturbation: the integrands are products of two eigenfunctions of the unperturbed Hamiltonian by a function describing the interaction.

Another important case is that

> The representation of the motions of all atoms of a molecule is the *direct product* of the *representation of equivalence of atomic sites* by the representation of the *vectors* of ordinary 3D space.

$$D(\text{tot})=D(\text{ve})\otimes D(\text{eas})$$
$$\text{eas: equivalence of atomic sites} \tag{6.10}$$

D(ve) describes how a *vector in ordinary 3D space* is transformed, that is how the three x, y, z functions are transformed. For the representation of *equivalence of atomic sites* D(eas), see Section 5.1.3.6. Equation (6.10) is obtained by formalizing the example of the 12D representation of the ammonia molecule presented in Section 5.1.3.8: see Section 5.1.5.2. The representation D(tot) can also be considered the representation of a product of operators because the transformation of a displacement vector relative to an atom comes from the product (in sequence) of two transformations, that of the atomic position D(eas) and that of a vector in ordinary 3D space D(ve).

Equation (6.10) is a fundamental step in the treatment of the vibrational degrees of freedom of a molecule (Section 6.2.4). Using the ammonia molecule as an example, in Section 5.1.3.8 we derive D(tot) and in Section 5.1.5.2 we show that Eq. (6.10) holds.

As already seen in Section 5.1.5.2, $D_{m\otimes n}(R)$ ***is not*** **necessarily an IR** *even if* D_m and D_n are. The problem is solved using the tools of representation and character theory.

The product of representations of **different groups** is also important but has different meanings and applications: see Section 6.3.

Still about products **of functions**, an important caveat is that (as we will see later) it is not always true that **all** IRs that are the direct product of IRs are pertinent to **possible states**. This is due to **other** symmetries that arise for systems of **indistinguishable particles**, namely the permutation symmetries that express the Pauli principle (symmetry for phonons, antisymmetry for electrons).

6.2 Building Symmetry Adapted Functions

While the eigenfunctions of the Hamiltonian form a basis for irreducible representations (IRs, one IR for each eigenvalue), this is *not generally* true for *other* sets of functions that are often used in applications, for example:

A. the AOs used to form MOs with the LCAO method,
B. the configurational coordinates, that is the displacements of atoms from their equilibrium positions, that are used for describing molecular vibrations, and

C. the sets of linear (x, y, z) or quadratic (x^2, y^2, z^2, xy, yz, zx) or higher-degree coordinate functions of ordinary 3D space that play a fundamental role when studying spectroscopic transitions.

Each of these *functions* is, in general, the *basis* of a *reducible representation* (RR) and group theory tells us how to *find linear combinations* of these functions corresponding to one or another IR contained in the RR. We say that these *linear combinations* are *symmetry adapted* (SALC), meaning that the symmetry operations transform a SALC into a linear combination of SALCs of the same IR or, in other words, the symmetry operations transform the SALCs of an IR *only between them*, as prescribed by Eq. (6.1), without mixture with functions belonging to other IRs.

The SALCs show orthogonality properties, analogous to those concerning the eigenfunctions of the Hamiltonian. In general:

> The symmetry adapted functions (SAFs) that belong to an IR are orthogonal to any other SAF that belongs to a different IR.

See notes (7.4) for the proof, which is based on the GOT (Eq. 5.16) and holds also for the **partners** of an IR:

> SAF of different partners (columns) of the *same* IR are orthogonal to each other.

Since the Hamiltonian is invariant under all the symmetry operations of the group, it transforms as identity so that the eigenfunctions of H share the property of a SAF:

> (Neglecting accidental degeneracy) the eigenfunctions of a given eigenvalue of the Hamiltonian belong to a single IR and are orthogonal to any SAF that belongs to a different IR.

Finally, we consider the linear combination of two or more functions that belongs to the same symmetry species, to obtain a function of the same symmetry species. This means that:

> If the eigenfunctions of a given level of the Hamiltonian are obtained by linear combination of SAF, both the SAF that are combined and the eigenfunctions belong to the same IR.

More specifically:

- a linear combination of two or more SAF of different symmetry species (IR) is not an eigenfunction of the Hamiltonian.
- When linearly combining various SAF to form an eigenfunction of the Hamiltonian of a given symmetry species, SAF of other species *cannot be used*, as they *cannot* give contributions to the linear combination.

6.2.1 Projection Operators: Definition and Use

Matrix representations and characters are the tools for generalizing to non-commutative groups the procedure already seen (Eq. 2.4) and used without proof in Section 2.2 for forming the SALC of commutative groups. Generally, the problem of forming SALC is solved by means of **projection operators**. When applied to a basis function, a projection operator, *by definition*, gives another basis function *of the same representation*. Let us consider a representation $\boldsymbol{D}_\alpha$ and a pair of its basis function, φ_m and φ_n. Then, the operator that projects φ_m onto φ_n, that is the operator that allows to obtain φ_n from φ_m, is abstractly defined as:

$$\hat{P}^{\alpha}_{nm}\varphi_m \stackrel{\text{def}}{=} \varphi_n \tag{6.11}$$

In detail, **two analytic forms for projection operators exist:** one is based on matrices, the other one is based on characters. It is worth noting that the properties of projection operators go far beyond what is strictly defined by Eq. (6.11), that is transforming the basis functions of a representation into each other. The most important property of projection operators in applications relies on the fact that they allow to solve the inverse problem with respect to that of paragraph 6.1.1. Here, we *are not* looking for the unitary matrices of a representation starting from a set of functions that are eigenfunction of the Hamiltonian of the system and therefore are surely symmetry adapted. On the contrary, we now *do have the unitary matrices* (or, at least, their characters) *of a given IR* and we want to use them:

- for **finding** a set **of functions that transform into each other as shown by these matrices,**
- and, more generally, for determining the proper linear combinations (SALC) of **a set of *arbitrary* functions,** to obtain **symmetry adapted** functions.

6.2.1.1 Matrix Form of the Projection Operators

The explicit definition of the operator transforming the m-th *basis function* of the α IR into the n-th *basis function* of the same IR is:

$$\hat{P}^{\alpha}_{nm} = \frac{g_\alpha}{g}\sum_R \left[\boldsymbol{D}_\alpha(R)\right]^* \hat{R} \tag{6.12}$$

Equation (6.12) can be obtained from the GOT, using the assumption that a projector operator can be written as a linear combination of symmetry operators. However, we are not interested in deriving Eq. (6.12), but in verifying that the operator defined by Eq. (6.12) satisfies Eq. (6.11), *i.e.*, that it is actually capable of transforming the m-th basis function of the α IR into the n-th basis function of the same IR. To this aim, given an IR (α) and any of its basis functions φ_m, let us use the definition of Eq. (6.12) for obtaining the n-th basis function. We need to consider that the dot products of the n-th basis function with any other basis function of the same IR are zero, while the dot product with itself is 1. Using Dirac notation

$$\left\langle n|\hat{P}^{\alpha}_{nm}|m\right\rangle = \left\langle n\left|\frac{g_\alpha}{g}\sum\nolimits_R [\boldsymbol{D}_\alpha(R)]^*_{mn}\cdot\hat{R}\right|m\right\rangle$$

$$= \frac{g_\alpha}{g}\sum\nolimits_R [\boldsymbol{D}_\alpha(R)]^*_{mn}\left\langle n|\hat{R}|m\right\rangle$$

Also, $\langle n|R|m\rangle = [\boldsymbol{D}_\alpha(R)]_{nm}$ as $\langle n|R|m\rangle$ is a matrix element of the symmetry operator in the basis of the IR. Then: $\left\langle n|\hat{P}^{\alpha}_{nm}|m\right\rangle = \frac{g_\alpha}{g}\sum_R [\boldsymbol{D}_\alpha(R)]^*_{mn}[\boldsymbol{D}_\alpha(R)]_{nm}$, which finally gives δ_{nm} using the GOT. This means that if n is not equal to m, the matrix element is zero, but this also imply that if $\hat{P}^{\alpha}_{nm} = \frac{g_\alpha}{g}\sum_R [\boldsymbol{D}_\alpha(R)]^*_{mn}\hat{R}$ acts on the m-th basis function, the result is the n-th basis function, which is the result we sought.

Moreover, the effect of the operator is to produce a SALC behaving as the n-th partner of the α IR **even when starting from an arbitrary function**. Let us consider an arbitrary function ψ, and write it as a linear combination of several SAF (φ) of different symmetry species:

$$\psi = \sum_\beta\sum_{m(\beta)} c^\beta_m \varphi^\beta_m$$

Here, β refers to the IR, the m index varies between 1 and the dimension of the IR, and the c are the coefficients of the combination.

With this linear combination, we use the projection operator as defined by Eq. (6.12):

$$\hat{P}^{\alpha}_{nm}\psi = \sum_\beta\sum_{m(\beta)} c^\beta_m\left[\frac{g_\alpha}{g}\sum_R [\boldsymbol{D}_\alpha(R)]^*_{mn}\hat{R}\left(\varphi^\beta_m\right)\right]$$

Taking the dot product by the φ^β_m function, in the Dirac notation $\left(\varphi^\alpha_k\right)^*\cdot\hat{R}\varphi^\beta_m \rightarrow \left\langle\alpha k\left|\hat{R}\right|\beta m\right\rangle$ and considering that the basis functions of different IR are orthogonal to each other (a property that is proved in Section 7.4): $\left\langle\alpha k\left|\hat{R}\right|\beta m\right\rangle = \delta_{\alpha\beta}\cdot\left\langle\alpha k\left|\hat{R}\right|am\right\rangle$. We are left with the summation:

$$\sum_{m(\alpha)} c^\alpha_m\left[\frac{g_\alpha}{g}\sum_R [\boldsymbol{D}_\alpha(R)]^*_{mn}\hat{R}\left(\varphi^\alpha_m\right)\right]$$

$$= \frac{g_\alpha}{g}\sum_R [\boldsymbol{D}_\alpha(R)]^*_{mn}\cdot\left\langle\alpha n|\hat{R}|\alpha m\right\rangle = c^\alpha_m\delta_{nm}$$

This shows that the projection operator $\hat{P}^\alpha_{mn}$ gives the coefficient (c^α_m) of n-th partner of the α IR in the linear combination $\sum_\beta\sum_{m(\beta)} c^\beta_m\varphi^\beta_m$, which is the desired property of a projection operator.

6.2.1.2 *An Example: the SALCs from the H-AOs of NH_3*

Let us consider the LCAO-MO problem of the ammonia molecule. Between the several AOs that can be used as basis functions for SALCs, let us analyse the 1s AO of the three H atoms. These AOs transform according to Table 4.22 and form a 3D RR that is the direct sum of $A_1 \oplus E$ of two IR of the group (the characters are in Table 4.20).

We directly use Eq. (5.27): $a(A_1)=(1\cdot3\cdot1+2\cdot0\cdot1+3\cdot1\cdot1)/6=6/6=1$; $a(A_2)=(1\cdot3\cdot1+2\cdot0\cdot1+3\cdot1\ (-1))/6=0$; $a(E)=(1\cdot3\cdot2+2\cdot0\cdot(-1)+3\cdot1\cdot0)/6=6/6=1$. To show how projection operators are used, let us start with the A_1 IR.

$$\hat{P}^{A1} = \frac{g_\alpha}{g}\sum_R \hat{R} \tag{6.13}$$

There is only one matrix element ($i=j=1$) and it is $=1$ for all R. The basis set is (1sH1, 1sH2 and 1sH3). Let us take 1sH1. Using Eq. (6.13), we get (1/6)[1sH1 + 1sH2 + 1sH3 + 1sH1 + 1sH3 + 1sH2] and, after normalization, [1sH1 + 1sH2 + 1sH3]/$\sqrt{3}$.

Looking at Eq. (6.13), we understand now that Eq. (2.4) is a *particular case* of Eq. (6.12): this justifies the use of Eq. (2.4) done in Sections 2.2.1 and 2.2.3.

Now, let us consider the (2D) E IR. We are looking for two SALCs that are transformed according to the matrices of this IR obtained by reducing the (3D) RR as done for Table 4.19:

TABLE 4.22
Transformations of the 1s AO for the three H of NH_3 and matrices of the related (3D) representation

E	C_3	C_{-3}	$\sigma^{(1)}$	$\sigma^{(2)}$	$\sigma^{(3)}$
a	b	c	a	c	b
b	c	a	c	b	a
c	a	b	b	a	c
$\begin{pmatrix}1&0&0\\0&1&0\\0&0&1\end{pmatrix}$	$\begin{pmatrix}0&0&1\\1&0&0\\0&1&0\end{pmatrix}$	$\begin{pmatrix}0&1&0\\0&0&1\\1&0&0\end{pmatrix}$	$\begin{pmatrix}1&0&0\\0&0&1\\0&1&0\end{pmatrix}$	$\begin{pmatrix}0&0&1\\0&1&0\\1&0&0\end{pmatrix}$	$\begin{pmatrix}0&1&0\\1&0&0\\0&0&1\end{pmatrix}$

The symbols a, b, c indicate the 1s AOs of the first, second and third H atoms.

E	C_3	C_{-3}	$\sigma^{(1)}$	$\sigma^{(2)}$	$\sigma^{(3)}$
$\begin{pmatrix} 1 & 0 \\ 0 & 1 \end{pmatrix}$	$\begin{pmatrix} \frac{-1}{2} & \frac{-\sqrt{3}}{2} \\ \frac{\sqrt{3}}{2} & \frac{-1}{2} \end{pmatrix}$	$\begin{pmatrix} \frac{-1}{2} & \frac{\sqrt{3}}{2} \\ \frac{-\sqrt{3}}{2} & \frac{-1}{2} \end{pmatrix}$	$\begin{pmatrix} \frac{-1}{2} & \frac{-\sqrt{3}}{2} \\ \frac{-\sqrt{3}}{2} & \frac{1}{2} \end{pmatrix}$	$\begin{pmatrix} 1 & 0 \\ 0 & -1 \end{pmatrix}$	$\begin{pmatrix} \frac{-1}{2} & \frac{\sqrt{3}}{2} \\ \frac{\sqrt{3}}{2} & \frac{1}{2} \end{pmatrix}$

Starting from 1sH1 and writing for short a, b, c to mean the 1s AOs of H1, H2, H3, we find (see Table 4.23) that there are only two independent solutions, one for each column of $\boldsymbol{D}$ (remind that the columns of the matrices are orthogonal to each other.

Other solutions are obtained with the cyclic permutations ($a \rightarrow b$, $b \rightarrow c$, $c \rightarrow a$) starting from another AO (1sH2 or 1sH3) and are equivalent to the solutions of the table above. Normalizing:

$$\left(\frac{1}{2}\right)a - b + \left(\frac{1}{2}\right)c \Rightarrow \frac{1}{\sqrt{6}}(a - 2b + c) \quad \text{and}$$

$$\left(\frac{-\sqrt{3}}{2}\right)a + \left(\frac{\sqrt{3}}{2}\right)c \Rightarrow \frac{1}{\sqrt{2}}(a - c)$$

So, in the $\begin{pmatrix} a\, b\, c \end{pmatrix}$ basis, the SALC are

$$\frac{1}{\sqrt{6}}\begin{pmatrix} 1 \\ -2 \\ 1 \end{pmatrix} \quad \text{and} \quad \frac{1}{\sqrt{2}}\begin{pmatrix} 1 \\ 0 \\ -1 \end{pmatrix} \tag{6.14}$$

Figure 4.31 shows the two SALCs and allows to verify visually the orthogonality of the two functions: the left SALC is even with respect to right/left reflection, the right SALC is odd. Analytically (using s to indicate the overlap integral between two 1s-H AOs) the dot product of the SALCs is:

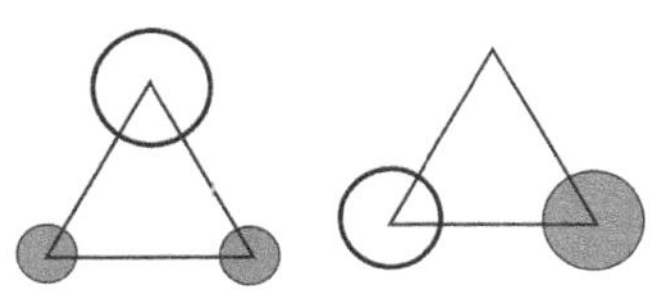

FIGURE 4.31 The two SALCs belonging to the E IR and based on the hydrogen atoms of NH_3.

$$\frac{1}{\sqrt{6}}\begin{pmatrix} 1 & -2 & 1 \end{pmatrix}\begin{pmatrix} 1 & s & s \\ s & 1 & s \\ s & s & 1 \end{pmatrix}\frac{1}{\sqrt{2}}\begin{pmatrix} 1 \\ 0 \\ -1 \end{pmatrix}$$

$$= \frac{1}{2\sqrt{3}}\begin{pmatrix} 1-s & 2s-2 & 1-s \end{pmatrix}\begin{pmatrix} 1 \\ 0 \\ -1 \end{pmatrix} = 0$$

The precise value of s is irrelevant, which clearly shows that the result is due to symmetry.

6.2.1.3 Projection Operators Using Characters

The alternative version of the projection operators uses characters instead of matrices:

$$\hat{P}^{\alpha} = \frac{g_{\alpha}}{g}\sum_{R}\left[\chi_{\alpha}(R)\right]^{*}\hat{R} \tag{6.15}$$

For the proof, let us consider the arbitrary function on which (as seen before) we operate with a projection operator according to Eq. (6.12):

$$\hat{P}^{\alpha}_{nm}\psi = \frac{g_{\alpha}}{g}\sum_{R}\left[\boldsymbol{D}_{\alpha}(R)\right]^{*}_{mn}\hat{R}\psi$$

We obtain the same c^{α}_{m} coefficient **multiplying the *n*th partner** of the α IR in the linear combination (on the various IRs and on their various partners) that gives the arbitrary function: $\psi = \sum_{\beta}\sum_{m(\beta)} c^{\beta}_{m}\varphi^{\beta}_{m}$. Consider now the case $m=n$ (that we will write s for the sake of better clarity) and sum over all the diagonal elements:

TABLE 4.23
Application of the projection operators to the E IR obtained by reducing the matrices of the (3D) RR of Table 4.22

i,j	φ	$[D(E)]_{ij}\cdot E$	$[D(C_3)]_{ij}\cdot C_3$	$[D(C_{-3})]_{ij}\cdot C_{-3}$	$[D(\sigma^{(1)})]_{ij}\cdot\sigma^{(1)}$	$[D(\sigma^{(2)})]_{ij}\cdot\sigma^{(2)}$	$[D(\sigma^{(3)})]_{ij}\cdot\sigma^{(3)}$
1,1	a	a	$-1/2b$	$-1/2c$	$-1/2a$	c	$-1/2b$
2,1	a	0	$\sqrt{3}/2b$	$-\sqrt{3}/2c$	$-\sqrt{3}/2a$	0	$\sqrt{3}/2b$
1,2	a	0	$-\sqrt{3}/2b$	$\sqrt{3}/2c$	$-\sqrt{3}/2a$	0	$\sqrt{3}/2b$
2,2	a	a	$-1/2b$	$-1/2c$	$1/2a$	$-c$	$1/2b$

TABLE 4.24
The Construction of the SALCs of the 1s AO of the Three Hs in the Character-Based Version of the Projection Operator

φ	$\chi{\cdot}E\varphi$	$\chi{\cdot}C_3\varphi$	$\chi{\cdot}C_{-3}\varphi$	$\chi{\cdot}\sigma^{(1)}\varphi$	$\chi{\cdot}\sigma^{(2)}\varphi$	$\chi{\cdot}\sigma^{(3)}\varphi$	Total
a	$2a$	$-b$	$-c$	0	0	0	$2a-b-c$
b	$2b$	$-c$	$-a$	0	0	0	$2b-c-a$
c	$2c$	$-a$	$-b$	0	0	0	$2c-a-b$

$$\sum_{ss}\hat{P}^{\alpha}_{ss}\psi = \frac{g_\alpha}{g}\sum_{s}\left[\sum_{R}[\boldsymbol{D}_\alpha(R)]^*_{ss}\,\hat{R}\psi\right]$$

$$= \frac{g_\alpha}{g}\sum_{R}\left[\sum_{s}[\boldsymbol{D}_\alpha(R)]^*_{ss}\,\hat{R}\psi\right] = \frac{g_\alpha}{g}\sum_{R}\chi_\alpha\cdot\hat{R}\psi$$

Note that the index for the basis function that is projected is lost: the projection is on the α IR, and not on a specific *partner*.

6.2.1.4 *Example: Again the SALCs from the H-AOs of NH_3*

As an example of application of Eq. (6.15), let us reconsider the IR E for the SALCs of the AO of the Hs in NH_3 by applying Eq. (6.15) to each of the AO: results are in Table 4.24.

It is easy to see from the sum (which gives zero) that these *three* SALCs are *linearly dependent*, in agreement with the fact that for a 2D IR we expect only *two linearly independent* SALCs. The three solutions are also obtained with cyclic substitutions ($a \to b,\ b \to c,\ c \to a$).

We then need to select **two orthogonal linear combinations**. The observation that the three solutions have a zero sum and the cyclic permutation relation between them, suggests the comparison with the geometric situation of three vectors arranged as the sides of an equilateral triangle and allows us to find two orthogonal vectors in a direct way: we choose the pair consisting of a vector selected among the three original ones and a second vector given by the difference of the other two, that is:

$$\left[\begin{pmatrix}-a\\2b\\-c\end{pmatrix}-\begin{pmatrix}-a\\-b\\2c\end{pmatrix}\right]=\left[\begin{pmatrix}0\\3b\\-3c\end{pmatrix}\right]$$

After normalization, the independent and orthogonal SALCs are: $\frac{1}{\sqrt{6}}\begin{pmatrix}2a\\-b\\-c\end{pmatrix}$ and $\frac{1}{\sqrt{2}}\begin{pmatrix}0\\b\\-c\end{pmatrix}$. Except for the order of the indices of the H atoms, this is the previous result (Eq. 6.14).

6.2.1.5 *Comparison between the Two Procedures*

On comparing the two projection procedures, we need to consider two aspects:

- for all symmetry groups and for all IRs, one can deem to *always* have the characters available in the tables, while the *complete matrices* of the various IRs are generally *not* available.
- The character version requires *orthogonalization*, while the matrix version directly provides orthogonal SALCs that only need to be *normalized*.

It is therefore clear that the matrix procedure is preferable, because it is faster, when the matrix representation of the IR of interest is already available. In the much more frequent case in which only the characters are available, it is unavoidable to resort to the characters and take on the subsequent orthogonalization.

Observations:

A. Both methods provide unnormalized SALCs.
B. For all the 1D representations, Eqs. (6.12) and (6.15) reduce to Eq. (2.4), which is then proved.
C. Applying Eq. (6.12) or Eq. (6.15) sometimes you get a null result. This means that a linear combination behaving like the nth partner of the α IR *does not exist*. We then say that the starting function is not a basis for the nth partner of the α IR. This corresponds to what is already seen applying Eq. (2.4) in Sections 2.2.1 and 2.2.2.

6.2.1.6 *SALCs from the $2p_z$ AOs of Benzene*

Let us now consider a more complex example of the procedure for obtaining SALCs of AOs. Once again, we use the π MOs benzene. A useful feature of this problem is that the SALCs of $2p_z$ AOs of equivalent C atoms coincide with π MOs that can be obtained with the simple Hückel approach. For this reason, the results obtained with a numerical procedure have been already discussed in Section 1.2.2.2 to introduce the difference between commutative and non-commutative groups and the dimensions of symmetry species and in Section 6.1.1.2 to compare orthogonal and non-orthogonal basis sets. We remind that the problem is characterized by the degeneracies 1, 2, 2, 1, as shown by Figure 4.11.

Let us now obtain the results using the group theoretical techniques of the previous sections.

The point group of benzene is D_{6h}: the multiplication table of the group is partially reported in Table 4.4 and the orientation of the symmetry elements is described in detail in the text concerning the same Table, while the following Table 4.31 lists the characters.

It is not easy here to use group theory, due to the complexity of group D_{6h} with 24 symmetry operations and 12 classes. To simplify the calculation, a "shortcut" is used. It simply involves using the character table of *another* group,

C_{6v}, having half of the symmetry elements (and classes) of the correct group, as shown in Table 4.25.

Using p1 to denote the $2p_z$ AO of C1, p2 to denote the $2p_z$ AO of C2 and so on, the symmetry operations transform the AOs as follows.

E	C_6	C_{-6}	C_3	C_{-3}	C_2	$\sigma_v^{(1)}$	$\sigma_v^{(2)}$	$\sigma_v^{(3)}$	$\sigma_d^{(1)}$	$\sigma_d^{(2)}$	$\sigma_d^{(3)}$
p1	p2	p6	p3	p5	p4	p1	p3	p5	p2	p4	p6
p2	p3	p1	p4	p6	p5	p6	p2	p4	p1	p3	p5

Using the first row (p1...p6) of the above table, the character-based projection operators (Eq. 6.13) give the following SALCs.

In the A_1 IR all AOs appear with the same coefficient, which is then normalised as $1/\sqrt{6}$: $A_1 \to (p1+p2+p3+p4+p5+p6)/\sqrt{6}$.

Analogously, $A_2 \to (p1+p2+p6+p3+p5+p4-p1-p3-p5-p2-p4-p6)=0$ and we say that there is no SALC with A_2 symmetry.

Also, $B_1 \to (p1-p2-p6+p3+p5-p4+p1+p3+p5-p2-p4-p6)=(2p1-2p2+2p3-2p4+2p5-2p6) \to (p1-p2+p3-p4+p5-p6)/\sqrt{6}$.

The last 1D IR (B_2) gives $(p1-p2-p6+p3+p5-p4-p1-p3-p5=0$ and $+p2+p4+p6)=0$ and we say that there is no SALC with B_2 symmetry.

TABLE 4.25
Characters for the C_{6v} Group

C_{6v}	E	$2C_6$	$2C_3$	C_2	$3\sigma_v$	$3\sigma_d$
A_1	+1	+1	+1	+1	+1	+1
A_2	+1	+1	+1	+1	−1	−1
B_1	+1	−1	+1	−1	+1	−1
B_2	+1	−1	+1	−1	−1	+1
E_1	+2	+1	−1	−2	0	0
E_2	+2	−1	−1	+2	0	0

Concerning the two 2D IRs, the method based on characters requires two AOs and therefore we use both rows (p1 ... p6 and p2 ... p5) of the table above.

The SALCs from E_1 are: $(2p1+p2+p6-p3-p5-2p4)/\sqrt{12}$ using p1 and $(2p2+p3+p1-p4-p6-2p5)/\sqrt{12}$ using p2.

The SALCs from E_2 are: $(2p1-p2-p6-p3-p5+2p4)/\sqrt{12}$ using p1 and $(2p2-p3-p1-p4-p6+2p5)/\sqrt{12}$ using p2.

The upper part of Figure 4.32 shows the non-orthogonal SALCs obtained with this procedure. The orthogonalization described in Section 3.2.5.3 of the Appendices gives the SALCs shown on the lower part of the figure.

It remains to justify the use of a point group *other than* the correct one: for this, see Section 6.3.3.

6.2.2 From SALCs to MOs

Once we have seen how to obtain the SALCs from AOs, we move on to examining the MOs, that is, the combination of SALCs. For this purpose, we can continue with the example of NH_3.

In Section 6.2.1.4 we already obtained the SALCs based on the H 1s AOs: one SALC of species A_1 and two SALCs of species E. Now we need to classify the AOs of the N atom in a similar way. Using a reduced basis with only 2s and 2p AOs, as done in Section 2.2.1 for the water molecule, it can be seen that the 2s AO belongs to the A_1 IR. For the 2p, the results are those already obtained when discussing the representations generated by the unit vectors of the ordinary 3D space and choosing the unit vector $\vec{z}$ and the others $(\vec{x}, \vec{y})$ in a plane perpendicular to the axis and parallel to that of the three H. In Section 5.1.3.2 it was found that $\vec{z}$ belongs to the A_1 IR, while the pair $(\vec{x}, \vec{y})$ belongs to the E IR. Substituting $2p_x$ to $\vec{x}$, $2p_y$ to $\vec{y}$ and $2p_z$ to $\vec{z}$, we have the result we need:

- the E pair of 2p-N AOs combines with the two SALCs E of the three H, forming two levels, one bonding and one antibonding, each level being twofold degenerate, while

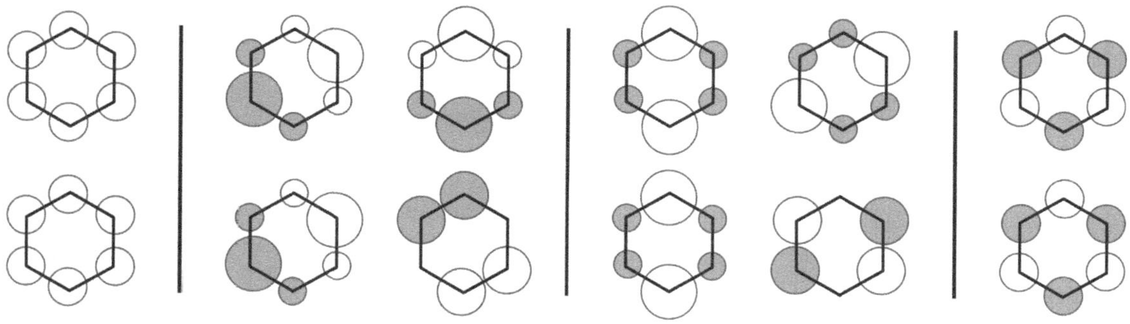

FIGURE 4.32 SALCs of the $2p_z$-C AOs in benzene. Upper part: SALCs obtained as explained in the text. Lower parts, SALCs of the same IRs after orthogonalization. The IRs are A_1, E_1, E_2, B_1, respectively, from left to right. The signs (positive or negative) of the lobes of the $2p_z$-C AOs placed above the molecular plane are indicated in light and dark, respectively. See also Figure 4.30 and Eq. (6.4).

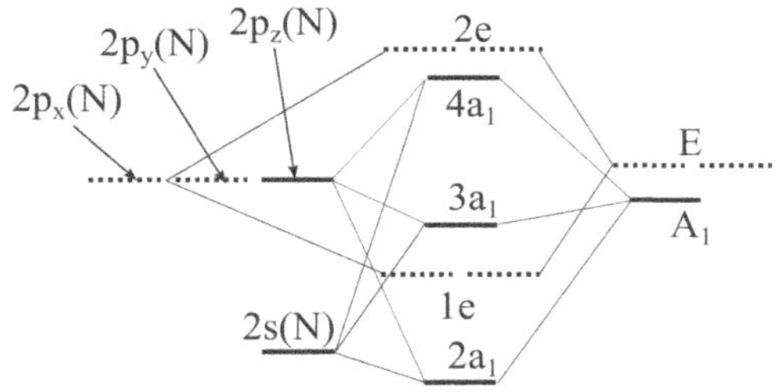

FIGURE 4.33 Scheme of the MOs in NH_3. The different line styles indicate different IR.

- the two AOs (2s and $2p_z$) of N (A_1 IR) combine with the SALC A_1 of the 1s AOs of the three H to give three non-degenerate MOs, one strongly bonding, one strongly antibonding and an intermediate MO that hosts the lone pair and has weakly bonding character.

NB: The usual nomenclature uses lowercase letters to specify the symmetry species of the energy levels of the MOs, preceded by an order number increasing with energy. Thus, these levels are normally indicated as $2a_1$, 1e, $3a_1$, $4a_1$, 2e while $1a_1$ is the MO that is essentially given by the 1s AO of N. Instead, uppercase letters are usually reserved for the energy levels of *configurations*, i.e., *multi-electronic states* (Figure 4.33).

It is worth noting from Figure 4.34, that each MO combines the SALCs of H and N belonging to the *same* partner within the *same* IR (E). This is made clear by the even/odd symmetries with respect to the right/left reflection, respectively shown by the first/second partners. On the other hand, the comparison of the bonding MO with the corresponding (same partner) antibonding MO exhibits a change of sign of one of the SALCs (of N or of the H atoms).

6.2.3 Many-Electron States

For the representations of polyelectronic states, the starting point is the consideration that **the representation of a product of SAF is the product of the respective representations**. The IR product is not necessarily an IR (Section 5.1.5.2). It can be equivalent to a direct sum of IRs, among which one must eliminate those prohibited by antisymmetrisation. As already shown (see Section 2.3.1), in commutative groups **any fully occupied non-degenerate electron shell belongs to the totally symmetric IR only**, and thus *it can be neglected in the determination of the symmetry species of the configuration as a whole*. Now we add that the result is also valid for non-commutative groups and therefore also for **degenerate levels**:

> Closed shells belong *only* to the totally symmetric IR also for degenerate levels.

The explanation given for non-degenerate levels does not hold here, but the result is obtained, albeit less directly: the problem is discussed in Section 7.5.

In addition, concerning the spin quantum number (S):

> Closed shells are singlets.

The effect of antisymmetry is particularly evident in closed shells. If one were to treat the problem with the general rule that the characters of a product of functions are the products of the characters of the factor functions (Section 6.1.2) and ignoring Pauli principle, one would obtain a *very* different result. For example, let us consider the full shell of the MOs of Figure 4.33 with symmetry 1e, that is the $(1e)^4$ configuration. The characters of the fundamental state (Table 4.20) are $2^4=16$ for identity, $(-1)^4=1$ for the rotation class, 0 for the reflection class, which is the E⊗E⊗E⊗E RR. The contributions of the different IR to this RR are $a(A_1)=[1\cdot16\cdot1+2\cdot1\cdot1+3\cdot0\cdot1]/6=3$; $a(A_2)=[1\cdot16\cdot1+2\cdot1\cdot1+3\cdot0\cdot(-1)]/6=3$; $a(E)=[1\cdot16\cdot2+2\cdot1\cdot(-1)+3\cdot0\cdot0]/6=5$. In conclusion, $E \otimes E \otimes E \otimes E = 3A_1 \oplus 3A_2 \oplus 5E$. Obviously, for this configuration, we have only one state, that belongs to the totally symmetric IR A_1: its spin state is a singlet ($S=0$, $2S+1=1$). The remarkable difference between the two results is obviously a consequence of the Pauli principle, which "eliminates" a large part (in this case 15 out of 16) of the apparently possible results.

For **open shells**, the situation is complicated, and it is worth to distinguish between:

1. the case of two or more different *open shells* with *one electron each* and
2. the case of (one or more) *open shells*, at least one of them containing *two or more electrons*.

The **first case** does not present anti-symmetry conflicts, *all the symmetries that are obtained by the product of the IR of each shell are allowed*, and any (spatial) IR can be associated to all the spin states (S) that are compatible. The reason is obviously that - with only one electron per shell - there

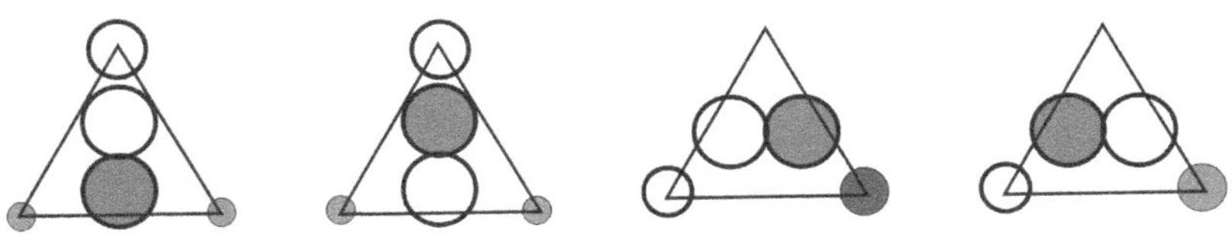

FIGURE 4.34 MOs of NH_3 for the two degenerate levels of E symmetry. From left to right: bonding MO, 1e, first partner; antibonding MO, 2e, first partner; bonding MO, 1e, second partner; antibonding MO, 2e, second partner. The weights for the AO of the H atoms *are not* in the same scale as the N 2s AO.

TABLE 4.26
Symmetry of Configuration $e^1e'^1$, group C_{3v} (see also Table 4.20)

	E	$2C_3$	$3\sigma_v$
A_1	1	1	1
A_2	1	1	−1
E	2	−1	0
$E \otimes E$	4	1	0
$A_1 \otimes A_2 \otimes E$	4	1	0

TABLE 4.27
IR Products of the C_{3v} Group

	A_1	A_2	E
A_1	A_1	A_2	E
A_2		A_1	E
E			A_1+[A_2]+E

[] indicates the IR that can be a **triplet.**

is no possibility of counting states corresponding to "two electrons in the same spin orbital (SO)". The simplest case is that of a configuration where there are two open shells and **both belong to a 1D IR**, for example the configuration $(b_1)^1(b_2)^1$ in a molecule with C_{2v} symmetry. Then $b_1 \otimes b_2 = A_2$ and both $^1A_2$ and $^3A_2$ states are allowed. We have *just one* SO to choose from two and *another* SO to choose from the other two, for a total of four different Slater determinants that combine linearly into one singlet state and three triplet states.

A slightly more complicated case is that of **one 1D and one 2D IR**: for example, the configuration $(a_1)^1(e)^1$ for a C_{3v} molecule as in Figure 4.33. We have $a_1 \otimes e = e$, and therefore: (i) two singlets ^1E and $^1E'$, with different spatial parts, E and E′ and (ii) two triplets ^3E and $^3E'$. There are then 2+6=8 states, corresponding to the 2×4 choices of SO to be combined.

Consider now the case of **two shells belonging to 2D** IRs, again referring to the C_{3v} group and to the configuration $e^1e'^1$, where with e and e′ we mean two different shells, with two different MO energies. Then, we have $E \otimes E = A_1 \oplus A_2 \oplus E$ (see Table 4.26). States can be singlets or triplets. So, we have four singlet states and 4×3=12 triplet states, for a total of 16 states, which corresponds to the 4×4 possible choices of an SO from each of the e and e′ shells.

The **second case** is much more intricate because the Pauli principle limits the number of possible states even with **only one open shell**. Consider the example of the configuration e^2 in the usual C_{3v} group. From Table 4.26 we have $E \otimes E = A_1 \oplus A_2 \oplus E$, but we cannot expect that all IRs are allowed both as singlets and as triplets; in fact, now only $\binom{4}{2} = 6$ states are possible, instead of $4^2 = 16$ states. On the other hand, Table 4.26 still holds for the spatial parts of two-electron functions. So, we can immediately say that the triplet state must be associated with a 1D IR, as otherwise we would only have a ^3E triplet, without singlets and 1D IRs. We are left with only two options: $^3A_1 \oplus {}^1A_2 \oplus {}^1E$ or $^1A_1 \oplus {}^3A_2 \oplus {}^1E$. The choice can be made considering that the three spin functions for a triplet are $\alpha(1)\alpha(2)$, $\beta(1)\beta(2)$ and the $[\alpha(1)\beta(2)+\beta(1)\alpha(2)]$ combination, while the singlet requires the $[\alpha(1)\beta(2)-\beta(1)\alpha(2)]$ combination. That is, we see that the two-electron spin functions of a triplet are symmetric with respect to the exchange of the two electrons, while the spin function of the singlet is antisymmetric. For the spatial parts, the opposite must be true, since the product of the spatial part and the spin part must be antisymmetric. As the spatial symmetric function is A_1 and the antisymmetric one is A_2, the correct results is $^1A_1 \oplus {}^3A_2 \oplus {}^1E$. The example also shows that it is not trivial to identify the polyelectronic states that are compatible with antisymmetry. The final scheme of the IR products is in Table 4.27.

Upon closer inspection, the situation is similar to that of the electronic states of atoms when classified according to the vector model. Here too, the conclusion can be drawn through a specialized treatment that goes beyond a basic approach, but appropriate **tables** with the results of the most frequent cases **are available**, and it is worth referring to these.[10] Essentially, for each direct product of the IRs of the individual partially occupied shells, the tables report which IRs are possible and with what spin multiplicity.

The comparison with the vector model of the atom is much more than an analogy: it is **actually the same problem** presented **without explicit reference to group theory**. In fact, the different values of the quantum number L correspond to the different species of the central field symmetry (the symmetry of the sphere).

A final observation, also in analogy to what is valid for the poly-electronic states of atoms, concerns the **hole formalism**, whereby the situation in which there are n (> 1) electrons in a shell that can contain *m electrons* ($m>n$, m even) has the same IR for the spatial part, and the same S of the configuration with $m-n$ electrons. Let's consider, for example, the partial configuration $(e)^3$: $(e)^1$ has four states belonging to the E IR, that are, obviously, ^1E and ^3E. Then, the configuration $(a_1)^1(e)^3$ is equivalent for the hole formalism to $(a_1)^1(e)^1$. As $a_1 \otimes e = E$, we can have two singlets, ^1E and $^1E'$, and two triplets, ^3E and $^3E'$, of a total of eight states (here the writings, with and without apex, indicate two different spatial parts of the *same* e *shell*.

6.2.4 Vibrational Degrees of Freedom of Molecules

The problem of the motions of the nuclei of a molecule can be treated with symmetry by following the steps indicated in Section 6.2.3 for the analogous electronic problem, but with the specification that it is preliminarily necessary to distinguish, within the context of the motions of the constituent atoms, between vibrational motions (those of interest here) and translational and rotational motions. For a molecule with n atoms, we have a total of $3n$ degrees of freedom, that yield a $\boldsymbol{D}$(tot) RR, given by:

$$\boldsymbol{D}(\text{tot}) = \boldsymbol{D}(\text{ve}) \otimes \boldsymbol{D}(\text{eas}) \qquad (6.10)$$

As already seen in Section 6.1.2, $\boldsymbol{D}$(ve) is the (3D) representation generated by a ***vector** of ordinary 3D space*, that is by the three x, y, z functions, while $\boldsymbol{D}$(eas) is the (n dimensional) representation of ***equivalence** of **atomic** sites* (see Section 5.1.3.6). The *characters* of the eas RR can also be obtained by counting the atoms that are *not modified* by the symmetry operations (see Section 5.3.3).

The $3n$-dimensional $\boldsymbol{D}$(tot) representation contains the contributions of all degrees of freedom relating to the motions of the atoms constituting a molecule and therefore:

$$\boldsymbol{D}(\text{tot}) = \boldsymbol{D}(\text{vib}) \otimes \boldsymbol{D}(\text{rot}) \otimes \boldsymbol{D}(\text{trans}) \qquad (6.16)$$

To obtain only the contributions of the vibrations, we need to eliminate the contributions of the translational and rotational degrees of freedom considering that the characters for rotations and translations are in the character tables (the x, y, z and R_x, R_y, R_z, functions, respectively).

6.2.4.1 An Example: The NH_3 Molecule

For the NH_3 molecule, the whole procedure is summarized in Table 4.28. $\boldsymbol{D}$(tot) is given by matrices of Section 5.1.5.2, which shows the direct product relationship $\boldsymbol{D}(\text{tot}) = \boldsymbol{D}(\text{ve}) \otimes \boldsymbol{D}(\text{eas})$. $\boldsymbol{D}(\text{ve})$ is given by the matrices in Section 5.1.3.1 or Section 5.1.3.2 (Eq. 5.5): the second case makes the direct sum $A_1 \oplus E$ apparent. $\boldsymbol{D}$(eas) is given by matrices in Section 5.1.3.6.

The characters of $\boldsymbol{D}$(tot) are given by the products of the characters of the factors: $\chi[\boldsymbol{D}(\text{tot})]=\chi[\boldsymbol{D}(\text{ve})]\cdot\chi[\boldsymbol{D}(\text{eas})]$ and $\boldsymbol{D}(\text{ve})=A_1 \oplus E$, $\boldsymbol{D}(\text{eas})=A_1 \oplus A_1 \oplus E$. The IR characters for translational and rotational degrees of freedom are in Table 4.20: for translations you need to look for the IRs to which the x, y, z functions belong; for rotations look for R_x, R_y, R_z.

To find how many times the IRs of the group are contained in $\boldsymbol{D}$(vib) we use Eq. (5.27):

$$a_\beta = \frac{1}{g}\sum_{k=1}^{h} N_k \left[\chi_\alpha(k)\right]^* \left[\chi_\beta(k)\right]$$

We find $a(A_1)=[1\cdot6\cdot1+2\cdot0\cdot1+3\cdot2\cdot1]/6=2$, $a(A_2)=[1\cdot6\cdot1+2\cdot0\cdot1+3\cdot2\cdot(-1)]/6=0$; $a(E)=[1\cdot6\cdot2+2\cdot0\cdot(-1)+3\cdot2\cdot0]/6=2$. Then $\boldsymbol{D}(\text{vib})=2A_1 \oplus E$. For confirmation, for the RR **D**(vib) one gets: $\chi(I)=1+1+2+2=6$; $\chi(C_3)=1+1-1-1=0$; $\chi(\sigma_v)=1+1+0+0=2$. There are therefore two totally symmetric modes and four modes belonging to two different frequencies.

Another way to obtain the same result (is to directly break down the $\boldsymbol{D}$(tot)] and to subtract the IR of translations , which are obtained from the species of the functions z, (x, y) and R_z, (R_x, R_y), respectively, and then $\boldsymbol{D}(\text{vib})=2A_1 \oplus E$.

TABLE 4.28
Steps to Obtain the Representation of the Normal Vibration Modes of the NH_3 Molecule

	E	$2C_3$	$3\sigma_v$
A_1	1	1	1
A_2	1	1	−1
E	2	−1	0
$\boldsymbol{D}$(ve)	3	0	1
$\boldsymbol{D}$(eas)	4	1	2
$\boldsymbol{D}$(tot)	12	0	2
(x, y)	2	−1	0
z	1	1	1
R_z	1	1	−1
(R_x, R_y)	2	−1	0
$\boldsymbol{D}$(vib)	6	0	2

6.2.4.2 Another Example: The H_2O Molecule

$$\boldsymbol{D}(\text{tot}) = \boldsymbol{D}(\text{ve}) \otimes \boldsymbol{D}(\text{eas})$$

It may be useful to compare the procedure using the characters with the procedure used previously to derive the symmetry species of the normal modes of the H_2O molecule (Section 2.2.3.1). Table 4.29 reports the characters for $\boldsymbol{D}(\text{ve})=A_1 \oplus B_1 \oplus B_2$ and $\boldsymbol{D}$(eas). $\boldsymbol{D}$(tot) has characters 9, −1, 3, 1; subtracting the translational and rotational contributions, we obtain $\boldsymbol{D}$(vib), with characters 3, 1, 3, 1. Therefore,

TABLE 4.29
The Characters of the Representations Involved in the Calculation of *D*(vib), the Representation of the Normal Vibration Modes of the H_2O Molecule

	E	C_2	σ	σ'
A_1	1	1	1	1
A_2	1	1	−1	−1
B_1	1	−1	1	−1
B_2	1	−1	−1	1
x	1	−1	1	−1
y	1	−1	−1	1
z	1	1	1	1
$\boldsymbol{D}$(trans) = $\boldsymbol{D}$(ve)	3	−1	1	1
$\boldsymbol{D}$(eas)	3	1	3	1
$\boldsymbol{D}$(tot)	9	−1	3	1
R_x	1	−1	−1	1
R_y	1	−1	1	−1
R_z	1	1	−1	−1
$\boldsymbol{D}$(rot)	3	−1	−1	−1
$\boldsymbol{D}$(vib)	3	1	3	1
$A_1 \otimes B_1$	3	1	3	1

The C_2 axis is z, the molecular plane (σ) is zx, σ' is yz (see Figure 4.10).

$\boldsymbol{D}$(vib)$=2A_1 \oplus B_1$, as obtained with the much more complex procedure of Section 2.2.3.1.

For check, considering that the classes are all made up of a single element, we have:

$a(A_1)=[3{\cdot}1+1{\cdot}1+3{\cdot}1+1{\cdot}1]/4=2$
$a(A_2)=[3{\cdot}1+1{\cdot}1+3{\cdot}(-1)+1{\cdot}(-1)]/4=0$
$a(B_1)=[3{\cdot}1+1{\cdot}(-1)+3{\cdot}1+1{\cdot}(-)1]/4=1$
$a(B_2)=[3{\cdot}1+1{\cdot}(-1)+3{\cdot}(-1)+1{\cdot}1]/4=0$

6.2.5 Vibrational Energy Levels and Degeneracies

We can now re-examine and partly clarify the considerations already made for the vibrations of molecules described by commutative symmetry groups (see Section 2.2.3.1) and discuss the limitations of the harmonic model (Section 2.3.2.1).

The existence of normal modes of vibration belonging to *non*-1D IRs requires considering the degeneracy of energy levels. Essentially, it is *not the states* (specified by quantum numbers, one for each normal mode) that need to be classified, but the energy levels, characterized by their IRs and already named in Section 2.2.2. An energy level is specified by the number of *phonons* - that is, of excitations – for each *frequency* (n_1, n_2, …, n_s, if the molecule has s distinct *vibrational frequencies*). If r indicates the total number of vibrational degrees of freedom, the presence of degenerate normal modes implies that $s<r$. In summary:

- The **ground state** of the molecule is that without phonons ($n_1=0$, $n_2=0$, …, $n_s=0$).
- In the so-called s **fundamental levels**, we have only one phonon: only one $n_i=1$, all the rest$=0$.
- In the different **overtones** we have more than one phonon with the *same* frequency: one $n_i>1$, all the others$=0$.
- Finally, there are the so-called **combination** levels in which there are phonons of at least two distinct frequencies.

To know the degeneracies and symmetries of the energy levels of the molecule, we first need to consider the possible energy levels for each single frequency and their degeneracies.

For any different frequency, regardless of the number of modes, the level corresponding to no phonon is a unique state, in which all the quantum numbers of the modes (of that frequency) are$=0$.

For excited levels, a distinction must be made; with non-degenerate frequencies (*i.e.* belonging to 1D IR) there is no degeneracy of the energy levels. Instead, for each frequency belonging to a mD ($m>1$) IR, the first excited level ($n_i=1$) has degeneracy equal to the dimension m of the corresponding IR, the other levels have larger degeneracy. In detail, for a **2D** IR, which has two normal modes with the same frequency ω, for the level $\hbar\omega$ ($n_i=1$) above the *ground state* we have two states where the quantum numbers of the degenerate modes are: (1,0) or (0,1); for the level $2\hbar\omega(n_i=2)$ we have three states: (2,0), (0,2) and (1,1); for the level $3\hbar\omega$ ($n_i=3$) four states: (3,0), (0,3), (2,1) and (1,2). Generalizing without proof, for 2D IR, for the level $n\hbar\omega$ we have $n+1$ states: (n,0), ($n-1$,1), …, (0, n).

If **three modes** have the same frequency, then the level $\hbar\omega$ ($n_i=1$) above the fundamental level has three states: (1,0,0), (0,1,0) and (0,0,1); the level $2\hbar\omega$ ($n_i=2$) has six states: (2,0,0), (0,2,0), (0,0,2), (1,1,0), (1,0,1) and (0,1,1); the level $3\hbar\omega$ ($n_i=3$) has ten states: (3,0,0), (0,3,0), (0,0,3), (2,1,0), (2,0,1), (1,2,0), (0,2,1), (1,0,2), (0,1,2), and (1,1,1). Generalizing, for 3D IR, the level $n\hbar\omega$ has $(n+1)(n+2)/2$ states.

Then, **for the molecule as a whole**:

- The ***ground state*** is **not** degenerate.
- Each **fundamental level** has degeneracy equal to the degeneracy of its frequency.
- For **overtones**: if the frequency belongs to a 1D IR, there is always only one state; for 2D or 3D IRs, each n_i gives (n_i+1) or $(n_i+1)(n_i+2)/2$ states, respectively.
- Finally, for the **combination** levels, the various distinct component frequencies are considered separately, and the relevant degeneracies are multiplied following what was seen for the previous cases.

For example, NH_3 has a total of four distinct characteristic frequencies and, in particular, two normal modes belonging to two distinct 1D IRs and the four modes belong to two distinct 2D IRs of species E, *i.e.* in total two distinct non-degenerate frequencies and two distinct frequencies, each doubly degenerate. So, considering the symmetry species in the order 1D, 1D, 2D, 2D, we have for example that the degeneracy of the combination level (1, 0, 1, 2), is 6 as (a) the fundamental level ($n_1=1$) of the first 1D IR has degeneracy 1, (b) the ground state ($n_2=0$) of the second 1D IR has degeneracy 1, (c) the fundamental level ($n_3=1$) of the first 2D IR has degeneracy 2, (d) the overtone $n_4=2$ of the second 2D IR has degeneracy 3, and then $1\times1\times2\times3=6$.

6.2.5.1 Anharmonicity

Vibrational spectroscopy **experimentally** shows, both in terms of energy levels and selection rules, that the effect of *anharmonicity* is **non**-negligible. Concerning *selection rules*, anharmonicity allows transitions in which two or more vibrational quanta of the same or different normal modes are formed (or destroyed): examples of this are *overtones* and *combination* transitions, which— not allowed in the harmonic approximation—often exhibit *significant intensities*. Regarding the *energy levels*, experimentally it is found, for example, that in absorption, *overtone* transitions *do not* occur at frequencies that are *exactly* multiples of the frequencies of the fundamental transitions, and that combination transitions *do not* occur at frequencies that are exactly the *integer combinations* of the frequencies of the fundamentals. Similar considerations can also be made by analysing the frequencies of *hot lines*.

Despite all these problems, the harmonic approximation maintains an essential role not only as a first approximation to the most accurate quantitative treatment, but above all for the fact - already anticipated in Section 2.3.2.1- that **the eigenfunctions of the harmonic oscillator represent the most appropriate basis set to account for anharmonicity** since:

> Correct vibrational eigenfunctions, approximate harmonic functions to be corrected and corrections for anharmonicity all belong to the same symmetry species.

It can be added that the eigenfunctions of the harmonic Hamiltonian constitute a complete basis set. Hence,

> Using the harmonic approximation one gets the correct information on the *symmetry species* of the vibrational functions.

6.2.5.2 The Symmetry of the Wavefunctions

Concerning the symmetry of the wave functions, it is convenient to first look at the *symmetries of the functions of a specific level of each frequency* and then derive the symmetry of the global wave function of the molecule as their product.

As already seen in Section 2.3.2.1, for **1D IR**, single-mode vibrational functions are totally symmetric if n_i is even (zero included) and belong to the IR of the normal coordinate if n_i is odd. For the m normal modes of an m-dimensional ($m>1$) IR we will consider the sum of the squares of the m coordinates. This sum is the harmonic approximation of the ground state and is invariant under the symmetry operations of the system, because the potential energy is given precisely by the sum of the squares of the coordinates: if the potential energy as a function of the coordinates changed as a result of a transformation of coordinates, the transformation would not be a symmetry operation of the system. Therefore, *the ground state single frequency wavefunction* (no phonons) ***is** totally symmetric,* even when the frequency does not correspond to a 1D IR. Therefore, the *ground state* of the molecule is always *totally symmetric***,** *even when there are degenerate normal modes***.** In fact, it is the product of only totally symmetric IRs. Furthermore, for all the other levels, it is necessary to examine *only the excited frequencies*, since the functions of the non-excited frequencies contribute a totally symmetric factor and therefore do not modify anything.

Therefore, for **fundamental levels,** where **only one phonon** exists, the representation of the molecular state is the IR of the only wave function with quantum number=1, *i.e.* the IR of the corresponding normal coordinate, whatever its dimension.

More difficulties are encountered with **overtone** levels. We need to make a distinction.

A. The case in which the frequency **is not** degenerate, *i.e.* it corresponds to a 1D IR, has already been seen: n_i even → totally symmetric molecular wavefunction; n_i odd → molecular wavefunction belonging to the IR of the excited normal mode. For example, the level (3 0 0 0) in NH_3 (three phonons at the frequency ω_1, no phonons for the other frequencies) belongs to the A_1 IR, because this is the IR of the relative normal coordinate, because it is an odd overtone and all vibrational functions of the other modes are totally symmetric because they have no phonons.
B. The case in which the frequency is degenerate, that is, it corresponds to a 2D or 3D IR, is much more complex and does not lead to a simple result but requires a *detailed and specific treatment* which is left to specialistic texts

For example, let's consider some overtones of one of the two twofold degenerate frequencies of NH_3. Without derivation, the overtone (0 0 2 0) with three states belongs to the $A_1 \oplus E$ RR; overtone (0 0 3 0) with four states belongs to the $A_1 \oplus A_2 \oplus E$ RR; overtone (0 0 4 0) with five states belongs to the $A_1 \oplus E \oplus E$ RR. Note for partial verification that the dimensions of the RR coincide with the numbers of states.

Finally, for **combination levels**, in which at least two different frequencies are excited, the level belongs to the representation that is the product of the IRs of the combining levels, regardless of whether (i) they involve two or more frequencies, (ii) whether these frequencies are degenerate or not, and (iii) whether they are fundamental levels or overtones. Naturally, if the combination includes overtone levels, calculating the contributions of these levels involves all the difficulties just described for overtones.

The simplest case is when all the excited frequencies are non-degenerate, and the level is non-degenerate. Another simple case is when the levels to be combined are only fundamental levels (no phonons or a single phonon for each different frequency).

As an example of a level **combining** two overtones, we can consider the level (0 0 2 2) of NH_3. We have already seen that overtone (0 0 2 0) with three states belongs to the $A_1 \oplus E$ RR ; this also holds for overtone (0 0 0 2). The direct product $(A_1 \oplus E) \otimes (A_1 \oplus E)$ gives $A_1 \otimes (A_1 \oplus E) \oplus E \otimes (A_1 \oplus E) \sim (A_1 \otimes A_1) \oplus (A_1 \otimes E) \oplus (E \otimes A_1) \oplus (E \otimes E)$. The direct product of A_1 for any IR X leaves the IR unchanged, while $(E \otimes E)$ gives $A_1 \oplus A_2 \oplus E$. Therefore, $(A_1 \oplus E) \otimes (A_1 \oplus E) \sim 2\,A_1 \oplus A_2 \oplus 3\,E$. The result is a 9D RR, in agreement with the threefold degeneracy of both the starting overtones.

For the application to vibrational transitions, see Section 6.5.1.

6.2.6 Final Remarks

We have already noted in Section 2.2.3, for commutative groups, that symmetry allows us to **characterize the solutions** of motion equations **without the need to solve the equations of motion**. Obviously, it is necessary to solve the equation of motion when the objective is the construction of MOs, a problem that is undoubtedly relevant to discussing many aspects of the chemistry of a molecular species. In this case, symmetry allows (i) to directly derive the relative contributions of OAs that are equivalent by symmetry and (ii) to simplify the solution by transforming the Hamiltonian matrix into a block diagonal matrix, a simplification that is essential in the case of translational symmetry, as seen in Section 3.2. However, when it comes to obtaining the properties of electronic or vibrational states, in order to deduce - in particular - which radiative transitions between these states are allowed, what is needed is to know the symmetry species, which can be obtained simply by processing the IRs (of the various sets of equivalent localized functions, of the MOs and normal modes, and finally of the polyelectronic or vibrational states) without the need to explicitly specify the various SALCs and the solutions to the equations of motion. To illustrate this type of use of symmetry, in the treatment of vibrational motions of Sections 6.2.4 and 6.2.5 only the symmetry species are involved. Note in particular that the whole treatment (i) is completely independent of the harmonic approximation and therefore is independent of its limits and (ii) is even more synthetic than what was seen for the analogous problem of the vibrations of the water molecule in the case of commutative groups (Section 2.2.3.1).

The case of the symmetry species of **open-shell polyelectronic functions** and of **vibrational functions with multiple excited normal modes** present remarkable similarities that can be easily grouped together considering that they are two cases of systems in which a wave function for *several indistinguishable particles* (electrons or phonons) is written as a product of single-particle functions. In both cases, the simple rule that the representation of the product is the product of the IRs of the factors *usually* applies, except when **single-particle degenerate levels** have occupancy > 1, *e.g.* the cases of a *degenerate MO level occupied by at least two electrons* or a *degenerate frequency with at least two phonons.*

The **vibrational functions of an entire molecule are symmetric under permutation of phonons**: this is somewhat hidden by the language most frequently used. Speaking explicitly of particles makes things simpler. For example, in the case already seen of the (0020) overtone with three states, we say that two phonons with the third frequency can occupy (i) both phonons the first state of the 2D IR, (ii) both phonons the second state of the IR, or (iii) one phonon - it does not matter which – is in the first state and one is in the second state. We can therefore see the indistinguishability of the two phonons in all three cases: it makes no sense to distinguish between the situations "phonon a in state 1 and phonon b in state 2" and "phonon a in state 2 and phonon b in state 1".

The difficulties encountered for the two different problems (electronic states and vibrations) are due to the same cause, the Pauli principle, which requires that the global function be antisymmetric (electrons) or symmetric (phonons), *limiting the number of acceptable products between the IRs of the factors.* There are no elementary solutions to these situations, and it is necessary to resort to specialized treatments.

6.3 Relations between Groups

6.3.1 IR of the Direct Product of Groups

In adding new symmetry operations to a given group, an important role is played by the **direct** product $P = G \otimes H$ of two groups (see Section 4.2):

> The direct product of two representations of different groups G and H is a representation of the direct product $P = G \otimes H$

Unlike what holds for the product of IRs of the *same* group (5.2):

> The direct product of two IRs of different groups, is an IR of the direct product of the groups.

In fact, a more general property also holds:

> The IRs of the direct product of two groups are **all** given by the direct matrix products of the IRs of the two groups

6.3.2 IRs of Groups and Subgroups

In many important cases, however, direct products are **not** involved, and less restrictive relations are used. In particular, remember Curie's principle (Section 1.4). A fundamental aspect is the group/subgroup relationship and therefore a first requirement (see Section 4.1.2) is that:

> The order of the subgroup must be a divisor of the order of the group.

A second important relationship is between the **IRs of the group and the subgroup**. Here, two cases can be considered.

A. The first case essentially has been already seen. It occurs when there is **no** reduction in degeneracy when moving from the group to the subgroup and the characters allows us to correlate the IRs of the group and those of the subgroup. A simple example is discussed in the following Section 6.3.3.
B. The most interesting case, however, is when an **IR of the group** has a **degeneracy** that **does not occur** for the **IRs of the subgroup** (*symmetry ascent/descent*). Then, the relation between the groups and their IRs is found:
   a. by forming an RR of the *subgroup* using the characters (of the group) only of the classes that are also present in the subgroup and then
   b. decomposing this RR into the sum of the IRs of the subgroup: the IRs resulting from the decomposition correspond to the IR of the subgroup.

TABLE 4.30
Character Tables for Relating the IRs of Group C_{4v} to the IRs of Its C_{2v} Subgroup

C_{4v}	E	C_2	$2\sigma_v$	$2C_4$	$2\sigma_d$	C_{2v}	E	C_2	σ_v	σ_v'
A_1	1	1	1	1	1	A_1	1	1	1	1
A_2	1	1	−1	1	−1	A_2	1	1	−1	−1
B_1	1	1	1	−1	−1	B_1	1	−1	1	−1
B_2	1	1	−1	−1	1	B_2	1	−1	−1	1
E	2	−2	0	0	0					
X	2	−2	0							

Here is an example for C_{4v} and its subgroup C_{2v}.

For a comparison we must limit ourselves to the first three classes of group C_{4v}, that is, those that contain the symmetry operations **also** present in the subgroup. Note that **one** class ($2\sigma_v$) of C_{4v} corresponds to **two** classes (σ_v and σ_v') of the C_{2v} subgroup. We see that the A_1 and A_2 IRs of C_{2v} have total correspondence with the IRs of the same name of C_{4v}. In both cases, the characters of the classes σ_v and σ_v' of C_{2v} are the same as the characters of the class $2\sigma_v$ of C_{4v}. The same A_1 and A_2 IRs of C_{2v} have total correspondence **also** with the B_1 and B_2 IR of C_{4v}. Group C_{4v}, which is *not* commutative, has a 2D IR (E in the Table 4.30). For this IR, limiting to the first three classes, the characters are 2, –2, 0, as shown by the row labelled X in the table. We use these characters of C_{4v}, and we treat them as a RR of subgroup C_{2v} assigning the value 0 to both classes of C_{2v} [$\{\sigma_v\}$ and $\{\sigma_v'\}$] corresponding to the class $\{2\sigma_v\}$ of C_{4v}. We find that this RR of C_{2v} is given by $B_1 \oplus B_2$. The result is obtained directly by adding the characters of B_1 and B_2 of C_{2v}, without using Eq. (5.27). The meaning is then that, in the *symmetry descent* ($C_{4v} \rightarrow C_{2v}$): $A_1 \rightarrow A_1 \quad A_2 \rightarrow A_2 \quad B_1 \rightarrow A_1 \quad B_2 \rightarrow A_2 \quad E \rightarrow B_1 \oplus B_2$.

In particular, the twofold degenerate level corresponding to the E IR of C_{4v}, is split into **two levels,** to which belong **functions of species B_1 and B_2**.

In the *symmetry ascent* ($C_{4v} \leftarrow C_{2v}$):

$$A_1, B_1 \leftarrow A_1 \quad A_2, B_2 \leftarrow A_2 \quad E \leftarrow B_1 \oplus B_2$$

Thus, the transition to a more symmetric situation keeps the symmetry states A_1 and A_2 non-degenerate, while states with symmetry B_1 and B_2 merge in a single twofold degenerate level.

An ample depiction of the *symmetry descent* from different point groups (and also from group SO(3), see Section 6.4.4) is available in an extensive compilation (see footnote 10).

6.3.3 Use of a Smaller Group

As seen in Section 6.2.1.6, the calculation of the SALCs of the $2p_z$ - C AOs of benzene can be simplified performing the calculations in C_{6v} (12 symmetry operations, see Table 4.25) and not in D_{6h} (which is the correct symmetry group and has 24 symmetry operations, see Table 4.31).

TABLE 4.31
Characters of $D_{6h} = C_{6v} \times C_i$

D_{6h}	E	$2C_6$	$2C_3$	C_2	$3\sigma_v$	$3\sigma_d$	i	$2S_3$	$2S_6$	σ_h	$3C_2'$	$3C_2''$
A_{1g}	+1	+1	+1	+1	+1	+1	+1	+1	+1	+1	+1	+1
A_{2g}	+1	+1	+1	+1	−1	−1	+1	+1	+1	+1	−1	−1
B_{1g}	+1	−1	+1	−1	−1	+1	+1	−1	+1	−1	+1	−1
B_{2g}	+1	−1	+1	−1	+1	−1	+1	−1	+1	−1	−1	+1
E_{1g}	+2	+1	−1	−2	0	0	+2	+1	−1	−2	0	0
E_{2g}	+2	−1	−1	+2	0	0	+2	−1	−1	+2	0	0
A_{1u}	+1	+1	+1	+1	−1	−1	−1	−1	−1	−1	+1	+1
A_{2u}	+1	+1	+1	+1	+1	+1	−1	−1	−1	−1	−1	−1
B_{1u}	+1	−1	+1	−1	+1	−1	−1	+1	−1	+1	+1	+1
B_{2u}	+1	−1	+1	−1	−1	+1	−1	+1	−1	+1	−1	+1
E_{1u}	+2	+1	−1	−2	0	0	−2	−1	+1	+2	0	0
E_{2u}	+2	−1	−1	+2	0	0	−2	+1	+1	−2	0	0

This can now be understood by exploiting the relation between D_{6h} and its subgroup C_{6v}.

Each IR of the least symmetric group (Table 4.25) corresponds – in the most symmetric group (Table 4.31) - to two IRs that are are different only in the character sign of the "added" operations, that is the operations given by the combination of C_i with each operation of C_{6v}. The sign is specified by the subscript g (*gerade*=even) or u (*ungerade*=odd). For example, the B_2 IR of C_{6v} corresponds to the B_{2g} and B_{2u} IR of D_{6h}: in B_{2g} and B_{2u}, all operations (classes) common to the two groups have the same characters, the classes pertinent only to D_{6h} have characters with opposite signs passing from B_{2g} to B_{2u}.

Each operation common to the two groups (R) has a corresponding one ($R'=iR$) among the operations specific to D_{6h}. In the SALC procedure based on characters (Eq. 6.15), each of the results ($R\phi$) of the applicaton of the operation R on an AO ϕ must be multiplied by the character: $\chi \cdot R\phi$. Now, the $2p_z$ AOs change in sign under i; so, each operation (R) of the left side of Table 4.31 produces the result of the corresponding operation (R') on the right side **with changed sign** due to the component i in $R'=R \cdot i(\psi=R\phi=-R'\phi=-\psi)$. Therefore, with all the IR with the g subscript, the sum of the various terms ($\chi_R \cdot R\phi$) provides an identically null result $\chi_R \cdot \psi - \chi_{R'} \cdot \psi = 0$ because the characters for R and for R' are equal.

To have a non-zero result starting from a $2p_z$ AO we need to use the IRs labelled with u, where the results of the right and left sides are added because the characters for R and R' have opposite signs: $\chi_R \cdot \psi - \chi_{R'} \cdot \psi = 2\chi_R \cdot \psi$. It is also seen - and this is the point of interest - that **using only the characters on the left side give the correct result.** The factor 2 is irrelevant because the SALCs must be normalized anyway. At this point, using only the characters on the left side makes no difference whether you refer to the original group or the subgroup C_{6v}. **The consideration can be easily extended to the various direct products of the previous examples in which the symmetry operations and the IRs are doubled in the product group.**

6.4 Continuous Groups

Point groups include **continuous** groups (see Section 1.1.5), such as the group $C_{\infty v}$ of linear molecules without inversion, the group $D_{\infty v}$ of linear molecules with inversion and reflection in the plane perpendicular to the molecular axis, and the group of rotations of a sphere, SO(3).

6.4.1 The $C_{\infty v}$ Group

The symmetry operations of group $C_{\infty v}$ are **rotations** C_n^k ($n=2, 3, \ldots$; $k=1, \ldots, n-1$) around the molecular axis and **reflections** with respect to any **vertical** plane σ_v. To get an idea of the group, it is useful to examine the case of groups C_{nv} (n finite, rotation of an angle $\varphi=2\pi/n$): there are n vertical planes, $(n-1)$ distinct rotations, and $2n$ symmetry operations, including identity. The group is *not* commutative. For what concerns classes, considering in particular an odd n, we have: a class of n vertical planes (labelled $n\sigma_v$), m classes [$m=(n-1)/2$] of two rotations each, one clockwise and one counterclockwise of an angle $j\varphi$; $j=1, \ldots, m$; the classes are labelled $2C_n$, $2(C_n)^2$, …, $2(C_n)^m$. Therefore $(n-1)/2+2=m+2$ classes, including the identity E that is in a class of its own. The IRs are A_1, A_2, and m 2D IR. Among these IRs, A_1 is totally symmetric. A_2 has characters 1 for E, 1 for $2C_n$, …, 1 for $2(C_n)^m$, −1 for $n\sigma_v$. E_1 has characters: 2 for E, $2\cos\varphi$ for $2C_n$, $2\cos2\varphi$ for $2(C_n)^2$, …, $2\cos(m\varphi)$ for $2(C_n)^m$, 0 for $n\sigma_v$. E_2 has characters: 2 for E, $2\cos2\varphi$ for $2C_n$, $2\cos3\varphi$ for $2(C_n)^2$, …, $2\cos(\varphi)$ for $2(C_n)^m$, 0 for $n\sigma_v$… E_m has characters: 2 for E, $2\cos(m\varphi)$ for $2C_n$, $2\cos\varphi$ for $2(C_n)^2$, …, $2\cos(m-1)\varphi$ for $2(C_n)^m$, 0 for $n\sigma_v$.

Note that $2\cos\varphi=e^{i\varphi}+e^{-i\varphi}$, as already seen for commutative groups C_n with complex eigenvalues (Section 2.1.4). However, IR pairs with conjugate eigenvalues in C_n are degenerate due to *time reversal* symmetry but in C_{nv} are degenerate due to point symmetry.

TABLE 4.32
The Character Table of Continuous Group $C_{\infty v}$

$C_{\infty v}$	I	$2C_\infty$	$2(C_\infty)^2$	…	$\infty\sigma_v$	Functions
$A_1=\Sigma^+$	1	1	1	1	1	z; (x^2+y^2); z^2; z^3; $z(x^2+y^2)$
$A_2=\Sigma^-$	1	1	−1	1	−1	R_z
$E_1=\Pi$	2	$2\cos\varphi$	$2\cos(2\varphi)$	…	0	(x, y); (R_x, R_y); (xz, yz); (xz^2, yz^2) $[x(x^2+y^2), y(x^2+y^2)]$
$E_2=\Delta$	2	$2\cos(2\varphi)$	$2\cos(4\varphi)$	…	0	(x^2-y^2, xy); $[xyz, z(x^2-y^2)]$
$E_3=\Phi$	2	$2\cos(3\varphi)$	$2\cos(6\varphi)$	…	0	$[y(3x^2-y^2), x(x^2-3y^2)]$
…	2	…	…	…	0	
E_n	2	$2\cos(n\varphi)$	$2\cos(2n\varphi)$	…	0	
…	2	…	…	…	0	

For $C_{\infty v}$, the classes are E (identity), $2C_\infty$, an infinite number of other classes $2(C_\infty)^2$, … and the class of vertical planes $\infty\sigma_v$. The IRs and the characters are (see Table 4.32): A_1 (or Σ^+) totally symmetric: 1 (all classes); A_2 (or Σ^-): 1(E), 1 (all classes of rotations), −1 ($\infty\sigma_v$); E_1 (or Π): 1(E), $2\cos\varphi$, $2\cos2\varphi$, … (all classes of rotations), 0 ($\infty\sigma_v$); E_2 (or Δ): 1(E), $2\cos2\varphi$, $2\cos4\varphi$, … (all classes of rotations), 0 ($\infty\sigma_v$); E_2 (or Φ): 1(E), $2\cos3\varphi$, $2\cos6\varphi$, … (all classes of rotations), 0 ($\infty\sigma_v$) and so on…

6.4.2 The $D_{\infty h}$ Group

In addition to the operations of $C_{\infty v}$, group, $D_{\infty h}$ contains inversion (i), improper rotations S_n and binary rotations C_2' around an axis in the horizontal plane (and therefore perpendicular to C_n and S_n). Table 4.33 reports the characters and shows that the relationship between $C_{\infty v}$ and $D_{\infty h}$ is

TABLE 4.33
The Character Table of Continuous Group $D_{\infty h}$

$D_{\infty h}$	E	$2C_\infty$	…	$\infty\sigma_v$	i	$2S_\infty$	…	$\infty C_2'$	Functions
$A_1=\Sigma_g^+$	1	1	…	1	1	1	…	1	(x^2+y^2); z^2;
$A_2=\Sigma_g^-$	1	1	…	−1	1	1	…	−1	R_z
$E_1=\Pi_g$	2	$2\cos\varphi$	…	0	2	$-2\cos\varphi$	…	0	(R_x, R_y); (xz, yz)
$E_2=\Delta_g$	2	$2\cos(2\varphi)$	…	0	2	$2\cos(2\varphi)$	…	0	(x^2-y^2, xy)
$E_3=\Phi_g$	2	$2\cos(3\varphi)$	…	0	2	$-2\cos(3\varphi)$	…	0	
…	2	…	…	0	2	…	…	0	
E_{ng}	2	$2\cos(n\varphi)$	…	0	2	$(-1)^n\cdot2\cos(n\varphi)$	…	0	
…	2	…	…	0	2	…	…	0	
$A_1=\Sigma_u^+$	1	1	…	1	−1	−1	…	−1	z; z^3; $z(x^2+y^2)$
$A_2=\Sigma_u^-$	1	1	…	−1	−1	−1	…	1	
$E_1=\Pi_u$	2	$2\cos\varphi$	…	0	−2	$2\cos\varphi$	…	0	(x, y); (xz^2, yz^2); $[x(x^2+y^2), y(x^2+y^2)]$
$E_2=\Delta_u$	2	$2\cos(2\varphi)$	…	0	−2	$-2\cos(2\varphi)$	…	0	$[xyz, z(x^2-y^2)]$
$E_3=\Phi_u$	2	$2\cos(3\varphi)$	…	0	−2	$2\cos(3\varphi)$	…	0	$[y(3x^2-y^2), x(x^2-3y^2)]$
…	2	…	…	0	−2	…	…	0	
E_n	2	$2\cos(n\varphi)$	…	0	−2	$-(-1)^n\cdot2\cos(n\varphi)$	…	0	
…	2	…	…	0	−2	…	…	0	

essentially the same as the relationship between the C_{6v} and $D_{6h} = C_{6v} \otimes C_i$ (Section 6.3.3).

The IRs of $C_{\infty v}$ and $D_{\infty h}$ are labelled with **capital Greek letters** ($\Sigma\ \Pi\ \Delta\ \Phi$, …) corresponding to the Latin letters of the hydrogenic AOs (s, p, d, f and then g, h, i, … in alphabetical order). The choice reflects the fact that the IRs include planar harmonic functions, just as the angular part of the AOs consists of spherical harmonics.

6.4.3 Symmetry Descent from $C_{\infty v}$ and $D_{\infty h}$ Groups

Despite their incomplete form, these character tables for groups $C_{\infty v}$ and $D_{\infty h}$ can be easily used in applications because in most cases only the first IRs have physical meaning, and it is convenient to refer to the *characteristic functions* to find the IRs of a finite subgroup that correspond to a given IR of group $C_{\infty v}$ or $D_{\infty h}$. For example, group C_{2v} (Table 4.21) is a subgroup of $C_{\infty v}$. Examining Table 4.32, let's consider various characteristic functions:

- z in Σ^+ of $C_{\infty v}$: in C_{2v} we have $z \rightarrow A_1$, therefore $\Sigma^+ \rightarrow A_1$;
- R_z in Σ^- of $C_{\infty v}$: in C_{2v} we have $R_z \rightarrow A_2$ therefore $\Sigma^- \rightarrow A_2$;
- (x, y) in Π of $C_{\infty v}$: in C_{2v} we have $x \rightarrow B_1$ and $y \rightarrow B_2$; the same result would be obtained by considering (R_x, R_y) or (xz, yz).
- (x^2-y^2, xy) in Δ of $C_{\infty v}$: in C_{2v} we have $x^2, y^2 \rightarrow A_1$ and $xy \rightarrow A_2$.

Thus, there is degeneracy removal for $\Pi \rightarrow B_1 \oplus B_2$ and $\Delta \rightarrow A_1 \oplus A_2$.

The case of $D_{\infty h}$ with the subgroup D_{2h} is analogous. D_{2h} is commutative and has 1D IR only, and therefore all the 2D IR of $D_{\infty h}$ are split; the correspondence is: $\Pi_g \rightarrow B_{2g} \oplus B_{3g}, \Pi_u \rightarrow B_{2u} \oplus B_{3u}, \Pi_g \rightarrow A_g \oplus B_{1g}, \Pi_u \rightarrow A_u \oplus B_{1u}$.

6.4.4 The SO(3) Group and Its Subgroups

Group SO(3) contains all the rotations by an angle $\varphi \neq 0$ ($\varphi = 2\pi/n$, n integer) around any axis passing through the origin of the spatial coordinates. The characters depend on the quantum number $l = 0, 1, 2, 3, \ldots$ of the angular momentum operator and are given by:

$$\chi^l(\varphi) = \frac{\sin\left[(l+1/2)\varphi\right]}{\sin(\varphi/2)} \quad \text{or} \quad 2l+1(\text{for identity}) \qquad (6.17)$$

The IRs of the group correspond to the possible values for l, are labelled with $S, P, D, F, G, H, I, \ldots$ and the degeneracies are 1, 3, 5, 7, 9, 11, 13, ….

Compared to the use of characteristic functions, Eq. (6.17) provides a more direct and convenient method for finding the correspondences between the IRs of SO(3) and those of a subgroup. As an example, consider the relationship with the octahedral group O (Table 4.34).

TABLE 4.34
Characters of Group O:

O	E	$8C_3$	$6C_2'$	$6C_4$	$3(C_4)^2$	Functions
A_1	1	1	1	1	1	$(x^2+y^2+z^2)$
A_2	1	1	−1	−1	1	
E	2	−1	0	0	2	$(x^2-y^2)\ (3z^2-r^2)$
T_1	3	0	−1	1	−1	$(x, y, z)\ (R_x, R_y, R_z)$
T_2	3	0	1	−1	−1	(xy, yz, zx)

In SO(3), the characters for quaternary, ternary and binary orbitals and rotations are:

AO $p(l = 1)$:
$E \rightarrow 2l+1 = 3 \qquad \rightarrow \chi = 3$
$C_4 \rightarrow \varphi = 2\pi/4, \sin(3\varphi/2) = \sin(\varphi/2) \qquad \rightarrow \chi = 1$
$C_3 \rightarrow \varphi = 2\pi/3, \sin(3\varphi/2) = 0 \qquad \rightarrow \chi = 1$
$C_2 \rightarrow \varphi = 2\pi/2, \sin(3\varphi/2) = -1, \sin(\varphi/2) = 1 \qquad \rightarrow \chi = 0$

These characters correspond to the T_1 IR.

AO $p(l = 2)$:
$E \rightarrow 2l+1 = 5 \qquad \rightarrow \chi = 5$
$C_4 \rightarrow \varphi = 2\pi/4, \sin(5\varphi/2) = -\sin(\varphi/2) \qquad \rightarrow \chi = -1$
$C_3 \rightarrow \varphi = 2\pi/3, \sin(5\varphi/2) = -\sin(\varphi/2) \qquad \rightarrow \chi = -1$
$C_2 \rightarrow \varphi = 2\pi/2, \sin(5\varphi/2) = \sin(\varphi/2) \qquad \rightarrow \chi = 1$

These characters correspond to $E \oplus T_2$.

Finally, for the f AO one finds that the 7D RR is given by $A_1 \oplus T_1 \oplus T_2$.

6.5 Spectroscopic Transitions in Molecules

The application of symmetry to spectroscopic transitions is now automatic, since it is simply a matter of drawing conclusions without any additional difficulty compared to what already seen with regard to the symmetry of electronic (Section 6.2.3) or vibrational (Section 6.2.5.2) states. Therefore, it is sufficient here to discuss some examples.

6.5.1 Combination Lines in Vibrational Absorption Spectroscopy

We have seen in Section 2.3.2 that the **harmonic oscillator model** only allows transitions in which the quantum number of **only one normal mode changes by 1**. It may therefore be interesting to derive selection rules exclusively based on the symmetry of a system and for **a transition forbidden by the harmonic oscillator rule**. Leaving aside symmetry, it is important to point out that the attention to the most excited levels is not pure curiosity: in fact, the transitions from the *ground state* to the levels with higher energies are those found at higher frequencies. Generally speaking, these transitions are visible in the *near*-IR, while the fundamentals, again in very general terms, are in the far end of the IR spectrum.

Let us consider, for example, for the *combination transition* of water, **from the ground state** to the level where there are **two phonons**, **one** *of* **symmetric stretching** (A_1) *and one of* **asymmetric stretching** (B_2). As we have seen in Section 6.2.5.2, the species of this doubly excited state is $A_1 \otimes B_2 = B_2$. We can then refer directly to Table 4.15 where, considering that the combination level belongs to the same symmetry species as the level in which only one asymmetric stretching phonon is excited, it follows directly that the transition is IR active. **This is, in fact, one of the strongest absorption lines in gas phase water.**

6.5.2 Electronic Transitions of a Monoatomic Ion in an Octahedral Field

Let us consider the ion Fe^{2+} with the $(3d)^6$ configuration. If the ion is embedded in a ligand field of octahedral symmetry, for example O_h, we have the *symmetry descent* to $E_g \oplus T_{2g}$ (see Section 6.4.3) The six electrons occupy the shell T_{2g} at lower energy, and the configuration $(t_{2g})^6$ consists of only the totally symmetric singlet state ${}^1A_1$.

Let's examine the d-d transition from the configuration $(t_{2g})^6$ to $(t_{2g})^5(e_g)^1$, having symmetry $T_{2g} \otimes E_g$. Considering Table 4.35, we have that $T_{2g} \otimes E_g = T_{1g} \oplus T_{2g}$. For confirmation, the product characters are 6, 0, 0, 0, −2, 6, 0, 0, −2, 0, the sum characters are 6, 0, 0, 0, −2, 6, 0, 0, −2, 0.

We have $6 \times 2 = 12$ states and must be a triplet and a singlet, both with spatial symmetry T.

The three vectors *x*, *y*, *z* belong to the T_{1u} IR. So the integrand of the electric dipole transition has possible symmetries given by the product of A_{1g} (initial state), $T_{1g} \oplus T_{2g}$ (final state) and T_{1u} (operator):

$A_{1g} \otimes T_{1g} \otimes T_{1u} = T_{1g} \otimes T_{1u} = A_{1u} \oplus E_u \oplus T_{1u} \oplus T_{2u}$ and $A_{1g} \otimes T_{2g} \otimes T_{1u} = A_{2u} \oplus E_u \oplus T_{1u} \oplus T_{2u}$

TABLE 4.35
Character Table for the Group O_h

O	E	$8C_3$	$6C_2'$	$6C_4$	$3(C_4)^2$	i	$6S_4$	$8S_6$	$3\sigma_h$	$6\sigma_d$	Functions
A_{1g}	1	1	1	1	1	1	1	1	1	1	$(x^2+y^2+z^2)$
A_{2g}	1	1	−1	−1	1	1	−1	1	1	−1	
E_g	2	−1	0	0	2	2	0	−1	2	0	$(2z^2-x^2-y^2)$ (x^2-y^2)
T_{1g}	3	0	−1	1	−1	3	1	0	−1	−1	(R_x, R_y, R_z)
T_{2g}	3	0	1	−1	−1	3	−1	0	−1	1	(xy, yz, zx)
A_{1u}	1	1	1	1	1	−1	−1	−1	−1	−1	
A_{2u}	1	1	−1	−1	1	−1	1	−1	−1	1	
E_u	2	−1	0	0	2	−2	0	1	−2	0	
T_{1u}	3	0	−1	1	−1	−3	−1	0	1	1	(x, y, z)
T_{2u}	3	0	1	−1	−1	−3	1	0	1	−1	

As a check, the characters of the first product are 9, 0, 1, 1, 1, −9, −1, 0, −1, −1; the characters of the first sum are 9, 0, 1, 1, 1, −9, −1, 0, −1, −1. The characters of the second product are: 9, 0, −1, −1, 1, −9, 1, 0, −1, 1; the characters of the second sum are 9, 0, −1, −1, 1, −9, 1, 0, −1, 1. Then, transitions from A_{1g} to T_{1g} or T_{2g} are forbidden by point symmetry because none of the possible products contains the totally symmetric IR.

6.5.3 Vibronic Transitions

It should not be surprising that d-d transitions are forbidden, because the result can be obtained directly from Laporte's rule (1.5). The peculiarity lies in the fact that, for example, the electronic transition from $(t_2)^6$ to $(t_2)^5(e)^1$ **is experimentally observed** (even if with significantly reduced intensity compared to *allowed* electronic transitions). The result can be explained by the coupling with a vibrational transition: this is called a *vibronic* transition. For these, we must take the ligands into account. To keep the problem at the simplest level, it is sufficient to add six identical monoatomic ligands, without the need to specify further. The complex ion therefore has seven atoms and $(7 \times 3) - 6 = 15$ vibrational degrees of freedom. The vibrational degrees of freedom are the direct sum: $D(\text{vibr}) \sim A_{1g} \oplus E_g \oplus 2T_{1u} \oplus T_{2g} \oplus T_{2u}$.

In the vibronic transition, the symmetry of the starting state can be ignored because it is totally symmetric. Final states are the product of the electronic part ($T_{2g} \otimes E_g \sim T_{1g} \oplus T_{2g}$) by the vibrational part, which in turn is given by the IR of each normal mode within *D*(vibration), assuming that the coupling with a fundamental vibrational transition is considered. Finally, it is necessary to multiply by the IR of the electric dipole operator (T_{1u}). This gives: $(T_{1g} \oplus T_{2g}) \otimes (A_{1g} \oplus E_g \oplus 2T_{1u} \oplus T_{2g} \oplus T_{2u}) \otimes T_{1u}$

The terms that multiply T_{1g} in the factor ($T_{1g} \oplus T_{2g}$) give: $(T_{1g} \oplus T_{2g}) \otimes (A_{1g} \oplus E_g \oplus 2T_{1u} \oplus T_{2g} \oplus T_{2u}) \otimes T_{1u}$, then $T_{1g} \otimes (A_{1g} \oplus E_g \oplus 2T_{1u} \oplus T_{2g} \oplus T_{2u}) \otimes T_{1u}$, then $\sim [(T_{1g}) \oplus (T_{1g} \otimes E_g) \oplus 2(T_{1g} \otimes T_{1u}) \oplus (T_{1g} \otimes T_{2g}) \oplus (T_{1g} \otimes T_{2u})] \otimes T_{1u}$, then this huge relation:

$$\sim \begin{bmatrix} (T_{1g}) \oplus (T_{1g} \oplus T_{2g}) \oplus 2(A_{1u} \oplus E_u \oplus T_{1u} \oplus T_{2u}) \oplus \\ (A_{2g} \oplus E_g \oplus T_{1g} \oplus T_{2g}) \oplus (A_{2u} \oplus E_u \oplus T_{1u} \oplus T_{2u}) \end{bmatrix} \otimes T_{1u}$$

By counting only once the IRs that appear more than once in the direct sums we have:

$$[A_{1u} \oplus E_g \oplus E_u \oplus T_{1g} \oplus T_{2g} \oplus T_{1u} \oplus T_{2u}] \otimes T_{1u}$$

We then need to consider only $T_{1u} \otimes T_{1u} \sim A_{1g} \oplus E_g \oplus T_{1g} \oplus T_{2g}$ to be able to say that the transition is allowed because its RR contains the totally symmetric IR (A_{1g}).

6.6 The Solid State beyond the Translational Symmetry

The symmetry of 3D periodic crystals combines (in a sense to be specified) *finite* point groups with the 3D translational symmetry discussed throughout Section 3, forming an algebraic structure known as the *space* group (see Section 6.6.4).

Combining, in the same symmetry group, point operations and translations is much more complex than combining only point operations, as is done, for example, when discussing the expansion of a point group. The important difference with respect to combining only point symmetries is that now the combination produces symmetry operations that *generally* do not leave any point in the 3D space unchanged.

To understand the results of the combination of point operations with translations, **three** specific aspects must be examined, namely the **constraints** on the operations of one type produced by the request for compatibility with the symmetry operations of the other type (similarly to what has already been seen for the constraints of point symmetries only) and the **new symmetry operations** originating from the combination.

6.6.1 Crystallographic Point Groups

A first aspect concerns the constraints that the presence of a **translation lattice** imposes on **point symmetries**. Starting with rotations, it is surely well known that rotations of **order 5, 7 and higher are incompatible** with translational symmetry.

A simple proof (Figure 4.35) considers two points (A and B, upper part of the figure) connected by a lattice vector $\vec{a}$. Rotating B by an angle $\alpha>0$ (counterclockwise) with centre A we obtain B′; rotating A by an angle $-\alpha$ (clockwise) with centre B we similarly obtain A′. If the rotations are symmetry operations of the group, the segment A′B′ is a lattice translation $\vec{l}$ parallel to $\vec{a}$ (= an integer number of $\vec{a}$ translations) and in ABA′B′ three sides (AB, AB′ and BA′) are $=|\vec{a}|$. We see that it can be $l=|\vec{l}|=a$ (and then B′ABA′ is a straight line), $l=2a$, $l=a$ (and B′ABA is a square) or finally $l=0$ (and the B′ABA is a triangle). In other words, $2\cos\alpha=|\vec{l}-\vec{a}|/|\vec{a}|$ must be an integer and vary between 2 and –1. Then, $2\cos\alpha=2$ gives $\alpha=180°$, $2\cos\alpha=1$ gives $\alpha=60°$, $2\cos\alpha=0$ gives $\alpha=90°$, $2\cos\alpha=-1$ gives $\alpha=120°$. These are the cases reported in the lower part of the figure and correspond to the rotations C_2, C_6, C_4 and C_3.

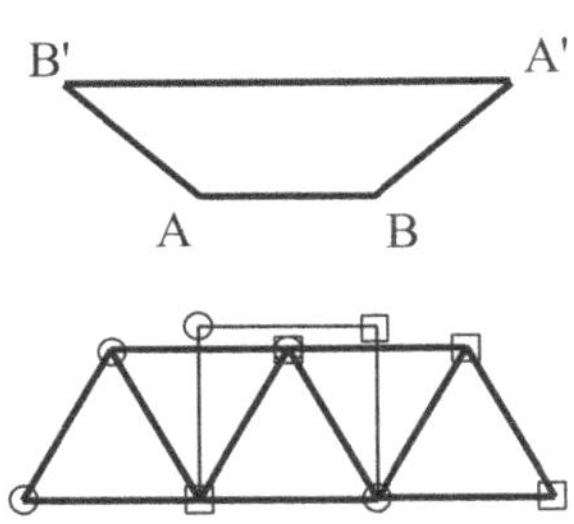

FIGURE 4.35 Compatibility of rotations with translation symmetry.

Concerning **inversion** and **reflection**, the constraints are only on the **position** of the symmetry elements (point and plane) within the unit cell. For example, an inversion centre is repeated by the translation $\vec{l}=m\vec{a}+n+o\vec{c}$, where m, n, o are positive, negative or null integers (Eq. 3.1), but also by the translations $1/2\vec{a}$ or $1/2\vec{b}$ or $1/2\vec{c}$. The common choice (but not the only one!) is to place inversion centres on lattice points and on the middle points of the sides of the unit cell.

In conclusion, among the point groups (Section 1.1.4) to be considered for a crystal structure, we find:

- 22 groups C_n, C_{nv}, C_{nh}, D_n, D_{nh} with $n=1, 2, 3, 4, 6$. We have 22 groups and not 25 as (i) C_{1v} coincides with C_{1h} (and both are preferentially referred to as C_s), (ii) D_1 coincides with C_2, (iii) D_{1h} coincides with C_{2v}.
- Two groups D_{2d}, D_{3d}. The D_{1d} group coincides with C_2, while D_{4d} and D_{6d} are to be rejected because they implicitly contain the improper rotations of order 8 and 12.
- Three groups S_2, S_4, S_6, (S_2 is preferentially referred to as C_i).
- Three tetrahedral groups T, T_h, T_d
- Two octahedral groups O, O_h.

The total amounts to 32 *crystallographic point groups*.

6.6.2 Bravais Lattices

The second point, on the contrary, concerns the **actual values** of the angles and the ratios between the lengths of the sides of the unit cell that **a lattice must present in order to be compatible with** the presence of **rotation** or **reflection** operations (inversion symmetry does not pose this problem). There are only 14 possible results, under a uniqueness constrain to be better explained later and are called **Bravais** lattice, and they are reported in all textbooks. It is noteworthy that Bravais lattices were historically obtained well

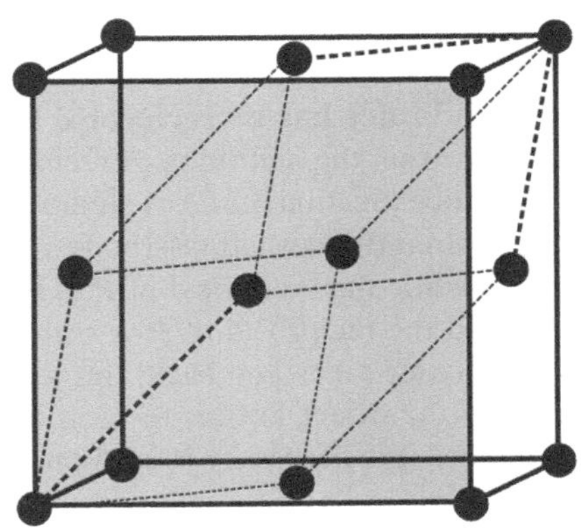

FIGURE 4.36 The F cubic lattice. The full circles are the lattice points. The cube is a fourfold cell, the rhombohedron is a primitive cell.

before the birth of structural diffraction and even before the complete acceptance of the atomic hypothesis.

The fact that the same technical expressions (face-centred cubic, body-centred tetragonal, …, primitive monoclinic) are used to specify both **cells** (primitive or multiple) and **lattices** requires due attention because it is open to misunderstandings. For example, the ***face-centred cubic* Bravais lattice** (F) is *usually* described by a unit cell, also called face-centred cubic, with edges $a=b=c$ and angles $\alpha=\beta=\gamma=90°$. As shown in Figure 4.36, using basis vectors $\vec{a}$, $\vec{b}$, $\vec{c}$ corresponding to the sides of this cell, the lattice points (full circles) are not *all* described by Eq. (3.1): $\vec{l} = m\vec{a} + n\vec{b} + o\vec{c}$, where m, n, o are three positive, negative or null numbers. We must add the vectors $\vec{A} = \left(\vec{b} + \vec{c}\right)/2$, $\vec{B} = \left(\vec{c} + \vec{a}\right)/2$ and $\vec{C} = \left(\vec{a} + \vec{b}\right)/2$ and we see that the cubic cell is quadruple. Alternatively, the **same face-centred cubic Bravais lattice** (F) *can* be described by a basic triplet consisting of the three vectors $\vec{A}$, $\vec{B}$, $\vec{C}$. The latter choice, $\vec{l} = m\vec{A} + n\vec{B} + o\vec{C}$, where m, n, o are all positive, negative or null integers, describes **all the lattice points**. On the other hand, the unit cell built using the basis vectors $\vec{A}$, $\vec{B}$, $\vec{C}$, is **primitive** and is not a cube (it *has a rhombohedral shape*, shown by the dotted edges in the figure). This *does not* make the symmetry of the lattice *immediately evident*: for example, there are three *equivalent* orthogonal directions, which are not along the vectors $\vec{A}$, $\vec{B}$, $\vec{C}$. So, the **Bravais** cubic face centred lattice (F) can be described by different choices of the unit cells, but all these choices describe **the same lattice**, and this is one of the 14 **Bravais lattices.** This is one example of the uniqueness constraint said above.

Note that the Wigner-Seitz cell (Section 3.1.2) does not show this kind of problems.

Let us now consider a *different case*. **For the primitive cubic lattice** (P) is *almost exclusively* used a cubic cell, with edge $a=b=c$ and angles $\alpha=\beta=\gamma=90°$. Nothing would prevent us from using a different primitive parallelepiped cell with different edges and/or angles other than 90° or a multiple cell. This is not *usually* done, because, apart from very specific technical problems, this choice does not present any advantage: the cubic symmetry of the lattice is well highlighted by the cubic cell and the lattice points are all described by Eq. (3.1), $\vec{l} = m\vec{a} + n\vec{b} + o\vec{c}$, with m, n, o are all positive, negative or null integers.

Every Bravais lattice has its reciprocal lattice, as it should be obvious from the definition of reciprocal triplet and reciprocal lattice (Section 3.2.2). A situation to note is the case of the centred Bravais lattices (body-centred I and face-centred F): while the reciprocal of P is P, the reciprocal of I is F and the reciprocal of F is I (in fact, going to the reciprocal twice takes you back). In general, using multiple cells for the direct lattice, to each lattice point with integer indices it is necessary **to add** one (if the cell is double), two (triple cell) or three (quadruple cell) vectors with fractional indices, as seen with the example of the cubic Bravais lattice F. Conversely, by constructing the reciprocal triplet of a multiple cell, one somewhat obtains the opposite result: that is, one has a lattice in which it is necessary **to remove** lattice points from the list of those described by the formula with integer coefficients, instead of adding them.

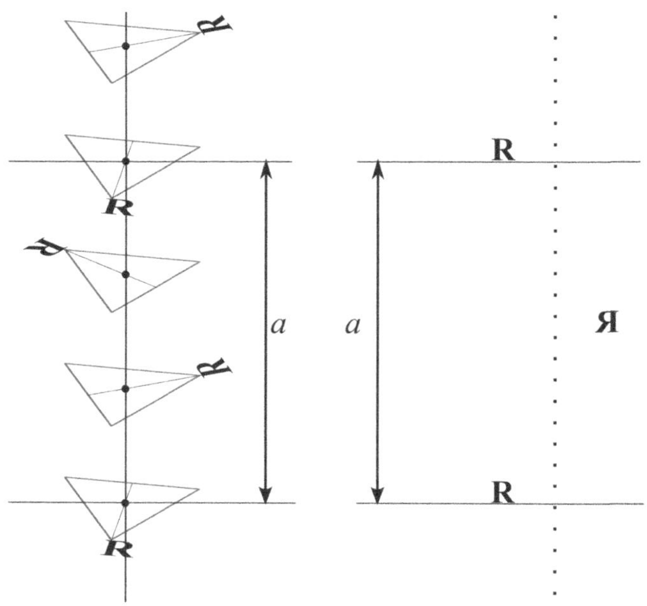

FIGURE 4.37 The screw axis 3_2 (left) and the a glide (right).

6.6.3 Glides and Screws

A third aspect to be considered is that the combination of translations with rotations or reflections can give origin to **a new kind of symmetry operations**. These are **screw axes** and **glide planes** (see Figure 4.37). These operations are obtained as a combination of rotations or reflections (R) with a translation ($\vec{s}$) which is a linear combination of the lattice vector $\vec{a}$, $\vec{b}$, $\vec{c}$, but with **fractional coefficients** (1/2, 1/3, 1/4, or 1/6 and multiples), and not with integer coefficients as in the definition of lattice by Eq. (3.1). It is worth noting that the two components (R and $\vec{s}$) - individually considered - *are not* symmetry operations of the space group: this is similar to what is already seen (Section 1.1.4) for improper rotations S_n (according to Schoenflies) where a rotation is combined with a reflection.

In the *Hermann-Maugin* notation, one has a 2_1 screw axis, which is a combination of a C_2 rotation with a 1/2 of a lattice translation along the axis direction, 3_1 and 3_2 screw axes, as combinations of anti-clockwise and clockwise C_3 rotations, respectively, combined with a one-third of a lattice translation along the axis direction, 4_1, 4_2 and 4_3, as C_4 rotations, the first anticlockwise and the last clockwise, combined with three translations by 1/4, 1/2 and ¼, and finally the 6_1, 6_2, 6_3, 6_4 and 6_5, as C_6 rotations, the first and second anticlockwise, penultimate and last clockwise, associated with translations of 1/6, 2/6, 3/6, 2/6 and 1/6. In the same notation, a glide plane is named a, b, or c when the translation component is parallel to one of the basis vectors of the lattice (with coefficient 1/2) or named n or d when the translation is along diagonal directions (with coefficients 1/2 and 1/2 or 1/4 and 1/4).

6.6.4 Space Groups

Space groups are the result of *all possible relations and compatibilities* between the set of point symmetry operations on one side and a translation lattice (Eq. 3.1), according to the three points discussed in the Sections 6.6.1, 6.6.2 and 6.6.3. They correspond to all pairs (crystallographic point group+Bravais lattice) of symmetries *compatible with each other* and - within each pair - to *all the distinct cases* originated by the replacement of rotations with screw axes and/or reflections with glide planes (if and when possible).

6.6.4.1 Seitz's Operators

More precisely and extending the definition already seen for the glide planes and screw axes, **each** symmetry operation in a space group has a point *component R* (inversion, proper or improper rotation, reflection, identity) and a translational *component* ($\vec{t}$) *referring to a* lattice, being a linear combination of the basis vectors (null vector included). The point operation R is performed first, and then the translation $\vec{t}$. The symmetry operation of the space group is the sequential execution of the two operations: as already said, *not necessarily each of the two operations is* by itself *a symmetry operation of the group.* The operator $\{R|\vec{t}\}$ performing the combined transformation is called **Seitz's operator.** With a Seitz's operator we can represent:

- Pure point transformations with $\{R|\vec{0}\}$, where $\vec{0}$ is the null vector,
- Pure translations with $\{E\,|\,\vec{l}\}$, where $\vec{l}$ is a lattice vector and E the identity,
- Screw axes and glide planes with $\{R|\vec{s}\}$, where $\vec{s}$ is a *fraction* of a lattice vector $\vec{l}$.
- All combinations of the above.

6.6.4.2 Crystal Classes

The $\{R|\vec{0}\}$ symmetry operations refer to a unique origin, keep the position of the origin unchanged and belong to one of the crystallographic point groups already seen (see Section 6.6.1). In crystallography, each crystallographic point group defines a **crystal class**. The importance of the class to which the space group belongs is that there is a **Neumann's Principle,** according to which the (tensorial) macroscopic properties of a crystal cannot have symmetry lower than that of the point group of the space group. For example, if the point group is centrosymmetric, the crystal cannot display a permanent electric dipole moment, and therefore it cannot be piezoelectric. Another example: a chiral crystal can have *proper* rotations only, screw axes included.

6.6.4.3 Symmorphic and Non-Symmorphic Space Group

The standard notation in crystallography is that of Hermann-Maugin. We can classify space groups into two different types. In **symmorphic space groups** *all* the symmetry operations have a *lattice vector* $\vec{l}$ as the translational component $\vec{t}$ in the Seitz's operator. Generating a symmorphic space group is relatively simple, and we can illustrate the procedure with an example. Suppose to consider a primitive orthorhombic lattice. The point symmetries compatible with an orthorhombic lattice are described by the groups 222 (D_2), *mm*2 (C_{2v}) and *mmm* (D_{2h}). If we now associate an object with point symmetry *mm*2 (C_{2v}) to each of the lattice points, the *Pmm*2 space group is generated. Similarly, the symmorphic space group *Cmm*2 can be obtained by associating to each point of a base-centred orthorhombic lattice *C* an object with symmetry *mm*2. It is important to note that the object is associated with *each point of the lattice*. In other words, in the case of the space group *Cmm*2, it will be necessary to specify two sets of point symmetry operations: one set is associated with the origin of the lattice, the other with the node of coordinates (½, ½, 0), which is at the centre of the *C* face.

The **non-symmorphic** space groups include symmetry operations (screw axes and glide planes) where the translational component is not a lattice translation but a **submultiple** of it, in the sense that *n* times the translation component - *n* integer - is a lattice vector. In this case, it is necessary to specify *at least one Seitz's operator* $\{R|\vec{t}\}$ where $\vec{t}$ is a fractional translation.

To obtain the non-symmorphic space groups, what we need are the 14 possible Bravais lattices and a set of symmetry operations that is isomorphic to one of the 32 crystalline classes. For example, given the crystalline class 4, the non-symmorphic space groups can have as symmetry operations only the screw axes 4_1, 4_2 and 4_3. Similarly, if in a class there are only symmetry planes, the corresponding non-symmorphic operations can only be glide planes.

Basically, we need to perform the combinations exhaustively and eliminate the redundant cases. Clearly this is a *lengthy but finite task*. For example, consider the crystal class 2 (monoclinic). We could then have either binary axes or 2_1 screw axes. Combining the two point groups with the primitive monoclinic lattice (*P*) or *B* centred monoclinic lattice (*B*), the *P*2 and *B*2 symmorphic space groups are obtained. By combining the 2_1 screw axis with *P* and *B*, we could obtain the $P2_1$ and $B2_1$ space groups. However, a detailed inspection of $B2_1$ will show that it is coincident with $P2_1$ by an origin shift. Therefore, only the three *P*2, *B*2 and $P2_1$ space groups are obtained.

Concluding,

- symmorphic space groups contain (i) lattice translation operations, (ii) point operations and (iii) their combinations.
- Non-symmorphic space groups contain (i) lattice translation operations, (ii) point operations, (iii) screw axes and/or glide planes, and (iv) combinations of all these.

For example, we have six *space groups* derived from the *point symmetry* C_{3v} indicated as C_{3v}^1, C_{3v}^2, C_{3v}^3, C_{3v}^4, C_{3v}^5, C_{3v}^6

in the Schoenflies notation, while the Hermann-Mauguin notation (*P*3*m*1, *P*31*m*, *P*3*c*1, *P*31*c*, *R*3*m* and *R*3*c*, respectively) specifies *both* the Bravais lattice and the *substitution* of reflection planes (*m*) with glide planes (*c*). Of course, we can *also* derive other space groups by substituting the C_3 axis with screw axes, but in this case the point group is different (it may be C_3 or D_3).

For the sake of completeness, it should be noted that it is often preferred to describe the Bravais lattice *R* using a *triple cell* (*H*) that has the same prismatic shape and the same constraints on sides and angles as the primitive cell of the (trigonal) lattice *P* but has two additional lattice nodes placed at 1/3 and 2/3 along the diagonal between the 0,0,0 and 1,1,1 points.

There are 230 space groups in total, of which 73 are symmorphic.

Some algebraic relations are worth mentioning. The symmetry operations of a space group *G* evidently also include pure lattice translations, and it can be seen that these themselves are a group *H*: technically, it is a *commutative and normal subgroup* (Section 4.1.3) of the space group, and it is called the **translation group**.

Translational symmetry replicates the symmetry operations present within a single unit cell in all primitive cells: for every non-translational operation in a given cell, we have a corresponding operation in every other cell. The so-called **quotient group** expresses this relationship: each element of this group is a **set** that contains all the non-translational symmetry operations corresponding to each other in different unit cells.

Symmorphic space groups are **semidirect products** (see Section 4.2) of the translation group and of the crystallographic group (Section 6.6.1). In fact, the operations of the two kinds do not commute in general.

Generally, a space group is ***non*-commutative**.

Even for space groups it is possible to have **matrix representations** in a similar way to what has been done for point groups. However, matrix representations of space groups are necessarily more complicated than those of point groups because of the need to combine point operations with *translations*.

For simplicity's sake, a less formal treatment is followed here. Essentially:

- the same concepts are used (unit cell, direct and reciprocal lattice, Bloch's theorem, dispersion relations for monoelectronic states and for vibrations, …) as developed for pure translational symmetry (see Section 3), and then
- are integrated with the consequences due to the additional symmetries and are applied to the various (Bravais) lattices compatible with these.

As found for commutative point groups, pure translational symmetry has only 1D IRs. In addition, IRs are defined by Bloch's vectors $\vec{k}$ (see Section 3.2) and the character (eigenvalue) for the translation $\vec{l}$ in the $\vec{k}$ IR is $\exp(i\vec{k} \cdot \vec{l})$. Finally, it should not be surprising that solving the different problems for non-symmorphic groups requires more complicated procedures (which we do not address).

6.6.5 Equivalent Positions and Multiplicities in Crystal Structures

The structural elements (and the electron density) are replicated by translational symmetry from a unit cell to another, and, within a unit cell, by *non-translational* symmetry operations. If, for example, at the centre of each unit cell there is an inversion centre, the position *x*, *y*, *z* of each cell is replicated at the position $1-x$, $1-y$, $1-z$ within the *same* unit cell and the **positions** (*x*, *y*, *z*) and ($1-x$, $1-y$, $1-z$) or ($-x$, $-y$, $-z$) are said to be **symmetry equivalent**. As said, *x*, *y*, *z* are fractional coordinates, meaning that they are fractions of the moduli of the $\vec{a}$, $\vec{b}$ and $\vec{c}$ unit vectors.

The minimal set of structural elements required to reproduce the *whole crystal structure* by ***all*** symmetry operations *is not* the whole unit cell but a fraction of it and has the technical name of **asymmetric unit**. The application of all the non-purely-translational symmetry operations, including screw axes and glide planes, to the asymmetric unit produces the *unit cell*.

> If a given *x*, *y*, *z*, position is *symmetry-equivalent to* $n-1$ different positions in the cell [(x_2, y_2, z_2), …, (x_n, y_n, z_n)], we say that position (*x*, *y*, *z*) has **multiplicity** *n*.

In the same space group, different *x*, *y*, *z* positions can have *different multiplicity*. The reason is simple. If, for example, there is a reflection plane (perpendicular to the drawing plane and indicated by the line in the figure) and a structural element (represented by A in Figure 4.38) is on the reflection plane, the reflection "replicates one half to the other" and the multiplicity is halved compared to that of another structural element (say, letter R) which is instead external to the reflection plane. For the same reason, the structural element on the plane (A) has the **reflection symmetry** that the other element (R) does not possess.

Similarly, for example, the multiplicity of a structural element located on a C_3 axis would be divided by three and the structural element would have symmetry C_3.

The **maximum multiplicity** occurs when a position *x*, *y*, *z* **is not** in a particular position with respect to the point symmetry elements and is therefore replicated by **all** the

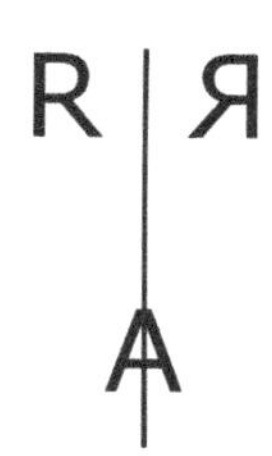

FIGURE 4.38 Letters R and A show a structural element that is (is not) replicated by reflection and is not (is) characterised by mirror symmetry.

symmetry operations of the unit cell. It is then said to be a **general position**. The multiplicity can reach up to 48 for the *primitive* unit cell of the spatial groups with the highest symmetry. The particular positions with respect to the symmetry elements are called **special positions**. For the special positions, the **multiplicity is reduced** with respect to the general positions and **jointly** a **particular symmetry** arises (this is the *symmetry of the position*):

Lesser multiplicity → higher symmetry.

A *special position* (and hence a specific *symmetry*) does not necessarily correspond to **isolated points** but may include **lines or surfaces** within the unit cell. An example with group *P*23 is discussed below.

These brief notes make it clear that the very *description* of the structure and its detailed analysis require a much more intensive use of symmetry than in the molecular case. The very well-known *International Tables for Crystallography* are the reference text, both for the in-depth analysis of the various topics and for the exhaustive list of the various cases. For each of the 230 space groups, the basic information found there, in addition to the conventional graphic description of all the symmetry operations, contains:

- the *complete* list of distinct sets of *equivalent positions*.
- the *multiplicity* of each set of equivalent positions,
- the *symmetry* of the equivalent positions in the set.

For illustration, consider the simplest of the cubic space groups, the group *P*23 (# 195 of the *International Tables*) which derives from the combination of point group 23 (or *T*, in the Schoenflies notation) with the *simple cubic* Bravais lattice, with primitive unit cell and no screw axes or glide planes. Every structural element in the general *x, y, z* position has multiplicity 12 (in short: the three cyclic permutations of the coordinates and for each permutation the four cases corresponding to all three positive values or only one positive value). Consider that ($y, -x, -z$) is equivalent to ($y, 1-x, 1-z$) by translational symmetry.

The group also presents nine special positions, with multiplicities 6 (four distinct cases), 4 (only one case), 3 (two cases), and 1 (two cases). One of these cases, for example, occurs when the three fractional coordinates are equal (x, x, x), that is, when the structural element is placed on the ternary axis passing through the (0,0,0) and (1,1,1) lattice points. The multiplicity is 4 and the equivalent positions are: (x, x, x), ($-x, -x, x$), ($-x, x, -x$) and ($x, -x, -x$), corresponding to points along each of the four ternary axes, which in turn are equivalent to each other (they are the four body diagonals of the primitive cubic cell). The symmetry of the position is obviously 3 (C_3 in Schoenflies notation). The two special positions with multiplicity 1 are the origin (0,0,0) and the centre (1/2, 1/2, 1/2) of the cell and have the full tetrahedral symmetry of the point group 23. The crystal structure is then obtained by specifying the fractional coordinates of all the atoms in the asymmetric unit. For example, tin iodide SnI_4 has been reported to crystallize within this space group. Sn is in the 0, 0, 0 position at the cell origin, with multiplicity 1, while I is on the ternary axis at 0.25, 0.25, 0.25, with multiplicity 4. Note that the ratio between multiplicities must reflect the stoichiometry.

6.6.6 Symmetry and Dispersion Relations

In Section 3.2 we saw that Bloch's vectors $\vec{k}$ express the eigenvalues of the translational symmetry operations. We now have to see how to combine translational symmetry with the other symmetries of the different space groups, regarding the dispersion relations of phonons and monoelectronic states. Since this is a specialist topic, we limit ourselves to general considerations and conclusions that can be drawn in a simple way.

1. Point symmetries act both on direct lattice vectors ($\vec{r}$) and on Bloch's vectors ($\vec{k}$) and indicate which $\vec{r}$ (and which $\vec{k}$) are symmetry equivalent and *how many* they are.
2. A given $\vec{k}$, which specifies the characters for translations, *does not* correspond to all the possible behaviours with respect to rotations, reflections, inversions. This fact should not be surprising: returning to the molecular symmetry groups and examining their character tables, one sees similarly that a specific character of a point operation *A*, is not necessarily associated with all the different characters that another symmetry operation (*B*) can present. For example, in the character table of point group D_{6h} (Table 4.31), characters −2, −1, +1, +2 are present for both C_2 and *i*, but the value −2 for C_2 is present with only −2 and +2 for *i*.
3. To clarify the previous point, the different $\vec{k}$ have different **multiplicity** and **symmetry**; in particular, the Block's vector $\vec{k} = \vec{0}$ (the **origin** of the reciprocal lattice) is transformed into *itself by all the point symmetry operations*. Other $\vec{k}(\neq \vec{0})$ are *generally* transformed into different points, consistently with non-translational symmetry. The situation is analogous to that already seen for the atomic sites within a unit cell of the direct lattice, in which, in addition to the general position characterized by low local symmetry and maximum multiplicity, there are special positions with lower multiplicity, and characterized by various degrees of symmetry up to the maximum symmetry possible for the space group.
4. With respect to the symmetries and multiplicities relating to the unit cell of the direct lattice, in the $\vec{k}$ space there is an **additional symmetry**. In Section 3.3.2 we saw the $\vec{k}\,/-\vec{k}$ symmetry as a consequence of the *time reversal* symmetry (see Section 2.1.3), in agreement with what was understood for

complex eigenvalues in commutative groups (see Section 1.3.2): a $\vec{k} \neq \vec{0}$ has always multiplicity≥2.
5. Again, because of time reversal symmetry, at the zone borders the dispersion relations have zero gradient in the $\vec{k}$ space (see Section 3.3.2).
6. Within the first Brillouin zone (FBZ) of the reciprocal lattice, singular points (*i.e.* $\vec{k}$ vectors) with high symmetry can be identified, and also lines and surfaces characterized by a given symmetry.

Based on what has been seen in detail in Section 3.3.2, the calculations of the dispersion relations are carried out by sampling the FBZ at appropriate intervals. So, based on the previous considerations, it can be seen immediately that

> There is no need to sample the whole first Brillouin zone (FBZ), but only a fraction of it. The dispersion relations show maximum symmetry at the origin of FBZ.

Dispersion relations are typically represented in 2D plots along appropriate lines within the FBZ (they thus appear similar to the 1D cases of Figures 4.23, 4.24, and 4.26): the lines are characterized by a particular symmetry, with the exception of singular points (for example $\vec{k} = \vec{0}$), with larger symmetry. A dispersion relation belongs to an IR with the symmetry of the line and along the whole line it maintains the degeneracy of the IR. The IR (and therefore the degeneracies) of the singular points can be determined on the basis of the relations between their respective symmetries. When two (or more) dispersion relations meet, they present a point, $\vec{k}$, where they have the same energy (degeneracy).

Then:

- If they belong to the same IR, then they can be combined, and the solutions of the electronic or vibrational problem are given by the combinations and produce a *repulsion* of the levels or - to put it another way - **crossing of the dispersion relations is avoided.**
- If they *do not* belong to the same IR, then they do not combine, but remain separate and show a simple level **crossing.**

It may be worth noting that there is **no degeneracy** (better: we do not say that there is degeneracy) when, for the dispersion relations E_1 and E_2, one finds $E_1(\vec{k}) = E_2(\vec{k'})$ for $\vec{k} \neq \vec{k'}$ (obviously excluding the case where they differ by a RL vector).

7 REMARKS

We report some proofs both for completeness and to facilitate the understanding of some more difficult derivations that are here presented in more details.

7.1 The Basic Property of the Cayley Table

We here show that each column of the table contains all the elements of a group in a different permutation (see Section 1.1.6). Consider a group consisting of *n distinct* elements, which we denote as $A_1, A_2, \ldots, A_n$ choosing the identity as the first element. Let us write them in this order as column headings and also in the first row, which has heading $I = A_1$. If A_k is the heading of *any other* row, then this row contains the products $A_kI = A_k, A_kA_2, \ldots, A_kA_n$ in this order. We must prove that (a) no element of the group is missing from this list and (b) no element appears twice.

a. Let us consider an element of the group that we may indicate with X and let us take the product $A_k^{-1}X$. This surely exists according to the group axioms (A_k surely has an inverse and the product of this inverse by X surely exists): let us call it A_j. Then the product $A_k A_j = A_k A_k^{-1} X = X$, that is, we surely obtain X in the row with heading A_k.
b. Let us suppose for the sake of argument that X appears twice, that is, as the product A_kA_j and as the product A_kA_r. By multiplying the two products from the left by A_k^{-1} we obtain $A_k^{-1} A_k A_j = A_j$ and $A_k^{-1} A_k A_r = A_r$, that is, $A_j = A_r$. Therefore, two elements of the group would be equal, contradicting the original statement. To complete, we note that the order in which the rows are arranged is completely arbitrary and the property just demonstrated remains valid even if it is changed. A similar proof is valid for the order of the columns.

7.2 The Regular Representation

The regular representation of a group allows us to demonstrate an important result (see Section 7.2.2).

7.2.1 Properties

Section 5.1.3.7 describes how the regular representation is constructed. Note that:

- the dimension of the matrices is equal to the order of the group g,
- for each symmetry element of the group, each row (or each column) of the matrix has only one element that is different from zero and that element is=1,
- all symmetry elements of the group have character=0, except the identity which has character=g.

It is necessary to first demonstrate that the matrices defined in Section 5.1.3.7 are a representation, *i.e.* that if a product of symmetry operations is $AB = C$ then $\boldsymbol{D}_{\text{reg}}(A) \cdot \boldsymbol{D}_{\text{reg}}(B) = \boldsymbol{D}_{\text{reg}}(C)$.

The proof is long and intricate and starts from the translation of the procedure (Section 5.1.3.7) into formulas. First, an order of enumeration of the elements of the group is established, say $A_1=E, A_2, \ldots, A_g$. Then it is seen that the elements ij of the matrices are given by:

$$\left[\boldsymbol{D}_{\text{reg}}(R)\right]_{i,j} = \begin{cases} 1 & \text{if}(A_i)^{-1} \cdot A_j = R \\ 0 & \text{otherwise} \end{cases}$$

Let us now consider two symmetry operations A_m and A_n and let $A_m \cdot A_n = A_o$: we must verify that $\boldsymbol{D}_{\text{reg}}(A_m) \cdot \boldsymbol{D}_{\text{reg}}(A_n) = \boldsymbol{D}_{\text{reg}}(A_o)$. Let us therefore examine the product of the i-th row of $\boldsymbol{D}_{\text{reg}}(A_m)$ by the j-th column of $\boldsymbol{D}_{\text{reg}}(A_n)$:

$$\sum_{k=1}^{g}\left[\boldsymbol{D}_{\text{reg}}(A_m)\right]_{i,k} \cdot \left[\boldsymbol{D}_{\text{reg}}(A_n)\right]_{k,j}$$

As said, the regular representation matrices are made up of rows in which all elements$=0$ except one element$=1$ (and similarly for columns). The only non-zero element in the i-th row of $\boldsymbol{D}_{\text{reg}}(A_m)$ is element k such that $(A_i)^{-1} \cdot A_k = A_m$. Similarly, the only non-zero element in the j-th column of $\boldsymbol{D}_{\text{reg}}(A_n)$ is element k' such that $(A_k)^{-1} \cdot A_j = A_n$. In the summation, *all* addends are *generally* zero, because one or the other of the two factors, or both, vanish. Therefore, the product of a row of $\boldsymbol{D}_{\text{reg}}(A_m)$ by a column of $\boldsymbol{D}_{\text{reg}}(A_n)$ is *generally* 0, in agreement with the fact that only g among the g^2 elements of a regular representation matrix are non-zero. The exception is the case in which – in correspondence with a particular value of j for a given i, or in correspondence with a particular value of i for a given j – the equality $k=k'$ occurs: then there is an addend that is not zero but 1, and consequently the element ij of the product is also 1. This case occurs when we simultaneously have $(A_i)^{-1} \cdot A_k = A_m$ and $(A_k c)^{-1} \cdot A_j = A_n$.

In conclusion, the element ij of $\boldsymbol{D}_{\text{reg}}(A_m) \cdot \boldsymbol{D}_{\text{reg}}(A_n)$ is 1 when $(A_i)^{-1} \cdot A_k = A_m$ and $(A_{k'})^{-1} \cdot A_j = A_n$ under the condition $k=k'$, that is, when the matrix product represents the operation $A_m \cdot A_n$.

Note that **the regular representation transforms a coordinate into (only) another coordinate, as a consequence of the fact that each row (each column) has only one element = 1 and all the others = 0**.

7.2.2 The IRs Contained in the Regular Representation

The result we are interested in is that the regular representation contains **all the IRs** of the group and contains them l_i times each (l_i is the dimension of the i-th IR).

The dimension of the regular representation is g (= order of the group) and we already know from 5.4.3 that $g^2 > g \geq \sum_i l_i^2$ if we exclude the trivial group with only one element. So, the regular representation is a RR.

The characters of the regular RR are then searched as a combination of the respective characters of the different IRs, with coefficient a_i for the i-th IR. For class k, the character in the regular RR is:

$$\chi_{\text{reg}}(C_k) = \sum_i a_i \chi_i(k)$$

An IR (say β) is contained in the regular RR (α) a_β times (see Eq. (5.27)):

$$a_\beta = \frac{1}{g}\sum_{k=1}^{h} N_k \left[\chi_\alpha(k)\right]^* \left[\chi_\beta(k)\right]$$

Now let's look at a particular class, the one that contains the identity and therefore $N_I=1$ (in any group the identity is a class in itself). Let us recall once again that, by construction (Section 5.1.3.7), in the regular representation only the identity has diagonal elements different from zero and they are all$=1$ and that the dimension of the regular representation is g. Therefore, the character of the identity is g and:

$$\begin{aligned} a_\beta &= \frac{1}{g}\sum_{k=1}^{h} N_k \left[\chi_\alpha(k)\right]^* \left[\chi_\beta(k)\right] = \frac{1}{g} N_I \left[\chi_\alpha(I)\right]^* \left[\chi_\beta(I)\right] \\ &= \frac{1}{g} \cdot 1 \cdot g \cdot \left[\chi_\beta(I)\right] = l_\beta \end{aligned} \tag{7.1}$$

Each IR is contained in the regular representation a number of times equal to the size of the IR itself. In particular, each IR is contained at least once.

Finally, concerning the relation

$$g^2 > g \geq \sum_i l_i^2 \tag{5.21}$$

To prove that the equality sign holds, we use again the properties of the *regular representation*, in particular:

a. that it contains l_i times every IR (i) of the group (Eq. 7.1),
b. that the character of the identity (E) is g and
c. that the characters of the other symmetry operations are zero.

Remembering that in any representation (i) the character of the identity is equal to the size l_i of the representation:

$$g = \chi_{\text{reg}}(E) = \sum_i a_i \chi_i(E) = \sum_i l_i l_i \tag{7.2}$$

This finally proves that in Eq. (5.22) the equality sign: $g = \sum_i l_i^2$.

7.3 Another Orthogonality Theorem for Characters

The starting point is Eq. (5.26): $\sum_{k=1}^{h} N_k [\chi_\alpha(k)]^* [\chi_\beta(k)] = g\delta_{\alpha\beta}$. The equal number ($h$) of IRs and classes makes possible to put characters into squares matrices where each column (l) is pertinent to a class and each row (i) is pertinent to an IR:

$$O_{ij} \overset{\text{def}}{=} \chi_i(C_j)$$

Let us write another matrix M defined as: $M_{ji} \overset{\text{def}}{=} \frac{N_j}{g}[\chi_i(C_j)]^*$ Therefore:

$$(MO)_{ij} = \sum_l M_{il} O_{lj} = \sum_l \frac{N_i}{g} \chi_l(C_j)[\chi_l(C_i)]^* = \delta_{ij}$$

Then:

$$\sum_{l=1}^{h} N_i \cdot \chi_l(C_j) \cdot [\chi_l(C_i)]^* = g\delta_{ij}, \text{which is Eq. (5.29)}$$

7.4 Orthogonality of SAF of Different IRs and Different Partners

The proof is based on the GOT. Let us consider a symmetry group (operations I, R, S, …) and an l_α-dimensional IR (α). An orthonormal basis set $(\varphi_{\alpha 1}, \varphi_{\alpha 2}, \ldots)$ of l_α linearly independent functions generates (see Sections 5.1 and 6.1.1.1) a matrix representation with unitary matrices $D_\alpha(E)$, $D_\alpha(R)$, $D_\alpha(S)$, …. For a generic function that belongs to the α IR, from Eq. (6.1):

$$\psi_{\alpha,k} = \hat{R}\varphi_{\alpha,k} = \sum_j [D_\alpha(R)]_{jk} \varphi_{a,j}; \quad j,k = 1,\ldots,l_\alpha$$

Let us also consider an l_β-dimensional IR (β) and a basis set $(\varphi_{\beta 1}, \varphi_{\beta 2}, \ldots)$ of l_β linearly independent functions that generates the matrix representation $D_\beta(I)$, $D_\beta(R)$, $D_\beta(S)$, …. For a generic function that belongs to β:

$$\psi_{\beta,m} = \hat{R}\varphi_{\beta,m} = \sum_w [D_\beta(R)]_{wm} \varphi_{\beta,w}; \quad w,m = 1,\ldots,l_\beta$$

so that: $\langle \psi_{\alpha,k} | \psi_{\beta,m} \rangle = \sum_j [D_\alpha(R)]_{jk} \varphi_{a,j} \Big| \sum_w [D_\beta(R)]_{wm} \varphi_{\beta,}$

Sections 2.2 and 3.2.3 of the Appendices discuss the linearity and anti-linearity properties of the scalar product. Accordingly:

$$\langle \varphi_{\alpha,k} | \varphi_{\beta,m} \rangle = \sum_j [D_\alpha(R)]^*_{jk} \cdot \sum_w [D_\beta(R)]_{wm} \cdot \langle \varphi_{a,j} | \varphi_{\beta,m} \rangle$$

This relation holds for any symmetry operation (R). Then, we can sum over the elements of the group and divide by g (the order of the group):

$$\langle \varphi_{\alpha,k} | \varphi_{\beta,m} \rangle = \frac{1}{g} \sum_R \sum_j \sum_w \sum_R [D_\alpha(R)]^*_{jk} \cdot \sum_w [D_\beta(R)]_{wm} \cdot \langle \varphi_{a,j} | \varphi_{\beta,m} \rangle$$

Let us now exchange the order of the sums:

$$\langle \varphi_{\alpha,k} | \varphi_{\beta,m} \rangle = \frac{1}{g} \sum_j \sum_w \sum_R [D_\alpha(R)]^*_{jk} \cdot [D_\beta(R)]_{wm} \cdot \langle \varphi_{a,j} | \varphi_{\beta,m} \rangle$$

Finally, let us replace with GOT (, (Eq. 5.16):

$$\langle \varphi_{\alpha,k} | \varphi_{\beta,m} \rangle = \frac{1}{g} \sum_j \sum_w \left(\frac{g}{g_\alpha} \right) \delta_{\alpha\beta} \delta_{jw} \delta_{km} \cdot \langle \varphi_{a,j} | \varphi_{\beta,m} \rangle$$

where g_α is the dimension of the α IR.

Now, if $\alpha = \beta$, and $k = m$,

$$\langle \varphi_{\alpha,k} | \varphi_{\beta,m} \rangle = \frac{1}{g_\alpha} \sum_j \sum_w \delta_{jw} \cdot \langle \varphi_{a,j} | \varphi_{\beta,m} \rangle$$

and we are left with only:

$$\frac{1}{g_\alpha} \sum_w \delta_{ww} \cdot \langle \varphi_{a,w} | \varphi_{\alpha,w} \rangle.$$

There are g_α integrals such as $\langle \varphi_{a,w} | \varphi_{\alpha,w} \rangle$ and all of them are = 1, because of the orthonormality of the basis set. Then:

$\langle \varphi_{\alpha,k} | \varphi_{\alpha,k} \rangle = 1$. In all other cases: $\langle \varphi_{\alpha,k} | \varphi_{\beta,m} \rangle = 0$

7.5 Filled Shells

For **non-degenerate orbital levels**, the complete occupancy is totally symmetric, as already seen in Section 2.3.1. For **degenerate levels**, the explanation does not hold, but the same result is reached, albeit less directly. As an original article on this problem reports,[11] "it is an often-stated theorem in quantum chemistry textbooks that closed electron shells are totally symmetric, and thus, for the purposes of determining the overall symmetry of a many-electron wave function, only the open shells need be examined" specifying, however, that the argument "… is not quite as trivial as the usual hand-waving arguments would indicate." We refer to this article for a complete analysis of the topic.

8 EXERCISES

Exercise 4.1

Show that there is only *one* abstract group of order 3.

(hint) Select at will (a) an order for the headings of columns and rows and (b) the identity. Write the row of the identity. Show that there is only one possibility for the second row and then for the last row. Is this group commutative? Is it cyclic? (see Section 1.1.3.7).

Exercise 4.2

There are two abstract groups of order 4 and one is isomorphous to group C_{2v} (see Table 4.2). Try to find the other one. Is it commutative? Is it cyclic? (hint: the difference concerns only the last two rows of the multiplication table).

Exercise 4.3

There is only one abstract group of order 5. Write its Cayley table knowing that it is cyclic. Is it commutative?

Exercise 4.4

There are two abstract groups of order 6 and one corresponds to group C_{3v} (see Table 4.3). Find the other one knowing that it is commutative and cyclic.

Exercise 4.5

Write the Cayley tables and find the eigenvalues of the symmetry operations and the symmetry species of groups D_2 and C_{2h}.

Exercise 4.6

Derive Eq. (2.3) from Eq. (2.2) using the assumption that the *Hamiltonian does not include time.*

Hint: start from the general form for the solution of the time-dependent Schrödinger equation as a series expansion on the Hamiltonian eigenfunctions with their temporal dependence and apply conjugation and change of sign of time.

Exercise 4.7

Prove the following properties of classes (see Section 4.1.3): (i) the identity always is a class in itself; (ii) the class of elements conjugated to any element A is never empty; (iii) if the group is commutative, each of its elements is a class in itself, (iv) the elements of a class all have the same order n, that is, for all of them the nth power equals the identity.

Exercise 4.8

Prove that the subgroup $\{E, C_3, C_{-3}\}$ of group C_{3v} is normal and that subgroups $\{E, \sigma^{(1)}\}$, $\{E, \sigma^{(2)}\}$ and $\{E, \sigma^{(3)}\}$ are not normal.

Exercise 4.9

Derive the right cosets of group C_{3v}. (hint: they are 2). Derive the left cosets of the group.

Exercise 4.10

Derive the following properties of left cosets: (1) if X belongs to the subgroup A that generates the coset, then the coset coincides with the underlying set of A; (2) if the group is commutative, then whatever the subgroup A and the element X, the right coset is equal to the left coset; (3) two right (left) cosets of a given subgroup A either contain exactly the same elements (and then we say that they are the same coset) or they are completely disjoint (see below); (4) all right (left) cosets contain the same number of elements, which equals the number of elements of the subgroup A that generates the cosets; (5) the number of right cosets is equal to the number of left cosets. This number is called the index of the subgroup and is obviously the ratio of the order of the group to the order of the subgroup.

Hint for (3): let us assume that two cosets AX and AY of subgroup A share one element $Z=A_rX=A_sY$. Then consider YX^{-1}. This is$=A_s^{-1}A_r$ and surely belongs to A because A is a group. What about YX^{-1} and the products of YX^{-1} by *any* element of A?

Exercise 4.11

Show that there are no elements of a group that are *not* contained in one of the cosets of a pre-assigned subgroup. In other words, show that for each element Y of the group it is possible to find a coset that contains Y: this can be done by selecting as element X that generates the coset all the elements of the group. (Hint: a subgroup always contains the identity).

Exercise 4.12

Using properties 3 and 6 of Exercise 4.10 show that the left (right) cosets of a subgroup are a partition of the underlying set of a group.

Exercise 4.13

Using property 4 of Exercise 4.10 show that the number of left (right) cosets of a subgroup is a divisor of the order g of the group.

Exercise 4.14

Assume that $\boldsymbol{D}_1$ and $\boldsymbol{D}_2$ are two representations of a group and that there is a set of matrices $\boldsymbol{D}_3$ such that for each element X of the group the matrices of this set are the direct products of the matrices of the two representations $\boldsymbol{D}_1$ and $\boldsymbol{D}_2$: $\boldsymbol{D}_3(X)=\boldsymbol{D}_1(X)\otimes\mathbf{D}_2(X)$. Show that if three elements of the group are related by $RS=T$ so that $\boldsymbol{D}_1(R)\cdot\boldsymbol{D}_1(S)=\boldsymbol{D}_1(T)$ and $\boldsymbol{D}_2(R)\cdot\boldsymbol{D}_2(S)=\mathbf{D}_2(T)$, then $\boldsymbol{D}_3(R)\cdot\boldsymbol{D}_3(S)=\boldsymbol{D}_3(T)$ using the suitable property of the direct product of matrices (Section 5.1.1.2).

Exercise 4.15

Using data for group C_{3v}, (matrix representation in Table 4.19, characters in Table 4.20) write the vectors $\boldsymbol{v}_s$ and $\boldsymbol{v}_\alpha$ (Eq. 5.18) and show the pairwise orthogonality between the three $\boldsymbol{v}_\alpha$ ($\alpha=\mathrm{A}_1$, A_2, E) and the four vectors $\boldsymbol{v}_s$ ($s=\{\mathrm{E}, 1, 1\}$, $s=\{\mathrm{E}, 1, 2\}$, $s=\{\mathrm{E}, 2, 1\}$, $s=\{\mathrm{E}, 2, 2\}$). Show that these vectors are normalized. Show that a $\boldsymbol{v}_s$ ($s=\{\mathrm{E}, i, j\}$) is orthogonal to both $\boldsymbol{v}_\alpha$ ($\alpha=\mathrm{A}_1$ or A_2).

Exercise 4.16

With reference to Eq. (5.28), write the vectors $\boldsymbol{w}_\alpha$ ($\alpha=\mathrm{A}_1$, A_2, E) for group C_{3v} (see Table 4.20 for the characters) and show that they are pairwise orthogonal.

Exercise 4.17

With reference to Section 6.1.1.2, using the eigenvectors of Eq. (6.2), and the rotation matrix $\boldsymbol{R}$ of Eq. (6.3), prove that $\boldsymbol{R}\cdot\boldsymbol{c}_2=\boldsymbol{c}_3$ and $\boldsymbol{R}\cdot\boldsymbol{c}_3=-\boldsymbol{c}_2-\boldsymbol{c}_3$.

NOTES

1 M.S. Dresselhaus, G. Dresselhaus, and A. Jorio. *Group Theory: Application to the Physics of Condensed Matter.* Springer-Verlag, Berlin, Heidelberg, 2008. https://doi.org/10.1007/978-3-540-32899-5.

2 The symbol ∀ is the so called *universal quantifier* and indicates that the statement is true for every possible value of the following variable(s). So, the axiom is: "For all elements *A, B, C* that belong to group *G*, *A*(*BC*) is equal to (*AB*)*C*".

3 A list of the point groups, including the character tables and examples of molecules, can be found on several websites, *e.g.* http://symmetry.jacobs-university.de/.

4 It should be noted that the clockwise and anticlockwise rotations of the same angle are different rotations. We must therefore define a convention for rotation directions. Although in the above we implicitly used the clockwise convention, this is not the only accepted. For example, in crystallography, the anticlockwise convention is used.

5 "Lattice" vibrations is of common but improper usage: vibrations regard atoms, not lattices.

6 Particular cases are *soft modes, i.e.* **optical modes** with frequency $\omega(\vec{q} \to 0) \sim 0$.

7 The reader who is not familiar with the topic can find a more detailed and complete presentation in the Appendix.

8 See Section 2.2 of the Appendices for the relation between scalar products and linear spaces and Section 3.2.3 for the properties of scalar products.

9 Some results of Sections 5.4.3 and 5.4.4 may appear quite abstract and therefore difficult to understand. For this reason, we frequently make examples, always using data for space group C_{3v} taken from Table 4.19 and Table 4.20. The reader is warned that the character tables derive from the results of this section, so that the examples should not be mistaken as part of the derivations, but as simple *illustrations* of an intricate topic.

10 P.W. Atkins, M.S. Child, and C.S.G. Phillips. *Tables for Group Theory.* Oxford University Press, Oxford, 1970.

11 H.F. Beebe Nelson. On the symmetry of filled shells. *Int. J. Quantum Chem.* **X**, 1007–1010, 1976.

5 Appendices

These appendices discuss various topics that are actually pertinent to other chapters but are not included there because not strictly necessary.

Section 1 supplements the first section of the chapter on Quantum Mechanics.

In Sections 2 and 3, the readers who have little familiarity with linear spaces and matrix calculus may find a very discursive presentation that develops the analogy with what every student should already know from previous courses about vector quantities.

In the final section, we offer short biographical notes of scientist that have been prominent in the development of the scientific topics that are presented in this book.

1 THE HAMILTONIAN VERSION OF CLASSICAL MECHANICS

Section 1 of the Chapter on Quantum Mechanics (QM) summarizes some very basic topics of Classical Mechanics (CM): the elementary approach based on Newton law, the use of generalized coordinates, and a very short account of Hamiltonian mechanics. The last topic is both a general method for writing the equation of motion with generalized coordinates and the version of CM suitable to introduce QM.

This Appendix gives a more detailed account of Hamiltonian mechanics and discusses the relation between the Newtonian and Hamiltonian approaches (1.1), solves the problem of two dimensionless particles bound by a central force (1.2) as an example of Hamilton's method, and finally extends the approach to the molecular degrees of freedom (1.3).

1.1 The Method

The *elementary* version of CM derives the total *energy* and its properties after having solved (when possible) Newton equations. *Hamilton's approach* is based on a *characteristic function* called Hamiltonian and derives the equations of motion (Hamilton equations) from this function.

1.1.1 The Hamiltonian Function

The Hamiltonian function corresponds to the total energy and is a function of n generalized coordinates $\{q_i\}$ and an equal number of generalized moments $\{p_i\}$. Hamilton equations are $2n$ first-order differential equations in the $2n$ variables$\{q_i, p_i\}$. Starting from the elementary description of the system based on the Cartesian coordinates of each point particle (Newton coordinates), the following steps are followed to write Hamilton function H.

A. The first step is to select the suitable generalized coordinates (q_i) and specify the direct transformations from Newton coordinates $\{x_i\}$ to generalized coordinates $\{q_i\}$, *i.e.* the relations expressing $\{x_i\}$ as functions of $\{q_i\}$:

$$\{x_i\} \rightarrow \{q_i\} = f\{x_i\} \tag{1.1}$$

and then the inverse transformations:

$$\{q_i\} \rightarrow \{x_i\} = g\{q_i\} \tag{1.2}$$

From the latter relations, we also obtain the transformations of the derivatives, that is the Newtonian velocities, as a function of both the generalized velocities $\mathrm{d}q_i/\mathrm{d}t$ and the $\{q_i\}$:

$$\{\mathrm{d}q_i / \mathrm{d}t\} \rightarrow \{\mathrm{d}x_i / \mathrm{d}t\} = h\{\mathrm{d}q_i / \mathrm{d}t\}$$

$$\text{or} \quad \{\dot{q}_i\} \rightarrow \{\dot{x}_i\} = h(\dot{q}_i, q_i) \tag{1.3}$$

where the superscript dot denotes the derivative: $\dot{q}_i \stackrel{\text{def}}{=} \frac{\mathrm{d}q_i}{\mathrm{d}t}$, and $\dot{x}_i \stackrel{\text{def}}{=} \frac{\mathrm{d}x_i}{\mathrm{d}t}$.

B. The potential energy V is then written as a function of the generalized coordinates: $V(\{q_i\})$ using the transformations of Eq. (1.2).

C. The kinetic energy T is written as a function of the generalized coordinates and velocities using Eq. (1.3).[1]

D. The moment conjugate to a generalized coordinate is obtained according to the definitions:

$$p_j \stackrel{\text{def}}{=} \frac{\partial T}{\partial \dot{q}_j} \tag{1.4}$$

When the potential energy depends on velocities as well as on coordinates, the definition to use is more precisely:

DOI: 10.1201/9781003644422-5

$$p_j \stackrel{\text{def}}{=} \frac{\partial(T-V)}{\partial \dot{q}_j} \qquad (1.5)$$

E. The kinetic energy T (see step C) is finally re-written as a function of moments, instead of generalized velocities. With non-conservative systems, also the potential energy V is re-written as a function of both generalized coordinates and generalized moments.

Now, we know how to write kinetic energy (T), potential energy (V) and therefore total energy ($E=T+V$) as functions of the generalized coordinates and their conjugated moments. This function is called the Hamilton's function or the *Hamiltonian*: $H=f(q_i, p_i)$.

1.1.2 Hamilton Equations

The fundamental equations of Hamilton's formulation of classical mechanics are:

$$\dot{q}_j = \frac{\partial H}{\partial p_j} \quad \text{and} \quad \dot{p}_j = \frac{-\partial H}{\partial q_j} \quad (\textbf{Hamilton equations}) \qquad (1.6)$$

Note A: Hamilton equations solve the problem of writing the *equations of motion* in generalized coordinates. It is worth remarking some formal differences between Newton equations and Hamilton equations. For each degree of freedom, the Newtonian formulation gives an equation of motion that is *a second-order* differential equation, while Hamilton's formulation gives *two first-order* differential equations. These are mathematically equivalent, because a second-order differential equation can be written as a system of two first-order differential equations. Furthermore, there is a difference in role between velocities and moments: while velocities have a direct dependence on the coordinates, being their total derivatives with respect to time, moments are now "promoted" to the rank of independent variables, being linked to the coordinates only through the equations of motion, just as the coordinates themselves are possibly linked to each other.

Note B: Hamilton's version allows ample freedom in selecting generalized coordinates and moments and also provides an easy way for the transformations between different sets of generalized coordinates.

Note C: As for the role of energy, one can say, figuratively, that Newton's version "discovers" that energy is conserved by solving the equations of motion, while Hamilton's version uses from the beginning the fundamental role of energy in determining the evolution in time of a mechanical system. The Hamilton function is the characteristic function of the system; it implicitly contains everything that is known about its time evolution, and the equations of motion are obtained by "operating" on the Hamiltonian function. This *point of view plays an important role also in QM.*

Note D: Hamilton equations of motion also provide insights on the choice of the most suitable generalized coordinates for solving a mechanical problem. For example, if one chooses the generalized coordinates so that H is independent of a particular q, then the moment conjugate to the q is conserved and the conservation law itself is part of the solution. Also, the fact that the energy of a mechanical system does *not* depend on some quantities (and that we can choose these as coordinates) often is easily derived from general considerations that do not require detailed calculations. These are essentially symmetry considerations and can appear obvious. For example, in the following sections we discuss a system in which the energy of a particle depends on the distance from a central point, but does not depend on the direction, *i.e.*, it shows a central (or spherical) symmetry. Hamilton equations then tell us that the motions of the system are decoupled if distance and angles are chosen as generalized coordinates.

Another strategy is to find coordinates that allow the potential energy V to be written as a sum of disjoint terms, that is as a sum of parts such that each part contains only some coordinates or moments, preferably a single coordinate or moment. Then, in each equation $\dot{q} = \partial H / \partial p$ (or $\dot{p} = -\partial H / \partial q$) the derivative on the right contains a single moment or coordinate, because the derivative of a sum is the sum of the derivatives, only one of these derivatives is nonzero and contains a single variable.

In the following, we will see some examples of coordinate transformations based on these considerations. **This strategy** of simplifying the equations of motion **is important also in QM**.

Note E: Two derivations are theoretically needed to justify the use of Hamilton's equations (H) of motion instead of Newton equations (N): (i) the derivation of Hamilton equations from Newton's equations and (ii) the derivation of Newton equations from Hamilton equations. The equivalence N ⇔ H requires both implications: N ⇒ H and H ⇒ N.

The assessment of Hamilton equations is very simple if we are only interested in proving that *everything that can be derived from Newton equations can be also derived from Hamilton equations.* In this case, we need to derive Newton equations from Hamilton equations: H ⇒ N. In fact, for any statement X that can be derived from N (that is, for any derivation N ⇒ X), H ⇒ X certainly holds, if it has already been proved that H ⇒ N, because H ⇒ N ⇒ X. Indeed, the proof of H ⇒ N is easy. Essentially, it consists in showing that Hamilton equations coincide with Newton equations if we use Newton coordinates in Hamilton's formulation. To that purpose, let's consider a conservative system of n point-like particles. Let's also use Newton coordinates as generalized coordinates, so the q_i are given by the x, y, or z of the particles, while the $\dot{q}$ are given by the x, y, or z components of the respective speeds. By hypothesis V is a function of the coordinates only, while T is given by:

$$T = \frac{1}{2}\sum_{i=1}^{n} m_i\left(v_i^2\right); \quad \left(v_i^2\right) = \left(v_{x_i}^2\right) + \left(v_{y_i}^2\right) + \left(v_{z_i}^2\right)$$

The moments are $p_j = \frac{\partial(T-V)}{\partial \dot{q}_j}$, that is: $p_j = \frac{\partial T}{\partial \dot{q}_j}$ (conservative system) and therefore: $p_j = m_j v_j$, where j specifies both a particle and a coordinate (x, y, or z) of the particle. The kinetic energy is: $\sum_{j=1}^{3n}\left(\frac{(p_j)^2}{2m_j}\right)$ and the first Hamilton equation becomes: $\dot{q}_j = v_j = \frac{\partial T}{\partial p_j} = \frac{p_j}{m_j}$, which simply redefines the conjugate moment. The second equation becomes: $\dot{p}_j = m_j \frac{\mathrm{d}v_j}{\mathrm{d}t} = \frac{-\partial V}{\partial q_j} \stackrel{\text{def}}{=} f_j$, which is *Newton equation for a cartesian coordinate of the particle* (index j).

On the other hand, if one is interested in proving that *Hamilton equations do not introduce results incompatible with Newton equations,* then one must demonstrate the opposite implication, that is, derive Hamilton equations from Newton equations. In formulas: for any statement Y that can be derived from H (that is, for any derivation H ⇒ Y), N ⇒ H ⇒ Y certainly holds, if it has already been proved that N ⇒ H. This result is much longer to derive, and we refer the interested reader to a standard book on Analytical Mechanics.

1.2 Central Field Motion

1.2.1 The Two Particles System

As an example of Hamilton's formulation of CM and an approach to problems of very general importance, let us consider the mechanical model of

a. *two* dimensionless particles with masses m_1 and m_2 moving in ordinary *3D space* and
b. linked by a force that *depends only on the distance between the two particles* (hence the name *central* field), that is a potential energy $V = V\left(\left|\vec{r}_2 - \vec{r}_1\right|\right)$.

The coordinates of the particles are $\vec{r}_1$, *i.e.*, x_1, y_1, z_1 and $\vec{r}_2$, *i.e.*, x_2, y_2, z_2.

Note: The dependence of V on $\left|\vec{r}_2 - \vec{r}_1\right|$ is not specified because we want to obtain the most general results. At the end, we will consider some examples, such as the harmonic interaction (in 3D), the gravitational interaction and the Coulomb interaction.

1.2.2 External Degrees of Freedom

There are 6 degrees of freedom in total. As a first step, we prove the exact separation between the **external** degrees of freedom, that is the three Cartesian components of the vector $\vec{R}$ that specifies the position of the centre of mass, and the **internal** degrees of freedom, currently identified by the three Cartesian components of the difference vector: $\vec{r} = \vec{r}_2 - \vec{r}_1$. We use vector notation to write formulas more compactly, remembering that when we use the product of vectors, we mean the scalar product, when we write the square of a vector, we mean the square of its modulus (= the sum of the squares of its components) and finally that the derivative (with respect to time) of a vector indicates the three derivatives of its components.

Let's follow the rules in Section 1.1.1. step by step. The transformation equations are:

$$\begin{pmatrix} \vec{R} = \frac{m_1}{m_1+m_2}\vec{r}_1 + \frac{m_2}{m_1+m_2}\vec{r}_2 \\ \vec{r} = \vec{r}_2 - \vec{r}_1 \end{pmatrix} \quad \text{and it is convenient to write:}$$

$$\begin{pmatrix} M\vec{R} = m_1\vec{r}_1 + m_2\vec{r}_2 \\ \vec{r} = \vec{r}_2 - \vec{r}_1 \end{pmatrix} \quad \text{with } M = m_1 + m_2.$$

The inverse transformations are:

$$\begin{pmatrix} \vec{r}_1 = \vec{R} - \frac{m_2}{M}\vec{r} \\ \vec{r}_2 = \vec{R} + \frac{m_1}{M}\vec{r} \end{pmatrix}, \quad \text{the first derivatives are:}$$

$$\begin{pmatrix} \frac{\mathrm{d}\vec{r}_1}{\mathrm{d}t} = \frac{\mathrm{d}\vec{R}}{\mathrm{d}t} - \frac{m_2}{M}\frac{\mathrm{d}\vec{r}}{\mathrm{d}t} \\ \frac{\mathrm{d}\vec{r}_2}{\mathrm{d}t} = \frac{\mathrm{d}\vec{R}}{\mathrm{d}t} + \frac{m_1}{M}\frac{\mathrm{d}\vec{r}}{\mathrm{d}t} \end{pmatrix}. \text{ The squares are:}$$

$$\begin{pmatrix} \left|\frac{\mathrm{d}\vec{r}_1}{\mathrm{d}t}\right|^2 = \left|\frac{\mathrm{d}\vec{R}}{\mathrm{d}t}\right|^2 + \frac{m_2}{M}\left|\frac{m_2}{M}\right|^2 - \frac{2m_2}{M}\left(\frac{\mathrm{d}\vec{R}}{\mathrm{d}t}\right)\cdot\left(\frac{\mathrm{d}\vec{r}}{\mathrm{d}t}\right) \\ \left|\frac{\mathrm{d}\vec{r}_2}{\mathrm{d}t}\right|^2 = \left|\frac{\mathrm{d}\vec{R}}{\mathrm{d}t}\right|^2 + \frac{m_1}{M}\left|\frac{m_2}{M}\right|^2 + \frac{2m_1}{M}\left(\frac{\mathrm{d}\vec{R}}{\mathrm{d}t}\right)\cdot\left(\frac{\mathrm{d}\vec{r}}{\mathrm{d}t}\right) \end{pmatrix}$$

Now, we write the kinetic energy:

$$T = \frac{1}{2}\left[m_1\left|\frac{\mathrm{d}\vec{r}_1}{\mathrm{d}t}\right|^2 + m_2\left|\frac{\mathrm{d}\vec{r}_2}{\mathrm{d}t}\right|^2\right] \tag{1.7}$$

as a function of the new coordinates and their derivatives, finally seeing that the terms containing the product of the derivatives of $\vec{R}$ and $\vec{r}$ cancel out and:

$$T = \frac{1}{2}M\left|\frac{\mathrm{d}\vec{R}}{\mathrm{d}t}\right|^2 + \frac{1}{2}\mu\left|\frac{\mathrm{d}\vec{r}}{\mathrm{d}t}\right|^2 \quad \text{where } \mu \stackrel{\text{def}}{=} \frac{m_1 m_2}{M} \text{ is the reduced mass}$$

Let us call X the x component of $\vec{R}$: its conjugate moment is:

$$p_X = \frac{\partial T}{\partial \dot{X}} = M\dot{X} \quad \text{which gives} \quad \vec{p_R} = M\frac{\mathrm{d}\vec{R}}{\mathrm{d}t} \tag{1.8}$$

Then, the part of the kinetic energy that depends on $\vec{R}$ and its derivatives is $T = \frac{(\vec{p_R})^2}{2M}$. This is also the part of H that depends on $\vec{R}$ and its derivatives, because the potential energy depends only on $\vec{r}$. The corresponding Hamilton equations become:

$$\dot{X} = \frac{\partial H}{\partial p_X} = \frac{\partial T}{\partial p_X} = \frac{p_X}{M}; \quad \dot{p}_X = \frac{-\partial H}{\partial X} = 0 \tag{1.9}$$

and analogously for Y and Z

The first equation essentially redefines p_X as the moment conjugated to X (the x component of the moment of the centre of mass is the product of total mass times velocity), while the second expresses the conservation of the same moment. From these and the analogous equations for the Y and Z components, we obtain that **the centre of mass moves with uniform rectilinear** (inertial) **motion**.

This result shows - as already said - that Hamilton equations *provide a criterion for identifying the most suitable generalized coordinates*. Essentially, the result comes from the symmetry of the system. In fact, it should appear obvious from the beginning that, since there are no external fields, H cannot change whether the system is globally "here" or "there" but depends only on how the two constituent particles are mutually placed, *i.e.* it depends only on $\vec{r}$. As H cannot depend on X, Y or Z, the conjugate moments of these generalized coordinates are conserved, and the centre of mass moves with inertial motion (p_X=constant, so that $\dot{X}$ = constant).

1.2.3 Internal Degrees of Freedom

The three internal degrees of freedom globally represented by $\vec{r}$ describe the position of one particle with respect to the other but can be visualized as the motion of a particle of mass μ subjected to a potential energy $V(|\vec{r}|)$ and placed at point $\vec{r}$ of the ordinary 3D space, if the origin is placed at the centre of mass of the *original* system of the two particles.[2] This point of view is convenient to simplify the language. Therefore, we directly discuss the motion of this fictitious particle, except to return to the original point of view once the solution has been found.

When the masses are very different from each other (for example, when $m_1 >> m_2$), the point of view of the motion of the particle of mass μ placed at the point $\vec{r}$ is not only a useful logical scheme, but also is a good approximation of the actual situation, because then the centre of mass of the system almost coincides with the position of the larger mass, μ almost coincides with the smaller mass m_2 and $\vec{r}$ represents not only the distance between the two particles, but also (with good approximation) the absolute position of the smaller particle when the larger mass is placed at the origin of the coordinates. This is, evidently, the case of the one-electron atom model: the larger mass is the nucleus, the smaller mass is the electron, the ratio is always > 1800. For this reason, the one-electron atom model is usually described by considering the nucleus as fixed instead of the centre of mass of the nucleus+electron system, as it should be.

In addition to the separation of internal and external degrees of freedom, the shape of potential energy suggests a separation between internal degrees of freedom. To better understand the topic, two preliminary considerations are useful.

First: if V depends only on the distance from the origin of the coordinates, the force acts in the same direction as the vector $\vec{r}$ wherever the "fictitious" particle of mass μ is placed. The equipotential surfaces (V=constant) are spheres, while the lines of force (perpendicular to the equipotential surfaces) are the rays passing through the origin.

From this we can directly deduce that **the motion occurs in a plane**.

Here is the logical line of the proof: let $\vec{r}(t)$ and $\vec{v}(t)$ indicate the position and velocity of the particle at time t. There are three cases: (a) $\vec{v}(t)$ is not parallel to $\vec{r}(t)$; (b) $\vec{v}(t)$ is parallel to $\vec{r}(t)$; (c) $\vec{v}(t)$ is zero (*i.e.*, its modulus = 0). In case (a) (see Figure 5.1) $\vec{v}(t)$ and $\vec{r}(t)$, not being parallel to each other, define a plane that we may call the initial plane. The variation in time of the velocity vector is specified by the acceleration ($\vec{a}$), which has the same direction as the force, that is, the same direction as $\vec{r}(t)$. Therefore, the acceleration also lies in the initial plane. Then, the velocity is modified only in a direction ($\vec{d}$, see Figure 5.1) which in turn lies in the plane and therefore the velocity remains in the initial plane. Since the velocity is the variation in time of the position vector, which in turn initially lies in the plane, the position vector remains in the initial plane forever. Case (b) is a special case of case (a), in the sense that the motion now occurs along the line on which the position and velocity initially lie. Finally, case (c), is a special case of case (b).

Second: the potential energy does not actually depend on the three Cartesian components of $\vec{r}$, but on its modulus $r = |\vec{r}|$. This suggests representing $\vec{r}$ in *polar coordinates*, so that V does not depend on the angles ϑ and φ. Moreover, the fact that the motion occurs in a plane allows us to use *only two* coordinates, that is the planar polar coordinates (r, φ) by orienting the coordinate axes so that the polar coordinate ϑ, which is the co-latitude, is initially zero and remains zero forever. It is easily verified that the choice is correct by counting the degrees of freedom and the initial conditions: in fact, to specify a plane in 3D space two numbers are needed, just as two conditions are required to specify the initial values of (i) the coordinate that is removed (ϑ_0) and (ii) its derivative $(\mathrm{d}\vartheta/\mathrm{d}t)_0$.

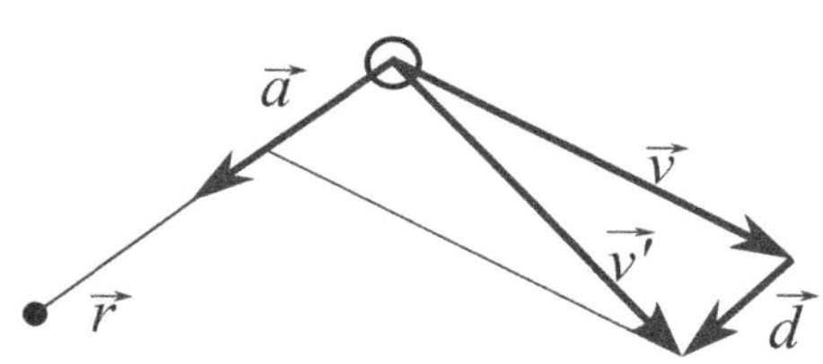

FIGURE 5.1 Central field motion. The large open circle is the particle, and the small circle is the centre of the force. Vectors $\vec{r}$, $\vec{v}$ and $\vec{a}$ are position, velocity and acceleration of the particle at time t, $\vec{v}'$ is the velocity at time $t+\delta t$ and $\vec{d} = \vec{v}' - \vec{v} \sim \vec{a}\delta t$.

1.2.4 The Radial and Angular Equations

The transformations made so far are exact, not approximate! For the final coordinates (r, φ) we now derive the equations of motion according to Hamilton's method and discuss the individual steps as an example of the whole procedure.

1. Let's start from the transformations between the plane polar coordinates and the Cartesian components (x, y) of $\vec{r}$ and, in particular, from the relations expressing the old coordinates (x, y) as a function of the new ones (r, φ):

$$\begin{pmatrix} x = r\cos\varphi \\ y = r\sin\varphi \end{pmatrix} \quad (1.10)$$

2. We derive the corresponding relations between the velocities, that is between the total derivatives with respect to time (specified by the superscript dot):

$$\begin{pmatrix} \dot{x} = \dot{r}\cos\varphi - r\dot{\varphi}\sin\varphi \\ \dot{y} = \dot{r}\sin\varphi + r\dot{\varphi}\cos\varphi \end{pmatrix} \quad (1.11)$$

3. In Cartesian coordinates the kinetic energy is:

$$T = \frac{1}{2}\mu\left[(\dot{x})^2 + (\dot{y})^2\right] \quad (1.12)$$

Substituting with Eq. (1.11) we get:

$$T = \frac{1}{2}\mu\left[(\dot{r}\cos\varphi)^2 + (r\dot{\varphi}\sin\varphi)^2 - 2(\dot{r}\cos\varphi)(r\dot{\varphi}\sin\varphi)\right]$$

$$+\frac{1}{2}\mu\left[(\dot{r}\sin\varphi)^2 + (r\dot{\varphi}\cos\varphi)^2 + 2(\dot{r}\sin\varphi)(r\dot{\varphi}\cos\varphi)\right]$$

and using $(\cos\varphi)^2 + (\sin\varphi)^2 = 1$ we finally have $T = \frac{1}{2}\mu\left\{(\dot{r})^2 + (r\dot{\varphi})^2\right\}$

4. Remembering that V is a function of r but does not depend on the velocities, the conjugate moments are:

$$\begin{pmatrix} p_r \overset{\text{def}}{=} \dfrac{\partial T}{\partial \dot{r}} = \mu\dot{r} \\ p_\varphi \overset{\text{def}}{=} \dfrac{\partial T}{\partial \dot{\varphi}} = \mu r^2\dot{\varphi} \end{pmatrix} \text{ and } \mu r^2\dot{\varphi} \overset{\text{def}}{=} l \text{ (l is the angular moment)} \quad (1.13)$$

5. Then, let's write the kinetic energy and the Hamiltonian as a function of the new coordinates and the conjugated momenta:

$$T = \frac{(\mu\dot{r})^2}{2\mu} + \frac{(\mu r^2\dot{\varphi})^2}{2\mu r^2} = \frac{(p_r)^2}{2\mu} + \frac{(l)^2}{2\mu r^2};$$

$$H = T + V = \frac{(p_r)^2}{2\mu} + \frac{(l)^2}{2I} + V(r) \quad (1.14)$$

The quantity $I = \mu r^2$ is the *moment of inertia*. Now, the kinetic energy is made up of **two uncoupled terms**, each of which is the square of a moment divided by twice the relative inertia coefficient, *i.e.* a mass and a moment of inertia.

6. Finally, Hamilton equations are:

$$\begin{pmatrix} \dot{r} = \dfrac{\partial H}{\partial p_r} = \dfrac{p_r}{\mu} & \dot{p}_r = \dfrac{-\partial H}{\partial r} = \dfrac{-\partial V}{\partial r} - \dfrac{(l)^2}{2\mu}\left(\dfrac{-2}{r^3}\right) \\ \dot{\varphi} = \dfrac{\partial H}{\partial l} = \dfrac{l}{\mu r^2} & \dot{l} = \dfrac{-\partial H}{\partial \varphi} = 0 \end{pmatrix} \text{ so that}$$

(1.15)

$$\begin{pmatrix} p_r = \mu\dot{r} & \mu\ddot{r} = \dfrac{-\partial V}{\partial r} + \dfrac{(l)^2}{\mu r^3} \\ l = \mu r^2\dot{\varphi} & \dot{l} = 0 \end{pmatrix}$$

The equations in the first column are immediately recognized as redefinitions of the two moments. From the last equation we find that l is conserved, and we substitute its constant value in the various equations. Then, the equations that remain to be solved are:

$$\begin{pmatrix} \mu\ddot{r} = \dfrac{-\partial V}{\partial r} + \dfrac{(l)^2}{\mu r^3} \\ \dot{\varphi} = \dfrac{l}{\mu r^2} \end{pmatrix} \quad \text{(l is constant)} \quad (1.16)$$

These two equations are only **partially decoupled**: the differential equation for r *does not* contain φ nor its derivatives, while the differential equation for φ *does* contain r. An obvious solution strategy would be to solve the first equation and insert the solution $r(t)$ thus obtained into the second equation. Of course, the actual possibility of having an analytic form for $r(t)$ and $\varphi(t)$ depends on the explicit expression of the potential energy $V(r)$.

1.2.4.1 Centrifugal Force and Centrifugal Potential

Before examining the solutions of the system of Eq. (1.16) for some interesting cases, we can continue a little further along the line followed so far to see what general conclusions can be obtained without specifying the potential energy $V(r)$. From this point of view, it is particularly interesting to look at the first Eq. (1.16):

$$\mu\ddot{r} = \frac{-\partial V}{\partial r} + \frac{(l)^2}{\mu r^3} \text{ and compare it with} \quad (1.16a)$$

Newton equation for 1D motion

$$ma = m\ddot{x} = F = -\frac{\partial V}{\partial x} \quad (1.17)$$

A few comments.

a. $\ddot{r}$ is the second total derivative with respect to time of the generalized coordinate r, while a in Eq. (1.17) is the second total derivative with respect to time of the Newton coordinate x. The variable r is defined on the real interval between 0 and ∞, while x is defined on the whole real axis.
b A force is generally derived from a potential energy V and the expression on the right of Eq. (1.16a) can also be written as the derivative of a function of r, $(V+V')$, which we may call the "effective potential energy" or effective potential. Indeed:

$$\frac{(l)^2}{\mu r^3} = \frac{-\partial V'}{\partial r} \quad \text{if} \quad V' \stackrel{\text{def}}{=} \frac{(l)^2}{2\mu r^2} \quad \text{so that} \quad (1.18)$$

$$\mu\ddot{r} = \frac{-\partial(V+V')}{\partial r} \quad (1.19)$$

$V' = (l)^2/(2\mu r^2)$ is known as centrifugal potential. Similarly, the additional term $(l)^2/(\mu r^3)$ in Eq. (1.16) is called *centrifugal force*.

The inclusion of the centrifugal potential or force can be considered as the "price we have to pay" for treating a non-Newtonian coordinate (r) as if it were Newtonian.

1.2.5 Synthetic Analysis

Now, we can use the similarity to Newton equation for 1D motion (see Eq. (1.17)). Using Eq. (1.19) as if r were a normal Newton coordinate, we are able to deduce some general features of the motion, *without having to derive a single formula*. For this discussion, of course, we cannot leave the form of the potential energy indeterminate, as we have done so far. We must consider specific cases: the most interesting are the r^{-1} attractive potential (Figures 5.2–5.4) and the r^2 (harmonic) potential (Figure 5.5).

Let us analyse the asymptotic behaviours (Figure 5.2) of the *attractive potential* (r^{-1}). As $r \to \infty$, $(V+V') \to 0$ with negative values because the negative part depending on (r^{-1}) prevails over the positive part depending on (r^{-2}). As $r \to 0$, $(V+V') \to +\infty$ because the (r^{-2}) part prevails over the (r^{-1}) part. We also see that there is a minimum and it is unique:

$$V_{\text{eff}} \stackrel{\text{def}}{=} (V+V') = \frac{-A}{r} + \frac{B}{r^2} \Rightarrow \frac{\partial V_{\text{eff}}}{\partial r} = \frac{A}{r^2} - \frac{2B}{r^3};\ \left.\frac{\partial V_{\text{eff}}}{\partial r}\right|_{r0} = 0$$

which leads to: $r_0 = \frac{2B}{A} = \frac{(l)^2}{A\mu}$ (A contains quantities that depend on the nature of the interaction, for example the charges and the dielectric constant for the Coulomb case, or the masses and the gravitational constant for the gravitational case).

It is easy to see that there can be both bounded and unbounded motions, depending on the initial (and constant) value of the total energy. If the total energy is positive, we see that the motion is unlimited on the right, (see Figure 5.3) and the coordinate is confined to the interval $r_{\text{min}} \leq r < \infty$, where r_{min} depends on the energy. If, instead, E is negative (Figure 5.4), the motion is confined to a circular crown ($r_{\text{min}} \leq r \leq r_{\text{max}}$).

In the case of the **3D harmonic potential**, looking at Figure 5.5 we see that only bounded motion is possible ($r_{\text{min}} \leq r \leq r_{\text{max}}$): r_{min} and r_{max} depend on both the energy and

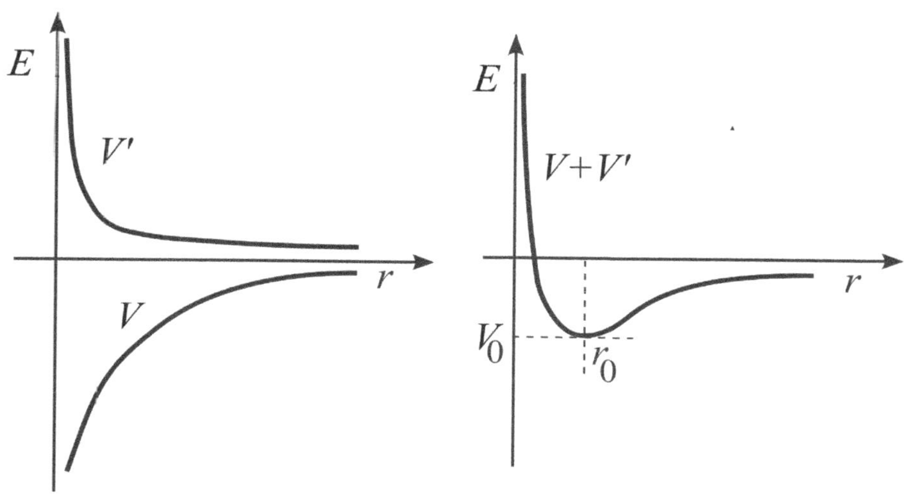

FIGURE 5.2 Attractive r^{-1} type interaction such as the Coulomb force between different charges, or the gravitational force. Left: the *assigned* potential V and the centrifugal contribution V'. Right: the *effective* potential $V+V'$.

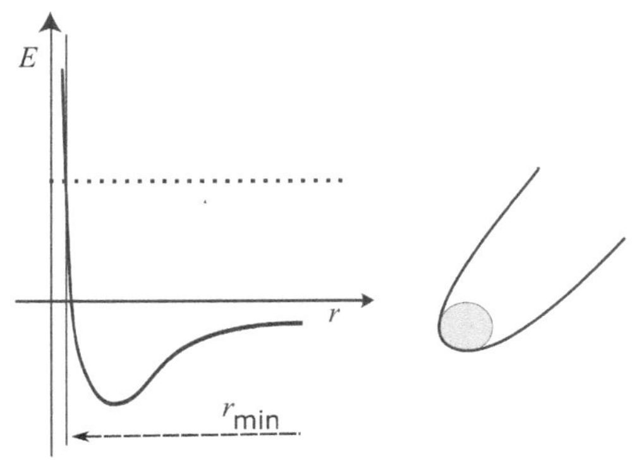

FIGURE 5.3 Unbounded motion due to attractive interaction of type r^{-1} (right part). Left: if the energy corresponds to the horizontal dotted line, the particle cannot approach the centre of force closer than the value r_{min} that is marked in the graph on the right by the enlarged grey circle. From the analytical solution, we also know that the trajectories are *parabolas*.

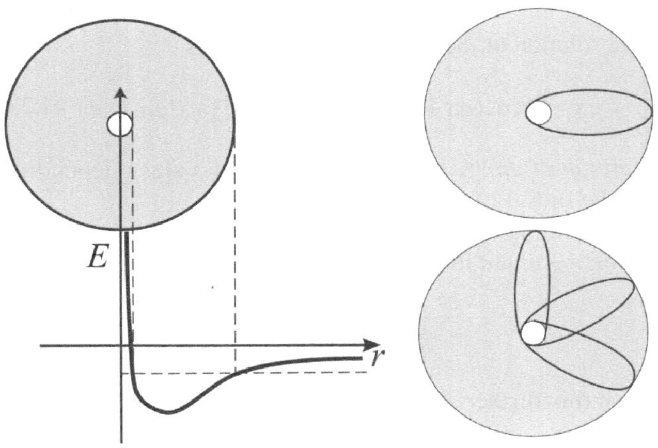

FIGURE 5.4 Bounded motion for attractive interaction of the r^{-1} type. On the right, there are two examples of a possible bounded motion: a trajectory that closes after a single turn and part of an open trajectory that does not return to the starting point or returns after many turns.

the angular momentum and the graph allows us to easily evaluate whether they increase or decrease as E or l change.

Of course, we cannot obtain from these simple considerations all the details of the motion: in the case of the attractive potential proportional to r^{-1}, for example, we cannot know what the trajectories of the bounded and unbounded motion are and, for the bounded motion, whether the trajectories are closed or open as schematized in the right part of Figure 5.4. To this aim the analytical solution of the equations of motion is needed. The problem of the attractive potential proportional to r^{-1}, as is well known, is the problem of planetary motion. The analytical solution provides Kepler's laws and, in particular, tells us that the trajectories of the unbounded motion are parabolas while the trajectories of the bounded motion are ellipses (or, as a special case, circles) and therefore the system returns to the top after each turn.

1.2.6 The Diatomic Molecule

Let's now apply the mechanical model to a diatomic molecule. The two particles are now the two atoms considered as point particles of mass m_1 and m_2. The results obtained so far tell us that the motion of the entire molecule can be separated into (i) three degrees of freedom for the inertial motion of the centre of mass with inertial coefficient $M=m_1+m_2$ and (ii) three internal degrees of freedom.

Concerning the internal degrees, the potential energy is due to the electrons, but these do not appear in the model. The chemical bond is a central field of force, that is, an interaction directed along the line connecting the two atoms. From a qualitative point of view, this is a bounded motion such as that of Figure 5.4, because this is our working hypothesis: we are interested in a *stable* molecule.

However, for the potential energy there is not a theoretical expression of general validity such as Coulomb law or Newton law of gravitation. It depends in a complicated way on the result of a calculation on electronic motions that must be carried out within the framework of QM. Of course, there *are* analytical expressions for the potential energy due to the chemical bond in a diatomic molecule, but they are essentially empirical in nature. The simplest and best-known example is Morse function: $V(r) \stackrel{\text{def}}{=} D_e\left(e^{-2a(r-r_0)} - 2e^{-a(r-r_0)}\right)$ where D_e, a and r_0 are suitable parameters that can be obtained from experiments or by a best fit of QM calculations. The Morse function corresponds to the qualitative plot of Figure 5.4 and models the potential energy as the sum of an attractive part

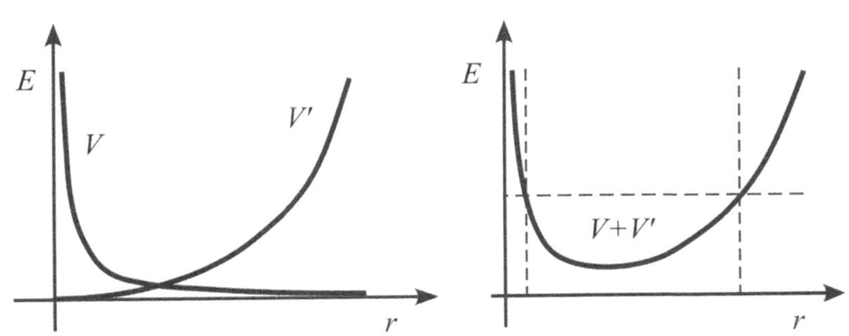

FIGURE 5.5 Attractive interaction of the type $1/r^2$. The motion is always bounded. The two vertical dashed lines indicate the radial limits r_{min} and r_{max} within which the motion is constrained.

$D_e\left(-2e^{-a(r-r_0)}\right) < 0$, which is predominant at long distances, and a repulsive part $D_e\left(e^{-2a(r-r_0)}\right) > 0$, which is predominant at short distances.

The lack of a generally valid analytical expression for the potential energy due to the chemical bond does not prevent us from achieving a solution, in particular when the system is "not much excited". This means that the total energy is only slightly higher than the minimum in Figure 5.4. Then, an approach is available which is based on the power series expansion and, although formally approximate, is in reality accurate enough for most applications. Under certain analytical conditions of continuity and derivability that we take here as verified, the unknown potential energy function can be written as:

$$\begin{aligned} V(r) = V(r)_{r=r_0} + (r-r_0)\left(\frac{\partial V(r)}{\partial r}\right)_{r=r_0} \\ +\frac{1}{2}(r-r_0)^2\left(\frac{\partial^2 V(r)}{\partial r^2}\right)_{r=r_0} + \cdots \\ +\frac{1}{n!}(r-r_0)^n\left(\frac{\partial^n V(r)}{\partial r^n}\right)_{r=r_0} + \cdots \end{aligned} \tag{1.20}$$

The right-hand side of the equation depends on the values of the function, the first partial derivative, the second partial derivative, …, the nth partial derivative, … at point r_0. These are no longer functions of r, but the whole expression on the right is still a function of r, through the various powers of $(r-r_0)$. Written as a series, Eq. (1.19) is exact. In applications, the series expansion is obviously truncated at an index n, depending on the desired accuracy and on the available information, *i.e.* on what is known about the value of the function and its derivatives at r_0, which is the *coordinate of the minimum of the binding energy*. Note that r_0 is the bond length and $(r-r_0)$ is its deformation. Let's discuss the coefficients in Eq. (1.20).

a. The first coefficient $\left[V(r)_{r=r_0}\right]$ is the ordinate of the minimum. Actually, this coefficient is not needed, because the potential energy essentially provides the force through its derivative, where it does not appear. In other words, moving the entire curve upwards or downwards does not change the *force* acting on the particle.
b. The second coefficient $\left[\left(\frac{\partial V(r)}{\partial r}\right)_{r=r_0}\right] = 0$ because at r_0 there is the minimum V.
c. The third coefficient $\left[\left(\frac{\partial^2 V(r)}{\partial r^2}\right)_{r=r_0}\right]$ is usually denoted by k and is positive since the curve is upwards concave in the minimum.

Therefore, the potential energy can be approximated at the *least accuracy* level by:

$$V(r) = \frac{1}{2}k(r-r_0)^2; \quad k \overset{\text{def}}{=} \left(\frac{\partial^2 V(r)}{\partial r^2}\right)_{r=r_0} \tag{1.21}$$

This is the *harmonic potential*. Considering that $V(r)$ is now the total potential, the equations obtained before can be rewritten as:

$$\mu\ddot{r} = \frac{-\partial(V)}{\partial r} \tag{1.16a}$$

$$\dot{\varphi} = \frac{l}{\mu r^2} \tag{1.16b}$$

with the hypotheses just explained, we can write:

$$\mu\ddot{r} \approx -k(r-r_0) \quad \text{or} \tag{1.22}$$

$\mu\ddot{x} \approx -kx; \quad x \overset{\text{def}}{=} (r-r_0)$ where x does *not* mean a Cartesian coordinate.

The solution of the differential equation is:

$$x = R\cos(\omega t + \tau) \text{ or } x = A\sin(\omega t) + B\cos(\omega t) \tag{1.23}$$

Here, $\omega = (k/\mu)^{1/2}$, and τ is, A and B are values depending on the initial conditions.

Assuming again that we are not too far from the minimum, $r \approx r_0$ and the other equation of motion becomes:

$$\dot{\varphi} = \frac{l}{\mu r^2} \approx \frac{l}{\mu r_0^2}$$

With this further approximation, the solution of Eq. (1.15b) is:

$$\varphi = \varphi_0 + \frac{l}{\mu r_0^2}t \tag{1.24}$$

This is a **uniform circular motion** and φ_0 is the value of the angular coordinate at time t=0.

Therefore, when the system is only *slightly excited the final result is* that *the motion* approximately **separates** into two very simple motions: **harmonic motion** along the radial coordinate and **uniform circular motion**. We have achieved what is known as the **roto-vibrational separation** and is the basis of the elementary treatment of di- or poly-atomic molecules also in QM.

1.2.6.1 Comments

A. Equation (1.21) shows that the *harmonic potential* finds its theoretical justification in the fact that it "grasps" at the simplest level *the essential aspects of a bounded motion*, that is an oscillation around a single minimum of potential energy. Higher order approximations are of course possible and provide better agreement with experimental data.
B. When the potential energy can be sampled numerically, more accurate expressions can be obtained

by fitting the sampled values with higher degree interpolation functions.

C. Concerning the simple harmonic case, it often happens that k is not known a priori: in many respects, this is not a serious shortcoming. At a basic level, in fact, it is important, essentially, to know the analytical expression of the solution and how it depends on k, which in turn can be determined through QM numerical calculations or from experiments. In QM, the harmonic model gives quantized, non-degenerate and equally spaced energy levels and their values depend on k according to: $E_n = (n + 1/2)\hbar\omega$ where $\omega = \sqrt{k/\mu}$ and $n=0, 1, 2, \ldots$. Since μ is known, one can then determine k spectroscopically by measuring the difference in energy between two adjacent levels: $E_{n+1} - E_n = \hbar\omega$.

1.3 Molecular Degrees of Freedom and Equipartition of Energy

The procedure of the previous section can be extended to all molecular systems. In this procedure, a molecule is considered as a set of massive points held together by chemical bonds. To avoid confusion, we here refer to the *positional* degrees of freedom of nuclei as motions of *atoms* rather than motions of *nuclei* as would be more correct because nuclei have other degrees of freedom. Examples are those concerning their dipole and quadrupole moments that are also the basis of very important spectroscopies in chemistry.

Of course, the contribution of electrons to the energy of the molecule is not ignored. It is not *explicitly* included in the dynamical problem, but its effects are *implicitly* included in the potential energy for atomic motions. In other words, the potential energy considers the interactions between nuclei and the centrifugal potential but also the *change of the energy of the electron system as the relative positions of the nuclei change.*

The approach is based on an approximate separation between the degrees of freedom of nuclei and electrons and is known as the **Born-Oppenheimer** (B-O) **separation**. The B-O approximation is in turn fundamentally justified by the value of the ratio between the mass of the electron and that of a nucleus, which is at least three orders of magnitude larger. According to this approximation, the QM equation of motion for the electrons can be adequately solved at fixed nuclei, *i.e.* for a specific configuration of the nuclei, which is not necessarily that of equilibrium. The electronic energy is then evaluated for various values of the nuclear coordinates and is added to the other interactions between the nuclei for specifying the potential energy that determines the atomic motions.

Then, a molecule with n atoms has $3n$ degrees of freedom. As already seen, three of these are *external* and concern the motion of the whole molecule as a particle of mass M equal to the sum of the masses of the atoms. The external degrees of freedom are treated in QM with the 3D version of the particle-in-a-box model.

The other $3n-3$ degrees of freedom are *internal* and concern the *relative* positions of the atoms. Extending again the considerations made for the diatomic molecule, we can distinguish the internal degrees of freedom into two sets.

The first set covers the **rotations**, in which the molecule behaves as a rigid body. A *rigid body*, whether it is given by a set of point masses or by a continuous distribution of mass, has *three* rotational degrees of freedom. This is easily explained by drawing a sphere centred on the centre of mass of the molecule and with radius given by the segment joining the centre of mass to a chosen point of the body. Two numbers, for example, a latitude and a longitude, are sufficient to specify the orientation of the segment, but a further number is needed to specify the rotation of the entire body around the segment. Linear molecules are an exception, because they have only two rotational degrees of freedom, those relative to the orientation of the molecular axis: the rotation around the axis is not a *molecular* degree of freedom.

The second set of internal degrees of freedom describes the **deformations** of the molecule, *i.e.* variations in bond lengths and/or bond angles and/or dihedral angles etc.... They are globally called **vibrations**.

Summarizing, the generalized coordinates are (in the case of **non-linear molecules**):

- *Three* coordinates of the centre of mass (*translational* degrees of freedom)
- *Three rotational* degrees of freedom (typically these are three angles that specify the orientation - with respect to an external frame - of three of non-coplanar axes integral with the rigid molecule),
- The remaining $(3n-6)$ degrees of freedom describe *vibrations*.

In the case of **linear molecules**, there are only *two* rotational degrees of freedom and an additional vibration so that the vibrational degrees of freedom are $3n-5$.

The separation between rotations and vibrations is not exact, as seen when discussing the diatomic molecule (Section 1.2.6). It derives from the assumption that the vibrations are little excited, that is, they have energies slightly higher than the minimum of potential energy (see Eq. (1.22)). This approach can be improved as needed but still has a fundamental conceptual relevance. It corresponds to the free rotor model (in 3D or 2D) and as many harmonic oscillators as needed.

There are several points to note here.

A. Rotations produce a centrifugal potential (see Eq. (1.15)) that adds to the electronic energy and nuclear (charge) repulsion in determining the potential energy of the vibrations, which therefore depends on the rotation energy.

B. Vibrations modify the moments of inertia of rotation. Finally,
C. in the potential energy there are cross terms between vibrations (see Section 1.3.2).

A *fundamental difference* between the translational and rotational degrees of freedom on the one hand and the vibrational degrees on the other is that the former ones have *only kinetic energy* (see Section 1.3.3)

1.3.1 Molecular Rotations

To treat the rotation of a molecule as a rigid (non-deformable) body *around its centre of mass*, the starting point is Eq. (1.24). For a diatomic molecule with reduced mass μ, fixed interatomic distance r_0 and constant angular momentum L the free rotor approximation gives:

$$\omega \overset{\text{def}}{=} \dot{\varphi} = \frac{L}{\mu r_0^2} \quad \text{or :} \qquad (1.25)$$

$$\omega = \frac{L}{I} \Rightarrow L = I\omega \qquad (1.26)$$

where $I = \mu r_0^2$ is the **moment of inertia**.

For generalizing to molecules with $n>2$ atoms, Eq. (1.25) is better interpreted as the equation of the rotation of *one* point particle of mass μ rotating with angular velocity ω around an axis placed at distance r_0. Then, it must be generalized in two respects: (i) consider n ($n>2$) point masses (m_1, m_i, …, m_n) placed at distances d_i from the rotation axis and (ii) consider nonlinear molecules.[3] Concerning the extension to more than two atoms, Eq. (1.26) can be directly generalized to the case of many masses by adding their relative angular momenta:

$$L = I\omega = \sum_{i=1}^{n} L_i = \left(\sum_{i=1}^{n} m_i d_i^2 \right) \cdot \omega \qquad (1.27)$$

For the second aspect, it is evident that the angular moment L (generally) changes when considering different orientations of the rotation axis with respect to the fixed molecule. A trivial example is the same diatomic molecule. If we orient the rotation axis along the molecular axis, the distances of the two atoms from the axis are null and $L=0$. For a given ω, L takes its maximum value when the rotation axis is perpendicular to the molecular axis, while intermediate orientations correspond to intermediate values of L. When there are no external fields (gravitational, electric, magnetic, …) that produce a torsional moment, the angular momentum is conserved.

Now, we introduce a set of Cartesian axes (x, y, z) integral with the molecule and an angular velocity vector $\vec{\omega}$: its module is the usual angular velocity ω and the direction is that of its rotation axis. The angular momentum vector ($\vec{L}$) is related to vector $\vec{\omega}$ by the generalization of the linear relation of Eq. (1.27):

$$\left[\begin{array}{l} L_x = I_{xx} \cdot \omega_x + I_{xy} \cdot \omega_y + I_{xz} \cdot \omega_z \\ L_y = I_{yx} \cdot \omega_x + I_{yy} \cdot \omega_y + I_{yz} \cdot \omega_z \\ L_z = I_{zx} \cdot \omega_x + I_{zy} \cdot \omega_y + I_{zz} \cdot \omega_z \end{array} \right. \quad \text{that is:} \qquad (1.28)$$

$$\boldsymbol{L} = \boldsymbol{I} \cdot \boldsymbol{\omega} \text{ which can be written } \vec{L} = \vec{\vec{I}} \cdot \vec{\omega} \qquad (1.29)$$

where $\vec{\vec{I}}$ is a real symmetric tensor, corresponding to the symmetric 3×3 real matrix:

$$\boldsymbol{I} = \begin{pmatrix} I_{xx} & I_{xy} & I_{xz} \\ I_{xy} & I_{yy} & I_{yz} \\ I_{xz} & I_{yz} & I_{zz} \end{pmatrix} \quad \text{and its components are:}$$

$$I_{xx} = \sum_{i=1}^{n} m_i \left(y_i^2 + z_i^2 \right); \quad I_{yy} = \sum_{i=1}^{n} m_i \left(x_i^2 + z_i^2 \right);$$

$$I_{zz} = \sum_{i=1}^{n} m_i \left(x_i^2 + y_i^2 \right)$$

$$I_{xy} = I_{yx} = -\sum_{i=1}^{n} m_i x_i y_i; \quad I_{yz} = I_{zy} = -\sum_{i=1}^{n} m_i y_i z_i;$$

$$I_{zx} = I_{xz} = -\sum_{i=1}^{n} m_i x_i z_i$$

From general considerations of linear algebra, we know that $\boldsymbol{I}$ can be diagonalized and has three *real positive* eigenvalues. There are three cases.

A. The eigenvalues are all different from each other; the associated eigenvectors are orthogonal to each other.
B. Two eigenvalues are equal; the relative eigenvectors can be chosen to be orthogonal to each other and are orthogonal to the eigenvector of the third eigenvalue.
C. All three eigenvalues are equal; the eigenvectors can be chosen to be orthogonal to each other.

Ordering the three eigenvalues according to $I_A \le I_B \le I_C$, we have the following cases.

A. Linear rotor: $0=I_A<I_B=I_C$.
B. Asymmetric rotor: $0<I_A<I_B<I_C$.
C. Prolate symmetric rotor: $0<I_A<I_B=I_C$.
D. Oblate symmetric rotor: $0<I_A=I_B<I_C$.
E. Spherical rotor: $0<I_A=I_B=I_C$.

A *prolate* symmetric rotor has an *elongated* shape (the shape of a rugby ball), while an *oblate* symmetric rotor has a *discoidal* shape. Examples are CH_3I (prolate) and benzene (oblate).

The symmetry of the rotor is higher than the symmetry of the molecule itself, because the inversion symmetry is added. It is also worth noting a possible confusion in the standard denomination: an *asymmetric* rotor is described by a *symmetric* tensor. The words symmetric/asymmetric are used in two different meanings.

Based on the assumptions made so far, the rotation is free, there is no potential energy of rotation and therefore the total (rotational) energy equals the kinetic energy K:

$$K = \frac{1}{2}\boldsymbol{L\omega} = \frac{1}{2}\boldsymbol{\omega}^{\mathrm{T}}\boldsymbol{I\omega} \quad K = \frac{1}{2}\vec{L}\cdot\vec{\omega} = \frac{1}{2}\vec{\omega}\cdot\vec{\vec{I}}\cdot\vec{\omega} \tag{1.30}$$

Here it is useful to highlight the formal analogy of these expressions with those of linear motion:

$$K = \frac{1}{2}\boldsymbol{p}\cdot\boldsymbol{v} = \frac{1}{2}\boldsymbol{v}^T\cdot m\cdot\boldsymbol{v} \tag{1.31}$$

If the moment of inertia is in diagonal form, Eqs. (1.29) and (1.30) become $\boldsymbol{L\cdot\omega} = L_x\omega_x + L_y\omega_y + L_z\omega_z$ and $\boldsymbol{\omega\cdot I\cdot\omega} = \omega_x I_{xx}\omega_x + \omega_y I_{yy}\omega_y + \omega_z I_{zz}\omega_z$. Hamiltonian and kinetic energy can be written:

$$H = K = \frac{L_x^2}{2I_{xx}} + \frac{L_y^2}{2I_{yy}} + \frac{L_z^2}{2I_{zz}} \tag{1.32}$$

which is the most useful form to shift to the QM treatment. The specific forms for the various cases then become as follows.

- Linear rotor ($I_A=0$, $I_B=I_C=I$) $H=L^2/2I$.
- Prolate symmetric rotor ($I_\parallel = I_A < I_B = I_C = I_\perp$) or oblate symmetric rotor ($I_\perp = I_A < I_B = I_C = I_\parallel$): in both cases: $H = \frac{L_\parallel^2}{2I_\parallel} + 2\frac{L_\perp^2}{2I_\perp^2}$, where $L_\perp$ indicates *one* of the two equal components of L in the plane of the two degenerate eigenvalues, while $L_\parallel$ is the component relative to the third eigenvalue.
- Spherical rotor ($I_A=I_B=I_C=I$) $H=L^2/2I$.

1.3.2 Molecular Vibrations

According to all the approximations previously described the potential energy of the vibrational degrees of freedom is independent of the rotational state of the molecule. There is *at least* one point of minimum energy in the $(3n-6)$- or $(3n-5)$-dimensional space of the vibrational coordinates, however these are defined. This is the *equilibrium configuration* of the molecule.

A potential energy *can* have more than one minimum, and the minima *can* have different values. We then speak of one or more *principal* minima at the lowest energy and *secondary* minima at higher energy. An example of three principal minima at the *same* energy is given by the three staggered configurations of the ethane molecule. A more complicated example is the 2-butene molecule, which has cis/trans isomerism with different energies and for each form has multiple minima, corresponding to its conformational isomers.[4]

The approximation that only considers *slightly excited* motions has another very important effect. The series is truncated at the quadratic term: this is the *harmonic approximation*, already seen in Eq. (1.21) for the diatomic molecule.

Before discussing the problem in general terms, it is worth considering discursively the nature of the most suitable coordinates. Let's analyse the potential energy of the vibrations of a three-atomic molecule as schematically shown by Figure 5.6 as a function of *only two* variables. The plot is for a symmetric AB_2 molecule, *i.e.* with an A atom between two B atoms. The coordinates are the *distances* between the middle A atom with one or the other B atom, d(A-B) or d(B-A), while the B–A–B bond angle is held fixed at its equilibrium value. The minimum of the potential energy is approximately at the centre of the innermost level curve.

Warning: the graph has two independent variables for visualization reasons only: a three-atomic molecule with 3 (or 4, if linear) vibrational degrees of freedom would require a $3+1=4$-dimensional graph (or $4+1=5$, if linear).

Let's also note a point to be discussed later: Figure 5.6 shows reflection symmetry with respect to the bisector of the angle between the two axes. The physical meaning is that the coordinates chosen for the graphical representation, *i.e.* the two bond distances, are equivalent by symmetry. We would not have had symmetry of the graph if we had

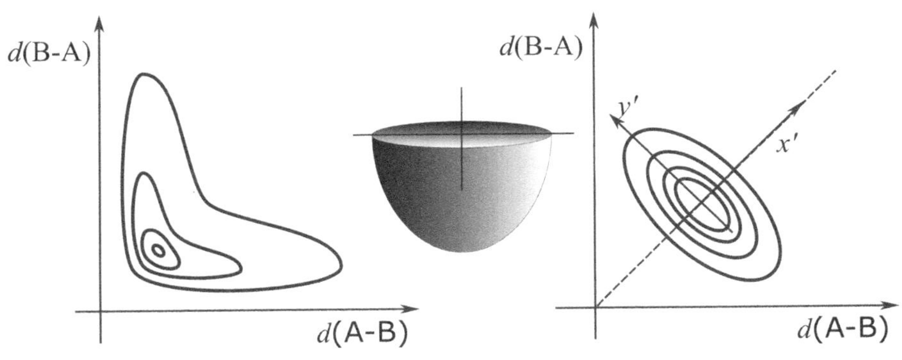

FIGURE 5.6 The potential energy for the vibrations of a triatomic B-A-B molecule. Left part: schematic plot near the minimum using only two coordinates (see text). Middle part: the harmonic paraboloid. Right part: *enlarged* plot of contour levels and normal modes (x' and y') in the harmonic approximation.

chosen to represent two other coordinates (*e.g.*: a bond distance and an angle) or to deal with an asymmetric molecule (*e.g.*: COS instead of CO_2). Figure 5.6 shows two essential aspects.

a. The equipotential lines do not generally have a simple shape but increasingly approach the elliptical shape when approaching the minimum.
b. The axes of these ellipses are not generally parallel to the coordinates of the plot.

The first point justifies the **harmonic approximation**: in a suitable neighbourhood of the minimum the series expansion of the potential energy can be truncated to the second power. This can be explained as follows. For simplicity, let's use again only two independent vibrational coordinates (as in Figure 5.6) instead of three or more, as is actually necessary for a molecule with three or more atoms. Then we have:

$$V(x,y) = A + B\cdot(x-x_0)^2 + C\cdot(y-y_0)^2 + D\cdot(x-x_0)\cdot(y-y_0) \tag{1.33}$$

(x_0, y_0) is the minimum and A, B, C, D are four parameters that indicate the energy of the minimum (A) and the second derivatives of the energy:

$$A = V(x,y)_{x=x_0,y=y_0}; \quad B = \left(\frac{\partial^2 V}{\partial x^2}\right)_{x=x_0,y=y_0};$$

$$C = \left(\frac{\partial^2 V}{\partial y^2}\right)_{x=x_0,y=y_0}; \quad D = \left(\frac{\partial^2 V}{\partial x\,\partial y}\right)_{x=x_0,y=y_0}$$

By the definition of minimum, B and C are strictly positive and $D \geq 0$. Equation (1.33) is both a generalization of Eq. (1.21) from one to two coordinates, and a truncation to the quadratic terms in the series expansion. From elementary analytical geometry it is known that Eq. (1.33) describes a paraboloid as shown by the middle part of Figure 5.6. The sections parallel to the (x, y) plane are ellipses, or circles as a special case, but the axes of the ellipses are not necessarily parallel to the axes (x and y) of the right part of Figure 5.6. They are parallel only if $D=0$ (see Eq. (1.33)). However, it is possible to obtain ellipses with parallel axes using a *linear* transformation of coordinates ($x, y \Rightarrow x', y'$). In the right part of Figure 5.6 the transformed coordinates are represented by the arrows at 45° with respects to the original axes. With the new coordinates, the potential energy is no longer expressed by Eq. (1.34), but by:

$$V(x,y) = A + B'\cdot(x'-x'_0)^2 + C'\cdot(y'-y'_0)^2 \tag{1.34}$$

with new coefficients (B' and C'). This achieves the separation between the vibrational degrees of freedom: the potential energy is now the sum of only pure quadratic terms without any cross terms such as the $D(x-x_0)(y-y_0)$ of Eq. (1.33). Thus, the Hamiltonian becomes the sum of disjoint parts and leads to distinct equations in which only one coordinate appears.[5]

The vibrational degrees of freedom separated by this transformation are called **normal modes**.

Notes:

- The normal modes depend on the coordinates of *more than one atom* and can be written as functions of geometric parameters such as bond lengths, bond angles, dihedral angles, ...
- The normal modes are *not* generally known *a priori* but are derived from the shape of the potential energy and the masses.
- However, there are significant symmetry constraints on the nature of the normal modes. The example of Figure 5.6 shows that the two normal modes are the sum and difference of the two original coordinates and are geometrically obtained with a rotation of 45°.
- The harmonic approximation fails when its basic requirement of small excitation no longer holds. It is indeed evident that the potential energy cannot increase indefinitely but must show one or more secondary minima corresponding to molecular fragments: the harmonic approximation cannot describe the dissociation of a molecule, and the normal modes lose their meaning at high energies.

In summary, the atomic motions of a polyatomic molecule can be described at the simplest level in classical mechanics using two simplifying assumptions, the Born-Oppenheimer separation and the harmonic approximation, which allow the potential energy to be written as:

$$V \cong \frac{1}{2}\sum_{ij} x_i \cdot x_j \cdot \frac{\partial^2 V}{\partial x_i \cdot \partial x_j} = \frac{1}{2}\cdot \boldsymbol{x}^{\mathrm{T}} \cdot \boldsymbol{V} \cdot \boldsymbol{x}; \quad (\boldsymbol{V})_{ij} \overset{\text{def}}{=} \frac{\partial^2 V}{\partial x_i \cdot \partial x_j} \tag{1.35}$$

In Eq. (1.35) the x are the Cartesian coordinates of the n atoms: $i = 1, 2, 3$ for the first atom, 4, 5, 6 for the second, ..., $3n-2$, $3n-1$, $3n$ for the last. The matrix $\boldsymbol{V}$ is symmetric: $\boldsymbol{V}_{ij} = \boldsymbol{V}_{ji}$ (the order of derivation is obviously irrelevant). The $\boldsymbol{V}_{ij}$ coefficients are known as *force constants*. For the kinetic energy is written:

$$K = \frac{1}{2}\cdot \dot{\boldsymbol{x}}^{\mathrm{T}} \cdot \boldsymbol{M} \cdot \dot{\boldsymbol{x}}; \quad \boldsymbol{M} = diag(m_1, m_1, m_1, m_2, \ldots) \tag{1.36}$$

using a superscript dot to denote total derivative with respect to time. The diagonal matrix $\boldsymbol{M}$ contains the masses ($\boldsymbol{M}_{11} = \boldsymbol{M}_{22} = \boldsymbol{M}_{33} = m_1$ is the mass of the first particle, and so on). Since we are using the Cartesian coordinates of the various atoms, we can directly write the matrix expression:

$$\begin{pmatrix} m_1 \ddot{x}_1 \\ m_1 \ddot{x}_2 \\ m_1 \ddot{x}_3 \\ \vdots \\ m_n \ddot{x}_{3n-2} \\ m_n \ddot{x}_{3n-1} \\ m_n \ddot{x}_{3n} \end{pmatrix} = \boldsymbol{M} \cdot \ddot{\boldsymbol{x}} \quad (1.37)$$

The force along each coordinate x_j is given by $F_j = \frac{-\partial V}{\partial x_j}$ and the set of forces is obtained by deriving Eq. (1.35). This gives $\boldsymbol{F} = -\boldsymbol{V} \cdot \boldsymbol{x}$ and the equations of motion are:

$$\boldsymbol{M} \cdot \ddot{\boldsymbol{x}} = -\boldsymbol{V} \cdot \boldsymbol{x} \quad \text{or} \quad \ddot{\boldsymbol{x}} = -\boldsymbol{M}^{-1} \cdot \boldsymbol{V} \cdot \boldsymbol{x}, \quad \text{and finally} \quad (1.38)$$

$$\ddot{\boldsymbol{x}} = \boldsymbol{D} \cdot \boldsymbol{x}; \quad \boldsymbol{D} = -\boldsymbol{M}^{-1} \cdot \boldsymbol{V}$$

is called the **dynamic *matrix*** of the system (1.39)

Equation (1.39) is a system of second order *homogeneous* ODEs (ordinary differential equations).

In this regard, mathematics tells us that:

i. the solution of the ODE system is a linear combination of independent motions, which are the normal modes discursively introduced in Figure 5.6:

$$\begin{pmatrix} Q_1 \\ \vdots \\ Q_{3s} \end{pmatrix} = \begin{pmatrix} c_{11} & \cdots & c_{1,3s} \\ \vdots & \ddots & \vdots \\ c_{3s,1} & \cdots & c_{3s,3s} \end{pmatrix} \cdot \begin{pmatrix} x_1 \\ \vdots \\ x_{3s} \end{pmatrix} \quad \text{or} \quad \boldsymbol{Q} = \boldsymbol{C} \cdot \boldsymbol{x} \quad (1.40)$$

The coefficients C_{ij} of column j of $\boldsymbol{C}$ tell us how much the original coordinate x_i contributes to the normal mode j.

ii. Each normal mode depends on time according to a harmonic law:

$$\ddot{Q}_i = -\omega_i^2 \cdot Q_i \Rightarrow Q_i = Q_{i,0} e^{i\omega_i t} \quad (1.41)$$

$$\begin{pmatrix} \ddot{Q}_1 \\ \vdots \\ \ddot{Q}_{3s} \end{pmatrix} = - \begin{pmatrix} \omega_1^2 & \cdots & 0 \\ \vdots & \ddots & \vdots \\ 0 & \cdots & \omega_{3s}^2 \end{pmatrix} \cdot \begin{pmatrix} Q_1 \\ \vdots \\ Q_{3s} \end{pmatrix} \quad (1.42)$$

and we write: $\ddot{\boldsymbol{Q}} = \boldsymbol{\Lambda} \cdot \boldsymbol{Q}$

This is obtained from $\ddot{\boldsymbol{Q}} = \boldsymbol{C} \cdot \ddot{\boldsymbol{x}} = \boldsymbol{C} \cdot \boldsymbol{D} \cdot \boldsymbol{x} = \boldsymbol{C} \cdot \boldsymbol{D} \cdot \boldsymbol{C}^{-1} \cdot \boldsymbol{Q}$ where the coefficients C_{ij} (the elements of $\boldsymbol{C}$) are obtained by diagonalizing the dynamic matrix $\boldsymbol{D}$:

$$\boldsymbol{\Lambda} = \boldsymbol{C} \cdot \boldsymbol{D} \cdot \boldsymbol{C}^{-1} \quad (1.43)$$

while $\boldsymbol{\Lambda}$ is the diagonal matrix of the eigenfrequencies ω. *In Λ, five or six elements are zero and obviously refer to translational and rotational motions.*

We see here why the motions are independent: while each $\ddot{x}$ depends linearly on several x (see Eq. (1.38)), each $\ddot{Q}$ depends only on the corresponding Q, as shown by Eq. (1.42). In Eq. (1.41), $Q_{i,0}$ is the complex amplitude of the i normal mode, a quantity that groups the true amplitude, *i.e.* the maximum extension of the oscillation, and the phase.

Furthermore, it can happen that two (or three) normal modes have the same frequency. Then we say that there is *degeneracy*, and the choice of normal modes is arbitrary, since any independent linear combination is still a normal mode with the same frequency. *Normally,* molecular vibrations are at most three times degenerate.[6]

1.3.3 Equipartition of Energy

According to a fundamental result of statistical thermodynamics, the *principle of equipartition of energy* (described in part 6.1 of the Chapter 3), the mean value of the contribution to the energy (kinetic or potential) of a separate degree of freedom is $kT/2$.

Let us consider an ideal gas. The degrees of freedom of different molecules are separated from each other by definition of ideal gas. The degrees of freedom of a single molecule are also separated from each other according to the discussion of the previous point. Then, we obtain that the contribution of a degree of freedom is $kT/2$ (for translations or rotations, because for these degrees of freedom we only have kinetic energy) or kT (for vibrations, which have both kinetic and potential energy). Then, the knowledge of the nature of the degrees of freedom that are separable at the molecular level allows us to count how many vibrational and rotational degrees the molecule has and to evaluate the thermodynamic functions.

Let us see some examples.

- CO_2 (linear molecule). There are 3 degrees of freedom (dof) for translations, 2 dof for rotations, then 9−5=4 dof for vibrations. Then: $E=(3+2+(4\times2))$ $kT/2=13$ $kT/2$ (one molecule) or 13 $RT/2$ ($R=kL$, L is Avogadro's constant): $C_v \overset{\text{def}}{=} \left(\frac{\partial E}{\partial T}\right)_V = 13R/2$.
- COS is an analogous case with the difference that - due to the different masses - the motions of O and S do *not* combine with the same weight in the normal modes of vibration, as instead happens for the two O of CO_2.

- H_2O (non-linear). There are 3 translational dof, 3 rotational dof and 9−6=3 vibrational dof. Then, $E=(3+3+(3\times2))\ kT/2=6\ kT$ (one molecule) or 6 RT (1 mol) and $C_v=6R$.
- C_2H_2 (ethyne or acetylene, linear molecule). There are 4×3=12 dof: 3 for translations, 2 for rotations, then 12−5=7 vibrational dof. Then, $E=(3+2+(7\times2))kT/2=19kT/2$ (one molecule) or $19RT/2$ (1 mol) and $C_v=19R/2$.
- H_2CO (methanal or formaldehyde, non-linear). There are 4×3=12 dof: 3 for translations, 3 for rotations, then 12−6=6 vibrational dof. Then, $E=(3+3+(6\times2))kT/2=9RT$ (1 mol) and $C_v=9R$.
- CH_4 (methane, non-linear). There are 5×3=15 dof: 3 for translations, 3 for rotations, then 15−6=9 vibrational dof. Then, $E=(3+3+(9\times2))$ $kT/2=12RT$ (1 mol) and $C_v=12R$.
- C_2H_6 (ethane, non-linear). There are 8×3=24 dof: 3 for translations, 3 for rotations, then 24−6=18 vibrational dof. Then, $E=(3+3+(18\times2))$ $kT/2=21RT$ (1 mol) and $C_v=21R$. One of the vibrational dof is the dihedral angle that describes the torsion of a methyl group with respect to the other and shows 3 equivalent minimums separated by maximums at not particularly high energy. When the temperature is sufficiently high, the motion relative to this coordinate becomes essentially free and therefore behaves like a rotation. It is called an **internal rotation** to distinguish it from the rotations of the entire molecule as a rigid body.

2 LINEAR SPACES (VECTOR SPACES)

A *linear space* (also called *vector space*) is an algebraic structure shared by several sets that have applications in many areas of science. Linear spaces are, for example:

- the vectors of ordinary three-dimensional (3D) space, here indicated by an upper arrow $\vec{r}$,
- the vectors (also indicated as $\vec{r}$) of the configuration space of a system of particles, in particular the configuration of a molecule when it is treated as a system of point-like atoms,
- the functions (now indicated by Greek letters) that are solutions of the Schrödinger equation of a QM problem.

The algebra of linear spaces deals with the relations between the elements (the vectors or wave functions) of these sets, studies the properties of these relations and derives the general consequences by abstracting from the specific features of the different sets or vectors.

Leaving the formal treatment to algebra textbooks or to specialized QM textbooks, here we focus on the points that are needed to understand the basic language of QM. For this limited purpose, and completely renouncing rigor in the exposition, we can follow a presentation of linear spaces based on the analogy with what is already known from previous courses on operations between *vector quantities, i.e.* quantities such as positions, forces, velocities, accelerations, …

2.1 Basic Operations

The most familiar case of linear space are surely the vectors of ordinary three-dimensional (3D) space, which are represented graphically by oriented segments and from now on are called *ordinary* vectors. A note: a vector is *not* simply an oriented segment, because two oriented segments that have the same direction, and the same length are considered the *same* vector even if they have a different starting point.

A linear space is first a set of vectors. The essential point of the algebraic structure is that on these vectors we can perform *two operations that define new vectors starting from given vectors* and that *these operations have specific properties*. The operations are:

- the **vector addition:** an operation that from two vectors ($\vec{r}$ and $\vec{s}$) produces a third vector $\vec{t}=\vec{r}+\vec{s}$
- and **the multiplication of a vector by a scalar**, that is by a number: with this operation, from a vector ($\vec{r}$) and a number (α) a second vector $\alpha\vec{r}$ is obtained.

For vectors in ordinary 3D space, these operations have a simple geometric meaning that we can describe in the well-known way:

- vector addition is defined by the parallelogram rule,
- multiplication of a vector by a scalar defines a new vector that has the same direction as the given vector, has length multiplied by the scalar and finally, if the scalar is negative, has the opposite direction.

The structure of linear space thus involves **two sets**: a *main set* (the set of vectors) and an *auxiliary set* of scalars (numbers).

In an equivalent way, we can say that we can make **linear combinations**: from two vectors ($\vec{r}$ and $\vec{s}$) and two numbers (α and β) a new vector $\vec{v}=\alpha\vec{r}+\beta\vec{s}$ is obtained.

These operations have a first and fundamental property, called *closure*. This means that

> Every linear combination of elements of a linear space is an element of the space.

We can also say that *with linear combinations we remain in the set* of vectors. To clarify, it may be useful to give an example of a set and an operation that do *not* have the closure property. Let us consider the set of natural numbers. Concerning the addition operation, on the one hand, we see that we can always add two natural numbers, and the result is always a natural number. So, the closure property holds for addition. On the other hand, it does not hold for division: if we divide two natural numbers, we can obtain a non-integer fraction, that is, we may end up outside the set of natural numbers. We can say, in an equivalent way, that *in the set of natural numbers* the division is not defined for each couple of numbers.

In the case of three-dimensional ordinary space and configuration spaces, the auxiliary set is the field of real numbers, while in the case of linear function spaces of QM it is the field of complex numbers. The word *field* has a precise technical meaning, because it in turn specifies a particular algebraic structure. We do not discuss the field structure, taking for granted that the properties of real numbers and complex numbers are known to the reader.

Moreover, the vector space structure implies that the linear combination has *other important properties*, in addition to the closure property. These properties appear obvious to us when we examine the case of ordinary 3D vectors. For example (the list is not complete):

- Vector addition is commutative: $\vec{a}+\vec{b}=\vec{b}+\vec{a}$,
- And associative: $(\vec{a}+\vec{b})+\vec{c}=\vec{a}+(\vec{b}+\vec{c})$,
- There is a (unique) null vector $\vec{0}$, such that: $\vec{a}+\vec{0}=\vec{0}+\vec{a}=\vec{a}$
- There is the opposite $-\vec{a}$ of every vector $\vec{a}$, and it is unique,
- Two distributive properties hold between vector addition and scalar multiplication: (i) $k(\vec{a}+\vec{b})=k\vec{a}+k\vec{b}$ and (ii) $(k+m)\vec{a}=k\vec{a}+m\vec{a}$.

These properties and their consequences specify at the **abstract** level a *linear space*. Adding two functions, obviously, is not the same thing as adding two ordinary vectors by applying the parallelogram rule, multiplying a function by a number is not the same thing as expanding or compressing an ordinary vector by multiplying it by a scalar, etc.... So, saying - for example - that the wave functions of a QM problem are a linear space means that the wave functions can be combined with each other with operations that have the same *formal properties* as the operations on the vectors of ordinary space even if they are different in detail. Some examples can clarify the meaning of the statements.

- The sum of two functions which are solutions of a Schrödinger equation is also a solution of the same Schrödinger equation, *i.e.* the closure property holds.
- The addition of functions is commutative and associative.
- There is a (unique) null function, *i.e.* a function that, when added to any other function of the set, leaves it unchanged: it is the function that is zero everywhere. From the postulates of QM, we also know that this function has no *physical* meaning.
- The distributive properties between the addition of functions and the multiplication of a function by a scalar hold.
- The opposite of every function exists, in the sense that if a function is a solution of a certain Schrödinger equation, then its opposite is also a solution of the same equation.

As we see, these are the same properties listed above for ordinary vectors and for this reason we say that ordinary vectors and QM wave functions have the same algebraic structure.

Furthermore, on a vector space, *other operations can be defined* with their relative properties. An important case is the *scalar product between two vectors* and the *orthogonality relation* that will be discussed in Section 2.2. These additional operations are *not necessary* for the algebraic structure of linear space and, when they are added, it is necessary to verify that they are compatible with it.

This presentation of linear spaces relies on the analogy with ordinary vectors, avoiding a rigorous treatment. Therefore, to become acquainted with the abstract structure of linear space and, in particular, the linear spaces of wave functions in QM, it is sufficient to compare the properties of ordinary vectors with the properties of wave functions, highlight the different meanings in the two contexts and discuss the consequences. The points to note are as follows:

a. The use of complex numbers instead of the real numbers (already seen),
b. The dimension of linear space,
c. The completeness of linear space,
d. The meaning of measures such as the length of a vector and the angle between two vectors (from which the orthogonality between vectors can be introduced).

2.1.1 The Dimension of a Linear Space

Ordinary space has dimension 3, a plane passing through the origin of ordinary 3D space has dimension 2, etc. .... This seems trivial but requires attention to correctly introduce abstraction.

Let's see what it means. Given one vector, we can talk about its linear combinations, that is, the set of all vectors obtained by multiplying it by all possible scalars. A line (a 1D linear space) is the set of all linear combinations of a given vector.

Given two non-collinear vectors ($\vec{v_1}$ and $\vec{v_2}$), the set of all their linear combinations such as $\vec{a}=a_1\vec{v_1}+a_2\vec{v_2}$ where a_1 and a_2 are arbitrary numbers is a plane, *i.e.* a 2D space. It is also said that $\vec{v_1}$ and $\vec{v_2}$ are a *basis of linear combinations*. The given vector $\vec{a}$ can be obtained as linear combination of two other non-collinear basis vectors in place of $\vec{v_1}$ and $\vec{v_2}$: just choose two vectors belonging to the same plane to which $\vec{v_1}$ and $\vec{v_2}$ belong. Of course, to obtain $\vec{a}$ you need different coefficients depending on the basis.

We can repeat the consideration for the entire space (now using three or more basis vectors).

All the above can be done with arbitrary vector bases, that is, with bases consisting not only of different vectors, but also of different *numbers* of vectors. For example, we can express the vectors of a plane or of ordinary 3D space as a linear combination of three, four, five, ... non-collinear basis vectors. Now, we can ask how many basis vectors we must choose in order to be able to represent all the vectors of a given linear space using all linear combinations.

For simplicity, let us consider a plane of ordinary space. If we choose a single basis vector, we know that with its linear combinations we cannot obtain all the vectors of the plane, because we obtain only the vectors that lie on the same line as the basis vector. With two basis vectors we reach all the points of the plane, provided that the second vector is *not* a simple multiple (*i.e.* a linear combination) of the first one. We say that it must be **linearly independent** of the first. We can equally reach the whole plane if we use three, four, ... basis vectors, but we understand that the third, fourth, ... basis vector can be written as a linear combination of the first two and therefore that *we have at most two linearly **independent** basis vectors.*

Moving on to ordinary 3D space, we know that we have *at most three linearly independent basis vectors.* In conclusion, we say that ordinary space has dimension 3 (or that it is three-dimensional: 3D), that a plane has dimension 2, ...

The dimension of a vector space is the *maximum number* of *linearly independent vectors* that can be chosen as *basis vectors.*

Warning:

The case in which this maximum number does **not** exist, makes sense and is important. It is then said that the *dimension is infinite.*

The difference between finite and infinite dimension is *not* trivial: infinite-dimensional linear spaces present additional mathematical difficulties compared to finite-dimensional vector spaces, as illustrated below.

2.1.2 Complete Linear Space

A linear space is said to be *complete* when it is possible to obtain, from linear combinations of suitable basis vectors, all the elements of the linear space. Completeness is a trivial property for finite-dimensional spaces. With a non-co-planar, *i.e.* linearly independent, set of three vectors we "reach" any point in the ordinary 3D space. With a non-co-linear pair of vectors, we "reach" any point in ordinary 2D plane. By multiplying any basis vector by an appropriate scalar, we "reach" any point on a straight line.

The property is no longer trivial when it concerns infinite-dimensional spaces because it is necessary to ensure that expressions of the type: $\psi_a = \sum_{i=1}^{\infty}(a_i\varphi_i)$ make sense for any choice of linearly independent basis vectors (φ_i). The series, in fact, *cannot* be briefly treated as a "sum of infinite terms", but it is the synthetic notation of a procedure consisting in the search for the limit of a sequence of partial sums. To study the existence of the limit, it is necessary to be able to demonstrate that "as k increases", the "difference" between $\psi_a \overset{\text{def}}{=} \lim_{n\to+\infty}\sum_{i=1}^{n}(a_i\varphi_i)$ and the partial sums $\sum_{i=1}^{k}(a_i\varphi_i)$, tends to zero, once the meaning of the expressions in quotation marks has been specified. Both each partial sum and the limit are vectors, and the difference is more precisely a distance between these two vectors. So, it is in fact necessary to be able to prove that the distance between limit and partial sum is less than any preselected real number, if an appropriate k is chosen.

Then, there is a connection between **completeness** and the possibility of defining a **distance between elements of a linear space**. This is an important aspect that requires the introduction of the additional notion of scalar product.

2.2 Components, Projections and Dot Product

When discussing ordinary 3D vectors, reference is often made to other quantities, operations and properties, in addition to vectors and scalars, to the operations of adding and multiplying by a scalar, and to the properties that derive from them. It is necessary to introduce, for example, the *projection* of a vector onto another vector, or it is necessary to consider the *length* of a vector, the *distance* and the *angle* between two vectors, as well as the fact that a vector has *length* = 1 or that two vectors are *orthogonal* (angle = 90°), that is, that two vectors have zero projection one along the direction of the other. It is essentially necessary to introduce the notions of length, distance and angle, that is (to put it concisely) to introduce a **metric**.

Considering the spaces of ordinary vectors, one can see that these additional quantities and operations are closely related to each other and can be derived from the single operation of *projecting* one vector onto another vector which, in turn, can be derived from the **scalar product of two vectors** also known as the **dot product**.

Beware of the confusion that names can create: the scalar product of two vectors is an operation that produces a number from two vectors. Among the operations that define the structure of linear space we had seen instead the product of a vector by a scalar, an operation that produces another vector from a vector and a scalar:

- Product of a vector by a scalar: (vector, number) ⇒ vector
- Dot (scalar) product of two vectors: (vector, vector) ⇒ number

The dot product of two vectors is different from the operations that define a linear space, even if it is compatible with them. To clarify that the dot product is an addition to the algebraic structure of a vector space, it may be useful to note that different scalar products can be defined on the same vector space.

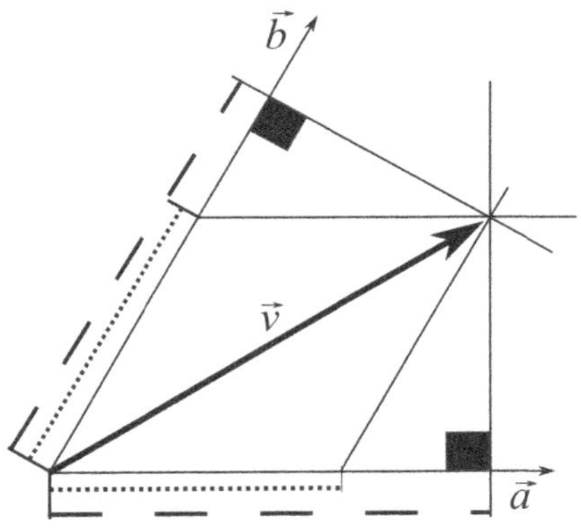

FIGURE 5.7 The basis vectors are $\vec{a}$ and $\vec{b}$. Consider the vector $\vec{v}$: its components along $\vec{a}$ and along $\vec{b}$ are represented by dotted lines, its projections by dashed lines and the black squares mark right angles. Components differ from projections because the basis vectors are *not* orthogonal to each other.

The scalar product essentially quantifies the projection of a vector on another vector. There can be confusion between two terms: component and projection (see Figure 5.7).

For a given vector ($\vec{v}$):

- We speak of a *component* (only) along one of the basis vectors, while we may want to determine the *projection* of $\vec{v}$ on *any* vector (even on $\vec{v}$ itself),
- The component along one of the basis vectors is the coefficient of the linear combination, *i.e.* a_r, if it is the component of vector $\vec{v} = \sum_{k=1} a_k \vec{v_k}$ along the basis vector $\vec{v_r}$;
- A component coincides with a projection if and only if the basis vectors are (two by two) orthogonal to each other.

Therefore, the meaning of a *component derives directly from the operations that define a vector space.* Instead, the *projection* requires another operation that must be *added explicitly.* This is the dot (scalar) product of two vectors.

The *elementary geometric definition* of the dot product $(\vec{a},\vec{b})$ of two vectors $\vec{a}$, $\vec{b}$ of ordinary 3D space requires the moduli (a and b) of the vectors and their angle $(\widehat{\vec{a},\vec{b}})$:

$$(\vec{a},\vec{b}) \overset{\text{def}}{=} |\vec{a}| \cdot |\vec{b}| \cdot \cos\left(\widehat{\vec{a},\vec{b}}\right)$$

$$\text{(a number that is a function of } \textit{two vectors}) \qquad (2.1)$$

Knowing the dot product and given the unit of length, we then obtain various related quantities, with their respective properties. The theoretical sequence is as follows.

- The **norm** of a vector $(\vec{a},\vec{a})$ is the dot product of a vector by itself. So, the norm is always a *positive real number*, it is null if and only if $\vec{a}$ is the null vector.
- The **length** $\sqrt{(\vec{a},\vec{a})}$ it is the square root of the norm. From the properties of the norm, it follows that also the length of a vector is always a positive real number but is zero if and only if $\vec{a}$ is the null vector.
- $\vec{a} / \sqrt{(\vec{a},\vec{a})}$ gives a vector with **unit length**.
- The **projection** of vector $\vec{a}$ on vector $\vec{b}$ is the *number* given by the dot product of the vectors divided by the length of vector $\vec{b}$:[7]
- $(\vec{a},\vec{b}) / b \overset{\text{def}}{=} (\vec{a},\vec{b}) / \sqrt{(\vec{b},\vec{b})}$
- The **distance** between two vectors is the *length* of the difference vector:

$$d(\vec{a},\vec{b}) \overset{\text{def}}{=} \sqrt{(\vec{b}-\vec{a})\cdot(\vec{b}-\vec{a})}$$

- The **angle between two vectors = dot product/ product of lengths:**

$$\left(\widehat{\vec{a},\vec{b}}\right) \overset{\text{def}}{=} \arccos \frac{(\vec{a},\vec{b})}{\sqrt{(\vec{a},\vec{a})} \cdot \sqrt{(\vec{b},\vec{b})}}$$

- **orthogonality between two vectors** means that the projection of one vector along the other is zero.

All these quantities have properties that derive from the properties of the scalar product, while the properties of the scalar product are in turn derived from the elementary geometric definition (Eq. 2.1). For example, based on the properties of the scalar product, one can derive these properties of distance:

- It is always a positive real number (or zero, when the two vectors are equal),
- It is symmetric: the distance between $\vec{a}$ and $\vec{b}$ is equal to the distance between $\vec{b}$ and $\vec{a}$,
- The triangle inequality holds: the distance between $\vec{a}$ and $\vec{c}$ is less than or equal to the distance between $\vec{a}$ and $\vec{b}$ plus the distance between $\vec{b}$ and $\vec{c}$: $d(\vec{a},\vec{c}) \le d(\vec{a},\vec{b}) + d(\vec{b},\vec{c})$

Now, we move to the abstract level by reversing the procedure. First, we *define* a quantity as a "scalar product", specifying only its formal properties as follows.

1. It is defined for every pair of elements of linear space.
2. It is symmetric: $(\vec{a},\vec{b}) = (\vec{b},\vec{a})$ (for a linear space on the real field) or it has *conjugate symmetry*:

$\left(\vec{a},\vec{b}\right)=\left(\vec{b},\vec{a}\right)^*$ (for a linear space on the complex field).

3. It is linear in the second argument: $\left(\vec{a},\left(\beta\cdot\vec{b}+\gamma\cdot\vec{c}\right)\right)=\beta\cdot\left(\vec{a},\vec{b}\right)+\gamma\cdot\left(\vec{a},\vec{c}\right)$ and antilinear in the first argument: $\left(\left(\beta\cdot\vec{b}+\gamma\cdot\vec{c}\right),\vec{a}\right)=\beta^*\cdot\left(\vec{b},\vec{a}\right)+\gamma^*\left(\vec{c},\vec{a}\right)$.
4. If $\vec{a}$ is *not* the null vector: $\left(\vec{a},\vec{a}\right)$ is a positive real numb; if $\vec{a}$ is the null vector: $\left(\vec{a},\vec{a}\right)=0$.

Note that the generalization to the abstract scalar product also considers the problems related to the *linear spaces over complex numbers* (rather than over real numbers). As an example, one can consider the property of the scalar product defined by Eq. (2.1) to be symmetric: $\left(\vec{a},\vec{b}\right)=\left(\vec{b},\vec{a}\right)$. If one instead considers the scalar product for a linear space over the complex field, the definition is: $\left(\vec{a},\vec{b}\right)=\left(\vec{b},\vec{a}\right)^*$. The generalized definitions are also applicable to infinite-dimensional linear spaces. Both these aspects are obviously essential for applications in QM.

Now, once defined the scalar product, the theory defines the derived quantities (norm, distance, …) as a function of the abstract scalar product and derives the abstract properties from the properties postulated for the scalar product.

In applications, if one wants to introduce a scalar product for a specific vector space, it is obviously defined using operations that make sense for the vectors of that space. Then, it is sufficient to verify that the proposed definition satisfies the properties (1)–(4). If so, also all the derived quantities are introduced, and it is guaranteed that they have the expected properties.

A linear (or vector) space over the complex field, with a scalar (dot) product and complete, is called a **Hilbert space**. The states of a QM problem are a Hilbert space.

3 MATRIX CALCULUS

3.1 Definitions

A matrix is a rectangular pattern of real or complex numbers on n rows and m columns ($n\times m$ matrix):

$$\begin{pmatrix} A_{11} & A_{12} & \cdots \\ A_{21} & A_{22} & \cdots \\ \vdots & \vdots & \ddots \end{pmatrix}$$

A_{ij} indicates the element in row i and column j. If necessary, A_{ij} can be used to define a matrix. For example, the notation $A_{ij}=2i+j$ defines the matrix $\begin{pmatrix} 3 & 4 & \cdots \\ 5 & 6 & \cdots \\ \vdots & \vdots & \ddots \end{pmatrix}$

3.1.1 Some Matrices

Important are **square matrices** ($n\times n$), **column matrices** ($n\times 1$) and **row matrices** ($1\times n$).[8]

Examples:

A square matrix (3×3): $\begin{pmatrix} 1.25 & \pi & 2.3+3.5\,i \\ -i & -0.23 & -1/2 \\ \mathrm{e} & \sqrt{2-2i} & 5.5 \end{pmatrix}$

A column matrix (3×1): $\begin{pmatrix} 7 \\ \mathrm{e} \\ 2.3 \end{pmatrix}$

And a row matrix (1×3): $\begin{pmatrix} 1.25 & \pi & 3.5\,i \end{pmatrix}$

To indicate a property of a matrix $\boldsymbol{A}$, one often indicates the property of the generic element $\boldsymbol{A}_{ij}$, meaning that the property is valid for every i and every j, as for example in the definition given below (see the following Eqs. (3.1), (3.2), (3.6), and (3.7)).

There are also the **diagonal matrices**: these are the square matrices that have non-zero elements only along the main diagonal, *i.e.* the elements $\boldsymbol{A}_{ii}$. Finally, among the diagonal matrices, in the **scalar matrices** all elements along the main diagonal equal.

A diagonal matrix (3×3): $\begin{pmatrix} 1.25 & 0 & 0 \\ 0 & -0.235 & 0 \\ 0 & 0 & 1.2+i \end{pmatrix}$

A scalar matrix (3×3): $\begin{pmatrix} 1.25 & 0 & 0 \\ 0 & 1.25 & 0 \\ 0 & 0 & 1.25 \end{pmatrix}$

The **transpose** of an $n\times m$ matrix $\boldsymbol{A}$ is the $m\times n$ matrix that contains the same elements as $\boldsymbol{A}$ rearranged by exchanging rows with columns. It is denoted with the **superscript T**:

$$\boldsymbol{A}^T:\ \left(\boldsymbol{A}^{\mathrm{T}}\right)_{ij}=\boldsymbol{A}_{ji} \tag{3.1}$$

Example: if $\boldsymbol{A}=\begin{pmatrix} 1.25 & \pi & 2.3+3.5\,i \\ -i & -0.32 & 1/2 \end{pmatrix}$

then $\boldsymbol{A}^{\mathrm{T}}=\begin{pmatrix} 1.25 & -i \\ \pi & -0.32 \\ 2.3+3.5\,i & 1/2 \end{pmatrix}$

When dealing with **complex** matrices, *i.e.* matrices whose elements are complex numbers, instead of the transpose we are typically interested in the **adjoint** or **Hermitian conjugate**, now indicated with the dagger (†) sign. This is the matrix obtained both with the row/column exchange *and* the transition to the complex conjugate:

$$\boldsymbol{A}^\dagger:\ \left(\boldsymbol{A}^\dagger\right)_{ij}=\boldsymbol{A}_{ji}^* \tag{3.2}$$

$$\text{Example: if } \boldsymbol{A} = \begin{pmatrix} 1.25 & \pi & 2.3+3.5\,i \\ -i & -0.32 & 1/2 \end{pmatrix}$$

$$\text{then} \quad \boldsymbol{A}^{\dagger} = \begin{pmatrix} 1.25 & +i \\ \pi & -0.32 \\ 2.3-3.5\,i & 1/2 \end{pmatrix}$$

It can be useful to know that the transpose of a sum is the sum of the transposes and that the adjoint of a sum is the sum of the adjoints:

$$\text{if } \boldsymbol{C} = \boldsymbol{A} + \boldsymbol{B} \quad \text{then} \quad \boldsymbol{C}^{\dagger} = \boldsymbol{A}^{\dagger} + \boldsymbol{B}^{\dagger} \tag{3.3}$$

The proof is trivial: just write explicitly the elements *ij* of the various matrices.

In the case of square matrices, it may happen that a matrix is equal to its transpose or its adjoint. These cases are important and deserve specific names: a matrix is said to be **symmetric** or **self-adjoint** or – more simply – **Hermitian** in the case of the adjoint matrix:

$$\text{Symmetric matrix: } \boldsymbol{A} = \boldsymbol{A}^{T}: \quad A_{ij} = A_{ji} \tag{3.4}$$

$$\text{Example: } \boldsymbol{A} = \begin{pmatrix} 1.25 & \pi & 2.3+3.5\,i \\ \pi & -0.32 & -1/2 \\ 2.3+3.5\,i & -1/2 & 5.5 \end{pmatrix}$$

$$\text{Hermitian matrix: } \boldsymbol{A} = \boldsymbol{A}^{\dagger}: \quad \boldsymbol{A}_{ij} = \boldsymbol{A}_{ji}^{*} \tag{3.5}$$

$$\text{Example: } \boldsymbol{A} = \begin{pmatrix} 1.25 & \pi & 2.3-3.5\,i \\ \pi & -0.32 & -1/2 \\ 2.3+3.5\,i & -1/2 & 5.5 \end{pmatrix}$$

Usually, symmetric matrices are important when dealing with real matrices, Hermitian matrices when dealing with complex matrices. Of course, a real Hermitian matrix is symmetric.

3.1.2 Linear Combination of Matrices

For matrices, various operations are defined, starting with the **addition** between two matrices and the **multiplication by a scalar**. These are simple and intuitive operations because they are performed element by element, but they are also fundamental, because, as we will see in Section 3.2.1, *they correspond directly to the operations that are the basis of the notion of linear space.*

You can **add** *two matrices only if they have the same dimensions:* you can add an $n \times m$ matrix only with another $n \times m$ matrix. The sum is a matrix with the same dimensions and the *element ij* of $\boldsymbol{C}=\boldsymbol{A}+\boldsymbol{B}$ is the sum of the *elements ij* of the matrices to add:

$$\boldsymbol{C} = \boldsymbol{A} + \boldsymbol{B}: \quad \boldsymbol{C}_{ij} = \boldsymbol{A}_{ji} + \boldsymbol{B}_{ij} \tag{3.6}$$

$$\text{For example: } \begin{pmatrix} 1 & 2 & 3 \\ 4 & 5 & 6 \end{pmatrix} + \begin{pmatrix} 2 & 4 & 6 \\ 6 & 4 & 2 \end{pmatrix} = \begin{pmatrix} 3 & 6 & 9 \\ 10 & 9 & 8 \end{pmatrix}$$

The **product** of a matrix **by a scalar** can always be done, no matter what the size of the matrix is, and is:

$$\boldsymbol{D} = k\boldsymbol{A}: \quad \boldsymbol{D}_{ij} = k\boldsymbol{A}_{ji} \tag{3.7}$$

$$\text{For example: } 3.5\begin{pmatrix} 1 & 2 & 3 \\ 4 & 5 & 6 \end{pmatrix} = \begin{pmatrix} 3.5 & 7 & 10.5 \\ 14 & 17.5 & 21 \end{pmatrix}$$

Among the $(n \times m)$ matrices there is a matrix that plays the role of neutral element with respect to the addition between matrices $(n \times m)$, and it is the **null matrix**, with all the elements$=0$:

$$\boldsymbol{0}_{ij} = 0: \forall i = 1, \ldots, n; \quad \forall j = 1, \ldots, m \tag{3.8}$$

$$\text{For example: the } 2 \times 3 \text{ null matrix is: } \begin{pmatrix} 0 & 0 & 0 \\ 0 & 0 & 0 \end{pmatrix}$$

For the null matrix *of order* $n \times m$ and for *any* matrix ($\boldsymbol{A}$) of the same order, therefore, it holds:

$$\boldsymbol{A} + \boldsymbol{0}_{nm} = \boldsymbol{0}_{nm} + \boldsymbol{A} = \boldsymbol{A}\ (\boldsymbol{A} \text{ is } n \times m) \tag{3.9}$$

In the particular case of square $(n \times n)$ matrices, the null matrix is simply said to be of order n: $\boldsymbol{0}_n = \boldsymbol{0}_{nn}$.

Then, it makes sense to define the **opposite** of a matrix (A) of dimensions $(n \times m)$:

$$\text{if } \boldsymbol{A} + \boldsymbol{B} = \boldsymbol{B} + \boldsymbol{A} = \boldsymbol{0}_{nm} \quad \text{we write } \boldsymbol{B} = -\boldsymbol{A} \quad \text{and} \quad \boldsymbol{A} = -\boldsymbol{B} \tag{3.10}$$

and it is said that $\boldsymbol{B}$ is the opposite of $\boldsymbol{A}$, and that $\boldsymbol{A}$ is the opposite of $\boldsymbol{B}$.

The corresponding elements in $\boldsymbol{A}$ and $\boldsymbol{B}$ are the opposite of each other: $\boldsymbol{A}_{ij} = -\boldsymbol{B}_{ij}$.

3.1.3 Matrix Product

Matrix-by-matrix multiplication is a *more complicated* operation than matrix addition and multiplication of a matrix by a scalar. First, *it is not always possible: for matrix-by-matrix multiplication to make sense, the second dimension of the first factor must be equal to the first dimension of the second factor.* An $n \times m$ matrix $\boldsymbol{A}$ and an $r \times s$ matrix $\boldsymbol{B}$ $(r \times s)$, can be multiplied together (in this order) only if $m = r$ and the result $\boldsymbol{C}=\boldsymbol{A}\cdot\boldsymbol{B}$ is an $n \times s$ matrix:

Congruity rule for multiplication between matrices:

$$(n \times m) \times (m \times s) \Rightarrow (m \times s)$$

For example, you can't multiply these matrices:

$$\begin{pmatrix} 1 & 2 & 3 \\ 4 & 5 & 6 \end{pmatrix} \times \begin{pmatrix} 2 & 4 & 6 \\ 6 & 4 & 2 \end{pmatrix}.$$

Instead, you can multiply the matrices $\begin{pmatrix} 1 & 2 & 3 \\ 4 & 5 & 6 \end{pmatrix}$ and $\begin{pmatrix} 2 & 6 \\ 4 & 4 \\ 6 & 2 \end{pmatrix}$. The result is a 2×2 matrix.

The algorithm that defines matrix x matrix multiplication is that each element of $\boldsymbol{C}=\boldsymbol{A}\cdot\boldsymbol{B}$ is obtained from a **row x column product** made according to Eq. (3.11) and the scheme of Figure 5.8:

Matrix-by-matrix product:

$$C_{ij} = \sum_{k=1}^{m} A_{ik} B_{kj} \tag{3.11}$$

Completing the previous example, the product

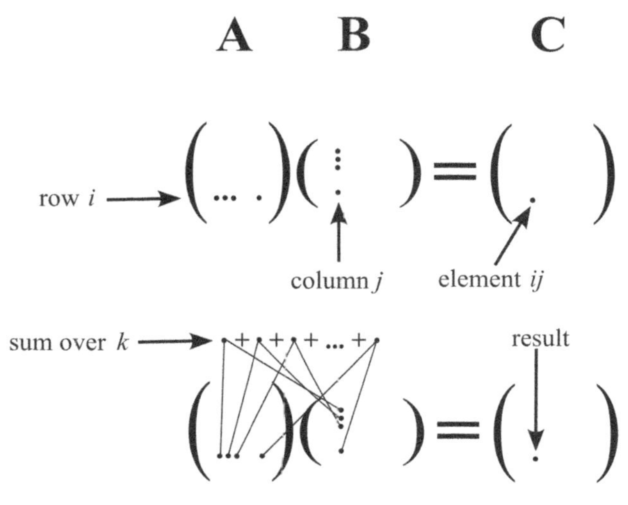

FIGURE 5.8 Scheme for the row-by-column product of matrices.

$$\boldsymbol{C} = \begin{pmatrix} 1 & 2 & 3 \\ 4 & 5 & 6 \end{pmatrix} \cdot \begin{pmatrix} 2 & 6 \\ 4 & 4 \\ 6 & 2 \end{pmatrix} \text{ is:}$$

$$C_{11} = 1\cdot 2 + 2\cdot 4 + 3\cdot 6 = 28 \qquad C_{12} = 1\cdot 6 + 2\cdot 4 + 3\cdot 2 = 20$$
$$C_{21} = 4\cdot 2 + 5\cdot 4 + 6\cdot 6 = 64 \qquad C_{22} = 4\cdot 6 + 5\cdot 4 + 6\cdot 2 = 56$$

So, we write: $\begin{pmatrix} 1 & 2 & 3 \\ 4 & 5 & 6 \end{pmatrix} \cdot \begin{pmatrix} 2 & 6 \\ 4 & 4 \\ 6 & 2 \end{pmatrix} = \begin{pmatrix} 28 & 20 \\ 64 & 56 \end{pmatrix}$.

To indicate the matrix x matrix product, we either use a dot ($\boldsymbol{A}\cdot\boldsymbol{B}$ or $\boldsymbol{A}\bullet\boldsymbol{B}$) or simply place the factors next to each other ($\boldsymbol{AB}$) as preferred.

Notes:

- When it is possible to calculate the product $\boldsymbol{AB}$, it is *not necessarily possible* to calculate the product $\boldsymbol{BA}$. It is possible *only when two square matrices with the same size* are multiplied
- The **two products $\boldsymbol{AB}$ and $\boldsymbol{BA}$** between two square matrices of the same size are *not necessarily* equal: matrix multiplication is **not necessarily commutative**.

A simple counterexample may suffice. If $\boldsymbol{A} = \begin{pmatrix} 1 & 2 & 3 \\ 2 & 3 & 4 \\ 3 & 4 & 5 \end{pmatrix}$ and $\boldsymbol{B} = \begin{pmatrix} 10 & 8 & 6 \\ 8 & 6 & 4 \\ 6 & 4 & 2 \end{pmatrix}$,

$$\text{then: } \boldsymbol{AB} = \begin{pmatrix} 1 & 2 & 3 \\ 2 & 3 & 4 \\ 3 & 4 & 5 \end{pmatrix} \begin{pmatrix} 10 & 8 & 6 \\ 8 & 6 & 4 \\ 6 & 4 & 2 \end{pmatrix} = \begin{pmatrix} 44 & 32 & 20 \\ 68 & 50 & 32 \\ 92 & 68 & 44 \end{pmatrix}$$

$$\text{and: } \boldsymbol{BA} = \begin{pmatrix} 10 & 8 & 6 \\ 8 & 6 & 4 \\ 6 & 4 & 2 \end{pmatrix} \begin{pmatrix} 1 & 2 & 3 \\ 2 & 3 & 4 \\ 3 & 4 & 5 \end{pmatrix} = \begin{pmatrix} 44 & 68 & 92 \\ 32 & 50 & 68 \\ 20 & 32 & 44 \end{pmatrix}$$

- Diagonal matrices are a special case in this respect, because they commute: if $\boldsymbol{A}$ and $\boldsymbol{B}$ are diagonal matrices, then $\boldsymbol{AB}=\boldsymbol{BA}$. Obviously, the result is valid also for scalar matrices.
- The *transpose* of a product is the product of the transposes performed in *reverse order*. Analogously (in the case of complex matrices) for the *adjoint* of a product:

$$\text{if } \boldsymbol{D} = \boldsymbol{C}\cdot\boldsymbol{B} \quad \text{then} \quad \boldsymbol{D}^T = \boldsymbol{B}^T \cdot \boldsymbol{C}^T \quad \text{and} \quad \boldsymbol{D}^\dagger = \boldsymbol{B}^\dagger \cdot \boldsymbol{C}^\dagger \tag{3.12}$$

The proofs are easily obtained by explicitly writing the products according to Eq. (3.11).

- There is *another product of matrices* which is more precisely indicated as the *direct product* (see Section 3.3.2), while the product we are talking about here - when it is necessary to specify - is indicated as a **row-by-column product**: these are completely different operations which must not be confused with each other.

3.1.3.1 Inverse of a Matrix

Among the square ($n\times n$) matrices there is one matrix that plays the role of unit with respect to multiplication. It is called the **unit matrix of order n** and is the scalar matrix with all non-zero elements = 1:

$$\mathbf{1}_n: \ \mathbf{1}_{ij} = \delta_{ij} \quad \forall i = 1, \ldots, n; \quad \forall j = 1, \ldots, n \tag{3.13}$$

Example: the unit matrix of the second order is $\begin{pmatrix} 1 & 0 \\ 0 & 1 \end{pmatrix}$

The symbol δ_{ij} (known as **Kronecker delta**) indicates a number that depends on two indices and is = 1 when the two indices are equal and = 0 when the two indices are different.

For the unit matrix of order n and for any $(n\times n)$ matrix $\boldsymbol{A}$:

$$\boldsymbol{A}\cdot\boldsymbol{1}_n=\boldsymbol{1}_n\cdot\boldsymbol{A}=\boldsymbol{A} \qquad (3.14)$$

The proof is a simple application of the definition of matrix product. Given the existence of a unit matrix, it makes sense to define the inverse of an $(n\times n)$ matrix $\boldsymbol{A}$:

$$\text{if } \boldsymbol{A}\cdot\boldsymbol{B}=\boldsymbol{B}\cdot\boldsymbol{A}=\boldsymbol{1}_n \quad \text{we write } \boldsymbol{B}=\boldsymbol{A}^{-1} \text{ and } \boldsymbol{A}=\boldsymbol{B}^{-1} \qquad (3.15)$$

We say that $\boldsymbol{B}$ is the inverse of $\boldsymbol{A}$ and $\boldsymbol{A}$ is the inverse of $\boldsymbol{B}$.

The inverse of a product is the product of the inverses performed in the other order:

$$\text{if } \boldsymbol{D}=\boldsymbol{C}\cdot\boldsymbol{B} \quad \text{then} \quad \boldsymbol{D}^{-1}=\boldsymbol{B}^{-1}\cdot\boldsymbol{C}^{-1} \qquad (3.16)$$

The proof can be easily obtained by multiplying from the left by the inverses:

$$\boldsymbol{D}=\boldsymbol{C}\cdot\boldsymbol{B}\Rightarrow\boldsymbol{C}^{-1}\cdot\boldsymbol{D}=\boldsymbol{C}^{-1}\cdot\boldsymbol{C}\cdot\boldsymbol{B}=\boldsymbol{B}\Rightarrow$$

$$\boldsymbol{B}^{-1}\cdot\left(\boldsymbol{C}^{-1}\cdot\boldsymbol{D}\right)=\boldsymbol{B}^{-1}\cdot\boldsymbol{B}=\boldsymbol{1}_n$$

$$\Rightarrow\left(\boldsymbol{B}^{-1}\cdot\boldsymbol{C}^{-1}\right)\cdot\boldsymbol{D}\Rightarrow(\boldsymbol{C}\cdot\boldsymbol{B})^{-1}=\boldsymbol{D}^{-1}=\left(\boldsymbol{B}^{-1}\cdot\boldsymbol{C}^{-1}\right)$$

Not all square $(n\times n)$ matrices can be inverted, *i.e.* **the inverse does not always exist**, but if it does, the inverse is **unique**.

3.1.3.2 Orthogonal Matrices and Unitary Matrices

It is easy to *verify* that a matrix ($\boldsymbol{R}$) is the inverse of a given matrix ($\boldsymbol{A}$): it is enough to calculate the product $\boldsymbol{RA}$ (or $\boldsymbol{AR}$) which must be equal to the unit matrix. On the contrary, *finding* the inverse usually requires long and complex calculations (not discussed here). By increasing the size of the matrix, in symbolic calculation one quickly arrives at formulas of intractable complexity. In numerical calculation the algorithms are computationally expensive and therefore in applications one usually avoids - when possible - the calculation of the inverse matrix. For example, a linear system $\boldsymbol{Ax}=\boldsymbol{b}$ is *not* solved by inverting $\boldsymbol{A}$ and writing $\boldsymbol{x}=\boldsymbol{A}^{-1}\boldsymbol{b}$.

For **diagonal** matrices it is easy to find the inverse, because $(\boldsymbol{A}^{-1})_{ii}=1/\boldsymbol{A}_{ii}$ and the off-diagonal elements are zero. An even more important case is given by **orthogonal** or **unitary** matrices. An **orthogonal** matrix is a real invertible matrix whose transpose is equal to the inverse.

$$\textbf{\textit{Orthogonal matrix}}: A^{-1}=\mathbf{A}^T:\ \left(A^{-1}\right)_{ij}=A_{ji} \qquad (3.17)$$

Example: $A=\begin{pmatrix}\sqrt{3}/2 & -1/2 & 0\\ 1/2 & \sqrt{3}/2 & 0\\ 0 & 0 & 1\end{pmatrix}$ is orthogonal.

In fact:

$$\begin{pmatrix}\sqrt{3}/2 & -1/2 & 0\\ 1/2 & \sqrt{3}/2 & 0\\ 0 & 0 & 1\end{pmatrix}\cdot\begin{pmatrix}\sqrt{3}/2 & 1/2 & 0\\ -1/2 & \sqrt{3}/2 & 0\\ 0 & 0 & 1\end{pmatrix}=\begin{pmatrix}1&0&0\\0&1&0\\0&0&1\end{pmatrix}.$$

It is easily verified that, for an orthogonal matrix: (a) the sum of the squares of the elements of a column (or a row) is = 1; (b) the sum of the products of corresponding elements in two different columns (or rows) is = 0.

For applications concerning linear spaces on the complex field, we are interested in **unitary** matrices, *i.e.* complex matrices whose adjoint (*i.e.* the transpose of the conjugate, cf. Section 3.1.1) is equal to the inverse.

$$\textbf{\textit{Unitary matrix}}: A^{-1}=A^{\dagger}:\ \left(A^{-1}\right)_{ij}=(A_{ji})^{*} \qquad (3.18)$$

Orthogonal matrices and unitary matrices are important for similarity transformations, *i.e.* for transformations of linear operators because of a change of the basis vectors (see Section 3.2.2.2).

3.2 Matrices and Linear Spaces

Matrices are the tool for expressing many operations on linear spaces in a concise and easy way, essentially by grouping the **components** of a vector with respect to a given basis into a single **column matrix** and by grouping the equations relating to all the components into a **single equation**. For example, if we need to indicate the *components* of the vector ($\vec{r}$) of the ordinary 3D space in the basis $\begin{pmatrix}\vec{e_1} & \vec{e_2} & \vec{e_3}\end{pmatrix}$, *i.e.* $\vec{r}=r_1\vec{e_1}+r_2\vec{e_2}+r_3\vec{e_3}$, we group the components r_1,r_2,r_3 in a column matrix:

$$\boldsymbol{r}=\begin{pmatrix}r_1\\ r_2\\ r_3\end{pmatrix}$$

If it is necessary to indicate the components of the function (φ_r) in the base (φ_1, φ_2,…), that is if $\varphi_r=r_1\varphi_1+r_2\varphi_2+\cdots$ then the components are grouped in a column matrix in a similar way:

$$\boldsymbol{r}=\begin{pmatrix}a_1\\ a_2\\ \vdots\end{pmatrix}$$

Usually, in matrix notation, basis vectors or basis functions do not appear. However, a notation that explicitly includes basis functions is possible. The starting point is the above formulas, which we can write as the product of a *row of basis elements* and a *column of components*:

$$\vec{r}=r_1\vec{e_1}+r_2\vec{e_2}+r_3\vec{e_3}=\begin{pmatrix}\vec{e_1} & \vec{e_2} & \vec{e_3}\end{pmatrix}\begin{pmatrix}r_1\\ r_2\\ r_3\end{pmatrix}=\boldsymbol{e}\cdot\boldsymbol{r} \quad \text{or}$$

$$\varphi_a = a_1\varphi_1 + a_2\varphi_2 + \cdots = \begin{pmatrix} \varphi_1 & \varphi_2 & \cdots \end{pmatrix} \begin{pmatrix} a_1 \\ a_2 \\ \vdots \end{pmatrix} = \boldsymbol{\varphi} \cdot \boldsymbol{a} \quad (3.19)$$

Note that the choice of writing *the vector basis as a row matrix* and *the components as a column matrix* is **conventional**. The choice followed here is the prevailing convention in the scientific literature. Ideally, the opposite choice (with change in the order of the factors to respect the matrix multiplication rule, Figure 5.8) is equally acceptable. However, once a convention has been chosen, the opposite choice, that is

$$\vec{r} = r_1\vec{e_1} + r_2\vec{e_2} + r_3\vec{e_3} = \begin{pmatrix} r_1 & r_2 & r_3 \end{pmatrix} \cdot \begin{pmatrix} \vec{e_1} \\ \vec{e_2} \\ \vec{e_3} \end{pmatrix},$$

should be carefully avoided

3.2.1 Linear Combinations

The fundamental operations of a linear space (addition of two vectors and multiplication by a scalar, *i.e.* linear combinations) can be written directly and synthetically using the corresponding operations on matrices:

if $\vec{t} = \alpha\vec{r} + \beta\vec{s}$ and

$\boldsymbol{r} = \begin{pmatrix} r_1 \\ r_2 \\ r_3 \end{pmatrix}$ and $\boldsymbol{s} = \begin{pmatrix} s_1 \\ s_2 \\ s_3 \end{pmatrix}$ are the components

of $\vec{r}$ and $\vec{s}$ in a given base,

$$\text{then } \boldsymbol{t} = \begin{pmatrix} t_1 \\ t_2 \\ t_3 \end{pmatrix} = \alpha \begin{pmatrix} r_1 \\ r_2 \\ r_3 \end{pmatrix} + \beta \begin{pmatrix} s_1 \\ s_2 \\ s_3 \end{pmatrix}$$

or, if $\varphi_t = \alpha \cdot \varphi_r + \beta\varphi_s$ and

$\boldsymbol{r} = \begin{pmatrix} r_1 \\ r_2 \\ \vdots \end{pmatrix}$ and $\boldsymbol{s} = \begin{pmatrix} s_1 \\ s_2 \\ \vdots \end{pmatrix}$ are the components

of φ_r and φ_s in a given base,

$$\text{then (analogously) } \boldsymbol{t} = \begin{pmatrix} t_1 \\ t_2 \\ \vdots \end{pmatrix} = \alpha \begin{pmatrix} r_1 \\ r_2 \\ \vdots \end{pmatrix} + \beta \begin{pmatrix} s_1 \\ s_2 \\ \vdots \end{pmatrix}$$

3.2.2 Linear Operators

Unlike matrix addition and scalar multiplication, the *matrix-by-matrix multiplication does not deal directly with the operations that define a vector space* (linear combinations between vectors), but *plays an important role in the applications of linear spaces, in particular for*:

a. linear **operators,**
b. **changes of basis**, *i.e.* transformations related to the passage from one set of basis vectors to another set of basis vectors, where the "new" basis vectors are linear combinations of the "old" ones or vice versa, and
c. the **scalar product between vectors.**

A **linear operator** transforms a given vector into *a vector of the same space*. Its applications concern both the observables of QM and the linear transformations of elementary geometry in ordinary 3D space with a particular reference to symmetry.

In matrix calculus, the **product of a square matrix and a column matrix** is used to obtain, from the components of the given vector, the components of the transformed vector (in the same basis). For example, the **linear transformation** $\hat{A}$ of the vector $\vec{r}$ into the vector $\vec{s}$ having components **r** and **s** in the same basis, *i.e.* the transformation $\vec{s} = \hat{A}\vec{r}$, means that each component of $\vec{s}$ is the linear combination of the components of $\vec{r}$. The linear transformation of $\vec{r} = r_1\vec{e_1} + r_2\vec{e_2} + r_3\vec{e_3}$ therefore corresponds to the matrix equation:

$$\boldsymbol{s} = \boldsymbol{A} \cdot \boldsymbol{r}, \text{ that is } \begin{pmatrix} s_1 \\ s_2 \\ s_3 \end{pmatrix} = \begin{pmatrix} a_{11} & a_{12} & a_{13} \\ a_{21} & a_{22} & a_{23} \\ a_{31} & a_{32} & a_{33} \end{pmatrix} \begin{pmatrix} r_1 \\ r_2 \\ r_3 \end{pmatrix} \quad (3.20)$$

The matrix $\boldsymbol{A}$ obviously depends on the transformation, but also on the basis chosen for $\vec{r}$ and $\vec{s}$. Equation (3.20) also holds in an analogous way for a function space.

It is useful to note the *geometric meaning* of a linear transformation $\hat{A}$, because it can be used to obtain the corresponding matrix $\boldsymbol{A}$. First of all, it is necessary to make the basis explicit:

$\boldsymbol{e} = \begin{pmatrix} \vec{e_1} & \vec{e_2} & \vec{e_3} \end{pmatrix}$ and then write:

$$\vec{r} = r_1\vec{e_1} + r_2\vec{e_2} + r_3\vec{e_3} = \begin{pmatrix} \vec{e_1} & \vec{e_2} & \vec{e_3} \end{pmatrix} \begin{pmatrix} r_1 \\ r_2 \\ r_3 \end{pmatrix} = \boldsymbol{e} \cdot \mathbf{r}$$

$$\textit{Similarly } \vec{s} = s_1\vec{e_1} + s_2\vec{e_2} + s_3\vec{e_3} = \begin{pmatrix} \vec{e_1} & \vec{e_2} & \vec{e_3} \end{pmatrix} \begin{pmatrix} s_1 \\ s_2 \\ s_3 \end{pmatrix} = \boldsymbol{e} \cdot \boldsymbol{s}$$

Then, Eq. (3.20) becomes $\vec{s} = \boldsymbol{e} \cdot \boldsymbol{A} \cdot \boldsymbol{r}$.

3.2.2.1 Basis Change

The linear operator $\hat{A}$ acts on all vectors and thus transforms the basis vectors into new vectors:

$$\boldsymbol{e}' = \begin{pmatrix} \vec{e_1'} & \vec{e_2'} & \vec{e_3'} \end{pmatrix} = \hat{A}\begin{pmatrix} \vec{e_1} & \vec{e_2} & \vec{e_3} \end{pmatrix}$$

These can be written as linear combinations of the original basis vectors:

$$\vec{e_1'} = u_{11}\vec{e_1} + u_{21}\vec{e_2} + u_{31}\vec{e_3}, \quad \vec{e_2'} = u_{12}\vec{e_1} + u_{22}\vec{e_2} + u_{32}\vec{e_3}$$

$$\text{and} \quad \vec{e_3'} = u_{13}\vec{e_1} + u_{23}\vec{e_2} + u_{23}\vec{e_3}$$

We can write synthetically in matrix notation by grouping the coefficients u_{jk} in the matrix $\boldsymbol{U}$:

$$\begin{pmatrix} \vec{e_1'} & \vec{e_2'} & \vec{e_3'} \end{pmatrix} = \begin{pmatrix} \vec{e_1} & \vec{e_2} & \vec{e_3} \end{pmatrix}\boldsymbol{U}$$

As the linear operator $\hat{A}$ acts on all vectors (the basis vectors as well as each particular vector such as $\vec{r}$), the components ($\boldsymbol{r}$) of $\vec{r}$ in the original basis must be equal to the components of $\vec{s}$ in the transformed basis, because this is the geometric meaning of the transformation: $\vec{s}$ is the vector $\vec{r}$ as transformed by $\hat{A}$. One can therefore write $\vec{s} = \boldsymbol{e}' \cdot \boldsymbol{r} = \boldsymbol{e} \cdot \boldsymbol{s}$ but also $= \boldsymbol{e} \cdot \boldsymbol{A} \cdot \boldsymbol{r}$ from which $\boldsymbol{e}' = \boldsymbol{e} \cdot \boldsymbol{A}$ and one sees that the matrix $\boldsymbol{U}$ in fact is the matrix $\boldsymbol{A}$ of Eq. (3.20) and therefore the basis transformation is:

$$\begin{pmatrix} \vec{e_1'} & \vec{e_2'} & \vec{e_3'} \end{pmatrix} = \begin{pmatrix} \vec{e_1} & \vec{e_2} & \vec{e_3} \end{pmatrix} \cdot \begin{pmatrix} a_{11} & a_{12} & a_{13} \\ a_{21} & a_{22} & a_{23} \\ a_{31} & a_{32} & a_{33} \end{pmatrix}$$

Note that the components ($\boldsymbol{r}$) of a vector $\vec{r}$ are multiplied *on the left* by the matrix $\boldsymbol{A}$ (*i.e.* $\boldsymbol{A} \bullet \boldsymbol{r}$), while the basis vectors $\boldsymbol{e} = \begin{pmatrix} \vec{e_1} & \vec{e_2} & \vec{e_3} \end{pmatrix}$ are multiplied *on the right* (*i.e.* $\boldsymbol{e} \bullet \boldsymbol{A}$).

In conclusion, to obtain the matrix $\boldsymbol{A}$ of a geometric transformation one *can* analyse the transformation of the basis vectors. For example, in ordinary 3D space, rotating an angle ϑ counterclockwise around the z-axis transforms the basis vectors $\boldsymbol{x}$ and $\boldsymbol{y}$ into ($\boldsymbol{x}\cos\vartheta + \boldsymbol{y}\sin\vartheta$) and ($\boldsymbol{x}\sin\vartheta - \boldsymbol{y}\cos\vartheta$) while $\boldsymbol{z}$ is unchanged. The transformation matrix is therefore:

$$\boldsymbol{A} = \begin{pmatrix} \cos\vartheta & \sin\vartheta & 0 \\ \sin\vartheta & -\cos\vartheta & 0 \\ 0 & 0 & 1 \end{pmatrix}$$

In the case of QM function spaces and operators corresponding to observables, however, the transformation matrix is preferably obtained by developing the functional notation of the operators (see Section 3.2.3 and Eq. (3.25)).

3.2.2.2 Changing Operators

Let us now consider a transformation of the basis vectors described by the linear relation:

$$\begin{pmatrix} \vec{e_1'} & \vec{e_2'} & \vec{e_3'} \end{pmatrix} = \begin{pmatrix} \vec{e_1} & \vec{e_2} & \vec{e_3} \end{pmatrix} \cdot \boldsymbol{T}$$

indicating with $\boldsymbol{T}$ the transformation matrix. Now, the components ($\boldsymbol{r}$ and $\boldsymbol{r}'$) of the same vector $\vec{r}$ are transformed by going **from one basis to another**:

$$\vec{r} = \begin{pmatrix} \vec{e_1} & \vec{e_2} & \vec{e_3} \end{pmatrix} \cdot \boldsymbol{r} = \begin{pmatrix} \vec{e_1'} & \vec{e_2'} & \vec{e_3'} \end{pmatrix} \cdot \boldsymbol{r}' \quad \text{so that:}$$

$$\begin{pmatrix} \vec{e_1} & \vec{e_2} & \vec{e_3} \end{pmatrix} \cdot \boldsymbol{r} = \begin{pmatrix} \vec{e_1} & \vec{e_2} & \vec{e_3} \end{pmatrix} \cdot \boldsymbol{T} \cdot \boldsymbol{r}'$$

Then:

$$\boldsymbol{r} = \boldsymbol{T} \cdot \boldsymbol{r}' \quad \text{or} \quad \boldsymbol{r}' = \boldsymbol{T}^{-1} \cdot \boldsymbol{r} \tag{3.21}$$

This means that the *columns* of $\boldsymbol{T}$ are the *components of the "new" basis vectors* with respect to the original basis vectors. From Eq. (3.21), we finally derive how the representative matrix $\boldsymbol{A}$ of a linear operator is transformed: $\boldsymbol{A}' \cdot \boldsymbol{r}' = \boldsymbol{s}' = \boldsymbol{T}^{-1} \cdot \boldsymbol{s} = \boldsymbol{T}^{-1} \cdot \boldsymbol{A} \cdot \boldsymbol{r} = \boldsymbol{T}^{-1} \cdot \boldsymbol{A} \cdot \boldsymbol{T} \cdot \boldsymbol{r}'$, and:

$$\boldsymbol{A}' = \boldsymbol{T}^{-1} \cdot \boldsymbol{A} \cdot \boldsymbol{T} \tag{3.22}$$

The transformation of operators according to Eq. (3.22) is known as **similarity transformation**.

3.2.3 Dot Product in Matrix Form

Once the scalar products between the *basis* vectors are known, it is also possible to obtain the scalar product between two *general* vectors of a linear space. We want to express the scalar product $(\vec{r},\vec{s})$ of the vectors (in the strict sense) $\vec{r} = \begin{pmatrix} \vec{e_1} & \vec{e_2} & \vec{e_3} \end{pmatrix} \cdot \boldsymbol{r}$ and $\vec{s} = \begin{pmatrix} \vec{e_1} & \vec{e_2} & \vec{e_3} \end{pmatrix} \cdot \boldsymbol{s}$ as a function of the scalar products $(\vec{e_i},\vec{e_j})$ and of the components $\boldsymbol{r}, \boldsymbol{s}$, or else the scalar product $\langle \chi_r / \chi_s \rangle$ of $\chi_r = \begin{pmatrix} \varphi_1 & \varphi_2 & \cdots \end{pmatrix} \cdot \boldsymbol{r}$ and $\chi_s = \begin{pmatrix} \varphi_1 & \varphi_2 & \cdots \end{pmatrix} \cdot \boldsymbol{s}$ as a function of the scalar products $\langle \varphi_i | \varphi_j \rangle$ and the $\boldsymbol{r}, \boldsymbol{s}$ components.

This is a direct application of the properties of the scalar product which is linear in the second term and antilinear in the first term (see Section 2.2). Considering the case of function spaces we have

$$\langle \chi_r | \chi_s \rangle = \langle \chi_r | s_1\varphi_1 + s_2\varphi_2 + \cdots \rangle = s_1 \langle \chi_r | \varphi_1 \rangle + s_2 \langle \chi_r | \varphi_2 \rangle + \cdots$$

For each of the scalar products $\langle \chi_r | \varphi_i \rangle$ the anti-linearity gives:

$$\langle \chi_r | \varphi_1 \rangle = \langle r_1\varphi_1 + r_2\varphi_2 + \cdots | \varphi_1 \rangle = r_1^* \langle \varphi_1 | \varphi_1 \rangle + r_2^* \langle \varphi_2 | \varphi_1 \rangle + \cdots$$

Finally,

$$\langle \chi_r | \chi_s \rangle = \sum_{i=1}^{\infty} \left(\sum_{j=1}^{\infty} r_i^* \cdot s_j \cdot \langle \varphi_i | \varphi_j \rangle \right) \quad \text{so that:}$$

$$\langle \chi_r | \chi_s \rangle = \boldsymbol{r}^\dagger \cdot \boldsymbol{S} \cdot \boldsymbol{s} \; \left(\boldsymbol{r}^\dagger \text{ is a row matrix and } S_{ij} = \langle \varphi_i | \varphi_j \rangle \right) \quad (3.23)$$

The particular case of linear spaces over **real numbers** is:

$$(\vec{r}, \vec{s}) = \boldsymbol{r}^T \cdot \boldsymbol{S} \cdot \boldsymbol{s} \left[\boldsymbol{r}^T \text{ is a row matrix and } S_{ij} \stackrel{\text{def}}{=} (\vec{e_i}, \vec{e_j}) \right] \quad (3.24)$$

The matrix $\boldsymbol{S}$ contains the scalar products between the basis vectors and therefore depends on the specific definition of the scalar product for the space considered. It is the tool for expressing in matrix notation lengths and angles of the vectors of the space and therefore is also known as the **metric** matrix.

3.2.3.1 Unitary Transformations

Particular importance is given to *basis changes* that do not modify the metric matrix, *i.e. that preserve the scalar product.* Considering function spaces, if all functions, including basis functions are modified by an operator $\hat{A}$ that transforms $\chi_r \rightarrow \hat{A}\chi_r = \chi_t$ and $\chi_s \rightarrow \hat{A}\chi_s = \chi_u$, the scalar product $\langle \chi_r | \chi_s \rangle$ given by Eq. (3.23) becomes $\langle \hat{A}\chi_r | \hat{A}\chi_s \rangle = \langle \chi_t | \chi_u \rangle$ and the requirement for preserving scalar products is:

$$\langle \chi_r | \chi_s \rangle = \langle \hat{A}\chi_r | \hat{A}\chi_s \rangle$$

The matrix notation: $\langle \chi_r | \chi_s \rangle = \boldsymbol{r}^\dagger \cdot \boldsymbol{S} \cdot \boldsymbol{s}$ and

$$\langle \hat{A}\chi_r | \hat{A}\chi_s \rangle = (\boldsymbol{A} \cdot \boldsymbol{r})^\dagger \cdot \boldsymbol{S} \cdot (\boldsymbol{A} \cdot \boldsymbol{s}) = \boldsymbol{r}^\dagger \cdot \boldsymbol{A}^\dagger \cdot \boldsymbol{S} \cdot \boldsymbol{A} \cdot \boldsymbol{s} \text{ that is}$$

$$\boldsymbol{S} \rightarrow \boldsymbol{S}' = \boldsymbol{A}^\dagger \cdot \boldsymbol{S} \cdot \boldsymbol{A} = \boldsymbol{S}$$

Now, if the **basis vectors are an orthonormal set**, $\boldsymbol{S} = 1 = \boldsymbol{A}^\dagger \cdot \boldsymbol{A}$ and $\boldsymbol{A}^{-1} = \boldsymbol{A}^\dagger$.

A matrix ($\boldsymbol{U}$) is said to be unitary if it is invertible and its inverse ($\boldsymbol{U}^{-1}$) equals the adjoint matrix $\boldsymbol{U}^\dagger$, that is: $\boldsymbol{U}^{-1} = \boldsymbol{U}^\dagger$. A **transformation** that preserves the scalar product is said to be *unitary.*

3.2.3.2 Quantum Mechanics Operators

For applications in QM, it is interesting to consider the integral of a generic operator ($\hat{G}$) between two generic states χ_a and χ_b:

$$\int_\Omega \chi_b^* \hat{G} \chi_a d\tau = \langle b | \hat{G} | a \rangle \quad \text{using Dirac notation}$$

We write the functions χ_a and χ_b as linear combinations of a suitable basis set $\{\varphi_j\}$: $\chi_a = \sum_j a_j \varphi_j$ and $\chi_b = \sum_j b_j \varphi_j$. Then:

$$\int_\Omega \chi_b^* \hat{G} \chi_a d\tau = \int_\Omega \chi_b^* \hat{G} \left(\sum_j a_j \varphi_j \right) d\tau$$

The linearity of QM operators gives:

$$\hat{G} \left(\sum_j a_j \varphi_j \right) = \sum_j a_j \cdot \hat{G} \varphi_j.$$

Combining with $\chi_b^* = \sum_i b_i^* \varphi_i^*$:

$$\langle b | \hat{G} | a \rangle = \sum_i \sum_j b_i^* \cdot a_j \cdot \int_\Omega \varphi_i^* \cdot \hat{G} \varphi_j d\tau = \sum_i \sum_j b_i^* \cdot a_j \cdot \langle \varphi_i^* | \hat{G} | \varphi_j \rangle$$

Finally, we denote the integrals $\langle \varphi_i^* | \hat{G} | \varphi_j \rangle$ as elements ij of a matrix $\boldsymbol{G}$, *i.e.* we define $G_{ij} = \langle \varphi_i^* | \hat{G} | \varphi_j \rangle$ and we have:

$$\langle b | \hat{G} | a \rangle = \boldsymbol{b}^\dagger \cdot \boldsymbol{G} \cdot \boldsymbol{a} \quad \text{with} \quad G_{ij} \stackrel{\text{def}}{=} \langle \varphi_i^* | \hat{G} | \varphi_j \rangle \quad (3.25)$$

From this we obtain in a different way Eq. (3.23) by considering the particular operator consisting of multiplication by 1.

Many matrix expressions take slightly different forms in the two cases:

- **Orthonormal basis**, where $\boldsymbol{S}$ is the identity matrix and can therefore be omitted and
- **Non-orthonormal basis**, where $\boldsymbol{S}$ must be inserted into the formulas.

3.2.4 Eigenvalue Equations

Let us now consider an eigenvalue equation: $\hat{G}\chi_i = \lambda_i \cdot \chi_i$. In the case of **NON-orthonormal basis** is:

$$\boldsymbol{G} \cdot \boldsymbol{a} = \lambda \cdot \boldsymbol{S} \cdot \boldsymbol{a} \quad \text{(non-orthonormal basis)} \quad (3.26)$$

The derivation starts from the relation (Eq. 3.25) $\langle b | \hat{G} | a \rangle = \boldsymbol{b}^\dagger \cdot \boldsymbol{G} \cdot \boldsymbol{a}$ by considering the particular case $b = a$:

$$\langle a | \hat{G} | a \rangle = \boldsymbol{a}^\dagger \cdot \boldsymbol{G} \cdot \boldsymbol{a}$$

Then, we assume that this χ_a is an eigenfunction of the operator, with eigenvalue λ_a. Then:

$$\int_\Omega \chi_a^* \hat{G} \chi_a d\tau = \int_\Omega \chi_a^* \cdot \lambda_a \cdot \chi_a d\tau = \lambda_a \cdot \int_\Omega \chi_a^* \cdot \chi_a d\tau$$

Using $\chi_a = \sum_j a_j \varphi_j$ we obtain $\langle a | \hat{G} | a \rangle = \lambda_a \cdot \sum_i \sum_j a_i^* \cdot a_j \cdot$

$$\int_\Omega \varphi_i^* \cdot \varphi_j d\tau = \lambda_a \cdot \boldsymbol{a}^\dagger \cdot \boldsymbol{S} \cdot \boldsymbol{a}$$

From the comparison, and considering that a scalar (λ_a) always commutes with matrices of any dimension we have

$\lambda_a \cdot \boldsymbol{a}^\dagger \cdot \boldsymbol{S} \cdot \boldsymbol{a} = \boldsymbol{a}^\dagger \cdot \lambda_a \cdot \boldsymbol{S} \cdot \boldsymbol{a} = \boldsymbol{a}^\dagger \cdot \boldsymbol{G} \cdot \boldsymbol{a}$, and we obtain Eq. (3.26) that in the case of orthonormal basis $\boldsymbol{S}_{ij} = \delta_{ij}$ is:

$$\boldsymbol{G} \cdot \boldsymbol{a} = \lambda \cdot \boldsymbol{a} \quad \text{(orthonormal basis)} \tag{3.27}$$

For the time-independent Schrödinger equation the analogous forms are:

$$\boldsymbol{H} \cdot \boldsymbol{c}_i = E_i \cdot \boldsymbol{S} \cdot \boldsymbol{c}_i \quad \text{(non-orthonormal basis)}$$
$$\text{or } \boldsymbol{H} \cdot \boldsymbol{c}_i = E_i \cdot \boldsymbol{c}_i \quad \text{(orthonormal basis)}$$

Now, it is convenient to collect in a single matrix the various columns corresponding to the various eigenvectors. If we indicate them with $\boldsymbol{a},\boldsymbol{b},\boldsymbol{c},\boldsymbol{d},\ldots$: $\boldsymbol{A} = \begin{pmatrix} a_1 & b_1 & \cdots \\ a_2 & b_2 & \cdots \\ \vdots & \cdots & \ddots \end{pmatrix}$

where a column corresponds to an eigenvector, and $\boldsymbol{A}_{ij}$ is element i of eigenvector j.

Similarly, we collect all the eigenvalues in a single diagonal matrix:

$$\lambda = \begin{pmatrix} \lambda_1 & 0 & \cdots \\ 0 & \lambda_2 & \cdots \\ \vdots & \cdots & \ddots \end{pmatrix} \quad (\text{element } ii \text{ is eigenvalue } \lambda_i)$$

Finally:

$$\boldsymbol{G} \cdot \boldsymbol{A} = \lambda \cdot \boldsymbol{S} \cdot \boldsymbol{A} \quad (\lambda \text{ is diagonal}) \tag{3.28}$$

When the basis is orthonormal, Eq. (3.28) becomes $\boldsymbol{G} \cdot \boldsymbol{A} = \lambda \cdot \boldsymbol{A}$ and

$$\boldsymbol{A}^{-1} \cdot \boldsymbol{G} \cdot \boldsymbol{A} = \lambda \quad (\lambda \text{ is diagonal}) \tag{3.29}$$

Looking back at Eq. (3.21), we understand now that

(In the orthonormal case)
The diagonalization of a matrix is a change of basis (a similarity transformation).

Finally, going to the more general case of a **non-orthonormal basis**, Eq. (3.28) ($\boldsymbol{G} \bullet \boldsymbol{A} = \lambda \bullet \boldsymbol{S} \bullet \boldsymbol{A}$) is solved as an extension of the methods used for the orthonormal basis case. In fact: $\boldsymbol{G} \cdot \boldsymbol{A} = \lambda \cdot \boldsymbol{S} \cdot \boldsymbol{A} \Rightarrow \boldsymbol{S}^{-1} \cdot \boldsymbol{G} \cdot \boldsymbol{A} = \boldsymbol{S}^{-1} \cdot \lambda \cdot \boldsymbol{S} \cdot \boldsymbol{A}$. Since diagonal matrices always commute, this is $\lambda \bullet \boldsymbol{S}^{-1}$ Ÿ$\boldsymbol{S} \bullet \boldsymbol{A} = \lambda \bullet \boldsymbol{A}$ so that $\boldsymbol{A}^{-1} \bullet (\boldsymbol{S}^{-1} \bullet \boldsymbol{G}) \bullet \boldsymbol{A} = \boldsymbol{A}^{-1} \bullet \lambda \bullet \boldsymbol{A} = \lambda$. In conclusion, we must diagonalize $(\boldsymbol{S}^{-1} \bullet \boldsymbol{G})$ instead of $\boldsymbol{G}$.

3.2.5 Diagonalizing Hermitian Matrices

The observables of QM correspond to Hermitian matrices. An important feature is that:

The eigenvalues of a Hermitian matrix are all real.

In fact, if $\boldsymbol{G}^\dagger = \boldsymbol{G}$ and λ is one of its eigenvalues, there is an eigenvector (let us say: $\boldsymbol{a}$) such that $\boldsymbol{G} \bullet \boldsymbol{a} = \lambda \boldsymbol{a}$. Let us multiply this on $\boldsymbol{a}^\dagger$ from the left:

$$\boldsymbol{a}^\dagger \cdot \boldsymbol{G} \cdot \boldsymbol{a} = \boldsymbol{a}^\dagger \cdot \lambda \boldsymbol{a} = \lambda \boldsymbol{a}^\dagger \cdot \boldsymbol{a} = \lambda |\boldsymbol{a}|^2$$

where $|\boldsymbol{a}|^2$ is a strictly positive real number. Now, let us write the adjoint of the whole equation. The adjoint of the left side is $(\boldsymbol{a}^\dagger \bullet \boldsymbol{G} \bullet \boldsymbol{a})^\dagger = \boldsymbol{a}^\dagger \bullet \boldsymbol{G}^\dagger \bullet (\boldsymbol{a}^\dagger)^\dagger = \boldsymbol{a}^\dagger \bullet \boldsymbol{G} \bullet \boldsymbol{a}$ because the adjoint of a product is the product of the adjoints in reverse order and $\boldsymbol{G}$ is Hermitian. Therefore, the left side is equal to the left side of the original equation. The adjoint of the right side is $\lambda^* |\boldsymbol{a}|^2$ because $|\boldsymbol{a}|^2$ is real. So, $\lambda |\boldsymbol{a}|^2 = \lambda^* |a|^2$ and $\lambda = \lambda^*$. This proves that λ is real.

Let us consider the relationship between the adjoints in Eq. (3.29). We have:

$$(\boldsymbol{A}^{-1} \cdot \boldsymbol{G} \cdot \boldsymbol{A})^\dagger = \lambda = (\boldsymbol{A})^\dagger \cdot (\boldsymbol{G})^\dagger \cdot (\boldsymbol{A}^{-1})^\dagger = \boldsymbol{A}^\dagger \cdot \boldsymbol{G} \cdot (\boldsymbol{A}^\dagger)^{-1} \tag{3.30}$$

where λ is a real diagonal matrix. The comparison with $\boldsymbol{A}^{-1} \bullet \boldsymbol{G} \bullet \boldsymbol{A} = \lambda$ (Eq. 3.29) shows that matrix $\boldsymbol{A}$ (which contains the eigenvectors of the problem as columns) has the property that its inverse is equal to the adjoint: $\boldsymbol{A}^{-1} = \boldsymbol{A}^\dagger$: it is a **unitary matrix** (see Section 3.1.3.2).

The diagonalization of a Hermitian matrix is a similarity transformation with unitary matrices.

There can be degeneracy: every case of degeneracy corresponds to a solution with multiplicity > 1 of the polynomial equation in the unknown λ (see Eq. (3.32)). Then, linear combinations of degenerate eigenvectors are still eigenvectors with the same eigenvalue.

Another important property, presented here *without proof*, is that:

The eigenfunctions of a Hermitian matrix are a complete and orthogonal set.

A **basis set** is **complete** if **any** state of the space can be represented as a linear combination of the functions of the basis set. As said, this property is important for infinite-dimensional linear spaces, where it means that any element of the space can be represented as a *series* expansion on the basis set, *i.e.* that *the series* always *converges*.

It is easy to see that the eigenvalue equations (3.28 and 3.29) are satisfied by both **normalized** eigenvectors and **non-normalized** eigenvectors (*i.e.* multiplied by an arbitrary complex factor).

Another important property is that

Eigenvectors of different eigenvalues are orthogonal.

Take two eigenvectors $\boldsymbol{x}$, $\boldsymbol{y}$ of the Hermitian matrix $\boldsymbol{A}$, with real eigenvalues μ and ν, respectively. Then: $\mu(\boldsymbol{x}, \boldsymbol{y}) = (\mu \cdot \boldsymbol{x}^\dagger \bullet \boldsymbol{y}) = ((\boldsymbol{A} \bullet \boldsymbol{x})^\dagger \bullet \boldsymbol{y}) = (\boldsymbol{x}^\dagger \bullet \boldsymbol{A}^\dagger \bullet \boldsymbol{y}) = (\boldsymbol{x}^\dagger \bullet \boldsymbol{A} \bullet \boldsymbol{y}) = (\boldsymbol{x}^\dagger \bullet \nu \cdot \boldsymbol{y}) = \nu(\boldsymbol{x}^\dagger \bullet \boldsymbol{y})$. Now: $(\mu - \nu) \cdot (\boldsymbol{x}^\dagger \bullet \boldsymbol{y}) = 0$ and therefore $(\boldsymbol{x}^\dagger \bullet \boldsymbol{y}) = 0$ because $\mu \neq \nu$. The property is also derived, in functional notation, in Section 5.1.3 of the Chapter 1.

3.2.5.1 The Analytical Solution

Considering the case of an **orthonormal basis**:

$$\boldsymbol{G}\cdot\boldsymbol{a}_i = \lambda_i \boldsymbol{a}_i, \quad \text{it can be written:} \tag{3.31}$$

$$(\boldsymbol{G}-\boldsymbol{\lambda}\cdot\boldsymbol{1})\cdot\boldsymbol{a}_i = 0 \quad \text{that is:} \tag{3.32}$$

$$\left[\begin{pmatrix} G_{11} & G_{12} & \cdots \\ G_{21} & G_{22} & \cdots \\ \vdots & \vdots & \ddots \end{pmatrix} - \begin{pmatrix} \lambda_1 & 0 & \cdots \\ 0 & \lambda_2 & \cdots \\ \vdots & \vdots & \ddots \end{pmatrix}\right]\cdot\begin{pmatrix} \vdots \\ a_i \\ \vdots \end{pmatrix} =$$

$$\begin{pmatrix} G_{11}-\lambda_1 & G_{12} & \cdots \\ G_{21} & G_{22}-\lambda_2 & \cdots \\ \vdots & \vdots & \ddots \end{pmatrix}\cdot\begin{pmatrix} \vdots \\ a_i \\ \vdots \end{pmatrix} = \begin{pmatrix} \vdots \\ 0 \\ \vdots \end{pmatrix}$$

To derive eigenvalues and eigenfunctions *analytically*, one starts noting that $|\boldsymbol{G}-\boldsymbol{\lambda}\cdot\boldsymbol{1}| = 0$, because the solution of Eq. (3.31) must be different from the trivial solution in which all coefficients of column $\boldsymbol{a}$ are zero. This leads to a polynomial equation of degree n in the unknown λ, where n is the dimension of the square matrices $\boldsymbol{G}$ and $\boldsymbol{1}$. By the fundamental theorem of algebra, the solution of the equation consists of n values of λ, not necessarily all distinct: if an eigenvalue is degenerate, it must be counted many times.

In general, an analytic solution is possible up to $n=4$. A special case is when the matrix to be diagonalized is **block-diagonal**. Then, each block can be treated separately and the limit $n=4$ regards each single block. As described in detail in the chapter on symmetry, the block-diagonal structure is a typical result of the use of symmetry: by using basis functions that are previously adapted to the symmetry, each block contains the matrix elements of the operator between (only) those basis functions that have the same behaviour with respect to the symmetry operations: the eigenvectors relating to the same block are said to belong to the same symmetry species.

Once we have obtained the eigenvalues with their degeneracies, Eq. (3.32) allows us to obtain the corresponding eigenvector. In the case of degeneracy, the substitution in Eq. (3.32) does not uniquely identify an eigenvector but a subspace of vectors, among which an orthonormal basis must be chosen.

3.2.5.2 The Numerical Solution

The numerical solution of an eigenvalue equation is a standard problem of numerical analysis. A widespread method essentially modifies the matrix to be diagonalized ($\boldsymbol{G}$) with a sequence of linear transformations, gradually bringing its off-diagonal elements towards zero within a predetermined tolerance. In parallel, the same transformations are performed on another matrix initially set equal to the unit matrix ($\boldsymbol{1}$) of the required order. When the predetermined tolerance has been reached and $\boldsymbol{G}$ has become close enough to a diagonal matrix, the eigenvectors are the columns of the matrix $\boldsymbol{1}$ so transformed and the eigenvalues are the diagonal elements of $\boldsymbol{G}$.

3.2.5.3 How to Select Eigenvectors of Degenerate Levels

For degenerate eigenvalues, the situation is complicated by the fact that *any linear combination* of eigenvectors relative to the degenerate eigenvalue *is an eigenvector with the same eigenvalue.* Hence:

1. There is ambiguity (or else freedom of choice) for the eigenvectors of a degenerate eigenvalue, and the different choices correspond to different independent linear combinations of the same set of eigenvectors that share the eigenvalue.
2. The numerical diagonalization procedures, having to provide a result in any case, necessarily introduce additional criteria that may be different from one procedure to another.
3. The choice made both in the analytical solution and as a result of a particular numerical procedure may not be the most suitable for what is needed.

On point 2, it should be clarified that the numerical procedures for diagonalizing Hermitian matrices may or may not also implement the **orthogonalization** of the degenerate eigenvectors. In the second case they provide eigenvectors that are only linearly independent of each other.

More typically, however, numerical diagonalization procedures incorporate an algorithm (called **Gram-Schmidt**) that also performs the orthogonalization starting from a set of functions that are not orthonormal, say the functions φ_1, …, φ_m for an m-times degenerate level. The algorithm starts by normalizing the first function of the set by dividing it by its normalization integral:

$$\chi_1 = \frac{\varphi_1}{\langle\varphi_1|\varphi_1\rangle}$$

This function becomes the first function of the required set of ortho-normalized functions. The algorithm then subtracts from the second function of the initial set (φ_2) its projection onto χ_1 and normalizes the difference:

$$\zeta_2 = \varphi_2 - \langle\varphi_2|\chi_1\rangle\cdot\chi_1; \quad \chi_2 = \frac{\zeta_2}{\langle\zeta_2|\zeta_2\rangle}$$

Then, we similarly subtract from the third function of the initial set its projections along the two new basis functions already found (χ_1, χ_2) and normalize the difference, … and so on until the functions of the initial set are exhausted.

The Gram-Schmidt procedure, therefore, also provides *a constructive proof* of the possibility of finding an ortho-normal set of eigenfunctions for a Hermitian matrix.

Note that the ortho-normalized set $\chi_1,\dots,\chi_m$ is still an arbitrary choice and hence presents the problem 4 above, because it derives from the initial set $(\varphi_1,\dots,\varphi_m)$, which in turn is a consequence of a procedure that can be implemented in different ways.

Point 4 requires care especially when the results need to be further elaborated. An example is when there is an additional interaction ("a perturbation") in the Hamiltonian operator. If the perturbation produces a "degeneracy removal", that is, it happens that the degenerate energy level not only shifts but is also "splits" into sublevels with different energies, it is necessary to choose appropriately the eigenvectors of the unperturbed degenerate case to correctly set up the calculation of the correction due to the perturbation. As said in the Chapter 1, the problem is essentially solved by finding the eigenvectors of the perturbation operator in the basis of the eigenvectors of the degenerate level of the unperturbed Hamiltonian.

3.3 Other Matrix Operations

In group theory, there are two other important matrix operations.

3.3.1 Direct Sum

The **direct sum** operates on two or more square matrices, not necessarily of equal size, and produces a square matrix that has size equal to the sum of the dimensions of the addend matrices. To be precise, given a matrix $\boldsymbol{A}$ $(n\times n)$ and a matrix $\boldsymbol{B}$ $(m\times m)$, their direct sum is a matrix $(n+m)\times(n+m)$ constructed by placing the elements of $\boldsymbol{A}$ in the upper left block and the elements of $\boldsymbol{B}$ in the lower right block:

$$\boldsymbol{A}\oplus\boldsymbol{B}=\begin{pmatrix} a_{11} & \cdots & a_{1n} & 0 & \cdots & 0 \\ \vdots & \ddots & \vdots & \vdots & \ddots & \vdots \\ a_{n1} & \cdots & a_{nn} & 0 & \cdots & 0 \\ 0 & \cdots & 0 & b_{11} & \cdots & b_{1m} \\ \vdots & \ddots & \vdots & \vdots & \ddots & \vdots \\ 0 & \cdots & 0 & b_{m1} & \cdots & b_{mm} \end{pmatrix},\quad \text{that is:}$$

$$\boldsymbol{A}\oplus\boldsymbol{B}=\begin{pmatrix} \boldsymbol{A} & \boldsymbol{0} \\ \boldsymbol{0}^T & \boldsymbol{B} \end{pmatrix}$$

where **0 and 0**^{T} are two matrices of **all elements = 0** that have dimensions $(n\times m)$ and $(m\times n)$ respectively. The direct sum can be extended to any number of matrices by operating in succession. The direct sum of the first two matrices is directly added to the third, and so on:

$$\boldsymbol{A}\oplus\boldsymbol{B}\oplus\boldsymbol{C}\oplus\cdots=\big((\boldsymbol{A}\oplus\boldsymbol{B})\oplus\boldsymbol{C}\big)\oplus\cdots$$

3.3.2 Direct Product

The **direct product** or **Kronecker product** (C) of two matrices $\boldsymbol{A}$ and $\boldsymbol{B}$ (in this order) is written $\boldsymbol{C}=\boldsymbol{A}\otimes\boldsymbol{B}$ and produces a matrix $\boldsymbol{C}$ in which each element of the first matrix $\boldsymbol{A}$ corresponds to a block of elements containing the entire second matrix $\boldsymbol{B}$ multiplied by the element of $\boldsymbol{A}$.

If $\boldsymbol{A}$ is an $n\times m$ matrix and $\boldsymbol{B}$ is a $p\times q$ matrix, the direct product can be schematized as follows:

$$\boldsymbol{A}\otimes\boldsymbol{B}=\begin{pmatrix} a_{11}\boldsymbol{B} & \cdots & a_{1m}\boldsymbol{B} \\ \vdots & \ddots & \vdots \\ a_{n1}\boldsymbol{B} & \cdots & a_{nm}\boldsymbol{B} \end{pmatrix}\quad \text{where } a_{jk}\boldsymbol{B} \text{ is a } p\times q \text{ block}$$

given by $\boldsymbol{B}$ multiplied by a_{jk}. In detail:

$$\boldsymbol{A}\otimes\boldsymbol{B}\begin{pmatrix} a_{11}b_{11} & a_{11}b_{12} & \cdots & a_{11}b_{1q} & \cdots & a_{1m}b_{11} & a_{1m}b_{12} & \cdots & a_{1m}b_{1q} \\ a_{11}b_{21} & a_{11}b_{22} & \cdots & a_{11}b_{2q} & \cdots & a_{1m}b_{21} & a_{1m}b_{22} & \cdots & a_{1m}b_{2q} \\ \vdots & \vdots & \ddots & \vdots & \cdots & \vdots & \cdots & \ddots & \vdots \\ a_{11}b_{p1} & a_{11}b_{p2} & \cdots & a_{11}b_{pq} & \cdots & a_{1m}b_{p1} & a_{1m}b_{p2} & \cdots & a_{1m}b_{pq} \\ \vdots & \vdots & \cdots & \vdots & \ddots & \vdots & \vdots & \vdots & \vdots \\ a_{n1}b_{11} & a_{11}b_{11} & \cdots & a_{n1}b_{1q} & \cdots & a_{nm}b_{11} & a_{nm}b_{21} & \cdots & a_{nm}b_{1q} \\ a_{n1}b_{21} & a_{11}b_{22} & \cdots & a_{n1}b_{2q} & \cdots & a_{nm}b_{21} & a_{nm}b_{21} & \cdots & a_{nm}b_{2q} \\ \vdots & \vdots & \ddots & \vdots & \cdots & \vdots & \vdots & \ddots & \vdots \\ a_{n1}b_{p1} & a_{n1}b_{p2} & \cdots & a_{n1}b_{pq} & \cdots & a_{nm}b_{p1} & a_{nm}b_{p2} & \cdots & a_{nm}b_{pq} \end{pmatrix}$$

The direct product is often denoted (or even *defined*) with a double index system:

$$\text{if } \boldsymbol{A}\otimes\boldsymbol{B}=\boldsymbol{C} \text{ then: } \boldsymbol{C}_{ij,kl}=\boldsymbol{A}_{ij}\boldsymbol{B}_{kl}$$

in the sense that the matrix element $\boldsymbol{C}_{ij,kl}$ is in position kl within the block (of equal shape to matrix $\boldsymbol{B}$) that occupies the partition ij of matrix $\boldsymbol{C}$. The double-index notation is important both in applications and in proofs.

Warning: do not confuse the double index notation with the multiplication of the indices $i{\cdot}j$.

In detail, given a matrix $(n\times m)$ $\boldsymbol{A}$ and a matrix $(p\times q)$ $\boldsymbol{B}$:

- The direct product $\boldsymbol{A}\otimes\boldsymbol{B}$ is a matrix of dimensions $(np)\times(mq)$.
- Each element of $\boldsymbol{C}$ is one of the $n{\cdot}p{\cdot}m{\cdot}q$ products of an element of $\boldsymbol{A}$ by an element of $\boldsymbol{B}$: these products are all contained in $\boldsymbol{C}$, each only once.
- The product $\boldsymbol{A}_{ij}\boldsymbol{B}_{kl}$ appears at row $[(i-1){\cdot}p+k]$ and at column $[(j-1){\cdot}q+l]$ of matrix $\boldsymbol{B}$.
- Reversing the relationship, the element $\boldsymbol{C}_{ij}$ is given by the product $\boldsymbol{A}_{rs}\boldsymbol{B}_{tu}$, where $r=(i-1)\backslash p+1$; $s=(j-1)\backslash q+1$; $t=i-p[(i-1)\backslash p]$; $u=j-q[(j-1)\backslash q]$.

For example, if $\boldsymbol{A}$ is a 2×3 matrix and $\boldsymbol{B}$ is a 4×5 matrix, then $\boldsymbol{C}$ is an 8×15 matrix.

The product $\boldsymbol{A}_{21}{\cdot}\boldsymbol{B}_{23}$ is in $\boldsymbol{C}_{63}$, because $(2-1)\times4+2=6$ and $(1-1)\times5+3=3$. In two-index notation, the element $\boldsymbol{C}_{63}$ is simply denoted as $\boldsymbol{C}_{21,23}$.

Conversely, $\boldsymbol{C}_{68}$ is the product $\boldsymbol{A}_{22}{\cdot}\boldsymbol{B}_{23}$, $r=(6-1)\backslash4+1=1+1=2$; $s=(8-1)\backslash5+1=1+1=2$; $t=6-4$

$[(6-1)\backslash 4]=6-4\times 1=2$; $u=8-5[(8-1)\backslash 5]=8-5\times 1=3$, where the\sign indicates the division truncated to the whole number (*e.g.*: $8 \backslash 3=2$). Therefore, C_{68} can be alternatively indicated as $C_{22,23}$.

For **this matrix product** there are **no constraints on the sizes of the matrices that can be multiplied** together, unlike the normal row×column multiplication which only makes sense when the number of columns of the first factor is equal to the number of rows of the second factor.

It is useful to report some properties of the direct product of two matrices.

- The direct product **is not commutative**: it is not necessarily true that $A \otimes B = B \otimes A$.
- An important relationship between **the direct product** and **the normal matrix product** (rows×columns) is that if two matrices (A and C) can be multiplied together with the row×column multiplication and if the same is true for two other matrices (B and D), then $(A \otimes B)(C \otimes D) = (AC) \otimes (BD)$.
- If two matrices A and B are invertible, then $(A \otimes B)^{-1} = (A^{-1}) \otimes (B^{-1})$.
- If A and B are square matrices, the eigenvalues of the direct product are all possible products of an eigenvalue of A and an eigenvalue of B.
- If A and B are square matrices, the **trace** (the sum of the diagonal elements) of the direct product is the product of the traces: $tr(A \otimes B) = tr(A)tr(B)$ and also $tr(A)tr(B) = tr(B)tr(A) = tr(B \otimes A)$.

Note that the *trace* of the direct product of two matrices is *independent of the order of the operation* $tr(A \otimes B) = tr(B \otimes A)$, even if the direct product is not. This property is the basis of using characters (matrix traces) when working on matrix representations of symmetry operations, a topic widely discussed in the Chapter 4.

4 BIOGRAPHICAL NOTES

We offer here short biographical notes of Scientist that have been prominent in the development of the topics that are presented in this book. This list is not pretending to be exhaustive in any way. Afterall, this is a textbook, and it is not about the history of Science. Anyway, we think that presenting the significant events in the life of Scientist helps the students to see the persons behind the scientific theories. We choose Clausius and Kelvin for their key role in establishing Classical Thermodynamics, Boltzmann and Gibbs for Statistical Thermodynamics, together with Plank, which also shared with Bohr the role in founding what is called the "Old Quantum Theory", which is a collection of results obtained by including quantization into Classical Mechanics. Schrödinger, Pauli, Heisenberg and Dirac then took the duty to develop Quantum Mechanics into a coherent theory. Finally, Wigner was the key figure in introducing the principles of Group Theory as an ideal tool in Quantum Mechanics. The biographies are short, and therefore we focused, mainly, on facts, rather than anecdotes, although some key narratives are presented to give some illustration of the personality of these Scientists. Portions of the biographical material were originally published in P. Ghigna, Le Forme della Simmetria. Quattro storie di Scienza, dalla Chiralità allo Spin, Mimesis Edizioni, Milan, 2026, and are reproduced here with permission.

Rudolf Julius Emanuel Clausius (2 January 1822 to 24 August 1888), was born in Köslin in Pomerania (now Koszalin, Poland). Clausius graduated at the Berlin University in 1844 and got his Doctorate in 1848 at the University of Halle. He was Professor at the ETH Zürich, then at the University of Würzburg and eventually, of Bonn. He is considered as one of the founding fathers of Thermodynamics. His most famous paper, *Über die bewegende Kraft der Wärme* ("On the Moving Force of Heat and the Laws of Heat which may be Deduced Therefrom") was published in 1850: in this work, starting from Carnot theorem, Clausius restated the First and Second Laws of Thermodynamics. In 1865, Clausius gave the first mathematical definition of entropy, and coined the same word from the Greek ἐν *en* "in" and τροπή *tropē* "transformation", meaning "content of transformation". He also summarized the contents of the First and Second Laws in a renowned sentence in an 1865 paper, stating *Die Energie der Welt ist konstant. Die Entropie der Welt strebt einem Maximum zu:* "The energy of the world is constant. Its entropy tends to a maximum." Clausius gave in addition important contributions to the kinetic theory of gases.

William Thomson, 1st Baron Kelvin (26 June 1824 to 17 December 1907), was born in Belfast, Ireland. From 1841 to 1845 he attended the University of Cambridge, where he graduated. In 1846, aged only 22, Thomson returned to Glasgow to become Professor of Natural Philosophy. He remained in this post for 53 years, despite numerous offers by many world-renowned Universities for prestigious positions. Together with Clausius, he is considered one of the founders of Thermodynamics, to which he gave two fundamental contributions: the establishment of the absolute temperature scale, the kelvin temperature scale, and his formulation of the Second Law, according to which heat cannot be entirely converted into work.

Thomson also worked on the transatlantic cable project, for which he was knighted in 1866, and then, in 1892, he was ennobled for his contributions to Thermodynamics to become the first Baron of Kelvin, from the name of river Kelvin, near the University of Glasgow. The Joule-Thomson effect is also named in his honour.

Ludwig Eduard Boltzmann (20 February 1844 to 5 September 1906) was born in Erdberg, a suburb of Vienna, Austria. He studied Physics at the University of Vienna where he received his doctorate in 1866. In 1865, he became Professor of Theoretical Physics at the University of Gratz, then he moved to Vienna to became Professor of Mathematics, where he eventually succeeded his mentor Joseph Stephan as Professor of Theoretical Physics. Boltzmann is considered as one of the greatest scientists of all time, being the founder of statistical mechanics, and for giving a statistical foundation of the Second Law, which is mainly formulated in a famous paper of 1877 *Über die Beziehung zwischen dem zweiten Hauptsatze der mechanischen Wärmetheorie und der Wahrscheinlichkeitsrechnung, respektive den Sätzen über das Wärmegleichgewicht*: "On the Relationship between the Second Fundamental Theorem of the Mechanical Theory of Heat and Probability Calculations Regarding the Conditions for Thermal Equilibrium". Boltzmann's belief of the ontological existence of atoms was quite contested by his contemporaries, especially by Ernst Mach. This may be one of the reasons behind his deteriorating mental conditions in his final years, which forced him to resign his position in 1906, and to commit suicide during a seaside vacation with his family in Duino, near Trieste (now in Italy, then in Austria). The statistical entropy definition $S = k \log W$ is engraved on his tombstone. Max Planck named k (or k_B) the Boltzmann constant.

Josiah Willard Gibbs (February 11, 1839 to April 28, 1903) was born in New Haven, Connecticut. He graduated from the Yale College in 1858, at the age of 19. In 1871, after a period as a tutor in Yale and some years spent in Europe, studying Mathematics and Chemistry, he was appointed as Professor of Mathematical Physics at Yale, the first such chair in the United States, and he remained at Yale till he passed away. With his works on Thermodynamics, Gibbs was instrumental in establishing Physical Chemistry as a rigorous discipline. Together with Boltzmann and Maxwell, he created statistical mechanics, a term he coined. He also introduced the notions of chemical potential and statistical ensemble, and was one of the discoverers of vector calculus, where he introduced the concept of reciprocal lattice. Most of Gibbs' papers appeared in the *Transactions of the Connecticut Academy of Science*, which was a quite little-known journal, and therefore had to be rediscovered later: for example, Einstein between 1902 and 1904 wrote three papers on statistical mechanics, but after reading Gibbs's textbook "Elementary Principles in Statistical Mechanics", which was translated into German in 1905, he acknowledged Gibbs's treatment as superior to his own, and declared that he had not written the papers if he were aware of Gibbs's work. Nonetheless, Gibbs was the first American theoretical scientist to obtain international recognition, and he died before to be awarded the Nobel prize. On the other hand, in 1901 he was honoured with what was considered the most important honour in Science, the Copley medal by the Royal Society of London "for his contributions to Mathematical Physics".

Max Karl Ernst Ludwig Planck (23 April 1858 to 4 October 1947) was born in Kiel, Germany. He graduated at the age of 17 at the University of Munich, where his father was Professor in Law. By 1880, at the age of 22 only, he obtained both a doctoral degree at the University Friederik Wilhelms of Berlin and the *Habilitation*, that in German-speaking countries is the first step into the academic career. He was then enrolled by the University of Kiel in 1885, and in 1888 by the University of Berlin, where he became full Professor in 1892. He held that position till he retired in 1926. Planck is prominent for his contributions to Thermodynamics, where he realized, contemporaneously to Gibbs but independently of him, that the condition of maximum entropy corresponds to the equilibrium state, and to the early quantum mechanics, when he solved the problem of the black body radiation in a series of papers published between 1899 and 1901. He introduced then the "quantum of action", which is now called the Planck constant, for which he was awarded the Nobel prize in Physics in 1908. Planck was quite an opponent of the Nazi regime in Germany, and in 1938 he protested the increasing interference of the regime on German science by resigning from the presidency of the Prussian Academy of Sciences. Planck was also president of the *Kaiser-Wilhelm-Gesellschaft zur Förderung der Wissenschaften*, renamed in 1948 as *Max-Planck-Gesellschaft* in his honour, which is widely recognised as one of the most important basic research institutions in the world.

Niels Henrik David Bohr (7 October 1885 to 18 November 1962), was born in Copenhagen, Denmark. Bohr studied at the University of Copenhagen, where he graduated in Mathematics in 1909, and obtained his Ph.D. in 1911. Bohr first became *Privatdozent* in Copenhagen, and in 1913 was fully enrolled in the University. In 1917 he obtained the support of the Danish government to found the Institute of Theoretical Physics (now Niels Bohr Institute), which opened in 1921, with Bohr as director. In 1943 he escaped Denmark, as the Nazi considered his family as Jewish, since his mother was Jewish. He then flew first to the UK, and then to the USA, where he paid some visits to Los Alamos, and made some occasional (but decisive) contributions to the Manhattan project. After the war, he returned to Copenhagen in 1945, and was honoured with the "Order of the Elephant" (a Danish order of chivalry: it is Denmark's highest-ranked honour, usually reserved to heads of states). Bohr was one of the founders of the so-called "old quantum theory", which is an (incomplete) extension of classical mechanics with some "ad hoc" postulates, accounting for the stability of electron states in atoms. For these contributions, Bohr received the Nobel prize in Physics in 1922. He also developed the concept of complementarity, and the interpretation of quantum mechanics which is known as "Copenhagen interpretation". Bohr helped refugees from Nazism to escape to Denmark, and, during the war, was also active in aiding the escape of Jews from the occupied Denmark to Sweden. After the war, Bohr demanded for international cooperation on nuclear research, and participated in establishing the CERN.

Erwin Rudolf Josef Alexander Schrödinger (12 August 1887 to 4 January 1961), was born in Vienna, Austria. He studied at the University of Vienna, where in 1910 he earned his doctorate. In 1927 he succeeded Max Planck in Berlin, but in 1933 he left Germany for Oxford after the seizure of power by the Nazi. Schrödinger's academic career was quite erratic, due to his personal lifestyle (he shared his house with both his wife and his mistress), and political inconsistency (his opposition to the Nazi persecutions of Jews was first recanted after the Anschluss in 1938, when he was Professor at the University of Gratz since 1936, and then he regretted doing so). At the end, he accepted an invitation to move to Ireland to found an Institute of Advanced Study in Dublin, where he remained until he retired in 1955. The main contribution of Schrödinger to the new quantum mechanics is his famous equation, for which he shared the Nobel prize in Physics with Paul Dirac in 1933. The equation was first formulated in the 1926 paper *Quantisierung als Eigenwertproblem:* "Quantization as an Eigenvalue problem"; a second paper, published just 4 weeks later, dealt with solvable problems like the harmonic oscillator and the rigid rotor. This was followed in the same year by other two papers, the first showing (although not rigorously) the equivalence of the wave mechanics with the matrix formulation by Heisenberg, and the final one dealing with time dependent problems. Schrödinger was also interested in Psychology and Philosophy, especially concerning Philosophy of Science: he devised the famous thought experiment known as "Schrödinger's cat" as a critique of the Copenhagen interpretation of quantum mechanics.

Wolfgang Ernst Pauli (25 April 1900 to 15 December 1958), was born in Vienna. Pauli studied at the University of Munich, where he received his Ph.D. in 1921. Pauli was first assistant to Born in Göttingen, and then to Bohr in Copenhagen. He was Professor at the University of Hamburg between 1923 and 1928, and then Professor of theoretical Physics at the ETH in Zürich. In 1940 Pauli relocated to the USA, where he became Professor of theoretical Physics at Institute for Advanced Study in Princeton. After the war, he was naturalized as an US citizen, but he eventually returned to Zürich to spend the rest of his life. Pauli contributions to quantum mechanics include the development of the *Aufbau* Principle along with Bohr, the exclusion principle (warranting him the Nobel prize in Physics in 1945), for which he devised a new two valued quantum number accounting for a new degree of freedom for the electron, later identified with the spin by Uhlenbeck and Goudsmit, the development of the Pauli matrices as basis for spin operators, and one of the first proofs, although not systematic, of the spin statistic theorem. Pauli rarely authored regular papers, and most of his contributions are in letters that circulated among his closest friends, such as Heisenberg and Dirac. The legendary "Pauli effect" was related to his seeming ability of breaking experimental devices by staying in their vicinity: Pauli was genuinely proud of that.

Werner Karl Heisenberg (5 December 1901 to 1 February 1976), was born in Würzburg, Germany. From 1920 to 1923 he studied Physics at the University of Munich, and he received his doctorate in 1923 under the supervision of Sommerfeld. In 1924 he was appointed as *Privatdozent* in Göttingen, and in the same year he obtained a fellowship from the Rockefeller Foundation for doing research in Copenhagen with Niels Bohr. During that period, he developed the main ideas leading to the seminal 1925 paper *Über quantentheoretische Umdeutung kinematischer und mechanischer Beziehungen*: "On the quantum-theoretical reinterpretation of kinematical and mechanical relationships", which eventually steered, with Born and Jordan, the formulation of matrix mechanics. Still in Copenhagen, in 1927 he formulated his Uncertainty Principle, and in the same year he was appointed as Professor of theoretical Physics at the University of Leipzig, being the youngest full Professor in all Germany. After the advent of the Nazi regime in 1933, Heisenberg was attacked by the propaganda as being a "white jew": at that time, supporters of the so-called Deutsche Physik, including eminent scientists like Philipp Lenhard and Johannes Stark, launched a campaign against theoretical physicist studying quantum mechanics and relativity, which were considered as "anti-Aryan" disciplines. This did not prevent Heisenberg from entering the German nuclear programme, although his role is controversial. In 1943 he was appointed to the chair of theoretical Physics at the University of Berlin, and eventually he became a Professor at the University of Munich in 1958. After the war, he was actively promoting scientific collaboration between countries through the *UNESCO* organization, and he was appointed to sign the convention that established *CERN* as the representative of the Federal Republic of Germany. In addition to his contributions to the foundation of the new quantum mechanics, for which he was awarded the Nobel prize in Physics in 1932, Heisenberg also made fundamental contributions to the theories of turbulent flow and ferromagnetism.

Paul Adrien Maurice Dirac (8 August 1902 to 20 October 1984), was born in Bristol, UK. He graduated in 1921 at the University of Bristol. He obtained his Ph.D. at the University of Cambridge in 1926, with the first ever thesis in quantum mechanics. Dirac held the *Lucasian Professorship* of Mathematics at Cambridge from 1932 till 1969, when he retired, and then moved to the USA, first at the University of Miami as visiting professor, and then (1970) at the Florida State University in Tallahassee. In 1928, in the famous paper "The quantum Theory of the electron", he proposed the Dirac equation as a relativistic equation for the wave function of the electron. This work led Dirac to predict the existence of the positron, the electron antiparticle, which he interpreted in terms of what came to be called the Dirac sea. The positron was observed by Carl Anderson in 1932. The Dirac equation also contributed to explaining the origin of the electron spin as a relativistic phenomenon. In 1933, he shared the Nobel prize in Physics with Schrödinger, for "the discovery of new productive forms of atomic theory". Dirac also made decisive contributions to quantum electrodynamics, coining the term. He was extremely gifted in Mathematics, and one of the principles he pursued in his research was that of mathematical beauty. Dirac also coined the terms fermion and boson, and, according to the Nobel laureate Abdus Salam "was undoubtedly one of the greatest physicists of this or any century ... No man except Einstein has had such a decisive influence, in so short a time, on the course of physics in this century".

Eugene Paul Wigner (November 17, 1902 to January 1, 1995), was born in Budapest, Austria-Hungary. In 1921 he enrolled at the Technische Hochschule Berlin, for studying chemical engineering. Wigner also worked at the Kaiser Wilhelm Institute for Physical Chemistry and Electrochemistry (now the Fritz Haber Institute), and there he met Michael Polanyi, who supervised his Ph.D. thesis. In 1926 he accepted a position at the Kaiser Wilhelm Institute, but shortly after moved to Göttingen as an assistant of David Hilbert. Wigner was disappointed with the aging Hilbert, whose interest had shifted to logic. Nonetheless, he worked independently, and in a series of papers between 1927 and 1931 he laid the foundations of the theory of symmetry in quantum mechanics. In 1929 he moved to Princeton, and in 1937 he also was naturalized with the US citizenship. During the war, he actively participated in the Manhattan project, leading a project for converting uranium into plutonium. After the war, he accepted a position as the director of research and development at the *Clinton Laboratory* (now the Oak Ridge National Laboratory) in Oak Ridge, Tennessee, but, feeling inapt for such a managerial role, in 1947 he returned to Princeton. In the postwar period, he served on a number of government bodies, including the National Bureau of Standards, the National Research Council, the National Science Foundation, and the Atomic Energy Commission. The interest of Wigner after that period shifted to the

Philosophy of Science and in 1960, he wrote the famous paper "The Unreasonable Effectiveness of Mathematics in the Natural Sciences", where he pointed out the predictive power of mathematical theories in the description of natural phenomena. He was awarded the Nobel prize in Physics in 1963 "for his contributions to the theory of the atomic nucleus and the elementary particles, particularly through the discovery and application of fundamental symmetry principles".

NOTES

1 In the case of non-conservative systems, the potential energy V also depends on the velocities dx_i/dt. Therefore, this step must be applied also to the potential energy.

2 The 1D version of this problem is discussed in point 1.3 of the Chapter 1.

3 To simplify the derivations and without loss of generality, it is convenient to place the origin of the coordinates and the rotation axis on the centre of mass.

4 Cis/trans isomers are usually treated as distinct molecules.

5 It can be shown that the kinetic energy (which was already separate) remains separate.

6 The result depends on the symmetry of the molecule and is explained in the specific chapter.

7 This is the elementary definition. Frequently, however, the projection of vector $\vec{a}$ on vector $\vec{b}$ is defined as the *unit vector* parallel to $\vec{b}$ *multiplied by* this length, that is as $\left[\left(\vec{a},\vec{b}\right)/\left(\vec{b},\vec{b}\right)\right]\cdot\vec{b}$

8 Bold italic uppercase is used for generic matrices ($n > 1$ and $m > 1$), bold italic lowercase for row or column matrices.

Index

A

Angular moment, 16, 22–26, 45–51, 70–74, 241
Antisymmetry, 28, 49, 53–57, 123, 217

B

Bloch's theorem, 182–188, 230
Boltzmann constant, 120, 142–143
Bose-Einstein statistics, 28, 40, 121–126
Bravais lattice, 227–231

C

Canonical ensemble, 89, 119–121, 124–127, 140–143
Character (symmetry), 166, 191–192, 201–206
Chemical potential, 84, 91–92, 97–101, 113–118, 120–12, 126–127, 131–132, 147
Commutator, 46–47
Coordinates, 2–6
Coordinates, generalized, 6, 8, 16, 237–238, 240, 245
Coulomb integral, 64

D

Degeneracy, 21, 36–38, 61–69, 73–76, 120–121, 128–129, 141, 147, 159–169, 186, 208, 220–222, 232, 249, 261–263
Degrees of freedom (dof), 2–7, 50, 58, 122–124, 128–129, 173–177, 180, 183, 187–190, 207, 211, 219–220, 226, 237, 239–240, 243, 245–249
Degrees of freedom, thermodynamic degrees of freedom, 103, 119

E

Eigenfunction, 18–23, 32, 35, 46, 54, 161, 163, 165–166, 168–169, 181–182, 208–209, 212–213
Eigenvalue, 18–26, 28–29, 32, 35, 45–46, 54, 74, 120, 159–170, 173, 175, 181–188, 200, 208–211, 246–247, 260–264
Electrochemistry, 93, 115–117
Ensemble, 88–89, 119–122, 137–143
Enthalpy, 93–95, 97, 100, 102, 112, 135
Entropy, 80, 86–88, 90–92, 94–97, 100, 102, 120–121, 132–134, 143, 145
Equilibrium conditions, 86, 90, 92, 95, 104, 112
Equilibrium constant, 110–114, 117, 132
Equipartition, 128, 131, 133, 137, 245, 249
Exchange energy, 68
Exchange integral, 64, 66, 68, 78
Exchange symmetry, 23, 28, 53–64
Exclusion principle, 56–57, 62, 123, 127, 134
Expectation value, 11–14, 18–21, 34, 59, 186

F

Fermi-Dirac statistic, 124, 126–128, 147
Fermi energy, 127
Ferromagnetism, 68, 122, 136

G

Gibbs energy, 80, 93, 95, 98, 100–108, 110, 112, 116, 121
Gibbs phase rule, 103–119
Glide, glide plane, 151, 228–231
Grand canonical ensemble, 121, 124–125
Group, algebraic group, 153, 154
Group, continuous group, 224
Group, point group, 155–157, 215–216, 227, 229–231
Group products, 192
Group, space group, 227–231
Group, symmetry group, 149–156

H

Hamiltonian, 6–8, 15–22, 24–26, 28–30, 35–36, 49–53, 59, 63–66, 68, 75, 77, 150, 159, 161–171, 181–182, 185–187, 207–212, 221–222, 235, 237–238, 247–248
Hamiltonian, function, 6, 128, 237–250
Hamiltonian, operator, 18, 22, 49
Hamiltonian, separable operator, 24–26, 51, 59
Hartree product, 52, 62, 66, 70, 79, 211
Heat, 89–91, 94, 96
Heat capacity, 81, 95
Helmholtz energy, 80, 93, 95, 98, 103, 109, 111, 120, 123–124, 134–135, 143, 145
Hermitian operator, 11, 15, 18, 32, 33, 45, 193, 254–255, 261
Homogeneous function, 83, 98, 140
Hückel HMO, 160–164, 170, 185–187, 209–210

I

Ideal gas, 81, 86–87, 90, 92, 94–97, 99–102, 122, 124, 129–131, 134, 143, 147, 249
Ideal mixture, 80, 99–101, 108, 111, 113, 134–136, 144
Indistinguishability, 28, 56, 57, 62–65, 121, 123–125, 131, 134, 141–146, 222
Internal energy, 86, 89–97, 111, 121–123, 142, 144–145
Inversion symmetry, 23, 37, 67, 165, 167–168
Irreducible representation, 200, 205–211
Irreversibility, 81–93

L

Lattice dynamics, 136
Lattice, translation lattice, 179–187, 227–232
LCAO, 29, 35, 66, 160–174, 185–187, 210–213
Legendre transformation, 92–93, 140
Linear operator, 12, 13, 22, 30, 34, 36, 194, 258–259

M

Many electron systems, 49, 54, 72, 124
Matrix, 16, 33–37, 193–207, 254–264
Matrix element, 40, 47, 55, 64, 69, 76, 78, 186, 192, 194, 213, 262–263
Maxwell-Boltzmann distribution, 115, 118, 126, 130, 131, 145–147
Micro-canonical ensemble, 89, 119–121, 138
Mirror symmetry, 152, 159, 230
Molecular dynamics, 136
Monte Carlo, 137
Multiplet, 75, 76

N

Nernst equation, 116
Normalization, 8, 10–11, 27, 29–30, 56

O

Operator, 11, 15, 16
Operator, Hermitian operator, 11–15, 18, 32, 33, 45, 193, 254–255, 261–262
Operator, linear operator, 12, 13, 22, 30, 34, 36, 194, 258–259
Orthogonality, 30, 35, 38, 64, 203–207, 234–235, 251, 253

P

Partial molar quantity, 83, 97–99
Partition function, 119–135, 140, 143–146
Perturbative method, 35–42, 164, 178, 263
Phase space, 7, 137–138
Polarizability polarization, 42–43, 178
Probability density, 8, 11, 65, 130
Projection operator, 212–216

Q

Quasi-chemical equilibria, 117–119, 122

R

Reducible representation, 200–212
Rotation symmetry, 150–151

S

SAF (symmetry adapted functions), 162, 169, 210–213, 217, 234
SALC (symmetry adapted linear combinations), 162, 169–176, 182, 185–189, 208–217, 222–223
Screw axis, 151, 228–231
Seitz operator, 229
Singlet, 64–71, 176, 217–218, 226
Slater determinant, 55–58, 62, 65, 70, 72
Spin, 19–28, 53–78, 123–127
Spin orbital, 52–54
Spin-orbit interaction, 50, 73–76
Standard state, 97, 99, 101–102, 112, 132

State function, 83–86, 89–98
Subgroup, 190–192, 222–225, 230, 235
Symmetry element, 149–151
Symmetry operation, 22, 149–183
Symmetry operator, 14, 22–23, 159–169, 208–213
System (thermodynamics), 81–83

T

Time dependent perturbation, 38
Time reversal symmetry, 151, 164, 168–169, 198, 203, 206, 224, 231–232
Translational symmetry, 178
Triplet, 64–65, 67–68, 218, 226
Turnover rule, 15, 44

U

Unit cell, 119, 133, 180–185, 227–231
Unitary matrix, 34–36, 198, 203–212, 257–261

V

Vector model, 69–73, 218
Vibration, 169–178

W

Wavefunction, 8–11, 15, 18–20, 32
Wigner-Seitz cell, 180–184
Work, 82, 84, 89–90, 93, 116

For Product Safety Concerns and Information please contact our EU representative GPSR@taylorandfrancis.com Taylor & Francis Verlag GmbH, Kaufingerstraße 24, 80331 München, Germany